Halophytes and Climate Change: Adaptive Mechanisms and Potential Uses

Halophytes and Climate Change: Adaptive Mechanisms and Potential Uses

Editors

Mirza Hasanuzzaman

Department of Agronomy, Faculty of Agriculture, Sher-e-Bangla Agricultural University, Sher-e-Bangla Nagar, Dhaka-1207, Bangladesh

Sergey Shabala

Stress Physiology Laboratory, Tasmanian Institute for Agriculture, University of Tasmania, Hobart, Tasmania 7001, Australia

Masayuki Fujita

Laboratory of Plant Stress Responses, Faculty of Agriculture, Kagawa University, Miki-cho, Kagawa 761-0795, Japan

CABI

CABI is a trading name of CAB International

CABI
Nosworthy Way
Wallingford
Oxfordshire OX10 8DE
UK

Tel: +44 (0)1491 832111
Fax: +44 (0)1491 833508
E-mail: info@cabi.org
Website: www.cabi.org

CABI
745 Atlantic Avenue
8th Floor
Boston, MA 02111
USA

Tel: +1 (617)682-9015
E-mail: cabi-nao@cabi.org

A catalogue record for this book is available from the British Library, London, UK.

Library of Congress Cataloging-in-Publication Data

Names: Hasanuzzaman, Mirza, editor. | Shabala, Sergey, editor. | Fujita, Masayuki, 1956- editor.
Title: Halophytes and climate change : adaptive mechanisms and potential uses / editors: Mirza Hasanuzzaman, Sergey Shabala, Masayuki Fujita.
Description: Boston, MA : CABI, 2018. | Includes bibliographical references and index.
Identifiers: LCCN 2018037171 | ISBN 9781786394330 (hardback) | ISBN 9781786394354 (epub)
Subjects: LCSH: Halophytes. | Salt-tolerant crops. | Plants--Adaptation. | Crops and climate.
Classification: LCC QK922 .H35 2018 | DDC 581.4/2--dc23 LC record available at https://lccn.loc.gov/2018037171

ISBN-13: 9781786394330 (hardback)
9781786394347 (ePDF)
9781786394354 (ePub)

Commissioning Editors: Rachael Russell/Rebecca Stubbs
Editorial Assistant: Tabitha Jay
Production Editor: Marta Patiño

Typeset by SPi, Pondicherry, India
Printed and bound in the UK by Severn, Gloucester

Contents

List of Contributors

Janine Barbara Adams, Nelson Mandela Metropolitan University, P.O. Box 77000, Port Elizabeth 6031, South Africa. E-mail: Janine.Adams@nmmu.ac.za

Svetlana Aćić, Applied Botany Department, Faculty of Agriculture, University of Belgrade, Nemanjina 6, 11080 Belgrade, Serbia. E-mail: acic@agrif.bg.ac.rs

Sunder Singh Arya, Department of Botany, Maharishi Dayanand University, Rohtak 124001, Haryana, India. E-mail: aryasunder.hau@gmail.com

Gary Bañuelos, USDA, Agricultural Research Service, San Joaquin Valley Agricultural Sciences Center, 9611 South Riverbend Avenue, Parlier, California 93648-9757, USA. E-mail: gary.banuelos@ars.usda.gov

Roberto Barbato, Department of Sciences and Technological Innovation, University of Eastern Piedmont, Viale T. Michel 11, I-15121 Alessandria, Italy. E-mail: roberto.barbato@uniupo.it

Isabel Caçador, MARE – Marine and Environmental Sciences Centre, Faculty of Sciences, University of Lisbon, Campo Grande, 1749-016 Lisbon, Portugal; and Department of Plant Biology, Faculty of Sciences, University of Lisbon, Campo Grande, 1749-016 Lisbon, Portugal. E-mail: micacador@fc.ul.pt

Romina Cannata, Department of Sciences and Technological Innovation, University of Eastern Piedmont, Viale T. Michel 11, I-15121 Alessandria, Italy. E-mail: romina.cannata@uniupo.it

Tiziana Centofanti, Central European University, Nador utca 9, 1051 Budapest, Hungary. E-mail: centofantit@spp.ceu.edu

Gurdev Chand, Sher-e-Kashmir University, Jammu, Srinagar, Jammu and Kashmir, India. E-mail: gurdev74@gmail.com

Doongar R. Chaudhary, Division of Biotechnology and Phycology, CSIR-Central Salt and Marine Chemicals Research Institute), Bhavnagar 364002, Gujarat, India. E-mail: drchaudhary@csmcri.res.in

Zora Dajić Stevanović, Applied Botany Department, Faculty of Agriculture, University of Belgrade, Nemanjina 6, 11080 Belgrade, Serbia. E-mail: dajic@agrif.bg.ac.rs

Sarita Devi, Chaudhary Charan Singh Haryana Agricultural University, Hisar 125004, Haryana, India. E-mail: devisaritaa@gmail.com

Bernardo Duarte, MARE (Marine and Environmental Sciences Centre), Faculty of Sciences, University of Lisbon, Campo Grande, 1749-016 Lisbon, Portugal. E-mail: baduarte@fc.ul.pt

Chloe M. Eggert, Department of Environmental Studies, Elon University, Elon, North Carolina, USA. E-mail: ceggert@elon.edu

José A. Fernández, Department of Plant Biology, University of Málaga, Málaga, Spain. E-mail: ja_fernandez@uma.es

Marius-Nicuşor Grigore, Faculty of Biology, Alexandru Ioan Cuza University, Carol 1, 20 A, 700505 Iaşi, Romania. E-mail: mariusgrigorepsyche@yahoo.com

Bilquees Gul, Institute of Sustainable Halophyte Utilization (ISHU), University of Karachi, Karachi 75270, Pakistan. E-mail: bilqueesgul@uok.edu.pk

Abdul Hameed, Institute of Sustainable Halophyte Utilization (ISHU), University of Karachi, Karachi 75270, Pakistan. E-mail: ahameed@uok.edu.pk

Pedja Janaćković, Institute of Botany and Botanical Garden 'Jevremovac', Faculty of Biology, University of Belgrade, Belgrade, Serbia. E-mail: pjanackovic@bio.bg.ac.rs

Bhavanath Jha, Division of Biotechnology and Phycology, Central Salt and Marine Chemicals Research Institute (CSIR-CSMCRI), G.B. Marg, Bhavnagar, Gujarat, India; and Academy of Scientific and Innovative Research, Ghaziabad, New Delhi, India. E-mail: jha.bhavanath@gmail.com

Rajesh Kumar Jha, Division of Biotechnology and Phycology, Central Salt and Marine Chemicals Research Institute (CSIR-CSMCRI), G.B. Marg, Bhavnagar, Gujarat, India; and Academy of Scientific and Innovative Research, Ghaziabad, India. E-mail: rajeshkumarjha2588@gmail.com

Adam Kapler, Polish Academy of Sciences – Botanical Garden in Powsin (PAS BG-CBDC), Prawdziwka 2, 02-973 Warsaw, Poland; and Centre for Wetland Conservation, Warsaw, Poland. E-mail: adam.kapler@obpan.pl

M. Ajmal Khan, Institute of Sustainable Halophyte Utilization (ISHU), University of Karachi, Karachi 75270, Pakistan. E-mail: majmalk@uok.edu.pk

Mariana K. Kneppers, Department of Environmental Studies, Elon University, Elon, North Carolina, USA. E-mail: mkneppers@elon.edu

Arvind Kumar, Central Soil Salinity Research Institute (ICAR), Karnal 132001, Haryana, India. E-mail: singh.ak92@gmail.com

Ashwani Kumar, Central Soil Salinity Research Institute (ICAR), Karnal 132001, Haryana, India. E-mail: ashwanisharma107@gmail.com

Neeraj Kumar, Chaudhary Charan Singh Haryana Agricultural University, Hisar 125004, Haryana, India. E-mail: neerajhau@yahoo.co.in

Pratima Kumar, Prannath Parnami Institute of Management and Technology, Hisar, Haryana, India. E-mail: pratimakumar66@gmail.com

Anita Kumari, Chaudhary Charan Singh Haryana Agricultural University, Hisar 125004, Haryana, India. E-mail: anitahsr@gmail.com

Asha Kumari, Division of Biotechnology and Phycology, CSRI-Central Salt and Marine Chemicals Research Institute (CSIR-CSMCRI), Gijubhai Badheka Marg, Bhavnagar-364002, Gujarat, India; and Academy of Scientific and Innovative Research, CSRI-Central Salt and Marine Chemicals Research Institute (CSIR-CSMCRI), Council of Scientific and Industrial Research (CSIR), Gijubhai Badheka Marg, Bhavnagar 364 002, Gujarat, India. E-mail: ashakyadav@gmail.com

Shubham Lamba, Department of Soil Science, CCS Haryana Agricultural University, Hisar 125004, Haryana, India. E-mail: shubham.hau@gmail.com

Dariusz Latowski, Faculty of Biochemistry, Biophysics and Biotechnology, Jagiellonian University, Gronostajowa 7, 30-387 Kraków, Poland. E-mail: dariusz.latowski@uj.edu.pl

Marta Libik-Konieczny, The Franciszek Górski Institute of Plant Physiology, Polish Academy of Sciences, Niezapominajek 21, 30-239 Kraków, Poland. E-mail: libik@ifr-pan.edu.pl

Analía Llanes, Plant Physiology Laboratory, Department of Natural Sciences, National University of Río Cuarto, Route 36 Km. 601, (CP X5804BYA) Río Cuarto-Córdoba, Argentina; and National Council for Science and Technology, Argentina. E-mail: allanes@exa.unrc.edu.ar

Milica Luković, Faculty of Hotel Management and Tourism, University of Kragujevac, Vrnjačka banja, Serbia. E-mail: mpetrovic.azs@gmail.com

Virginia Luna, Plant Physiology Laboratory, Department of Natural Sciences, National University of Río Cuarto, Route 36 Km. 601, (CP X5804BYA) Río Cuarto-Córdoba, Argentina; and National Council for Science and Technology, Argentina. E-mail: vluna@exa.unrc.edu.ar

Ulrich Lüttge, Institute of Botany, Technical University of Darmstadt, Schnittspahnstrasse, Darmstadt, Germany. E-mail: luettge@bio.tu-darmstadt.de

Anita Mann, Central Soil Salinity Research Institute (ICAR), Karnal 132001, Haryana, India. E-mail: anitadgr13@gmail.com

João Carlos Marques, Marine and Environmental Sciences Centre (MARE), c/o DCV, Faculty of Sciences and Technology, University of Coimbra, Coimbra, Portugal. E-mail: jcmimar@ci.uc.pt

Enrique Mateos-Naranjo, Department of Plant Biology and Ecology, Biology Faculty, University of Seville, Campus Reina Mercedes, 41012 Seville, Spain. E-mail: emana@us.es

Ana Rita Matos, Biosystems and Integrative Sciences Institute (BioISI), Plant Functional Genomics Group, Plant Biology Department, Faculty of Sciences, University of Lisbon, Campo Grande, 1749-016 Lisbon, Portugal. E-mail: armatos@fc.ul.pt

Jennifer Mesa-Marín, Department of Plant Biology and Ecology, Biology Faculty, University of Seville, Campus Reina Mercedes, 41012 Sevilla, Spain. E-mail: jmesam@us.es

Avinash Mishra, Division of Biotechnology and Phycology, Central Salt and Marine Chemicals Research Institute (CSIR-CSMCRI), G.B. Marg, Bhavnagar, Gujarat, India; and Academy of Scientific and Innovative Research, Ghaziabad, India. E-mail: avinash@csmcri.res.in

Zbigniew Miszalski, The Franciszek Górski Institute of Plant Physiology, Polish Academy of Sciences, Niezapominajek 21, 30-239 Kraków, Poland. E-mail: miszalski@ifr-pan.edu.pl

Farah Nisar, Institute of Sustainable Halophyte Utilization (ISHU), University of Karachi, Karachi 75270, Pakistan. E-mail: farah.ishu@hotmail.com

Eloísa Pajuelo, Department of Microbiology and Parasitology, Faculty of Pharmacy, University of Seville, Calle Profesor García González 2, 41012 Seville, Spain. E-mail: epajuelo@us.es

Asish Kumar Parida, Division of Biotechnology and Phycology, CSRI-Central Salt and Marine Chemicals Research Institute (CSIR-CSMCRI), Gijubhai Badheka Marg, Bhavnagar-364002, Gujarat, India; and Academy of Scientific and Innovative Research, CSRI-Central Salt and Marine Chemicals Research Institute (CSIR-CSMCRI), Council of Scientific and Industrial Research (CSIR), Gijubhai Badheka Marg, Bhavnagar 364 002, Gujarat, India. E-mail: asishparida@csmcri.org

Jaykumar Patel, Division of Biotechnology and Phycology, Central Salt and Marine Chemicals Research Institute (CSIR-CSMCRI), G.B. Marg, Bhavnagar, Gujarat, India; and Academy of Scientific and Innovative Research, Ghaziabad, New Delhi, India. E-mail: jaypatelbt14@gmail.com

Monika Patel, Division of Biotechnology and Phycology, CSRI-Central Salt and Marine Chemicals Research Institute (CSIR-CSMCRI), Gijubhai Badheka Marg, Bhavnagar-364002, Gujarat, India; and Academy of Scientific and Innovative Research, CSRI-Central Salt and Marine Chemicals Research Institute (CSIR-CSMCRI), Council of Scientific and Industrial Research (CSIR), Gijubhai Badheka Marg, Bhavnagar 364 002, Gujarat, India. E-mail: patelmonika0909@gmail.com

Ilinka Pećinar, Applied Botany Department, Faculty of Agriculture, University of Belgrade, Nemanjina 6, 11080 Belgrade, Serbia. E-mail: ilinka@agrif.bg.ac.rs

Sílvia Pedro, MARE – Marine and Environmental Sciences Centre, Faculty of Sciences, University of Lisbon, Campo Grande, 1749-016 Lisbon, Portugal. E-mail: sspedro@fc.ul.pt

Pooja, Sugarcane Breeding Institute, ICAR, Regional Centre, Karnal 132001, Haryana, India. E-mail: poojadhansu@gmail.com

Dragana Rančić, Applied Botany Department, Faculty of Agriculture, University of Belgrade, Nemanjina 6, 11080 Belgrade, Serbia. E-mail: rancicd@agrif.bg.ac.rs

Babita Rani, Chaudhary Charan Singh Haryana Agricultural University, Hisar 125004, Haryana, India. E-mail: babitachahalkharb@gmail.com

Jaykumar Rangani, Division of Biotechnology and Phycology, CSRI-Central Salt and Marine Chemicals Research Institute (CSIR-CSMCRI), Gijubhai Badheka Marg, Bhavnagar-364002, Gujarat, India; and Academy of Scientific and Innovative Research, CSRI-Central Salt and Marine Chemicals Research Institute (CSIR-CSMCRI), Council of Scientific and Industrial Research (CSIR), Gijubhai Badheka Marg, Bhavnagar 364 002, Gujarat, India. E-mail: jayrangani99@yahoo.com

Aysha Rasheed, Institute of Sustainable Halophyte Utilization (ISHU), University of Karachi, Karachi 75270, Pakistan. E-mail: halophyte_aysha@yahoo.com

Susana Redondo-Gómez, Department of Plant Biology and Ecology, Biology Faculty, University of Seville, Campus Reina Mercedes, 41012 Seville, Spain. E-mail: susana@us.es

Mariana Reginato, Plant Physiology Laboratory, Department of Natural Sciences, National University of Río Cuarto, Route 36 Km. 601, (CP X5804BYA) Río Cuarto-Córdoba, Argentina; and National Council for Science and Technology, Argentina. E-mail: mreginato@exa.unrc.edu.ar

Ignacio D. Rodríguez-Llorente, Department of Microbiology and Parasitology, Faculty of Pharmacy, University of Seville, Calle Profesor García González 2, 41012 Seville, Spain. irodri@us.es

Lourdes Rubio, Department of Plant Biology, University of Málaga, Málaga, Spain. lrubio@uma.es

Dhouha Belhadj Sghaier, UR MaNE, Faculty of Sciences, Bizerte, University of Carthage, Bizerte, Tunisia. E-mail: dhouhasghaier@hotmail.fr

Urban Šilc, Scientific Research Centre of the Slovenian Academy of Sciences and Arts (ZRC SAZU), Institute of Biology, Ljubljana, Slovenia. E-mail: urban@zrc-sazu.si

Noomene Sleimi, UR MaNE, Faculty of Sciences, Bizerte, University of Carthage, Bizerte, Tunisia. E-mail: noomene.sleimi@gmail.com

Jovana Stanković, Institute of Chemistry, Technology and Metallurgy, University of Belgrade, Belgrade, Serbia. E-mail: jovanas@chem.bg.ac.rs

Milan S. Stanković, Department of Biology and Ecology, Faculty of Sciences, University of Kragujevac, Kragujevac, Serbia. E-mail: mstankovic@kg.ac.rs

Miroslava Stanković, Nuclear Facilities of Serbia, Mike Petrovića Alasa 12-14, 11351 Vinča, Serbia. E-mail: miroslava.stankovic@nuklearniobjekti.rs

Danijela Stešević, Biotechnical Faculty, University of Podgorica, Podgorica, Montenegro. E-mail: danijela.denist@gmail.com

Ewa Surówka, The Franciszek Górski Institute of Plant Physiology, Polish Academy of Sciences, Niezapominajek 21, 30-239 Kraków, Poland. E-mail: ewa.surowka@gmail.com

Pascal Tabi Tabot, Nelson Mandela Metropolitan University, Port Elizabeth 6031, South Africa; and Department of Agriculture, Higher Technical Teachers' Training College Kumba, University of Buea, P.O. Box 63, Buea, Cameroon. E-mail: ttabot@yahoo.com

Brant W. Touchette, Biology and Environmental Studies, Elon University, Elon, North Carolina 27244, USA. E-mail: btouchette@elon.edu

Preface

The world's food production will need to increase dramatically to match the predicted population growth. Achieving this goal will be very challenging because of the decreased availability of arable land, resulting from urbanization and land degradation caused by salinity (Panta *et al.*, 2014). The economic impact of soil salinity on agricultural crop production exceeds US$27 billion p.a. (Qadir *et al.*, 2014) and is expected to rise, given present trends in climate change. Currently, soil salinity is claiming about 3 ha of arable land from conventional crop farming every minute (Shabala *et al.*, 2014), forcing agricultural production into marginal areas and making the use of traditional staple crops more and more challenging and less efficient. The physiological and genetic complexity of the salt tolerance trait means that real progress in crop breeding for higher salinity stress tolerance may be achieved only by a painfully slow pyramiding of essential physiological traits (Shabala, 2013; Flowers and Colmer, 2015). Given the time constraints, a safer solution to meet the 2050 challenge may be to find alternative crop and forage species for farming in salt-affected conditions and to restore salt-affected areas (Rozema and Schat, 2013; Panta *et al.*, 2014). In this context, halophytes 'tick all the boxes' and promise to become important components of 21st-century farming systems.

The term 'halophyte' entered plant biology literature at the beginning of the 18th century, and these plants have been fascinating generations of scientists ever since. Halophytes thrive in conditions that kill 99% of conventional crops and, over the course of evolution, have developed highly efficient mechanisms to cope with the salt load. What are these mechanisms? Are they unique to the halophytes, or are they simply underutilized by conventional crops? Can we make the latter more efficient? Some answers to these questions are given in this volume. The range of topics covered includes the patterns of ion accumulation in halophytes; specific anatomical traits such as leaf or stem succulence or secretory glands/salt bladders, conferring their superior salinity tolerance; and mechanisms of reactive oxygen species (ROS) detoxification. Several chapters also investigate specific features of interaction between halophytes and soil microorganisms.

While the number of halophytic species is relatively low (< 1%), they are present in about half of the higher plant families (Flowers and Colmer, 2008; Flowers *et al.*, 2010). Some orders contain many species of halophytes, but many orders only have a few (Flowers *et al.*, 2010; Shabala and Mackay, 2011). A similarly uneven distribution is also characteristic of the presence of specific physiological traits between orders (e.g. salt secretion through the glands; Dassanayake and Larkin, 2017). Why? How does the habitat affect mechanisms employed by halophytes to deal with the salt load? Several chapters in this volume address this question, dealing with various aspects of halophyte ecophysiology.

In addition to their potential use as a 'deciphering tool' to understand mechanisms of salinity stress tolerance to boost current breeding programmes for this trait, halophytes are already being used directly for human food, for forage and animal feeds, as oilseed and energy crops, and for desalination and phytoremediation purposes. Other chapters provide a comprehensive update on these topics.

The rapid progress in molecular biology and the recent genome sequencing of several halophyte species (Oh *et al.*, 2012; Wu *et al.*, 2012; Yang *et al.*, 2013; Zou *et al.*, 2017) will undoubtedly provide an additional boost to halophyte research. However, plant organisms are not merely a number of genes put together. Further progress in the field may only be achieved when various omics tools are intrinsically interspersed with the understanding of plant *function* and put into an environmental context. We hope this volume will contribute to this purpose. Happy reading!

Mirza Hasanuzzaman, Sergey Shabala and Masayuki Fujita
1 January 2019

References

Dassanayake, M. and Larkin, J.C. (2017) Making plants break a sweat: the structure, function, and evolution of plant salt glands. *Frontiers in Plant Science* 8, doi:10.3389/fpls.2017.00406.

Flowers, T.J. and Colmer, T.D. (2008) Salinity tolerance in halophytes. *New Phytologist* 179, 945–963.

Flowers, T.J. and Colmer, T.D. (2015) Plant salt tolerance: adaptations in halophytes. *Annals of Botany* 115, 327–331.

Flowers, T.J., Galal, H.K. and Bromham, L. (2010) Evolution of halophytes: multiple origins of salt tolerance in land plants. *Functional Plant Biology* 37, 604–612.

Oh, D.H., Dassanayake, M., Bohnert, H.J. and Cheeseman, J.M. (2012) Life at the extreme: lessons from the genome. *Genome Biology* 13, doi:10.1186/gb-2012-13-3-241.

Panta, S., Flowers, T., Lane, P., Doyle, R., Haros, G. and Shabala, S. *et al.* (2014) Halophyte agriculture: Success stories. *Environmental and Experimental Botany* 107, 71–83.

Qadir, M., Quillérou, E., Nangia, V., Murtaza, G., Singh, M. *et al.* (2014) Economics of salt-induced land degradation and restoration. *Natural Research Forum* 38, 282–295.

Rozema, J. and Schat, H. (2013) Salt tolerance of halophytes, research questions reviewed in the perspective of saline agriculture. *Environmental and Experimental Botany* 92, 83–95.

Shabala, S. (2013) Learning from halophytes: physiological basis and strategies to improve abiotic stress tolerance in crops. *Annals of Botany* 112, 1209–1221.

Shabala, S. and Mackay, A. (2011) Ion transport in halophytes. *Advances in Botanical Research* 57, 151–199.

Shabala, S., Bose, J. and Hedrich, R. (2014) Salt bladders: do they matter? *Trends in Plant Science* 19, 687–691.

Wu, H.J., Zhang, Z.H., Wang, J.Y., Oh, D.H., Dassanayake, M. *et al.* (2012) Insights into salt tolerance from the genome of *Thellungiella salsuginea*. *Proceedings of the National Academy of Science USA* 109, 12219–12224.

Yang, R.L., Jarvis, D.E., Chen, H., Beilstein, M.A., Grimwood, J. *et al.* (2013) The reference genome of the halophytic plant *Eutrema salsugineum*. *Frontiers in Plant Science* 4, doi:10.3389/fpls.2013.00046.

Zou, C.S., Chen, A.J., Xiao, L.H., Muller, H.M., Ache, P. *et al.* (2017) A high-quality genome assembly of quinoa provides insights into the molecular basis of salt bladder-based salinity tolerance and the exceptional nutritional value. *Cell Research* 27, 1327–1340.

Section I

Ecophysiology

1 Defining Halophytes: a Conceptual and Historical Approach in an Ecological Frame

Marius-Nicuşor Grigore*
Alexandru Ioan Cuza University, Iaşi, Romania

Abstract
Halophytes have certainly been recognized since the beginning of the 18th century, but subtle earlier allusions to salt-tolerant plants can be found back to 1500. By the time of the French *Encyclopédie* (1751–1765), and Goethe (1786), consistent data about halophytes had already been accumulated. Halophytes were brought to scientific attention through the classic and iconic papers of the German botanist Schimper, and especially those of the Danish plant ecologist Warming. However, their definitions remain controversial; the lack of a unique definition is related mainly to the complex features of this ecological group of plants. There are many definitions of halophytes, and some reflect the scientific background of the researchers who defined them. The 'historical' evolution of a definition can be delineated, and it especially reflects the progress of accumulating knowledge about halophyte biology. The definition of halophytes is thus manifold. The criteria and the historical context in which their definitions occurred and developed are discussed in detail in this chapter. Special attention is given to textual analysis that reveals the convergent approaches of researchers who used different languages. Ecological perspectives in defining halophytes suggest that, although they in fact represent a large ecological group of plants, their common xeromorphic nature must be universally recognized.

Keywords: Ecology; Halophytes; Saline environments; Xerophytes

1.1 Historical and Semantic Evolution of Terms

Many definitions of halophytes now exist. Halophytes are so diverse from taxonomical and ecological points of view that there is no consensus on a unique definition of the concept of 'halophyte' (Grigore, 2008, 2012; Grigore *et al.*, 2010; Grigore and Toma, 2017). Several studies deal with this issue (Grigore *et al.*, 2010, 2014; Grigore, 2012) and in Table 1.1 new definitions are mentioned and discussed in a conceptual frame, with emphasis on the historical and linguistic evolution of basic terms.

It is well known that the history of botany is a part of the history of science and reflects the cultural evolution of humanity (Von Sachs, 1890; Green, 1914; Arber, 1953; Isley, 1994; Stevens, 1994; Grigore, 2016; Hardy and Totelin, 2016). Table 1.2 shows a list of terms, extracted from the languages most used in relation to halophyte semantics, with special attention to Latin, which was a rooting language for French and Romanian. Latin was the universal

* Corresponding author e-mail: mariusgrigorepsyche@yahoo.com

Table 1.1. A review of halophyte definitions (adapted from Grigore *et al*., 2010, 2014; Grigore, 2012).

Definition or descriptions related to halophytes	References	Comments
Plants of the seashore	Willdenow (1805)	First observations of halophytes were made of plants from seashore – regarded as the 'absolute' case of plants growing in saline environments. Later, plants from inland saline habitats were also recognized as halophytes; some botanists referred sometimes to 'maritime' and 'terrestrial' forms of the same species
Plants that vegetate in saline soils	Bischoff (1833)	An ecological definition
Saline plants are those found in maritime situations, or near salt lakes, and which seem to require much soda in their composition, and which have been called Halophytes	Balfour (1862)	The 'obligatory' character of halophytes is suggested
Plants which inhabit salt marshes, and by burning yield barilla	Fowler (1875)	An ecological definition
A plant containing a large quantity of common salt in its composition, and which thrives best in salty places	Crozier (1892)	Despite its earlier character, this definition is interesting because it suggests the capacity of halophytes to accumulate salt in large amounts. We now know that this is a group of halophytes accumulating salts, in contrast to those secreting it
Salt-loving plants (are in the most of their characters, strikingly similar to the xerophytes)	Barnes (1898)	Many plant ecologists consider halophytes a particular case of xerophytes (see further comments in this table)
Species of saline and alkaline soils (salt plants)	Clements (1907)	Saline and/or alkaline soils are terms more precise than other words designating saline environments
A certain amount of soluble salts must be present before halophytic vegetation is called into existence	Warming (1909)	How precise is the term 'certain'?
Plants which grow where the water contains salt; the effect upon them is seen in their fleshy habit	Bower (1911)	In fact, the soil solution always contains 'salt'; the issue is concentration. Not all halophytes display a fleshy tissue
Strand plants, or Halophytes, living along the margin of salt water, and therefore condensed and otherwise adapted to the difficult absorption thereof	Ganong (1913)	We must discriminate that not all halophytes are strand plants; they can also appear in inland salt marshes/areas
Halo-philous/phytes, plants of sea-coasts and salt-steppes, where the presence of salt, by checking absorption, compels a reduction of transpiration	Willis (1919)	Here we can notice the introduction of 'physiological drought' hypothesis characterizing saline soils. This is 'famous' for a certain period of plant ecology (see Grigore and Toma, 2010)
Plants which at any stage of their life are subjected to a concentration of salt, which is more than 'normal' glycophytic plants can bear without dying	Stocker (1928)	The salt concept is an ambiguous one (see the discussions above). It is difficult to establish if the plants are exposed all the time to salt, at any stage of their life cycle

Continued

Table 1.1. Continued.

Definition or descriptions related to halophytes	References	Comments
Salt plants; Typical halophytes; true halophytes; *absolute halophytes**; the obligate halophytes are plants which for their normal development need certain ions of the alkali metals and halogens, and which, therefore, can exist and bear seed only in soils containing salt	Braun-Blanquet (1932)	A good definition of obligate halophytes
Plants that grow in saline soil or in salty water are called halophytes and they are strikingly xeric	McDougall (1941)	An interesting definition stating that halophytes are a peculiar case of xerophytes (for extensive comments, see Grigore and Toma, 2010)
All plants that are capable of growing in an environment where there is more than 0.5 per cent sodium chloride	Chapman (1942)	Chapman's comments: 'its use [i.e. the definition of 'halophytes'] will not imply that the species is either common or rare in such habitats nor will the term involve the assumption that a plant cannot grow under any other conditions'. Salinity is a very changeable ecological factor: choosing a number for drawing a line between two different plant groups could be hazardous
Plants that can tolerate the concentrations of salts found in saline soils are termed halophytes	Oosting (1948)	
Plants tolerant of various mineral salt in the soil solution, usually sodium chloride	Lawrence (1951)	
Plants growing on salinized media	Bucur *et al.* (1957)	
Plant that grow exclusively on salt soil	Dansereau (1957)	'Exclusively' could also suggest that Dansereau was thinking only of euhalophytes
Plants growing in saline soils	Fernald (1950)	
Salt-tolerant plants	Chapman (1960)	Neither salt nor tolerant are well defined
[..] the extremely saline soils which are inhabited only by specially adapted plants (halophytes); plants which habitually grow in very salty soils – halophytes, or at least *can* grow in such soils (facultative halophytes); Halophytes are plants which can tolerate a considerable degree of salinity	Polunin (1960)	A good definition of euhalophytes; however, growing does not necessarily imply plant reproducing
Plants of salty or alkaline soils	Correl and Johnston (1970)	
1. Plants which grow and complete their life cycle in habitats with a high salt content 2. Usually, the term is reserved only for plants which appear in salty habitats constantly and specifically	Waisel (1972)	Very difficult to say precisely what 'high salt content' represents. This remark of Waisel suggests that the term be applied only to euhalophytes ('true halophytes')

Continued

Table 1.1. Continued.

Definition or descriptions related to halophytes	References	Comments
Plants that can tolerate sea water, pure or diluted	Duncan (1974)	The sea water concentration is not a universal standard, so 'pure' or 'diluted' could be regarded as relative terms
Plants of salty environments; plants adapted to live in a saline environment, be it seawater, a salt-water marsh, or a salt-desert. Plants found growing under naturally saline conditions; for terrestrial plants, this means a minimum salt concentration of about 100 mM in the soil solution. Plants adapted to complete their life cycles in salinities about that of seawater	Flowers *et al.* (1986)	This is perhaps among the first physiological definition of halophytes
The term halophyte literally means salt plants, but is used specifically for plants that can grow in the presence of high concentrations of Na salts	Sharma and Gupta (1986)	Perhaps referring also to the character of euhalophytes
Those species for which salt marsh is a major and, in many cases, only habitat.	Adam (1990)	A good ecological definition
Plants that grow in saline conditions	Ingrouille (1992)	
Plant species with a set of ecological and physiological characteristics allowing growth and reproduction in a saline environment. *Arbitrarily* a salinity of 0.5% NaCl in soil water should be tolerated by halophytic plants	Gorham (1995) [cited by Rozema, 1996]	Some authors are aware of this arbitrariness
Halophytes are defined as those plants which grow and complete their entire life-cycle in saline habitats. Coping with salinity needs adaptations on all levels from the autecological, the tissue and cellular level to subcellular and biochemical adaptations	Breckle (1995)	'Entire' means inclusively producing seeds to assur plant survival, colonization and stabilization in any habitat A holistic definition
Plants that occur naturally on soils or in water too salty for the average plants are usually designated as halophytes	Dagar (1995)	
[The growth] of halophytes is optimal at relatively high levels of NaCl, a response which can be explained only in part by the role of sodium as a mineral nutrient in these species	Marschner (1995)	This is an example of an indirect definition of euhalophytes
Halophytes are adapted to survive in a range of saline environments	Weber (1995)	
Halophyte species are those occurring in naturally saline conditions *only*	Aronson and Le Floc'h (1996)	Also suggesting the 'obligate' character of (some) halophytes
The vegetation of saline habitats is designated 'halophytic'	Poljakoff-Mayber and Lerner (1999)	Saline habitats are defined by these authors as those whose soils contain a high percentage of soluble salts, and one or more of these salt components is usually in excess

Continued

Table 1.1. Continued.

Definition or descriptions related to halophytes	References	Comments
Salt tolerant plants (halophytes, including salt marsh and mangrove plants) are highly evolved and specialized organisms with well-adapted morphological and physiological characteristics allowing them to proliferate in the soils possessing high salt concentrations	Khan and Duke (2001)	A good holistic definition
Plants that can grow on soils with a high salt content are termed halophytes	Fitter and Hay (2002)	
Plants that can survive in or benefit from an environment with a high level of salt (i.e. sodium chloride), as in saline soils and seawater	Mooney and Canadell (2002)	
Halophytes are salt-resistant or salt-tolerant plants that thrive and complete their life cycles in soils or waters containing high salt concentration	Ness (2003)	
Halophytes are able to adapt faster and to tolerate extreme salinity	Schulze *et al.* (2005)	A deeper physiological definition
Plants that are able to grow on mildly to strongly saline soils (halobiomes). Halophytes which tolerate or endure high levels of salt are known as euhalophytes	Ingrouille and Eddie (2006)	'Mildly', 'strongly', 'high levels' are not well-defined terms. However, these authors are among the only ones distinguishing between 'halophytes and salt-tolerant plants', a very subtle but pertinent remark in the context of our previous discussions on semantics
Plants that survive to reproduce in environments where the salt concentration is around 200 mM NaCl or more	Flowers and Colmer (2008)	
Halophytes grow naturally in very salty soils; they still have not lost their resistance mechanisms to salt-stress conditions	Koyro *et al.* (2008)	
Plants of saline habitats	Holzapfel (2009)	
Plants able to complete their life cycle on saline substrates	Koyro *et al.* (2009)	
Plants that are tolerant of excess salt	Quinn (2009)	

*The term *absolute halophytes* was solely found in Braun-Blanquet (1932).

language for natural sciences from the time of ancient history, during the Middle Ages and towards the beginning of the 19th century. In parallel, however, some people started to use their national languages in scientific works from the 17th century.

1.2 Halophytes: Starting to Define Them as an Ecological Group of Plants

Table 1.2 shows the earlier references to halophytes in Latin language, which may be further

Table 1.2. Terms used in different languages in relation to halophytes – an historical overview.

Language	Time period	Terms and references	Remarks
Latin	1700–1799	*Plantae salsae* (Leonhard, 1753; Georgi, 1777) *Herbae salsae* (Henckel, 1722, 1755) *Marinae plantae* (Oeder, 1764; Nemnich, 1798) *Plantae marinae* (Boerhaave, 1724; Brückmann, 1749; Khell, 1755)	An interesting classification is given by Oeder (1764): under salt (marine) water ecosystem, he includes '*marinae plantae*', which 'nourish without roots, being subjected to wave's fluctuations'; he may be referring to algae or perhaps *Zostera* species. Under 'sea shore', he includes 'maritime (littoral) plants [*littoraceae maritimae*], which often taste salty, are succulent and they are found even in continental saline environments'. Note that Latin has two different terms related to plant succulence: '*succulentae*' and '*carnosae*'
	1800–1899	*Halophyta* (Pallas, 1803) *Plantae salsae*; *plantae salsuginosae* (Bischoff, 1833)	Latin was used almost exclusively as the botanical language until the first decades of the 19th century when it started to be abandoned, to be replaced by national languages
French	1700–1799	*Plantes marines* (Prevost d'Exile, 1777); *Plantes maritimes* (du Monceau, 1755; Macquer, 1756, 1778; Roux, 1765; Rozier, 1781; Miller, 1788) *Plantes salines* (de Haller, 1776; de la Peyronie, 1793)	Sometimes, there is a substantial difference between *plante marine* and *plante maritime* in the French language: for instance, Roux (1765) refers to the first as organisms that belong today to animals, while *plantes maritimes* are 'plants that grow on seashore like the soude' (a species that corresponds to *Salsola* species)
	1800–1899	*Halophytes* (Meyen, 1840; Bortschof, 1859; Grisebach, 1875); *Plantes salicoles* (Gadaceau, 1881)	
	1900–Present	*Halophytes*	
English	1700–1799	*Marine plants* (Short, 1746) *Plants as love the sea-side* (Tournefort, 1741)	
	1800–1899	*Maritime plants* (Willdenow, 1805) *Halophytes* (Balfour, 1862; Sullivan and O'Reilly, 1863; Grisebach, 1864; Hoblyn, 1865; Brown, 1874; von Hellwald, 1874)	
	1900–Present	*Halophytes; salt tolerant plants; salt plants; high salinity tolerant plants; salt loving plants; halophylous plants; halophytic plants; maritime plants* (Grigore, 2012)	

Continued

Table 1.2. Continued.

Language	Time period	Terms and references	Remarks
German	1700–1799	*Meere pflanzen* (Willdenow, 1792) *Salzpflanzen* (von Lehman, 1791; Hochheimer, 1793)	
	1800–1899	*Halophyten* (Göbel,1838)	
	1900–Present		
Romanian	1800–1899	*Plante halofile, Halofite* (Grecescu, 1898) *Specii litorale* (Brândza, 1880)	
	1900–Present	*Halofite; plante de sărătură; plante halofile; plante iubitoare de săruri; plante de locuri sărate* (Grigore, 2012)	

found in the specific etymology of several modern languages. Thus, *plantae (herbae) salsae* literally means *salt(y) plants*, and the ecological affinity of this group is clearly correlated with saline environments. This is because the references to halophytes were to plants that grow in *locis salsuginosis* (salty habitats), a description of habitats largely widespread in the 18–19th centuries. Many salt plants (*plantae salsae*, as they are called) also taste salty, and this purely sensorial observation is of great importance in the 18th century, since it is also related to succulence and maritime habitats. This simple but striking finding most likely led to major research in the chemistry of salt – typical in the context of the increasing modernity of chemical and mineral sciences from the Age of Enlightenment (Hutchings, 1969; Conlon, 1991). Halophytes were recognized from that time as natural resources of salt – *soda* (English), *sel de soude (soude)* (French). An extensive literature is dedicated to this issue, and especially to the tradition and practice of obtaining salts to be used in manufacturing, especially that of glass. However, most likely, saline environments were recognized much earlier than the plants living on them. This is a simple but important aspect, because it can explain how a particular type of habitat was delineated in relation to a dominant (ecological) factor such as the salt. For instance, early foreign travellers in the Romanian Principalities mentioned salty environments within their observations: Paul de Alep (1637–1667) referred to *sărătura* (salty area) and noticed that *all the surface area of this land* ('this land' refers to Ocnele Mari) *is salty* (Alexandrescu-Dersca Bulgaru, 1976).

The monumental French work, *L'Encyclopédie* (17 volumes, 1751–1765) contains many references about or related to halophytes as a salt resource. Thus, the contributor Jaucourt dealt with the terms *Salicornia, Statice, Kali* and *soude (sel de soude)* (1765); and Venel with *sel marin* (sea salt) (1765). *Soude* is defined as 'the ash of many plants that contain marine salt and which grow mostly on seashores – despite some of them are also found in the interior (saltmarshes)' (Jaucourt, 1765). This is a good implicit definition of halophytes. The French language uses the term *soude* for the ash of halophytes and for species of the actual genus *Salsola*. The fact that halophytes can also grow in inland saline environments is also a good observation from an ecological point of view and for the French language, which also used the term *plantes maritimes* for halophytes. Detailed information about the tradition of extracting this *soude* is also given, particularly in two iconic works from that period: de Jussieu (1717) and Henckel (1760). The French author de Jussieu gave an account of a halophyte species, *Kali d'Alicante*, which is most likely a species of *Salsola* (*kali*). He offers much accurate information about this species (having 'cylindrical and succulent leaves') in respect of the extraction of salt and its use; and, finally, he concludes that 'the history of other species that provide these types of salt, the particular and physical observation of each species of *Salsola* [soude, in the original French] ... are so extended, that they would offer material for another work' (de Jussieu, 1717: p. 78).

Henckel (1760) elaborated a massive treatise on mineralogy (*Pyritologia*, 1725; 1755, German editions) and a curious, interesting work – *Flora Saturnizans* (1722). The 1760 French edition of the latter includes an explanation of this title, 'where the author shows the alliance between plants and minerals', and he added two long chapters to its supplement. These chapters are very well documented: 'Historical description and chemical analysis of *Kali geniculatum* and Chemical analysis of *Kali geniculatum*'. These chapters may be regarded as genuine parts of a modern treatise on plant biochemistry, and particularly on the 'mineral nutrition' of halophytes.

The French contributor Venel (1765), who was responsible for the *sel marin* (marine salt) entry in the *Encyclopédie*, underlines the point that plants from seashores represent a natural resource for salt: 'some plants contain *perfect* (pure) *marine salt*, while others, *potassium*'. Finally, a simple circuit of marine salt is recognized. Salt passes through plants that are nourished by it and, with their decomposition, salt moves through the soil.

However, we found a much earlier record, most likely in a relation to a salt plant, in a Latin translation of the medical works of the famous Persian Muhammad al-Rāzī (Rhazes, Rasis – Latinized names, 854–925), from 1500 (Rhazis *et al.*, 1500). In a section of this translated work, *Synonima Rafis*, there is an entry on *cali cinis cribellatus* (filtered ash of c(k)ali) with *herba salsa* within its short explanation.

Pallas's iconic work from 1803 is an important moment in the evolution of concepts related to halophytes. His contribution, *Illustrationes plantarum imperfect vel nondum cognitarum, cum centuria iconum. De halophytis, seu plantis apetalis kalicis generatim*, is perhaps the first that introduces the term *Halophyta*. Despite his work being written in Latin, Pallas preferred the Greek-based term *Halophyta*. His choice could have been motivated by the fact that he included under this concept a distinct botanical family, in order to deal with these problematic species. He says that '[I] understand under this family species gathered based on natural affinity: Salicornia, Anabasis, Salsola, Suaeda, Polycnemum, Camphorosma which correspond today to Chenopodiaceae (or Amaranthaceae)'. The indissoluble relationship with the salinity factor is clearly delineated and underlined in the context of defining halophytes: 'plants that prefer salinity [...], living in salty and maritime environments, which are preferentially loved by these plants'. Therefore, Pallas created the term *Halophyta* for these striking halophytic chenopod species, mostly for taxonomic purposes, but also from an ecological perspective. Since the use of terms derived from *halophyta* had not been recorded in any modern languages before his work, we may consider that it was a starting point for creating *halophyta*-derived terms in French (*halophytes*), English (*halophytes*), German (*halophyten*), Romanian (*halofite*) and Danish (*halofyt*). However, the Greek etymology is clear and logical and we cannot exclude the possibility that it will have been foreseen by botanists before 1800. Pallas's contribution has been extensively discussed and commented on (Schrader, 1810) and, in this context, some authors refined the definition of *halophyta*. Schrader (1810) said in a footnote 'better to nominate this family as Halophyta, where to include all the plants that love the salt or soda'.

Before using actual names (derived from *halophyta* – as described by Pallas or elsewhere), other relevant terms have been used in different languages (Table 1.2); most suggest the influence of maritime (marine) salt on plants and, consequently, on their origin. As already underlined, even from the time of the French *Encyclopédie*, botanists realized that halophytes also grow in inland saline environments.

Therefore, the origin of the first references to halophytes and their subsequent refinements and extensions are of an ecological nature. This observation is very important because it can explain the logical evolution of concepts and the associated fields of research to be conducted by future botanists. Actually, this explains why many studies from the 19th century – especially conducted in the frame of German, Danish and French botany schools – emphasized the xeromorphic nature of many halophytes and its close correlation with the action of salt water or salt spray upon these plants.

Goethe made an interesting observation during his stay in Italy, in 1786. He later wrote:

> On the sea-coast I found also several plants whose characters similar to others I already knew, enabled me to recognize pretty well their properties. They are all alike, fat and strong-full of sap and clammy,- and it is evident that the old salt of the sandy soil, but still more the saline atmosphere gives them these properties. Like aquatic plants they abound in sap and are fleshy and tough like mountainous ones: those whose leaves shew a tendency to put forth prickles after the manlier of thistles have them extremely sharp and strong.
>
> (Goethe, 1885)

Goethe thus recognized the xeromorphic nature of halophyte adaptations and clearly indicated that salinity is responsible for characteristics like succulence.

Poiret (1820) described *plantes maritimes* or *salines* as plants that – without growing immersed in salt water and without floating on its surface – need to live near to salt water so they can absorb the necessary amount in their nutrition. They, like *Salicornia* species, grow in salt marshes and seem to absorb saline compounds through their roots and leaves. Others, like *Eryngium campestre*, do not require (need) salt water but live on seashore like others because they are robust enough to withstand the action of salt water. Poiret clearly suggested that some halophytic species are 'obligatory'; he foresaw that these kinds of species uptake salt water and it is necessary for their growth, while others are 'accidental': they do not need salt, but simply resist its action.

1.3 Halophytes and Physiological Drought in Soil

Halophytes have been perceived as a heterogeneous ecological group of plants, and are recognized and discussed as such today (Grigore, 2008, 2012; Grigore and Toma, 2010, 2011, 2017). The works of Battandier (1887) and Contejean (1855, 1874), Brick (1888 and references therein) demonstrate that this ecological vision of halophytes will have been delineating within plants, especially those growing on the seashore. However, under the great influence of Thurmann's work (1849) on the dispersion of plants under the influence of adjacent rocks, many botanists felt that – in relation to maritime flora – an ecological approach is logically needed in order to consider also the chemical nature of the soil with respect to halophyte distribution. Thus, Contejean (1874) first distinguished between *flore maritime* (terrestrial species that grow in free-air spaces and shores) and *flore marine* (species that grow exclusively immersed under saline waters – mostly the case of algae). He then emphasized that all halophytes contain consistent amounts of sodic salts and their existence is closely related to the presence of NaCl in the interior of continents, next to the sea and around salty efflorescences and saline sources. Further, he realized that within halophytes (maritime flora) the hygrophytes and xerophytes must be recognized. The first seek humidity and deep soils and the second seek drought and, consequently, superficial soils. He also subdivided hygrophytes into *péliques* (growing on a clay soil), *psammiques* (preferring sandy soils) and *pélopsammiques* (species growing in both clay and sandy environments).

Nevertheless, with respect to this point, we can consider that the halophytes have been clearly defined as an ecological group of plants, and plant ecologists such as Schimper and Warming have consequently refined this vision in terms of the adopted physiological drought concept (see Grigore and Toma, 2011).

This is the moment when halophytes – based on their anatomical and morphological features and characteristics of saline environments – started to be considered as a (special) type of xerophyte.

Schimper (1898, 1903) emphasized that, apart from the structural features of halophytes or xerophytes, environmental characteristics (especially of the water and soil) are also of great importance for establishing the nature of adaptations in plants. A distinction should, therefore, be made between physical and physiological dryness and between physical and physiological moistness. He pointed out that only the physiological characteristics need to be considered in plant life and in geographical botany. Hygrophilous vegetation corresponds to physiological moistness, and xerophylous vegetation to physiological dryness.

Schimper (1903) concluded that physiological drought is caused by external factors that either reduce water uptake or favour transpiration; frequently, there is a combination of these influences (Grigore and Toma, 2011, 2017; Grigore *et al.*, 2014). Among the factors that reduce water uptake, an abundance of soluble salts in the soil must be mentioned for the halophytes. As Schimper shows, a small quantity of salt favours absorption, while a large quantity impedes it. The degree of concentration at which the retardation commences varies with the species of plant, but rarely exceeds 0.5%. Mixtures of salts impede absorption more than pure salts, and certain types (for example NaCl) act more energetically than others.

The most important natural regions and habitats where physiological dryness prevails and only xerophytes therefore thrive are grouped according to their physical characters:

> 1. Deserts, steppes, and other districts with a dry substratum and dry air, occasional or persistent great heat, and intense illumination [Fig. 1.1];

Fig. 1.1. *Zygophyllum cornutum*; xerophilous structure; hot, dry, desert climate (Engler, 1896).

> 2. The bark of trees, rocks where there is rapid drying up of the substratum, owing to deficient depth [Fig. 1.2];
> 3. Sandy soil, gravel, and the like, on account of the rapid drying up of the substratum owing to its great permeability;
> 4. Sea-shores, solfataras, which have an abundance of soluble salts in the soil [Fig. 1.3];
> 5. Peat-bogs, because of the humic acids in the soil;
> 6. Polar zones, vicinity of glaciers in high mountains, where the temperature of the soil is low;
> 7. Alpine highlands, which are under rarefied air and strong insolation characteristic of the alpine climate.
>
> (Schimper, 1903)

All these types of habitats have been mentioned to better understand how one common factor (physiological drought) found in various environments can induce similar xeromorphic traits in plants growing in habitats that seem to be so different (Grigore *et al.*, 2014). Based on these details, together with some experiments aimed to show plant behaviour in saline conditions, Schimper concluded that the morphological characteristics of halophytes are consistent with those of xerophytes.

Therefore, it would be correct to consider a halophyte to be a special form of xerophyte (Wiesner, 1889; Warming, 1909; Grigore and Toma, 2010, 2011). This statement is supported both by analysis of habitat characteristics (physiological drought-affected), and especially by histo-anatomical features of halophytes. In this context, it seems that considering halophytes as a distinct ecological group (in relation to xerophytes and hydrophytes) can be questioned (Kearney, 1904). Kearney (1904) believed that even halophytes should be regarded as a subdivision of xerophytes (referring especially to succulence, Kearney considered that, in halophytes, this feature is of a xerophytic nature). He stated that, in using the term 'halophyte', it should not be forgotten that halophytes do not constitute an ecological class coordinated with hydrophytes and xerophytes, but are properly only a subdivision of the xerophytes. Like other xerophytes, they are generally characterized by having various adaptations for reducing transpiration, as promoters of physiological drought (Warming 1895, 1897; Schimper, 1903) have pointed out. Kearney (1904) underlined that halophytes are xerophytes owing to their need for a reduction in transpiration largely owing to the presence in the soil of an excessive amount of readily soluble salts. Despite Kearney believing that the character of the soil is the final criterion governing whether the vegetation should be called halophytic or non-halophytic, halophytes and xerophytes may actually occupy similar habitats. According to Kearney (1904), no ecological characters can be cited as exclusively halophytic, although halophytes seem to show a more marked tendency than other xerophytes towards development of water-storage tissue. The well-marked succulence in halophytes has been discussed in detail in their ecological context (Grigore, 2008; Grigore *et al.*, 2014; Grigore and Toma, 2017). In halophytes, the xerophytic habit is often developed

Fig. 1.2. Xerophilous structure – dry substratum of tree bark and rocks. *Octomeria* ssp. (left) and *Cattleya bicolor* (right) (Schimper, 1903).

to an extreme; this may be explained by the fact that, in saline environments, a combination of stressing factors occur, and drought and high salinity in the soil are synergic factors that induce the xeromorphic character of anatomical adaptations in halophytes.

Henslow (1895), in his important work on plant adaptations to living environments, also stressed the similarities between the anatomical characteristics of halophytes and xerophytes. Particularly taking into account plant succulence and pubescence, he concluded that there are certain peculiarities in common between desert plants and alpine and sub-alpine species. Henslow (1895) believed that resemblances in plants from widely different localities such as seashores, alpine regions and desert areas can be explained by the fact that certain elements in the environments are found to excess. Through the simplest deduction, he stated that these elements act upon the plants, and they respond accordingly, so that much the same results occur. For instance, the succulence of desert plants, especially where the soil is saline, arises from identically the same cause as that of plants from seashores. He concluded correctly that, since the moisture of the air and soil is charged with saline

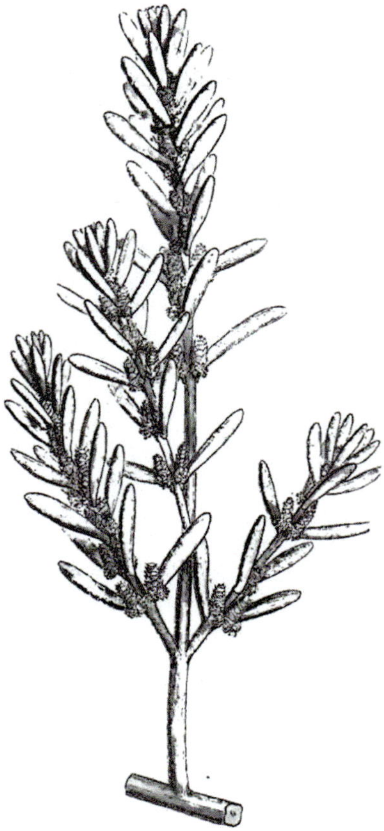

Fig. 1.3. *Batis maritima*; xerophilous structure; wet, saline soil of tropical shores. Plant with succulent leaves (Dammer, 1893).

matter in maritime regions and salt marshes, the presence of salts is to some extent responsible for succulence in many halophytes.

This concept of physiological dryness (drought) was further adopted by several researchers who studied plant ecology. McDougall (1941) described halophytes as 'plants of habitats that are physiologically dry'. Plants that grow in saline soil or in salty water are called halophytes, and they are strikingly xeric. Perhaps the most characteristic feature of halophytes as a group is their succulence, which is accompanied by very high osmotic pressure. Oosting (1948) defined halophytes as plants that can tolerate the concentrations of salts found in saline soils. If not actually dry, these saline habitats may be termed 'physiologically dry' because of the high concentrations of salts, which limit osmotic activity and, consequently, absorption of water by ordinary plants.

The introduction and operational use of physiological drought complicated the ecological classification of plants, and especially of halophytes, to some extent (Warming, 1895, 1909), Schimper (1898, 1903) and Clements (1920).

A typical example relates to the iconic works of the Danish pioneer of plant ecology, Warming, who in 1895 published a book on plant ecology in Danish. This was subsequently translated into other languages: German (1896, 1902, 1914) and English (1909). These translations are also *revised* editions of the original book. This is a very important point, because the evolution of ecological classifications of plants can be observed during the period. The 'volatile' position of halophytes is also noticeable within these editions. Thus, until the 1909 English edition, the classification system in ecological classes (*vereinsklasse, formationklasse* in German; *(o)ecological class* or *classes* in English – terms used here to mean an ecological group of plants – contained the following plant communities: hydrophytes, xerophytes, halophytes and mesophytes. Schimper (1903) divided the plants into terrestrial and aquatic, with terrestrial plants including xerophytes, hygrophytes and tropophytes.

In the English (and comprehensively revised) edition, Warming (1909) emphasized the role of physiological drought when choosing the place of halophytes within the xerophyte group. The soil is physiologically dry when it contains a considerable amount of water which, nevertheless, is available to the plant only to a slight extent or can be absorbed only with difficulty, either because the soil holds firmly to a large quantity of water or because the osmotic force of the root is inadequate to overcome that of the concentrated salt solution in the soil. This may be the case when:

1. The soil is rich in free humic acids, or in chemical bodies that by their peculiar action on the plant evoke xerophily; there result those formations that grow on acid soil;
2. The soil is rich in soluble salts, usually common salt, which brings into existence the form of xerophily we see in halophilous formations. Consequently, a halophyte is, in fact, a special form of xerophyte.

Based on these considerations, Warming (1909) included halophytes in the class subordinated to formations where the soil is physiologically dry (the soil contains water which is available to the plant only to a slight extent). This class is, therefore, composed of xerophilous species. Next to the halophytes he placed oxylophytes (formations on acid soil) and psychrophytes (formations on cold soil).

The soil can be also physically dry, and its slight power of retaining water determines the vegetation, the climate being of secondary import; the formations are therefore likewise xerophilous.

It can be seen, as emphasized in the paragraphs above, that in contrast to previous editions of Warming's work, halophytes are not still treated as a principal group of the plant in the English translation (1909). This distinction in plant classification into ecological groups was motivated explicitly by Warming, in the foreword to the English edition (1909), having taken into account the soil and climatic differences involved in shaping plant adaptations.

Clements (1907) included halophytes in the category of xerophytes growing in conditions where the chresard is low, the soil is physically or physiologically dry, the climate is usually dry or variable and transpiration is low. Halophytes grow in environments where the chresard is low because of an excess of soil salts.

1.4 Conclusion

Semantic analysis of the works consulted revealed the natural etymology of terms related to halophytes. Many of the associated terms have origins in Latin and Greek. Halophytes have been correctly defined and described as an ecological group of plants since the early years of their formal study. The specific characteristic of saline environments – those affected by physiological drought – was a fair criterion for considering halophytes as a special set of xerophytes. Further investigation and critical analysis of literature will contribute to identifying new criteria for the definition (and especially the classification) of halophytes; biochemical and molecular investigations should clarify if there is the possibility of clustering them according to their ability to synthesize and accumulate specific biochemical compounds in relation to soil salinity. In parallel, literature surveys and textual analysis are expected to bring new and challenging data about cultural and historical implications of salt from an ecological perspective.

References

Adam, P. (1990) *Saltmarsh Ecology*. Cambridge University Press, Cambridge, New York.

Alexandrescu-Dersca Bulgaru, M.M. (1976) *Călători străini despre Țările Române, vol. VI, Partea I. Paul de Alep*, Edit. Științifică și Enciclopedică, București.

Arber, A. (1953) *Herbals: Their Origin and Evolution: A Chapter in the History of Botany, 1470–1670*, 2nd edn. Cambridge University Press, Cambridge.

Aronson, J. and Le Floc'H, E. (1996) Restoration ecology of salt-affected, arid and semi-arid lands. In: Choukr-Allah, R., Malcolm, C.V. and Hamdy, A. (eds) *Halophytes and Biosaline Agriculture*. Marcel Dekker, Inc., New York, pp. 55–72.

Balfour, J.H. (1862) *Outlines of Botany: Designed for Schools and Colleges*. 2nd edn. Adam and Charles Black, North Bridge, Edinburgh.

Barnes, C.R. (1898) *Plant Life Considered with Special Reference to Form and Function*. Henry Colt Company, New York.

Battandier, J.A. (1887) Quelques mots sur les causes de la localisation des espèces d'une région. *Buletin de la Societé Botanique de France* 34, 189–195.

Bischoff, G.W. (1833) *Handbuch der botanischen Terminologie und Systemkunde*. Vol 1. Verlag von Johann Leonard Schrag, Nürnberg, Germany.

Boerhaave, H. (1724) *Institutiones et experimenta chemiae*. Tomus secundus. Paris.

Bortschof, E. (1859) Compte rendu général sur les résultats botaniques, obtenus pendant un voyage dans les régions Aralo-Caspiennes en 1857 et 1858. *Bulletin de la classe physico-mathématique de l'Académie Impériale des Sciences de Saint-Pétersbourg* 17(30), 471–479.

Bower, F.O. (1911) *Plant-Life on Land Considered in Some of its Biological Aspects*. University Press, Cambridge, UK.

Brândza, D. (1880) *Despre vegetațiunea României și exploratorii ei. Date asupra climei și a regiunilor botanice. Discurs de recepție în Academia Română*. Tipografia Academiei Române, București.

Braun-Blanquet, J. (1932) *Plant Sociology. The Study of Plant Communities*. 1st edn. George D. Fuller and Henry S. Conard, USA.

Breckle, S.-W. (1995) How do halophytes overcome salinity? In: Khan, M.A. and Ungar, I.A. (eds) *Biology of Salt Tolerant Plants*. Department of Botany, University of Karachi, Pakistan, pp. 199–213.

Brick, C. (1888) Beiträge zur biologie und vergleichenden anatomie der baltischen strandpflanzen. *Schriften der Naturforschenden Gesselschaft in Danzig, neue folge* 7(1–2), 108–156.

Brown, R. (1874) *A Manual of Botany: Anatomical and Physiological for the Use of Students*. William Blackwood and Sons, Edinburgh.

Brückmann, F.E. (1749) *Centuria secunda Epistolarum itinerariarum accedit Museum Closterianum*, Wolffenbüttelae.

Bucur, N., Dobrescu, C., Turcu, Gh., Lixandru, Gh., Teşu, C. *et al.* (1957) Contribuţii la studiul halofiliei plantelor din păşuni şi fâneţe de sărătură din Depresiunea Jijia-Bahlui (partea a I-a). *Studii şi Cercetări (Biologie şi Ştiinţe Agricole), Academia R. P. Române., filiala Iaşi* 8(2), 277–317.

Chapman, V.J. (1942) The new perspective in the halophytes. *The Quarterly Review of Biology* 17(4), 291–311.

Chapman, V.J. (1960) *Salt Marshes and Salt Deserts of the World*. Plant Science Monographs. Interscience, Inc., New York.

Clements, F.E. (1907) *Plant Physiology and Ecology*. Henry Holt and Company, New York.

Clements, F.E. (1920) *Plant Indicators. The Relation of Plant Communities to Process and Practice*. Camegie Institution of Washington, Washington, DC.

Conlon, P.M. (1991) *Le siècle des Lumières: Bibliographie Chronologique, Vol. 8 (1757–1760)*. Librairie Droz S.A., Geneva, Switzerland.

Contejean, C. (1855) Remarques sur la dispersion des plantes vasculaires relativement aux roches subjacentes dans les environs de Montbéliard. *Actes de la Societé Helvétiques des Sciences Naturelles*. Session 38, 189–201.

Contejean, C. (1874) De l'influence du terrain sur la végétation. *Annales des Sciences Naturelles, 5-ème sér., Botanique* 20, 266–304.

Correll, D.S. and Johnston, M.C. (1970) *Manual of the Vascular Plants of Texas*. Texas Research Foundation, Renner, Texas.

Crozier, A.A. (1892) *A Dictionary of Botanical Terms*. Henry Holt and Company, New York.

Dagar, J.C. (1995) Characteristics of halophytic vegetation in India. In: Khan, M.A. and Ungar, I.A. (eds) *Biology of Salt Tolerant Plants*. Department of Botany, University of Karachi, Karachi, Pakistan, pp. 255–276.

Dammer, U. (1893) Batidaceae. In: Engler, A. and Prantl, K. (eds) *Die natürlichen pflanzenfamilien*. Verlag von Wilhelm Engelmann, Leipzig, Germany. 3(1a), pp. 118–120.
Dansereau, P. (1957) *Biogeography, an Ecological Perspective*. Ronald Press, New York.
de Haller, A. (1776) *Description courte et abrégée des salines du gouvernement d'Aigle*. De l'imprimerie de la Soc. Litt. & Typog, Verdon.
de Jussieu, M. (1717) Histoire du Kali d'Alicante. *Memoires de Mathematique et Physique, Academie Royale des Sciences*, 73–78.
de la Peyronie, G. (1793) *Voyages de M. P. S. Pallas en différentes provinces de l'Empire de Russie, et dans l'Asie septentrionale, tome troisième*. Maradan, Paris.
du Monceau, D. (1755) *Traité des arbres et arbustes que se cultivent en France en pleine terre*. H.L. Guerin & L.F. Delatour, Paris.
Duncan, W.H. (1974) Vascular halophytes of the Atlantic and Gulf coasts of North America North Mexico. In: Reimold, R.J. and Queen, W.H. (eds) *Ecology of Halophytes*. Academic Press, New York, pp. 23–50.
Engler, A. (1896) Zygophyllaceae. In: Engler, A. and Prantl, K. (eds) *Die natürlichen pflanzenfamilien*. Verlag von Wilhelm Engelmann, Leipzig, Germany. 3(4), pp. 74–93.
Fernald, M.L. (1950) *Gray's Manual of Botany*. 8th edn. American Book Company, New York.
Fitter, A.H. and Hay, R.K.M. (2002) *Environmental Physiology of Plants*. 3rd edn. Academic Press, San Diego, San Francisco, New York.
Flowers, T.J. and Colmer, T.D. (2008) Salinity tolerance in halophytes. *New Phytologist* 179, 945–963.
Flowers, T.J., Hajibagheri, M.A. and Clipson, N.J.W. (1986) Halophytes. *The Quarterly Review of Biology* 61, 313–337.
Fowler, R. (1875) *The Medical Vocabulary Containing a Concise Explanation of the Terms Used in Medicine and its Accessory Sciences*. Henry Renshaw, London.
Gadaceau, E. (1881) Note sur la découverte de deux plantes salicoles spontanées a Soulvache (Loire-Inférieure). Excursion botanique de Rougé a Basin de Brétagne. *Bulletin de la Societé Botanique de France* 28, 250–256.
Ganong, W.F. (1913) *The Living Plant. A Description and Interpretation of its Functions and Structure*. Henry Holt and Company, New York.
Georgi, I.G. (1777) De natro ruthenico observationes. *Acta Academiae Scientiarum Imperislis Petropolitanae* 1, 197–212.
Göbel, F. (1838) *Reise in die Steppen des südlichen Russlands*. Vol. 2. Verlage bei C.A. Kluge, Dorpat.
Goethe, J.W. (1885) *Goethe's Travels in Italy (translated from the German)*. George Bell and Sons, London.
Grecescu, D. (1898) *Conspectul florei României*, Tipografia Dreptatea, București.
Green, J.R. (1914) *A History of Botany in the United Kingdom from the Earliest Times to the End of the 19th Century*. J.M. Dent & Sons Ltd, E.P. Dutton & Company, New York.
Grigore, M.-N. (2008) *Introducere în Halofitologie. Elemente de anatomie integrativă*. Edit. Pim, Iași, Romania.
Grigore, M.-N. (2012) *Romanian Salt Tolerant Plants. Taxonomy and Ecology*. Edit. Tehnopress, Iasi, Romania.
Grigore, M.-N. (2016) Rediscovering the first monograph on plant anatomy - *Anatome Plantarum* (1675) by Marcello Malpighi. *The Biologist (Lima)* 14 (2), 155–170.
Grigore, M.-N. and Toma, C. (2010) *Halofitele. Aspecte de anatomie ecologică*. Editura Universității Alexandru Ioan Cuza. Iași, Romania.
Grigore, M.-N. and Toma, C. (2011) Halofitele, o categorie ecologică polimorfă. Între seceta fiziologică a solului și stresul salin. *Revista Botanică (Chișinău)* 2(3), 38–46.
Grigore, M.-N. and Toma, C. (2017) *Anatomical Adaptations of Halophytes. A Review of Classic Literature and Recent Findings*. Springer, Cham, Swizerland.
Grigore, M.-N., Toma, C. and Boscaiu, M. (2010) Dealing with halophytes: an old problem, the same continuous exciting challenge. *Analele Științifice ale Universității „Alexandru Ioan Cuza" din Iași, s. II.a (Biologie Vegetală)* 56 (1), 21–32.
Grigore, M.-N., Ivănescu, L. and Toma C. (2014) *Halophytes. An Integrative Anatomical Study*. Springer, Cham, Switzerland.
Grisebach, A.H.R. (1864) *Flora of the British West Indian Islands*. Lovell Reeve & Co., London.
Grisebach, A. (1875) *La végétation du Globe*. Tome 1. L. Guérin et Cie, Paris.
Hardy, G. and Totelin, L. (2016) *Ancient Botany*. Routledge, London.
Henckel, J.F. (1722) *Flora saturnizans*. Verlegts Johann Christian Martini, Leipzig, Germany.
Henckel, J.F. (1755) *Flora saturnizans*. Grossischen Handlung, Leipzig, Germany.
Henckel, J.F. (1760) *Pyritologie, ou histoire naturelle de la pyrite, on y a joint le Flora saturnisans*. J.-T. Herissant, Paris.

Henslow, G. (1895) *The origin of plant-structures by self-adaptation to the environment*. Kegan Paul, Trench, Trubner, & Co Ltd, London.

Hoblyn, R.D. (1865) *A Dictionary of Terms Used in Medicine and the Collateral Sciences*. Henry C. Lea, Philadelphia, PA, USA.

Hochheimer, C.F.A. (1793) *Sammlung auserlesener Abhandlungen über die interessantesten Gegenstände der Chemie*. A.F. Böhme, Leipzig, Germany.

Holzapfel, C. (2009) Deserts. In: Jørgensen, S.V. (ed.) *Ecosystem Ecology*. Elsevier B.V., Amsterdam, the Netherlands, pp. 222–240.

Hutchings, D. (1969) *Late Seventeenth Century Scientists*. Pergamon Press, Oxford, UK.

Ingrouille, M. (1992) *Diversity and Evolution of Land Plants*. Chapman & Hall, London, New York.

Ingrouille, M. and Eddie, B. (2006) *Plants. Diversity and Evolution*. Cambridge University Press, Cambridge, UK.

Isley, D. (1994) *One Hundred and One Botanists*. Iowa State University Press, Ames, IA, USA.

Jaucourt, D. (1765) In: *Encyclopedie, ou Dictionnaire raisonné des sciences, des arts et des métieres*. Vols 9, 14, 15. Samuel Faulche & Compagnie, Neufchastel.

Kearney, T.H. (1904) Are plants of sea and dunes true halophytes? *Botanical Gazette* 37, 424–436.

Khan, M.A. and Duke, N.C. (2001) Halophytes – a resource for the future. *Wetland Ecology Management* 6, 455–456.

Khell, J. (1755) *Physica ex recentiorum observationibus accomodata usibus academicis, t.2, editio altera*. Officina Trattneriana, Vienna.

Koyro, H.-W., Geissler, N., Hussin, S. and Huchzermeyer, B. (2008) Life strategies of halophytes – the long way from system ecology, whole plant physiology, cell biochemistry and molecular aspects back to sustainable utilization at field sites. In: Abdely, D., Ozturk, M., Ashraf, M. and Grignon, F.C. (eds) *Biosaline Agriculture and High Salinity Tolerance*. Birkhäuser, Basel, Boston, Berlin, pp. 1–20.

Koyro, H.-W., Geissler, N. and Hussin, S. (2009) Survival at extreme locations: life strategies of halophytes. In: Ashraf, M., Ozturk, M. and Athar, H.R. (eds) *Salinity and Water Stress. Improving Crop Efficiency*. Springer, New York, pp. 167–177.

Lawrence, G.H.M. (1951) *Taxonomy of Vascular Plants*. MacMilian Co., New York.

Leonhard, P.C. (1753) *De novo aquae salsae fonte detecto et experimentis confirmato*. Typis Pavli Cristoph. Hageri, Gottingae.

Macquer, M. (1756) *Elemens de chymie-théorique*. P.F. Didot le Jeune, Paris.

Macquer, M. (1778) *Dictionnaire de chymie contenant la théorie et la pratique de cette science*. Tome 1. P.F. Didot le Jeune, Paris.

Marschner, H. (1995) *Mineral Nutrition of Higher Plants*. 2nd edn. Academic Press, Amsterdam, The Netherlands.

McDougall, W.B. (1941) *Plant Ecology*. 3rd edn. Lea & Febiger, Philadelphia, PA, USA.

Meyen, M. (1840) Recherches chimiques sur les substances alcalines continues dans les plantes aux diverses periods de leur accroissement, par M.F. Goebel. *Annales des Sciences Naturelles, seconde série* 14, 162–164.

Miller, P. (1788) *Dictionnaire des jardiniers et des cultivateurs*. Volume 4. Benoit le Franco, Brussels.

Mooney, H.A. and Canadell, J.G. (2002) The earth system: biological and ecological dimensions of global environmental change. In: Munn, T. (ed.) *Encyclopedia of Global Environmental Change*. Wiley & Sons, Chichester, UK, p. 339.

Nemnich, P.A. (1798) *Wörterbücher der Naturgeschichte*. Licentiat Nemnich und Leipzig bey Adam Friedrich Bökme, Hamburg, Germany.

Ness, B.D. (2003) Magill's encyclopedia of science. In: Moose, C.J. (ed.) *Plant Life, Vol. 2: DNA Replication-Metabolites: Primary vs. Secondary*. Salern Press, Inc., Pasadena, CA, USA, Hackkensack, NJ, USA.

Oeder, G.C. (1764) *Elementa Botanicae*. Typis Fratrum Philibert, Hafinae, p.1.

Oosting, H.J. (1948) *The study of Plant Communities*. W.H. Freeman and Co., San Francisco, CA, USA.

Pallas, P.S. (1803) *Illustrationes plantarum imperfecte vel nondum cognitarum, cum centuria iconum. De Halophytis, seu plantis apetalis kalicis generatim*. Sumtibus Godofredi Martini, Lipsiae, Germany.

Poiret, J.L.M. (1820) *Géographie Botanique. Dictionnaire des Sciences Naturelles*. Vol. 18. F.G. Levrault, Strasbourg, Le Normant, Paris, pp. 359–439.

Poljakoff-Mayber, A. and Lerner, H.R. (1999) Plants in saline environments. In: Pessarakli, M. (ed.) *Handbook of Plant and Crop Stress*. 2nd edn. Marcel Dekker, Inc., New York, Basel, pp. 125–152.

Polunin, N. (1960) *Introduction to Plant Geography*. McGraw-Hill Book Company, Inc., New York.

Prevost d'Exile, A.F. (1777) *Histoire générale des voyages*. Tome 22. E. van Harrevelt, D.J. Changuion, Amsterdam.

Quinn, J.A. (2009) *Desert biomes*. In: Woodward S.L. (ed.) *Greenwood Guides to Biomes of the World*. Greenwood Press, Westport, CT, USA.

Rhazis, M. *et al*. (1500) *Opera Medica*. Johannes Hamman.

Roux, A. (1765) *Dictionnaire domestique portatif, tome troisième*. Vincent, Paris.

Rozema, J. (1996) Biology of halophytes. In: Choukr-Allah, R., Malcolm, C.V. and Hamdy, A. (eds) *Halophytes and Biosaline Agriculture*. Marcel Dekker, Inc., New York, pp. 17–30.

Rozier, A. (1781) *Cours complet d'agriculture théorique, pratique, économique, et de médicine rurale et vétérinaire*. Tome 1. Rue et Hôtel Serpente, Paris.

Schimper, A.F.W. (1898) *Pflanzengeographie auf physiologischer Grundlage*. G. Fischer Verlag, Jena, Germany.

Schimper, A.F.W. (1903) *Plant Geography Upon a Physiological Basis*. Clarendon Press, Oxford, UK.

Schrader, H.A. (1810) *De Halophytis Pallasii, respectu imprimis ad Salsolam et Suaedam habito commentatio*. Henricum Dietrich, Gottingae.

Schulze, E.-D., Beck, E. and Müller-Hohenstein, K. (2005) *Plant Ecology*. Springer Verlag, Berlin, Heidelberg, Germany.

Sharma, S.K. and Gupta, I.C. (1986) *Saline Environment and Plant Growth*. Agro Botanical Publishers, Bilkaner, India.

Short, T. (1746) *Medicina Britannica*. R. Manby and H. Shute Cox, London.

Stevens, P.F. (1994) *The Development of Biological Systematics: Antoine-Laurent de Jussieu, Nature, and the Natural System*. Columbia University Press, New York.

Stocker, O. (1928) Das Halophytenproblem. *Ergebnisse der Biologie* 3, 265–353.

Sullivan, K. and O'Reilly, J.P. (1863) *Notes on the Geology and Mineralogy of the Spanish Provinces of Santander and Madrid*. Williams and Norgate, London.

Thurmann, J. (1849) *Essai de phytostatique appliqué à la chaine du Jura et aux contrées voisines*. Jent et Gassmann, Libraires, Berne.

Tournefort, J.P. (1741) *A Voyage into the Levant: Performed by Command of the Late French King*. Vol. 3. D. Midwinter, London.

Venel, G.-F. (1765) In: *Encyclopedie, ou Dictionnaire raisonné des sciences, des arts et des métieres, vol.* 14, Samuel Faulche & Compagnie, Neufchastel.

Von Hellwald, F. (1874) *The Russians in Central Asia, a Critical Examination of the Geography Down to the Present Time of the Geography and History of Central Asia*. Henry S. King & Co, London.

Von Lehmann, J.C. (1791) *Grundsätze der Mineralogie*. Varrentrapp und Wenner, Frankfurt am main, Germany.

Von Sachs, J. (1890) *History of Botany (1530–1860)*. Clarendon Press, Oxford, UK.

Waisel, Y. (1972) *Biology of Halophytes*. Academic Press, New York, London.

Warming, E. (1895) *Plantesamfund - Grundtræk af den økologiske Plantegeografi*. P.G. Philipsens Forlag, Kjøbenhavn.

Warming, E. (1896) *Lehrbuch der Ökologischen Pflanzengeographie - Eine Einführung in die Kenntnis der Pflanzenverenie*. Translated by Emil Knoblauch. Gebrüder Borntraeger, Berlin.

Warming, E. (1897) Halophyt-studier. *Det Kongelige Danske videnskabernes selskabs skrifter 6, Raekke, naturvidenskabeling og mathematisk afdeling* 8(4), 173–272.

Warming, E. (1902) *Lehrbuch der Ökologischen Pflanzengeographie - Eine Einführung in die Kenntnis der Pflanzenverenie*. Translated by Emil Knoblauch. Gebrüder Borntraeger, Berlin.

Warming, E. (1909) *Oecology of Plants. An Introduction to the Study of Plant-Communities*. Clarendon Press, Oxford, UK.

Warming, E. and Graebner, P. (1914) Eug. Warming's Lehrbuch der Ökologischen Pflanzengeographie. Gebrüder Borntraeger, Berlin.

Weber, D.J. (1995) Mechanism and reactions of halophytes to water and salt stress. In: Khan, M.A. and Ungar, I.A. (eds) *Biology of Salt Tolerant Plants*. Department of Botany, University of Karachi, Pakistan, pp. 170–182.

Wiesner, J. (1899) Uber die Formen der Anpassung der Blatter an die Lichtstarke. *Biologisches Centralblatt* 19, 1–14.

Willdenow, C.L. (1792) *Grundriss der Kräuterkunde*. Haude und Spener, Berlin.

Willdenow, C.L. (1805) *The Principles of Botany, and of Vegetable Physiology* (translated from the German). University Press, Edinburgh.

Willis, J.C. (1919) *A Dictionary of the Flowering Plants and Ferns*. 4th edn. Cambridge University Press, Cambridge, UK.

2 Habitats of Halophytes

Adam Kapler*
Polish Academy of Sciences – Botanical Garden in Powsin (PAS BG-CBDC), Warsaw, Poland; Centre for Wetland Conservation, Warsaw, Poland

Abstract
Salt-tolerant plants occur all over the world in a number of different ecosystems, ranging from pristine alkaline semi-deserts and mangrove forests; through semi-natural meadows and pastures; to man-made habitats such as the environs of graduation towers; over irrigated arable lands with poor drainage in the tropics; and to city lawns in the boreo-temperate zone polluted with NaCl and $CaCl_2$ during deicing. Natural habitats disappear because of urbanization, tourism and agriculture intensification. Since 1980 one-fifth of the Earth's mangrove biome has disappeared as well as more than one-half of alkaline steppes and nearly all Earth's coastal and inland salt meadows, glassworts and other annual communities of muds and sands, Mediterranean and warm Atlantic halophilous scrubs, vegetated sea cliffs and machairs. At the same time halophytes colonize new, man-made habitats, becoming dominant or even the sole species there. Some salt-resistant species, such as *Rhizophora mangle* in Hawaii and *Spartina anglica* in the UK, become dangerous invasive species. Mangrove swamps deserve more efficient conservation and restoration efforts since they shelter coasts from erosion, tsunami and storm surge; trap a wide variety of heavy metals; and provide habitats for juvenile fish, oysters and crustaceans. In the temperate and boreal zones the traditional land use of saline meadows and pastures needs to be maintained to preserve the original biodiversity and ecosystem services. Further halophyte domestication will lead to establishment of completely new, artificial agro-ecosystems to yield food, fodder and fuel, as well as fibre and phytoremediation, for rapidly expanding human populations. A range of halophyte crop cultivation systems can help to reduce damage caused by salinization of soils and freshwater, increase food production up to 70% by 2050 and combat coastal erosion in the era of climate change and global pollinator crisis. At the same time we need to eradicate some monospecific thickets built by invasive, alien halophytes to restore primeval, species-rich communities in areas of naturally high salinity.

Keywords: Halophyte-based agriculture; Halophytes; Mangroves; Ruderal communities; Salt deserts; Salt marshes; Salt steppes; Sargasso Sea

I can only compare these great aquatic forests... with the terrestrial ones in the intertropical regions. Yet if in any country a forest was destroyed, I do not believe so nearly so many species of animals would perish as would here, from the destruction of kelp. Amidst the leaves of this plant numerous species of fish live, which nowhere else could find food or shelter; with their destruction the many cormorants and other fishing birds, the otters, seals and porpoise,

* Corresponding author e-mail: adam.kapler@obpan.pl

> would soon perish also; and lastly, the Fuegian... would...decrease in numbers and perhaps cease to exist.
>
> (Darwin, c. 1909)

2.1 Introduction

Halophytes occur all over the world, often possessing pantropical or cosmopolitan distribution, in a number of different ecosystems ranging from pristine alkaline semi-deserts, steppes and mangrove forests; through half-natural pastures and meadows; to man-made habitats such as the environs of graduation towers. A good example of a halophyte with an extremely wide ecological niche is *Spartina anglica*. It dominates marshes, grows in irrigated arable lands with poor drainage in the tropics, and in city lawns in the boreo-temperate zone that are contaminated by salts during periods when the roads are cleared of ice and snow.

In the 20th century kelp forests, mangroves, salt marshes and alkali steppes – previously treated like wasteland, a kind of no-man's-land – have attracted some attention and respect due to their ability to stabilize shorelines and protect coastal communities. Although this paradigm has recently been called into question by small-scale experiments, the shoreline safeguarding theory still stands, but gaps still remain in our knowledge of the context-dependent aspects of coastline protection mentioned above (Gedan *et al.*, 2011; Millennium Ecosystem Assessment, 2005).

2.1.1 Mangroves

The mangrove forest (mangal) is a distinct environment found in tropical and subtropical tidal areas (shorelines and estuaries) chiefly between latitudes 25° N and 25° S, in nearly 120 countries and territories. It is found where fine, usually organic sediments gather and plants developed special adaptations to the stressful marine conditions: root anoxia, fluctuating but high salinity (from 1% up to 9%) and intertidal growth (Teas, 1983; Giri *et al.*, 2011; Spalding *et al.*, 2010; Tomlinson, 2016; Fig. 2.1). Such evolutionary traits remain uncommon in tropical flora, thus mangrove forests are dominated by only a few tree species, even in New Guinea, Malaysia or the Caribbean. This contrasts with inland rainforests composed of thousands of woody species from hundreds of families (Hogarth, 1999). True mangroves (species from Table 2.1 and around 50 other species in 20 genera from 16 families) originated in the Paleogene from inland ancestors (Plaziat, 2001;

Fig. 2.1. Mangrove forests on the eastern coast of Australia (photo courtesy of B. Sudnik-Wójcikowska).

Table 2.1. Major components of mangrove forests.

Family	Genus	Number of species	Vernacular name
Rhizophoraceae	*Bruguiera*	7	Red mangrove
	Ceriops	5	
	Kandelia	2	
	Rhizophora	8	
Avicenniaceae	*Avicennia*	9	Black mangrove
Lythraceae	*Sonneratia*	5	Mangrove apple
Combretaceae	*Conocarpus*	1	White mangrove, buttonwood
	Laguncularia	1	
	Lumnitzera	3	
Meliaceae	*Xylocarpus*	1	Cannonball mangrove
Arecaceae	*Nypa*	1	Mangrove palm

Ricklefs *et al.*, 2006) and are good examples of convergent evolution to solve the same ecophysiological challenges (for more details see Table 2.2). True mangrove forests occupy two distinct, hemispheric regions: 'the Old World'/eastern hemisphere including South-east Asia, India, eastern Africa and Australia; and the 'New World'/ western hemisphere including western Africa, the Caribbean, Florida and Central America, as well as the Pacific coasts of North and South America. Mangroves are more abundant and diverse in the Old World than in the New World (Tomlinson, 2016).

Mangroves protect the coast from tsunami, storm surge and erosion (Dahdouh-Guebas *et al.*, 2005). They form indispensable shelter for algae and animals, especially those requiring a hard surface for anchoring (Skov and Hartnoll, 2002). Mangrove trees themselves have numerous uses in the traditional economy and medicine (Węglarski and Węglarska, 2008). Nearly 35% of the world's mangrove swamps have disappeared through man-made habitat loss, chiefly shrimp farming and extension of tourist infrastructure (Bowen *et al.*, 2001; Hamilton, 2013). Fortunately, the swift rate of this loss has decreased since 2000 (Hamilton, 2013).

Table 2.2. Mangrove ecophysiological adaptations to extreme conditions.

Extreme condition	Adaptation
Salt intake	Suberized roots, storing salt in vacuoles
Low oxygen availability	Pneumatophore roots
Fresh water loss	Closing stomata, changing the orientation of the leaves, growing new leaves after rainstorm
Limited nutrient uptake	Pneumatophore roots
Threats to offspring	Vivipary within the fruit (*Avicennia*, *Aegialitis*) or through the fruit (*Nypa*, *Rhizophora*, *Ceriops*)

2.1.2 Sand and cliff shorelines in the tropics

Tropical beaches subjected to tidal waves, with a sandy, weakly compact and very unstable substrate, grow by 'pes-caprae' formation, and plants often form vast mats on dunes just above the highest tide line (Fig. 2.2). Its main components are perennial herbs and non-woody, trailing vines, chiefly *Ipomoea pes-caprae* (which gives its name to the entire ecosystem, and incidentally is a famous antidote to jellyfish, sea urchins and fish venoms), *Vignea* and *Canavalia*, as well as the succulent halophyte *Sesuvium portulacastrum* and thorny, hedgehog-like *Spinifex* grasses. All these plant species have pantropical (nearly cosmopolitan) ranges, and are widely dispersed by waves, sea currents, birds and fish (Devall, 1992). Most of these species are not only halophytes but also calcium tolerant, because tropical sands are from corals, coral-like red algae and seashells, not from the siliceous rocks like the sands in boreal and Atlantic zones (Węglarski and Węglarska, 2008).

Fig. 2.2. The *Ipomoea pes-caprae* community in Laguna de Cayuca, west coast of Mexico (photo courtesy of B. Sudnik-Wójcikowska).

More stabilized, 'grey' dunes and rocky cliffs have more diverse vegetation, composed of lianas (wood vines), epilithic forbs and small trees with small epiphytes. Most of them are adapted to extremely dry and barren soils and have many features of succulents including narrow, leathery, evergreen leaves (e.g. *Cycas*, *Dracaena*, *Pandanus*, etc.), reduced leaves and photosynthesis carried by stems (e.g. *Casuarina*; succulent, cacti-like spurges).

The above plant communities are not as important and well studied as mangals but carry out similar ecologic services, protect the beach from erosion, and have many uses in traditional medicine and economies (Węglarski and Węglarska, 2008).

2.1.3 Salt deserts and semi-deserts

Deserts are defined as barren areas of landscape where little or no precipitation occurs, and they have limited or no vegetation. About one-third of the land surface of the world is arid or semi-arid, often in the form of salted deserts and semi-deserts. Many former prairies, evergreen oakwoods in the USA and savannah in the Sahel zone changed into semi-deserts as a result of bad irrigation. Both in North America and in Central Asia the alcaline sunk vegetation is composed mostly of woody *Chenopodiaceae* such as:

- Shadscale (*Atriplex confertifolia*)
- Four-wing salt bush (*A. canescens*)
- All-scale (*A. polycarpa*)

in the Mojave Desert (the saltbush scrub). At higher elevations than the saltbush scrub, another similar community occurs: the shadscale scrub. This shares many plant species from the Goosefoot family, dominated by *A. confertifolia* and accompanied by:

- Hop-sage (*Grayia spinosa*)
- Winter-fat (*Krascheninnikovia lanata*)
- Inyo County star tulip (*Calochortus excavatus*)

While a dry lake is itself typically devoid of vegetation, it is commonly ringed by shadscale, saltbush and other salt-tolerant plants that provide critical winter fodder for livestock and other herbivores (Knapp, 1965).

2.1.4 Sargasso Sea

The Sargasso Sea is the only sea not adjacent to any land. It is bounded by four currents (the North Atlantic Current in the north, the Canary Current in the east, the North Atlantic Equatorial Current in the south and the Gulf Stream in the west). Its unique littoral-like flora and fauna floating in pelagic waters can be called one of the weirdest, but hence most well-studied regions of the global ocean (Kingsford, 1995; Friedland *et al.*, 2007). It is named after the brown algae *Sargassum natans* and *S. fluitans*, which cover its surface and shelter numerous turtles, fish and invertebrates. The vast, sustained biomass of pelagic *Sargassum* was observed by early seafarers such as Himilco the Navigator (late 6th century BC) and Columbus (1492) (Murray, 1893; Cunliffe, 2017). It forms a 'floating, golden rainforest' in nutrient-poor waters and has been one of the most fascinating puzzles in world oceanography and ecology (Ryther, 1956). Very poor productivity and low growth rates of both *S. fluitans* and *S. natans* have been associated with high C:P and C:N ratios in oceanic populations in the oligotrophic gyre of the Sargasso Sea. These findings suggest that new production of *Sargassum* can occur in the littoral Gulf of Mexico, as well as in the north-west Atlantic, where mutualistic relationships with fishes contribute to significant N and P supplies (Lapointe *et al.*, 2014). Economically, the most important species are the larvae of American and European eels hatching in the sea and later migrating to estuaries and inland waters (Bonhommeau *et al.*, 2008). Another key animal spawning there is the loggerhead sea turtle (Putman *et al.*, 2012). The previously unknown, 'hidden' biodiversity of Sargasso Sea bacterioflora has been studied by Venter *et al.* (2004). The Sargasso Sea is threatened by a range of adverse human impacts. As an open ocean, it is part of the High Seas and beyond the jurisdiction of any national government, hence almost without protection. It has become more and more polluted and overfished. To secure its unique biodiversity, especially eel and turtle hatching places, the Sargasso Sea Alliance (under the leadership of the Government of Bermuda) was established in 2010 (Laffoley *et al.*, 2011).

2.1.5 Temperate and boreal zones

2.1.5.1 Mudflats (tidal and wind flats) and salt marshes

Mudflats (wind flats and tidal flats) are low grasslands growing on muds (marine algae and animal detritus, estuarine/river clays and silts) accumulated by sea tide and/or rivers, typical of sheltered areas (estuaries and lagoons) in boreal, temperate and subtropical climates. As mudflats occur within the intertidal zone, they are exposed and submerged around twice a day. In the past, tidal flats were considered unhealthy (malaria, tuberculosis, fluke infections), and economically unimportant 'wastelands'. This is why wind and tidal flats have been dredged and changed into agricultural land (polders). In fact the agriculture revolution, leading to the development of the modern crop rotation (which followed the earlier three-field system), began in the Wadden Sea, Netherlands (Roding and van Voss, 1996). Intertidal salt marshes (Fig. 2.3) and mudflats (tidal and wind flats; Fig. 2.4) are overgrown in Europe by *Salicornia europaea*, *Suaeda maritima*, *Spartina anglica*, *Juncus balticus* spp. *littoralis*, *Glaux maritima*, *Puccinellia fasciculata*, *Triglochin maritima*. At the Atlantic coasts of Northern America we find *Salicornia bigelovii*, *Aster tenuifolius*, *Puccinellia pumila*, *Plantago eriopoda*, *Solidago sempervirens*. Collectively, these are indispensable feeding and resting sites for myriads of migratory shorebirds travelling each year from their breeding sites in the northern hemisphere to moulting areas in the south (Knapp, 1965; Burger *et al.*, 1997). Although tidal flats and salt marshes are necessary to prevent coastal erosion and to safeguard the populations of fish, molluscs, crabs and birds, they are still disappearing worldwide because of dredging (for shipping purposes), land reclamation for development and chemical pollution (Murray *et al.*, 2014; Lazarus and Wszałek-Rożek, 2016). Recently, botanists have discovered the previously overlooked diversity of neoendemic, cryptic species that have recently evolved in these salt-affected habitats (Kadereit *et al.*, 2012; Bosiacka *et al.*, 2016).

2.1.5.2 Kelp forests and beds

Underwater areas of shallow, nutrient-rich sea water near the shore with a high density of kelp are known as kelp forests (larger areas) and kelp beds (smaller areas). Aside from coral reefs, the Sargasso Sea and mangrove kelp forests are the most productive and dynamic marine ecosystems on Earth. Large kelps occur in dense 'stands', resembling forests on land. They also provide food and shelter for a vast diversity of molluscs, crustaceans, fish, turtles, marine birds and mammal species. Dense canopies of brown algae depend on cool, nearshore waters where high levels of both nutrients and light are available. Kelp forests can change rapidly over time (Jackson and Winant, 1983; Dayton, 1985). Kelp forests not only provide many ecosystem services, but for more than 120 years have been recognized as 'natural laboratories' where we have learned about numerous ecological processes, mostly in trophic ecology. Those unique 'sea forests' still provoke important ideas (Steneck et al., 2002). The architecture of kelp forests and beds, based on their physical structure, is very complicated. It can include as many as three guilds of kelp, and two additional guilds occupied by green and red algae:

1. Sparsely aggregated, canopy kelps with the largest, dominant species extending up to the sea surface (chiefly *Alaria* and *Macrocystis*).
2. Densely aggregated, stipitate brown algae extending only few metres above the ocean floor (mostly *Ecklonia* and *Eisenia*).
3. Prostrate kelps lying along the ocean floor, such as *Laminaria*.

Fig. 2.3. Marshes on the North Sea coast, showing an *Artemisia maritima–Statice–Limonium* community (photo courtesy of B. Sudnik-Wójcikowska).

4. Layers 2 and 3 together form the understory canopy. Foliose and filamentous brown, red and green algae and coral-like red algae known as corallines, as well as sessile animals bryozoans, sea anemones, sea lilies).
5. Coralline red algae directly covering the floor substrate (Knapp, 1965).

Overfishing in nearshore waters can cause a sudden kelp decline, when invertebrate herbivores become released from the regulation of their normal population and overgraze the brown and red algae. The next step is a sudden transition into barren 'desert' where relatively few ubiquitous species remain (Nordenhaug and Christie, 2009).

2.1.5.3 Salt lakes and salt steppes of the Pannonian Region (Hungarian: Puszta (Pannonian Steppe), Alföld (Great Hungarian Plain))

In Europe, salt-tolerant species and salt-dependent ecosystems can be found not only on the seashore

Fig. 2.4. Intertidal mudflats ('watts') on the shore of the North Sea, Netherlands (photo courtesy of B. Sudnik-Wójcikowska).

but also in the Pannonian Biogeographic Region (Puszta, the Pannonian Steppe), covering the Great Hungarian Plain in present-day Hungary and the neighbouring regions of Slovakia, Serbia, Romania, Bulgaria and Austria (Valko *et al.*, 2014; Török *et al.*, 2016). Salt lakes (e.g. Neusiedler See, Kelemen; Fig. 2.5) and the salt steppes of Puszta, spanning a south–north gradient of at least 4° latitude, formed after the total, irreversible drying of a vast, landlocked water body (the Central Paratethys Sea or Great Pannonian Lake, a northern appendage of the Miocene Tethys Sea – the early Mediterranean Sea plus Sargasso Sea) (Magyar *et al.*, 1999; Harzhauser and Piller, 2007). The Pannon landscape is now anthropogenic, with intensively cultivated arable fields, vineyards and pastures, and the primeval landscape of ancient Puszta survives only locally in a few nature sanctuaries such as the Hortobágy National Park (Hungary) and Čenkovská Steppe near Mužla (Slovakia) (Zlinszky *et al.*, 2015). Pannonian endemics are safeguarded *ex situ* in the Pannon Seed Bank (Török *et al.*, 2016). The Pannonic salt marshes and salt steppes (Natura 2000 habitat code 1530) (European Commission, 1992) are dominated by specialist salt-tolerant herbs, sedges and grasses that are adapted to alkaline soils and the extreme continental climate with hot, dry summers and cold winters. The key driver of the Puszta vegetation pattern is the general availability of seasonal water. Microtopography decides local water levels and salt accumulation, explaining why the Pannonian Puszta is a mosaic of numerous vegetation types. The drought-tolerant Pannonian Steppe is short grassland occurring in moderately alkaline soils dominated by *Festuca pseudovina*. Alkali meadows (flower steppes) are tall grasslands located in the lower depressions, also on slightly salty soils. The Pannonian Salt Marshes and Pannonian Salt Steppes, an isolated 'Steppe Island' in forested and urbanized Europe, host a number of endemic species (e.g. *Artemisia pancicii*, *Crepis*

Fig. 2.5. Plant communities on the shore of Kelemen-szék salt lake, Kiskunsag National Park (Hungarian: Kiskunsági Nemzeti Park), Pannonian Biogeographic Region, Hungary. (Photo courtesy of B. Sudnik-Wójcikowska.)

pannonica, Ferula sadleriana, Thlaspi kovatsii) and endemic habitats, unique among mostly cosmopolitan halophytic ecosystems (Deak *et al.*, 2014). They are, therefore, listed as priority habitats in Annex 1 of the Habitats Directive (Council Directive 92/43/EEC) (European Commission, 1992). Major threats to these grasslands are: soil disturbance, fragmentation, improper grazing and – most threatening – the decrease in the water table, which is causing significant, irreversible loss of salinity and invasion of generalist, glycophytic competitors (Kelemen *et al.*, 2013; Zlinszky *et al.*, 2015).

2.1.5.4 Wash fringes (wash fringe or wash margin; German: Spülsaum, Strandline; Polish: kidzina)

These are floristically meagre but zoologically plenteous – often Natura 2000 Bird areas

(European Commission 1979) – areas of the seashore where washed material is deposited or washed up from the sandy beach, and from the front of white dunes, running along the water margin. In Europe such communities are highly endangered, even in national parks and nature reserves, because of growing tourist activities (and in the Baltic region also from amber collecting). Washed material is formed mostly from seaweed remains from the kelp forests and *Zostera–Characeae* sea meadows, often enriched with dead jellyfish, fish, crustaceans, shells, inland trash and – in the Baltic – also amber. Due to its stressful marine conditions, it can be colonized only by a few specialized plant species. These are nitrophilous, psammophilous and halophilous. *Atriplex calotheca, A. littoralis, A. prostrata* var. *salina, Cakile maritima, Matricaria maritima* ssp. *maritima* and *Salsola kali* ssp. *kali* form the wash fringe communities in nutrient-rich places near to the sea. *Honckenya peploides* at the Baltic shore and *Elymus farctus* at the North Sea form another community, in nutrient-poor white dunes, a bit further from the sea than the aforementioned *A. calotheca, C. maritima, S. kali* wash fringe community. Although wash fringes on sandy coasts are natural habitats included in the Natura 2000 network, on recreational beaches seaweed debris forming this habitat is treated as 'waste', and removed. Consequently, the number of wintering birds has decreased and 40–50 vascular plant species strongly associated with wash margins have become extinct or become critically endangered (Sudnik-Wójcikowska and Krzyk, 2015).

2.1.5.5 Isolated inland saline grasslands outside the Pannon and Pontic regions

Small patches of halophilous and subhalophilous grasslands, floristically resembling alkali steppes and semi-deserts, can be found outside the Pannonian, Pontic and Tropic regions. They form 'ecological islands' surrounded by an 'ocean' of deciduous or taiga forests, 'glycophytic' steppes, farms, roads and urbanized areas. Good examples are the Beka, Owczary and Ciechocinek nature reserves in Poland, central Europe (a deciduous forest ecozone). Beka Nature Reserve is a bird sanctuary established in 1988. It covers an area of 193 ha on Puck Bay at the mouth of the Reda River, which flows into the Baltic Sea. Increased water levels and soil salinity result not only from the proximity of the sea, but also from local topography: organic soils accumulating salt. The reserve secures salt meadows, non-salt wetland meadows and white dunes. There are numerous halophilous species there, which are rare, protected or endangered throughout the Baltic region: *Glaux maritima, Aster tripolium, Juncus gerardii* and *Triglochin maritima* (Figs 2.6 and 2.7). The Beka area is home to many rare bird species which are protected by an EU Bird Directive. Until recently there was also one of the largest populations of *Calidris alpina* ssp. *schinzii* (up to 50 breeding pairs). However, as a result of the culling of cattle in the reserve, vast areas of salty meadows have been dominated by densely growing glycophytes, mostly the reed *Phragmites australis*, resulting in the almost complete disappearance of this bird species in the local breeding grounds. In 1999 there were only 1–2 nesting pairs and now only single, wandering individuals are generally seen. *Philomachus pugnax* has also ceased nesting (Piotrowska, 1974, 1976; Rąkowski *et al*., 2005; Skóra, 2014). The Owczary nature reserve, in the village of Owczary (central–southern Poland), is the only salt reserve in central Poland and one of only a few in continental Europe outside the Pannonian Biogeographic Region. In its north-western corner there is a distinctive salty precipice beneath a steep slope, with *Zostera* beds and mud flat-like vegetation, around 600 km from the Baltic Sea and 1200 km from the Black Sea. Species found there include *Ruppia maritima, Zannichellia palustris* ssp. *pedicellata* and *Spergularia salina*. Rare fauna is represented by the beetle *Pogonus persicus*; this site, isolated from the Pannonian region, is the only record of its species in central–eastern Europe (Piernik, 2005; Rąkowski *et al*., 2007). The Ciechocinek nature reserve (covering 188 ha) has an anthropogenic, but unique, salt vegetation, formed at a retrograde shaft near graduation tower no. 3 in Ciechocinek Spa. The local environment is continuously enriched in NaCl in two ways: through its settling from the air in fine crystals around the graduation tower, and through the brining process. One can observe characteristic zonation of vegetation along watercourses from the one-species aggregations of obligatory halophytes like *Salicornia europaea* sensu lato and *Spergularia salina* to salt meadows with *Glaux maritima, Salsola kali* (recently not observed in Ciechocinek), *Aster tripolium, Melilotus dentatus*, etc. (Rąkowski *et al*., 2005; Fig. 2.8).

Fig. 2.6. Halophilous grasslands in Beka reserve, Poland (photo courtesy of B. Sudnik-Wójcikowska).

2.1.6 Anthropogenic Salinization

2.1.6.1 Ruderal communities of salt-tolerant species in industrial regions

Industrial and highly urbanized areas (roadsides, gas and trans-shipment stations, some factories) in boreal and temperate climate zones need deicing every winter. In many areas NaCl and $CaCl_2$ are still the most popular anti-slipping agents, because urea and glycol, as well as calcium and magnesium formates and acetates, are too expensive for the company or local administration. Due to frequent extreme disturbances (such as the entire natural topsoil being covered with a foreign substance) few ornamental woody species and ruderal forbs can form saline-resistant plant cover there. Trees and shrubs resistant to growing salt concentrations are mostly multipurpose Eurasian ornamentals: *Ailanthus altissima*, *Elaeagnus* spp., *Caragana arborescens*, *Ulmus pumila* and *Tamarix* spp. North American plants seem to be slightly less halotolerant but *Robinia pseudoacacia*, *Amorpha fruticosa*, *Ribes aureum* and *Rhus typhina* can hold out against urban salinity. Some can escape cultivation and form their own communities. Halophyte herbs for cut flowers and flowerbeds include *Limonium vulgare* and *Armeria maritima* ssp. *maritima*. True halophytes such as *Cochlearia danica*, *Lepidium latifolium* or *Lactuca tatarica* can naturalize but remain uncommon in hyper-urbanized areas. Far more abundant in urban salt ecosystems are subhalophilous annuals (e.g. *Kochia scoparia*, *Atriplex tatarica* and *Polygonum aviculare*) and ruderal perennials (*Elymus repens*, *Puccinellia distans*) with extremely broad ecological amplitudes (Sudnik-Wójcikowska, 2011; Sudnik-Wójcikowska and Krzyk, 2015).

2.2 Halophyte-based Agriculture, Silviculture and Mariculture

Salt-tolerant species have not been consumed in a large amount by humans for millennia, and

Fig. 2.7. Education board near the entrance to the Beka reserve, Poland (photo courtesy of B. Sudnik-Wójcikowska).

neither agriculture nor forestry based solely on true halophytes has been successfully developed. There are no cultivation protocols for future salt-dependent crops (Ventura *et al.*, 2014). Watering edible, industrial and/or decorative crops with seawater remains one of the greatest dreams of plant researchers and farmers, especially in hot and dry climates. Most domestic plants are all highly sensitive to even minor amounts of NaCl and $CaCl_2$, even those whose ancestors were halophytes or halotolerant, such as *Beta vulgaris*. Soil salinity equal just to 2 dS m^{-1} results in vast yield reduction in maize (*Zea mays*, 12% per dS m^{-1}), potato (*Solanum tuberosum*, 12% per dS m^{-1}) and beans (*Phaseolus vulgaris*, 19% per dS m^{-1}) (Maas 2004). As a result of single gene transfer some enhancement in crop yields in saline soils can be obtained; but real, necessary progress in yield may be achieved only through a time-consuming 'pyramiding' of the major, physiological traits. A better solution to meet the 2050 climate change and desertification challenge will be domestication and popularization of alternative forage and crop

Fig. 2.8. Ciechocinek inland halophyte reserve, Poland (photo courtesy of B. Sudnik-Wójcikowska).

plants for farming and restoration of salt-affected lands (Panta *et al.*, 2014). Hybridization of key crops with their salt-tolerant congeners do not help to solve the problem, and so domestication of naturally salt-tolerant plants with economic potential seems reasonable. There are several dozen salt-resistant, wild vegetables (e.g. *Salicornia* spp. (Fig. 2.9) and multipurpose plant species in the wild (e.g. numerous mangrove tree species, *Casuarina* spp. (Fig. 2.10) (see Table 2.3). These could be grown in marginal soils and watered with seawater. Both the soils and seawater are useless for typical plant production) (Ventura *et al.*, 2014). The virtually infinite availability of marine water highlights the importance of new, salt-dependent and salt-tolerant crops as a source of food, fodder, medicine, decoration and – last but not least – renewable energy. Their greatest advantage is the lack of competition with 'traditional' alimentary and ornamentals for fresh water and the lack of competition with wildlife for space. Increasing research is targeting the improvement in halophyte agricultural traits (nutritional values, ease of harvesting, yield). A fine example of a domesticated halophyte is the Polynesian ironwood *Casuarina excelsa*, grown for construction wood and fuel; as a bonsai, a pollution-resistant and easily trained urban tree; and also as a windbreak and source of shade in nutmeg plantations

Fig. 2.9. Halophytes (chiefly *Salicornia europaea*) on the bank of the channel carrying salt water from the Ciechocinek salt basin, Poland (photo courtesy of B. Sudnik-Wójcikowska).

Fig. 2.10. *Casuarina* thickets on Fraser Island, east coast of Australia (photo courtesy of B. Sudnik-Wójcikowska).

(Węglarski and Węglarska, 2008). Another promising possibility is the use of salt-dependent (*Spartina anglica*, *Salicornia* spp.) and salt-tolerant species (e.g. *Phragmites australis*, *Elaeagnus angustifolia*, *Amorpha fruticosa*) for phytostabilization and/or phytoextraction of heavy metal-polluted soils and sea muds, and also of heavy metal-polluted soils affected by artificial

Table 2.3. Salt-tolerant lost or underutilized crops.

Botanical name	Vernacular names	Salt tolerance	Family	Origin	Uses
Crithmum maritimum	Sea fennel, samphire	Very high	Apiaceae	Mediterranean, Black Sea, Atlantic Europe	Ornamental, food, edible oil, fodder, biofuel
Portulaca oleracea	Little hogweed, common purslane, verdolaga, red root	Very high	Portulacaceae	Native in the Old World and Australasia, uncertain in the New World	Food, fodder
Aster tripolium	Seashore aster, Pannon aster	Very high	Asteraceae	Eurasia and N. Africa, mostly Pannon Biogeographic Region and Atlantic shores	Ornamental, food, fodder
Limbarda crithmoides	Golden samphire	Very high	Asteraceae	Mediterranean, Black Sea, Atlantic Europe	Food, fodder
Bruguiera gymnorrhiza	Burma mangrove, Kankra	Very high	Rhizophoraceae	Australasia, Oceania, E. Africa, Madagascar	Fuel, charcoal, construction, medicine, food, fodder
Rhizophora mucronata	Asiatic mangrove	Very high	Rhizophoraceae	Australasia, Oceania, E. Africa, Madagascar	Construction, especially for aquaculture, food, fodder, fuel, medicine
Sonneratia alba	Crabapple mangrove	Very high	Sonneratiaceae	Australasia, Oceania, E. Africa	Food, fodder, fuel, roots used as buoys
Xylocarpus granatum	Cannonball mangrove	Very high	Meliaceae	Pantropical	Medicinal. Seed oils used to make candles, soap, amadou
Mesembryanthemum crystallinum	Common ice plant	High	Aizoaceae	Native to Mediterranean and Sinai Peninsula, invasive in the New World and Australia	Ornamental, food, fodder
Sesuvium portulacastrum	Shoreline purslane, dampalit	High	Aizoaceae	Pantropical	Food, fodder
Cochlearia danica	Danish scurvy-grass	Moderate	Brassicaceae	Atlantic Europe, now invasive along motorways in Eurasia	Food, spice
Glaux maritima	Black saltwort, sea milkweed	Very high	Primulaceae	Native to Old and New World	Medicine
Salicornia europaea agg. and *Sarcocornia*	Glasswort, pickleweeds, sea asparagus, sea beans, picklegrass, samphire greens	Very high	Amaranthaceae	*S. europea* native to Old World, many species cosmopolitan	Probiotic, fodder, biofuel, phytoremediation, ornamental
Atriplex and *Chenopodium*, mostly *A. hortensis*	Orach, saltbush and goosefoots	Moderate to very high	Amaranthaceae	Cosmopolitan, uncertain in many places	Food, fodder, spice, biofuel

Ipomoea pes-caprae	Katang, goat's foot, beach morning glory, bayhops	High	Convolvulaceae	Pantropical	Medicinal, fodder, green manure
Althaea officinalis	Common marshmallow	Moderate	Malvaceae	Mediterranean, invasive in other regions	Medicine, phytoremediator, ornamental
Derris heterophylla	Poison tree, flame tree	Very high	Fabaceae	Pantropical	Medicine, natural insecticide and fish poison for fishing and biomanipulation
Tetragonolobus and *Lotus*	Deer vetches, bacon-and-eggs, bird's foot trefoils	Moderate	Fabaceae	Native in the Old World, invasive in the New World	Fodder, phytoremediation
Pandanus tectorius, P. odoratissimus	Walking tree, Hala screw pine	High	Pandanaceae	Australasia, Oceania	Food, fodder, fibre, medicinal, perfume
Nypa fruticans	Mangrove palm, water coconut, Nipah	High	Arecaceae	Australasia, Oceania	Construction and fibre material, source of edible salt and sugar
Oncosperma tigillarium	Nibung	High	Arecaceae	Australasia	Prized wood, food, construction (traditional spears)
Eugeissonia insignis	Wild sago palm	High	Arecaceae	Sarawak endemic	Food, fodder
Phoenix paludosa	Mangrove date palm, hental	High	Arecaceae	Australasia, Oceania	Food, fodder
Pennisetum clandestinum (syn. *Sporobolus virginicus*)	Sand couch, coastal rat-tail, nioaka, seashore dropseed	High	Poaceae	Pantropical	Fodder, biomass, groundcover
Ammophila arenaria	Marram grass, European beasch grass	High	Poaceae	Cosmopolitan, uncertain in many places	Fodder, biomass, groundcover
Spinifex spp.	Sea urchin plant, hedgehog	Moderate	Poaceae	Australasia, Oceania	Groundcover
Triglochin maritima	Shore arrowgrass, seaside arrowgrass	Very high	Juncaginaceae	Cosmopolitan, uncertain in many places	Food, fodder, biofuel
Acrostichum aureum	Golden leather fern, mangrove fern	High	Polypodiaceae	Pteridaceae	Ornamental, medicinal

salinity (Nehring and Adsersen 2006; Mesa *et al.*, 2015).

Although mariculture has been subject to controversy because of its environmental and social impacts, it has rapidly expanded over the last four decades. It owes its rise to numerous factors: (i) greater demand for seafood products; (ii) new technologies in water quality and fodders; (iii) better understanding of the biology of farmed species; (iv) technological progress; and (v) private and national investment. Most detrimental effects on the environment of marine farms, chiefly affecting mangroves and salt marshes, are:

- habitat loss and alteration;
- waste from cage culture;
- farm escapes of invasive species and/or sick specimens transferring parasites and diseases (Jennings *et al.*, 2001; Read and Fernandes, 2003; Molnar *et al.* 2008; Rosenberg, 2008; Pfeiffer, 2010).

2.3 Habitats of Halophytes as Natural Laboratories for Ecologists and Physiologists

Saline ecosystems have been important for mankind for countless years. Indeed, many archeologists now theorize that the first colonization of the New World was due to Pre-Aleute fisher-gatherers swimming along the Pacific 'kelp highway' during the last Ice Age (Pringle, 2008). Classic studies on the ecology of brown algae beds and forest ecology have helped us to better understand food webs and the regulation mechanism of top-down trophic processes (Graham, 2004). Bottom-up regulation has been studied in ephemeral alkaline lakes in salt deserts (Gliwicz, 2003). We would not know so much about allelopathy if we had not studied the ecology of the competitively strongest bushes in salt semi-deserts (Woodell *et al.*, 1969; Charley and West, 1975). Grime's CSR theory, which claims that every plant species has a blend of three life strategies: C (competitiveness), S (stress tolerance) and R (rapid propagation), would not have been formulated without having observed the extreme halophytes as model S-strategists (Grime and Pierce, 2012). Many key, classical observations on plant demography and zonation were performed in industrial salt marshes dominated by *Salicornia europaea* agg. (Wilkoń-Michalska, 1976; Fig. 2.9). Mangrove species have now been found to be good models for studying convergent evolution in plants. The multiple, evolutionary origins of salt tolerance in flowering plants is a marvellous case study in the ecology and evolution of a complex physiological trait. The amazing diversity of cell and tissue mechanisms used by halophytes, based on ecophysiologic processes common to all vascular plants, sheds new light on adaptations to extreme environments. A better understanding of the phylogenetic origins of the present diversity in halophytes may be critical for obtaining novel organisms for land reclamation, phytoremediation and salt pan-based agriculture (Flowers *et al.*, 2010). Thanks to genetic analyses, and more thorough morphological research, we are discovering the hidden species diversity of the halophilous and subhalophilous coastal and inland grasslands (Bosiacka *et al.*, 2016).

Acknowledgements

I would like to thank Professor Barbara Sudnik-Wójcikowska (Faculty of Biology, University of Warsaw) for her critical comments on an earlier version of the manuscript, which greatly helped in improving the content of this chapter. I am also grateful to her for kindly sharing unpublished photographs of halophytes and their habitats.

References

Bonhommeau, S., Chassot, E. and Rivot, E. (2008) Fluctuations in European eel (*Anguilla anguilla*) recruitment resulting from environmental changes in the Sargasso Sea. *Fisheries Oceanography* 17(1), 32–44.

Bosiacka, B., Kull, T., Więcław, H., Marciniuk, P. and Podlasiński, M. (2016) Habitat requirements of marsh dandelions (*Taraxacum*) in Polish and Estonian coastal grasslands. *Polish Journal of Ecology* 64(2), 213–230.

Bowen, J., Valiela, I. and York, J. (2001) Mangrove forests: One of the world's threatened major tropical environments. *Bio Science* 51, 807–815.
Burger, J., Niles, L. and Clark, K.E. (1997) Importance of beach, mudflat and marsh habitats to migrant shorebirds on Delaware Bay. *Biological Conservation* 79(2), 283–289.
Charley, J.L. and West, N.E. (1975) Plant-induced soil chemical patterns in some shrub-dominated semi-desert ecosystems of Utah. *The Journal of Ecology* 9, 945–963.
Cunliffe, B. (2017) *On the Ocean: The Mediterranean and the Atlantic from prehistory to AD 1500*. Oxford University Press, Oxford, UK.
Dahdouh-Guebas, F., Jayatissa, L.P., Di Nitto, D., Bosire, J.O., Seen, D.L. *et al.* (2005) How effective were mangroves as a defence against the recent tsunami? *Current Biology* 15(12), 443–447.
Darwin, C. 1909. *The Voyage of the Beagle*. The Harvard Classics Volume 29. P.F. Collier & Son Company, New York.
Dayton, P.K. (1985) Ecology of kelp communities. *Annual Review of Ecology and Systematics* 16, 215–245.
Devall, M.S. (1992) The biological flora of coastal dunes and wetlands. 2. *Ipomoea pes-caprae* (L.) Roth. *Journal of Coastal Research* 8(2), 442–456.
European Commission (1979) Directive 2009/147/EC of the European Parliament and of the Council of 30 November 2009 on the conservation of wild birds. *Official Journal of the European Communities* 1979, 1–103.
European Commission (1992) Council Directive 92/43/EEC of 21 May 1992 on the conservation of natural habitats and of wild fauna and flora. *Official Journal of the European Union* 206, 7–50.
Flowers, T.J., Galal, H.K. and Bromham, L. (2010) Evolution of halophytes: Multiple origins of salt tolerance in land plants. *Functional Plant Biology* 37(7), 604–612.
Friedland, K.D., Miller, M.J. and Knights, B. (2007) Oceanic changes in the Sargasso Sea and declines in recruitment of the European eel. *ICES Journal of Marine Science* 64(3), 519–530.
Gedan, K.B., Kirwan, M.L., Wolanski, E., Barbier, E.B. and Silliman, B.R. (2011) The present and future role of coastal wetland vegetation in protecting shorelines: answering recent challenges to the paradigm. *Climatic Change* 106(1), 7–29.
Giri, C., Ochieng, E., Tieszen, L.L., Zhu, Z., Singh, A. *et al.* (2011) Status and distribution of mangrove forests of the world using earth observation satellite data. *Global Ecology and Biogeography* 20(1), 154–159.
Gliwicz, Z.M. (2003) *Between Hazards of Starvation and risk of Predation: The Ecology of Offshore Animals* (Vol. 12). International Ecology Institute, Oldendorf/Luhe, Germany.
Graham, M.H. (2004) Effects of local deforestation on the diversity and structure of Southern California giant kelp forest food webs. *Ecosystems* 7, 341–357.
Grime, J.P. and Pierce, S. (2012) *The Evolutionary Strategies that Shape Ecosystems*. Wiley-Blackwell, London.
Hamilton, S. (2013) Assessing the role of commercial aquaculture in displacing mangrove forest. *Bulletin of Marine Science* 89(2), 585–601.
Harzhauser, M. and Piller, W.E. (2007) Benchmark data of a changing sea - palaeogeography, palaeobiogeography and events in the Central Paratethys during the Miocene. *Palaeogeography, Palaeoclimatology, Palaeoecology* 253(1), 8–31.
Hogarth, P.J. (1999) *The Biology of Mangroves*. Oxford University Press, Oxford, UK.
Jackson, G.A. and Winant, C.D. (1983) Effect of a kelp forest on coastal currents. *Continental Shelf Report* 2, 75–80.
Jennings, S., Kaiser, M.J. and Reynolds, J.D. (2001) *Marine Fisheries Ecology*. Blackwell, Victoria, Australia.
Jin-Eong, O. (2004) The ecology of mangrove conservation and management. *Hydrobiologia* 295, 343–351.
Kadereit, G., Piirainen, M., Lambinon, J. and Vanderpoorten, A. (2012) Cryptic taxa should have names: Reflections in the glasswort genus *Salicornia* (Amaranthaceae). *Taxon* 61(6), 1227–1239.
Kelemen, A., Török, P., Valkó, O., Miglécz, T. and Tóthmérész, B. (2013) Mechanisms shaping plant biomass and species richness: Plant strategies and litter effect in alkali and loess grasslands. *Journal of Vegetation Science* 24, 1195–1203.
Kingsford, M.J. (1995) Drift algae: A contribution to near-shore habitat complexity in the pelagic environment and an attractant for fish. *Marine Ecology Progress* 116(1), 297–301.
Knapp, R. (1965) Die Vegetation von Nord- und Mittelamerika und der Hawaii-Inseln. Gustav Fisher Verlag, Stuttgart, Germany.

Laffoley, D.D.A., Roe, H.S.J., Angel, M.V., Ardron, J., Bates, N.R. *et al.* (2011) *The Protection and Management of the Sargasso Sea: The Golden Floating Rainforest of the Atlantic Ocean*. Summary Science and Supporting Evidence Case. Sargasso Sea Alliance, Bermuda, p. 71.

Lapointe, B.E., West, L.E., Sutton, T.T. and Hu, C. (2014) Ryther revisited: Nutrient excretions by fishes enhance productivity of pelagic *Sargassum* in the western north Atlantic Ocean. *Journal of Experimental Marine Biology and Ecology* 458, 46–56.

Lazarus, M. and Wszałek-Rożek, K. (2016) Two rare halophyte species: *Aster tripolium* L. and *Plantago maritima* L. on the Baltic coast in Poland – their resources, distribution and implications for conservation management. *Biodiversity Research and Conservation* 41(1), 51–60.

Maas, E.V. (1990) Crop salt tolerance. In: Tanji, K.K. (ed.) *Agricultural Salinity Assessment and Management*. ASCE Manuals and Reports on Engineering Practice 71, 262–304.

Magyar, I., Geary, D.H. and Müller, P. (1999) Paleogeographic evolution of the Late Miocene Lake Pannon in Central Europe. *Palaeogeography, Palaeoclimatology, Palaeoecology* 147(3), 151–167.

Mesa, J., Rodríguez-Llorente, I.D., Pajuelo, E., Piedras, J.M.B., Caviedes, M.A. *et al.* (2015) Moving closer towards restoration of contaminated estuaries: Bioaugmentation with autochthonous rhizobacteria improves metal rhizoaccumulation in native *Spartina maritima*. *Journal of Hazardous Materials* 300, 263–271.

Millennium Ecosystem Assessment (2005) *Ecosystems and Human Well-being: Synthesis*. Island Press, World Resources Institute, Washington, DC, p. 2.

Murray, J. (1893) The discovery of America by Columbus: The Influences which led up to that Great Event, and its Effect on the Development of Oceanographical Knowledge. *Scottish Geographical Magazine* 9(11), 561–586.

Murray, N.J., Clemens, R.S., Phinn, S.R., Possingham, H.P. and Fuller, R.A. (2014) Tracking the rapid loss of tidal wetlands in the Yellow Sea. *Frontiers in Ecology and the Environment* 12(5), 267–272.

Nehring, S. and Adsersen, H. (2006) NOBANIS – Invasive Alien Species Fact Sheet – *Spartina anglica*. Online Database of the European Network on Invasive Alien Species. NOBANIS. https://www.nobanis.org/search/?q=Spartina+anglica, accessed 14 March 2018.

Norderhaug, K.M. and Christie, H. (2009) Sea urchin grazing and kelp revegetation in the NE Atlantic. *Marine Biology Research* 5, 515–528.

Panta, S., Flowers, T., Lane, P., Doyle, R., Haros, G. *et al.* (2014) Halophyte agriculture: Success stories. *Environmental and Experimental Botany* 107, 71–83.

Pfeiffer, T. (2010) Recirculation technology: the future of aquaculture. *Resource* 17(3), 7–9.

Piernik, A. (2005) Vegetation-environment relations on inland saline habitats in Central Poland. *Phytocoenologia* 35(1), 19–38.

Piotrowska, H. (1974) Nadmorskie zespoły solniskowe w Polsce i problemy ich ochrony. *Ochrona Przyrody* 39, 7–63.

Piotrowska, H. (1976) Przyczyny i skutki regresywnych zmian w nadmorskiej florze halofitów. *Phytocoenosis* 5, 3–4.

Plaziat, J.C., Cavagnetto, C., Koeniguer, J.C. and Baltzer, F. (2001) History and biogeography of the mangrove ecosystem, based on a critical reassessment of the paleontological record. *Wetlands Ecology and Management* 9(3), 161–179.

Pringle, H. (2008) Did humans colonize the world by boat? http://discovermagazine.com/2008/jun/20-did-humans-colonize-the-world-by-boat, accessed 10 October 2017.

Putman, N.F., Verley, P., Shay, T.J., and Lohmann, K.J. (2012) Simulating transoceanic migrations of young loggerhead sea turtles: Merging magnetic navigation behavior with an ocean circulation model. *Journal of Experimental Biology* 215(11), 1863–1870.

Rąkowski, G., Wójcik, J., Walczak, M., Smogorzewska, M. and Brodowska, M. (2005) *Rezerwaty przyrody w Polsce Północnej*. Instytut Ochrony Ś rodowiska, Warszawa.

Rąkowski, G., Walczak, M. and Smogorzewska, M. (2007) *Rezerwaty przyrody w Polsce Południowej*. Instytut Ochrony Ś rodowiska, Warszawa.

Read, P. and Fernandes, T. (2003) Management of environmental impacts of marine aquaculture in Europe. *Aquaculture* 226, 139–163.

Ricklefs, R.E., Schwarzbach, A. and Renner, S.S. (2006) Rate of lineage origin explains the diversity anomaly in the world's mangrove vegetation. *American Naturalist* 168(6), 805–810.

Roding, J. and van Voss, L.H. (eds) (1996) *The North Sea and Culture (1550–1800): Proceedings of the International Conference Held at Leiden 21–22 April 1995*. Uitgeverij Verloren, Leiden, Netherlands.

Rosenberg, A.A. (2008) Aquaculture: The price of lice. *Nature* 451(7174), 23–24.

Ryther, J.H. (1956) The Sargasso Sea. *Scientific American* 194, 98–104.
Skora, M.E. (2014) Nowe stanowisko piskorza Misgurnus fossilis w dorzeczu Redy i propozycje jego ochrony. *Chrońmy Przyrodę Ojczystą* 2(70), 178–184.
Skov, M.W. and Hartnoll, R.G. (2002) Paradoxical selective feeding on a low-nutrient diet: Why do mangrove crabs eat leaves? *Oecologia* 131(1), 1–7.
Steneck, R.S., Graham, M.H., Bourque, B.J., Corbett, D., Erlandson, J.M. *et al.* (2002) Kelp forest ecosystems: biodiversity, stability, resilience and future. *Environmental Conservation* 29, 436–459.
Sudnik-Wójcikowska, B. (2011) *Rośliny synantropijne*. Oficyna Wydawnicza MULTICO, Warszawa, Poland.
Sudnik-Wójcikowska, B., and Krzyk, A. (2015) *Rośliny wydm, klifów, solnisk i aluwiów*. Oficyna Wydawnicza MULTICO, Warszawa, Poland.
Teas, H.J. (1983) *Biology and Ecology of Mangroves*. W. Junk Publishers, The Hague, Netherlands.
Tomlinson, P.B. (2016) *The Botany of Mangroves*. Cambridge University Press, Cambridge, UK.
Török, K., Szilágyi, K., Halász, K., Zsigmond, V., Kósa, G. *et al.* (2016) Seed collection data encompassing half of the vascular flora of the Pannonian Ecoregion stored by the Pannon Seed Bank. *Acta Botanica Hungarica* 58(3–4), 435–445.
Venter, J.C., Remington, K., Heidelberg, J.F., Halpern, A.L., Rusch, D. *et al.* (2004) Environmental genome shotgun sequencing of the Sargasso Sea. *Science* 304(5667), 66–74.
Ventura, Y., Eshel, A., Pasternak, D. and Sagi, M. (2014) The development of halophyte-based agriculture: Past and present. *Annals of Botany* 115(3), 529–540.
Węglarski, K. and Jańczak-Węglarska, J. (2008) Użyteczne rośliny tropików: szkice etnobotaniczne. Bogucki Wydawnictwo Naukowe, Poznań, Polska.
Wilkoń-Michalska, J. (1976) Struktura i dynamika populacji *Salicornia patula* Duval-Jouve. *Rozprawy Uniwersytetu im. Mikołaja Kopernika, Folia Sozologica* 3, 123–129.
Woodell, S.R., Mooney, H.A. and Hill, A.J. (1969) The behaviour of *Larrea divaricata* (creosote bush) in response to rainfall in California. *The Journal of Ecology* 57(1), 37–44.
Zlinszky, A., Deák, B., Kania, A., Schroiff, A. and Pfeifer, N. (2015) Mapping Natura 2000 habitat conservation status in a pannonic salt steppe with airborne laser scanning. *Remote Sensing* 7(3), 2991–3019.

3 Intra-habitat Variability of Halophytic Flora of North-west India

Sarita Devi[1]*, Ashwani Kumar[2], Anita Mann[2], Sunder Singh Arya[3], Gurdev Chand[4], Neeraj Kumar[1], Anita Kumari[1], Pooja[5], Babita Rani[1] and Arvind Kumar[2]

[1]Chaudhary Charan Singh Haryana Agricultural University, Hisar, India; [2]ICAR – Central Soil Salinity Research Institute, Karnal, India; [3]Maharishi Dayanand University, Rohtak, India; [4]Sher-e-Kashmir University, Jammu, Chatha, India; [5]ICAR – Sugarcane Breeding Institute, Karnal, India

Abstract

The growth and ion-accumulating characteristics of different plant species native to saline-arid areas *vis-à-vis* the ionic status of their rhizospheric soils are important factors to consider when studying the survival of plant species under harsh conditions. A survey of saline areas in Haryana (CCS Haryana Agricultural University and Central Institute for Research on Buffaloes at Hisar) and Rajasthan (Lunkaransar in Bikaner, Sambhar Lake, Jaipur and Gangani-Kaparda at Jodhpur) was done to explore the variability in halophytic species in different arid and saline regions. It was observed that 44 species spread over 16 families of Angiosperms (i.e. Chenopodiaceae, Mimosaceae, Poaceae, Capparidaceae, Portulacaceae, Tamaricaceae, Fabaceae, Caesalpiniaceae, Aizoceae, Asteraceae, Salvadoraceae, Asclepiadaceae, Boraginaceae, Solanaceae, Amaranthaceae and Cyperaceae) were growing across these locations with most species from the Chenopodiaceae family. Among these highly flourishing species were *Salsola baryosma*, *Suaeda fruticosa*, *S. nudiflora* and *Saccharum munja*, producing the highest biomass per unit area of land. The rhizospheric soil of different locations was saline to highly saline. While exploring the ionic homeostasis, it was found that most ions (e.g. Na^+, K^+, Ca^{2+}, Mg^{2+}, Cl^- and SO_4^{2-}) were accumulating more in their leaves than in their stems, which may be a survival mechanism of these halophytic species under worse conditions of salinity. In terms of soil habitat across these surveyed regions, the rhizospheric salinity is patchy and supports both facultative halophytes and glycophytes.

Keywords: Biodiversity; Growth; Habitats; Halophytes; Mineral ions

3.1 Introduction

The build-up of soil salinity is one of the world's oldest and most serious agricultural problems in arid and semi-arid regions. About 7.0 million ha of agricultural land worldwide is affected by salinity and the area is expanding. In Haryana state, northern India, about one-third of problem soils are saline, and the chlorides and sulfates of sodium, calcium and magnesium are the dominant soluble salts in them. The existing technologies used in farm salinity management that work well include surface and sub-surface drainage. These are basically civil engineering technologies and are costly to install, difficult to maintain and have the problem of saline effluent

* Corresponding author e-mail: devisaritaa@gmail.com

management. Apart from that, under Indian conditions and with fragmented land holdings a wide application of such technology seems utopian. Soil remediation technologies can, first, be an *ex situ* technique in which remediation of the excavated contaminated soils by various processes like thermal treatment, chemical extraction and solidification prior to disposal is carried out. Second, an *in situ* technique where remediation is done without excavation of the contaminated soil (e.g. by volatilization via air venting, leaching with a surfactant, vitrification, isolation and containment with physical barriers). *Ex situ* treatment is very costly as it involves returning the treated soils to the restored site. *In situ* techniques are favoured over the *ex situ* technique because of their low cost and reduced impact on the ecosystem (Burns *et al.*, 1996).

South-west Haryana, parts of Rajasthan and adjoining areas of Pakistan comprise the 'Thar Desert' (Fig. 3.1). This is a vast arid zone interspersed with saline domains. The native flora of saline soils has a unique biology that can be exploited through a variety of phytoremediation and biotechnological interventions (Vera-Estrella *et al.*, 2005; Angrish and Devi, 2014). Characterization of the eco-physiological characteristics of the plants growing in the saline niches of this arid zone is of fundamental interest. The present studies were, therefore, carried out to characterize the growth magnitude and mineral ion accumulation characteristics of the plants found in selected pockets of the semi-arid zones of the states of Haryana and Rajasthan.

3.2 Methodology

3.2.1 Floristic surveys

We investigated the growth and ion-accumulating characteristics of different plant species *vis-à-vis* the ionic status of the saline soils that form their habitat. To do this, we explored the wild vegetation native to saline soils in arid zones

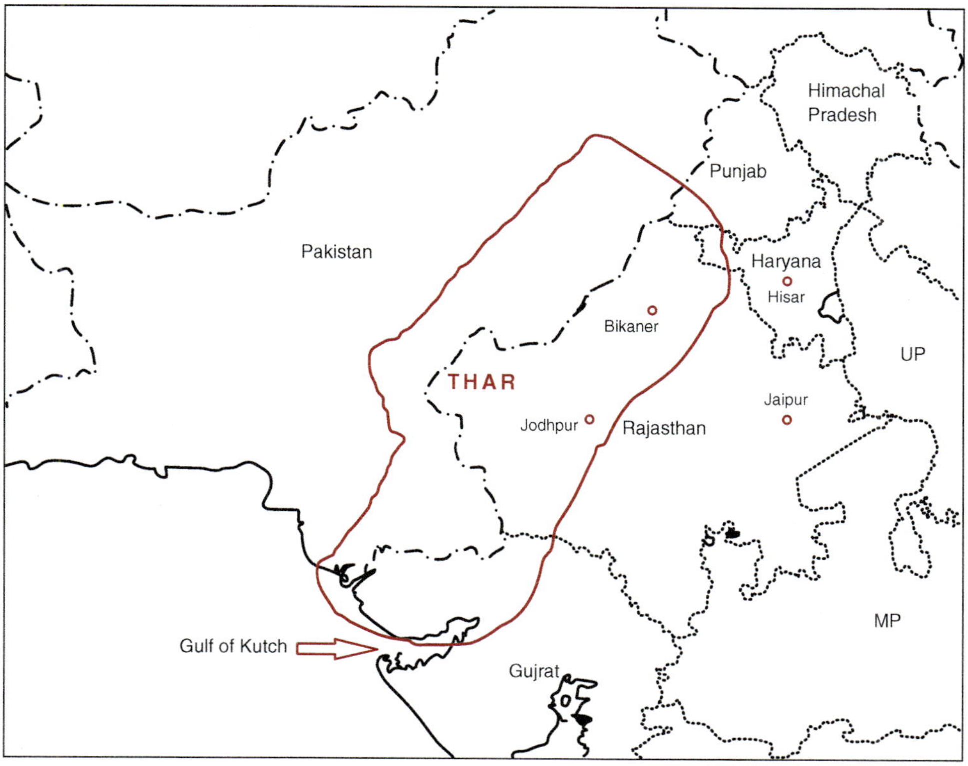

Fig. 3.1. Illustration of sampling area in Haryana and Rajasthan.

of north-west India. In general, growth of such vegetation is initiated in February–March. It may subside in the dry, hot period of late May and June. However, it picks up again in response to the monsoon in July–August and reaches a peak towards the end September or early October. In all, five locations were explored. These were (latitude and longitude):

- Location 1: Fallow saline land at Haryana Agricultural University, Hisar (29.121362 and 75.725646).
- Location 2: Wild area west of Sector 14 and behind the Central Institute of Research on Buffalo (CIRB), Hisar (29.182237 and 75.703688).
- Location 3: Lunkaransar, Bikaner (28.497998 and 73.747076).
- Location 4: Sambhar Lake, Jaipur (26.904152 and 75.179712).
- Location 5: Gangani and Kaparda, Jodhpur (26.503581 and 73.216756).

Plant sampling was done towards September at flower initiation stage in plants exhibiting flowering, and at maximum vegetative stage where flowering was not observed. Sampling was done by removing the aboveground biomass of the plants in the wild from a unit (preferably 1 m^2) land area. However, as salt uptake per unit land area for a particular species was the prime objective, mixed population areas were avoided, so the sampling was done in areas dominated by the individual species in question. Total fresh and dry weight of all plants was calculated and expressed in kg m^{-2}. Rhizospheric soil sampling from 0–30 cm depth was collected from the spot where the aboveground mass was harvested. The parameters of sodium adsorption ratio (SAR), soil pH and ECe, were determined by using standard procedures. The mineral ions (Na^+, K^+, Ca^{2+}, Mg^{2+}, Cl^-, and SO_4^{2-}), total dissolved solids (TDS), ash content per unit dry mass of various plant organs (*viz*. stem and leaves) were determined from the oven-dried material. Standard analytical methods for mineral ion determination were as enumerated in Richards (1954) and Kalra (1997).

3.2.2 Digestion

We placed 500 mg oven-dried and well-ground material in a 50 ml conical flask to which 5 ml of 9:1 H_2SO_4 (97%–100%) and $HClO_4$ (60%) mixture was added. The flasks were heated gently on a hotplate until dense white fumes formed. When fumes reduced and subsided, heating was increased and digestion was continued for another 25–30 min to obtain a colourless digest. The digest was cooled and diluted to 50 ml with 50 ml deionised water. This acid digest was used for the estimation of sodium, potassium, calcium and magnesium.

3.2.3 Sodium (Na^+) and potassium (K^+) estimation

Na^+ and K^+ contents were determined in the above acid digest with a flame photometer (Elico CL 378, Hyderabad, India) using standard KCl for K^+ and NaCl for Na^+. The values were calculated and expressed as mg g^{-1} tissue dry weight.

3.2.4 Calcium (Ca^{2+}) and magnesium (Mg^{2+})

Ca^{2+} and Mg^{2+} were estimated by EDTA titration according to the method given in the USDA Handbook (Richards, 1954) as detailed below.

3.2.4.1 Reagents

1. N/100 EDTA solution 1.86 of EDTA (di-sodium salt) was dissolved in water and volume increased to 1 l with distilled water.
2. Erichrome Black T indicator (EBT) 4.5 g of hydroxylamine hydrochloride and 0.5 g EBT were dissolved in ethanol and volume was increased to 100 ml with double distilled water. This solution was prepared freshly before use.
3. NH_4OH-NH_4Cl buffer 67.5 g of NH_4Cl was dissolved in small volume of distilled water and 570 ml concentrated NH_4OH was added and volume was made to 1 l with distilled water.
4. Carbamate crystals.

3.2.4.2 Procedure

We placed 1 ml aliquot in a china dish and added 5 ml distilled water. The contents were stirred with a magnetic stirrer on slow speed. Five crystals of carbamate were added to the china dish and then 2 ml NH_4OH-NH_4Cl buffer and 4 drops of EBT indicator were added to it. The contents of the china dish were titrated with 0.1 N EDTA until the appearance of a bluish-green colour (end-point). The volume of EDTA solution used was noted and the following calculations were made:

$$\mathrm{Ca+Mg\ (meq/ml)} = \frac{\mathrm{EDTA\ used\ in\ titration\ (ml)} \times \mathrm{Normality\ of\ EDTA}}{\mathrm{Aliquot\ taken\ (ml)}}$$

3.2.4.3 Determination of calcium

REAGENTS

1. N/100 EDTA (di-sodium salt) solution.
2. Calcon indicator: 0.5 g calcon in 100 ml methanol.
3. 4N NaOH solution: 16 g NaOH was dissolved in 100 ml distilled water.
4. 1% PVA solution.
5. 5% Hydroxylamine hydrochloride solution.
6. Tri ethanolamine.

PROCEDURE. We placed 1 ml of aliquot in a china dish and added 5 ml distilled water to it. After stirring, 1 ml 5% hydroxylamine hydrochloride solution, 10 drops of triethanolamine, 2 ml 4N NaOH, 1 ml PVA solution and 4 drops of calcon indicator were added and mixed. The content was titrated against N/100 EDTA until a bluish-green colour (i.e. end-point) appeared. The volume of EDTA used was recorded.

3.2.4.4 Calculation for calcium

$$\mathrm{Ca\ (meq/ml)} = \frac{\mathrm{ml\ of\ EDTA\ used} \times \mathrm{Normality\ of\ EDTA}}{\mathrm{Aliquot\ taken\ (ml)}}$$

3.2.4.5 Calculation for magnesium

$$\mathrm{Mg\ (meq/ml)} = \left[\mathrm{Ca+Mg\ (meq/ml)}\right] - \mathrm{Ca\ (meq/ml)}$$

The data for Ca^{2+} and Mg^{2+} were expressed as mg g^{-1} tissue dry weight.

3.2.5 Sulfate

The SO_4^{2-} was estimated by the turbidimetric method, as suggested by Chesnin and Yien (1950).

3.2.5.1 Digestion for sulfate

We placed 500 mg of oven-dried and well-ground material in a 50 ml conical flask to which 5 ml HNO_3:$HClO_4$ (4:1) diacid mixture was added. The flasks were heated gently on a hotplate until dense white fumes formed. At the stage at which fumes reduced and subsided, heating was increased and digestion was continued for another 25–30 min to obtain a colourless digest. The digest thus obtained was cooled and diluted to 50 ml with distilled water.

3.2.5.2 Reagent

1. Gum acacia solution of 0.25% is prepared. Dissolve 25 g of gum acacia in distilled water and dilute to 100 ml.
2. Barium chloride: Grind $BaCl_2.2H_2O$ crystals in a mortar until they pass through a 20–30 mm mesh sieve, but are retained on a 60 mm mesh sieve.
3. Standard SO_4 solution: Dissolve 0.1815 g of reagent grade K_2SO_4 in 1 l distilled water. This is 100 mg/l stock solution of SO_4^{2-}. Transfer 1.25, 2.50, 5.0, 7.5, 10.0, 12.5 and 15.0 of the 100 mg/l SO_4^{2-}, stock solution in a series of 25 ml volumetric flasks to obtain 5, 10, 20, 30, 40, 50 and 60 mg/l SO_4^{2-}, respectively. Prepare a standard curve by plotting percentage transmittance (T) on Y-axis and concentration on X-axis on a graph paper. There should be a straight line relationship between concentration and transmittance.

3.2.5.3 Procedure

We transferred 6 ml aliquot of digest to a 25 ml volumetric flask, 1 ml of gum acacia solution was added to make the volume up to the mask of volumetric flask (50 ml) and the flask was shaken for 1 min. A further 1.0 g of sieved $BaCl_2$ crystals was added and the mixture shaken for 1 min. Turbidity was measured after 25–30 min, after adding $BaCl_2$ crystals, as absorbance on an atomic absorption spectrophotometer (Shimadzu corporation, Japan) using a blue filter at a wavelength of 420 nm. Simultaneously, a blank (without sample) was carried out. Data were expressed as mg g^{-1} tissue dry weight.

3.2.5.4 Calculation

$$C \times \text{Dilution factor}$$

$$C = SO_4^{2-} \text{ content } (\text{mg l}^{-1}) \text{ from the standard curve against T}$$

$$T = \text{Transmittance } (\%T) \text{ as read from spectrophotometer}$$

3.2.6 Chloride (Cl^-)

Cl^- was estimated by the ion analyser (Orion, USA), using a chloride-specific electrode.

3.2.6.1 Digestion

We placed 500 mg of well-ground material in a 50 ml conical flask to which 5 ml (1M HNO_3-KNO_3) solution was added. The solution was stirred for 30 min with a magnetic stirrer and then filtered. The filtrate was used for chloride estimation.

3.2.7 TDS

We placed 1.0 g dry weight of material in a 100 ml conical flask and 100 ml of distilled water was added. The solution was boiled and filtered. The volume was made up to 100 ml with distilled water and TDS was recorded using a conductivity meter (Systronic-308, Gujarat, India).

3.2.8 Ash content

We weighed an empty crucible and then added 1.0 g dry weight of material into the crucible. The crucible was heated with the material in a muffle furnace at 550°C for 30 min. After cooling, the crucible was weighed again to get the final weight.

$$\text{Ash content/g dry weight} = \text{Final weight} - \text{Crucible weight}$$

3.2.9 SAR

The amount of adsorbed sodium is usually expressed as percentage of the cation exchange capacity.

3.2.9.1 Calculation

$$SAR = \frac{Na^+}{\sqrt{(Ca + Mg)/2}}$$

3.2.10 pH

pH was determined with an Elico glass electrode (Elico LI 120, Hyderabad, India) on a pH meter at the soil saturation extract.

3.2.11 ECe

Electrical conductivity of soil saturation extract (ECe) was measured by using a conductivity bridge and expressed as dSm^{-1} at 25°C.

3.3 Results

Table 3.1 shows that a total of 44 naturally occurring plant species were collected, belonging to 16 families of Angiosperms from saline locations in Haryana and Rajasthan. A maximum number of species encountered belong to the family Chenopodiaceae (*Atriplex amnicola*, *A. lentiformis*, *A. nummularia*, *Chenopodium album*, *C. ambrosioides*, *C. murale*, *Haloxylon recurvum*, *Salsola baryosma*, *Suaeda fruticosa* and *S. nudiflora*) followed by four each in Mimosaceae (*Acacia nilotica*, *A. ampliceps*, *A. colei* and *Prosopis juliflora*) and Poaceae (*Cynodon dactylon*, *Arundo donax*, *Setaria glauca* and *Saccharum munja*), Capparidaceae (*Capparis aphylla*), Portulacaceae (*Portulaca oleracea*), Tamaricaceae (*Tamarix dioica*), Fabaceae (*Sesbania sesban*), Caesalpiniaceae (*Parkinsonia aculeata*), Aizoceae (*Trianthema portulacastrum*),

Table 3.1. Taxonomic details of species collected from different saline locations.

Sl. No.	Family	Species: Botanical name	Species: Vernacular name
1.	Capparidaceae	*Capparis aphylla*	Kair
2.	Portulacaceae	*Portulaca oleracea*	Kulfa
3.	Tamaricaceae	*Tamarix dioica*	Morpankhi
4.	Fabaceae	*Sesbania sesban*	Rawasan
5.	Caesalpiniaceae	*Parkinsonia aculeata*	Vilaytikikar
6.	Mimosaceae	*Acacia nilotica*	Kikar
		Acacia ampliceps	
		Acacia colei	
		Prosopis juliflora	
7.	Aizoaceae	*Trianthema portulacastrum*	Santhi
8.	Asteraceae	*Xanthium strumarium*	Bhangra
9.	Salvadoraceae	*Salvadora persica*	Pilu
10.	Asclepiadaceae	*Calotropis procera*	Ak
11.	Boraginaceae	*Heliotropium ramosissimum*	
12.	Solanaceae	*Solanum xanthocarpum*	Berkateli
		Physalis longifolia	Solanaceae
13.	Amaranthaceae	*Achyranthes aspera*	Puthkunda
		Aerva tomentosa	Dholimundi
14.	Chenopodiaceae	*Salsola baryosma*	Bui
		Chenopodium ambrosioides	Khatua
		Chenopodium murale	Khartua
		Suaeda fruticosa	Bui (Lonia)
		Suaeda nudiflora	
		Chenopodium album	Bathua
		Atriplex nummularia	Stocksii
		Atriplex amnicola	
		Atriplex lentiformis	
		Haloxylon recurvum	
15.	Cyperaceae	*Cyperus rotundus*	Motha
16.	Poaceae	*Cyanodon dactylon*	Doob
		Arundo donax	Narhal
		Setaria glauca	Bandarighas
		Saccharum munja	Munj

Asteraceae (*Xanthium strumarium*), Salvadoraceae (*Salvadora persica*), Asclepiadaceae (*Calotropis procera*), Boraginaceae (*Heliotropium ramosissimum*), Cyperaceae (*Cyperus rotundus*), Solanaceae (*Solanum xanthocarpum* and *Physalis longifolia*); and Amaranthaceae (*Achyranthes aspera and Aerva tomentosa*) had one or two representatives each.

3.3.1 Magnitude of plant growth

Table 3.2 shows that the highest biomass per unit land area was accumulated in *S. baryosma* (5170 g m^{-2}). This was followed by *S. fruticosa* (5145 g m^{-2}), *S. nudiflora* (4213 g m^{-2}) and *S. munja* (4145 g m^{-2}). *A. aspera*, *C. dactylon*, *H. ramosissimum*, *S. xanthocarpum* and *S. glauca*, on the other hand, accumulated low biomass per unit land area and ranged from 906.2 to 1078.7 g m^{-2}. The biomass of species like *P. longifolia* and *P. juliflora*, with tree-like habits, could not be meaningfully quantified and their data are not included in the table. Dry biomass per unit area followed the same trend as that of fresh mass and was, in general, about 10% of the fresh mass.

In arid and saline environments germplasm diversity is of paramount significance for species to exploit different physiological and biotechnological interventions. A survey of the five locations showed that as many as 40 species

Table 3.2. Growth characteristics of plants at different saline locations.

Location	Species	Fresh biomass (g m^{-2})	Dry biomass (g m^{-2})	Water content (g m^{-2})
Location 1	*Acacia nilotica*	n.d.	n.d.	n.d.
	Achyranthes aspera	915.0	93.1	821.9
	Calotropis procera	2118.4	203.1	1915.4
	Capparis aphylla	1078.7	113.8	964.9
	Cyanodon dactylon	1115.7	98.8	1016.9
	Cyperus rotundus	1339.1	135.8	1203.2
	Physalis longifolia	n.d.	n.d.	n.d.
	Prosopis juliflora	n.d.	n.d.	n.d.
	Saccharum munja	4145.0	501.0	3643.9
	Salsola baryosma	4226.1	499.3	3726.8
	Sesbania sesban	3333.2	300.2	3033.1
	Setaria glauca	906.2	90.0	816.2
	Solanum xanthocarpum	990.9	93.0	897.9
	Suaeda fruticosa	5145.1	535.2	4609.9
	Tamarix dioica	n.d.	n.d.	n.d.
	Trianthema portulacastrum	845.1	92.2	752.9
	Mean	2105.6	229.6	1725.3
	CD (0.05) species	90.1	8.2	83.5
Location 2	*Aerva tomentosa*	2349.4	205.2	2144.2
	Arundo donax	3113.3	292.1	2821.2
	Chenopodium ambrosioides	1572.5	142.9	1429.6
	Chenopodium murale	3080.0	300.1	2779.9
	Cyanodon dactylon	900.3	92.1	821.2
	Heliotropium ramosissimum	994.0	90.4	903.6
	Portulaca oleracea	1003.0	91.2	912.2
	Salsola baryosma	5170.0	546.4	4623.7
	Setaria glauca	993.3	90.3	903.0
	Suaeda fruticosa	4440.0	505.5	3934.5
	Trianthema portulacastrum	991.7	90.5	901.9
	Mean	2237.0	22.4	1942.3
	CD (0.05) species	89.4	8.1	81.4
Location 3	*Atriplex amnicola*	2225.5	195.1	2030.4
	Atriplex nummularia	2315.4	200.5	2114.9
	Cyanodon dactylon	1005.0	93.5	911.9
	Salvadora persica	n.d.	n.d.	n.d.
	Suaeda fruticosa	3345.0	320.5	3024.6
	Mean	2222.7	202.8	2020.5
	CD (0.05) species	102.5	19.6	97.3
Location 4	*Heliotropium ramosissimum*	1000.0	96.9	903.1
	Salsola baryosma	5000.4	495.8	4504.7
	Solanum xanthocarpum	990.5	92.0	898.5
	Suaeda fruticosa	4348.8	510.0	3838.8
	Mean	2834.9	298.7	2536.8
	CD (0.05) species	115.6	11.7	109.3
Location 5	*Acacia ampliceps*	n.d.	n.d.	n.d.
	Acacia colei	n.d.	n.d.	n.d.
	Atriplex amnicola	2230.2	197.0	2033.2
	Atriplex lentiformis	2110.5	190.1	1920.4
	Atriplex nummularia	2328.8	203.8	2125.6
	Colophospermum mopane	2115.9	192.5	1923.7
	Salvadora persica	n.d.	n.d.	n.d.

Continued

Table 3.2. Continued.

Location	Species	Fresh biomass (g m^{-2})	Dry biomass (g m^{-2})	Water content (g m^{-2})
	Suaeda nudiflora	4213.8	480.8	3733.0
	Mean	2599.8	252.6	2347.8
	CD (0.05) species	117.5	15.3	114.7

representing 17 angiospermic families were growing, showing good species diversity in the arid–saline zones (Table 3.1). Different workers (Dagar, 2005; Shah, 2017) also observed considerable plant diversity in the different habitats they surveyed. It is interesting to note that a total of 10 species out of the 44 encountered belonged to the family Chenopodiaceae. Sen *et al.* (1982) also encountered considerable representation of the family Chenopodiaceae at the locations they surveyed.

A species growing in a saline-arid habitat may be very tolerant but slow growing and have much less biomass productivity. As research interest in our laboratory is towards phytoremediation (Angrish and Devi, 2014; Devi *et al.*, 2016), species that could yield high biomass were of primary interest. Mahmood *et al.* (1994) observed that *L. fusca* (Kallar grass) accumulated 20 t dry matter ha^{-1} from 4–5 cuts per year, and also successfully improved the soil conditions of the existing saline sodic soils, showing increased vegetation growth after a 5-year period. Similarly, quinoa (*Chenopodium quinoa*), a South American seed crop widely grown in Bolivia, has low grain yields (< 0.5 t ha^{-1}) but the potential yield may reach 3–4 t ha^{-1} through optimizing cultivation conditions (Gomez *et al.*, 2011; Adolf *et al.*, 2013).

3.3.2 Composition of the rhizospheric soil

An important feature of our investigation was that the rhizospheric soil of the area where the plants were harvested was also analysed for its EC, pH and salt quantity and quality. Results presented in Table 3.3 show that the rhizospheric soil of the different locations was saline to highly saline. Thus, the ECe of Location 1 was in the range of 9.62–60.80, whereas the pH range was 7.36–9.42. The corresponding variation in these parameters at other locations was: Location 2: ECe 11.99–12.34, pH 7.23–7.78; Location 3: ECe 12.47–20.65, pH 8.77–9.21; Location 4: ECe 14.90–18.60, pH 8.20–9.50; and Location 5: ECe 5.19–10.85, pH 7.90–8.95. Table 3.3 also shows that Na^+ was the predominant cation as compared to K^+, Ca^{2+} and Mg^{2+}, and that Cl^- was the predominant anion as compared to SO_4^{2-} at Location 1.

This indicated that Location 1 had a Cl^--dominated salinity. Location 2 had an even more uniformly higher soil Na^+ and Cl^-, again indicating the presence of Cl^-- dominating salinity. Examination of ionic content data of Locations 3, 4 and 5 also indicates a Cl^-- dominating salinity spread, though of a much lesser magnitude. The perusal of pH data at various locations indicated that the soil varied from neutral to slightly alkaline. The composition of rhizospheric soil *vis-à-vis* the growth and ionic characteristics of aboveground biomass is of vital interest. It was observed that within a marked site the salinity was variable or patchy. Thus, for example, at Location 1 (fallow saline land at Haryana Agricultural University, Hisar) the ground salinity varied from 9.62 to 60.80 (Table 3.3). This point highlights that the mere existence of a plant in an overall saline habitat can be due to its rhizospheric conditions being non-saline. Ionic composition of soil showed that chloride-dominated salinity was prevalent at almost all the locations, as there was a preponderance of Cl^- over SO_4^{2-} ions. Our studies are in accord with reports in literature showing that the saline soils of north-west India mostly have chloride-dominated salinity over sulfate-dominated salinity. Chloride-dominated salinity is more toxic than sulfate-dominated salinity (Mor *et al.*, 1992).

3.3.3 Ionic accumulation in plants

Tables 3.4 and 3.5 show that salt accumulation indicative parameters of ash content and TDS

Table 3.3. Rhizospheric soil characteristics of soil from different saline locations.

Location	Species	ECe	pH	Ions (mg g^{-1})						
				Na^+	K^+	Ca^{2+}	Mg^{2+}	Cl^-	SO_4^{2-}	Total
Location 1	*Acacia nilotica*	40.40	8.77	0.307	0.030	0.022	0.042	0.729	0.480	1.611
	Achyranthes aspera	16.65	8.62	0.285	0.028	0.022	0.043	0.623	0.400	1.403
	Calotropis procera	9.62	8.02	0.081	0.018	0.010	0.019	0.096	0.035	2.061
	Capparis aphylla	13.32	8.07	0.139	0.007	0.022	0.042	0.325	0.188	7.025
	Cyanodon dactylon	12.52	8.40	0.091	0.010	0.024	0.045	0.277	0.150	6.001
	Cyperus rotundus	54.71	9.10	0.383	0.030	0.046	0.060	0.643	0.049	1.213
	Physalis longifolia	17.80	8.96	0.196	0.013	0.039	0.055	0.469	0.295	1.069
	Prosopis juliflora	23.50	8.65	0.295	0.010	0.034	0.053	0.562	0.378	1.334
	Saccharum munja	19.67	7.49	0.151	0.014	0.026	0.031	0.449	0.298	9.071
	Salsola baryosma	27.60	8.50	0.236	0.023	0.026	0.051	0.585	0.378	1.300
	Sesbania sesban	27.90	8.87	0.388	0.058	0.060	0.065	0.655	0.500	1.728
	Setaria glauca	21.80	8.98	0.256	0.013	0.025	0.053	0.567	0.490	1.406
	Solanum xanthocarpum	23.60	8.32	0.252	0.009	0.029	0.057	0.579	0.478	1.407
	Suaeda fruticosa	17.26	9.42	0.273	0.014	0.022	0.043	0.549	0.475	1.379
	Tamarix dioica	60.80	7.36	0.316	0.022	0.050	0.071	0.679	0.480	1.621
	Trianthema portulacastrum	10.15	7.93	0.085	0.026	0.010	0.017	0.130	0.075	3.044
	Mean	24.83	8.47	0.233	0.020	0.029	0.047	0.495	0.322	
	CD (0.05) species	1.31	0.11	2.28	1.03	1.00	1.10	3.89	3.20	
Location 2	*Aerva tomentosa*	12.30	7.38	0.345	0.058	0.086	0.152	0.735	0.480	
	Arundo donax	12.05	7.27	0.383	0.056	0.091	0.152	0.769	0.468	1921.4
	Chenopodium ambrosioides	12.25	7.65	0.392	0.059	0.090	0.155	0.749	0.490	1937.7
	Chenopodium murale	12.28	7.37	0.386	0.056	0.091	0.151	0.731	0.476	1893.3
	Cyanodon dactylon	12.28	7.37	0.350	0.062	0.009	0.158	0.752	0.490	1904.8
	Heliotropium ramosissimum	12.14	7.35	0.394	0.060	0.091	0.153	0.751	0.465	1917.2
	Portulaca oleracea	12.04	7.23	0.408	0.058	0.090	0.150	0.759	0.463	1931.5
	Salsola baryosma	12.27	7.65	0.384	0.064	0.084	0.154	0.757	0.475	1921.8
	Setaria glauca	12.34	7.78	0.400	0.061	0.091	0.149	0.778	0.498	1978.8
	Suaeda fruticosa	11.99	7.47	0.378	0.060	0.090	0.148	0.780	0.475	1933.9
	Trianthema portulacastrum	12.15	7.26	0.378	0.057	0.090	0.150	0.766	0.478	1920.9
	Mean	12.19	7.43	0.382	0.059	0.082	0.152	0.757	0.478	
	CD (0.05) species	0.11	0.13	2.42	1.16	2.36	2.02	4.02	3.95	

Location 3	*Atriplex amnicola*	15.54	8.77	0.256	0.014	0.036	0.066	0.427	0.367	1.168
	Atriplex nummularia	13.17	8.94	0.276	0.013	0.029	0.051	0.342	0.290	1.003
	Cyanodon dactylon	20.65	8.86	0.357	0.021	0.046	0.081	0.641	0.370	1.518
	Salvadora persica	12.47	9.21	0.218	0.013	0.029	0.048	0.372	0.245	9.027
	Suaeda fruticosa	13.68	8.91	0.303	0.013	0.034	0.054	0.342	0.398	1.146
	Mean	15.10	8.94	0.282	0.015	0.035	0.060	0.425	0.334	
	CD (0.05) species	0.31	0.22	2.15	1.00	1.07	1.18	3.80	3.35	
Location 4	*Heliotropium ramosissimum*	17.60	8.20	0.208	0.012	0.017	0.028	0.325	0.268	8.060
	Salsola baryosma	18.60	8.80	0.211	0.013	0.020	0.026	0.268	0.225	7.065
	Solanum xanthocarpum	17.50	9.50	0.218	0.011	0.017	0.025	0.416	0.300	9.089
	Suaeda fruticosa	14.90	9.30	0.206	0.010	0.018	0.024	0.324	0.278	8.061
	Mean	17.15	8.95	0.211	0.012	0.018	0.026	0.333	0.268	
	CD (0.05) species	0.21	0.20	2.10	1.01	1.03	1.03	3.50	3.00	
Location 5	*Acacia ampliceps*	5.19	7.95	0.106	0.009	0.018	0.030	0.216	0.170	5.052
	Acacia colei	6.97	8.69	0.176	0.009	0.022	0.034	0.318	0.215	7.076
	Atriplex amnicola	5.31	8.32	0.116	0.008	0.018	0.026	0.294	0.200	6.063
	Atriplex lentiformis	8.75	8.70	0.176	0.008	0.025	0.035	0.440	0.275	9.062
	Atriplex nummularia	10.85	8.06	0.220	0.010	0.020	0.034	0.330	0.275	8.092
	Colophospermum mopane	6.10	8.95	0.124	0.007	0.013	0.026	0.266	0.215	6.053
	Salvadora persica	7.34	8.06	0.168	0.011	0.019	0.029	0.306	0.230	7.065
	Suaeda nudiflora	6.94	7.90	0.142	0.010	0.016	0.031	0.248	0.205	6.053
	Mean	7.18	8.33	0.153	0.009	0.019	0.031	0.302	0.223	
	CD (0.05) species	0.25	0.21	1.89	0.99	1.00	1.10	3.06	2.75	

Table 3.4. Mineral ion accumulation characteristics of stem of plants at different locations.

Location	Species	Ash content (mg g^{-1} dry mass)	TDS (mg l^{-1})	Ions (mg g^{-1})						
				Na^+	K^+	Ca^{2+}	Mg^{2+}	Cl^-	SO_4^{2-}	Total
Location 1	*Acacia nilotica*	80	0.95	61.80	12.10	6.30	21.70	56.30	15.45	173.65
	Achyranthes aspera	135	1.78	63.20	12.90	7.60	25.60	59.70	20.88	189.88
	Calotropis procera	130	0.80	61.20	11.70	7.60	20.40	48.30	18.15	167.35
	Capparis aphylla	104	0.99	76.70	13.60	6.50	23.30	69.20	18.75	208.05
	Cyanodon dactylon	90	0.89	36.20	12.70	8.10	25.60	41.30	15.00	138.90
	Cyperus rotundus	93	0.98	34.10	13.30	8.90	26.70	35.30	16.15	134.45
	Physalis longifolia	99	0.95	76.30	9.40	6.00	19.80	70.10	10.80	192.40
	Prosopis juliflora	70	0.99	29.60	11.50	8.70	26.60	31.70	14.89	122.99
	Saccharum munja	99	0.99	50.10	13.80	9.20	25.40	36.30	18.35	153.15
	Salsola baryosma	296	1.90	61.30	9.50	6.10	19.90	56.30	19.10	172.20
	Sesbania sesban	84	0.73	51.70	12.60	8.50	25.70	41.80	15.75	156.05
	Setaria glauca	110	0.95	64.90	15.60	7.30	24.40	59.30	19.25	190.75
	Solanum xanthocarpum	142	1.02	56.50	14.20	7.30	26.60	53.10	20.18	177.88
	Suaeda fruticosa	295	3.15	81.30	10.20	5.80	17.60	71.30	18.98	205.18
	Tamarix dioica	100	1.00	31.60	9.60	6.30	23.60	42.60	16.78	130.48
	Trianthema portulacastrum	115	1.85	36.20	15.70	11.11	27.30	29.20	18.15	137.66
	Mean	127.63	1.25	54.54	12.40	7.58	23.76	50.11	17.29	
	CD (0.05) species	4.01	0.12	0.70	0.73	1.01	1.04	3.01	1.03	
Location 2	*Aerva tomentosa*	125	1.85	75.75	7.25	4.60	14.00	59.15	13.62	29.06
	Arundo donax	105	1.45	65.01	6.45	3.65	13.14	46.35	10.86	145.46
	Chenopodium ambrosioides	199	0.86	25.28	6.39	3.06	9.09	31.56	8.95	84.33
	Chenopodium murale	230	1.38	44.96	5.86	3.35	12.06	41.62	10.56	118.41
	Cyanodon dactylon	114	0.93	27.89	6.44	3.13	12.07	34.54	12.04	96.11
	Heliotropium ramosissimum	270	0.95	26.23	4.48	3.00	8.94	27.08	8.08	77.81
	Portulaca oleracea	233	1.86	82.55	7.61	4.13	13.80	60.58	15.06	183.73
	Salsola baryosma	295	1.88	78.76	7.38	4.36	14.15	60.11	14.61	179.37
	Setaria glauca	112	0.90	27.85	3.69	2.00	7.45	23.58	6.55	71.12
	Suaeda fruticosa	297	3.11	81.82	13.27	5.98	22.56	80.26	26.52	230.41
	Trianthema portulacastrum	113	1.78	86.92	7.48	3.60	14.06	59.16	14.09	185.30
	Mean	19.27	1.54	56.64	6.94	3.71	12.85	47.64	12.81	
	CD (0.05) species	4.21	0.13	0.75	0.81	0.93	0.97	3.15	0.98	

Location 3	*Atriplex amnicola*	110	1.82	44.60	21.50	13.20	27.60	38.30	22.45	167.65
	Atriplex nummularia	147	1.93	56.70	20.00	12.80	26.78	37.18	23.15	176.61
	Cyanodon dactylon	118	1.40	43.10	21.10	24.20	27.10	37.60	20.00	173.10
	Salvadora persica	105	1.07	48.30	18.70	13.60	25.20	39.40	18.15	163.35
	Suaeda fruticosa	301	3.15	87.30	16.40	18.30	21.70	92.30	25.18	261.18
	Mean	156.2	1.87	56.00	19.54	16.42	25.68	48.96	21.79	
	CD (0.05) species	3.98	0.11	0.86	0.70	1.02	0.98	3.40	1.10	
Location 4	*Heliotropium ramosissimum*	275	0.98	51.60	16.50	15.70	24.20	57.90	20.10	186.00
	Salsola baryosma	296	2.00	59.10	15.40	14.70	23.10	81.80	15.15	209.25
	Solanum xanthocarpum	145	0.98	21.20	19.70	15.30	26.20	22.30	20.01	124.71
	Suaeda fruticosa	300	3.15	92.60	15.70	14.30	21.90	103.30	17.80	265.60
	Mean	254	1.78	56.12	16.83	15.00	23.85	66.33	18.27	
	CD (0.05) species	4.26	0.15	0.99	0.79	0.95	0.95	3.35	1.15	
Location 5	*Acacia ampliceps*	090	1.00	34.30	19.40	20.60	24.30	42.40	21.05	162.05
	Acacia colei	093	1.75	64.60	18.90	26.30	22.70	72.60	19.98	225.08
	Atriplex amnicola	092	1.01	38.60	23.40	19.60	23.40	46.40	20.25	171.65
	Atriplex lentiformis	130	1.97	70.10	20.30	24.60	27.60	68.90	23.45	234.95
	Atriplex nummularia	145	1.89	61.70	16.20	23.90	25.10	59.30	20.78	206.98
	Colophospermum mopane	093	1.00	18.40	18.10	20.40	20.90	17.45	17.45	112.70
	Salvadora persica	102	1.05	18.90	16.60	24.40	16.90	17.00	17.00	110.80
	Suaeda nudiflora	280	3.01	54.10	18.50	13.40	25.10	30.45	30.45	172.00
	Mean	128.13	1.59	45.09	18.93	21.65	23.25	44.31	21.30	
	CD (0.05) species	4.25	0.14	0.90	0.75	1.05	1.15	3.38	1.10	

Table 3.5. Mineral ion accumulation characteristics of leaf of plants at different locations.

Location	Species	Ash content (mg g^{-1} dry mass)	TDS (mg l^{-1})	Ions (mg g^{-1})						
				Na^+	K^+	Ca^{2+}	Mg^{2+}	Cl^-	SO_4^{2-}	Total
Location 1	*Acacia nilotica*	94	1.00	91.10	10.60	11.30	24.20	89.60	15.20	242.00
	Achyranthes aspera	198	2.00	71.30	11.70	9.30	21.30	68.70	18.45	200.75
	Calotropis procera	202	0.86	118.30	13.40	9.60	21.40	106.70	17.75	287.15
	Capparis aphylla	199	1.02	109.40	11.10	5.30	26.70	116.50	18.15	287.15
	Cyanodon dactylon	125	1.27	44.20	13.10	7.20	19.30	55.60	15.30	154.70
	Cyperus rotundus	129	1.02	74.60	13.30	6.70	14.40	62.30	12.78	184.08
	Physalis longifolia	140	1.00	168.30	10.40	9.40	20.30	141.70	16.45	366.55
	Prosopis juliflora	95	1.05	83.60	10.40	9.20	21.40	71.40	17.78	213.78
	Saccharum munja	113	1.03	77.30	12.70	9.70	20.10	79.40	15.20	214.40
	Salsola baryosma	335	2.25	149.30	10.30	6.90	21.40	131.30	20.35	339.55
	Sesbania sesban	115	0.98	66.70	12.30	6.40	21.10	79.60	13.80	199.90
	Setaria glauca	215	1.02	85.40	15.60	8.40	20.30	71.30	18.78	219.78
	Solanum xanthocarpum	170	1.00	111.30	12.40	5.10	29.30	101.30	20.15	279.55
	Suaeda fruticosa	336	3.35	141.20	13.40	9.40	21.30	121.30	20.15	326.75
	Tamarix dioica	190	1.20	68.40	9.60	5.30	20.10	69.40	15.20	188.00
	Trianthema portulacastrum	125	1.99	68.40	11.30	13.60	20.30	61.30	16.80	191.70
	Mean	173.81	1.38	95.55	11.98	8.30	21.43	89.21	17.02	
	CD (0.05) species	2.99	0.10	2.00	0.17	0.89	1.00	1.20	0.79	
Location 2	*Aerva tomentosa*	340	2.15	155.00	16.00	9.55	27.00	145.02	35.55	388.12
	Arundo donax	158	1.57	180.57	8.69	5.86	19.06	45.51	20.25	279.94
	Chenopodium ambrosioides	335	0.98	56.30	7.88	5.96	17.41	43.99	18.47	150.01
	Chenopodium murale	265	1.49	78.93	7.89	5.35	18.13	50.61	18.31	167.22
	Cyanodon dactylon	363	1.22	58.48	8.12	6.07	21.03	46.45	21.57	161.72
	Heliotropium ramosissimum	285	0.99	71.37	5.15	4.88	16.41	40.46	16.07	154.34
	Portulaca oleracea	298	2.13	148.30	13.49	7.62	23.07	97.53	25.46	315.47
	Salsola baryosma	332	2.14	140.59	12.55	7.65	17.35	98.23	22.11	298.48
	Setaria glauca	219	0.92	98.23	5.60	3.09	16.11	36.38	14.29	113.70
	Suaeda fruticosa	340	3.29	161.80	16.43	9.67	27.08	148.08	37.59	400.65
	Trianthema portulacastrum	121	1.98	148.00	12.39	7.12	21.10	103.86	22.38	314.85
	Mean	277.82	1.71	112.51	10.38	6.62	20.34	77.83	22.91	
	CD (0.05) species	3.23	0.11	2.02	0.80	0.91	1.01	1.42	0.97	

Location 3	*Atriplex amnicola*	189	2.23	79.50	16.20	7.40	22.90	71.60	23.30	220.90
	Atriplex nummularia	269	2.32	85.70	15.50	7.12	21.15	68.69	25.00	223.16
	Cyanodon dactylon	268	1.50	71.30	14.80	18.90	24.90	63.10	21.00	214.00
	Salvadora persica	129	1.15	86.40	12.30	7.60	23.50	72.70	20.01	222.51
	Suaeda fruticosa	345	3.33	156.70	9.10	5.70	19.30	161.90	28.09	380.79
	Mean	240	2.11	95.92	13.58	9.34	22.35	87.60	23.48	
	CD (0.05) species	3.00	0.12	1.90	0.20	1.01	0.99	1.23	0.99	
Location 4	*Heliotropium ramosissimum*	295	1.02	96.60	11.50	7.30	19.90	103.80	16.20	255.30
	Salsola baryosma	335	2.18	127.30	9.80	6.40	20.20	136.20	10.80	310.70
	Solanum xanthocarpum	178	1.08	49.70	16.40	7.90	22.30	53.80	18.15	168.25
	Suaeda fruticosa	345	3.32	154.10	9.30	6.10	19.70	176.50	15.75	381.45
	Mean	288.25	1.90	106.93	11.75	6.93	20.53	117.58	15.23	
	CD (0.05) species	3.20	0.11	1.99	0.25	1.00	1.02	1.35	1.00	
Location 5	*Acacia ampliceps*	97	1.02	80.90	16.30	19.80	24.90	86.10	21.25	249.25
	Acacia colei	180	2.15	102.10	13.20	18.40	23.90	104.60	20.48	282.68
	Atriplex amnicola	96	1.03	76.70	15.40	19.30	13.70	81.20	15.45	221.75
	Atriplex lentiformis	265	2.31	106.70	12.60	17.90	23.20	108.30	16.48	285.18
	Atriplex nummularia	267	2.26	109.40	14.30	18.60	24.20	91.70	27.45	285.65
	Colophospermum mopane	97	1.03	19.70	33.80	62.60	19.40	26.70	18.30	180.50
	Salvadora persica	120	1.10	35.30	19.60	48.80	15.40	70.20	20.05	209.35
	Suaeda nudiflora	320	3.20	113.70	13.90	9.30	23.60	119.50	22.15	302.15
	Mean	180.25	1.76	80.56	17.39	26.84	21.04	86.04	20.20	
	CD (0.05) species	3.15	0.12	2.00	0.19	1.05	1.03	1.25	1.00	

were higher in *S. fruticosa*, *S. baryosma*, *Aerva tomentosa* and *Chenopodium murale*. Further, in general, ash content and TDS were higher in leaf as compared to stem. Data on ionic composition of stem and leaf presented in Tables 3.4 and 3.5, respectively, also show that Na^+, K^+, Ca^{2+}, Mg^{2+}, Cl^- and SO_4^{2-} (as well as total ionic accumulation) were lower in stem as compared to leaves. It was interesting to note that, in general, Na^+ was the predominant cation as compared to K^+, Ca^{2+} and Mg^{2+}, and Cl^- was the predominant anion as compared to SO_4^{2-} in both tissues. Higher ionic levels, particularly leaf Na^+ and Cl^-, were observed in *S. fruticosa*, *A. tomentosa*, *S. baryosma* and *C. murale*. Tables 3.4 and 3.5 show that plants like *C. dactylon*, *C. rotundus*, *P. juliflora* and *S. glauca* growing in soil with high rhizospheric salinity had, in general, lower ionic accumulation as indicated by their low ash content and tissue TDS. Their ionic accumulation (e.g. Na^+ and Cl^-), particularly in the leaves, was also low.

Gonzalez *et al.* (2005), Karimi *et al.* (2005) and Kumar *et al.* (2016) reported that high leaf Na was a characteristic feature of leaves of halophytes, and that it correlated well with the overall ionic accumulation as well as ash content of the leaves (shoots). Results presented in Table 3.5 also shows that high leaf Na and ash content was reflected in ionic content and ash content. Thus, *S. fruticosa* had a leaf Na^+ content of 161.8 mg g^{-1} and 345 mg g^{-1} dry mass ash content, whereas *C. dactylon* had 44.2 mg g^{-1} Na^+ and a low ash content of 125 mg g^{-1} dry mass. It is further concluded that species like *S. fruticosa* were typical halophytes whereas *Cyanodon* etc., are salt-avoiding plants that share the overall saline habitat but confine their roots to non-saline domains of the rhizosphere. It is concluded that simple and straightforward parameters like ash content can be handy tools for rapid screening of the ion hyperaccumulation and halophytic nature of plants.

3.3.4 Species selection for control condition studies

Using the data pertaining to plant biomass accumulated on saline soils (Table 3.2) and on saline soil ionic accumulation per unit leaf (mg g^{-1}), the plants (Table 3.5) of different locations were shortlisted for subsequent control condition experiments. We concluded that the ten promising species were: *S. fruticosa*, *H. recurvum*, *S. baryosma*, *P. oleracea*, *S. nudiflora*, *A. amnicola*, *A. nummularia*, *A. lentiformis*, *A. donax* and *H. ramosissimum*.

Further close examination of the data (Table 3.3) suggested that plant species like *S. fruticosa*, *S. nudiflora*, *S. baryosma*, *A. nummularia* and *A. amnicola* grow *in situ* under comparatively high saline areas (ECe ± 18 dS m^{-1}). On the other hand, species like *A. tomentosa*, *P. oleracea*, *T. portulacastrum*, *A. donax* and *S. persica* grew widely in marginal saline areas (ECe ± 12 dS m^{-1}); and *C. procera*, *S. xanthocarpum*, *S. glauca* and *S. munja* grew under arid conditions but in the least saline patches (ECe ± 9 dS m^{-1}). While these results indicated that members of the Chenopodiaceae and Portulacaceae grew under comparatively more saline conditions; that members of the Salvadoraceae and Amaranthaceae grew in marginal saline lands; and that those belonging to Poaceae, Asclepiadaceae and Solanaceae were restricted to arid and low-saline areas. Sen *et al.* (1982) have also categorized such plants on the basis of physiographic ionic accumulation in both plant and soil into true halophytes, facultative halophytes, transitional halophytes, marshy halophytes and true glycophytes. Similarly, Salvi *et al.* (2017) has described the existence of alpha-, beta- and gamma-biodiversity of the arid zone flora of the Kuchh region of Gujarat. These workers have also analysed the biomass of plants native to saline lands concomitant with the analysis of rhizospheric soil.

3.4 Concluding Remarks

The studies discussed in section 3.3.4 provide useful information for the scientists working in the field of salinity, enabling them to instantly consider the flora of such extreme arid and saline-arid climates *vis-à-vis* their endogenous ionic and edaphic conditions and to investigate further. Using such data the present investigators (Devi *et al.*, 2008; Devi *et al.*, 2016) identified and exploited salt hyperaccumulator plants like *S. fruticosa*, *S. baryosma* and *S. nudiflora* for salinity phytoremediation. Further, we are of the opinion that the application of modern plant biology techniques on the membrane physiology of these plants can help to identify various molecular biology traits (2D-PAGE, DNA Finger Printing, QTLs) for biotechnological interventions.

Acknowledgements

We are grateful to Dr Rajiv Angrish, ex. Principal Scientist, Department of Botany and Plant Physiology, CCS Haryana Agricultural University, Hisar, for his valuable suggestions, extensive discussions and critical reading of this chapter.

References

Adolf, V.I., Jacobsen, S.E. and Shabala, S. (2013) Salt tolerance mechanisms in quinoa (*Chenopodium quinoa* Willd.). *Environmental and Experimental Botany* 92, 43–54.

Angrish, R. and Devi, S. (2014) Potential of salt hyperaccumulation plants in salinity phytoremediation. *Advanced Plant Physiology* 15, 307–323.

Burns, R.G., Rogers, S. and Meghee, I. (1996) Remediation of inorganics and organics in industrial and urban contaminated soil. In: Naidu, R., Kookana, R.S., Oliver, D.P., Rogers, S. and Mclanghin, M.J. (eds) *Contaminants and the Soil Environment in the Australasia – Pacific Region*. Kluwer Academic Publishers, London, pp. 125–181.

Chesnin, L. and Yien, C.H. (1950) Turbidimetric determination of available sulphur. *Soil Science Society of America, Proceedings* 15, 149–151.

Dagar, J.C. (2005) Ecology, management and utilization of halophytes. *Bulletin of the National Institute of Ecology* 15, 81–97.

Devi, S., Rani, C., Datta, K.S., Bishnoi, S.K., Mahala, S.C. *et al.* (2008) Phytoremediation of soil salinity using salt hyperaccumulator plants. *Indian Journal of Plant Physiology* 13, 347–356.

Devi, S., Nandwal, A.S., Angrish, R., Arya, S.S., Kumar, N. *et al.* (2016) Phytoremediation potential of some halophytic species for soil salinity. *International Journal of Phytology* 18, 693–696.

Gomez, M.B., Castro, P.A., Mignone, C. and Bertero, H.D. (2011) Can yield potential be increased by manipulation of reproductive partitioning in quinoa (*Chenopodium quinoa*)? Evidence from gibberellic acid synthesis inhibition using paclabutrazol. *Functional Plant Biology* 38, 420–430.

Gonzalez, M.B., Fournier, J.M., Ramos, J. and Benlloch, M. (2005) Strategies underlying salt tolerance in halophytes are present in *Cynara cardunculus*. *Plant Science* 168, 653–659.

Kalra, Y.P. (1998) *Handbook of Reference Methods for Plant Analysis*. Soil and Plant Analysis Council, Inc. CRC press, Boca Raton; Boston; London.

Karimi, G., Ghrobanli, M., Heidari, H., Nejad, R.A.K. and Assareh, M.H. (2005) The effect of NaCl on growth, water relation osmolytes and ion content in *Kochia prostrata*. *Biologia Plantarum* 49, 301–304.

Kumar, A., Kumar, A., Lata, C. and Kumar, S. (2016) Eco-physiological responses of *Aeluropus lagopoides* (grass halophyte) and *Suaeda nudiflora* (non-grass halophyte) under individual and interactive sodic and salt stress. *South African Journal of Botany* 105, 36–44.

Mahmood, K., Malik, K.A., Lodhi, M.A.K. and Sheikh, K.K. (1994) Soil-plant relationships in saline wastelands: Vegetation, soils and succession changes during biological amelioration. *Environment Conservation* 21, 236–241.

Mor, R.P. and Manchanda, H.R. (1992) Influence of phosphorous on the tolerance of table pea to chloride and sulphate salinity in sandy soil. *Arid Land Research* 6, 41–52.

Salvi, H., Das, L., Brahmbhatt, B., Vagheta, N. and Kamboj, R.D. (2017) Diversity of halophytes in gulf of kachchh Gujarat. *International Journal of Life Science and Scientific Research* 3(3), 995–1002.

Sen, D.N. and Rajpurohit, K.S. (1982) *Contributions to the Ecology of Halophytes*. Dr. W. Junk Publishers, The Hague.

Sen, D.N, Rajpurohit, K.S. and Wissing, F.W. (1982) *Survey and Adaptive Biology of Halophytes in Western Rajasthan, India*. Department of Botany and Geography, University of Jodhpur. Vol. 2. Dr. W. Junk Publishers, The Hague; London.

Shah, J.P. (2017) Halophytes a resource for the future in Arid regions of Kachchh. *Current Trends in Biomedical Engineering & Biosciences* 4(1), 1–2.

Richards, L.A. (ed.) (1954) *Diagnosis and Improvement of Saline and Alkali Soils*. USDA Handbook No. 60. USDA, Oxford; and IBH, New Delhi.

Vera-Estrella, R., Barkla, J.B. and Pantoja, O. (2005) Salt stress in *Thellungiella halophylla* activates Na^+ transport mechanism required for salinity tolerance. *Plant Physiology* 139, 1507–1517.

4 Halophytic Vegetation in South-east Europe: Classification, Conservation and Ecogeographical Patterns

Zora Dajić Stevanović[1,*], Svetlana Aćić[1], Danijela Stešević[2], Milica Luković[3] and Urban Šilc[4]

[1]University of Belgrade, Faculty of Agriculture, Belgrade, Serbia; [2]Biotechnical Faculty, University of Podgorica, Podgorica, Montenegro; [3]University of Kragujevac, Vrnjačka banja, Serbia; [4]Scientific Research Centre of the Slovenian Academy of Sciences and Arts (ZRC SAZU), Institute of Biology, Ljubljana, Slovenia

Abstract

Halophytic vegetation is very complex and diverse, and the main factors for its development are salinity and moisture. Investigations into different types of halophytic vegetation have been compiled in vegetation databases as scientific tools for classification analyses and interpretation of European saline habitats. Two major types of halophytic habitats exist in Europe: inland (continental) and coastal (maritime), with different halophytic plant communities. Inland halophytic vegetation is classified into two main classes: Thero-Salicornietea: annual succulent vegetation of extreme salt-rich soils, and Festuco-Puccinellietea: saline steppe grasslands. Coastal halophytic vegetation is represented by five classes: Saginetea maritimae: ephemeral vegetation, Crithmo-Staticetea: vegetation of salt-sprayed coastal cliffs, Cakiletea maritimae: pioneer vegetation of sandy and shingle beaches, Ammophiletea: tall-grass vegetation on mobile coastal dunes and Helichryso-Crucianelletea maritimae: dwarf shrub and grassland vegetation on stabilized dunes. Saline habitats are globally endangered; major threats are thought to be ploughing and melioration, as well as abandonment of traditional management (grazing and mowing), eutrophication and ruderalization. Several species of continental and maritime halophytic vegetation are listed in Annex I of the Habitats Directive, and saline habitats are part of the Natura 2000 network (Council Directive 92/43/EoEC 1992).

Keywords: Coastal ecosystem; Eutrophication; Grassland vegetation; Inland salinity; Salinity

4.1 Introduction

Salt-affected soils, essential elements for the formation of halophytic and sub-halophytic vegetation, occur particularly in regions where the strong concentration of salts in the soil is caused by high evaporation of water in the summer (negative balance of precipitation) and specific geological and geomorphological conditions of the area. Naturally occurring salt-affected soils cover about 1 billion ha. Main regions of appearance of this kind of habitat are arid areas (like deserts, semi-deserts, chernozem soil and forest–steppe), which are considered to occupy about 36% of total world territory (Vasin, 2009). It is estimated that saline habitats are distributed

* Corresponding author e-mail: dajic@agrif.bg.ac.rs

across the USA, Russia, western Siberia and Central Asia, eastern Mongolia, southern Ukraine, Hungary and Romania west of the Carpathian Basin; whereas only isolated fragments are found along the lower Morava River, in the Viennese Basin and Marchfeld, China, the central Eurasian continent, South-east Asia, Egypt, South Africa and Australia (Vasin, 2009; Borhidi *et al.*, 2012; Janssen *et al.*, 2016).

There are many definitions of halophytes (Grigore *et al.*, 2000), but phytosociologists generally use this term for plants growing naturally in salt-affected environments. Such environments are widespread and diverse, and include coastal habitats, land within the tidal zone, salt marshes, inland salt lakes and saline deserts (Flowers and Colmer, 2015). In order to survive in such an extreme ambience, plants have developed a variety of salt-tolerance mechanisms at molecular, cellular and tissue levels, as well as on a whole-plant level (e.g. Flowers *et al.*, 1986; Hameed *et al.*, 2010). Halophytic flora and vegetation represent a key denominator of saline habitats. Approximately 2% of terrestrial plant species are salt tolerant (Glenn *et al.*, 1999). Halophytes are considered a native flora of saline habitats, able to grow successfully in conditions of 2%–6% salts (Liang *et al.*, 2018), and of withstanding even 20% salt in the soil. Salt-tolerant plants are the most pronounced feature of saline habitats, where different species have evolved a range of adaptive strategies and mechanisms to survive the different levels of salt stress (e.g. Dajić, 2006; Hasegawa *et al.*, 2000). Only based on salt tolerance, the plants are divided into the euhalophytes (tolerant) and glycohalophytes (sensitive), with several transitional types such as salt-excreting plants and salt-avoiding plants, for example (Voronkova *et al.*, 2008). Many different types of salt-tolerant plants can be found within the same habitat, being spatially and temporally distributed upon local and specific environmental conditions, which can vary on a very narrow scale, depending first on the composition and concentration of salts in the soil. However, different sets of soil properties influence the distribution pattern of species and communities (Castañeda *et al.*, 2013).

Halophytic vegetation is influenced by several stress factors such as high osmotic pressure, ion toxicity, unfavourable soil structure, suboptimal soil pH and nutrient deficiency (e.g. Füzy *et al.*, 2010). The important factors influencing the vegetation composition in salt-affected communities are soil pH and elevation (Wanner *et al.*, 2014). Halophytic vegetation is very diverse, complex, fragile and changeable; important factors in its development include changes in intensity of salinity and moisture. Despite many years of research, there are controversial concepts in halophytic vegetation classification, depending on the classification criteria, geographical scale and research 'school'.

4.2 Review of Existing Databases of Halophytic Vegetation

During the last decade, development in the compilation of vegetation plot databases have led to an effort to create classification systems valid across large areas. These could serve, for example, as a basis for interpretation of the habitats of European conservation interests within the Natura 2000 network (European Commission, 1992; Eliáš, 2012). So far, these studies have been strongly biased towards central Europe, where the most complete vegetation databases exist. In addition, some threatened habitats, such as inland halophytic wetlands (including those of central Europe), have not yet been included in transnational vegetation surveys.

The Global Index of Vegetation-Plot Databases (GIVD) is an overview of existing vegetation data worldwide, where 259 databases with 3,226,037 vegetation plots are registered (Dengler *et al.*, 2011).

The list of European vegetation databases with *relevés* of halophytic vegetation only is as follows:

- Halophytic Vegetation in Serbia (GIVD No. EU-RS-001), with more than 2000 *relevés* of halophytic vegetation (Dajić Stevanović *et al.*, 2012).
- Halophytic and coastal vegetation database of Ukraine (GIVD No. EU-UA-005), with more than 4600 *relevés* of halophytic vegetation. The database contains *relevés* of halophytic vegetation of Ukraine (Thero-Salicornietea, Salicornietea fruticosae, Juncetea maritimi, Festuco-Puccinellietea, Crypsidetea aculeatae, Bolboschoenetea maritimi), and also of the Black and Azov Sea coast vegetation including the estuary regions of large rivers

(Cakiletea maritimae, Ammophiletea, Crithmo-Staticetea) (http://www.givd.info/ID/EU-UA-005).
- Coastal Vegetation Database of North-Western Seas of Russia (GIVD No. EU-RU-004), with 1530 *relevés* of halophytic vegetation from coasts of the Barents Sea, White Sea and Baltic Sea (http://www.givd.info/ID/EU-RU-004) (Sorokin *et al.*, 2012a).
- Coastal Vegetation Database of Southern Seas of Russia (GIVD No. EU-RU-005), with about 2000 *relevés* of halophytic vegetation from coasts of the Black Sea and Sea of Azov (Sorokin *et al.*, 2012b).
- Mediterranean Ammophiletea database (GIVD No. EU-UA-005), with more than 8000 *relevés* of halophytic vegetation of Ammophiletea class of West Mediterranean, North Africa, Middle East and Black Sea (http://www.givd.info/ID/EU-00-016).
- The European Coastal Vegetation Database (GIVD No. EU-00-017), with 26,000 *relevés* from beaches, dunes, salt marshes and cliffs in Europe (http://www.givd.info/ID/EU-00-017).
- Coastal Vegetation Germany (GIVD No. EU-DE-035), with 10,772 *relevés* of halophytic vegetation from coastal areas in Germany (http://www.givd.info/ID/EU-DE-035).
- The database VegDunes (GIVD No. EU-IT-005), with phytosociological *relevés* of more than 2800 Italian coastal dunes (http://www.givd.info/ID/EU-IT-005).
- The database RanVegDunes (GIVD No. EU-IT-020) with phytosociological *relevés* of about 1000 Italian coastal dunes (http://www.givd.info/ID/EU-IT-020).

Many *relevés* of halophytic vegetation are part of national vegetation databases such as: Austrian Vegetation Database (GIVD No. EU-AT-001), Bulgarian Vegetation Database (GIVD No. EU-BG-001), Czech National Phytosociological Database (GIVD No. EU-CZ-001), German Grassland Vegetation Database (GrassVeg.DE, GIVD No. EU-DE-020), Romanian Grassland Database (GIVD No. EU-RO-008), Vegetation Database of Slovenia (GIVD No. EU-SI-001) and the Slovak Vegetation Database (EU-SK-001).

Records of halophytic vegetation are parts of vegetation databases of specific regions or grasslands such as:

- Lower Volga Valley Phytosociological Database (EU-RU-002), with *relevés* of Crypsidetea aculeatae, Thero-Salicornietea, Salicornietea fruticosae and Artemisietea lerchianae classes from the Volga delta (Golub *et al.*, 2009).
- Vegetation Database of the Volga and the Ural Rivers Basins (GIVD No. EU-RU-003), with *relevés* of halophytic vegetation from classes Thero-Salicornietea and Festuco-Puccinellietea (Lysenko *et al.*, 2012).
- Phytosociological Database of Non-Forest Vegetation in Croatia (GIVD No. EU- HR-001), with 483 *relevés* of halophilous coastal vegetation from Croatia (http://www.givd.info/ID/EU-HR-001).
- Vegetation Database of Ukraine and Adjacent Parts of Russia (GIVD No. EU-UA-006), with *relevés* of halophytic vegetation from classes Ammophiletea, Thero-Salicornietea, Salicornietea fruticosae and Festuco-Puccinellietea from the Azov Sea region on the territories of Ukraine and Russia (http://www.givd.info/ID/EU-UA-006).

4.3 Main Saline Habitats in Europe

In Europe, two major types of halophytic habitats exist: the continental (inland) and coastal (maritime), comprising a range of specific, different and unique vegetation units: that is, halophytic plant communities.

Inland halophytic vegetation represents a relevant type of intra-zonal vegetation whose development depends on local soil and hydrological characteristics (Vuckovic, 1982). The inland alkaline and saline vegetation is characterized by high landscape-scale patchiness and rapid shifts among vegetation types due to uneven salt accumulation in the soil and water supply, various tolerance responses of plant species to salt stress and different human disturbances (Kelemen *et al.*, 2013).

The origin and development of the inland halophytic vegetation of continental Eurasia greatly depend on climate. This type of vegetation develops on saline and alkaline soils and, owing to long-lasting summer aridity and evaporation, the groundwater with dissolved salts moves upwards in the soil and the salts are left behind in the topsoil, or form the crystals on the soil surface (Borhidi *et al.*, 2012). European inland saline habitats mainly belong to the Pannonian biogeographic region, characterized by solonetz soils. These are saturated, even shallowly flooded locally, by soluble carbonates in spring, then dry in summer with polygonal surface cracking. Solonetz soils are characterized by high concentrations of easily soluble salts, especially sodium and potassium carbonates (Dajić, 2006). These soils are wet in spring and can be locally shallowly flooded; however, they are dry in summer when polygonal cracks often appear on the surface. The columnar structure of the solonetz soils and erosion by floods causes the development of complex microtopography of the salt steppe. According to variations in salinity, slope and erosion by spring floods, the steppe habitats are a complex mosaic of dry grasslands and herbaceous halophytic communities, often rich in endemic species and endemic plant communities (Janssen et al., 2016). Salt steppes (also called alkali steppes) occur on plains in the Eurasian steppe and forest-steppe zones from the Great Hungarian Plain and adjacent areas through the Danube Lowland in Serbia, Romania and Bulgaria to Ukraine, Russia, Kazakhstan and Mongolia (Eliáš *et al.*, 2013).

Inland, highly salinized ecosystems are characterized by herbaceous vegetation with a limited species pool, where floristic biodiversity is negatively correlated with increases in soil salinity and soil pH (Dajić Stevanović *et al.*, 2008). They develop under the influence of increased concentration of sodium salts originating from decomposition of particular soil sediments, saline underground waters or accumulation of salts, usually in shallow soil sinks (Zedler *et al.*, 2008). Such salt-affected soil type is known as solonchak; it is often characterized by spring floods of saline water, and a remarkable white soil crust on the surface during the dry summer season. Continental solonchak habitats frequently occur around salt lakes and salt marshes. Worldwide, they occupy a total of about 260 million ha. However, according to data from the European soil atlas (Jones *et al.*, 2005), solonchaks represent a relatively small area in Europe (less than 1% of a whole territory). In south-east Europe (the territory of Serbia and the former Yugoslav republics) they are distributed on about 70,000 ha, but locally they are very important (Vasin, 2009). These habitats mostly occur in the form of a mosaic of plots within larger non- or less-salinized soils. The vegetation of continental saline habitats is well pronounced in the area of the Pannonian Plain; while, at its south, the occurrence of solonchak soils – and therefore the appearance of euhalophytic vegetation – decreases (Zlatkovic *et al.*, 2005).

Recent studies on the inland halophytic vegetation of central and southern Europe are focused on alkaline wetlands and salt marshes (Wanner *et al.*, 2014), dry alkaline grasslands (Kelemen *et al.*, 2013; Willner *et al.*, 2013; Valkó *et al.*, 2014; Dajić Stevanović *et al.*, 2016), saline meadows (Dítě *et al.*, 2009) and the management of various types of vegetation of salt-affected soils (Dajić Stevanović *et al.*, 2008). Communities of continental halophytic vegetation mainly appear within complexes consisting of several different plant associations. There are two basic types of the association's layout within the complex. One is a zonal type on the edges of the saline lake; the other is mosaic and appears in the conditions of a complex micro-relief (Vuckovic, 1986).

From a general point of view, coastal vegetation comprises marine algal vegetation of the littoral and sub-littoral, the phanerogamic and algal vegetation of salt and brackish marshes, the vegetation of sand dunes together with that of their 'slacks', specialized vegetation associated with the driftline, the vegetation of shingle beaches and plants found on coastal cliffs and mangrove vegetation (Chapman, 1976).

Saline soils on the seashore develop as a result of regular flooding at high tide. In these soils, salt concentration decreases proportional to the distance from the waterline. Associations of halophytic species differing in degree of salinity tolerance are, accordingly, arranged in strips running parallel to the seashore. Exceptions to this pattern are the concentric vegetation belts at the shores of salt- or brackish-water lagoons cut off from the open sea by maritime dunes, which are relatively similar to the vegetation zones around ordinary lakes (Borhidi *et al.*, 2012).

4.4 Basic Characteristics and Classification of European Halophytic Vegetation

According to recent research into vegetation units in Europe proposed by Mucina *et al.* (2016), halophytic vegetation is organized into 19 classes, while a previous study (Rodwell *et al.*, 2002) listed only 14 classes. Some problems in classification of halophytic vegetation are generally the same as for all types of vegetation, but there are specific issues for halophytic plant communities due to extreme environmental conditions, sometimes causing the appearance of a vegetation unit comprising only few or even one single species.

Fig. 4.1. Halophytic species *Salicornia prostrata* on solonchak, Vojvodina, Serbia, August 2012.

4.4.1 Inland halophytic vegetation: examples of central and south-east European saline habitats

Traditional phytosociological classification of halophytic vegetation is very inconsistent among specific smaller regions, such as for countries of central and south-eastern Europe (e.g. Kojić *et al.*, 1998; Sanda *et al.*, 2008; Tzonev *et al.*, 2008; Chytrý, 2010, Borhidi *et al.*, 2012; Dajić Stevanović *et al.*, 2016), as well within large-scale analyses (Golub *et al.*, 2005; Eliáš *et al.*, 2013). Halophytic vegetation is most commonly classified into two main classes: Thero-Salicornietea and Festuco-Puccinellietea. Some authors distinguish classes Crypsietea aculeatae and Scorzonero-Juncetea gerardii, while plant communities of less-salinized soils are classified within Phragmito-Magnocaricetea, Molinio-Arrhenatheretea and Plantaginetea majoris classes and are transitional towards other vegetation types (e.g. meadows, wetlands and trampled vegetation). South-eastern Europe and the Balkan Peninsula, with its long biogeographical gradient and altitudinal shift, represent a special challenge for vegetation classification (Dajić Stevanović *et al.*, 2016).

4.4.1.1 Annual succulent vegetation of extreme salt-rich soils (Thero-Salicornietea)

This vegetation includes open, species-poor communities of annual succulent halophytes of the genera *Salicornia* and *Suaeda* in both maritime and inland salt marshes (Fig. 4.1). This type of vegetation is distributed on extreme salt-rich wet soils (Dajić Stevanović *et al.*, 2016). The soils are periodically flooded – or at least wet – in spring, and dry out in summer. The habitats of this vegetation have been drained and subsequently converted to agricultural land in central Europe and thus are disappearing (Chytrý, 2010).

1. Pannonian vegetation of annual succulent halophytes on solonchak soils (Salicornion prostratae): This type of vegetation includes continental salt marshes with domination of annual succulent halophytes such as *Salicornia prostrata*, *Suaeda prostrata* and *S. pannonica*. It occurs on strongly saline, loamy or clayey soils. This alliance is distributed in Austria, Hungary, Romania, Serbia, Bulgaria and Ukraine (Borhidi *et al.*, 2012; Dajić Stevanović *et al.*, 2016). It was found in dry areas of southern Moravia, but today this vegetation has been destroyed in the Czech Republic (Chytrý, 2010).

2. Vegetation of annual graminoids of Crypsietalia aculeatae: This type of vegetation includes species-poor stands of pioneer ephemeral halophytic communities with *Crypsis aculeata* (Fig. 4.2). These communities are distributed in south-east Europe and Pannonia lowland. They develop on the shores of salt pools or at the bottom of small dried depressions. Habitats are very salinized, with salt crusts on the soil surface (Dajić Stevanović *et al.*, 2016).

3. Alliance Cypero-Spergularion salinae (Slavnić, 1948): This vegetation includes saline annual grasslands developed on shallow depressions

Fig. 4.2. Mosaic complex of halophytic vegetation communities dominated by *Scirpus maritimus, Puccinellia limosa* and *Crypsis aculeata,* south Serbia, July 2013.

with domination of *Crypsis aculeata, Cyperus pannonicus* and *Spergularia salina* (Borhidi *et al.*, 2012). The north-western distribution of this alliance is in the southern Czech Republic (Chytrý, 2010).

4.4.1.2 Saline steppe grasslands (Festuco-Puccinellietea)

The class Festuco-Puccinelietea includes saline steppe grasslands of the continental regions of Europe. These perennial grasslands develop on continental saline soils influenced by trampling or grazing. The habitat is slightly less salinized and drier than habitats of the annual halophytic vegetation of Thero-Salicornietea and Crypsietalia aculeatae. These halophytic grasslands spread from western to south-eastern Europe and continue to Asia (Eliáš *et al.*, 2013; Dajić Stevanović, 2016).

1. Saline steppe pastures on solonetz soils (Festucion pseudovinae): Alliance F. pseudovinae includes Pontic-Pannonian saline steppic grasslands dominated by the grass *Festuca pseudovina* and obligate halophytes such as *Artemisia santonicum, Limonium gmelinii* and *Plantago schwarzenbergiana.*

2. Pannonian open saline grasslands on solonetz soils (Puccinellion limosae): This alliance includes saline grasslands dominated by halophytic grasses of the genus *Puccinellia*, occurring on drier habitats on both solonetz and solonchak soils. It is distributed in Pannonian dry lowlands in contact with steppe vegetation (Dajić Stevanović *et al.*, 2010; Borhidi *et al.*, 2012).

3. South Balkan relict saline grasslands (Puccinellion convolutae): South Balkan halophytic vegetation develops in the lowest parts of the flat depressions in the form of patchy mosaics (Fig. 4.3). The dominant species are the halophytic grass *Puccinellia festuciformis* subsp. *convoluta, Spergularia marina* and annuals such as *Plantago coronopus* and *Camphorosma monspeliaca.* Alliance Pucinellion convolutae is distributed in southern Serbia, the Republic of Macedonia and Bulgaria (Eliáš *et al.*, 2013; Dajić Stevanović *et al.*, 2016).

Fig. 4.3. Mosaic of communities dominated by *Camphorosma monspeliaca* and *Puccinellia convoluta*, south Serbia, July 2013.

4. Wet subsaline meadows and pastures (Juncion gerardii): This alliance includes wet saline meadows developed on solonchak soils dominated by *Carex distans*, *C. divisa* and *Juncus gerardi*.

5. Pannonian wet subsaline meadows (Beckmannion eruciformis): This alliance develops on slightly saline and periodically flooded soils with dominance of *Agrostis stolonifera* and *Beckmannia eruciformis* (Dítě *et al.*, 2012). According to Deák *et al.* (2014) this halophytic vegetation is among the best-preserved saline habitats in Europe, harbouring a unique flora including several endemics such as *Cirsium brachycephalum* and *Puccinellia distans* subsp. *limosa*.

4.4.1.3 Vegetation of slightly salt-affected soils (Molinio-Arrhenatheretea and Phragmito-Magnocaricetea)

Other types of halophytic vegetation developed on the slightly salinized soils of south-eastern Europe are classified into the wet meadows of Molinio-Arrhenatheretea or the swamps of the Phragmito-Magnocaricetea class.

1. Pannonian saline meadows (Trifolio-Ranunculion pedati): The alliance Trifolio-Ranunculion pedati includes the mesophilous grassland vegetation of Pannonian plain that thrives in sites with high variability of soil moisture and slight salinity (Šilc *et al.*, 2014; Mucina *et al.*, 2016).

2. Balkan wet meadows (Trifolion resupinati): This alliance includes the vegetation of wet meadows of the sub-arid continental regions of the southern Balkans (Eliáš *et al.*, 2013; Šilc *et al.*, 2014; Dajić Stevanović *et al.*, 2016).

3. Tall-rush subsaline reed communities (Meliloto dentati-Bolboschoenion): This alliance includes tall, wet communities (Fig. 4.4) developed on moderately saline swamp habitats with *Scirpus maritimus*, *Cyperus pannonicus*, *Eleocharis palustris* and *S. lacustris* subsp. *tabernaemontani* of the continental regions of the Iberian Peninsula and the Pannonian Basin (Hroudová *et al.*, 2009; Dajić Stevanović *et al.*, 2016).

4.4.2 Coastal halophytic vegetation: examples of south-east European and Adriatic maritime habitats

According to the recent study of Mucina *et al.* (2016), the vascular vegetation of the European

Fig. 4.4. *Meliloto dentati-Bolboschoenion*, south Serbia, July 2013.

coastline is represented by the following five classes:

1. Saginetea maritimae: Atlantic-Mediterranean and Macaronesian ephemeral winter-annual vegetation in disturbed saline habitats and inland saline badlands. The plant communities of Saginetea maritimae typically occupy small disturbed sites within large stands of Juncetea maritimi (Fig. 4.5). The class includes 2 orders, 2 groups of alliances and 11 alliances.
2. Crithmo-Staticetea: Rupicolous vegetation of salt-sprayed coastal cliffs of the Atlantic and Mediterranean seaboards of Europe (also found in North Africa and the Middle East). The plant communities of Crithmo-Staticetea are classified into 4 orders, 5 groups of alliances and 18 alliances.
3. Cakiletea maritimae: Pioneer halo-nitrophilous short-lived vegetation in strandlines of sandy and shingle beaches of the coasts of the North Atlantic and Arctic Oceans, the Mediterranean and the Black Sea. Class Cakiletea maritimae comprises 4 orders and 7 alliances.
4. Ammophiletea: Tall-grass perennial swards on mobile coastal dunes of the seaboards of Europe, but also in North America, Greenland, North Africa, the Middle East and the Caspian Sea. Class Ammophiletea includes 3 orders and 6 alliances.
5. Helichryso-Crucianelletea maritimae: Atlantic, Mediterranean and also Euxinian dwarf shrubs and grasslands on stabilized coastal grey hind dunes. Class Helichryso-Crucianelletea maritimae includes 3 orders and 10 alliances.

Owing to the huge heterogeneity of the halophytic vegetation of Europe, in this contribution we will focus on the psammophytic plant communities of the Balkan Peninsula and its Adriatic coast.

Sand dunes occur all along the Mediterranean coast (Fig. 4.6) but, because of specific geomorphology, they are very rare along the Adriatic coast of the Balkan Peninsula, in contrast to the Adriatic coasts of Italy and the Aegean coasts of the Balkan Peninsula (Trinajstić, 1974, Simeoni *et al.*, 1997). According to some general patterns of zonation of the psammophytic vegetation along the gradient perpendicular to the shoreline, the following more or less distinct zones are distinguished:

i) zone without vegetation (aphytoic zone);
ii) zone of organic/sand deposition (annual vegetation on drift lines, EU habitat type 1210);
iii) embryonic dunes (EU habitat type 2110);
iv) mobile (white) dunes (shifting dunes along the shoreline with *Ammophila arenaria*- EU habitat 2120); and
v) stabilized (grey) dunes (Crucianellion maritimae fixed beach dunes, EU habitat type 2210 and fixed coastal dunes with herbaceous vegetation, EU habitat type 2130, endemic for the northern Adriatic) (Biondi, 2007).

Nevertheless, such complete zonation is not uniformly developed along the entire coast. Sometimes the reason is the small width of a dune (e.g. Croatia; Alegro *et al.*, 2004), but more frequently natural or human stressors (Šilc *et al.*, 2016a). In zones (ii) to (v) the following plant communities are mainly observed: (ii): Cakilo-Xanthietum strumarii, Salsolo kali-Cakiletum maritimae; (iii): Sporoboletum, Echinophoro-Elymetum farctii, Echinophoro-Elymetum farcti othanthetosum, Eryngio-Sporoboletum virginici, Euphorbio paraliae-Agropyretum junceiformis, Cypero mucronati-Agropyretum juncei; (iv): Ammophiletum australis, Medicagini marinae-Ammophiletum australis, Echinophoro-Ammophiletum australis, Sileno-Vulpietum; and (v) Ephedra distachya, Euphorbio-Silenetum nicaeensis, Ephedro distachyae-Silenetum subconicae, Crucianellietum maritimae, Artemisio variabilis-Ephedretum distachyae, Plantagini albicantis-Scabiosetum albae and Tortulo-Scabiosetum albae (Šilc *et al.*, 2016a). Not all listed plant communities are present at the same locality, and differences between the regions are relatively obvious.

General features considering the common species are: (i) among the species characteristic

Fig. 4.5. Mediterranean salt meadows (Juncetalia maritimi), Montenegro, June 2013.

Fig. 4.6. Foredune, Montenegro, June 2013.

for the class Cakiletea maritimae (sensu Mucina, 1997) most common are *Xanthium orientale* ssp. *italicum* and *Cakile maritima*; and (ii) among the common species of the Ammophiletea are *Elytrigia juncea*, *Echinophora spinosa*, *Eryngium maritimum*, *Euphorbia paralias* and *Medicago marina*. There are, however, some differences. Along the Italian part of the Adriatic coast the vegetation of a drift line mainly corresponds to the association Salsolo kali-Cakiletum maritimae, although it occurs very discontinuously, because of the impact of touristic activities (Veronico *et al.*, 2017).

In the Balkan part of the Adriatic coast the vegetation of a drift line is ascribed to ass. Cakilo-Xanthietum strumarii (Šilc *et al.*, 2016a), while in Greece both associations occur (Sýkora *et al.*, 2003). Vegetation of embryonic dunes and white dunes (Fig 4.7) is more or less uniform and represented by ass. Echinophoro-Elymetum farctii and Medicagini marinae-Ammophiletum australis (Korica and Lovrić 1979; Trinajstić, 2008; Pirone, 2014; Šilc *et al.*, 2016a). Vegetation of grey dunes, alliance Crucianellion maritimae, that occurs in the back dunes or interdunes of mobile dunes (Stanisci *et al.*, 2004), is common in Greece, on the northern coasts and in the Peloponnese (Sýkora *et al.*, 2003), while on the Adriatic coast it is known in Puglia (Pirone, 2014) and Albania (Šilc *et al.*, 2016a). Thus, it does not occur in the eastern part of the Adriatic coast from Montenegro to Croatia. The grey dune vegetation of the Malcolmietalia, which has western and southern Mediterranean distribution and comprises ephemeral therophytic dune vegetation, is present only on the Italian coast (Pirone, 2014). As a specificum of Montenegrin and Albanian dunes, Šilc *et al.* (2016a) pointed out the hygrophilous communities in interdunal depressions: Juncetum maritimo-acuti, Eriantho-Schoenetum nigricantis and Holoschoenetum romani (Šilc *et al.*, 2016a). Vegetation of eastern Adriatic sandy shores is represented by six plant communities.

Fig. 4.7. White dune, Montenegro, June 2013.

4.5 Main Threats and Conservation of Saline Habitats

Saline habitats are fragile and highly endangered. Generally, they are specific ecosystems that characterize the presence of characteristic flora and vegetation. In addition to fragility, the other important trait of these habitats is fragmentation. Saline vegetation is highly endangered and under extreme human pressure because of efforts to increase the area of fertile arable land, as well as to place ponds, infrastructure or building sites (Zedler *et al.*, 2008, Dajić Stevanović *et al.*, 2016).

The major threat that led to the disappearance of these habitats in central Europe was agriculture, especially through melioration and ploughing, and the drainage which lowered the groundwater level (Šefferová Stanová *et al.*, 2008). In addition, abandonment of traditional management (grazing and mowing), eutrophication and ruderalization have changed halophytic vegetation: rare halophytic species have become extinct and invasive species are spreading (e.g. *Phragmites australis* and *Elaeagnus angustifolia*). We urgently need to consider adequate conservation measures to maintain such unique and rare habitats, because of the permanent human impact on biodiversity in agricultural and rural areas of central and south-eastern Europe.

Continental saline wet and solonetz meadows are characterized by several species listed in Annex I of the Habitats Directive (Eliáš *et al.*, 2013). Halophytic habitats are, therefore, of European conservation interests and are listed within the Natura 2000 network (Council Directive 92/43/EEC 1992). These habitats are classified as 1310 *Salicornia* and other annuals colonizing mud and sand, 1340 Inland salt meadows and 1530 Pannonian salt marshes.

According to the Habitats Directive (92/43/EEC), coastal habitats are classified as 'Annual vegetation of drift lines', habitat code 1210, the 'Embryonic shifting dunes', habitat code 2110, the 'Shifting dunes along the shoreline with *Ammophila arenaria*', habitat code 2120, 'Vegetated sea cliffs of the Mediterranean coasts with endemic *Limonium* spp.', habitat code 1240 etc. Recently, the conservation value of coastal and halophytic ecosystems has been stressed in the EU project 'Red List of Habitats' (Janssen *et al.*, 2016). According to this publication, coastal habitats are among the most threatened groups of habitats in the EU.

Five habitat types are assessed as Endangered:

1. A2.5b Baltic coastal meadow.
2. B1.4b Mediterranean and Macaronesian coastal dune grassland (grey dune).
3. B1.4c Black Sea coastal dune grassland (grey dune).
4. B1.6c Macaronesian coastal dune scrub.
5. B1.7b Black Sea broad-leaved coastal dune woodland.

Eight habitat types are assessed as vulnerable:

1. A2.5c Atlantic coastal salt marsh.
2. B1.1a Atlantic, Baltic and Arctic sand beach.
3. B1.3b Mediterranean and Black Sea shifting coastal dune.
4. B1.4a Atlantic and Baltic coastal dune grassland (grey dune).
5. B1.5a Atlantic and Baltic coastal Empetrum heath.
6. B1.6b Mediterranean and Black Sea coastal dune scrub.
7. B1.7c Baltic coniferous coastal dune woodland.
8. B1.8a Atlantic and Baltic moist and wet dune slack.

Urbanization is recognized as the most important threat to all coastal habitats, followed by recreational (over-)use by people. Invasive non-native species form the next important threat in coastal habitats, both in open habitat types such as beaches, white dunes and cliffs, and in more stabilized ecosystems, like grey dunes and forests. The main threats to inland European halophytic plant communities are melioration, over-grazing, construction of fish ponds and – in some cases – pollution and improper waste management. The impact of climate change is as yet hard to assess, but some changes are clear and probably increasing, particularly with higher temperatures in the Arctic and at high altitudes (Jansen *et al.*, 2016). Because of very specific flora and vegetation and numbers of rare (e.g. *Salsola soda*, *Lepidium cartilagineum*, *Limonium gmelinii*) and/or endemic plants (e.g. *Plantago schwarzenbergiana*, *Rorippa kerneri*, *Suaeda pannonica*), such habitats need further conservation attention.

References

Alegro, A., Biljaković, M., Bogdanović, S. and Boršić, I. (2004) Psammo-halophytic vegetation on the largest sand area on the Croatian coast: The island of Mljet, southern Adriatic. *Biologia* 59, 435–445.

Biondi, E. (2007) Thoughts on the ecology and syntaxonomy of some vegetation typologies of the Mediterranean coast. *Fitosociologia* 44, 3–10.

Borhidi, A., Kevey, B. and Lendvai, G. (2012) *Plant Communities of Hungary*. Akadémiai Kiadó, Budapest.

Castañeda, C., Herrero, J. and Conesa, J.A. (2013) Distribution, morphology and habitats of saline wetlands: a case study from Monegros, Spain. *Geologica Acta* 11, 371–388.

Chapman, V.J. (1976) *Coastal Vegetation*. 2nd edn. Pergamon Press Ltd, Oxford, UK.

Chytrý, M. (ed.) (2010) *Vegetace České republiky 1. Travinná a keříčková vegetace.* [*Vegetation of the Czech Republic 1. Grassland and heathland vegetation*]. 2nd edn. Academia, Prague.

Dajić Stevanović, Z., Pecinar, I., Kresovic, M., Vrbnicanin, S. and Tomovic, L. (2008) Biodiversity, utilization and management of grasslands of salt affected soils in Serbia. *Community Ecology* 9(Suppl.1), 107–114.

Dajić Stevanović, Z., Kresović, M., Pećinar, I., Aćić, S., Obratov-Petković, D. *et al.* (2010) Distribution of the halophytic grass *Puccinellia limosa* (Schur.) Holomb. on salt affected soils in Serbia in relation to its main adaptive responses to salinity. *Ekológia* 29, 258–268.

Dajić Stevanović, Z., Petrović, M., Šilc, U. and Aćić, S. (2012) Database of halophytic vegetation in Serbia. *Biodiversity and Ecology* 4, 417–417.

Dajić Stevanović, Z., Aćić, S., Luković, M., Zlatković, I., Vasin, J., *et al.* (2016) Classification of continental halophytic grassland vegetation of Southeastern Europe. *Phytocoenologia* 46(3), 317–331.

Dajic, Z. (2006) Salt stress. In: Madhava Rao K., Raghavendra A., Janardhan Reddy K. (eds) *Salinity and Tolerance Mechanisms in Plants. Physiology and Molecular Biology of Stress Tolerance in Plants.* Springer, Dordrecht, pp. 41–99.

Deák, B., Valkó, O., Török, P. and Tóthmérész, B. (2014) Solonetz meadow vegetation (*Beckmannion eruciformis*) in East-Hungary – an alliance driven by moisture and salinity. *Tuexenia* 34, 187–203.

Dengler, J., Jansen, F., Glöckler, F., Peet, R.K., De Cáceres, M. *et al.* (2011) The global index of vegetation-plot databases (GIVD): A new resource for vegetation science. *Journal of Vegetation Science* 22, 582–597.

Dítě, D., Eliáš, P. and Šuvada, R. (2009) The current distribution and status of community *Puccinellietum limosae* in Slovakia. *Thaiszia* 19, 63–70.

Dítě, D., Hrivnák, R., Melečková, Z., Eliáš, P. and Dajić-Stevanović, Z. (2012) *Beckmannia eruciformis* vegetation in the Pannonian Basin (Central and South-Eastern Europe). *Phyton* 52, 177–194.

Eliáš, P., Sopotlieva, D., Dítě, D., Hájková, P., Apostolova, I., *et al.* (2013) Vegetation diversity of salt-rich grasslands in southeast Europe. *Applied Vegetation Science* 16, 521–537.

European Commission (1992) Council Directive 92/43/EEC of 21 May 1992 on the conservation of natural habitats and of wild fauna and flora. Available at: https://eur-lex.europa.eu/eli/dir/1992/43/oj (accessed 26 September 2018).

Flowers, T.J. and Colmer, T.D. (2015) Plant salt tolerance: Adaptations in halophytes. *Annals of Botany* 115(3), 327–331.

Flowers, T.J., Hajibagheri, M.A. and Clipson, N.J.W. (1986) Halophytes. *The Quarterly Review of Biology* 61, 313–337.

Füzy, A., Bíró, B. and Tóth, T. (2010) Effect of saline soil parameterson endomycorrhizal colonisation of dominant halophytes in four Hungarian sites. *Spanish Journal of Agricultural Research* 8, 144–148.

Glenn, E.P., Brown, J.J., and Blumwald, E. (1999) Salt tolerance and crop potential of halophytes. *Critical Reviews in Plant Sciences* 18(2), 227–255.

Golub, V.B., Sorokin, A.N., Ivakhnova, T.L., Starichkova, K.A., Nikolaychuk, L.F., *et al.* (2005) Soobshchestva klassa Festuco-Puccinellietea Soó ex Vicherek 1973 na territorii Evrazii. [Communities of the class Festuco-Puccinellietea Soó ex Vicherek 1973 in Eurasia]. *Rastitel'nost' Rossii* 7, 59–75.

Golub, V.B., Sorokin, A.N., Ivakhnova, T.L., Starichkova, K.A., Nikolaychuk, L.F., *et al.* (2009) Geobotanicheskaja baza dannykh doliny Nizhney Volgi [Lower Volga Valley Phytosociological Database]. – *Izvestiya Samarskogo nauchnogo tzentra RAN* 11, N 1(4), 577–582.

Grigore, M.N., Toma, C. and Boşcaiu, M. (2010) Dealing with halophytes: An old problem, the same continuous exciting challenge. *Analele ştiinţifice ale Universităţii "Al. I. Cuza" Iaşi Tomul LVI, fasc. 1, s. II a. Biologie vegetală* 56, 21–32.

Hameed, M., Ashraf, M., Ahmad, M.S.A. and Naz, N. (2010) Structural and functional adaptations in plants for salinity tolerance. In: Ashraf, M., Ozturk, M. and Ahmad, M. (eds) *Plant Adaptation and Phytoremediation*. Springer Science and Business Media, Berlin, pp.151–170.
Hasegawa, P.M., Bressan, R.A., Zhu, J.-K. and Bohnert, H.J. (2000) Plant cellular and molecular responses to high salinity. *Annual Review of Plant Physiology and Plant Molecular Biology* 51, 463–499.
Hroudová, Z., Hrivnák, R. and Chytrý, M. (2009) Classification of inland *Bolboschoenus* dominated vegetation in Central Europe. *Phytocoenologia* 39, 205–215.
Janssen, J.A.M., Rodwell, J.S., Criado, M.G., Gubbay, S., Haynes, T., *et al*. (2016) European Red List of Habitats, Part 2. Terrestrial and Freshwater Habitats, Luxembourg Publications Office of the European Union, Luxembourg.
Jones, A., Montanarella, L., and Jones, R. (2005) *Soil Atlas of Europe*. European Commission Brussels, Belgium.
Kelemen, A., Török, P., Valkó, O., Miglécz, T. and Tóthmérész, B. (2013) Mechanisms shaping plant biomass and species richness: Plant strategies and litter effect in alkali and loess grasslands. *Journal of Vegetation Science* 24, 1195–1203.
Kojić, M., Mrfat-Vukelić, S., Dajić, Z. and Đorđević-Milošević, S. (1998) *Syntaxonomic review of the vegetation of Serbia*. Institute for Biological Research "Sinisa Stankovic", Beograd, R. Serbia. [In Serbian].
Korica, B. and Lovrić, A.Ž. (1979) Ecology and conservation of coastal psammophytes of NE Adriatic. In: Rauš, Đ. (ed.) Second Conference of Ecologists of Yugoslavia. Alliance of Ecology Associations, Zadar-Plitvice, Croatia, pp. 205–217. [In Croatian]
Liang, W., Ma, X., Wan, P. and Liu, L. (2018) Plant salt-tolerance mechanism: A review. *Biochemical and Biophysical Research Communications* 495, 286–291.
Lysenko, T., Kalmykova, O. and Mitroshenkova, A. (2012) Vegetation database of the Volga and the Ural rivers basins. *Biodiversity & Ecology* 4, 420–421.
Mucina, L. (1997) Conspectus of classes of European vegetation. *Folia Geobotanica et Phytotaxonomica* 32, 117–172.
Mucina, L., Bultmann, H., Dierßen, K., Theurillat, J.P., Raus, T., *et al*. (2016) Vegetation of Europe: Hierarchical floristic classification system of vascular plant, bryophyte, lichen, and algal communities. *Applied Vegetation Science* 19(Suppl. 1), 3–264.
Pirone, G. (2014) Notes on the vegetation diversity on the Adriatic and Ionian Italian coasts: The dunes and cliffs. *Plant Sociology* 51, 7–18.
Rodwell, J.S., Schaminée, J.H.J., Mucina, L., Pignatti, S., Dring, J., *et al*. (2002) The diversity of European vegetation. An overview of phytosociological alliances and their relationships to EUNIS habitats. Report EC-LNV, 54. National Reference Centre for Agriculture, Nature and Fisheries, Wageningen, Netherlands.
Sanda, V., Öllerer, K. and Burescu, P. (2008) *Fitocenozele din România – sintaxonomie, structură, dinamică şi evolutie* [Phytocoenology of Romania – syntaxonomy, structure, dynamics and evolution]. Ars Docendi. Universitatea din Bucureşti, Bucarest.
Šefferová Stanová, V., Janák, M. and Ripka, J. (2008) *Management of Natura 2000 habitats. 1530 *Pannonic salt steppes and salt marshes*. Technical Report 200803/24. European Commission Brussels, Belgium.
Šilc, U., Aćić, S., Škvorc, Ž., Krstonošić, D., and Franjić, J. (2014) Grassland vegetation of the Molinio-Arrhenatheretea class in the NW Balkan Peninsula. *Applied Vegetation Science* 17(3), 591–603.
Šilc, U., Mullaj, A., Alegro, A., Ibraliu, A., Dajić Stevanović, Z., *et al*. (2016a) Sand dune vegetation along the eastern Adriatic coast. *Phytocoenologia* 46, 339–355.
Šilc, U., Stevanović, Z.D., Ibraliu, A., Luković, M. and Stešević, D. (2016b) Human impact on sandy beach vegetation along the southeastern Adriatic coast. *Biologia* 71(8), 865–874.
Simeoni, U., Pano, N. and Ciavola, P. (1997) The coastline of Albania: Morphology, evolution and coastal management issues. *Bulletin de l'Institut Oceanographique* 18, 151–168.
Slavnić, Ž. (1948) Halophytic vegetation of Vojvodina. *Archives for Agricultural Sciences and Technique* 3, 1–80. [In Serbian]
Sorokin, A., Golub, V., Grechushkina, N., Nikolaychuk, L. and Bondareva, V. (2012a) Coastal vegetation database of north-western seas of Russia. *Biodiversity & Ecology* 4, 422–422.
Sorokin, G., Grechushkina, C. and Nikolaychuk, B. (2012b) Coastal vegetation database of the southern seas of Russia. *Biodiversity & Ecology* 4, 423–423.
Stanisci, A., Acosta, A., Ercole, S. and Blasi, C. (2004) Plant communities on coastal dunes in Lazio (Italy). *Annali di Botanica* 4, 115–128.

Sýkora, K.V., Babalonas, D. and Papastergiadou, E.S. (2003) Strandline and sand-dune vegetation of coasts of Greece and of some other Aegean countries. *Phytocoenologia* 33, 409–446.

Trinajstić, I. (1974) Ass. Agropyretum mediterraneum (Kühn.) Br.-Bl. 1933 in the vegetation of south Dalmatian island Korčula. *Glasnik Republičkog Zavoda za Zaštitu Prirode Titograd* 6, 71–76. [In Serbian]

Trinajstić, I. (2008) *Plant Communities of Republic of Croatia*. Academy of Forest Sciences, Zagreb, Croatia. [In Croatian]

Tzonev, R., Lysenko, T., Gusev, C. and Zhelev, P. (2008) The halophytic vegetation in south-east Bulgaria and along the Black Sea coast. *Hacquetia* 7, 95–121.

Valkó, O., Tóthmérész, B., Kelemen, A., Simon, E., Miglécz, T., *et al.* (2014) Environmental factors driving seed bank diversity in alkali grasslands. *Agriculture, Ecosystems & Environment* 182, 80–87.

Vasin, J. (2009) Solonchak soils of Vojvodina – characteristics and modern classification. PhD thesis, University of Novi Sad, Serbia. [In Serbian]

Veronico, G., Sciandrello, S., Medagli, P. and Tomaselli, V. (2017) Vegetation survey and plant landscape mapping of the SCI IT9140002 "Litorale Brindisino" (Puglia, Southern Italy). *Plant Sociology* 54, 89–106.

Voronkova, N.M., Burkovskaya, E.V., Bezdeleva, T.A. and Burundukova, O.L. (2008) Morphological and biological features of plants related to their adaptation to coastal habitats. *Russian Journal of Ecology* 39, 1–7.

Vuckovic, R. (1982) The novel association of the alliance Festucion pseudovinae Soo 1933. *Acta Biologica Yugoslavia, Ekologija* 17, 15–23. [In Serbian]

Vuckovic, R. (1986) Association complex of the halophytic vegetation on locality Deracka bara in the middle Banat. *Proceedings from the Meeting:"Plant and Human"*. Matica srpska, Novi Sad, Serbia, pp. 469–476.

Wanner, A., Suchrow, S., Kiehl, K., Meyer, W., Pohlmann, N. *et al.* (2014) Scale matters: Impact of management regime on plant species richness and vegetation type diversity in wadden sea salt marshes. *Agriculture, Ecosystems & Environment* 182, 69–79.

Willner, W., Sauberer, N., Staudinger, M. and Schratt-Ehrendorfer, L. (2013) Syntaxonomic revision of the Pannonian grasslands of Austria – Part I: Introduction and general overview. *Tuexenia* 33, 399–420.

Zedler, J.B., Bonnin, C.L., Larkin, D.J. and Varty, A. (2008) Salt marshes. In: Jorgensen, S.E. (ed.) *Ecosystem Ecology*. Elsevier, Amsterdam, pp. 384–393.

Zlatkovic, B., Randjelovic, V. and Amidzic, L. (2005) *Flora and vegetation of the central and south Serbia and its valorization from the environment protection view*. Institute for Nature Protection, Nis, Serbia, pp. 1–63. [In Serbian]

5 South African Salt Marshes: Ecophysiology and Ecology in the Context of Climate Change

Pascal Tabi Tabot[1,2,*] and Janine Barbara Adams[1]
[1]*Nelson Mandela Metropolitan University, Port Elizabeth, South Africa;* [2]*University of Buea, Buea, Cameroon*

Abstract
This paper reviews the distribution, zonation and ecophysiological tolerances of salt marsh plants in South Africa. In South Africa salt marsh covers 11,788.8 ha with the largest areas occurring in the Berg, Olifants, Orange and Langebaan Estuaries. The changing climate necessitates in-depth studies on the responses of these plants so that we can understand and predict future changes in their abundance and diversity. Responses of salt marsh plants to abiotic stressors are reviewed and it is shown that, on a local ecosystem scale, distribution of species is driven by moisture/inundation levels and salinity gradients. Ecophysiology studies have been completed on at least 15 salt marsh species in South Africa, focusing on the responses of salt marsh plants to various factorial combinations of abiotic stressors. Studies also exist on characterization and zonation of salt marsh habitats and have significantly affected salt marsh management policy. Salt marsh plants respond to osmotic and ionic stress through reduction in growth and increase in root:shoot ratio, ion sequestration and accumulation of ionic and compatible organic solutes for osmoregulation. These responses are similar, whether elicited by soil moisture deficit or salinity, but may vary in intensity. Submergence results in increased cytoplasmic acidosis, membrane lipid peroxidation and a rapid increase in reactive oxygen species. Physiologically, tolerant species respond to submergence by down-regulating metabolism, increased synthesis of antioxidant enzymes and up-regulating compatible osmolytes. This translates into increased species growth. These responses collectively determine salt marsh zonation and are important, especially in the context of climate change.

Keywords: Ecophysiology; Halophytes; Osmolytes; Reactive oxygen species; Salinity

5.1 Introduction

Salt marshes are transitional habitats between land and sea where herbaceous halophytes in the form of low-growing shrubs, herbs and grasses are dominant. Salt marshes occur predominantly in arctic regions, through to temperate regions where they appear to be less constrained by climate compared to mangroves. They are also found in the subtropics and tropics in localized areas where mangrove development is precluded, or as a component of a salt marsh–mangrove ecotone (Adam, 1990). Perhaps the most commonly used (or referenced) definition is:

> areas, vegetated by herbs, grasses or low shrubs, bordering saline water bodies. Although such areas are exposed to the air for the majority of the time, they are subjected to periodic flooding

* Corresponding author e-mail: ttabot@yahoo.com

as a result of fluctuations (tidal or non-tidal) in the level of the adjacent water body.
(Adam *et al.*, 1990)

Some of the global definitions of salt marsh only consider tidal systems (e.g. Best *et al.*, 2007; Weis and Butler, 2009). In South Africa we have included the seldom-flooded supratidal habitat with halophytic species as salt marsh, as this is a component of the estuary. The 5 m topographical contour is currently used to demarcate the South African estuarine functional zone (EFZ) and the lateral boundaries. Supratidal salt marsh occurs at > 1.5 m above mean sea level (amsl) and an ecotone with terrestrial species can occur from > 2.5 m amsl (Veldkornet *et al.*, 2015a).

Thus, in this review, salt marshes are associated with estuaries and do not include the saline wetlands of the arid inland Northern Cape and Free State provinces of South Africa. In these systems there is no tidal inundation and the chemistry of the dominant salts is derived from geological sources (Seaman *et al.*, 1991; Eallonardo and Leopold, 2014). Here, salt-tolerant lawn grasses and halophytic shrubs are found with some species in common with estuary salt marshes, such as *Cyperus laevigatus* and *Sporobolus virginicus* (Sieben *et al.*, 2015, 2016). In addition, the abiotic drivers of soil moisture and electrical conductivity are similar.

In the vegetation map of South Africa, Mucina and Rutherford (2006) identified Arid Estuarine Salt Marshes found in the cool temperate region with a mean annual temperature of 18°C and mean annual rainfall ranging from 45 mm at the Orange River Estuary to 150 mm at the Olifants Estuary. Cape Estuarine Salt Marshes occurred where there was greater variability in rainfall and climate.

Coastal salt marshes provide a number of ecosystem services such as nursery functions, carbon sequestration, flood regulation, storm protection and food. They are sensitive to physicochemical changes, which influence the abundance and distribution of species. As a result, the changing climate necessitates in-depth studies on the responses of these plants, and how their zonation, abundance and diversity would be affected by climate change. This study reviews salinity, inundation and drought-tolerance mechanisms of salt marsh plants. It reviews the ecophysiological responses of salt marsh plants to climate change, and examines how these responses have influenced species distribution in specific South African salt marshes.

5.2 Distribution of Salt Marsh in South Africa

Salt marsh covers an area of 11,788.8 ha in South Africa and is the dominant habitat type in estuaries of the temperate region, whereas mangroves, reeds and sedges are dominant in subtropical estuaries. It is interesting to note that the five estuaries with the largest salt marsh areas in the country (Table 5.1) are found in different estuary types, namely permanently open estuaries (Berg and Olifants), river mouth (Orange), estuarine bay (Knysna) and a coastal embayment (Langebaan). In the Langebaan and Knysna systems the intertidal salt marshes cover a larger area compared to the supratidal salt marshes, which reflects the large tidal prism and gentle slopes in these estuaries. Although salt marshes are better developed in estuaries with regular tidal exchange and gentle topographic gradients, they may still occur in temporarily open/closed estuaries (TOCEs) which are at times closed to the sea. Here, the frequency and duration of open mouth conditions determines the extent of the salt marsh and the species found typically form a subset of the full suite of species occurring in permanently open estuaries. High salinity conditions as a result of barrier overwash resulted in the growth of large salt marsh areas in the TOCEs in the former Ciskei region along the southern east coast of South Africa (Colloty *et al.*, 2002). In comparison, high rainfall and leaching of salts favour the development of brackish reeds, sedges and grasses in the TOCEs of the subtropical province of KwaZulu Natal, South Africa.

Ecological importance of the intertidal salt marsh is more obvious than that of the supratidal salt marsh, as it stabilizes banks and provides shelter and habitat for fish and invertebrates. Intertidal salt marsh is fairly evenly distributed between the cool temperate (2740.8 ha) and warm temperate (2529.8 ha) regions, whereas supratidal salt marsh is common to the dry cool temperate coast (4568.6 ha) compared to the warm temperate (1297.8 ha) region (Adams *et al.*, 2016). Although the Berg Estuary has the largest intertidal and supratidal salt marsh areas

Table 5.1. Estuaries with the largest salt marsh areas (ha) in South Africa. Extracted from estuary botanical database described in Adams *et al.* (2016).

Estuary	Estuary type	Intertidal	Supratidal	Total (ha)
Berg	Permanently open	1 667	2 545	4212
Langebaan	Coastal embayment	791.7	523.7	1315.4
Olifants	Permanently open	91.94	849.1	941.04
Orange	River mouth	144	602	746
Knysna	Estuarine bay	551	–	550
Klein Brak	Temporarily open/closed	–	494	494
St Lucia	Estuarine lake	414.7	–	414.7
Keiskamma	Permanently open	210.37	91.26	301.63
Heuningnes	Permanently open	5.53	292.51	298.51
Great Fish	Permanently open	133	65	198
Gamtoos	Permanently open	92.92	80.84	173.76
Swartkops	Permanently open	165	5	170
Klein	Estuarine lake	9.7	160	169.7

nationally, the Olifants, Orange and Klein Brak estuaries also have large supratidal salt marsh areas (Table 5.1). Langebaan (coastal embayment) and Knynsa (estuarine bay) have the most extensive intertidal salt marsh areas because of the large tidal prism and gentle slopes. Langebaan is not an estuary as it does not have any riverine freshwater input (Whitfield, 2005). Groundwater input maintains the extensive rush (*Juncus* spp. ~400 ha) stands. The salt marsh species and zonation are similar to other estuaries, indicating the importance of elevation and tides in influencing floristic structure. Besides the salt marshes, the Berg Estuary has a diversity of large freshwater wetlands worthy of conservation. The total estuarine habitat consists of halophytic floodplain (1521 ha), open pan (1159 ha), sedge pan (975 ha) and xeric floodplain (919 ha). Intertidal mudflats with eelgrass (*Zostera capensis*) occupy 206 ha, intertidal salt marsh (499 ha) and reeds and sedges (587 ha) (Adams, 2010).

By virtue of its size, St Lucia Estuary can have a large intertidal salt marsh area (Table 5.1). However, this is only extensive when the water level is low. The area shown in Table 5.1 was mapped for 2008 when succulent species such as *Salicornia meyeriana* and *Sarcocornia* spp. colonized exposed saline soils in False Bay and in the mudflats of North Lake. When the water level rises and inundates these plants they die back as they are not tolerant to long periods of inundation. Rautenbach (2015) showed that the biggest change in vegetation composition between 2008 and 2013 in St Lucia was the overall decrease in salt marsh by 57% and increase in submerged macrophytes by 96%. After the drought, water levels rose rapidly as rainfall returned to normal and the Mfolozi River connected to the sea and St Lucia Estuary. This caused flooding and inundation of the salt marsh habitat, reducing the area covered by these plants.

5.3 Salt Marsh Zonation in South Africa

Salt marsh plants show distinct zonation patterns along tidal inundation (Table 5.2) and salinity gradients. Salt marshes that have a small tidal influence tend to have vegetation arranged in a mosaic pattern and not in distinct zones. In South Africa, salt marsh zonation is well developed in estuaries with a large tidal range (e.g. Berg, Knysna and Swartkops). Intertidal salt marsh occurs below mean high water spring tide, and supratidal salt marsh is found above this. The supratidal salt marsh may only be flooded twice a year during exceptional spring tide events (Adams *et al.*, 1999). *S. pillansii* is common in this supratidal zone and large stands can occur in estuaries such as the Orange, Olifants, Berg, Gouritz and Klein Brak estuaries. Veldkornet *et al.* (2015b) showed that sediment organic content, electrical conductivity and redox potential increased with elevation, while pH decreased. Intertidal species such as *Spartina maritima*, *Triglochin buchenaui* and *T. striata* were associated with low elevation, high sediment moisture content and high sediment electrical conductivity.

Table 5.2. Zonation of salt marsh species in South Africa across an inundation gradient.

Salt marsh zone and description	Dominant species	Source
Submerged (mean high water neap to below sea level) 90–80 cm amsl[a]	*Zostera capensis*	Adams *et al.* (1999)
Lower intertidal (mean high water spring tide to mean high water neap tide level) msl – 100 cm amsl[b]	*Sarcocornia tegetaria, Triglochin bulbosa, T. buchenaui, Cotula coronopifolia*	Day (1981); O'Callaghan (1994); Adams *et al.* (1999) in: Allanson and Baird (1999); Köcke *et al.* (2010)
Upper intertidal (extreme high water spring tide to mean high water spring tide level) 100–130 cm amsl	*Sarcocornia decumbens, Limonium scabrum, Bassia diffusa*	O'Callaghan (1994)
Supratidal (above normal spring tide level) 130–170 cm amsl	*Stenotaphrum secundatum, Cynodon dactylon, Sporobolus virginicus, Sarcocornia pillansii, Disphyma crassifolium, Plantago crassifolia*	Day (1981); O'Callaghan (1994); Bezuidenhout (2011)

[a]amsl = above mean sea level; [b]msl = mean sea level

Zonation patterns across salt marshes have been attributed to both abiotic factors such as waterlogging, salinity gradients, siltation and tidal action, and biotic factors such as competition (Ungar, 1998; Carter and Ungar, 2003). Authors agree that the lower limits of these zones are set primarily by tolerance to physical stress and the upper limits by competition (Snow and Vince, 1984; Bertness and Ellison, 1987; Davy *et al.*, 2001; Emery *et al.*, 2001; Pennings *et al.*, 2005; Cacador *et al.*, 2007; Bornman *et al.*, 2008; Kim *et al.*, 2013; Weilhoefer *et al.*, 2013; González-Alcaraz *et al.*, 2014). Davy *et al.* (2001) found that few halophytes are found naturally in non-saline environments, demonstrating that halophytes have lost their competitive ability through energy and resource expenditure on salt tolerance. This was confirmed by Veldkornet *et al.* (2015a). The zonation patterns of salt marshes are well studied (Weilhoefer *et al.*, 2013; González-Alcaraz *et al.*, 2014; Kim, 2014); however, studies focusing on the salt marsh–terrestrial boundaries are rare. Veldkornet *et al.* (2015a) recent studies have provided some data on these boundary/ecotone zones for South African systems. Ecological thresholds were found for terrestrial vegetation but not for salt marshes; sediment electrical conductivity (threshold ~20–30 mS/cm), groundwater electrical conductivity (~60–80 mS/cm), groundwater salinity (~20–40 ppt) and sediment moisture content (~20%–25%). These results suggest that in the absence of competition and disturbance at the salt marsh–terrestrial boundary, salt marshes will be unaffected by sea level rise and will migrate landwards. This will, of course, depend on available habitat, and Veldkornet *et al.* (2015b) showed that most lateral boundaries of estuaries are disturbed by development. Understanding zonation as a direct response to sea level rise is important, as this will alter habitat conditions at the intertidal and supratidal zones. Critically, with sea level rise, upper intertidal ranges become more submerged and soils more anoxic, and the responses of the species to these changes influence the future zonation of salt marshes.

5.4 Climate Change and Predicted Variability in Abiotic Stress

By their very nature, plants live in constantly varying environments, yet this variability has been accentuated by anthropogenic climate change. Greenhouse gases from deforestation, slash and burn agriculture, burning of fossil fuels, fossil fuel exploration, cultivation, breeding large herds of cattle, disposing of organic matter in landfills and industrial processes – among others – result in increased temperatures, with a ripple effect on other environmental parameters such as rainfall variability and increase in sea levels (IPCC, 2001; Simas *et al.*, 2001; IPCC, 2007). In southern Africa, the resulting temperature rise is associated with rainfall variability and increased drought in some regions (Mason and Jury, 1997; Kruger and Shongwe, 2004). In the central parts of southern Africa, significant increases in rainfall are predicted (Mason and Joubert, 1997), while in eastern South Africa there would be a decrease in expected rainfall and an increase in drought events (Mason and Joubert, 1996). By 2050, rainfall is predicted to decrease in parts of South Africa by 10% (Watson *et al.*, 2001). In Limpopo Province, for example, Thomas *et al.* (2007) found that drought frequency and intensity have increased in the last two decades, concurrent with more rainfall variability in the north-west KwaZulu Natal Province. Drought would result in low freshwater flow and possible mouth closure in TOCEs, resulting in submerged or dry conditions depending on the extent of the drought. In all estuary types, drought events could result in prolonged dry conditions in salt marsh, which could be detrimental to plant growth. Besides the increase in rainfall variability and drought, predicted increase in sea levels would influence survival of salt marsh species. Hernes *et al.* (1995), cited in Watson *et al.* (2001), projected an increase in sea level of 25 cm for Africa. Along the east coast of South Africa, Mather (2007) found an increase in mean monthly and mean annual sea level of 2.7 ± 0.05 and 2.4 ± 0.29 mm year^{-1} respectively, for a 1970–2003 data set. The most current estimates of sea level rise for South Africa (1.5–2.7 mm yr^{-1}) are reported by Mather *et al.* (2009). This increase in sea levels would shift the tidal front and result in larger submerged areas, and habitat alteration beyond the areas directly under tidal inundation (see Table 5.2 in Tabot and Adams, 2013c). For salt marsh plants already living under variable abiotic stress, these conditions could impact species diversity, zonation and survival, thus understanding their responses is fundamental to their conservation. Van Niekerk *et al.* (2011) explored the susceptibility of salt marshes to climate variability. Typically, climate change will alter the key abiotic stressors in salt marshes – changing inundation patterns, changes in salinity and increased frequency of unpredictable storm events. In practice, the effect of these changes can be assessed as impacts of salinity and inundation levels based on predicted climate change conditions. Tabot and Adams (2012, 2013a,b,c, 2014) conducted such a study, using *Triglochin buchenaui*, *Bassia diffusa* and *Limonium linifolium* as model plants. The results of this and other similar studies are presented in the subsequent section.

5.5 Drought Tolerance

Drought stress is an important stressor to consider in salt marsh ecophysiology and the response of salt marshes to climate change. Two categories of salt marsh plants exist: those tolerant to drought and those that are susceptible. Most salt marsh species will tolerate some degree of drought, except perhaps the submerged *Zostera capensis* and *Ruppia cirrhosa*. The extent of drought tolerance depends on mechanisms that maintain water uptake and growth through solute potential (Ψs) adjustments. Some response mechanisms to drought are presented in Table 5.3. Such mechanisms include uptake of inorganic ions and/or synthesis of organic osmolytes (Ashraf and Foolad, 2007; Lee *et al.*, 2007; Chen and Murata, 2008; Naidoo *et al.*, 2008; Chen and Jiang, 2010). While inorganic ions function mainly through ion transport systems, compatible organic solutes balance the osmotic potential due to increasing inter- and intracellular ion build-up, in addition to stabilizing enzymes and proteins, protecting membrane integrity and detoxifying reactive oxygen species (ROS) (Ashraf and Foolad, 2007; Chen and Jiang, 2010). Among the most important compatible osmolytes is proline which is effective in *Limonium vulgare*, *Armeria maritima*, *Puccinellia maritima*, *Triglochin maritima* and *Triglochin buchenaui* (Storey *et al.*, 1977; Tabot and Adams 2012), *Limonium linifolium* (Tabot and Adams, 2014), *T. bulbosa* and *T. striata* (Naidoo and Naicker, 1992; Naidoo, 1994; Naidoo and Naidoo, 2001). Members of the genus *Limonium* have been reported to accumulate large quantities of proline and can convert this to more effective osmoprotectants such as proline betaine and hydroxyproline betaine (Hanson *et al.*, 1994). Glycinebetaine is important in *Atriplex portulacoides*, *Spartina townsendii* and *Aster tripolium* (Storey *et al.*, 1977).

Table 5.3. Tolerance mechanisms used by selected salt marsh species.

Species	Tolerance mechanism under drought and salinity	Compatible solutes/ inorganic ions	Reference
Bassia diffusa	Increase in anthocyanin pigments; lower growth rate	Oxalate	Tabot and Adams (2013a)
Triglochin buchenaui	Reduced growth, roots modify to rhizomes	Proline concentration increase	Tabot and Adams (2012)
Limonium linifolium	Salt secretion, increased branching	Proline concentration increase	Tabot and Adams (2014)
Genus Limonium		Proline and prolinebetaine	Hanson *et al.* (1994)
Triglochin bulbosa and *T. striata*		Proline, ion accumulation	Naidoo (1994)
Juncus kraussii	Growth reduction	Proline and ion accumulation	Naidoo and Kift (2006)

Accumulation of inorganic ions is reported in salt marsh species like *T. bulbosa* and *T. striata*, in which inorganic ions accounted for 72% and 83% of the osmotic adjustment, respectively. Sodium accounted for 22% and 41% of these inorganic ions (in *T. bulbosa* and *T. striata*, respectively) (Naidoo, 1994); in *Bassia diffusa*, the main ionic solute identified was oxalate (Tabot and Adams, 2013a). In the course of taking up salty water, excess salts need to be excreted from the plant to avoid ion toxicity. Salt secretion in salt marsh plants is typically through salt glands, and this process has been reported in *Limonium linifolium* (Tabot and Adams, 2014).

The response to abiotic stress by plants is not only through physiological mechanisms, but also through morphological adaptations. These determine the survival strategies of most species under different stressors. Typically, growth of salt marsh plants is reduced under water stress, and root:shoot ratios increase (Naidoo and Kift, 2006; Brown and Pezeshki, 2007; Tabot and Adams, 2012, 2013a,b, 2014). This increase is driven by an increase in abscisic acid concentration (Munns and Sharp, 1993).

On an ecosystem scale, it has been shown that some species possess the ability to drastically increase root proliferation to groundwater level. The availability of such water drives the distribution and abundance of the species. It was shown, for example, that the distribution of *Sarcocornia pillansii* in the Olifants Estuary in South Africa was determined by the depth to (and salinity of) groundwater, and the ability of the plants to access such water enabled them to survive prolonged drought. Vegetation cover was significantly reduced where the water table was greater than 1.5 m, and where the groundwater electrical conductivity was greater than 80 mS cm^{-1} (Bornman *et al.*, 2004, 2008). This dependence on groundwater has significant implications for salt marshes, as drought conditions have been predicted to increase either in intensity or frequency. Low sediment moisture content in salt marshes invariably leads to high salinity concentration in the sediment, with its accompanying effects.

5.6 Submergence Tolerance in Salt Marsh Plants

Flooding and waterlogging are constant features in intertidal ranges of salt marshes, and are predicted to increase in intensity and frequency as a result of climate change. Documented effects of flooding are presented in Table 5.4. Flooding and waterlogging might reduce growth and photosynthesis in salt marsh plants (Adkins *et al.*, 1990; Naidoo and Naidoo, 1992; Naidoo and Mundree, 1993; Weisner *et al.*, 1993; Colmer and Flowers, 2008). Submergence reduces plant metabolism and decreases available ATP (Voesenek *et al.*, 2006; Bailey-Serres and Voesenek, 2008). It accelerates chlorophyll degradation and membrane lipid peroxidation (Sarkar *et al.*, 2006 and references therein). Under submergence, the enzymes alcohol dehydrogenase (ADH) and

Table 5.4. Documented water level and inundation tolerances for different species.

		Biotic responses to abiotic drivers		
Species	Driver	Physiological response	Morphological response	Reference
Bassia diffusa	Inundation	• reduces plant metabolism • decreases available ATP • accelerates chlorophyll degradation and membrane lipid peroxidation • ADH and PDC are upregulated • accumulation of ROS (Naidoo and Mundree, 1993; Sarkar *et al.*, 2006; Voeseneck *et al.*, 2006; Bailey-Serres and Voesenek, 2008)	Adult plants did not survive 2 weeks of complete submergence	Tabot and Adams (2013b)
Triglochin buchenaui			Growth ceased for 3 months of submergence	Tabot and Adams (2012)
Limonium linifolium			Adult plants died within 2 weeks of complete submergence	Tabot and Adams (2014)
Cynodon dactylon			Tolerates 28 days of flooding	Furness and Breen (1985)
Juncus kraussii			Flood-tolerant species – growth in both drained and flooded conditions	Naidoo and Kift (2006)
Sporobolus virginicus			Tolerates 42 days flooding	Naidoo and Mundree (1993)
Sporobolus virginicus seedlings ≥ 3 months			Growth improved by inundation of 3 cm for 38 days	Breen *et al.* (1977)
Sporobolus virginicus			Maintained growth for 42 days of submergence by re-allocating resources to shoot relative to root growth	Naidoo and Naidoo (1992)
Sarcocornia natalensis			Endures submergence for ≤ 3 months without negative effects	Tölken (1967)
Sarcocornia perennis (now *S. tegetaria*)			≥ 35 PSU and complete submergence over 2 weeks reduced growth	Adams and Bate (1994)
Sarcocornia perennis (now *S. tegetaria*)			Inundation with 5 cm water increased/ stimulated growth	Jackson and Drew (1984)
Spartina maritima			Survives long periods of inundation, therefore occupies lower intertidal regions of salt marsh. Tidal flooding necessary for growth	Adams and Bate (1994, 1995)

pyruvate decarboxylase (PDC) are upregulated for glycolytic respiration (Naidoo and Mundree, 1993; Sarkar *et al.*, 2006 and references therein). Because of the light and oxygen limitation, there is accumulation of ROS which are toxic to membranes, proteins and DNA (Bohnert and Jensen, 1996; Jithesh *et al.*, 2006), and the ability to detoxify and scavenge these free radicals through enzymatic and non-enzymatic pathways (Apel and Hirt, 2004; Miller *et al.*, 2010, Ramel *et al.*, 2012) determines tolerance and survival of species under submergence.

Morphologically, submergence influences growth of species, and results in two sets of responses for tolerant species: escape and quiescent. For example, Cooper (1982) found that upper salt marsh species, but not lower salt marsh species, exhibited severe growth reduction under waterlogged conditions. Table 5.4 shows some of the responses of South African salt marsh species. In a tidal tank experiment, it was shown that *Bassia diffusa* would not survive beyond 2 weeks of complete submergence (Tabot and Adams, 2013b). Susceptibility is marked by a drastic drop in chlorophyll *a* concentration within the first 6 h, and a corresponding increase in the concentration of carotenoids. Oxalate was upregulated to 1.7 g g^{-1}DM in 4 days, with sharp increases in membrane damage after Day 7 (Tabot and Adams, 2013b). This susceptibility to submergence was also observed in the lower intertidal *T. buchenaui* (Tabot and Adams, 2012) and the supratidal *L. linifolium* (Tabot and Adams, 2014).

Species that exhibit such responses possess high concentrations of ABA under submergence. High concentrations of ABA imply low ethylene synthesis needed to stimulate synthesis of gibberelic acid for stem and petiole elongation (Benschop *et al.*, 2005; Jackson, 2008). On the other hand, tolerant species would exhibit rapid growth upon submergence due to ABA down-regulation and reciprocal ethylene up-regulation, the so-called escape response (Voesenek *et al.*, 2003; Benschop *et al.*, 2005; Bailey-Serres and Voesenek, 2008; Jackson, 2008). The species *Sporobolus virginicus*, *Spartina alterniflora* and *Spartina patens* exhibited increase in growth under waterlogged conditions (Naidoo and Naidoo, 1992; Naidoo *et al.*, 1992; Naidoo and Mundree, 1993). Such plants tolerate low oxygen via a suite of metabolic and morphologic adjustments that include increased photosynthetic rates, increased ADH activity counteracted by increased aerenchyma for elimination of hypoxia, increased aboveground growth and reduction in root:shoot ratios.

Beside growth responses, plants that survive submergence possess the capacity for aerenchyma and adventitious root formation (Naidoo and Naidoo, 1992; Naidoo and Mundree, 1993; Colmer, 2003; Colmer and Flowers, 2008), down-regulation of metabolism (Colmer and Flowers, 2008), compatible osmolytes for ROS scavenging (Apel and Hirt, 2004; Ashour and Mekki, 2006; Jithesh *et al.*, 2006; Naidoo and Kift, 2006; Voesenek *et al.*, 2006) and the ability to take up gases and nutrients from the water. Under submergence plants could develop or increase aerenchyma formation and develop thinner leaves and barriers to oxygen diffusion. Studies have reported on adventitious root formation in *Sporobolus virginicus* under submergence (Naidoo and Naidoo, 1992; Naidoo and Mundree, 1993). It has been reported that adventitious root formation is stimulated by accumulation of ethylene during submergence (Voesenek *et al.*, 2006). The development of aerenchyma is a major anoxia avoidance mechanism in plants adapted to submerged conditions, as it enhances the movement of gases within the roots (Colmer, 2003).

For species that exhibit escape responses, rapid internode and petiole elongation and re-orientation towards the light is evident; this facilitates emergence of the plants from complete submergence (Naidoo and Naidoo, 1992; Naidoo and Mundree, 1993; Voesenek *et al.*, 2006). In such species (for example as shown in *Sporobolus virginicus*), there would be a shift in resource re-allocation to aboveground parts (Naidoo and Naidoo, 1992; Naidoo and Mundree, 1993).

Salt marsh species are regarded as terrestrial plants that have evolved tolerance mechanisms to survive in saline waterlogged environments. As with salinity adaptation, different species cope with different degrees of waterlogging. Tolerance to waterlogging conditions involves an approach that waits for optimum conditions for growth. This can be seen with plant species that take on dormancy during severe waterlogged stages and includes seed dormancy (Colmer and Voesenek, 2009). This form of adaptation enables plants to survive short periods of inundation (Bailey-Serres and Voesenek, 2008).

In South Africa, the impacts of submergence have been investigated on an ecosystem scale. In the East Kleinemonde TOCE, the cover abundance of submerged and intertidal macrophytes was shown to be in anti-phase, depending on the mouth status and water levels. There was complete dieback of intertidal macrophytes, and flourishing of submerged macrophytes; the latter die back following subsidence of water, resulting in exposed mud banks which are then repopulated by intertidal macrophytes (Riddin and Adams, 2008). It has been shown (Riddin and Adams, 2009) that such vegetation regeneration is highly dependent on the natural seed bank, but that the size of the seed bank cannot always be correlated with extant populations. This raises important questions: *would the seed bank in these estuaries sustain persistent interrupted lifecycles, or is there a need for artificially maintained seed banks? If prolonged submergence persists for longer periods, would a new guild of species tolerant to these conditions emerge, or would the landscape become void of salt marsh halophytes as some numerical models predict?* Tabot and Adams (2013c) investigated these scenarios under predicted conditions in South African salt marshes, and showed that whereas the lower intertidal species can survive conditions typical of upper intertidal ranges in salt marshes, the reverse was not true of upper intertidal species; thus, predicted climate change conditions would result in depletion of salt marsh species in an already species-poor ecosystem.

5.7 Salinity Tolerance in Salt Marsh Plants

Although salt marshes are naturally saline environments, salinity is augmented by increased seawater intrusion and drought, which concentrates sediment salinity. Salinity responses of salt marsh plants are presented in Table 5.5. Saline soils inhibit the ability of plants to take up water and nutrients, and also result in high internal sodium concentrations that can reach toxic levels (Munns and Tester, 2008 and references therein). Salt stress is manifested as osmotic stress in the root zone, and ionic stress within the plant. Within plants, salt stress is observed as a decrease in new growth of young leaves, shoots and branches due to osmotic stress, and accelerated senescence of older leaves due to ionic stress resulting from high internal Na^+ concentrations. This has been shown for *Spartina maritima* where growth reduces as salinity increases beyond 20% seawater (Naidoo *et al.*, 2012). If the rate of senescence exceeds that of new leaf formation, photosynthesis would be drastically reduced (Munns and Tester, 2008 and references therein; Naidoo *et al.*, 2012). Salt stress also results in reduced stomatal conductance and hence the rate of photosynthesis (Redondo-Gómez *et al.*, 2007). The effects of salt stress are not limited to the single species, but extend to the entire ecosystem. In the Orange River Estuary in South Africa, salt marsh degradation was linked to high sediment electrical conductivity (120.9 ± 4.7 mS cm^{-1} in 2005 (dry year) to 52.5 ± 4.8 mS cm^{-1} (wet year)) as a result of prolonged drought, and the increased salinity reduced seed germination, requiring freshwater for the recovery of the species (Shaw *et al.*, 2008).

Most halophytes are facultative and would grow better under non-saline conditions. For example, *Sarcocornia natalensis* has a tolerance range of 0–17.5 ppt (Table 5.5), surviving such high salinity through ion accumulation, increased succulence and dilution through growth (Naidoo and Rughunanan, 1990), while *Odyssea paucinervis* shows over 50% growth at salinity of 30.4 ppt, but with a decrease in physiological processes as salinity is increased (Naidoo *et al.*, 2008). In *Sporobolus virginicus*, salinity up to 8.8 ppt NaCl increased shoot growth while root growth was improved at salinity of up to 26.3 ppt NaCl, resulting in high root:shoot ratios (Marcum and Murdoch, 1992). In facultative halophytes, therefore, a decrease in growth as salinity is increased is the norm. Obligate halophytes exhibit increased growth as salinity is increased, and would not grow well under non-saline conditions. In these extreme halophytes the thresholds for tolerance and stimulation of growth are much higher. Growth of *Arthrocnemum macrostachyum* is stimulated up to salinity of 29.8 ppt and the increased leaf and shoot area results in dilution and increased tolerance (Khan *et al.*, 2005; Redondo-Gómez *et al.*, 2010). Increase in salt concentration in NaCl-treated grasses was prevented by the diluting effects of continuous growth and production of new leaves in plants like *Aster tripolium* and *Triglochin maritima*, and constitutes a 'growing away from the salt' phenomenon. This is consistent with findings of other authors (Ashour and Mekki, 2006 on *Leptochloa fusca*; and Hamed *et al.*, 2008 on *Batis maritima*).

Less-tolerant plants, on the other hand, would exhibit growth reduction with increasing external salinity (Vincente *et al.*, 2004; Naidoo and Kift, 2006 on *J. kraussii*; Tabot and Adams, 2012 on *T. buchenaui;* Tabot and Adams, 2013a on *B. diffusa*; and Tabot and Adams, 2014 on *L. linifolium*). Naidoo and Kift (2006) studied the tolerance to salinity of *J. kraussii*, and reported a decrease in both aboveground and belowground biomass, as well as in stomatal conductance and the rate of photosynthesis. This is consistent with findings of other authors (Bernstein *et al.*, 1974; Cooper, 1982; Adams and Bate, 1995; Pennings *et al.*, 2005; Naidoo *et al.*, 2008; Marín-Guirao *et al.*, 2011). This

Table 5.5. Optimum salinity range for growth, survival and seed germination of some salt marsh species. Shaded rows indicate international studies included for comparative purposes.

Species	Driver	Tolerance range (ppt)	Biotic response to abiotic drivers	Source
Arthrocnemum macrostachyum	Salinity	10–30	Growth is optimum within this range	Redondo-Gómez *et al*. (2010)
Bolboschoenus maritimus		7–12		Bally *et al*. (1985)
Bolboschoenus maritimus		0.5–5		Lubke and Van Wijk (1988)
Distichlis spicata		0–15		Hansen *et al*. (1976)
Juncus bulbosus		0.3–1.5		Spence (1982)
Juncus kraussii		0–15		Heinsohn and Cunningham (1991)
Juncus kraussii		0–20		Adams *et al*. (1999)
Juncus kraussii		0–24.5		Naidoo and Kift (2006)
Odyssea paucinervis		0.08–30.4		Naidoo *et al*. (2008)
Phragmites australis		0–25		Starfield *et al*. (1989)
Potamogeton pectinatus		5–15		Howard-Williams and Liptrot (1980)
Potamogeton pectinatus		0–19		Ward (1976)
Ruppia cirrhosa		2–40		Verhoeven (1975)
Ruppia cirrhosa		0–50		Ward (1976)
Ruppia cirrhosa		0–55		Adams and Bate (1994)
Ruppia maritima		2–18		Verhoeven (1975)
Salicornia virginica		0–45		Pearcy and Ustin (1984)
Sarcocornia fruticosa		0–30		Redondo-Gómez *et al*. (2006)
Sarcocornia natalensis		0–18		Naidoo and Rughunanan (1990)
Sarcocornia perennis		12–42	Growth is optimum within this range	Adams and Bate (1994)
Sarcocornia pillansii		0–35		Bornman *et al*. (2002)
Spartina maritima		0–35		Adams and Bate (1995)
Spartina patens		63–93		Hester *et al*. (1996)
Sporobolus virginicus		1–13		Breen *et al*. (1977); Marcum and Murdoch (1992); Naidoo and Naidoo (1998)
Tecticornia pergranulata		0–47		Colmer *et al*. (2009)
Triglochin bulbosa		6–12		Naidoo and Naicker (1992); Naidoo (1994)
Triglochin striata		6–12		Naidoo and Naicker (1992); Naidoo (1994)
Zostera capensis		5–40		Day (1981)
Zostera capensis		0–45		Ward (1976)
Zostera capensis		10–46		Adams and Bate (1994)
		Limit of tolerance beyond which growth is inhibited		
Arthrocnemum macrostachyum	Salinity	<58	Growth is inhibited beyond this range	Khan *et al*. (2005)
Atriplex portulacoides		<49		Redondo-Gómez *et al*. (2007)
Sarcocornia natalensis		≤140		O'Callaghan (1992)
Spartina alterniflora		<45		Linthurst and Seneca (1981)
Sporobolus virginicus		<30		Breen *et al*. (1977)
Triglochin bulbosa		<23		Naidoo and Naicker (1992); Naidoo (1994)
Triglochin striata		<15		Naidoo and Naicker (1992); Naidoo (1994)

Continued

Table 5.5. Continued.

Species	Driver	Tolerance range (ppt)	Biotic response to abiotic drivers	Source
			Tolerance range for seed germination (all international studies)	
Aster tripolium		≥53*	At this salinity seeds do not germinate	Woodell (1985)
Atriplex triangularis		≥15*		Khan and Ungar (1984)
Chenopodium glaucum		≥23*		Duan *et al.* (2004)
Juncus maritimus		≥35*		Woodell (1985)
Limonium emarginatum		0–20‡		Redondo-Gómez *et al.* (2008)
Limonium bedilifolium	Salinity	≥17.5*	At this salinity seeds do not germinate	Woodell (1985)
Limonium humile		≥35*		Woodell (1985)
Plantago maritima		≥17.5*		Woodell (1985)
Salicornia fruticosa, Salicornia perennis		≥20‡	At this salinity seeds do not germinate	Redondo-Gómez *et al.* (2004)
		≥20‡		Redondo-Gómez *et al.* (2004)
		≥60*		Redondo-Gómez *et al.* (2004)
Salicornia rubra		≥58*		Khan *et al.* (2000)
Suaeda salsa		≥6‡		Duan *et al.* (2007)
Suaeda salsa		≥30*		Duan *et al.* (2007)
Suaeda vera		≥17.5*		Woodell (1985)
Triglochin maritima		≥35*		Woodell (1985)

Adapted from Adams *et al.* (1999) in: Allanson and Baird (1999); Whitfield and Bate (2007). *Salinity at which germination is inhibited. ‡Optimum range for seed germination.

reduction in plant growth is often through reducing the water potential (Naidoo *et al.*, 2008) or by interfering with nutrient uptake (Awad *et al.*, 1990, Lee *et al.*, 2007). However, salinity-susceptible plants can grow in saline soil by utilizing groundwater in areas where the water table is shallow and less saline, and where there is freshwater seepage. This is the case for *Phragmites australis* stands in the Goukou and Keurbooms estuaries in South Africa (Adams and Bate, 1999). Other plants respond by sequestering Na^+ ions in roots and excluding it in leaf blades and salt glands such that they do not reach toxic levels, or by compartmentalization of Na^+ in vacuoles and intercellular cell walls (Redondo-Gómez *et al.*, 2007; Munns and Tester, 2008; Redondo-Gómez *et al.*, 2010). According to Naidoo and Kift (2006), *J. kraussii* tolerates salinity of 70% seawater by compartmentalizing Na^+ and Cl^- ions in vacuoles of leaves as well as ion sequestration in culms. *Limonium linifolium* survives excess salinity through salt secretion (Tabot and Adams, 2014).

What happens when high salinity is combined with submergence? For some species, growth is stimulated, and decreases as salinity is increased; hence salinity and submergence function synergistically. For example, combined salinity and inundation have been shown to significantly influence growth of *Spartina maritima*; there was greater growth in the submerged treatments compared to dry treatments, with the typical decrease in growth as salinity increased (Adams and Bate, 1995). For salinity-susceptible species such as *J. kraussii*, rates of photosynthesis, stomatal conductance, growth and seed germination would decrease as salinity increases irrespective of the water state (Naidoo and Kift, 2006). It is also possible that some species are more sensitive to submergence, leading to a cessation of growth upon submergence, irrespective of the salinity. In a long-term study of the East Kleinemonde Estuary, Riddin and Adams (2008) found that salinity had little effect on cover abundance of the macrophytes; this was rather determined by water levels.

The above abiotic responses are also mediated by compatible osmotic solutes. Compatible osmotic solutes have the ability to accumulate in the plant cell's cytoplasm at high, yet non-toxic concentrations. *Spartina alterniflora* and *S. patens* accumulate high concentrations of proline as salinity is increased, if this increase is not coupled with submergence (Naidoo *et al.*, 1992). High concentrations of glycinebetaine, proline, amino acids and soluble sugars in leaves of salt marsh plants result in a much more negative water potential than that of the roots and soil (Silveira *et al.*, 2009). Organic solutes like proline accumulate in the cytoplasm and organelles of plants to balance the low osmotic potential in the vacuole (Naidoo and Kift, 2006; Ashour and Mekki, 2006; Tabot and Adams, 2012). Brock (1981) studied the accumulation of proline in members of the genus *Ruppia* under salt stress, and estimated that these plants accumulate proline to a concentration of about 50% of the solute concentration of the substrate. Increase in proline with salinity has also been reported for *R. cirrhosa* and *Zostera capensis*, with higher concentrations in *R. cirrhosa* which was more tolerant to salinity (Adams and Bate, 1994).

Proline and glycinebetaine are thought to preserve protein structure and activity, and deactivate hydroxyl radicals and reactive chemical species (Naidoo and Kift, 2006), and hence protect photosystems.

Another important implication of salinity is the reduction of nutrient uptake by salt marsh plants. High Na^+ concentrations in soils result in replacement of essential macronutrient cations (Ca^{2+} and Mg^{2+}) on clays by Na^+ and inhibit establishment of intolerant species by both specific ion toxicity and nutrient deficiencies (Ungar, 1998). Thus, as Table 5.5 shows, the optimum growth of intertidal species is below 25 ppt, and only submerged salt marshes and some supratidal species thrive at salinity above 35 ppt. At higher salinities, a higher fraction of nitrogen absorbed is diverted to the production of glycinebetaine and/or proline, compatible organic solutes associated with osmotic regulation (Storey *et al.*, 1977; Naidoo and Naidoo, 2001). Another physiological implication of high salinity is its impact on seed germination. Salinity induces physiological dormancy on seeds of salt marsh plants. There was a marked decrease in germination of *Triglochin bulbosa* and *T. striata* seeds (Naidoo and Naiker, 1992), *Juncus kraussii* seeds (Naidoo and Kift, 2006) and a general decline for seed banks of the East and West Kleinemonde TOCEs under different salinity treatments (Riddin and Adams, 2009). Data in Table 5.5 show that the response to salinity varies with species, making salinity a dominant factor influencing zonation in salt marshes. Salinity determines the range of survival of different species, and so halophytes occupy different tidal ranges depending on their salinity tolerances (Table 5.5).

5.8 Episodic Events as a Window into the Future

Episodic events offer unique opportunities for the study of certain phenomena and extremes. The Orange River Estuary, an important RAMSAR site, has been experiencing hypersaline conditions and progressive degradation due to a variety of pressures. Bornman and Adams (2010) found that a flood event was important in creating uniform groundwater salinity across the soil profile (60–90 mS cm^{-1}), the high rainfall created niches for seed germination and the floods increased the tidal reach for salt flushing, but on the whole the hypersalinity was not ameliorated because a causeway prevented the floods from reaching the degraded salt marsh area to reduce salinity. Human intervention was thus suggested for rehabilitation of the salt marsh (Bornman and Adams, 2010). In the East Kleinemonde Estuary, a sea storm surge resulted in high levels of saline water above historical levels in the estuary. This significantly altered vegetation composition: submerged macrophyte cover was reduced by 38%, there was an 8% increase in macroalgae cover; and a significant reduction in cover of supratidal vegetation, reeds and sedges, all due to prolonged submergence under highly saline (31 ± 0.9 ppt) conditions (Riddin and Adams, 2010). This provided an insight into the probable transformation of TOCEs under predicted climate change conditions in which such events would increase in frequency.

An understanding of the effect of environmental factors on plant growth and survival is required to predict, manage and adapt to global change. Completed research on the ecophysiology of estuarine macrophytes has contributed to an understanding of the responses to changes in freshwater inflow and future climate change. The stress responses of 38 species have been studied. Results from laboratory studies on species representative of different tidal ranges (Tabot and Adams, 2012; Tabot and Adams, 2013a,

2013c; Tabot and Adams, 2014) showed that the two upper intertidal salt marsh species were sensitive to waterlogging and would not survive complete submergence, whereas the lower intertidal species could (in addition to its natural range), thrive in conditions typical of the upper intertidal range, namely prolonged dry conditions and high sediment salinity. These results have important implications for the future management of salt marshes under predicted climate change conditions. Field studies have complemented laboratory research to provide a comprehensive understanding of responses of estuarine macrophytes to stress. Detailed studies in the Olifants (Bornman *et al.*, 2008) and Orange estuaries (Shaw *et al.*, 2008, Bornman and Adams, 2010) provided an understanding of the responses of salt marsh to changes in freshwater inflow and extreme saline conditions. Overabstraction of freshwater has resulted in an increase in the duration and frequency of mouth closure in TOCE. The majority of South Africa's estuaries are closed to the sea, and research has contributed to an understanding of the structure and function of these systems locally and globally (Whitfield *et al.*, 2008, 2012). The dynamics of macrophytes including salt marsh was investigated in the East Kleinemonde Estuary and showed that water level rather than salinity was most important in influencing macrophyte responses (Riddin and Adams, 2008).

5.9 A Closer Look at South African Salt Marshes

In South African salt marshes, research has shown that on an ecosystem scale, soil moisture and salinity help shape vegetation distribution. Soil moisture functions by its own merit as one of the requirements for growth, but also mediates sediment salinity. For example, a mapping approach was used to determine four main habitats with unique vegetation types in the St Lucia Estuary, South Africa (Taylor *et al.*, 2006). Table 5.6 shows that, in the intertidal shorelines, the main

Table 5.6. Drivers of vegetation distribution in the St Lucia Estuary.

Species	Driver	Biotic response to abiotic drivers
		Primary Habitat 2: Intertidal shoreline
Avicennia marina, *Bruguiera gymnorrhiza*, *Sporobolus virginicus*, *Paspalum vaginatum*, *Sarcocornia natalensis*, *Salicornia meyeriana*	Salinity	As salinity rises, the mangroves die, then the succulent salt marsh, leaving mud flats where salt forms a surface crust
	Tidal exchange	Salinity, temperature and inundation extremes are controlled by tidal exchange, which imposes a regular periodicity on variability. During mouth closure, sediments may become too dry or, if flooded, too anoxic to support plant growth
	Inundation	Gas exchange is limited if mangrove pneumatophores are covered
		Primary habitat 3: Dry shorelines, islands and estuarine floodplains
Salicornia meyeriana, *Sarcocornia natalensis*, *Sporobolus virginicus*, *Paspalum vaginatum*, *Stenotaphrum secundatum*, *Phragmites australis*	Salinity	A surface crust of salt prevents seedling establishment. This may result in a bare mud flat, or else plants will take a 'window of opportunity' after a rainfall event or freshwater flooding for seedling germination
	Inundation	Flooding kills succulent salt marsh when water levels increase after mouth closure
	Desiccation	Desiccation stress can occur, as the water potential of the saline sediment is too low for plant uptake
	Temperature	High temperatures inhibit seedling establishment. If plants can establish, then their shading effect reduces the heat

Adapted from Taylor *et al.* (2006)

drivers of the distribution of *Juncus kraussii*, mangroves and *Phragmites australis*, were fluctuations in inundation, tidal exchange, salinity and temperature, while distribution of *Sarcocornia natalensis* and *S. meyeriana* was driven by rainfall, drought and grazing (Taylor *et al.*, 2006). *Phragmites australis* stands thrived in highly saline soils in the Goukou and Keurbooms estuaries in South Africa by proliferating roots into freshwater seepage sites (Adams and Bate, 1999). Similar studies in the Olifant estuary showed that the depth-to-ground water was essential in the distribution of *S. pillansii*, which either proliferated roots to the water table to benefit from low-saline water, or grew in hummocks with less salinity (Bornman *et al.*, 2004). In the Orange Estuary, Shaw *et al.* (2008) showed that the regeneration of degraded salt marsh was mainly hampered by hypersaline conditions accentuated by low freshwater input. Across all South African salt marshes, two distinct vegetation assembly patterns can be discerned. Firstly, there is clear-cut zonation with submerged, lower intertidal, upper intertidal and supratidal ranges inhabited by unique species in estuaries with a large tidal range and clear inundation and/or salinity gradients as is the case with the Swartkops Estuary. Secondly, in estuaries where there are heterogeneous moisture and salinity conditions, salt marsh vegetation exists in mosaics, with different species occupying different niches based on local salinity and moisture combinations. Here, one would expect to find traditional upper intertidal species co-existing with lower intertidal species or vice versa. Irrespective of the vegetation patterns, however, a study of the local conditions would reveal that salinity and inundation are the main determinants of species distribution in South African estuaries. This is important because predicted changes in climate would mainly manifest in salt marshes as variation in moisture and inundation levels, with corresponding changes in salinity. The effects of these predicted changes have been studied (Tabot and Adams, 2013c), using an ecophysiological predictive methodology. This study assumed that erosion and accretion were equal.

However, for accurate measurements, Rod Surface Elevation Table (RSET) stations are necessary; but these have only been set up in three estuaries in South Africa. Marker horizons were also established at these sites in 2010 and are used to measure the contribution of vertical accretion (a surface process) in the root zone compared to changes in the overall surface elevation as measured by the RSET (Cahoon *et al.*, 2002, 2006). Schmidt's (2013) preliminary results from this show that the salt marsh surface elevation of the lower and middle sections of the Swartkops, Kromme and Knysna estuaries are generally declining relative to current sea level rise. In the Knysna Estuary, development will prevent a landward transgression of the salt marsh, forming an artificial coastal squeeze. Geomorphological limitations (steep hills adjacent to the salt marsh) will prevent any transgression in the upper reaches of the Kromme Estuary. Only supratidal habitat in the upper reaches of the Swartkops Estuary will (if undeveloped) provide a viable habitat for the salt marsh to migrate into, given sufficient surface elevation. This estuary has already lost 88% of supratidal salt marsh area to development (Colloty *et al.*, 2000).

5.10 Conclusion, Knowledge Gaps and Research Needed

An understanding of the distribution and status of salt marshes is important, as we need to prioritize these habitats for biodiversity conservation. Salt marshes are threatened by sea level rise at the sea interface and development at the land interface. Disturbance at the landward margin of salt marshes has already resulted in changes in species composition and altered environmental conditions (Veldkornet *et al.*, 2015b). Kirwan and Megonigal (2013) stated that multidisciplinary studies on salt marsh dynamics should focus on the factors that control the lateral position of salt marsh boundaries. Intact ecotones need to be identified and conserved as these are important buffer zones.

The ecophysiological tolerances of the plants at the sea–land margin need to be investigated. Knowledge of the stress responses of estuarine plants underpins understanding of resilience to global change. Future sea level rise translates into a modification of salinity and inundation effects, and an understanding of the effect of these environmental factors on plant growth and survival is required to predict, manage and adapt to climate change. We suggest that the salt tolerance of invasive plants such as *Acacia* spp. and

Tamarisk spp. is studied, as these plants are increasingly found inhabiting the ecotone habitat. Long-term monitoring plots should be used to measure the expansion of different zones and track changes in the plant and sediment characteristics of this ecotone habitat. Continual monitoring of elevation changes in salt marshes is also needed, as we are unsure whether these habitats are subsiding or accreting. Salt marshes will disappear if they are not able to accrete at a rate such that surface elevation gain is sufficient to offset the rate of sea level rise. More severe wave action has been predicted from simulation for the South African coastline (Mori *et al*., 2010) and this will potentially remove salt marsh vegetation. However, van Niekerk *et al*. (2011) noted that the South African wave record is too short to make any firm conclusions at this stage. Research of this nature is needed as we are still uncertain of the fate of coastal salt marshes in response to global change (i.e. sea level rise, increase in storminess and CO_2, and human-mediated disturbances).

Acknowledgements

We thank the South African National Research Foundation and the Nelson Mandela Metropolitan University for funding specific studies (Tabot and Adams, 2012–2014) that constituted part of this review.

References

Adam, P. (1990) *Salt Marsh Ecology*. Cambridge University Press, Cambridge, UK.

Adams, J.B. (2010) Macrophytes of the Berg Estuary. Specialist report prepared for Anchor Environmental. Department of Botany, Nelson Mandela Metropolitan University, South Africa.

Adams, J.B. and Bate, G.C. (1994) The ecological implications of tolerance to salinity by *Ruppia cirrhosa* (Petagna) Grande and *Zostera capensis* Setchell. *Botanica Marina* 37, 449–456, doi: 10.1515/botm.1994.37.5.449.

Adams, J.B. and Bate, G.C. (1995) Ecological implications of tolerance to salinity and inundation by *Spartina maritima*. *Aquatic Botany* 52, 183–191.

Adams, J.B. and Bate, G.C. (1999) Growth and photosynthetic performance of *Phragmites australis* in estuarine waters: A field and experimental evaluation. *Aquatic Botany* 64, 359–367.

Adams, J.B., Bate, G.C. and O'Callaghan, M. (1999) Primary producers. In: Allanson, B.R. and Baird, D. (eds) *Estuaries of South Africa*. Cambridge University Press, Cambridge, UK.

Adams, J.B., Veldkornet, D. and Tabot, P.T. (2016) Distribution of macrophyte species and habitats in South African estuaries. *South African Journal of Botany* 107, 5–11.

Adkins, S.W., Shiraishi, T. and McComb, J.A. (1990) Submergence tolerance of rice - a new glass house method for the experimental submergence of plants. *Physiologia Plantarum* 80, 642–646.

Allanson, B. and Baird, D. (1999) *Estuaries of South Africa*. Cambridge University Press, Cambridge, UK.

Apel, K. and Hirt, H. (2004) Reactive oxygen species: Metabolism, oxidative stress, and signal transduction. *Annual Review of Plant Biology* 55, 373–399.

Ashour, N.I. and Mekki, B.B. (2006) Halophytic forage plant response to irrigation with diluted seawater. *World Journal of Agricultural Sciences* 2, 169–173.

Ashraf, M. and Foolad, M.R. (2007) Role of glycinebetaine and proline in improving plant abiotic stress resistance. *Environmental and Experimental Botany* 59, 206–216.

Awad, A.S., Edwards, D.G. and Campbell, L.C. (1990) Phosphorus enhancement of salt tolerance of tomato. *Crop Science* 30, 123–128.

Bailey-Serres, J. and Voesenek, L.A.C.J. (2008) Flooding stress: Acclimations and genetic diversity. *Annual Review of Plant Biology* 59, 313–339.

Bally, R., McQuaid, C.D. and Pierce, S.M. (1985) Primary productivity of the Bot River estuary, South Africa. *Transactions of the Royal Society of South Africa* 45, 333–345.

Benschop, J.J., Jackson, M.B., Guhl, K., Vreeburg, R.A.M., Croker, S.J., Peeters, A.J.M. and Voesenek, L.A.C.J. (2005) Contrasting interactions between ethylene and abscisic acid in *Rumex* species differing in submergence tolerance. *The Plant Journal* 44, 756–768.

Bernstein, L., Francois, L.E. and Clark, R.A. (1974) Interactive effects of salinity and fertility on yields of grains and vegetables. *Agronomy Journal* 66, 412–421.

Bertness, M.D. and Ellison, A.M. (1987) Determinants of pattern in a new England salt marsh community. *Ecological Monographs* 57(2), 129–147.

Best, M., Massey, A., Prior, A. (2007) Developing a saltmarsh classification tool for the European water framework directive. *Marine Pollution Bulletin* 55(1-6), 205–214.

Bezuidenhout, C. (2011) Macrophytes as indicators of physico-chemical factors in South African Estuaries. PhD thesis. Nelson Mandela Metropolitan University, Port Elizabeth, South Africa.

Bohnert, H.J. and Jensen, R.G. (1996) Strategies for engineering water stress tolerance in plants. *Trends in Biotechnology* 14, 89–97.

Bornman, T.G. and Adams, J.B. (2010) Response of a hypersaline salt marsh to a large flood and rainfall event along the west coast of southern Africa. *Estuarine, Coastal and Shelf Science* 87, 378–386.

Bornman, T.G., Adams, J.B. and Bate, G.C. (2002) Freshwater requirements of a semi-arid supratidal and floodplain saltmarsh. *Estuaries* 25, 1394–1405.

Bornman, T.G., Adams, J.B. and Bate, G.C. (2004) The influence of floodplain geohydrology on the distribution of *Sarcocornia pillansii* in the Olifants Estuary on the West Coast, South Africa. *Journal of Arid Environments* 56, 603–625.

Bornman, T.G., Adams, J.B. and Bate, G.C. (2008) Environmental factors controlling the vegetation zonation patterns and distribution of vegetation types in the Olifants Estuary, South Africa. *South African Journal of Botany* 74, 585–695.

Breen, C.M., Everson, C. and Rogers, K. (1977) Ecological studies on *Sporobolus virginicus* (L.) Kunth with particular reference to salinity and inundation. *Hydrobiologia* 54, 135–140.

Brock, M.A. (1981) Accumulation of proline in a submerged aquatic halophyte *Ruppia* L. *Oecologia* 51, 217–219.

Brown, C.E. and Pezeshki, R.S. (2007) Threshold for recovery in the marsh halophyte *Spartina alterniflora* grown under the combined effects of salinity and soil drying. *Journal of Plant Physiology* 164, 274–282.

Cacador, I., Tiberio, S. and Cabral, H.N. (2007) Species zonation in Corroios salt marsh in the Tagus estuary (Portugal) and its dynamics in the past fifty years. *Hydrobiologia* 597, 205–211.

Cahoon, D.R., Lynch, J.C., Perez, B.C., Segura, B., Holland, R., *et al.* (2002) High precision measurement of wetland sediment elevation: II. The rod surface elevation table. *Journal of Sedimentary Research* 72, 734–739.

Cahoon, D.R., Hensel, P., Spencer, T., Reed, D.J., McKee, K.L., *et al.* (2006) Coastal wetland vulnerability to relative sea-level rise: wetland elevation trends and process controls. In: Verhoeven, J.T.A., Beltman, B., Bobbink, R. and Whigham, D. (eds) *Wetlands and Natural Resource Management*. Springer, Berlin, Germany.

Carter, C.T. and Ungar, I.A. (2003) Germination response of dimorphic seeds of two halophyte species to environmentally controlled and natural conditions. *Canadian Journal of Botany* 81, 918–926.

Chen, H. and Jiang, J. (2010) Osmotic adjustment and plant adaptation to environmental changes related to drought and salinity. *Environmental Review* 18, 309–319.

Chen, T.H.H. and Murata, N. (2008) Glycinebetaine: An effective protectant against abiotic stress in plants. *Trends in Plant Science* 13, 499–505.

Colloty, B.M., Adams, J.B. and Bate, G.C. (2000) The use of the botanical importance rating to assess changes in the flora of the Swartkops Estuary over time. *Water SA* 26, 171–180.

Colloty, B.M., Adams, J.B. and Bate, G.C. (2002) Classification of Estuaries in the Ciskei and Transkei regions based on physical and botanical characteristics. *South African Journal of Botany* 68, 312–321.

Colmer, T.D. (2003) Long-distance transport of gases in plants: A perspective on internal aeration and radial oxygen loss from roots. *Plant, Cell and Environment* 26, 17–36.

Colmer, T.D. and Flowers, T.J. (2008) Flooding tolerance in halophytes. *New Phytologist* 179, 964–974.

Colmer, T.D. and Voesenek, L.A.C.J. (2009) Flooding tolerance: Suites of plant traits in variable environments. *Functional Plant Biology* 36, 665–681.

Cooper, A. (1982) The effects of salinity and waterlogging on the growth and cation uptake of salt marsh plants. *New Phytologist* 90, 263–275.

Davy, A.J., Bishop, G.F. and Costa, C.S.B. (2001) Biological flora of the British Isles: *Salicornia* L. (*Salicornia pusilla* J. Woods, *S. ramosissima* J. Woods, *S. europaea* L., *S. obscura* P.W. Ball & Tutin, *S. nitens* P.W. Ball & Tutin, *S. fragilis* P.W. Ball & Tutin and *S. dolichostachya* Moss). *Journal of Ecology* 89, 681–707.

Day, J.H. (1981) The estuarine flora. In: Day, J.H. (ed.) *Estuarine Ecology*. AA Balkema, Cape Town, South Africa.

Duan, D., Liu, X., Khan, M.A. and Gul, B. (2004) Effects of salt and water stress on the germination of *Chenopodium glaucum* L. seed. *Pakistan Journal of Botany* 36, 793–800.

Duan, D., Li, W., Liu, X., Ouyang, H. and An, P. 2007. Seed germination and seedling growth of Suaeda salsa. *Annales Botanici Fennici* 44, 161–169.

Eallonardo Jr, A.S. and Leopold, D.J. (2014) Inland Salt Marshes of the Northeastern United States: Stress, Disturbance and Compositional Stability. *Wetlands* 34, 155–166.

Emery, N.C., Ewanchuk, P.J. and Bertness, M.D. (2001) Competition and salt marsh plant zonation: Stress tolerators may be dominant competitors. *Ecology* 82(9), 2471–2485.

Furness, H.D. and Breen, C.M. (1985) Interactions between period of exposure, grazing and crop growth rate of *Cynodon dactylon* (L.) Pers. in seasonally flooded areas of the Pongolo River floodplain. *Hydrobiologia* 126(1), 65–73.

González-Alcaraz, M.N., Jiménez-Cárceles, F.J., Álvarez, Y., Álvarez-Rogel, J. (2014) Gradients of soil salinity and moisture, and plant distribution, in a Mediterranean semiarid saline watershed: A model of soil-plant relationships for contributing to the management. *Catena* 115, 150–158.

Hamed, K.B., Messedi, D., Ranieri, A. and Abdelli, C. (2008) Diversity in the response of two potential halophytes (*Batis maritima* and *Crithmum maritimum*) to salt stress. In: Abdelly, C., Öztürk, M., Ashraf, M., Grignon, C. (eds) *Biosaline Agriculture and High Salinity Tolerance*. Birkhäuser, Berlin, Germany.

Hansen, D.J., Dayanandan, P., Kaufman, P.B. and Brotherson, J.D. (1976) Ecological adaptations of salt marsh grass, *Distichlis spicata* (Gramineae), and environmental factors affecting its growth and distribution. *American Journal of Botany* 63, 635–650.

Hanson, A.D., Rathinasabapathi, B., Rivoal, J., Burnet, M., Dillon, M.O. *et al.* (1994) Osmoprotective compounds in the Plumbaginaceae: A natural experiment in metabolic engineering of stress tolerance. *Proceedings of the National Academy of Science of USA* 91, 306–310.

Heinsohn, R.D. and Cunningham, A.B. (1991) Utilization and potential cultivation of salt marsh rush *Juncus kraussii*. *South African Journal of Botany* 57, 1–5.

Hernes, H., Dalfelt, A., Berntsen, T., Holtsmark, B., Naess, L.O. *et al.* (1995) Climate strategy for Africa. Cicero, Oslo.

Hester, M.W., Mendelssohn, I.A. and McKee, K.L. (1996) Intraspecific variation in salt tolerance and morphology in the coastal grass *Spartina patens*. *American Journal of Botany* 83, 1521–1527.

Howard-Williams, C. and Liptrot, M.R.M. (1980) Submerged macrophyte communities in a brackish south African estuarine lake system. *Aquatic Botany* 9, 101–116.

IPCC (2001) *The Regional Impacts of Climate Change: An Assessment of Variability*. In: Watson, R.T., Zinyowera, M.T., Moss, R.H. and Jokken, D.J. (eds) Cambridge University Press, Cambridge, UK and New York.

IPCC (2007) Summary for Policymakers. In: Solomon, S., Qin, D., Manning, M., Chen, Z., Marquis, M., Averyt, K.B., Tignor, M. and Miller, H.L. (eds) *Climate Change 2007: The Physical Science Basis. Contribution of Working Group I to the Fourth Assessment Report of the Intergovernmental Panel on Climate Change*. Cambridge University Press, Cambridge, UK and New York.

Jackson, M.B. (2008) Ethylene-promoted elongation: An adaptation to submergence stress. *Annals of Botany* 101, 229–248.

Jackson, M.B. and Drew, M.C. (1984) Effects of flooding on herbaceous plants. In: Kozlowski, T.T. (ed.) *Flooding and Plant Growth*. Academic Press, New York.

Jithesh, M.N., Preashanth, S.R., Sivaprakash, K.R. and Parida, A.K. (2006) Antioxidant response mechanisms in halophytes: Their role in stress defence. *Journal of Genetics* 85(3), 237–254.

Khan, M.A., Gul, B. and Weber, D.J. (2000) Germination responses of *Salicornia rubra* to temperature and salinity. *Journal of Arid Environments* 45, 207–214.

Khan, M.A., Ungar, I.A. and Showalter, A.M. (2005) Salt stimulation and tolerance in an intertidal stem-succulent halophyte. *Journal of Plant Nutrition* 28, 1365–1374.

Kim, D. (2014) Rates of Vegetation Dynamics Along Elevation Gradients in a Backbarrier Salt Marsh of the Danish Wadden Sea. *Estuaries and Coasts* 37(3), 610–620.

Kim, D., Cairns, D.M. and Bartholdy, J. (2013) Tidal creek morphology and sediment type influence spatial trends in salt marsh vegetation. *Professional Geographer* 65, 544–560.

Kirwan, M.L. and Megonigal, P. (2013) Tidal wetland stability in the face of human impacts and sea-level rise. *Nature* 504, 53–60.

Köcke, A.V., von Mering, S., Mucina, L. and Kadereit, J.W. (2010) Revision of the Mediterranean and Southern African *Triglochin bulbosa* complex (Juncaginaceae). *Edinburgh Journal of Botany* 67, 1–47.

Kruger, A.C. and Shongwe, S. (2004) Temperature trends in South Africa: 1960–2003. *International Journal of Climatology* 24(15), 1929–1945.

Lee, G.J., Duncan, R.R. and Lee, R.N. (2007) Nutrient uptake responses and inorganic ion contribution to solute potential under salinity stress in halophytic seashore paspalums. *Crop Science* 47, 2410–2419.

Linthurst, R.A. and Seneca, E.D. (1981) Aeration, nitrogen and salinity as determinants of *Spartina alterniflora* Loisel. growth response. *Estuaries* 4, 53–63.

Lubke, R.A. and Van Wijk, Y. (1988) Estuarine plants. In: Lubke, R.A., Gess, F.W. and Brutton, M.N. (eds) *A Field Guide to the Eastern Cape*. The Grahamstown Centre of the Wildlife Society of South Africa, Grahamstown, South Africa.

Marcum, K.B. and Murdoch, C.L. (1992) Salt tolerance of the coastal salt marsh grass, *Sporobolus virginicus* (L.) Kunth. *New Phytologist* 120, 281–288.

Marín-Guirao, L., Sandoval-Gil, J.M., Ruíz, J.M. and Sánchez-Lizaso, J.L. (2011) Photosynthesis, growth and survival of the Mediterranean seagrass *Posidonia oceanica* in response to simulated salinity increases in a laboratory mesocosm system. *Estuarine, Coastal and Shelf Science* 92, 286–296.

Mason, S.J. and Joubert, A.M. (1996) Droughts over southern Africa in a doubled CO_2 climate. *International Journal of Climatology* 16, 1146–1156.

Mason, S.J. and Joubert, A.M. (1997) Simulated changes in extreme rainfall over southern Africa. *International Journal of Climatology* 17, 291–301.

Mason, S.J. and Jury, M.R. (1997) Climatic variability and change over southern Africa: A reflection on underlying processes. *Progress in Physical Geography* 21, 23–50.

Mather, A.A. (2007) Linear and non-linear sea level changes at Durban, South Africa. *South African Journal of Science* 103, 509–512.

Mather, A.A., Garland, G.G. and Stretch, D.D. (2009) Southern African sea levels: Corrections, influences and trends. *African Journal of Marine Science* 31, 145–156.

Miller, G., Suzuki, N., Ciftci-Yilmaz, S. and Mittler, R. (2010) Reactive oxygen species homeostasis and signalling during drought and salinity stresses. *Plant, Cell and Environment* 33, 453–467.

Mori, N., Yasuda, T., Mase, H., Tom, T. and Oku, Y. (2010) Projection of extreme wave climate change under global warming. *Hydrology Research Letters* 4, 15–19.

Mucina, L. and Rutherford, M.C. (eds) (2006) *The vegetation of South Africa, Lesotho and Swaziland*. SANBI, Pretoria, South Africa.

Munns, R. and Sharp, R.E. (1993) Involvement of abscisic acid in controlling plant growth in soils of low water potential. *Australian Journal of Plant Physiology* 20, 425–437.

Munns, R. and Tester, M. (2008) Mechanisms of salinity tolerance. *Annual Review of Plant Biology* 59, 651–681.

Naidoo, G. (1994) Growth, water and ion relationships in the coastal halophytes *Triglochin bulbosa* and *T. striata*. *Environmental and Experimental Botany* 34, 419–426.

Naidoo, G. and Kift, J. (2006) Responses of the salt marsh rush *Juncus kraussii* to salinity and waterlogging. *Aquatic Botany* 84, 217–225.

Naidoo, G. and Mundree, S.G. (1993) Relationship between physiological and morphological responses to waterlogging and salinity in *Sporobolus virginicus* (L.) Kunth. *Oecologia* 93, 360–366.

Naidoo, G. and Naicker, K. (1992) Seed germination in the coastal halophytes *Triglochin bulbosa* and *Triglochin striata*. *Aquatic Botany* 42, 217–229.

Naidoo, G. and Naidoo, S. (1992) Waterlogging responses of *Sporobolus virginicus* (L.) Kunth. *Oecologia* 90, 445–450.

Naidoo, G. and Naidoo, Y. (2001) Effects of salinity and nitrogen on growth, ion relations and proline accumulation in *Triglochin bulbosa*. *Wetlands Ecology and Management* 9, 491–497.

Naidoo, G. and Rughunanan, R. (1990) Salt tolerance in the succulent, coastal halophyte, *Sarcocornia natalensis*. *Journal of Experimental Botany* 41, 497–502.

Naidoo, G., McKee, K.L. and Mendelssohn, I.A. (1992) Anatomical and metabolic responses to waterlogging and salinity in *Spartina alterniflora* and *S. patens* (Poaceae). *American Journal of Botany* 79, 765–770.

Naidoo, G., Somaru, R. and Achar, P. (2008) Morphological and physiological responses of the halophyte *Odyssea paucinervis* (Staph) (Poaceae), to salinity. *Flora* 203, 437–447.

Naidoo, G., Naidoo, Y. and Achar, P. (2012) Ecophysiological responses of the salt marsh grass *Spartina maritima* to salinity. *African Journal of Aquatic Science* 37, 81–88.

O'Callaghan, M. (1994) The marsh vegetation of Kleinmond lagoon. *Bothalia* 24, 235–240.

Pearcy, R.W. and Ustin, S.I. (1984) Effects of salinity on growth and photosynthesis of three California tidal marsh species. *Oecologia* 62, 68–73.

Pennings, S.C., Grant, M. and Bertness, M.D. (2005) Plant zonation in low-latitude salt marshes: Disentangling the roles of flooding, salinity and competition. *Journal of Ecology* 93, 159–167.

Ramel, F., Birtic, S., Cuiné, S., Triantaphylidès, C., Ravanat, J. *et al* (2012) Chemical quenching of singlet oxygen by carotenoids in plants. *Plant Physiology* 158, 1267–1278.

Rautenbach, K. (2015) Present state of macrophytes and responses to management scenarios at the St Lucia and Mfolozi estuaries. MSc Thesis. Nelson Mandela Metropolitan University, Port Elizabeth, South Africa.

Redondo-Gómez, S., Rubio-Casal, A.E., Castillo, J.M., Luque, C.J., Alvarez, A.A. *et al.* (2004) Influences of salinity and light on germination of three *Sarcocornia* taxa with contrasted habitats. *Aquatic Botany* 78, 255–264.

Redondo-Gómez, S., Mateos-Naranjo, E., Davy, A.J., Fernandez-Munoz, F., Castellanos, E.M. *et al.* (2007) Growth and photosynthetic responses to salinity of the salt-marsh shrub *Atriplex portulacoides*. *Annals of Botany* 100, 555–563.

Redondo-Gómez, S., Mateos-Naranjo, E., Figueroa, M.E. and Davy, A.J. (2010) Salt stimulation of growth and photosynthesis in an extreme halophyte *Arthrocnemum macrostachyum*. *Plant Biology* 12, 79–87.

Riddin, T. and Adams, J.B. (2008) Influence of mouth status and water level on the macrophytes in a small temporarily open/closed estuary. *Estuarine, Coastal and Shelf Science* 79, 86–92.

Riddin, T. and Adams, J.B. (2009) The seed banks of two temporarily open/closed estuaries in South Africa. *Aquatic Botany* 90, 328–332.

Riddin, T. and Adams, J.B. (2010) The effect of a storm surge event on the macrophytes of a temporarily open/closed estuary, South Africa. *Estuarine, Coastal and Shelf Science* 89, 119–123.

Sarkar, R.K., Reddy, J.N., Sharma, S.G. and Ismail, A.M. (2006) Physiological basis of submergence tolerance in rice and implications for crop improvement. *Current Science* 91, 899–906.

Schmidt, J. (2013) Influence of predicted sea-level rise on the salt marsh of the Swartkops, Kromme and Knysna estuaries. MSc Thesis, Nelson Mandela Metropolitan University, Port Elizabeth, South Africa.

Seaman, M.T., Ashton, P.J. and Williams, W.D. (1991) Inland salt waters of southern Africa. *Hydrobiologia* 210, 75.

Shaw, G.A., Adams, J.B. and Bornman, T.G. (2008) Sediment characteristics and vegetation dynamics as indicators for the potential rehabilitation of an estuary salt marsh on the arid west coast of south Africa. *Journal of Arid Environments* 72, 1097–1109.

Sieben, E.J.J., Ellery, W.N., Dullo, B.W. and Grootjans, A.P. (2015) Impact of marine inundation after a period of drought on the lakeshore vegetation of Lake St. Lucia, South Africa: resilience of estuarine vegetation. *African Journal of Aquatic Science* 1–11.

Sieben, E.J.J., Collins, N.B., Mtshali, H. and Venter, C.E. (2016) The vegetation of inland wetlands with salt-tolerant vegetation in South Africa: Description, classification and explanatory environmental factors. *South African Journal of Botany* 104, 199–207.

Silveira, J.A.G., Araujo, S.A.M., Lima, J.P.M.S. and Viégas, R.A. (2009) Roots and leaves display contrasting osmotic adjustment mechanisms in response to NaCl-salinity in *Atriplex nummularia*. *Environmental and Experimental Botany* 66, 1–8.

Simas, T., Nunes, J.P. and Ferreira, G.J. (2001) Effect of global climate change on coastal salt marshes. *Ecological Modelling* 139, 1–15.

Snow, A. and Vince, S.W. (1984). Plant zonation in an Alaskan salt marsh. II. An experimental study of the role of edaphic conditions. *Journal of Ecology* 71, 669–684.

Spence, D.H.N. (1982) The zonation of plants in freshwater lakes. *Advances in Ecological Research* 12, 37–125.

Starfield, A.M., Farm, B.P. and Taylor, R.H. (1989) A rule-based ecological model for the management of an estuarine lake. *Ecological Modelling* 46, 107–119.

Storey, R., Ahmad, N. and Jones, R.G.W. (1977) Taxonomic and ecological aspects of the distribution of glycinebetaine and related compounds in plants. *Oecologia* 27, 319–332.

Tabot, P.T. and Adams, J.B. (2012) Morphological and physiological responses of *Triglochin buchenaui* Köcke, Mering & Kadereit to various combinations of water and salinity: Implications for resilience to climate change. *Wetlands Ecology and Management* 20, 373–388.

Tabot, P.T. and Adams, J.B. (2013a) Responses of the salt marsh succulent *Bassia diffusa* (Thunb.) Kuntze to salinity and inundation. *Wetlands* 33, 787–797.

Tabot, P.T. and Adams, J.B. (2013b) Early responses of *Bassia diffusa* (Thunb.) Kuntze to submergence for different salinity treatments. *South African Journal of Botany* 84, 19–29.

Tabot, P.T. and Adams, J.B. (2013c) Ecophysiology of salt marsh plants and predicted responses to climate change in South Africa. *Oceans and Coastal Management* 80, 89–99.

Tabot, P.T. and Adams, J.B. (2014) Salt secretion, proline accumulation and increased branching confer tolerance to drought and salinity in the endemic halophyte *Limonium linifolium*. *South African Journal of Botany* 94, 64–73.

Taylor, R., Adams, J.B. and Haldorsen, S. (2006) Primary habitats of the St Lucia estuarine system, south Africa, and their responses to mouth management. *African Journal of Aquatic Science* 31, 31–41.

Thomas, D.S.G., Twyman, C., Osbahr, H. and Hewitson, B. (2007) Adaptation to climate change and variability: Farmer responses to intra-seasonal precipitation trends in South Africa. *Climatic Change* 83, 301–322.

Tolken, H.R. (1967) The species of Arthrocnemum and Salicornia (Chenopodiaceae) in Southern Africa. *Bothalia* 9, 255–307.

Ungar, I.J. (1998) Are biotic factors significant in influencing the distribution of halophytes in saline habitat? *The Botanical Review* 64, 176–199.

van Niekerk, L., Engelbrecht, F., James, N., Meyer, A. and Theron, A. (2011) How vulnerable are South Africa's estuaries to climate change. In: van Niekerk, L. and Turpie, J.K. (eds) *South African National Biodiversity Assessment 2011: Technical Report. Estuary Component*. Council for Scientific and Industrial Research, Stellenbosch, South Africa.

Veldkornet, D., Adams, J.B. and Potts, A. (2015a) Where do you draw the line? Determining the transition thresholds between estuarine salt marshes and terrestrial vegetation. *South African Journal of Botany* 101, 153–159.

Veldkornet, D.A., Adams, J.B. and van Niekerk, L. (2015b) Characteristics and landcover of estuarine boundaries: implications for the delineation of the South African estuarine functional zone. *Southern African Journal of Marine Science* 37, 313–323.

Verhoven, J.T.A. (1975) *Ruppia*-dominated communities in the Camargue, France. Distribution and structure in relation to salinity and salinity fluctuations. *Aquatic Botany* 1, 217–241.

Vincente, O., Boscaiu, M., Narangio, M.A., Estrelle, E., Bellés, J.M. *et al.* (2004) Responses to salt stress in the halophyte *Plantago crassifolia* (Plantaginaceae). *Journal of Arid Environments* 58, 463–481.

Voesenek, L.A.C.J., Benschop, J.J., Bou, J., Cox, M.C.H., Groeneveld, H.W. *et al.* (2003). Interactions between plant hormones regulate submergence-induced shoot elongation in the flooding-tolerant dicot *Rumex palustris*. *Annals of Botany* 91, 205–211.

Voesenek, L.A.C.J., Colmer, T.D., Pierik, R., Millenaar, F.F. and Peeters, A.J.M. (2006) How plants cope with complete submergence. *New Phytologist* 170, 213–226.

Ward, C.J. (1976) Aspects of the ecology and distribution of submerged macrophytes and shoreline vegetation of Lake St Lucia. In: Heydon, A.E.F. (ed.) *St Lucia Scientific and Advisory Council workshop, Charters Creek*. Natal Parks Board, Pietermaritzburg, South Africa.

Watson, R.T., Zinyowera, M.T., Moss, R.H., Jokken, D.J. (eds) (2001) *IPCC Special Reports on the Regional Impacts of Climate Change: An Assessment of Variability*. GRID-Arendal, New York.

Weilhoefer, C.L., Nelson, W.G., Clinton, P. and Beugli, D.M. (2013) Environmental determinants of emergent macrophyte vegetation in Pacific Northwest estuarine tidal wetlands. *Estuaries and Coasts* 36, 377–389.

Weis, J.S. and Butler, C.A. (2009) *Salt Marshes: A Natural and Unnatural History*. Rutgers University Press, New Brunswick, New Jersey.

Weisner, S.E.B., Graneli, W. and Ekstam, B. (1993) Influence of submergence on growth of seedlings of *Scirpus lacustris* and *Phragmites australis*. *Freshwater Biology* 29, 371–375.

Whitfield, A.K. (2005) Langebaan – A new type of estuary? *African Journal of Aquatic Science* 30(2), 207–209.

Whitfield, A.K. and Bate, G. (eds) (2007) A review of information on temporarily open/closed estuaries in warm and cool temperate biogeographic regions of South Africa, with a particular emphasis on the influence of river flow on these systems. Water Research Commission, Pretoria, South Africa.

Whitfield, A.K., Adams, J.B., Bate, G.C., Bezuidenhout, K., Bornman, T.G. *et al.* (2008) A multidisciplinary study of a small, temporarily open/closed South African estuary, with particular emphasis on the influence of mouth state on the ecology of the system. *African Journal of Marine Science* 30, 453–473.

Whitfield, A.K., Bate, G.C., Adams, J.B., Cowley, P.D., Froneman, P.W. *et al.* (2012) A review of the ecology and management of temporarily open/closed estuaries in South Africa, with particular emphasis on river flow and mouth state as primary drivers of these systems. *African Journal of Marine Science* 34(2), 163–180.

Woodell, S.R.J. (1985) Salinity and seed germination patterns in coastal plants. *Vegetatio* 61, 223–229.

6 Seagrasses, the Unique Adaptation of Angiosperms to the Marine Environment: Effect of High Carbon and Ocean Acidification on Energetics and Ion Homeostasis

Lourdes Rubio* and José A. Fernández
University of Málaga, Málaga, Spain

Abstract

As a functional group, seagrasses form highly productive ecosystems present along the coasts of all continents (except Antarctica), where they rival tropical rainforests and coral reefs in ecosystem services (Costanza *et al.*, 1997; Fourqurean *et al.*, 2012). Unfortunately, seagrasses are diminishing worldwide and several studies confirm a lack of appreciation for the value of these systems (Cullen-Unsworth *et al.*, 2014). Since the last century, the effects of climate change on natural and agricultural terrestrial plant communities have already received significant attention, but relatively little emphasis has been given to aquatic plant communities, including seagrasses (Koch *et al.*, 2013). Here we analyse the potential impact of global atmospheric CO_2 increase on the adaptation mechanisms of these vascular plants to marine environments, highlighting the effects on membrane energization, nutrient uptake and cytosolic ion homeostasis.

Keywords: Climate change; Halophytes; Ion homeostasis; Sodium exclusion; Nutrient uptake

6.1 Introduction: True Marine Higher Plants

Seagrasses are unique angiosperms that evolved from land plants to live submerged in the sea, possibly the most severe habitat change ever completed by flowering plants. Originating in the Cretaceous (around 140 million years ago), seagrasses colonized the sea on at least three independent occasions to form the basis of one of the highest productive and widespread coastal ecosystems on the planet (Larkum *et al.*, 2006). The group, which includes around 72 species, is a polyphyletic assemblage of basal monocots belonging to four families in the Alismatales (Les *et al.*, 1997). Three of the four families consist exclusively of seagrasses (Zosteraceae, Cymodoceae and Posidoniaceae), classified in the past as subfamilies of the Potamogetonaceae (den Hartog, 1970). The fourth family, the Hydrocharitaceae, contains three genera of seagrasses, although 14 genera are freshwater plants (den Hartog and Kuo, 2006).

These angiosperms fulfil the requirements to be considered true marine water plants, accordingly to Arber's (1920) properties: (i) seagrasses

* Corresponding author e-mail: lrubio@uma.es

are adapted to live in a saline medium; (ii) they are able to grow when fully submerged; (iii) they have a secure anchoring system; and (iv) they have a hydrophilous pollination mechanism. In contrast to eurysaline plants (those able to withstand waters with unstable salinity, from mixo- and hyperhaline brackish waters to continental salt waters), seagrasses are characteristic for homoiohaline marine habitats. These angiosperms are able to compete successfully with other organisms in the marine environment, extending from 90 m depth (Duarte, 1991) and to the middle of the intertidal zone, where some members of the genera *Zostera*, *Cymodocea*, *Halodule* and *Halophyla* are found (den Hartog and Kuo, 2006).

The fact that seagrasses occupy relatively stable environmental conditions suggests that these plants have developed special physiological mechanisms, firmly fixed in their genome, to cope with adaptation to the marine environment. The first marine angiosperm fully sequenced was *Zostera marina* (Olsen *et al.*, 2016), the most widespread species throughout the temperate northern hemisphere. A month later the genome of the southern hemisphere temperate species, *Z. muelleri*, was also published (Lee *et al.*, 2016). The analysis of these genomes shows losses and gains involved in accomplishing morphological and physiological adaptations associated with the 'back to the sea' reverse evolutionary trajectory that has occurred in this angiosperm lineage (Williams, 2016). Key land angiosperm innovations that are lost in seagrasses include the entire collection of genes involved in stomata differentiation, genes related to the synthesis and sensing of terpenoids and other volatile substances, genes for ultraviolet protection and phytochromes for far-red sensing (Olsen *et al.*, 2016). On the other hand, seagrasses seems to have re-evolved mechanisms enabling them to adjust to high salinity and the scarce occurrence of essential nutrients such as nitrate or phosphate in seawater. One of the six H^+-ATPase (*AHA*) genes found in *Z. marina* is strongly expressed in vegetative tissue and encodes a salt-tolerant H^+-ATPase (Muramatsu *et al.*, 2002). In addition, the *Z. marina* genome shows three *AHA* genes in a cluster unique to alismatids (Olsen *et al.*, 2016). However, comparison with land plant species indicates that less than 20% of the gene families seem to be specific in the genome of seagrasses (Lee *et al.*, 2016; Olsen *et al.*, 2016). Thus, despite their unique habitat and morphology, early divergence and marine adaptation of the seagrasses did not involve a major increase in the abundance of novel gene families (Lee *et al.*, 2016).

Hence, changes in expression patterns and/or regulation of the same genes expressed in land angiosperms could be on the basis of secondary marine adaptation. As we discuss below, the impact of climate change on the marine environment is complex, prompting alterations not only in water temperature, but also in dissolved inorganic carbon forms (DIC) and water pH. Therefore, changes in speciation of inorganic carbon and critical nutrients could lead to alterations in marine primary production, bearing in mind that seagrasses may account for as much as 30% of marine net primary productivity (Duarte and Cebrián, 1996). A combination of the physiological and genome resources of seagrasses should promote a wide range of functional ecological studies of adaptation to the marine environment under the conditions anticipated by climate change. These should aim to clarify the mechanisms of photosynthetic inorganic carbon use, osmoregulation and nutrient uptake under high salinities, and so increase our understanding of the evolution of salt tolerance in crop plants facing climatic stress.

6.2 Plasma Membrane Energization and Na^+ Permeability in Seagrasses

The maintenance of a highly negative membrane potential (Em) appears to be crucial for salinity tolerance in halophytes (Shabala and Mackay, 2011). In natural seawater, the green cells of mature seagrass leaves exhibit Em ranging from −160 mV in *Z. marina* (Fernández *et al.*, 1999) and *Cymodocea nodosa* (our data, unpublished) to −174 mV in the case of *Posidonia oceanica* (Rubio *et al.*, 2017), values relatively negative compared to those reported for most marine algae (Gutknecht and Dainty, 1968; Gradmann and Boyd, 1995) or in root cells of terrestrial halophytes such as *Atriplex hastata* (between −130 and −140 mV; Anderson *et al.*, 1977), *Atriplex lentiformis* and *Chenopodium quinoa* (−140 mV in plants grown in the absence of Na^+, Bose *et al.*, 2015) or *Halimione (= Atriplex) portulacoides* (−140 mV in plants grown in full strength K^+, Ait-Hammou *et al.*, 2014).

In vascular plants the primary pump responsible for energizing the plasma membrane is a P-type H^+-ATPase. This PM-ATPase is an electrogenic enzyme that extrudes positive charges (H^+) and consequently generates a negative membrane potential (Sze *et al.*, 1999; Palmgren, 2001; Gaxiola *et al.*, 2007). As flowering plants lacking Na^+ pumps (Pedersen *et al.*, 2012), marine angiosperms also use P-type H^+-ATPases to energize the plasma membrane (Fernández *et al.*, 1999; Rubio *et al.*, 2017).

Plasma membrane H^+-ATPases are irreversibly activated by the phytotoxin fusicoccin, a response that has been used to identify this enzyme in plants (Marrè, 1979). In relation to seagrasses, it has been shown that the addition of 10 μM of this phytotoxin to the seawater evokes a rapid hyperpolarization of the plasma membrane in mesophyll leaf cells of *Z. marina* and the acidification of the external medium (Fernández *et al.*, 1999). These responses have been also found in *P. oceanica* (Rubio *et al.*, 2017) and *Cymodocea nodosa* (our data, unpublished), proving that the primary pump of the plasma membrane of these seagrasses is a P-type H^+-ATPase. Since the phytotoxin has no effect on fungi, the irreversible activation of the H^+-ATPases of seagrasses may play an important role in the infection mechanisms of fungal endophytes in seagrasses (Panno *et al.*, 2011; Supaphon *et al.*, 2017).

In the presence of respiratory inhibitors, the plasma membrane of ATP-depleted cells depolarize until the so-called diffusion potential is reached (Felle, 1981). This value, defined by the Goldman–Hodgking–Katz equation, represents the passive ionic distribution of K^+, Na^+ and Cl^- on the two sides of the plasma membrane. This depends on the ion concentration gradients and also on the ease with which those ions cross the membrane (i.e. on their permeability coefficients) (Nobel, 2009). In plant cells, K^+ is the most permeable ion, followed by Na^+ and Cl^-, whose permeability is as much as 1%–0.1% lower than that of K^+. In the case of seagrasses, patch clamp studies show that the plasma membrane cation channels are highly selective for K^+ over Na^+ (Tyerman, 1989; Carpaneto *et al.*, 2004). Furthermore, despite the presence of 0.5 M Na^+, the diffusion potential is strongly responsive to changes in external K^+ (Fernández *et al.*, 1999), resulting in a particularly low Na^+ plasma membrane permeability compared with terrestrial halophyte plants (Table 6.1).

Downregulation of Na^+ permeability, consistent with a high K^+/Na^+ selectivity ratio is assumed to be an avoidance strategy to prevent high passive cytosolic accumulation of Na^+ in terrestrial halophytes, classified as excluders (Maathuisand Prins, 1990; Volkov and Amtmann, 2006; Shabala and Mackay, 2011). The bulk of terrestrial halophytes show much stronger accumulation of Na^+ in their shoots compared with glycophytes (Flowers and Colmer, 2008). Living in seawater and afforded relatively negative resting membrane potentials (lower than −160 mV), seagrasses face the problem of accumulating physiologically harmful concentrations of sodium. Their cytosolic concentrations are lower than 20 mM Na^+, similar values to those measured in terrestrial glycophytes (Table 6.2) but much lower than the proposed values for terrestrial halophytes, for which predictions are 70 and 200 mM Na^+ in plants grown in the presence of 50 and 400 mM Na^+, respectively (Flowers *et al.*, 2015).

Maintenance of the low cytosolic Na^+ concentrations found in seagrasses seems to be the

Table 6.1. Plasma membrane Na^+/K^+ permeability ratios (P_{Na^+}/P_{K^+}) in seagrasses and terrestrial plants.

Species	P_{Na^+}/P_{K^+}	Reference
Zostera marina	0.003 (Mesophyll leaf cells)	Fernández *et al.* (1999)
Posidonia oceanica	0.005 (Mesophyll leaf cells)	Rubio and Fernández, unpublished data
Cymodocea nodosa	0.015 (Mesophyll leaf cells)	Rubio and Fernández, unpublished data
Sarcocornia perennis	0.009 (Epidermal root cells)	Rubio and Fernández, unpublished data
Thellungiela halophila	0.12 (Root protoplasts)	Demidchik and Tester (2002) Volkov *et al.* (2003)
Halimione portulacoides	0.19 (Epidermal root cells)	Rubio *et al.* (2009)
Arabidopsis thaliana	0.67 (Root protoplasts)	Demidchik and Tester (2002) Volkov *et al.* (2003)

result of both the very low plasma membrane Na^+ permeability and the operation of an active Na^+ efflux mechanism (Fernández *et al.*, 1999). Contrary to expectations, the secondary adaptation to the marine environment did not result in the achievement of Na^+-ATPases in seagrasses (Garciadeblás *et al.*, 2001). In *Z. marina*, the salt-tolerant H^+-ATPases maintain a membrane potential of around −160 mV (Fernández *et al.*, 1999), generating a H^+ electrochemical potential gradient sufficient to fuel the active extrusion of Na^+ from the cytosol through a plasma membrane Na^+/H^+ exchanger (Rubio *et al.*, 2011). This system seems to be encoded by the *SOS1* gene, cloned in *C. nodosa* (*CnSOS1A* and *CnSOS1B*; Garciadeblás *et al.*, 2007) and found in the genome of *Z. marina* (Zosma28g00970.1; Olsen *et al.*, 2016) using sequence searching in the *Phytozome* platform (Goodstein *et al.*, 2012).

Apart from avoiding toxic implications, the preservation of a low cytosolic Na^+ concentration guarantees the maintenance of a high inwardly directed Na^+ electrochemical gradient, an important property in seagrasses for the colonization of the marine environment, and for essential nutrient uptake in the diluted and alkaline medium of seawater (Fernández *et al.*, 1999).

With respect to salt tolerance, leaf production and differentiation patterns seem to be a clue to the success of seagrasses. Thus, young leaves growing in the internal part of the shoot are protected from direct contact with seawater by a hyaline sheath that surrounds external mature leaves. In fact, the non-spherical protoplasts obtained from external (mature) leaves are highly resistant to a wide range of osmotic potentials and salinities; in contrast, spherical protoplasts isolated from meristematic and immature leaf tissues, protected by closely enveloping sheaths, are more sensitive to salinity (Arai *et al.*, 1991).

Thus, seagrasses seem to have acquired different mechanisms for salt tolerance from those of most terrestrial halophytes, such as a high invagination of the plasma membrane and high activity of the salt tolerant H^+-ATPase observed in mature leaves of *Z. marina* (Fukuhara *et al.*, 1996; Muramatsu *et al.*, 2002). Neither feature was detected in the epidermal cells of the immature (salt sensitive) leaves of the seagrasses (Pak *et al.*, 1995). These observations suggest that morphological and functional changes in the plasma membrane of seagrass leaf cells and the plasma membrane H^+-ATPase play an important role in their survival and growth in seawater. To overcome the osmotic stress in marine environments, leaf cells of several species of seagrasses regulate turgor and exchange K^+, Na^+ and Cl^- when exposed to salinity changes (Tyerman, 1989). Cytosolic accumulation of organic solutes, such as sucrose and proline, was also found in different seagrass species (Tyerman, 1989), whose osmotic equilibrium was further achieved in combination with a small cytoplasm:vacuole volume ratio, lower than 10% (Fukuhara *et al.*, 1996).

6.3 Nutrient Uptake in Seagrasses: Energization and Kinetics

In contrast to terrestrial plants, seagrasses can uptake mineral nutrients through roots and also through leaves. Some authors consider the contribution of root tissues more important, often citing it as an explanation of their success in colonizing even oligotrophic waters (Romero *et al.*, 2006). However, there is substantial evidence that leaf tissues have higher affinity for nutrient uptake and can substantially contribute to total

Table 6.2. Cytosolic Na^+ concentration (Na^+c, mM) measured with intracellular Na^+- selective microelectrodes in seagrasses and barley.

Species	Na^+c (mM)	Reference
Zostera marina	13.2 ± 2.8 (leaf cells)	Rubio *et al.* (2011)
	10.7 ± 3.3 (root cells)	Rubio *et al.* (2005)
Posidonia oceanica	16 ± 1 (leaf cells)	Rubio *et al.* (2017)
Hordeum vulgare		
Na⁺-sensitive variety	19 (root cells)	Carden *et al.* (2003)
Na⁺-tolerant variety	28 (root cells)	Carden *et al.* (2003)

nutrient acquisition (Burkholder *et al.*, 2007 and references therein). Consequently, seagrass leaf tissues have developed a new characteristic to manage nutrient uptake in a liquid and alkaline medium (from 7.9 to 8.2) that contains a high salt concentration (0.5 M NaCl). In addition, essential nutrients, such as nitrate or phosphate, usually occur at concentrations below 10 μM (Romero *et al.*, 2006). Furthermore, at the seawater pH the concentration of dissolved CO_2 is low while HCO_3^- is the more abundant dissolved inorganic carbon species in marine environments.

The preservation of a high negative membrane potential in cells exposed to high Na^+ concentrations provides the maintenance of an elevated driving force for nutrient uptake (Volkov and Amtmann, 2006). In the case of seagrasses, membrane potentials below −160 mV and the maintenance of cytosolic Na^+ less than 20 mM (Table 6.2), living in the presence of 500 mM Na^+, guarantees an inwardly directed Na^+ motive force higher than for H^+ at seawater pH. Contrary to studies in vascular plants and algae, where NO_3^- uptake is powered by the electrochemical potential gradient for protons (reviewed in Wang *et al.*, 2012), the high electrochemical Na^+ potential gradient in seagrasses has been proposed to be the driving force for the high-affinity uptake of nitrate (García-Sánchez *et al.*, 2000) and also for phosphate (Rubio *et al.*, 2005), in both epidermal root and mesophyll leaf cells (Fig. 6.1). These Na^+-dependent transport systems show saturation kinetics at the micromolar range, rendering lower semi-saturation constants (Km) than occur in terrestrial angiosperms (Table 6.3). The values found in seagrasses are in the range of those reported for marine algae (2-13 μM NO_3^-; De Boer, 1985) but are much lower than those reported for the high-affinity NO_3^- uptake in terrestrial plants, close to 60 μM NO_3^- in barley (Glass *et al.*, 1992) or *Arabidopsis* (Okamoto *et al.*, 2006; reviewed in Wang *et al.*, 2012). Before being found in marine angiosperms, Na^+- coupled transport had been demonstrated in marine diatoms (Rees *et al.*, 1980), cyanobacteria (Rodríguez *et al.*, 1992; Lara *et al.*, 1993; Rodríguez *et al.*, 1994; Shibata *et al.*, 2002) and charophyte algae (Walker and Sanders, 1991). These mechanisms, as was found in *Z. marina* (Table 6.3), exhibited a Km for Na^+ in a low millimolar range, illustrating that Na^+-dependent high affinity NO_3^- transporters will be functioning to saturate Na^+ concentrations in the marine environment (García-Sánchez *et al.*, 2000).

Genome analyses indicate that only one gene encoding for the high-affinity nitrate transporter NRT2.1 is present in *Z. marina* (quoted as Zosma70g00300.1 in *Phytozome* platform). In terrestrial plants, this high-affinity nitrate transporter functions as a NO_3^-/H^+ symporter and requires the co-expression of NAR2 protein (Zhou *et al.*, 2000; Okamoto *et al.*, 2006). Thus, this transporter could have evolved to use Na^+ as a driving ion in the case of seagrasses (Fig. 6.1). Although those mechanisms could be a feature of nutrient uptake in seagrasses, further molecular and physiological analyses are needed to test this hypothesis in other seagrass species.

In addition to high affinity NO_3^- uptake, seagrasses are also able to uptake NH_4^+ that is usually present at lower concentrations in the water column of seagrass beds (≈ 3.1 μM NH_4^+, Hemminga and Duarte, 2000). However, these plants show a higher affinity for ammonium than for nitrate, as well as higher uptake rates for ammonium (reviewed in Touchette and Burkholder, 2000). However, in contrast to Na^+- coupled transport systems of nitrate and phosphate, high-affinity ammonium uptake seems not to be dependent on Na^+ or H^+ (Rubio *et al.*, 2007) suggesting that ammonium transport in seagrasses could be mediated by a uniport (Fig. 6.1), as has been proposed for terrestrial plants (Ludewig *et al.*, 2002; Wood *et al.*, 2006).

Apart from the low occurrence of nitrogen forms, seagrasses also face low phosphate availability. Phosphate levels in seagrass meadows typically range from 0.1 to 1.7 μM in the water column, with higher concentrations in the sediment pore water (0.3–20 μM PO_4^{-3}), reviewed in Touchette and Burkholder (2000). Despite semi-saturation constant (Km) values found in seagrass literature being lower in roots than in leaves (Table 6.3), a high nutrient uptake efficiency (higher uptake rates and/or lower Km values) is observed in leaves than in roots, consistent with nutrient availability in the surrounding water (Romero *et al.*, 2006). In addition, reported phosphate Km values in seagrasses (Table 6.3) are slightly lower than the range of high-affinity values quoted for phosphate transporters in terrestrial vascular plants (Km ≈ 10 μM;

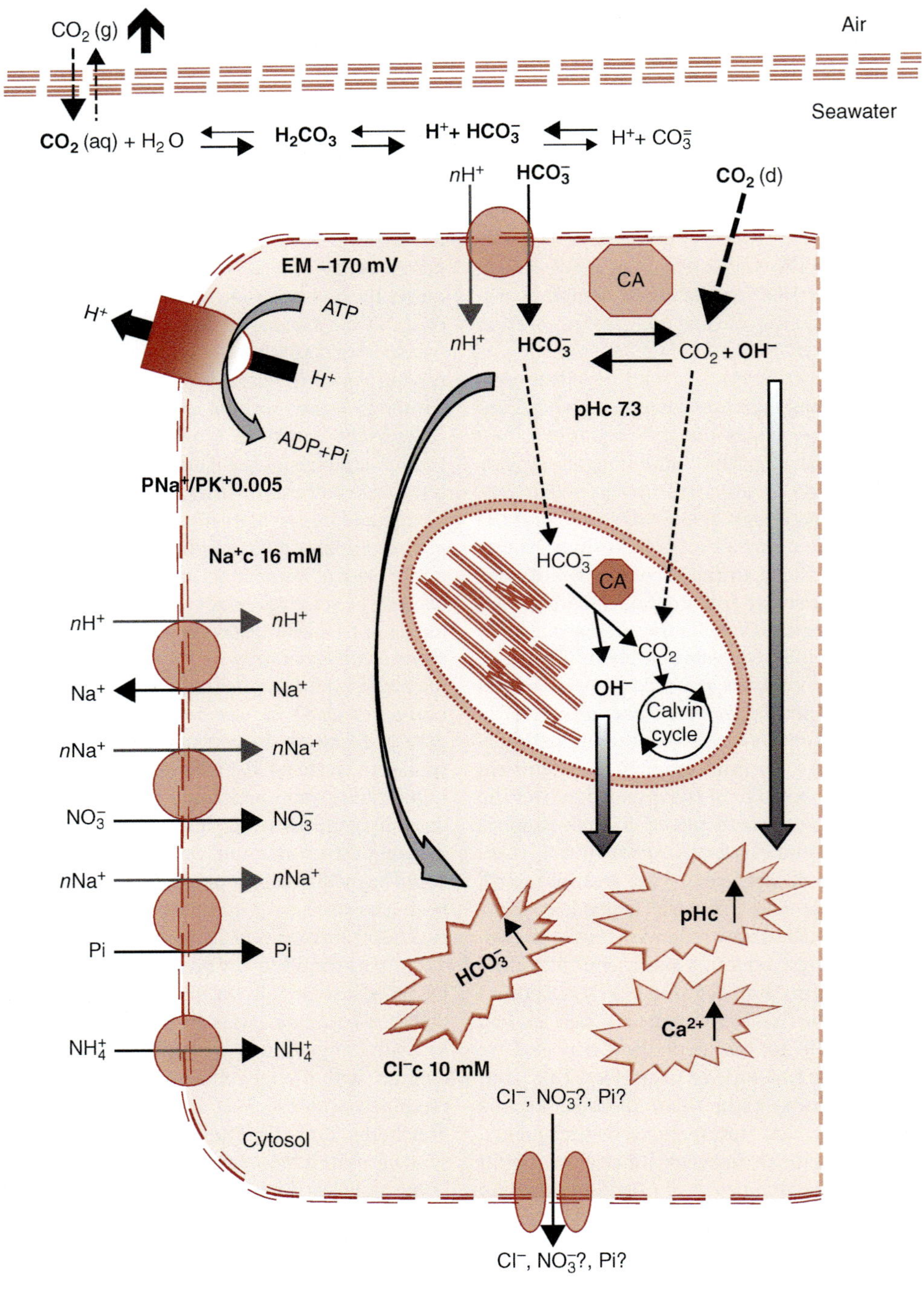

Fig. 6.1. Model of seagrass leaf cell plasma membrane energization, sodium homeostasis, high-affinity NO_3^-, NH_4^+ and Pi, and direct uptake of HCO_3^- transport system. The implications of atmospheric CO_2 increase on water dissolved inorganic carbon species, CO_2 diffusion and HCO_3^- uptake and its consequences on photosynthetic carbon use, cytosolic pH and anion homeostasis is included. Membrane potential (Em), Na^+/K^+ plasma membrane permeability P_{Na^+}/P_{K^+}) and cytosolic ion concentration values are referenced in the text. g: gas; aq: aqueous, c: cytosolic, CA: Carbonic anhydrase.

Table 6.3. Nitrate and phosphate semi-saturation constants (Km, µM) for high-affinity transport systems reported in seagrasses. Semi-saturation constant for Na^+, the driving ion of the Na^+-dependent NO_3^- uptake in *Z. marina* is included in brackets.

Species	Organ	Km	Reference
Zostera marina	Leaves	2.3 µM NO_3^- (720 µM Na^+)	García-Sánchez *et al.* (2000)
	Roots	8.9 µM NO_3^- 1.5 µM Pi	Rubio *et al.* (2005)
Pyllospadix torreyi	Leaves	8.7 µM NO_3^-	Terrados and Williams (1997)
Thalasia testudinum	Leaves	2.2 µM NO_3^-	Lee and Dunton (1996)
Zostera noltii	Leaves	12.1 µM Pi	Pérez Llorens and Niell (1995)
Thalassia hemprichii	Leaves	7.7 µM Pi	Stapel *et al.* (1996)
Thalassia testudinum	Leaves	11.9 µM Pi	Gras *et al.* (2003)
	Roots	3.7 µM Pi	

Mitsukawa *et al.*, 1997; Mimura, 2001; Rae *et al.*, 2003), in which it is assumed that the driving force for phosphate influx is the proton gradient generated by the H^+-ATPase (Ullrich-Eberius *et al.*, 1984; Schachtman *et al.*, 1998; Bucher, 2007). As discussed below, in the case of *Z. marina*, high-affinity phosphate uptake is meditated by a Na^+-coupled transport system (Rubio *et al.*, 2005), which had not been previously characterized in other vascular plants (Rausch and Bucher, 2002), despite several genes of putatively Na^+-coupled Pi being identified (Smith *et al.*, 2003). In terrestrial vascular plants, high-affinity phosphate uptake seems to be mediated by PTH1 transporters belonging to the *PTH1* family comprising nine members (*PTH1;1* through *PTH1;9*) in *Arabidopsis* (reviewed in Nussaume *et al.*, 2011). It is suggested that these systems, related to the fungal transporter, use the H^+-gradient at the plasma membrane to drive the H^+/Pi symport activity (Rausch and Bucher, 2002). Sequence searching in the *Phytozome* platform also renders nine genes in the genome of *Z. marina* that are classified as phosphate transporters. Further molecular studies are needed to reveal the molecular evolution of this putative high-affinity phosphate transport system to use Na^+ as the driving ion in seagrasses.

In addition, apart from the use of Na^+ as the driving ion, seagrasses also conserve the use of H^+ to fuel ion uptake (Fig. 6.1). Recently, it has been reported that in *P. oceanica* H^+ is used as the driving ion for the direct uptake of HCO_3^- through a H^+/HCO_3^- symporter in leaf cells (Rubio *et al.*, 2017). This mechanism shows similar semi-saturation kinetics constant (Km = 0.36 mM HCO_3^-) than the half saturation constant of photosynthesis (S0.5 = 0.35 mM HCO_3^-), both processes being almost saturated at the current HCO_3^- concentration of the seawater, 2 mM HCO_3^- (Rubio *et al.*, 2017). The plasma membrane H^+ economy in seagrasses appears to be important for photosynthesis, not only to fuel the direct uptake of HCO_3^- but also for the apoplastic carbonic anhydrase-mediated dehydration of HCO_3^-, the second model of the photosynthetic use of HCO_3^- proposed for these species (James and Larkum, 1996; Björk *et al.*, 1997; Invers *et al.*, 1997, 1999; Beer *et al.*, 2002). In any case, the existence of acid zones in seagrasses is a matter for discussion, since there has been no direct demonstration of their occurrence, which is expected to occupy small areas (Raven and Beardall, 2016). This situation could be altered as a consequence of ocean acidification in a climate change context (discussed further on).

6.4 Ocean Acidification: Nutrient Speciation and Energetic Consequences for Seagrasses

Atmospheric CO_2 has increased over the last 200 years (IPCC, 2001). Current observations indicate that it has reached a 50% higher value than preceding observations. That increment amounts to about 3 parts per million of CO_2 per year, or 6.3 GTons of carbon. In recent years, the average annual increase had been closer to 4 GTons of carbon per year. This record increment occurred even though emissions from human activities in

2015–2016 are estimated to have remained roughly the same as they were prior to the El Niño, which is a cyclical warming pattern of ocean circulation in the central and eastern tropical Pacific Ocean that can affect weather worldwide, and increased the temperature in 2016. Consequently, the CO_2 concentration recorded at Mauna Loa observatory remained for the first time above 400 ppm all year and perhaps will remain so for our lifetimes (Betts *et al.*, 2016).

The increase in the concentration of CO_2 in the atmosphere has other implications besides the observable increase of the average of the planet. The oceans have been a sink of 30% of CO_2 released during the industrial era, at a rate of 3.8 GTons/year^{-1}, compared to 1.8 GTons/year^{-1} fixed by photosynthesis and 2 GTons/year^{-1} removed by abiotic absorption (Behrenfeld *et al.*, 2002). The addition of such an amount of CO_2 to the oceans has led to a decrease in the pH of seawater (Doney *et al.*, 2009), an event described by Kerr (2010) as one of the most notable cataclysms in the history of the Earth. This acidification, although small in terms of pH units, is associated with a significant decrease in the amount of $CO_3^=$, it means that much of the acidity caused by the CO_2 increase has been buffered by the equilibrium $H_2CO_3/CO_3^=$ (Hoegh-Guldberg and Bruno, 2010).

Although there is no doubt that oceans sequester large amounts of anthropogenic CO_2 (Sabine and Feely, 2007), there is controversy over the potential rates of increase, owing to variable estimates of future burning of fossil fuels, and strong positive and negative feedbacks (Koch *et al.*, 2013). Predictions of future increments vary, but it is estimated that in the next 100 years the current concentrations will increase by two to three times (Beardall *et al.*, 1998). An increase in the amount of CO_2 in the gas phase from 370 to 1000 ppm will generate, in full equilibrium, an increase in the concentration of CO_2 dissolved in seawater from 14.2 to 38.4 μM (Beardall and Raven, 2004). It is estimated that this increase will acidify seawater over the same period by about 0.5 pH units (Beardall and Raven, 2004), which will alter the speciation of DIC. Thus, in addition to the increase in dissolved CO_2, the concentration of HCO_3^- will rise slightly from 2000 to 2200 μM, while $CO_3^=$ concentration will decrease from 200 to 100 μM (Beardall *et al.*, 1998).

The decrease in the pH of seawater by 0.5 units will alter the speciation of other dissolved ions. Although the most conspicuous case, as a limiting nutrient for marine phytoplankton, is iron (Shi *et al.*, 2010), speciation of both ammonium and phosphate will also change. A pH change from 8.1 to 7.8 evokes a decrease in the amount of NH_3 in the NH_4^+/NH_3 ratio (Raven, 1986), whereas the amount of NO_3^- would not be affected (Zeebe and Wolf-Gladrow, 2001). Similarly, a decrease in pH will cause a decrease in PO_4^{-3} concentration to the same extent as the $H_2PO_4^-$ form, which appears to be the inorganic phosphate form energetically more feasible for incorporation by seagrasses (Rubio *et al.*, 2005), whereas the $HPO_4^=$ will vary slightly (Zeebe and Wolf-Gladrow, 2001). These changes in speciation would be extremely relevant to marine plants, disturbing nutrient uptake mechanisms due to changes in essential nutrient availability, in addition to the energy cost for nutrient transport.

Acidification of seawater affects the energetic load of the plasma membrane and may have an impact on the uptake of nutrients. Because of their lineage, seagrasses are expected to use H^+-ATPases to energize the plasma membrane (Pedersen *et al.*, 2012). In *Z. marina* and *P. oceanica* the data from Em and cytosolic pH (pHc) of mesophyll leaf cells allows us to estimate the impact of external acidification on the H^+ motive force. In the case of *Z. marina*, and assuming a constant cytosolic pH (7.3) and Em (−160 mV) (Fernández *et al.*, 1999), the electrochemical potential gradient for the protons ($\Delta\mu H^+/F$) at external pH of 8.2 is −108 mV. The same calculations can be obtained for *P. oceanica*, in which the Em value is −174 mV and pHc is 7.3 (Rubio *et al.*, 2017), rendering a $\Delta\mu H^+/F$ of −122 mV at external pH 8.2. A decrease of 0.5 units in the external pH at constant temperature renders an increase of this proton motive force of −23 mV. This amount represents more than a 20% increase in the proton-associated motive force (directed into the cell), which could facilitate the operation of H^+-dependent transport systems in seagrasses.

Apparently, ocean acidification could alter both the H^+ gradient and the electrochemical Na^+ gradient due to changes in the activity of the plasma membrane Na^+/H^+ antiporter. The activity of this antiporter is sensitive to external pH, increasing at acid external pH, therefore a high efflux of Na^+ would be expected in acid seawater (Rubio *et al.*, 2011). Consequently, a diminution in cytosolic Na^+ could be expected in response to ocean acidification. Thus, an intensification of the inwardly directed Na^+ motive force could be also expected, favouring the Na^+-dependent high-affinity transport of nutrients. However, at least in the case of NO_3^-, it was demonstrated that the decrease in one pH unit of seawater promoted, in the short term, a reduction of around 30% in NO_3^- uptake (García-Sánchez *et al.*, 2000). The long-term response of seagrasses to the effects of seawater acidification are unknown, although changes in cytosolic Na^+ homeostasis and nutrient uptake should not be ruled out.

6.5 Effects of Inorganic Carbon Increase on Seagrass Ion Homeostasis

Living submerged, seagrasses have developed new combinations of structural characters related to the cell wall. The latter contains the polysaccharides typical of land plants, but also polyanionic, low-methylated pectins and sulfated galactans (Aquino *et al.*, 2005), a feature shared with the cell walls of all macroalgae (Popper *et al.*, 2011). Considering the diffusion limitation in aquatic ecosystems (Beer *et al.*, 1980), these achievements seem to be important for ion homeostasis, nutrient uptake and O_2/CO_2 exchange through leaf epidermal cells in marine environments (Olsen *et al.*, 2016).

Compared to land plants, the photosynthesis of most of seagrass species seems to be not saturated by the present concentration of DIC (Beer *et al.*, 2002). Consequently, most authors hypothesize a possible increase in inorganic carbon assimilation by seagrasses at high atmospheric CO_2 (Borum *et al.*, 2016). This increase could be different due to the inorganic carbon species that crosses the plasma membrane (CO_2 or HCO_3^-).

Although seagrasses are able to use HCO_3^-, most species do so inefficiently (Doney *et al.*, 2009). In several species tested, light-saturated photosynthetic rates increased dramatically with increased dissolved CO_2 concentration (Zimmerman *et al.*, 1997, Short and Neckles 1999; Invers *et al.*, 2001). In the case of *Z. marina*, an increase in biomass and reproductive output under high-CO_2 conditions was observed (Palacios and Zimmerman, 2007), suggesting potentially higher productivity of seagrasses, which should be beneficial for many fish and invertebrate communities. Nevertheless, CO_2-enriched seawater benefits could be limited due to the negative effects of increased temperature on vegetative growth (Koch *et al.*, 2013 and references therein).

Seawater contains 2.0–2.5mM HCO_3^- and 10–15 µM CO_2. Assuming equilibrium between the atmosphere and seawater, an increase in atmospheric CO_2 will cause a rise in the concentration of dissolved CO_2. Such an increase will produce a slight shift in the DIC species equilibrium, resulting in a greater concentration of undissociated H_2CO_3 and the concomitant reduction of pH discussed previously.

Apparently, increasing CO_2 atmospheric levels up to 800 ppm does not evoke a photosynthetic rise in seagrasses, as has been found in nine species from south-western Australia (Borum *et al.*, 2016) or in the case of the Mediterranean endemic *P. oceanica* exposed to 1000 ppm CO_2 (Rubio *et al.*, 2017). By contrast, *P. oceanica* shows a high photosynthesis increase in natural seawater supplemented by 3 mM HCO_3^- (around 50%), supporting the reliance on HCO_3^- uptake in the photosynthetic activity of this species (Rubio *et al.*, 2017).

Apart from photosynthetic effects, in the case of *P. oceanica* depending on the inorganic carbon species with which the seawater is enriched, membrane potential and cytosolic ion responses are different. External CO_2 increase (1000 ppm) causes a slight hyperpolarization (less than 5 mV) and a slow cytosolic acidification, around 0.1 pH unit. In contrast, addition of 3 mM HCO_3^- evokes an initial depolarization, turns into a hyperpolarization and is followed by a depolarization that lasts in the presence of high HCO_3^-. Simultaneously, an initial cytosolic acidification (from 7.3 to 7.2) followed by an alkalinization, levelling off at 7.5 in the presence of high HCO_3^- was observed (Rubio *et al.*, 2017). Similar responses have been found in *Z. marina* (our data, unpublished), denoting that the rise in external HCO_3^- alters cytosolic ion homeostasis and plasma membrane energetics wider than increasing CO_2.

Simultaneous to membrane depolarization induced by high HCO_3^-, leaf cells of *P. oceanica* show a diminution in cytosolic Cl^-, from 9.5 to

6.3 mM (Rubio *et al.*, 2017). Thus, the plasma membrane depolarization observed in *P. oceanica* leaf cells exposed to high HCO_3^- has been associated with the operation of slow (S-type) anion channels (Rubio *et al.*, 2017) which are activated when cells depolarize (Roberts, 2006; Roelfsema *et al.*, 2012). The situation in *P. oceanica* leaf cells submitted to high HCO_3^- also matches the Cl^- efflux mediated by the activation of S-type channels in stomata of terrestrial angiosperms, which takes place when the HCO_3^- concentration increases and the cytosolic pH is alkaline (Xue *et al.*, 2010). Intriguingly, these S-type channels are also permeable to NO_3^- (Schmidt and Schroeder, 1994). Accordingly, the Cl^- efflux observed in *P. oceanica* leaf cells subjected to high HCO_3^- agrees with the prediction that HCO_3^- enrichment not only evokes Cl^- efflux but may also cause NO_3^- efflux from leaf cells (Rubio *et al.*, 2017). Therefore, one of the consequences of the DIC rise due to climate change could be the decrease in cytosolic anion concentrations in seagrass cells (Fig. 6.1).

The high negative membrane potential (lower than −160 mV) pushes the anion flux outward in seagrass cells. The energy associated with the efflux of each anion, however, depends on its concentration gradients, which in the case of seagrass cells is much higher for NO_3^- than for Cl^-, since the NO_3^- concentration in seawater is much lower than the Cl^- (≈5 μM NO_3^- versus 500 mM Cl^-). Thus, in *P. oceanica*, the electrochemical potential gradient for Cl^- results in +71 mV, considering a membrane potential of −170 mV and a cytosolic Cl^- of 10 mM (Rubio *et al.*, 2017). The outwardly directed driving force for NO_3^- is almost fivefold higher (+326 mV), considering a cytosolic NO_3^- concentration as 2.5 mM (value in the range of that reported for *Arabidopsis* leaf cells) (Cookson *et al.*, 2005). However, plasma membrane permeability could be different for NO_3^- and Cl^-, limiting the efflux of NO_3^- over Cl^- in seagrasses.

In terrestrial vascular plants, NO_3^- efflux at the plasma membrane takes place through SLAH3 (Zheng *et al.*, 2015), whose activation depends on its phosphorylation by the CPK21 protein (Wang *et al.*, 2012). In comparison to its homologue SLAC1, a channel involved in anion homeostasis in guard cells, SLAH3 exhibits a higher NO_3^- conductance, having a NO_3^-/Cl^- permeability ratio of 10 for SLAC1 (Geiger *et al.*, 2009) and of 20 for SLAH3 (Geiger *et al.*, 2011). Examination of the *Arabidopsis* homologues SLAC/SLAH sequences in the *Z. marina* genome revealed two S-type anion channels, SLAH1 (Zosma91g00860.1) and SLAC1 (Zosma76g00610.1). These could be good candidates to encode the channels that mediate the NO_3^- efflux in response to high HCO_3^-.

6.6 Conclusions and Perspectives

Two major consequences of climate change are faced by marine angiosperms: the rise of DIC species (CO_2 and HCO_3^-) and the consequent water acidification. The predicted water pH decrease will increase the proton motive force generated by the H^+-ATPase operating at the plasma membrane of seagrasses (Fig. 6.1), and will probably change cytosolic sodium homeostasis owing to the promotion of Na^+ efflux through the plasma membrane Na^+/H^+ antiporter (Fig. 6.1). Thus, the electrochemical Na^+ gradient could also be altered because of water acidification. Apart from promotion of salinity tolerance, changes in cytosolic sodium and the predicted alteration of the Na^+ motive force will impact on Na^+-dependent high-affinity transport systems functioning in the plasma membrane (Fig. 6.1), which could modify nitrate and phosphate uptake rates in seagrasses. In addition, although high-affinity NH_4^+ transport seems to be meditated by a uniport system, water acidification could also change seagrass NH_4^+ uptake owing to changes in NH_4^+/NH_3 equilibrium (Raven, 1986).

Most seagrasses show a non-saturated C3 photosynthetic pathway and are able to use HCO_3^- as an inorganic carbon source (reviewed in Koch *et al.*, 2013). Accordingly, higher photosynthetic activity should be expected due to the rise of both CO_2 and HCO_3^- in seawater. However, the impact of the increase of these DIC species promotes not only high photosynthesis, but also important consequences for anion homeostasis. As in the model recently proposed for *P. oceanica* (Rubio *et al.*, 2017), while CO_2 increase has no significant effects on the photosynthetic rate, seawater HCO_3^- enrichment enhances photosynthesis and leads the efflux of chloride from the cytosol (probably through S-type anion channels), which could also allow the efflux of nitrate and, probably, phosphate. Thus, as has been predicted for plant tissues under high CO_2 (Taub and Wang, 2008), the unpredicted effect of HCO_3^- enrichment on the seagrass environment may be a key consideration in the context of climate change, in which nutrient biomass dilution of seagrasses should be expected.

References

Ait-Hammou, K., Rubio, L., Fernández, J.A. and García-Sánchez, M.J. (2014) Potassium uptake in the halophyte *Halimione portulacoides* L. Aellen. *Environmental and Experimental Botany* 107, 15–24.

Anderson, W.P., Willcocks, D.A. and Wright, B.J. (1977) Electrophysiological measurements on the root of *Atriplex hastata*. *Environmental and Experimental Botany* 28, 894–901.

Aquino, R.S., Landeira-Fernández, A.M., Valente, A.P., Andrade, L.R. and Mourao, P.A.S. (2005) Occurrence of sulfated galactans in marine angiosperms, evolutionary implications. *Glycobiology* 15, 11–20.

Arai, M., Pak, J.Y., Nomura, K. and Nitta, T. (1991) Seawater-resistant, non-spherical protoplast from seagrass leaves. *Physiologia Plantarum* 83, 551–559.

Arber, A., (1920) *Water Plants: A Study of Aquatic Angiosperms*. Cambridge University Press, Cambridge.

Beardall, J. and Raven, J.A. (2004) The potential effects of global change on microalgal photosynthesis, growth and ecology. *Phycologia* 43, 26–40.

Beardall, J., Beer, S. and Raven, J.A. (1998) Biodiversity of marine plants in an era of climate change, some predictions on the basis of physiological performance. *Botanica Marina* 41,113–123.

Beer, S., Shomer-Ilan, A. and Waisel, Y. (1980) Carbon metabolism in seagrasses. II. Patterns of photosynthetic CO_2 incorporation. *Journal of Experimental Botany* 31, 1019–1026.

Beer, S., Bjork, M., Hellblom, F. and Axelsson, L. (2002) Inorganic carbon utilization by marine angiosperms (seagrasses). *Functional Plant Biology* 29, 349–354.

Behrenfeld, M.J., Esaias, W.E. and Turpie, K.R. (2002) Assessment of the primary production at the global scale. In: Williams, P.J. le B., Thomas, D.N. and Reynolds, C.S. (eds) *Phytoplankton Productivity. Carbon Assimilation in Marine and Freshwater Ecosystems*. Blackwell Science, Oxford, UK, pp. 156–186.

Betts, R.A., Jones, C.D., Knight, J.R., Keeling, R.F. and Kennedy, J.J. (2016) El Niño and a record CO_2 rise. *Nature Climate Change* 6, 806–810.

Björk, M., Weil, A., Semesi, S. and Beer, S. (1997) Photosynthetic utilisation of inorganic carbon by seagrasses from Zanzibar, East Africa. *Marine Biology* 129, 363–366.

Borum, J., Pedersen, O., Kotula, L., Fraser, M.W., Statton, J., *et al.* (2016) Photosynthetic response to globally increasing CO_2 of co-occurring temperate seagrass species. *Plant, Cell and Environment* 39, 1240–1250.

Bose, J., Rodrigo-Moreno, A., Lai, D., Xie, Y., Shen, W., *et al.* (2015) Rapid regulation of the plasma membrane H^+-ATPase activity is essential to salinity tolerance in two halophyte species, *Atriplex lentiformis* and *Chenopodium quinoa*. *Annals of Botany* 115(3), 481–494.

Bucher, M. (2007) Functional biology of plant phosphate uptake at root and mycorrhiza interfaces. *New Phytologist* 173, 11–26.

Burkholder, J.M., Tomasko, D.A. and Touchette, B.W. (2007) Seagrasses and eutrophication. *Journal of Experimental Marine Biology and Ecology* 350, 46–72.

Carden, D., Walker, D., Flowers, T.J. and Miller, A.J. (2003) Single–cell measurements of the contributions of cytosolic Na^+ and K^+ to salt tolerance. *Plant Physiology* 131, 676–683.

Carpaneto, A., Naso, A., Paganetto, A., Cornara, L., Pesce, E-R., *et al.* (2004) Properties of ion channels in the protoplast of the Mediterranean seagrass *Posidonia oceanica*. *Plant, Cell and Environment* 27, 279–292.

Cookson, S.J., Williams, L.E. and Miller, A.J. (2005) Light–dark changes in cytosolic nitrate pools depend on nitrate reductase activity in Arabidopsis leaf cells. *Plant Physiology* 138, 1097–1105.

Costanza, R., D'Arge, R., de Groot, R., Farber, S., Grasso, M., *et al.* (1997) The value of the world's ecosystem services and natural capital. *Nature* 387, 253–260.

Cullen-Unsworth, L.C., Nordlund, L.M., Paddock, J., Baker, S., McKenzie, L.J. *et al.* (2014) Seagrass meadows globally as a coupled social-ecological system: Implications for human wellbeing. *Marine Pollution Bulletin* 83, 387–397.

De Boer, J.A. (1985) Nutrients. In: Lobban, C.S., Harrison, P.J., Duncan, J.J. (eds) *The Physiological Ecology of Seaweeds*. Cambridge University Press, Cambridge, UK, pp. 356–392.

Demidchik, V. and Tester, M. (2002) Sodium fluxes through nonselective cation channels in the plasma membrane of protoplast from Arabidopsis roots. *Plant Physiology* 128, 379–387.

Den Hartog, C. (1970) *The Seagrasses of the World*. North-Holland Publishing Company, Amsterdam and London.

Den Hartog, C. and Kuo, J. (2006) Taxonomy and Biogeography of Seagrass. In: Larkum, A.W.D., Orth, R.J. and Duarte, C.M. (eds) *Seagrasses: Biology, Ecology and Conservation*. Springer, Netherlands, pp. 1–23.

Doney, S.C., Fabry, V.J., Feely, R.A. and Kleypas, J.A. (2009) Ocean acidification: The other CO_2 problem. *Annual Review of Marine Sciences* 1, 169–192.

Duarte, C.M. (1991) Seagrass depth limits. *Aquatic Botany* 40, 363–377.

Duarte, C.M. and Cebrián, J. (1996) The fate of marine autotrophic production. *Limnology and Oceanography* 41, 1758–1766.

Felle, H. (1981) A study of the current-voltage relationships of electrogenic and passive membrane elements in *Riccia fluitans*. *Biochimica et Biophysica Acta* 646, 151–160.

Fernández, J.A., García-Sánchez, M.J. and Felle, H. (1999) Physiological evidence from a proton pump at the plasma membrane of the marine angiosperm *Zostera marina* L. *Journal of Experimental Botany* 50, 1763–1768.

Flowers, T.J. and Colmer, T.D. (2008) Salinity tolerance in halophytes. *New Phytologist* 179, 945–963.

Flowers, T.J., Munns, R. and Colmer, T.D. (2015) Sodium chloride toxicity and the cellular basis of salt tolerance in halophytes. *Annals of Botany* 115, 419–431.

Fourqurean, J.W., Duarte, C.M., Kennedy, H., Marbà, N., Holmer, M., *et al.* (2012) Seagrass ecosystems as a globally significant carbon stock. *Nature Geoscience* 5, 505–509.

Fukuhara, T., Pak, J.Y., Ohwaki, Y., Tsujimura, H. and Nitta, T. (1996) Tissue-specific expression of the gene for a putative plasma membrane H^+-ATPase in a seagrass. *Plant Physiology* 110, 35–42.

Garciadeblás, B., Benito, B. and Rodríguez-Navarro, A. (2001) Plant cells express several stress calcium ATPases but apparently no sodium ATPase. *Plant and Soil* 235, 181–192.

García-Sánchez, M.J., Jaime, M.P., Ramos, A., Sanders, D. and Fernández, J.A. (2000) Sodium-dependent nitrate transport at the plasma membrane of leaf cells of the marine higher plant *Zostera marina* L. *Plant Physiology* 122, 879–885.

Gaxiola, R.A., Palmgren, M.G. and Schumacher, K. (2007) Plant proton pumps. *FEBS Letters* 581, 2204–2214.

Geiger, D., Scherzer, S., Mumm, P., Stange, A., Marten, I., *et al.* (2009) Activity of guard cell anion channel SLAC1 is controlled by drought stress signalling kinase-phosphatase pair. *Proceedings of the National Academy of Sciences of the United States of America* 106, 21425–21430.

Geiger, D., Maierhofer, T., Al-Rasheid, K.A.S., Scherzer, S., Mumm, P., *et al.* (2011) Stomatal closure by fast abscisic acid signaling is mediated by the guard cell anion channel SLAH3 and the receptor RCAR1. *Science Signaling* 4. doi: 10.1126/scisignal.2001346.

Glass, A.D.M., Shaff, J. and Kochian, L.V. (1992) Studies of the uptake of nitrate in barley. IV Electrophysiology. *Plant Physiology* 99, 456–463.

Goodstein, D.M., Shu, S., Howson, R., Neupane, R., Hayes, R.D. *et al.* (2012) Phytozome: A comparative platform for green plant genomics. *Nucleic Acid Research* 40, 1178–1186.

Gradmann, D. and Boyd, C.M. (1995) Membrane voltage of marine phytoplankton, measured in the diatom *Coscinodiscus radiates*. *Marine Biology* 123, 645–650.

Gras, A.F., Koch, M.S. and Madden, C.J. (2003) Phosphorus uptake kinetics of a dominant tropical seagrass *Thalassia testudinum*. *Aquatic Botany* 76, 299–315.

Gutknecht, J. and Dainty, J. (1968) Ionic relations of marine algae. *Oceanography and Marine Biology Annual Review* 6, 163–200.

Hemminga, M.A. and Duarte, C.M. (2000) *Seagrass Ecology*. Cambridge University Press, Cambridge, UK.

Hoegh-Guldberg, O. and Bruno, J.F. (2010) The impact of climate change on the world´s marine ecosystems. *Science* 328, 1523–1528.

Invers, O., Pérez, M. and Romero, J. (1997) Effects of pH on seagrass photosynthesis: A laboratory and field assessment. *Aquatic Botany* 59, 185–194.

Invers, O., Pérez, M. and Romero, J. (1999) Bicarbonate utilization in seagrass photosynthesis: role of carbonic anhydrase in *Posidonia oceanica* (L.) Delile and *Cymodocea nodosa* (Ucria) Ascherson. *Journal of Experimental Marine Biology and Ecology* 235, 125–133.

Invers, O., Zimmerman, R.C., Alberte, R.S., Pérez, M., and Romero, J. (2001) Inorganic carbon sources for seagrass photosynthesis: An experimental evaluation of bicarbonate use in species inhabiting temperate waters. *Journal of Experimental Marine Biology and Ecology* 265, 203–217.

IPCC (2001) Climate change 2001: The scientific basis. Contribution of working group I to the third assessment report of the intergovernmental panel on climate change. Cambridge University Press, Cambridge, UK.

James, P.L. and Larkum, A.W.D. (1996) Photosynthetic inorganic carbon acquisition of *Posidonia australis*. *Aquatic Botany* 55, 149–157.

Kerr, R.A. (2010) Ocean acidification unprecedented, unsettling. *Science* 238, 1500–1501. DOI: 10.1126/science.328.5985.1500

Koch, M., Bowes, G., Ross, C. and Zhang, X.H. (2013) Climate change and ocean acidification effects on seagrasses and marine macroalgae. *Global Change Biology* 19, 103–132.

Lara, C., Rodríguez, R. and Guerrero, M.G. (1993) Sodium-dependent nitrate transport and energetics in cyanobacteria. *Journal of Phycology* 29, 389–395.

Larkum, W.D., Orth, R.J. and Duarte, C.M. (2006) *Seagrasses: Biology, Ecology and Conservation*. Springer, Netherlands.

Lee, H., Golicz, A.A., Bayer, P.E., Jiao, Y., Tang, H., *et al.* (2016) The Genome of a southern hemisphere seagrass species (*Zostera muelleri*). *Plant Physiology* 172, 272–283.

Lee, K.S. and Dunton, K.H. (1996) Inorganic nitrogen acquisition in the seagrass *Thalassia testudinum:* Development of a whole-plant nitrogen budget. *Limnology and Oceanography* 44, 1204–1215.

Les, D.H., Cleland, M.A. and Waycott, M. (1997) Phylogenetic studies in Alismatidae, II: Evolution of marine angiosperms (seagrasses) and hydrophily. *Systematic Botany* 22, 443–463.

Ludewig, U., von Wiren, N. and Frommer, W.B. (2002) Uniport of NH_4^+ by the root hair plasma membrane ammonium transporter LeAMT1;1. *Journal of Biological Chemistry* 277, 13548–13555.

Maathuis, F.J.M. and Prins, H.B.A. (1990) Electrophysiological membrane characteristics of the salt tolerant *Plantago maritima* and the sensitive *Plantago media*. *Plant and Soil* 123, 233–238.

Marrè, E. (1979) Fusicoccin: A tool in plant physiology. *Annual Review of Plant Physiology* 30, 273–288.

Mimura, T. (2001) Physiological control of phosphate uptake and phosphate homeostasis in plant cells. *Australian Journal of Plant Physiology* 28, 653–658.

Mitsukawa, N., Okumura, S., Shirano, Y., Sato, S., Kato, T. *et al.* (1997) Overexpression of an Arabidopsis thaliana high-affinity phosphate transporter in tobacco cultured cells grown under phosphate-limited conditions. *Proceedings of the National Academy of Sciences, USA* 94, 7098–7102.

Muramatsu, Y., Harada, A., Ohwaki, Y., Kasahara, Y., Takagi, S., *et al.* (2002) Salt-tolerance ATPase activity in the plasma membrane of the marine angiosperm *Zostera marina* L. *Plant and Cell Physiology* 43, 1137–1145.

Nobel, P.S. (2009) *Physicochemical and Environmental Plant Physiology*. 4th edn. Elsevier, Oxford, UK.

Nussaume, L., Kanno, S., Javot, H., Marin, E., Pochon, N., *et al.* (2011) Phosphate import in plants: Focus on the PTH1 trasnporters. *Frontiers in Plant Science* 2. doi: 10.3389/fpls.2011.00083.

Okamoto, M., Kumar, A., Li, W., Wang, Y., Siddiqi, M.Y., *et al.* (2006) High-affinity nitrate transport in roots of Arabidopsis depends on expression of the NAR2-Like gene *AtNRT3.11*. *Plant Physiology* 140, 1036–1046.

Olsen, J.L., Rouzé, P., Verhelst, B., Lin, Y.C., Bayer, T., *et al.* (2016) The genome of the seagrass *Zostera marina* reveals angiosperm adaptation to the sea. *Nature* 530, 331–335.

Pak, J.Y., Fukuhara, T. and Nitta, T. (1995) Discrete subcellular localization of membrane-bound ATPase activity in marine angiosperm and marine algae. *Planta* 196, 15–22.

Palacios, S.L. and Zimmerman, R.C. (2007) Response of eelgrass *Zostera marina* to CO_2 enrichment: possible impacts of climate change and potential for remediation of coastal habitats. *Marine Ecology Progress Series* 344, 1–13. DOI: 10.3354/meps07084

Palmgren, M.G. (2001) Plant plasma membrane H^+-ATPases: Powerhouses for nutrient uptake. *Annual Review of Plant Molecular Biology* 52, 817–845.

Panno, L., Voyron, S., Anastasi, A., Mussat Sartor, R. and Varese, G.C. (2011) Biodiversity of marine fungi associated with the seagrass *Posidonia oceanica*: An ecological and biotechnological perspective. *Biologia Marina Mediterranea* 18, 85–88.

Pedersen, C.N.S., Axelsen, K.B., Harper, J.F. and Palmgren, M.G. (2012) Evolution of plant P-Type ATPases. *Frontiers in Plant Science* 3. doi.org/10.3389/fpls.2012.00031.

Pérez-Lloréns, J.L. and Niell, F.X. (1995) Short-term phosphate uptake kinetics in *Zostera noltii* Hornem: A comparison between excised leaves and sediment-rooted plants. *Hydrobiologia* 297, 17–27.

Popper, Z.A., Michel, G., Hervé, C., Domozych, D.S., Willats, W.G.T., *et al.* (2011) Evolution and diversity of plant cell walls: From algae to flowering plants. *Annual Review of Plant Biology* 62, 567–590.

Rae, A.L., Cybinski, D.H., Jarmey, J.M. and Smith, F.W. (2003) Characterization of two phosphate transporters from barley: Evidence for diverse function and kinetic properties among members of the Pht1 family. *Plant Molecular Biology* 53, 27–36.

Rausch, C. and Bucher, M. (2002) Molecular mechanisms of phosphate transport in plants. *Planta* 216, 23–37.

Raven, J.A. (1986) Physiological consequences of extremely small size for autotrophic organisms in the sea. In: Platt, T. and Li, W.K.W. (eds) *Photosynthetic Picoplankton*. Canadian Bulletin of Fisheries and Aquatic Sciences, Ottawa, Canada, pp. 1–70.

Raven, J.A. and Beardall, J. (2016) The ins and outs of CO_2. *Journal of Experimental Botany* 67, 1–13.

Rees, T.A., Cresswell, R.C. and Syrett, P.J. (1980) Sodium-dependent uptake of nitrate and urea by a marine diatom. *Biochimica et Biophysica Acta* 596, 141–144.

Roberts, S.K. (2006) Plasma membrane anion channels in higher plants and their putative functions in roots. *New Phytologist* 169, 647–666.

Rodríguez, R., Lara, C. and Guerrero, M.G. (1992) Nitrate transport in the cyanobacterium *Anacystis nidulans* R2: kinetic and energetic aspects. *Biochemical Journal* 282, 639–643.

Rodríguez, R., Guerrero, M.G. and Lara, C. (1994) Mechanism of sodium/nitrate symport in *Anacystis nidulans* R2. *Biochimica et Biophysica Acta* 1187, 250–254.

Roelfsema, M.R.G., Hedrich, R. and Geiger, D. (2012) Anion channels: Master switches of stress responses. *Trends in Plant Science* 17, 221–229.

Romero, J., Lee, K.S., Pérez, M.A. and Alcoverro, T. (2006) Nutrients dynamics. In: Larkum, A.W.D., Orth, R.J. and Duarte, C.M. (eds) *Seagrasses: Biology, Ecology and Conservation*. Springer, Netherlands, pp. 227–254.

Rubio, L., Linares-Rueda, A., García-Sánchez, M.J. and Fernández, J.A. (2005) Physiological evidence for a sodium-dependent high-affinity phosphate and nitrate transport at the plasma membrane of leaf and root cells of *Zostera marina* L. *Journal of Experimental Botany* 56, 613–622.

Rubio, L., Linares-Rueda, A., García-Sánchez, M.J. and Fernández, J.A. (2007) Ammonium uptake kineticts in root and leaf cells of *Zostera marina* L. *Journal of Experimental Marine Biology and Ecology* 352, 271–279.

Rubio, L., Hammou, K.A., Elena, B., García-Sánchez, M.J. and Fernández, J.A. (2009) Na^+/K^+ permeability at different salt growth regimes and Na^+ and K^+ uptake kinetics in the halophyte *Halimione portulacoides* (L.) Aellen. In: *Comparative Biochemistry and Physiology Part A: Molecular & Integrative Physiology* 153(2), Supplement, pp. S193. doi:10.1016/j.cbpa.2009.04.433.

Rubio, L., Belver, A., Venema, K., García-Sánchez, M.J. and Fernández, J.A. (2011) Evidence for a sodium efflux mechanism in the leaf cells of the seagrass *Zostera marina* L. *Journal of Experimental Marine Biology and Ecology* 402, 56–64.

Rubio, L., García, D., García-Sánchez, M.J., Niell, F.X., Felle, H.H., *et al.* (2017) Direct uptake of HCO_3^- in the marine angiosperm *Posidonia oceanica* (L.) Delile driven by a plasma membrane H^+ economy. *Plant Cell and Environment* 40, 2820–2830.

Sabine, C.L. and Feely, R.A. (2007) The oceanic sink for carbon dioxide. In: Reay, D., Hewitt, N., Gracem, J., and Smith, K. (eds) *Greenhouse Gas Sinks*. CABI Publishing, Wallingford, UK, pp. 31–49.

Schachtman, D.P., Reid, R.J. and Ayling, S.M. (1998) Phosphorus uptake by plants: From soil to cell. *Plant Physiology* 116, 447–453.

Schmidt, C. and Schroeder, J.I. (1994) Anion selectivity of slow anion channels in the plasma membrane of guard cells (large nitrate permeability). *Plant Physiology* 106, 383–391.

Shabala, S. and Mackay, A. (2011) Ion transport in halophytes. In: Turkan, I. (ed.) *Plant Responses to Drought and Salinity Stress – Developments in a Post-Genomic Era, Advances in Botanical Research*. Academic Press, Amsterdam, pp. 151–199.

Shi, D., Xu, Y., Hopkinson, B.M. and Morel, F.M.M. (2010) Effect of ocean acidification on iron availability to marine phytoplankton. *Science* 327, 676–679. DOI: 10.1126/science.1183517

Shibata, M., Katoh, H., Sonoda, M., Ohkawa, H., Shimoyama, M. *et al.* (2002) Genes essential to sodium-dependent bicarbonate transport in cyanobacteria: Function and phylogenetic analysis. *Journal of Biological Chemistry* 277, 18658–18664.

Short, F.T. and Neckles, H.A. (1999) The effects of global climate change on seagrasses. *Aquatic Botany* 63,169–196.

Smith, F.W., Mudge, S.R., Rage, A.L. and Glassop, D. (2003) Phosphate transport in plants. *Plant and Soil* 248, 71–83.

Stapel, J., Aarts, T.L., van Duynhoven, B.H.M., de Groot, J.D., van den Hoogen, P.H.W., *et al.* (1996) Nutrient uptake by leaves and roots of the seagrass *Thalassia hemprichii* in the Spermonde Archipelago, Indonesia. *Marine Ecology Progress Series* 134, 195–206.

Supaphon, P., Phongpaichit, S., Sakayaroj, J., Rukachaisirikul, V., Kobmoo, N., *et al.* (2017) Phylogenetic community structure of fungal endophytes in seagrass species. *Botanica Marina* 60, 489–501.

Sze, H., Li, H.X. and Palmgren, M.G. (1999) Energetization of plant cell membranes by H^+ - pumping ATPases: Regulation and biosynthesis. *The Plant Cell* 11, 677–689.

Taub, D.R. and Wang, X. (2008) Why are nitrogen concentrations in plant tissues lower under elevated CO_2? A critical examination of the hypotheses. *Journal of Integrative Plant Biology* 50, 1365–1374.

Terrados, J. and Williams, S.L. (1997) Leaf versus root nitrogen uptake by the surfgrass *Phyllospadix torreyi*. *Marine Ecology Progress Series* 149, 267–277.

Touchette, B.W. and Burkholder, J.M. (2000) Review of nitrogen and phosphorus metabolism in seagrasses. *Journal of Experimental Marine Biology and Ecology* 250, 133–167.

Tyerman, S.D. (1989) Solute and water relationship of seagrasses. In: Larkum, A.W.D., McComb, A.J., Shepherd, S.A., (eds) *Biology of Seagrasses: A Treatise on the Biology of Seagrasses with Special Reference to the Australian Region*. Elsevier Science Publications, Amsterdam, pp. 723–759.

Ullrich-Eberius, C.I., Novacky, A. and van Bel, A.J.E. (1984) Phosphate uptake in *Lemna gibba* G1: Energetics and kinetics. *Planta* 161, 46–52.

Volkov, V. and Amtmann, A. (2006) *Thellungiella halophila*, a salt-tolerant relative of *Arabidopsis thaliana*, has specific root ion-channel features supporting K^+/Na^+ homeostasis under salinity stress. *Plant Journal* 48, 342–353.

Volkov, V., Wang, B., Dominy, P.J., Fricke, W. and Amtmann, A. (2003) *Thellungiella halophila*, a salt-tolerant relative of *Arabidopsis thaliana*, possesses effective mechanisms to discriminate between potassium and sodium. *Plant, Cell and Environment* 27, 1–14.

Walker, N.A. and Sanders, D. (1991) Sodium-coupled solute transport in charophyte algae: A general mechanism for transport energization in plant cell? *Planta* 185, 443–445.

Wang, Y.Y., Hsu, P.K. and Tsay, Y.F. (2012) Uptake, allocation and signalling of nitrate. *Trends in Plant Science* 17, 458–467.

Williams, S.L. (2016) Genomics: From sea to sea. *Nature* 530, 290–291.

Wood, C., Poree, F., Dreyer, I., Koehler, G. and Udvardi, M. (2006) Mechanisms of ammonium transport, accumulation, and retention in oocytes and yeast cells expressing Arabidopsis *AtAMT1;1*. *FEBS Letters* 580, 3931–3936.

Xue, S., Hu, H., Ries, A., Merilo, E., Kollist, H., *et al.* (2011) Central functions of bicarbonate in S-type anion channel activation and OST1 protein kinase in CO_2 signal transduction in guard cell. *The EMBO Journal* 30, 1645–1658.

Zeebe, R.E. and Wolf-Gladrow, D. (2001) *CO_2 in Seawater: Equilibrium, Kinetics, Isotopes*. 1st edn. Elsevier, Amsterdam.

Zheng, X., He, K., Kleist, T., Chen, F. and Luan, S. (2015) Anion channel SLAH3 functions in nitrate dependent alleviation of ammonium toxicity in Arabidopsis. *Plant Cell and Environment* 38, 474–486.

Zhou, J.J., Fernández, E., Galván, A. and Miller, A.J. (2000) A high affinity nitrate transport system from *Chlamydomonas* requires two gene products. *FEBS Letters* 466, 225–227.

Zimmerman, R., Kohrs, D.G., Steller, D.L. and Alberte, R. (1997) Impacts of CO_2 enrichment on productivity and light requirements of eelgrass. *Plant Physiology* 115, 599–607.

7 Ecophysiology of Seed Heteromorphism in Halophytes: an Overview

Aysha Rasheed, Farah Nisar, Bilquees Gul, M. Ajmal Khan and Abdul Hameed*
University of Karachi, Karachi, Pakistan

Abstract

Seed heteromorphism is the production of two or more types of seeds by a single individual of many plant species. On the basis of an extensive literature survey, this chapter aims to provide an overview of the research on seed heteromorphism, especially in the context of halophytes, which are emerging as non-conventional crops for arid saline lands. We cover the occurrence, ecophysiological significance and different biochemical aspects of seed heteromorphism. Dormancy, and the responses of heteromorphic seeds of halophytes to salinity, temperature and photoperiod are discussed. We also provide an overview of the carryover effects of heteromorphic seeds to progeny. This article also attempts to highlight the gaps in existing knowledge about seed heteromorphism in halophytes.

Keywords: Dormancy; Germination; Halophyte; Photoperiod; Salinity; Seed heteromorphism

7.1 Introduction

Seeds are reproductive units that play a key role in the continuity and distribution of plants on spatial and temporal scales (Bewley *et al.*, 2013; Penfield, 2017). Plants, being sessile organisms, depend completely on their seeds to escape stress, and so seeds have been the core survival strategy for most plants, especially annuals and plants of stressful habitats (Baskin and Baskin, 1998; Gul *et al.*, 2013). The evolution of seed in the Devonian period is, therefore, considered a big leap in the success of higher plants, which enabled them to colonize diverse habitats on Earth (Sabelli and Larkins, 2015). Environment and genotype of the mother plants are key factors governing seed attributes such as morphology (size, number, shape and colour, etc.), germination and dormancy characteristics, which in turn affect establishment and vigour of the seedlings (Bewley *et al.*, 2013). Hence, seeds act as a link between two generations and are also considered important in population dynamics (Gul *et al.*, 2013; Yang *et al.*, 2017). During the course of evolution, plants evolved diverse dormancy and germination strategies which ensured the formation of new seedlings in suitable places at appropriate times (Baskin and Baskin, 1998; Bewley *et al.*, 2013). Production of heteromorphic or morphologically different seeds, which in turn differ in germination and dormancy responses, is also an adaptive strategy of many plants of arid and

* Corresponding author e-mail: ahameed@uok.edu.pk

saline areas, which confers a selective advantage to plants in extreme and fluctuating environments (Harper, 1977; Baskin and Baskin, 2014). A large of number of studies have described occurrence (Harper, 1977; Gul *et al.*, 2013; Baskin and Baskin, 2014), dormancy and germination responses (Song *et al.*, 2008; H.L. Wang *et al.*, 2008; Yao *et al.*, 2010; Cao *et al.*, 2012; Gul *et al.*, 2013), spatial/geographic variations (Yang *et al.*, 2017), carryover effects on progeny (Ruiz de Clavijo, 2005; Mandák and Pyšek, 2005; Wang *et al.*, 2012; Baskin and Baskin, 2014) and physio-chemical variations (Siddiqui and Khan, 2010, 2011) of the heteromorphic seeds of different species. Here, we aim to present a comprehensive overview of research into seed heteromorphism, especially in the context of halophytes (i.e. plants of saline habitats), which are emerging as non-conventional crops for the arid saline lands.

7.2 Seed Heteromorphism: Definition and Occurrence

Seed heteromorphism is the phenomenon of producing different type of seeds in many angiospermic plant species (Harper, 1977; Matilla *et al.*, 2005; Gul *et al.*, 2013). In the past, a number of terms were coined to describe this phenomenon. For example, Harper (1977) used 'somatic polymorphism', but this term was already in use for the differentiation among individuals, especially in the animal kingdom. Hannan (1980) coined two different expressions: 'heterospermy' for heterogeneity among seeds and 'heterocarpy' to differentiate among morphs of different fruits. Later, Venable (1985) suggested the term 'heteromorphism' for the production of seeds of different forms or behaviours by single individuals.

Heteromorphic seeds are produced on the same individual at the same time and often differ in colour, size, shape, seed-coat texture and/or surface secondary structures (Williams and Harper, 1965; McGinley, 1989; Imbert, 2002; Matilla *et al.*, 2005; Gul *et al.*, 2013; Bhatt and Santo, 2016). Two (Ameixa *et al.*, 2016; Wang *et al.*, 2017) or more (Gross, 1984; Khan and Ungar, 1984) morphs of the seeds can be seen (Table 7.1). Currently, about 200 plant species from 18 families are known to produce heteromorphic seeds, with high incidence of seed heteromorphism in genera within Amaranthaceae, Brassicaceae, Caryophyllaceae, Asteraceae and Poaceae (Imbert, 2002; Matilla *et al.*, 2005; Mira *et al.*, 2011; Gul *et al.*, 2013). The phenomenon of seed heteromorphism is also common in many halophyte genera such as *Arthrocnemum* (Khan and Gul, 1998), *Atriplex* (Li *et al.*, 2008; Bhatt and Santo, 2016), *Cakile* (Zhang, 1993), *Chenopodium* (Yao *et al.*, 2010), *Salicornia* (Ungar, 1979; Ameixa *et al.*, 2016), *Salsola* (Wei *et al.*, 2008), *Spergularia* (Telenius and Torstensson, 1989), *Suaeda* (Khan *et al.*, 2001; Wang *et al.*, 2017) and *Trianthema* (Mohammed and Sen, 1988).

7.3 Seed Heteromorphism: Basis and Nature

Our understanding about seed heteromorphism is far from complete. A number of factors such as position of seeds in the inflorescence or on the plant, and/or distance from vascular tissues may contribute towards production of heteromorphic seeds (Gutterman, 1993; Susko and Lovett-Doust, 2000). Often such positional effects lead to differential grain-filling, hormonal gradient and/or testa structure and composition, which in turn affect germination and dormancy characteristics of the heteromorphic seeds (Gutterman, 1993; Susko and Lovett-Doust, 2000). Hence, most studies indicate that the phenomenon of seed heteromorphism is a variation within an individual (i.e. intra-individual variation) and probably has a physio-chemical but not a genetic basis, as heteromorphic seeds originate from single individuals (Imbert, 1999, 2002; Mandák and Pyšek, 2005).

Differences in morphology of heteromorphic seeds often lead to differences in their dispersal, germination and dormancy responses, seed bank characteristics and seedling attributes (Khan and Ungar, 1984; Puga-Hermida *et al.*, 2003; Brändel, 2004; Yao *et al.*, 2010; El-Keblawy and Bhatt, 2015; Koen *et al.*, 2017; Wang *et al.*, 2017). Seed heteromorphism is hence an adaptation of many plants to maximize fitness under harsh and unpredictable environments, whereby two/more seed morphs would germinate under different sets of environmental conditions (Khan and Ungar, 1984; Imbert, 2002; Song *et al.*, 2008; Yao *et al.*, 2010; El-Keblawy and Bhatt, 2015).

Table 7.1. Examples of halophytic plants with reported seed heteromorphism.

Trait	Morph	Species	References
Colour	Black and brown seeds	*Arthrocnemum indicum*	Khan and Gul (1998)
		Atriplex canescens	Bhatt and Santo (2016)
		Atriplex inflata	Atia *et al.* (2011)
		Atriplex rosea	Khan *et al.* (2004)
		Chenopodium album	Yao *et al.* (2010)
		Halopyrum mucronatum	Khan and Ungar (2001)
		Suaeda aralocaspica	L. Wang *et al.* (2012)
		Suaeda acuminata	H.L. Wang *et al.* (2012)
		Suaeda moquinii	Khan *et al.* (2001)
		Suaeda salsa	Song *et al.* (2014)
		Suaeda splendens	Redondo-Gómez *et al.* (2008)
Colour with other traits	Flat brown, flat black, humped brown and humped black seeds	*Atriplex centralasiatica*	Li *et al.* (2008)
	Light and dark achenes	*Anthemis chrysantha*	Aguado *et al.* (2011)
	Large brown and small black seeds	*Atriplex prostrata*	Carter *et al.* (2003)
	Small black achene with small bracteoles, large brown achene with extended bracteoles	*Atriplex tatarica*	Kochánková and Mandák (2009)
	Black elongated achenes, brown triangular achenes	*Bidens pilosa*	Forsyth and Brown (1982)
	Yellow and red perianth	*Salsola rubescens*	El-Keblawy *et al.* (2013)
	Dark curved seeds, pale straight seeds	*Tragopogon pratensis*	van Mölken *et al.* (2005)
Size	Large, medium and small seeds	*Atriplex sagittata*	Mandák and Pyšek (2005)
		Atriplex triangularis	Khan and Ungar (1984)
	Large and small seeds	*Cakile edentula*	Zhang (1993)
		Salicornia europea	Ungar (1979)
Size with other traits	Outer bracts short and oval, inner long and lanceolate	*Crepis sancta*	Imbert (1999)
	Central seeds and lateral seeds	*Salicornia ramosissima*	Ameixa *et al.* (2016)
	Long winged and short winged	*Salsola komarovii*	Yamaguchi *et al.* (1990)
	Winged and un-winged seeds	*Spergularia marina*	Telenius and Torstensson (1989)
	Spring and summer seeds	*Tamarix ramosissima*	Yan *et al.* (2011)

Seed heteromorphism of size often accompanies differences in the food/nutrient reserves of the seed morphs, generally with higher amounts in larger as compared to smaller seeds. For example, large seeds had higher food reserves than small seeds in *C. album* (Yao *et al.*, 2010) and *Halopyrum mucronatum* (Siddiqui and Khan, 2011). Generally, larger seeds show higher or faster germination and greater stress tolerance, as reported in *Cakile edentula* (Zhang, 1993), *Atriplex sagittata* (Mandák and Pyšek, 2005) and *Salsola komarovii* (Takeno and Yamaguchi, 1991). In addition, seed size also plays an important role in dispersal and persistence in the soil seed bank (Imbert, 2002). Often smaller seeds, being lighter in weight, tend to disperse further than the larger seeds and may also occupy deeper soil layers that may lead to less predation (Fenner, 1985; Bakker *et al.*, 1996; Imbert, 2002; Gul *et al.*, 2013).

Heteromorphism of seed colour is reportedly linked to variations in the chemical composition of the seed coat/testa. For example, black seeds had higher total phenols and tannins than the brown seeds of *H. mucronatum* (Siddiqui and Khan, 2011). Similarly, the seed coat of beans in the genus *Faba* showed high amount of tannins in pigmented cultivars as compared to low amounts in unpigmented cultivars (Bos and Jetten, 1989). Composition of phenols may also vary between seed morphs of different colour. For instance, in *H. mucronatum* black seeds had catechol and brown seeds had vanillic acid (Siddiqui and Khan, 2011). Gul *et al.* (2013) reviewed that generally brown seeds germinated quickly and to higher percentages than the black seeds. This difference could be linked to differential phenol levels/composition, as many phenols are inhibitors of enzymes involved in certain stages of germination. For instance, phenolic compounds inhibited seed germination of *Pinus laricio* by inhibiting activities of enzymes of glycolysis and the oxidative pentose phosphate pathway (Muscolo *et al.*, 2001). Likewise, exogenous application of phenols such as gentisic, caffeic and 2-hydroxy-5 methoxy benzoic acids and protocatechol inhibited seed germination of *Atriplex triangularis* (Khan and Ungar, 1986). Seed coat coloration may also impact the ability of the seed to absorb water. Permeability of seed coats in dark coloured seeds can be reduced because of the number of layers making up the coat, cell density and/or certain chemicals (e.g. phenolics) in comparison to light coloured seeds in which permeability is quite high (Matilla *et al.*, 2005; Siddiqui and Khan, 2011). Heteromorphic seeds differ in many other biochemical aspects; for example, levels of endogenous abscisic acid (ABA) and polyamines, and the activity of the final step of the ethylene pathway (reviewed by Matilla *et al.*, 2005).

7.4 Seed Heteromorphism in Dicotyledonous Halophytes

The phenomenon of seed heteromorphism is more frequent in the dicotyledonous (dicot) than the monocotyledonous (monocot) families, as 16 of 18 families having plants with heteromorphic seeds are dicots (Imbert, 2002). According to the data reviewed by Imbert (2002), 63% of the plants with heteromorphic seeds are from Asteraceae and 8% belong to Chenopodiaceae (now part of Amaranthaceae). Amaranthaceae is the largest (~24 of total halophytes) family of halophytes (eHaloph Version 3.15) and 18 species from 10 genera are known to produce heteromorphic seeds (Imbert, 2002). Heteromorphism of seeds is common in many halophyte genera such as *Arthrocnemum*, *Atriplex*, *Cakile*, *Chenopodium*, *Salicornia*, *Salsola*, *Spergularia*, *Suaeda* and *Trianthema* (Ungar 1978; Matilla *et al.*, 2005; Yao *et al.*, 2010; Gul *et al.*, 2013; Bhatt and Santo, 2016; Wang *et al.*, 2017). Mohammed and Sen (1991) reported seed heteromorphism in a number of dicot halophytes such as *Cressa cretica*, *Salsola baryosma*, *Sesuvium sesuvioides*, *Suaeda fruticosa*, *Trianthema triquetra* and *Zygophyllum simplex*. Gul *et al.* (2013) also enlisted seed heteromorphism in 18 dicot halophytes in their review. A number of species from the halophyte genus *Suaeda*, such as *S. depressa* (Williams and Ungar, 1972), *S. maritima* (Boucaud and Ungar, 1976), *S. japonica* (Yokoishi and Tanimoto, 1994) and *S. moquinii* (Khan *et al.*, 2001; Li *et al.*, 2005) reportedly display seed heteromorphism.

7.5 Seed Heteromorphism in Monocotyledonous Halophytes

Heteromorphism of seeds is less common in monocots than in dicot families, as only two monocot families compared to 16 dicot families have plants with heteromorphic seeds (Imbert, 2002). Heteromorphic seeds are reported in seven species (from seven genera) of Poaceae and one species (only one genus) of Commelinaceae (Imbert, 2002). Hence, seed heteromorphism is infrequent in monocots, as only 1% of total genera (668) and 0.07% of total species (9500) of Poaceae exhibit seed heteromorphism, as compared to 9.7% of total genera (103) and 1.38% of total species (1300) of Chenopodiaceae. Generally, about 8% of the total genera (99) and 3.6% of the total species (218) of plants with reported seed heteromorphism are monocot, as compared to the 92% and 96.4% contributions, respectively, of dicots (Imbert, 2002). *Halopyrum mucronatum* (Noor and Khan, 1995) and *Cenchrus longispinus* (Twentyman, 1974) are common halophytic grasses which produce heteromorphic seeds. *Halopyrum mucronatum* is a perennial grass commonly found on the coastal dunes of Pakistan and adjacent countries, and produces black and brown seeds; which not only differ in their seed germination and stress tolerance attributes but also in their biochemical composition (Noor and Khan, 1995; Siddiqui and Khan, 2011). Generally, brown seeds are characterized by more food reserves, fewer phenols and higher germination as compared to their black counterparts (Siddiqui and Khan, 2011).

7.6 Seed Heteromorphism in Different Habitats

Plants that produce heteromorphic seeds are common in habitats where temporal environmental heterogeneity is very frequent, so that no single ecological response can ensure fitness/survival; or where there is a need to ensure less competition of siblings (Harper 1977; Mandák, 1997; Imbert, 2002; Gul *et al.*, 2013). Hence, production of seeds with two or more types with differing germination and dormancy responses would act as a bet-hedging strategy, thereby increasing fitness in a fluctuating environment (Ungar, 1987; Gul *et al.*, 2013).

Coastal sand dunes and marshes display frequent fluctuations in moisture and salinity (Ungar, 1987; Gulzar and Khan, 1998; Khan and Gul, 1998; Gulzar *et al.*, 2014). *Arthrocnemum macrostachyum* is a stem-succulent halophyte commonly distributed in coastal salt marshes of the Arabian Sea, including Pakistan, and produces dimorphic (i.e. brown and black) seeds (Khan and Gul, 1998). Similarly, *H. mucronatum* is a perennial grass of the coastal dunes found along Arabian Sea coasts, including Pakistan, and produces black and brown seeds (Noor and Khan, 1995). Gul *et al.* (2013) reviewed that seed heteromorphism is common in the halophytes of temperate as compared to subtropical areas.

Salt flats and desert are also characterized by large variations and spatial heterogeneity in soil salinity, moisture and temperatures (Gulzar and Khan, 1998; Khan and Qaiser, 2006). Hence, many species such as *S. aralocaspica* (Wang *et al.*, 2008) and *Salsola affinis* (Wei *et al.*, 2008) with heteromorphic seeds are also reported from these habitats.

7.7 Seed Heteromorphism: Germination and Dormancy

Heteromorphic seeds may display differing germination and dormancy responses (Gul *et al.*, 2013). Mean final germination percentages of brown seeds of *A. macrostachyum* (Khan and Gul, 1998), *Atriplex centralasiatica* (Li *et al.*, 2008) and *A. inflata* (Atia *et al.*, 2011) were higher than those of their black seeds. Similarly, large seeds of *A. sagittata* (Mandák and Pyšek, 2005) and *Cakile edentula* (Zhang, 1993) germinated to higher percentages in distilled water than did their small seeds. However, dimorphic seeds of *A. rosea* germinated to similar percentages under non-saline conditions (Khan *et al.*, 2004).

Heteromorphic seeds may display different combinations of dormancy levels/types. The presence of dormancy was detected in black but not brown seeds of many halophytes such as *A. canescens* (Bhatt and Santo, 2016), *Suaeda aralocaspica* (Wang *et al.*, 2008) and *S. acuminata* (H.L. Wang *et al.*, 2012). However, black seeds of *S. acuminata* had intermediate physiological dormancy (H.L. Wang *et al.*, 2012), whereas black seeds of *S. aralocaspica* had non-deep type-2 physiological dormancy (L. Wang *et al.*, 2012). Thus, there is diversity in the dormancy combinations of heteromorphic seeds. Similarly, some halophytes with heteromorphic seeds may display annual dormancy cycles. For example, the presence of an annual dormancy cycle in black but not in brown seeds of *S. corniculata* has been reported (Cao *et al.*, 2012).

Variations in germination and dormancy responses of heteromorphic seeds may be related to their endogenous chemical environment, which is seldom studied. Short-winged fruits of *Salsola komarovii* have more abscisic acid than long-winged seeds and this difference is reflected in their germination rates (Takeno and Yamaguchi, 1991). Heteromorphic seeds may also vary in their longevity. For instance, dark seeds of *A. semibaccata* remain viable for a longer period than the light coloured seeds (Beadle, 1952). Similarly, peripheral achenes of *Crepis sancta* have greater longevity than the central achenes (Imbert, 1999).

7.8 Responses of Heteromorphic Seeds to Salinity

Heteromorphic seeds of halophytes may vary in their salinity tolerance (Gul *et al.*, 2013). For instance, the salinity tolerance limit of brown seeds of *Suaeda aralocaspica* was higher (1400 mM NaCl) than its black (400 mM NaCl) seeds (Wang *et al.*, 2008). Similarly, brown seeds of *A. centralasiatica* were more tolerant (480 mM NaCl) than the black (240) seeds (Li *et al.*, 2008). Similar results were obtained for *Arthrocnemum macrostachyum* (Khan and Gul, 1998), *Chenopodium album* (Yao *et al.*, 2010), *S. salsa* (Li *et al.*, 2005), *S. moquinii* (Khan *et al.*, 2001) and *S. splendens* (Redondo-Gómez *et al.*, 2008). Hence, it appears that the brown seeds display greater salinity tolerance than the black seeds. However, brown and black seeds of *Atriplex rosea* (Khan *et al.*, 2004) and *Halopyrum mucronatum* (Khan and Ungar, 2001) showed comparable salinity tolerance limits.

The salinity tolerance of seeds varying in size may also differ in many species. For example, large seeds of *A. triangularis* had higher (856 mM NaCl) salinity tolerance limit than the medium (685 mM NaCl) and small (342 mM NaCl) seeds (Khan and Ungar, 1984). Likewise, large seeds of *Salicornia europaea* could germinate in 856 mM NaCl while small seeds germinated in a solution of only 514 mM NaCl (Ungar, 1979). Thus, large seeds seem to have greater salinity tolerance than small seeds.

7.9 Responses of Heteromorphic Seeds to Temperature

In some cases, heteromorphic seeds may require a different temperature (or thermoperiod) for optimum germination (Gul *et al.*, 2013). For example, brown seeds of *H. mucronatum* germinated optimally at 20/30°C, while optimum thermoperiod for the germination of its black seeds was 25/35°C (Khan and Ungar, 2001). Similarly, large seeds of *A. prostrata* germinated optimally at 20/35°C, whereas its small seeds germinated optimally at 5/25°C (Carter *et al.*, 2003). However, for a number of halophytes, the temperature requirements of the heteromorphic seeds are similar. For example, dimorphic seeds of *Arthrocnemum macrostachyum* (Khan and Gul, 1998), *Atriplex centralasiatica* (Li *et al.*, 2008), *Suaeda splendens* (Redondo-Gómez *et al.*, 2008) and polymorphic seeds of *A. sagittata* (Mandák and Pyšek, 2005) showed similar temperature optima for germination.

7.10 Responses of Heteromorphic Seeds to Light/Photoperiod

Light (photoperiod) acts as a regulatory cue for the seed germination of most halophytes (Khan and Ungar, 1996; Gul *et al.*, 2013). Often a positive link between light requirement and small seed size is reported (Grime *et al.*, 1981), which is probably needed to prevent germination of small seeds when buried too deep, as small seeds possess less food reserves to support seedling emergence from deeper soil layers (Bewley and Black, 1994; Khan and Gul, 2006). However, information about the light requirements of heteromorphic seeds is both scanty and inconclusive. Brown seeds of *S. acuminata* showed 95% and 7% germination in light and dark, respectively; while its black seeds displayed 96% and 1% germination in light and dark, respectively (H.L. Wang *et al.*, 2012). Similarly, light enhanced the germination of black seeds (both unstratified and stratified) but not of brown (stratified) seeds of *A. patula* (Nurse *et al.*, 2008). Black seeds of

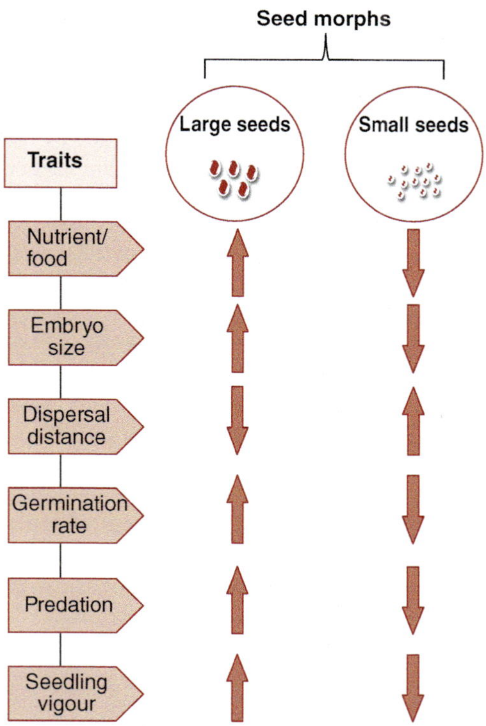

Fig. 7.1. Comparison of the characteristics of large and small seed morphs based on literature survey.

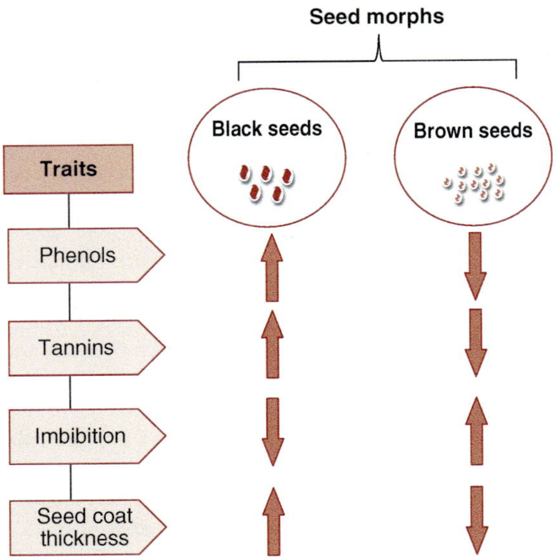

Fig. 7.2. Comparison of the characteristics of black and brown seed morphs based on literature survey.

S. corniculata need light to germinate, while brown seeds do not (Cao *et al.*, 2012).

7.11 Carryover Effects of Heteromorphic Seeds to Progeny

Often heteromorphic seeds display differences in their germination and dormancy responses, stress tolerance and longevity (Imbert, 2002; Gul *et al.*, 2013). Generally it is believed that the heteromorphism provides advantage only during the early life history of plants, and subsequent seedlings are quite similar (Harper, 1977; Mandák, 1997; Imbert, 2002; Gul *et al.*, 2013). However, evidence is accumulating in support of carryover effects of heteromorphy to the progeny (Bendall, 1973; Cheplick and Quinn, 1982; Xu *et al.*, 2011; Nisar and Hameed, unpublished data). For example, Redondo-Gómez *et al.* (2008) reported carryover of differential salt tolerance in plants of *S. splendens* grown from dimorphic seeds. Xu *et al.* (2011) also reported that the seedlings grown from yellow seeds of *A. centralasiatica* produced greater amounts of nitric oxide as compared to the seedlings from brown seeds. Similarly, seedlings produced from subterranean seeds of *Emex spinosa* had larger leaf area and stem as compared to the seedlings produced from aerial achenes in mixed cultures (Weiss, 1980). However, plants grown from the dimorphic seeds of *S. aralocaspica* showed similar responses to salinity and nutrient availability. Hence, information about carryover effects of heteromorphic seeds to the progeny is inconclusive and needs further investigation at ecophysiological, biochemical and molecular levels.

7.12 Conclusions

Many plants of heterogeneous and fluctuating environments such as coastal dunes, marshes, salt flats and deserts produce heteromorphic or morphologically different seeds, which vary in their dispersal capacity, germination and dormancy responses, longevity and/or stress tolerance. These seeds constitute a bet-hedging strategy for continuity of lineage (Figs 7.1 and 7.2). Seed heteromorphism, being an intra-individual variation, should be differentiated from intra- or inter-population variations in seed size and characteristics. Despite an increasing number of studies on seed heteromorphism, in general there is a dearth of knowledge about the mechanisms underlying seed heteromorphism. Further studies are also required on aspects such as maternal effects on heteromorphic seeds, differences in germination metabolism of contrasting seed morphs, relative contribution of heteromorphic seeds to progeny population and carryover effects to progeny. There is also a shortage of information about specific gene(s) for production of heteromorphic seeds.

References

Aguado, M., Martínez-Sánchez, J.J., Reig-Arminana, J., García-Breijo, F.J. and Franco, J.A. *et al.* (2011) Morphology, anatomy and germination response of heteromorphic achenes of *Anthemis chrysantha* J. Gay (Asteraceae), a critically endangered species. *Seed Science Research* 21, 283–294.

Ameixa, O.M., Marques, B., Fernandes, V.S., Soares, A.M. and Calado, R. *et al.* (2016) Dimorphic seeds of *Salicornia ramosissima* display contrasting germination responses under different salinities. *Ecological Engineering* 87, 120–123.

Atia, A., Rabhi, M., Debez, A., Barhoumi, Z. and Abdelly, C. *et al.* (2011) Factors controlling germination and dormancy processes in dimorphic fruits of *Atriplex inflata* (Chenopodiaceae). *Plant Ecology and Evolution* 144, 307–312.

Bakker, J.P., Poschlod, P., Strykstra, R.J., Bekker, R.M. and Thompson, K. (1996) Seed banks and seed dispersal: important topics in restoration ecology. *Acta Botanica Neerlandica* 45, 461–490.

Baskin, C.C. and Baskin, J.M. (1998) *Seeds: Ecology, Biogeography and Evolution of Dormancy and Germination*. Elsevier, London.

Baskin, J.M. and Baskin, C.C. (2014) What kind of seed dormancy might palms have? *Seed Science Research* 24, 17–22.

Beadle, N.C.W. (1952) Studies in Halophytes. I. The Germination of the Seed and Establishment of the Seedlings of Five Species of *Atriplex* in Australia. *Ecology* 33(1), 49–62.

Bendall, G.M. (1973) Some aspects of the biology, ecology and control of slender thistle, *Carduus pycnocephalus* L. and *C. tenuiflorus* Curt: (Compositae) in Tasmania. PhD thesis. University of Tasmania, Tasmania.

Bewley, J.D. and Black, M. (1994) *Seeds*. 2nd edn. Springer, Boston, MA, USA, pp. 1–33.

Bewley, J.D., Bradford, K. and Hilhorst, H. (2013) *Seeds: Physiology of Development, Germination and Dormancy*. 3rd edn. Springer, New York.

Bhatt, A. and Santo, A. (2016) Germination and recovery of heteromorphic seeds of *Atriplex canescens* (Amaranthaceae) under increasing salinity. *Plant Ecology* 217, 1069–1079.

Bos, K.D. and Jetten, J. (1989) Determination of tannins in faba beans. In: Huisman, J., Van der Poel, T.F.B. and Liener, I.E. (eds) *Recent Advances of Research in Antinutritional Factors in Legume Seeds*. Pudoc, Wageningen, Netherlands, pp. 285–296.

Boucaud, J. and Ungar, I.A. (1976) Hormonal control of germination under saline conditions of three halophytic taxa in the genus *Suaeda*. *Physiologia Plantarum* 37, 143–148.

Brändel, M. (2004) Dormancy and germination of heteromorphic achenes of *Bidens frondosa*. *Flora-Morphology, Distribution, Functional Ecology of Plants* 199, 228–233.

Cao, D., Baskin, C.C., Baskin, J.M., Yang, F. and Huang, Z. (2012) Comparison of germination and seed bank dynamics of dimorphic seeds of the cold desert halophyte *Suaeda corniculata* subsp. *mongolica*. *Annals of Botany* 110, 1545–1558.

Carter, C.T., Brown, L.S. and Ungar, I.A. (2003) Effect of temperature regimes on germination of dimorphic seeds of *Atriplex prostrata*. *Biologia Plantarum* 47, 269–272.

Cheplick, G.P. and Quinn, J.A. (1982) *Amphicarpum purshii* and the "pessimistic strategy" in amphicarpic annuals with subterranean fruit. *Oecologia* 52, 327–332.

eHaloph Version 3.15. https://www.sussex.ac.uk/affiliates/halophytes/index.php, accessed 30 September 2017.

El-Keblawy, A.A. and Bhatt, A. (2015) Aerial seed bank affects germination in two small-seeded halophytes in Arab Gulf desert. *Journal of Arid Environments* 117, 10–17.

El-Keblawy, A.A., Bhatt, A. and Gairola, S. (2013) Perianth colour affects germination behaviour in wind-pollinated *Salsola rubescens* in Arabian deserts. *Botany* 92, 69–75.

Fenner, M. (1985) *Seed Ecology*. Chapman and Hall, London.

Forsyth, C. and Brown, N.A.C. (1982) Germination of the dimorphic fruits of *Bidens pilosa* L. *New Phytologist* 90, 151–164.

Grime, J.P., Mason, G., Curtis, A.V., Rodman, J. and Band, S.R. (1981) A comparative study of germination characteristics in a local flora. *The Journal of Ecology* 69(3), 1017–1059.

Gross, K.L. (1984) Effects of seed size and growth form on seedling establishment of six monocarpic perennial plants. *The Journal of Ecology* 72(2), 369–387.

Gul, B., Ansari, R., Flowers, T.J. and Khan, M.A. (2013) Germination strategies of halophyte seeds under salinity. *Environmental and Experimental Botany* 92, 4–18.

Gulzar, S. and Khan, M.A. (1998) Diurnal water relations of inland and coastal halophytic populations from Pakistan. *Journal of Arid Environments* 40, 295–305.

Gulzar, S., Hameed, A., Ahmed, M.Z. and Khan, M.A. (2014) Is soil heterogeneity the major factor influencing vegetation zonation at Karachi coast? In: Khan, M.A., Böer, B., Öztürk, M., Al Abdessalaam, T.Z., Clüsener-Godt, M. and Gul, B. (eds) *Sabkha Ecosystems*. Springer, Dordrecht, Netherlands, pp.199–207.

Gutterman, Y. (1993) *Seed Germination in Desert Plants*. Springer-Verlag, Heidelberg, New York.

Hannan, G.L. (1980) Heteromericarpy and dual seed germination modes in *Platystemon californicus* (Papaveraceae). *Madroño* 27(4), 164–170.

Harper, J.L. (1977) *Population Biology of Plants*. Vol. 892. Academic Press, London.

Imbert, E. (1999) The effects of achene dimorphism on the dispersal in time and space in *Crepis sancta* (Asteraceae). *Canadian Journal of Botany* 77, 508–513.

Imbert, E. (2002) Ecological consequences and ontogeny of seed heteromorphism. *Perspectives in Plant Ecology, Evolution and Systematics* 5, 13–36.

Khan, M.A. and Gul, B. (1998) High salt tolerance in germinating dimorphic seeds of *Arthrocnemum indicum*. *International Journal of Plant Sciences* 159, 826–832.

Khan, M.A. and Gul, B. (2006) Halophyte seed germination. *Ecophysiology of High Salinity Tolerant Plants*. Springer, Dordrecht, Netherlands, pp. 11–30.

Khan, M.A. and Qaiser, M. (2006) Halophytes of Pakistan: characteristics, distribution and potential economic usages. In: *Sabkha Ecosystems Volume II: West and Central Asia*. Springer, Dordrecht, Netherlands, pp. 129–153.

Khan, M.A. and Ungar, I.A. (1984) Seed polymorphism and germination responses to salinity stress in *Atriplex triangularis* Willd. *Botanical Gazette* 145, 487–494.

Khan, M.A. and Ungar, I.A. (1986) Inhibition of germination in *Atriplex triangularis* seeds by application of phenols and reversal of inhibition by growth regulators. *Botanical Gazette* 147, 148–151.

Khan, M.A. and Ungar, I.A. (1996) Influence of salinity and temperature on the germination of *Haloxylon recurvum*. *Annals of Botany* 78, 547–551.

Khan, M.A. and Ungar, I.A. (2001) Alleviation of salinity stress and the response to temperature in two seed morphs of *Halopyrum mucronatum* (Poaceae). *Australian Journal of Botany* 49, 777–783.

Khan, M.A., Gul, B. and Weber, D.J. (2001) Germination of dimorphic seeds of *Suaeda moquinii* under high salinity stress. *Australian Journal of Botany* 49, 185–192.

Khan, M.A., Gul, B. and Weber, D.J. (2004) Temperature and high salinity effects in germinating dimorphic seeds of *Atriplex rosea*. *Western North American Naturalist* 46(2), 193–201.

Kochánková, J. and Mandák, B. (2009) How do population genetic parameters affect germination of the heterocarpic species *Atriplex tatarica* (Amaranthaceae). *Annals of Botany* 103, 1303–1313.

Koen, J., Slabbert, M.M., Bester, C. and Bierman, F. (2017) Germination characteristics of dimorphic honeybush (*Cyclopia* spp.) seed. *South African Journal of Botany* 110, 68–74.

Li, W., Liu, X., Khan, M.A. and Yamaguchi, S. (2005) The effect of plant growth regulators, nitric oxide, nitrate, nitrite and light on the germination of dimorphic seeds of *Suaeda salsa* under saline conditions. *Journal of Plant Research* 118, 207–214.

Li, W., An, P., Liu, X., Khan, M.A., Tsuji, W. *et al.* (2008) The effect of light, temperature and bracteoles on germination of polymorphic seeds of *Atriplex centralasiatica* Iljin under saline conditions. *Seed Science and Technology* 36, 325–338.

Mandák, B. (1997) Seed heteromorphism and the life cycle of plants: A literature review. *Preslia, Praha* 69, 129–159.

Mandák, B. and Pyšek, P. (2005) How does seed heteromorphism influence the life history stages of *Atriplex sagittata* (Chenopodiaceae)? *Flora-Morphology, Distribution, Functional Ecology of Plants* 200, 516–526.

Matilla, A., Gallardo, M. and Puga-Hermida, M.I. (2005) Structural, physiological and molecular aspects of heterogeneity in seeds: a review. *Seed Science Research* 15, 63–76.

McGinley, M.A. (1989) The influence of a positive correlation between clutch size and offspring fitness on the optimal offspring size. *Evolutionary Ecology* 3, 150–156.

Mira, S., González-Benito, M.E., Ibars, A.M. and Estrelles, E. (2011) Dormancy release and seed ageing in the endangered species *Silene diclinis*. *Biodiversity and Conservation* 20, 345–358.

Mohammed, S. and Sen, D.N. (1991) Polymorphism as an adaptation in seeds of halophytes in Indian desert. In: Sen, D.N. and Mohammad, S. (eds) *Marvels of Seed*. University of Joudpur, India, pp. 207–212.

Mohammed, S.H.E.R. and Sen, D.N. (1988) A report on polymorphic seeds in halophytes 1. *Trianthema triquetra* L. in Indian desert. *Current Science Bangalore*, 57, 616–617.

Muscolo, A., Panuccio, M.R. and Sidari, M. (2001) The effect of phenols on respiratory enzymes in seed germination. *Plant Growth Regulation* 35, 31–35.

Noor, M.E.H.E.R. and Khan, M.A. (1995) Factors affecting germination of summer and winter seeds of *Halopyrum mucronatum* under salt stress. In: Khan, M.A. and Ungar, I.A. (eds) *Biology of Salt Tolerant Plants*. Book Crafters, Michigan, USA, pp. 51–58.

Nurse, R.E., Reynolds, W.D., Doucet, C. and Weaver, S.E. (2008) Germination characteristics of the dimorphic seeds of spreading orach (*Atriplex patula*). *Weed Science* 56, 216–223.

Penfield, S. (2017) Seed dormancy and germination. *Current Biology* 27, R874–R878.

Puga-Hermida, M.I., Gallardo, M., del Carmen Rodríguez-Gacio, M. and Matilla, A.J. (2003) The heterogeneity of turnip-tops (*Brassica rapa*) seeds inside the silique affects germination, the activity of the final step of the ethylene pathway, and abscisic acid and polyamine content. *Functional Plant Biology* 30, 767–775.

Redondo-Gómez, S., Mateos-Naranjo, E., Cambrollé, J., Luque, T., Figueroa, M.E. *et al.* (2008) Carry-over of differential salt tolerance in plants grown from dimorphic seeds of *Suaeda splendens*. *Annals of Botany* 102, 103–112.

Ruiz De Clavijo, E. (2005) The reproductive strategies of the heterocarpic annual *Calendula arvensis* (Asteraceae). *Acta Oecologica* 28, 119–126.

Sabelli, P.A. and Larkins, B.A. (2015) New insights into how seeds are made. *Frontiers in Plant Science* 6, 196. doi: 10.3389/fpls.2015.00196.

Siddiqui, Z.S. and Khan, M.A. (2010) The role of seed coat phenolics on water uptake and early protein synthesis during germination of dimorphic seeds of *Halopyrum mucronatum* (L.) Staph. *Pakistan Journal of Botany* 42, 227–238.

Siddiqui, Z.S. and Khan, M.A. (2011) The role of enzyme amylase in two germinating seed morphs of *Halopyrum mucronatum* (L.) Stapf. in saline and non-saline environment. *Acta Physiologiae Plantarum* 33, 1185–1197.

Song, J. and Wang, B. (2014) Using euhalophytes to understand salt tolerance and to develop saline agriculture: *Suaeda salsa* as a promising model. *Annals of Botany* 115, 541–553.

Song, J., Fan, H., Zhao, Y., Jia, Y., Du, X. *et al.* (2008) Effect of salinity on germination, seedling emergence, seedling growth and ion accumulation of a euhalophyte *Suaeda salsa* in an intertidal zone and on saline inland. *Aquatic Botany* 88, 331–337.

Susko, D.J. and Lovett-Doust, L. (2000) Patterns of seed mass variation and their effects on seedling traits in *Alliaria petiolata* (Brassicaceae). *American Journal of Botany* 87, 56–66.

Takeno, K. and Yamaguchi, H. (1991) Diversity in seed germination behavior in relation to heterocarpy in *Salsola komarovii* Iljin. *Journal of Plant Research* 104, 207–215.

Telenius, A. and Torstensson, P. (1989) The seed dimorphism of *Spergularia marina* in relation to dispersal by wind and water. *Oecologia* 80, 206–210.

Twentyman, J. (1974) Environmental control of dormancy and germination in the seeds of *Cenchrus longispinus* (Hack.) Fern. *Weed Research* 14, 1–11.

Ungar, I.A. (1978) Halophyte seed germination. *The Botanical Review* 44, 233–264.

Ungar, I.A. (1979) Seed dimorphism in *Salicornia europaea* L. *Botanical Gazette* 140, 102–108.

Ungar, I.A. (1987) Population ecology of halophyte seeds. *The Botanical Review* 53, 301–334.

van Mölken, T., Jorritsma-Wienk, L.D., van Hoek, P.H. and de Kroon, H. (2005) Only seed size matters for germination in different populations of the dimorphic *Tragopogon pratensis* subsp. *pratensis* (Asteraceae). *American Journal of Botany* 92, 432–437.

Venable, D.L. (1985) The evolutionary ecology of seed heteromorphism. *The American Naturalist* 126, 577–595.

Wang, H.L., Wang, L., Tian, C.Y. and Huang, Z.Y. (2012) Germination dimorphism in *Suaeda acuminata*: A new combination of dormancy types for heteromorphic seeds *South African Journal of Botany* 78, 270–275.

Wang, H.L., Tian, C.Y. and Wang, L. (2017) Germination of dimorphic seeds of *Suaeda aralocaspica* in response to light and salinity conditions during and after cold stratification. *PeerJ 5*, p.e3671. doi: 10.7717/peerj.3671

Wang, L., Huang, Z., Baskin, C.C., Baskin, J.M. and Dong, M. (2008) Germination of dimorphic seeds of the desert annual halophyte *Suaeda aralocaspica* (Chenopodiaceae), a C_4 plant without Kranz anatomy. *Annals of Botany* 102, 757–769.

Wang, L., Baskin, J.M., Baskin, C.C., Cornelissen, J.H.C., Dong, M. *et al.* (2012) Seed dimorphism, nutrients and salinity differentially affect seed traits of the desert halophyte *Suaeda aralocaspica* via multiple maternal effects. *BMC Plant Biology* 12, 170. doi: 10.1186/1471-2229-12-170.

Wei, Y., Dong, M., Huang, Z.Y. and Tan, D.Y. (2008) Factors influencing seed germination of *Salsola affinis* (Chenopodiaceae), a dominant annual halophyte inhabiting the deserts of Xinjiang, China. *Flora - Morphology Distribution Functional Ecology of Plants* 203, 134–140.

Weiss, P.W. (1980) Germination, reproduction and interference in the amphicarpic annual *Emex spinosa* (L.) Campd. *Oecologia* 45, 244–251.

Williams, J.T. and Harper, J.L. (1965) Seed polymorphism and germination. *Weed Research* 5, 141–150.

Williams, M.D. and Ungar, I.A. (1972) The effect of environmental parameters on the germination, growth, and development of *Suaeda depressa* (Pursh) Wats. *American Journal of Botany* 912–918.

Xu, J., Yin, H., Yang, L., Xie, Z. and Liu, X. (2011) Differential salt tolerance in seedlings derived from dimorphic seeds of *Atriplex centralasiatica*: From physiology to molecular analysis. *Planta* 233, 859–871.

Yamaguchi, H., Ichihara, K., Takeno, K., Hori, Y. and Saito, T. (1990) Diversities in morphological characteristics and seed germination behavior in fruits of *Salsola komarovii* Iljin. *Journal of Plant Research* 103, 177–190.

Yan, C., Wei, Y. and Yang, M. (2011) Comparative germination of *Tamarix ramosissima* spring and summer seeds. *EXCLI Journal* 10, 198–204.

Yang, F., Baskin, J.M., Baskin, C.C., Yang, X., Cao, D., *et al.* (2017) Divergence in life history traits between two populations of a seed-dimorphic halophyte in response to soil salinity. *Frontiers in Plant Science* 8, 1028. doi: 10.3389/fpls.2017.01028.

Yao, S., Lan, H. and Zhang, F. (2010) Variation of seed heteromorphism in *Chenopodium album* and the effect of salinity stress on the descendants. *Annals of Botany* 105, 1015–1025.

Yokoishi, T. and Tanimoto, S. (1994) Seed germination of the halophyte *Suaeda japonica* under salt stress. *Journal of Plant Research* 107, 385–388.

Zhang, J. (1993) Seed dimorphism in relation to germination and growth of *Cakile edentula*. *Canadian Journal of Botany* 71, 1231–1235.

8 Salt Marsh Plants: Biological Overview and Vulnerability to Climate Change

Brant W. Touchette*, Mariana K. Kneppers and Chloe M. Eggert
Elon University, Elon, USA

Abstract

Salt marshes are vegetative ecosystems that occupy intertidal zones in estuaries or behind coastal barrier islands where there is some degree of protection from strong wave energy. Physiochemical properties of these marshes are stressful for most angiosperms, as only a few species can tolerate the anoxic, chemically reduced, high-saline soils typical of these ecosystems. Despite the dynamic properties that make salt marshes inhospitable to most plants, they maintain some of the highest levels of biological productivity observed in nature. The vast majority of salt marsh halophytes do not require environmental salts to grow or reproduce, and only a few species are restricted to salinities greater than 0.5%. For salt marsh plants, tolerance to high salinities may involve multiple physiological strategies including ion compartmentalization, synthesis of compatible solutes, changes in membrane and/or cell wall properties, and/or salt exudation by way of salt glands or bladders. Even though salt marsh halophytes are well adapted to highly dynamic and, often, stressful environmental conditions, it is generally recognized that climate change will result in a global net loss of these ecologically and economically important ecosystems. Sea-level rise may decrease overall plant diversity by selecting species that are more tolerant to sustained flooding, or through the loss of mid- and high-marsh species that are less competitive to changing conditions. In areas where soil accretion fails to keep pace with rising waters and/or where landward migration is impeded, rising sea levels are likely to promote the conversion of marshes into unvegetated open water systems. Soil and water hypersalinity may also foster salt marsh die-offs in areas that are expected to experience seasonal declines in precipitation. Nevertheless, because of the natural complexity of these systems, the degree of salt marsh loss remains uncertain. Localized differences in climate, geology, hydrology and topography, along with biological and anthropogenic interactions, are likely to determine which marshes withstand the challenges of climate change and which marshes will be lost.

Keywords: Halophytes; Saline soil; Seagrass; Species diversity; Waterlogging

8.1 Introduction

Salt marshes are complex saline–coastal ecosystems found in waterlogged and flooded areas up to and beyond the limits of extreme high tides. While salt marshes and mangroves share similar physiographic features, marshes are typically dominated by herbaceous graminoids with relativity few woody plant species. Well-established marshes often maintain gently sloping topographies

* Corresponding author e-mail: btouchette@elon.edu

that permit inundation of large areas with saline waters. These vegetative intertidal (and supratidal) zones usually occur in areas with low wave energy, such as behind barrier islands or in estuaries where there is protection from more intense waves (Wiegert, 1979; Mitsch and Gosselink, 2015). Although marshes occasionally appear in areas that are more open or along exposed coastlines, these areas tend to maintain either minimal wave action or have extremely shallow topographies minimizing the development of destructive wave forces (Chapman, 1960; Mitsch and Gosselink, 2015).

Physiochemical properties of coastal salt marshes are considered highly stressful for most flowering plants. Studies have shown that only a few species can tolerate the anoxic, chemically reduced, saline soils typical of these systems (Kramer and Boyer, 1995; Munns, 2002). In most regions, lower marsh plants growing within the intertidal zone will receive semi-diurnal tidal inundations. This may cause permanently waterlogged substratums that undergo prolonged anaerobiosis, fostering redoximorphic features (e.g. depleted matrix, low chroma, or the presence of soil nodules/concretions) typical of many wetland soils. While most angiosperms are intolerant of reduced-flooded soils, arguably it is the presence of environmental salts that restrict these systems to a relatively small number of genera. Although the majority of salt marsh plants are perennials (Chapman, 1977), some genera (including *Salicornia*) appear to have both annual and perennial species (Wiegert, 1979).

There have been a number of early attempts to classify aquatic halophytes (Chapman, 1960). Iversen's (1936) classification of halophytes based on range of tolerance to saline pore-waters is still widely used today. These ranges include oligohaline (0.01%–0.1% NaCl), mesohaline (0.1%–1% NaCl) and polyhaline (>1%) species. Extending beyond these three classes, plants with wider tolerances (e.g. between 0.01% and 1% NaCl) may be considered as glyco-mesohaline; and plants overlapping all three classes are referred to as euryhaline (Grigore *et al.*, 2014). In salt marsh systems, oligohaline species are often restricted to the most upland areas, whereas polyhaline species are more common along the lower reaches, where they are in direct contact with intertidal saline waters.

Salt marshes are commonly found along intertidal waters in mid- and high latitudes, frequently in localized areas where sediment accumulation is equal to or greater than the rate of land subsidence, and where there is sufficient protection from wave and storm disruptions. Despite the dynamic abiotic forces that make these systems inhospitable to most plant species, salt marshes are among the most biologically productive areas in the world. In marshes along the Atlantic coast of North America, for example, plant productivity can approach 8 kg $m^{-2}yr^{-1}$ (Mitch and Gosselink, 2000). As much as half of this productivity can be transported to adjacent coastal systems, where it supports diverse coastal food webs, including those involving commercially valuable finfish and shellfish. Because of the ecological significance of salt marshes and – by way of extension – the importance of plants in these systems, a number of studies have considered the interactions between marsh halophytes and environmental salts. Therefore, the purpose of this chapter is to characterize fundamental concepts that influence and shape important biological features of salt marshes and resident halophytes with respect to the salt environment. This includes factors that promote species distributions within a marsh system; physiological adaptations in plants that foster salt tolerance and salt avoidance; and how a changing climate may disrupt important marsh functions and, in turn, promote a global decline in marsh quality and quantity.

8.2 Plant Zonations Within Salt Marshes

Considerable attention has been directed towards understanding the often observed vegetative–zonation patterns in coastal marshes. These zones are typically presented as unique plant assemblages that change with marsh elevation, resulting in the occurrence of specific plant communities within the low-to-high-marsh continuum (Partridge and Wilson, 1987; Fig. 8.1). These patterns are likely driven by variations in stress tolerance. Low marsh plants must contend with periodic tidal inundations, water-saturated substratum, reduced anoxic soils, erosion and deposition of sediment, and direct exposure to environmental

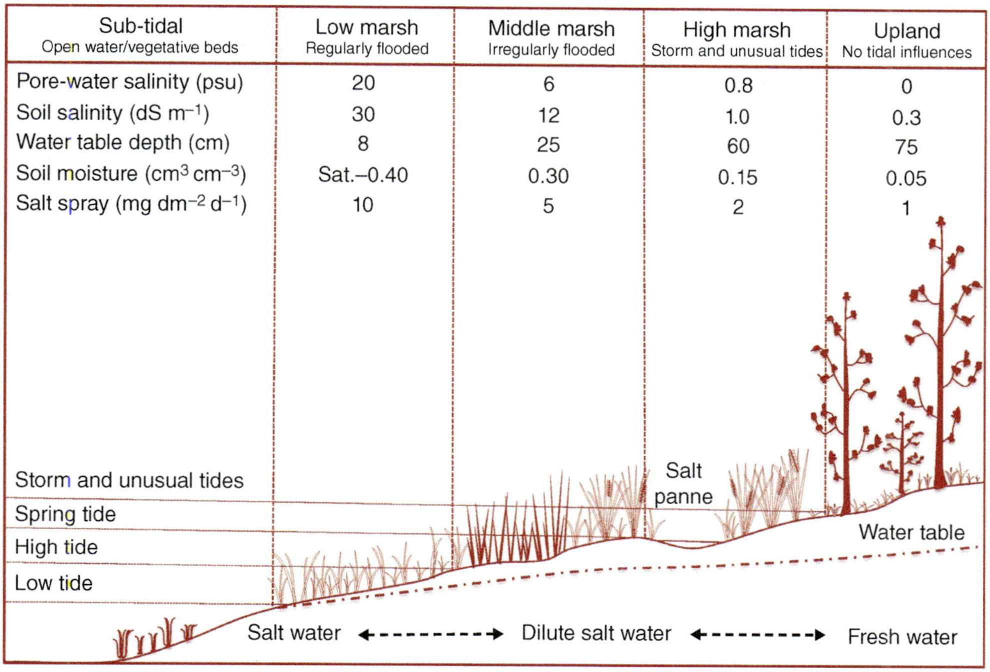

Fig. 8.1. Diagram illustrating general characteristics of different marsh zones (low, middle and high). Tidal ranges are depicted along the lower left and illustrate the important role of tides in shaping different marsh zones and plant communities. Values for pore-water salinity, soil salinity, water table depth, soil moisture and salt spray are used to illustrate the unique salt and hydrological characteristics for each zone. Comparative values were recorded in summer during low tide with moderate winds in a mid-Atlantic North American salt marsh.

salts. While any one of these components could preclude the growth and development of most species, the presence of seawater is thought to have a greater contribution in determining which plants become fully established. That is, in regions with moderate to high rainfall, species that are most tolerant to environmental salts reside in the lowermost zones, and there is a progressive decrease in salt tolerance as one moves up in elevation towards areas with characteristically lower salinities (Partridge and Wilson, 1987). In contrast, regions with low or seasonal rainfall often maintain different patterns of salinity tolerance, and more salt-tolerant species can be found within high marsh areas. This reversal in salinity tolerance is attributed to the accumulation of soil salts through evaporative processes in regions that are infrequently inundated by saltwater, along with limited rainfall that would otherwise dilute soil salinities (Partridge and Wilson, 1987; Rogel *et al.*, 2000).

While the role of tides as a source of environmental salts in coastal marshes is evident, there are additional mechanisms that foster salt exposure in resident flora. As early as 1904, Kearney recognized that edaphic factors such as groundwater salinity may influence species composition in coastal marshes and dunes. He observed, for example, that marsh systems exposed to normal tidal oscillations maintained groundwater salinities between 2.9 and 26 (practical salinity units; psu) at soil depths of 40 cm. Indeed, soil salinity has received more attention than any other parameter as an explanatory factor for plant zonation in salt marshes (Mahall and Park, 1976).

Tidal salt marshes in many parts of Australia, Europe and North America maintain increasing soil salinities with increasing elevation, up to or just above mean high water. Afterwards there is a prominent decrease in soil salinity as one continues to move landward (Mahall and Park, 1976; Adam, 1990; Silvestri *et al.*, 2005). In contrast, other studies along North America's Atlantic coast have demonstrated a gradual decline in salinity from the seaward edge of the marsh to the marsh's upper limits (Reed, 1947; Mahall and Park, 1976; Touchette, 2006). For example, in a *Juncus roemerianus* system, groundwater salts ranged between 0.8 and 18 psu, with decreasing water table depth and salinity at higher elevations (i.e., lowest soil water salinities occurred at 80 m from shoreline at 1 m depth; Touchette, 2006). It is possible that other factors, such as the degree of drainage within the substratum between tides and/or the presence of a landward freshwater lens that flows into these areas, may to some extent foster decreasing salinity gradients at higher marsh elevations (Cooper, 1982; Burdick *et al.*, 2001; Touchette, 2006).

Temporal and spatial patterns in soil salinity are rather complex and involve both abiotic and biotic components. Some prominent abiotic factors that control soil salinities in salt marshes include soil moisture, soil texture, evaporation, duration of tidal flooding, frequency and amount of precipitation, and salinity of tidal waters (Mahall and Park, 1976; Rogel *et al.*, 2000; González-Alcaraz *et al.*, 2014). Biotic components that influence soil salinities tend to be more complex in nature, but often involve the presence or absence of vegetation. In wet areas, vegetation appears to influence salinity by lowering water loss through evaporative processes and thus supporting more diluted soil waters (Bertness *et al.*, 1992; Srivastava and Jefferies, 1995; Di Bella *et al.*, 2014). A study on geese foraging responses in Hudson Bay coastal marshes, for example, resulted in a mosaic of soil patches with differing salinities (Srivastava and Jefferies, 1995). In these marshes, heavy grazing produced small areas devoid of vegetation. These bare areas typically had higher soil salinities due to elevated evaporation that, in turn, promoted the upward movement of saline waters from deeper horizons. This resulted in soils with elevated salinities but with similar moisture content (Srivastava and Jefferies, 1995). In Argentina, cattle grazing promoted higher soil salinities in mid-marsh regions, once again due to increased evaporation following vegetation removal (Di Bella *et al.*, 2014). But, unlike the Hudson Bay marshes, it was primarily evaporative processes that led to higher soil salinities following grazing.

Under certain circumstances, such as high marsh systems in temperate climates with ample rainfall, direct exposure to seawater (from surface water and/or groundwater) can be minimal and approach conditions more favourable to glycophytes. Nevertheless, these plants are still susceptible to environmental salts through salt spray. Indeed, for some high marsh plants (or even nearshore upland species), salt deposition may represent a significant abiotic stressor that extends well beyond areas inundated with tides. Depending on ambient weather conditions (e.g. wind and rain) and local topography, appreciable salt deposition has been observed up to 2 km from the shoreline (Rozema *et al.*, 1983). Therefore, plants residing in a high marsh system, well beyond the reaches of saline surface water and/or groundwater, may still be susceptible to abiotic challenges from salt (Ahmad and Wainwright, 1976; Rozema *et al.*, 1985; Touchette *et al.*, 2009a).

Salt spray tends to accumulate on coastal plants as dry deposition on aboveground structures, and the degree of accumulation is often correlated with wind velocity and inversely correlated with distance from shore (Gustafsson and Franzén, 1996; Griffiths, 2006). As with surface-water salinity, salt spray can strongly influence plant survival and species composition, fostering unique plant assemblages and zonation patterns consistent with species tolerance to environmental salts (Wilson and Sykes, 1999; Griffiths and Orians, 2003; Griffiths, 2006). For many species, salt spray can disrupt plant–water relations, promote tissue necrosis and leaf loss, and/or reduce growth in exposed plants (Sykes and Wilson, 1988; Cheplick and Demetri, 1999; Griffiths and Orians, 2003; Touchette *et al.*, 2009a). Plants that reside in areas where contact with other forms of salts are unavoidable tend to be more resistant to salt spray (Barbour, 1978). Moreover, high soil moisture levels seem to mitigate the damaging effects of salt spray in some species, allowing more salt-sensitive hydrophytes to grow in areas with high salt spray (Griffiths and Orians, 2003; Griffiths, 2006). Notwithstanding, a few coastal dune species with notable resistance to high soil salinities appear to be less tolerant to foliar applications of seawater (e.g. *Senecio elegans* and *Austrofestuca littoralis*; Sykes and Wilson, 1988); while other species such as *Lupinus arboreus* are intolerant to soil salinities, but unaffected by aerial salt exposure (Sykes and Wilson, 1988).

8.3 Growth and Survival of Salt Marsh Halophytes

The vast majority of salt marsh halophytes do not require salt to survive and reproduce. Indeed, comprehensive evaluations of coastal angiosperms suggest that very few species are restricted to environmental salinities greater than 0.5% (Barbour and Davis, 1970; Partridge and Wilson, 1987). Therefore, similar to glycophytes, many coastal halophytes grow best with no extraneous environmental salts (Fig. 8.2). This tendency to perform better in freshwater systems suggest that most salt marsh and coastal plant species would not satisfy the more restrictive definitions of halophytes, which include the ability to thrive in and/or require comparatively high saline conditions to complete their life cycle (Marschner, 1995; Grigore *et al.*, 2014). For salt marsh plants in general, a broader definition of halophyte is necessary to include all plants observed throughout the marsh. In this case, definitions that consider halophytes as plants adapted to (and capable of) surviving in saline environments (Grigore *et al.*, 2014; Flowers and Colmer, 2015) are more inclusive and consider high marsh plant species that: (1) experience intermittent or infrequent high surface water salinities attributed to unusually high tides or storm surges; (2) reside in soils that receive predominately freshwater with occasional salt exposure by subsurface waters through unusual events such as long-term droughts that would diminish the diluting effects of a freshwater lens; or (3) typically reside in low-saline soils but must contend with appreciable deposition by salt spray and/or seawater aerosols. For some high marsh plants, exposure to saline soils or surface waters may be sporadic and ephemeral; therefore, it can be challenging to determine if certain species have meaningful halophytic attributes. Unique adaptations may further complicate this distinction if based strictly on occurrence within the marsh. For example, salinity is a well-known stressor of *Phragmites australis*, although it may occur in mesohaline estuaries. Its survival is predicated on its ability to tolerate saline surface waters in favour of fresh groundwater; thus the success of *P. australis* depends not on the salinity of surface waters, but on the presence of deeper interstitial freshwaters (Adams and Bate, 1999; Burdick *et al.*, 2001).

Most salt marsh plants can sustain growth with salinities well above 15 psu (Table 8.1). Members of the grass and amaranth families (Poaceae and Amaranthaceae, respectively) are

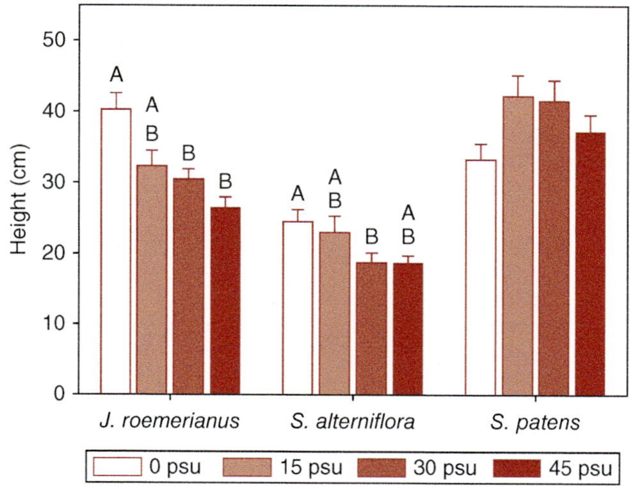

Fig. 8.2. Final plant height (cm) of three salt marsh plants, *Juncus roemerianus, Spartina alterniflora* and *S. patens*, grown in different salinities (0, 15, 30 and 45 psu) for 10 weeks. Letters above the bars indicate significant differences; the same letter indicates no significance ($p < 0.05$; Kruskal–Wallis one–way ANOVA on ranks). Data are presented as means ± 1 SE (n = 20).

Table 8.1. Growth responses of different salt marsh species to experimental and/or prevailing salinities. Species are arranged within taxonomic families and include growth response (or growing conditions), duration of growth, typical marsh zone where they occur, optimal salinity range for plant growth and reference. The table is not exhaustive (as information is lacking for many species and/or data are not readily available) but illustrates a range of tolerance in salt marsh halophytes and the broad taxonomic breadth of marsh plants. Information that was not reported or was not relevant to the referenced study is labelled NR.

Family (species)	Response or conditions	Duration	Marsh zone	Salinity (psu)	Reference
Amaranthaceae					
Arthrocnemum subterminale	Max. growth at high salinities	3 months	High	34[d]	Kuhn and Zedler (1997)
Arthrocnemum fruticosum	Natural range (soil salinity)	NR	Low–high	10–95[e]	Zörb *et al.* (2013)
Atriplex prostrata	Sustained maximum growth	> 5 weeks	High	1–15[a]	Partridge and Wilson (1987)
Halimione portulacoides	Natural range (soil salinity)	NR	Middle–high	0–75[e]	Zörb *et al.* (2013)
Halimione portulacoides	Low growth at high salinity	2.5 months	Middle–high	0 and 22[f]	Rozema *et al.* (1985)
Halocnemum strobilaceum	Natural range (soil salinity)	NR	Middle–high	10–115[e]	Zörb *et al.* (2013)
Petrosimonia spp.	Natural range (soil salinity)	NR	Middle–high	0–45[e]	Zörb *et al.* (2013)
Salicornia depressa	Sustained growth	41 days	Low–high	0.01–0.05[b]	Mahall and Park (1976)
Salicornia europaea	Max. growth at high salinities	2 months	Low	20[b]	Cooper (1982)
Salicornia europaea	Sustained maximum growth	2.5 months	Low	0 and 22[f]	Rozema *et al.* (1985)
Salicornia fragilis	Natural range (soil salinity)	NR	Low–high	3–94[e]	Zörb *et al.* (2013)
Salsola kali	Natural range (soil salinity)	NR	Middle–high	5–45[e]	Zörb *et al.* (2013)
Sarcocornia quinqueflora	Positive growth	3 months	Low–middle	0–35[c]	Clark and Hannon (1970)
Sarcocornia quinqueflora	Sustained maximum growth	> 5 weeks	Low–middle	8–35[a]	Partridge and Wilson (1987)
Suaeda australis	Positive growth	3 months	Low–middle	0–35[c]	Clarke and Hannon (1970)
Suaeda novae-zelandiae	Sustained maximum growth	5 months	Middle–high	16–42[a]	Partridge and Wilson (1987)
Suaeda prostrate	Natural range (soil salinity)	NR	Middle–high	10–95[e]	Zörb *et al.* (2013)
Apiaceae					
Apium prostratum	Sustained maximum growth	> 5 weeks	High	0–8[a]	Partridge and Wilson (1987)
Asteraceae					
Artemisia santonicum	Natural range (soil salinity)	NR	High	0–15[e]	Zörb *et al.* (2013)
Cotula coronopifolia	Sustained maximum growth	> 5 weeks	High	8–19[a]	Partridge and Wilson (1987)

Continued

Table 8.1. Continued.

Family (species)	Response or conditions	Duration	Marsh zone	Salinity (psu)	Reference
Leptinella dioica	Sustained maximum growth	> 5 weeks	High	3–15[a]	Partridge and Wilson (1987)
Tripolium pannonicum	Positive growth	2 months	High	20[b]	Cooper (1982)
Tripolium pannonicum	Sustained maximum growth	2.5 months	High	0 and 20[f]	Rozema *et al.* (1985)
Caryophyllaceae					
Spergularia maritima	Sustained maximum growth	> 5 weeks	Middle–high	5–22[a]	Partridge and Wilson (1987)
Spergularia sp.	Sustained maximum growth	2.5 months	NR	0 and 22[f]	Rozema *et al.* (1985)
Cyperaceae					
Carex flagellifera	Sustained maximum growth	> 5 weeks	High	1–9[a]	Partridge and Wilson (1987)
Isolepis cernua	Sustained maximum growth	> 5 weeks	High	2–10[a]	Partridge and Wilson (1987)
Schoenoplectus pungens	Sustained maximum growth	> 5 weeks	Low–middle	5–31[a]	Partridge and Wilson (1987)
Schoenus nitens	Sustained maximum growth	> 5 weeks	Middle	2–23[a]	Partridge and Wilson (1987)
Goodeniaceae					
Selliera radicans	Sustained maximum growth	> 5 weeks	Middle–high	0–18[a]	Partridge and Wilson (1987)
Juncaceae					
Juncus gerardii	Low growth at high salinities	2 months	Middle–high	20[b]	Cooper (1982)
Juncus gerardii	Low growth at high salinities	2.5 months	Middle–high	0 and 22[f]	Rozema *et al.* (1985)
Juncus maritimus	Sustained maximum growth	> 5 weeks	Middle–high	5–32[a]	Partridge and Wilson (1987)
Juncus maritimus	Sustained growth	3 months	Middle–high	0–25[c]	Clark and Hannon (1970)
Juncus maritimus	Low growth at high salinity	2.5 months	Middle–high	0 and 22[f]	Rozema *et al.* (1985)
Juncus maritimus	Natural range (soil salinity)	NR	Middle–high	0–34[e]	Zörb *et al.* (2013)
Juncus roemerianus	Sustained maximum growth	8 weeks	Middle–high	0–30	Touchette *et al.* (2009b)
Juncus roemerianus	Natural range (soil salinity)	NR	Middle–high	0–18	Touchette (2006)

Continued

Table 8.1. Continued.

Family (species)	Response or conditions	Duration	Marsh zone	Salinity (psu)	Reference
Juncaginaceae					
Triglochin striatum	Sustained growth	3 months	Low–middle	0–32[c]	Clark and Hannon (1970)
Triglochin striatum	Sustained maximum growth	> 5 weeks	Low–middle	5–25[a]	Partridge and Wilson (1987)
Triglochin maritima	Sustained growth	2 months	Low	20[b]	Cooper (1982)
Triglochin maritima	Low growth at high salinity	2.5 months	Low	0 and 22[f]	Rozema *et al.* (1985)
Malvaceae					
Plagianthus divaricatus	Sustained maximum growth	> 5 weeks	Middle–high	0–6[a]	Partridge and Wilson (1987)
Phrymaceae					
Mimulus repens	Sustained maximum growth	> 5 weeks	Middle–high	0–5[a]	Partridge and Wilson (1987)
Plantaginaceae					
Plantago coronopus	Sustained maximum growth	> 5 weeks	High	3–18[a]	Partridge and Wilson (1987)
Plantago maritima	Low growth at high salinities	2 months	High	20[b]	Cooper (1982)
Plantago maritima	Low growth at high salinity	2.5 months	High	0 and 22[f]	Rozema *et al.* (1985)
Plumbaginaceae					
Armeria maritima	Sustained growth	2 months	Middle–high	20[b]	Cooper (1982)
Armeria sp.	Low growth at high salinity	2.5 months	NR	0 and 22[f]	Rozema *et al.* (1985)
Limonium graecum	Natural range (soil salinity)	NR	High	0–40[e]	Zörb *et al.* (2013)
Limonium sp.	Low growth at high salinity	2.5 months	High	0 and 22[f]	Rozema *et al.* (1985)
Poaceae					
Aeluropus littoralis	Natural range (soil salinity)	NR	High	5–25[e]	Zörb *et al.* (2013)
Agrostis stolonifera	Sustained maximum growth	5 weeks	High	0–5[a]	Partridge and Wilson (1987)
Elymus pycnanthus	Sustained maximum growth	> 5 weeks	High	5–25[a]	Partridge and Wilson (1987)
Elytrigia pungens	No growth at high salinities	2.5 weeks	High	0 and 22[f]	Rosema *et al.* (1985)
Festuca rubra	Low growth at high salinities	2 months	High	20[b]	Cooper (1982)
Festuca rubra	No growth at high salinities	2.5 months	High	0 and 22[f]	Rosema *et al.* (1985)
Lachnagrostis filiformis	Sustained maximum growth	> 5 weeks	High	10–23[a]	Partridge and Wilson (1987)
Poa laevis	Sustained maximum growth	> 5 weeks	High	0–9[a]	Partridge and Wilson (1987)

Continued

Table 8.1. Continued.

Family (species)	Response or conditions	Duration	Marsh zone	Salinity (psu)	Reference
Polypogon monspeliensis	Low growth at high salinities	3 months	Middle–high	11[d]	Kuhn and Zedler (1997)
Polypogon monspeliensis	Sustained maximum growth	> 5 weeks	Middle–high	5–19[a]	Partridge and Wilson (1987)
Puccinellia fasciculata	Sustained maximum growth	> 5 weeks	Low–middle	5–15[a]	Partridge and Wilson (1987)
Puccinellia maritima	Low growth at high salinities	2 months	Low–high	20[b]	Cooper (1982)
Puccinellia maritima	Sustained maximum growth	2.5 months	Low–high	0 and 22[f]	Rozema *et al.* (1985)
Puccinellia walker	Sustained maximum growth	> 5 weeks	Middle–high	8–25[a]	Partridge and Wilson (1987)
Puccinellia stricta	Sustained maximum growth	> 5 weeks	High	1–20[a]	Partridge and Wilson (1987)
Schedonorus arundinaceus	Sustained maximum growth	> 5 weeks	High	0–6[a]	Partridge and Wilson (1987)
Spartina alterniflora	Sustained maximum growth	8 weeks	Low	0–30	Touchette *et al.* (2009b)
Spartina anglica	Sustained maximum growth	2.5 months	Low	0 and 22[f]	Rozema *et al.* (1985)
Spartina anglica	Sustained maximum growth	> 5 weeks	Low	8–29[a]	Partridge and Wilson (1987)
Spartina foliosa	Sustained maximum growth	2 months	Low	0–27[c]	Phleger (1971)
Spartina foliosa	Sustained maximum growth	41 days	Low	0.04[b]	Mahall and Park (1976)
Sporobolus virginicus	Sustained maximum growth	3 months	Low–middle	0–18[b]	Marcum and Murdoch (1992)
Urochondra setulosa	Sustained growth	6 weeks	High	58[b]	Gulzar *et al.* (2003)
Primulaceae					
Lysimachia maritima	Low growth at high salinities	2.5 months	NR	0 and 22[f]	Rozema *et al.* (1985)
Samolus repens	Sustained maximum growth	> 5 weeks	High	5–25[a]	Partridge and Wilson (1987)

Note: seawater is 35 practical salinity units (psu) or 3.5% NaCl; a = converted from % NaCl; b = converted from mM NaCl (where 600 mM NaCl is full-strength seawater); c = converted from % seawater; d = converted from g/l NaCl; e = converted from electrical conductivity (assuming room temp ~23°C); f = converted from water potential (where –0.8 MPa is ~ 10 psu).

among the most common and can be observed in all three plant zones. Collectively, there appears to be no distinction between salt tolerance and marsh zone (Fig. 8.3). Based on 68 species, there were no significant distinctions between the plants' ability to tolerate salt and their position within the marsh ($p = 0.32$). Moreover, while some plants are fairly productive in low marsh systems, only a few plant families have species that can sustain maximum growth at salinities equal to or greater than seawater (≥ 35 psu; Fig. 8.4). Indeed, most studies on saltmarsh halophytes observed sustained positive growth at salinities far below 35 psu (Table 8.1). This suggests that, in many cases, salt marshes are most productive in areas with dilute surface waters (e.g. estuaries) or in areas where groundwater mixing lowers overall soil salinity. It is not that salt marsh halophytes cannot grow and reproduce in hypersaline conditions; rather, overall productivity appears to be diminished at higher salinities for most species. Exceptions to this general trend are members of the amaranth family (Amaranthaceae) that appear to sustain positive growth at salinities well above 35 psu (Clarke and Hannon, 1970; Partridge and Wilson, 1987).

Growth rates in response to changing environmental conditions can vary among species. In one study, aboveground productivity was nearly twofold higher in *Juncus roemerianus* when salinities alternated weekly from 0 to 30 psu over the course of 8 weeks, in comparison to plants maintained in constant salinities (0 or 30 psu; Touchette *et al.*, 2009b). This response, however, was not observed in *Spartina alterniflora* grown under similar conditions. Moreover, plant growth responses may be influenced by prior salt exposures and/or plant age (Hwang and Morris, 1994; Touchette *et al.*, 2012). Growth following sudden increases in salinity (30 psu) for *J. roemerianus* was low in 5-year-old plants previously acclimated only to freshwater (Touchette *et al.*, 2012). Younger plants (6- and 24-months old), however, were more physiologically plastic and able to sustain high growth rates comparable to control plants in freshwater. This suggests that, while plasticity is important for coastal plants residing in remarkably dynamic environmental salinities, there are – at least for some species – fixed modifications in plant tissues to prevailing salinities after maturation (Richards *et al.*, 2010; Touchette *et al.*, 2012).

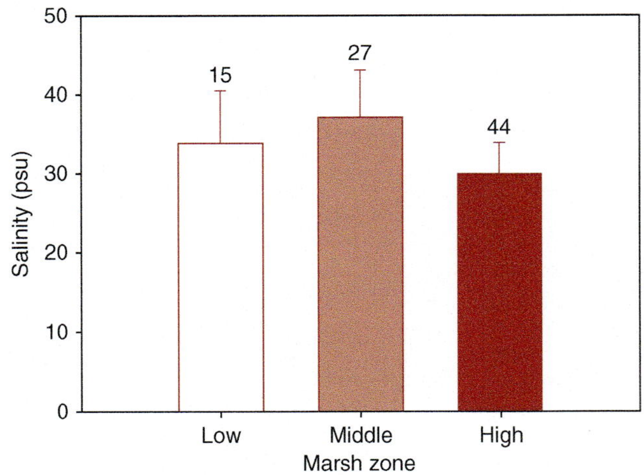

Fig. 8.3. Maximum salinity tolerance (practical salinity units; psu) for species present in each marsh zone (low, middle and high). Values were compiled from 68 species (Table 8.1). When species occurred in multiple zones, their tolerance values were counted towards each zone. When multiple data sources were available for the same species, the highest value was selected. No significant difference in tolerance was observed ($p = 0.32$; Kruskal–Wallis one-way ANOVA on ranks) among the three zones. Nevertheless, it is important to note that this figure includes species from both temperate–wet systems and arid systems, where differences in salinity tolerance with respect to zone are expected. Data are presented as means ± 1 SE with the number of species considered for each zone above the bars.

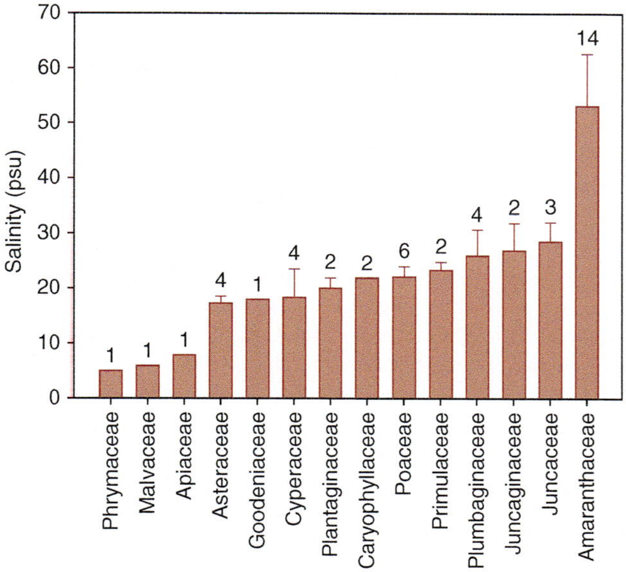

Fig. 8.4. Maximum salinity tolerance (practical salinity units; psu) for fourteen plant families (Table 8.1). When multiple data sources were available for the same species, the highest value was selected. Values presented here are not exhaustive, as many families have not been fully evaluated and data is not readily available. Data is presented as means ± 1 SE with the number of species considered above each bar.

8.4 Physiological Adaptations to Saline Environments

Full-strength seawater is more than 3% salt, with molarities of Na^+ and Cl^- at 460 mM and 540 mM, respectively. At these levels most angiosperms would experience hyperionic and hyperosmotic stress with concomitant decreases in growth and increases in mortality (Mahajan and Tuteja, 2005). High salinities promote low soil-water potentials, making it increasingly difficult for plants to acquire water and nutrients from soils and surface waters, and fostering what has been termed 'physiological drought' (Munns, 2002; Larcher, 2003; Touchette, 2006). It is important to note, however, that this term can misrepresent subtle responses at the cellular level. While both high salinity and desiccation will increase cellular ion concentrations, the proportions of ions seldom change during dehydration but will change drastically during salinity stress due to selective ion uptake (Kirst, 1989; Touchette, 2007).

In halophytes, high intercellular osmotic pressure can be generated with vacuolar ion sequestration and through accumulation of osmotically active solutes (osmolytes; Flowers *et al.*, 1977; Vasquez *et al.*, 2006). Compartmentalizing ions in vacuoles probably involves ion transporters along the tonoplast, as well as vacuolar Na^+/H^+ exchangers (e.g. NHX; Flowers and Colmer, 2015; Yuan *et al.*, 2015). Thus, salt marsh plants must be able to balance the flux of Na^+ and Cl^- needed for osmotic equilibrium with the deleterious accumulation of these metabolically disruptive ions within the cytosol. With proper ion and osmolyte balance, plants residing in saline environments can maintain tissues that are hyperosmotic relative to their environment (Vasquez *et al.*, 2006; Touchette, 2007). For many plants, tolerance to saline environments may involve multiple physiological strategies including ion compartmentalization, reductions in water and osmotic potentials, ion exclusion from belowground structures, ion partitioning among different organs, synthesis of compatible solutes or osmolytes, changes in membrane and/or cell wall structure, induction of antioxidative enzymes, and salt exudation by way of salt glands or bladders (Parida and Das, 2005; Baisakh *et al.*, 2006; Touchette, 2007; Flowers and Colmer, 2015).

Changes in plant–water status, through sudden fluctuations in environmental salinity, can alter cellular turgor pressure as water moves in the direction of the osmotic gradient. The degree of water flux depends on a variety of factors including the hydraulic conductance of membranes, intracellular osmotic pressure, cell volume and cell wall elasticity (Tyerman, 1982; Kramer and Boyer, 1995; Touchette *et al.*, 2014).

Bulk modulus of elasticity (ε) varies among species, and may be modified by individual plants when experiencing changing water status (e.g. drought or high salinities). When the freshwater plant *Alternanthera philoxeroides* was exposed to higher salinities, for example, there was a 30-fold increase in ε along with a concomitant decrease in osmotic potential (Ψ_π) attributed to osmolyte accumulation (Bolaños and Longstreth, 1984). Observed ε values across different species illustrate both highly flexible cell walls (e.g. 0.06 MPa in *Halicystis parvula*) and extremely rigid walls (e.g. 70 MPa in *Chara corallina*), wherein a higher ε represents greater wall rigidity (Daintly *et al.*, 1974; Graves and Gutknecht, 1976). Salt marsh plants exposed to high environmental salinities tend to maintain disproportionately higher tissue ε with lower Ψ_π than glycophytes found in other systems (Touchette *et al.*, 2014). Rigid cell walls can restrict overall water flux and minimize changes in cell volume. This helps maintain optimal metabolic function (e.g. maintaining cellular pH or avoiding 'salting out' of important metabolites) by supporting appropriate intermolecular distances within cell fluids (Clifford *et al.*, 1998). Indeed, ε and Ψ_π are closely tied to cell volume (i.e., $\Delta V/\Delta\Psi = V/(\varepsilon+\Psi_\pi)$; Tyerman, 1982), and it is likely that rigid cell walls can minimize dehydration as high ε and low Ψ_π together are effective means of maintaining optimal cellular water content (Bartlett *et al.*, 2012; Touchette *et al.*, 2014). This notion that both ε and Ψ_π work in concert to maintain cellular water volume, referred to as the 'cell water conservation hypothesis', may play an important role in marsh plants that accumulate ions and other solutes in response to elevated environmental salts.

Once salt enters the plant, modifications may be necessary to prevent accumulation of excess ions within sensitive structures. Among the more direct responses is to isolate salt in specific tissues, such as older leaves, which may then be discarded by the plant (Fig. 8.5). Alternatively, some plants known as recretohalophytes (representing approximately 370 species) possess specialized cells or structures that actively secrete ions. These specialized structures include salt bladders and salt glands, and appear to be quite diverse in structure and complexity; they are probably products of multiple evolutionary origins (Flowers *et al.*, 2010; Dassanayake and Larkin, 2017). Salt-secreting structures appear to have originated from the epidermal tissues and are generally considered to be modified tricombs (Yuan *et al.*, 2016; Dassanayake and Larkin, 2017). Salt bladders are composed of a single bladder cell that, in some instances, is supported by a short stalk consisting of one or more cells (Yuan *et al.*, 2016). These endo-recretohalophytes accumulate salts in large vacuoles of swollen bladder cells until they eventually rupture, depositing their contents on the plant's surface (Dassanayake and Larkin, 2017). In contrast, exo-recretohalophytes possess more structurally stable salt glands that accumulate and secrete ions directly from the plant and, depending on species, may or may not increase in abundance with increasing salinity (Barhoumi *et al.*, 2007). These glands can comprise a single secretory hair (observed in *Porteresia* spp.), bicellular structures including a basal cell and a cap cell (common among Poaceae) or multicellular structures (between 4 and 40 cells) differentiated into collecting and secretory cells (typical of many dicots; Kobayashi, 2008; Yuan *et al.*, 2016; Dassanayake and Larkin, 2017). Regardless of the gland's complexity, ion secretion is energy consumptive, comprising cells with highly developed mitochondria that deliver the necessary energy to transport ions against large concentration gradients (Kobayashi, 2008; Yuan *et al.*, 2016).

8.5 Sea-level Rise and Salt Marshes

Current IPCC models suggest that the rate of global sea-level rise may range between 7 and 16 mm yr^{-1} by the end of the century (IPCC, 2013). Even at the lower projected rates, it is generally recognized that there will be some salt marsh decline. The extent and magnitude of marsh loss, nevertheless, remains uncertain. Regional rates of sea-level rise may be considerably different, with some areas receiving effectively higher or lower changes in sea-level due to variations in coastal subsidence, seasonal precipitation/drought events, groundwater withdrawal, isostatic rebound and other factors (Crosby *et al.*, 2016; Raposa *et al.*, 2017). Therefore, some areas may experience very little change, whereas other regions may undergo accelerated rates of sea-level rise.

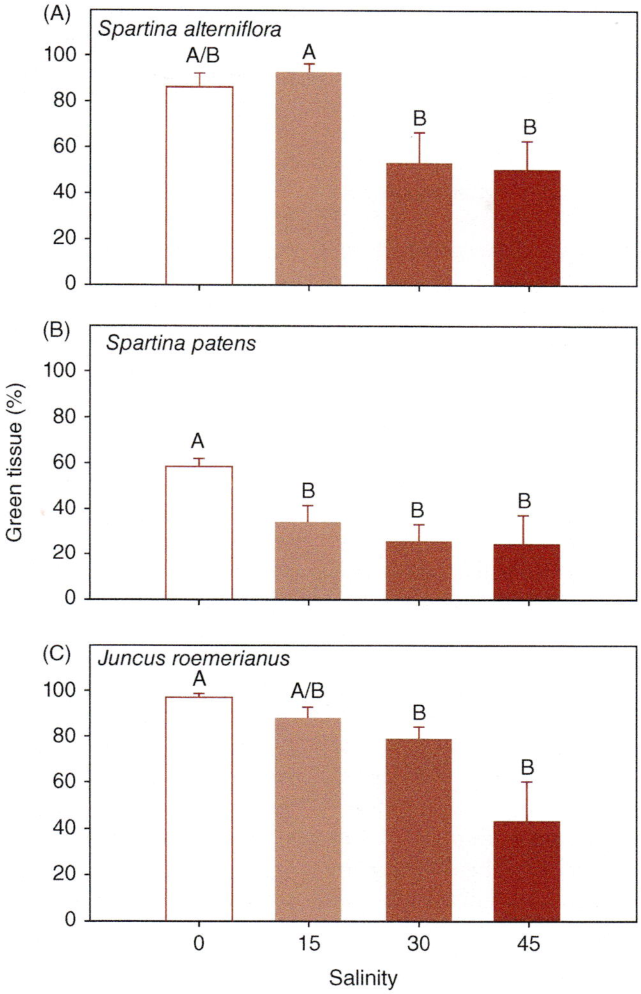

Fig. 8.5. Amount of green tissue (%) along the vertical axis of salt marsh halophytes, *Spartina alterniflora* (Panel A), *S. patens* (Panel B) and *Juncus roemerianus* (Panel C) grown in different salinities (0, 15, 30, 45) over 10 weeks. Some salt marsh plants respond to elevated environmental salts with increased chlorosis through: (1) natural tissue decline due to increased environmental stress; and (2) accumulation of salts in certain leaves which promotes tissue dieback and removal of excess salts once the leaf is discarded. Letters above the bars indicate significant differences; the same letter indicates no significance ($p < 0.05$; Kruskal–Wallis one–way ANOVA on ranks). Data are presented as means ± 1 SE ($n = 20$)

Globally, more than 50% of all wetlands have been lost, mostly due to anthropogenic activities; and coastal wetland loss is occurring at faster rates than that of inland wetlands during the 20th century (Davidson, 2014). Some studies suggest that an additional 20%–50% of salt-marsh habitat will transition into open waters by the end of the century because of sea-level rise (Craft *et al.*, 2009; Brisson *et al.*, 2014). A model developed by Spencer *et al.* (2016), using a sea-level rise scenario of 29 cm by 2100, estimated coastal wetland losses of between 23% and 28% by the 2050s, and 32%–40% by the 2080s. However, many models are not able to assess feedback modifications such as plant physiological responses to warmer temperatures or higher atmospheric CO_2 levels, and future changes in system hydrology (Spencer *et al.*, 2016). Indeed, some researchers argue that, in many instances, salt marshes have the capacity to maintain elevations with rising sea levels, and that marsh vulnerability is overestimated as these systems maintain biophysical feedback mechanisms that can accelerate soil accumulation (Kirwan and Megonigal, 2013; Kirwan *et al.*, 2016).

Salt marshes are highly dynamic systems and have been capable of balancing elevations with sea levels for the past 4000 years (Redfield, 1972). To avoid inundation and drowning during sea-level rise, salt marshes must either retreat to higher elevations or increase elevation through accumulation of mineral sediments and/or organic matter (Redfield, 1972; Morris *et al.*, 2002). Upland migration with rising sea levels, however, may be limited by available space; as human populations continue to expand into coastal areas, there has been a steady increase in physical barriers along shorelines that may prevent further landward movement (Torio and Chmura, 2013; Crosby *et al.*, 2016). Even in the absence of human barriers, local topographies may hinder landward migrations, as areas with more pronounced slopes may possess natural resistance to marsh encroachment.

While it remains uncertain whether elevation gains can keep pace with projected rates of sea-level rise for many coastal wetlands, marsh loss due in part to rising seas in the past 30 years has been observed in the Mississippi River Delta, USA, mid-Atlantic estuaries in the USA; and along the coast of the Yellow Sea, including the coast of China (Hartig *et al.*, 2002; Kearney *et al.*, 2002; Blum and Roberts, 2009; Murray *et al.*, 2014). Marshes with unusually low elevations and/or with accelerated rates of sea-level rise appear most vulnerable. Moreover, systems with low sediment supply are similarly at risk of decline, and it appears that sediment loads are diminishing in many coastal areas due to hydrological modifications by humans, including damming and changes in land-use practices (Murray *et al.*, 2014; Weston, 2014). Although climate change may increase storm frequencies in certain regions, thereby enhancing sediment loads to coastal areas, these storms can also promote marsh erosion and cause an overall decline in distribution (Crosby *et al.*, 2016). Likewise, elevated temperatures and atmospheric CO_2 levels may enhance plant productivity and provide much-needed organic matter to increase marsh elevation. The increased rates of productivity may not be sufficient to match sea-level rise in some areas, and warmer temperatures could also accelerate decomposition within organic soils, potentially slowing the rate of soil accumulation (Touchette *et al.*, 2010; Crosby *et al.*, 2016). Clearly there is a high degree of uncertainty about how coastal marshes will respond to rising sea levels on a global scale. Further studies that encompass unique biophysical and topographical features are necessary to identify local and regional responses to specific marsh systems.

For marshes that experience rising sea levels, responses may vary and can include shifts in dominant vegetation, plant dieback, surface ponding and/or soil subsidence (Raposa *et al.*, 2017). In New England, for example, sea-level rise has resulted in a loss of a high marsh species, *Spartina patens*, in favour of a more flood-tolerant species, *S. alterniflora* (Warren and Niering, 1993; Donnelly and Bertness, 2001). It is suspected that this replacement by *S. alterniflora* may be attributed to a slow landward encroachment of *S. alterniflora* into *S. patens*-dominant areas, and through the formation of episodic dieback patches that were subsequently recolonized by *S. alterniflora* (Raposa *et al.*, 2017). In the latter case, the occurrence of a high number of dieback patches coincided with a 2-year period of unusually high water levels that promoted waterlogged soils in high marsh regions. Once formed, these open areas were often repopulated by *Salicornia* spp. within 1 year, and were mostly composed of *S. alterniflora* after 3 years (Raposa *et al.*, 2017). In other areas, a more extreme response has occurred involving the conversion of coastal wetlands and vegetative flats into open water systems that lack the habitat and productivity necessary to support dynamic, shallow coastal ecosystems (Kearney *et al.*, 2002; Couvillion *et al.*, 2011; Murray *et al.*, 2014).

8.6 Macroclimatic Influences on Coastal Salt Marshes

While sea-level rise alone may result in significant declines in coastal wetlands, macroclimatic changes (e.g. temperature and precipitation) are also essential drivers influencing coastal wetland vulnerability to climate change (Osland *et al.*,

2016). Coastal vegetation assemblages often reflect regional climate. For example, graminoids tend to dominate in temperate–wet conditions, whereas succulent plants are more common in arid and semi-arid climates. Although it is uncertain how changes in precipitation and temperature will alter many of these systems, evidence points to an increase in unvegetative pannes and mud-flats, succulent-based marshes and mangroves, at the expense of graminoid-based systems (Osland *et al.*, 2016).

Winter air temperatures are important in determining the boundary between salt marshes and mangroves (Saintilan *et al.*, 2014). In the south-eastern USA, transitional zones between marshes and mangroves occur where minimal air temperatures fall below -7.0°C (Osland *et al.*, 2013). With the last major freezing event occurring over 25 years ago, mangroves in this region are expanding into areas previously occupied by salt marsh graminoids. A global poleward expansion of mangroves is expected to continue with rising global temperatures (Saintilan *et al.*, 2014).

Little is known about how salt marsh plants will respond to elevated temperatures. Studies suggest that, based on photosynthetic outputs, plants do have the capacity to acclimate to higher temperatures; however, unusually high temperatures can result in lower photosynthetic outputs and a concomitant decline in primary productivity (Yamori *et al.*, 2014). This, coupled with hypersaline conditions (attributed to increased evaporation) may impose even greater physiological stress on resident plants. Nevertheless, an interesting interaction between temperature stress and salinity has been observed in some halophytes. Some species acclimated to high salinities appear to be more tolerant (as indicated by photosynthetic quantum yields) to heat shock than plants grown in freshwater (Table 8.2). It is possible that preacclimation to salt may provide some degree of thermal protection to salt marsh halophytes. Furthermore, while not well documented in salt marsh plants, seed germination in halophytes may be influenced by interactions between soil salinity and thermoperiods (Khan and Ungar, 1997). In some species, germination is inhibited by warmer temperatures (Badger and Unger, 1989), while in others germination is enhanced (Khan *et al.*, 2000). Interactions may also exist where, in subtropical maritime desert plants for example, germination rates at high salinities decline as temperatures rise (Khan *et al.*, 2000).

In more arid systems, the amount of rainfall and access to freshwater are important drivers in determining wetland plant distribution and abundance. Osland *et al.* (2016) suggest that a critical tipping point for precipitation is approximately 99.5 cm yr^{-1}. In areas that receive less rainfall, small changes in freshwater availability may result in large changes in wetland plant distribution and abundance. In contrast, areas with greater rainfall will probably maintain more stable plant populations. Nevertheless, this threshold could vary considerably depending on rain periodicity (evenly distributed rainfall versus wet–dry climates) and interactions between temperature and precipitation. In wet–dry climates, for example, vegetation decline due to hypersaline soils could occur during dry seasons even though annual rainfall may exceed 99.5 cm yr^{-1} (Osland *et al.*, 2016).

The notion that lower freshwater inputs (from precipitation, surface-water runoff and/or groundwater) can adversely impact coastal marshes is exemplified by the ongoing acute marsh dieback along the Atlantic and Gulf coasts of North America. Multiple hypotheses have been developed to explain the sudden die-off of marsh grasses, including herbivory, *Fusarium*-fungi infections, pore-water geochemical changes and pollution (Bertness *et al.*, 2014). While these influences may explain localized decline in marsh area in some regions, the bulk of the dieback in the south-eastern USA is thought to be related to drought and low freshwater inputs (Alber *et al.*, 2008). It is likely that evapotranspirative processes fostered elevated pore-water salinities when freshwater inputs were diminished (Alber *et al.*, 2008; Hughes *et al.*, 2012). While marsh plants are tolerant of high salinities, short periods of hypersalinity may impact the osmoregulatory capacity of resident plants, promoting stress that could directly or indirectly (via secondary infections and/or additional stressors) cause plant death. Current climate projections suggest that these regions will experience rising temperatures and prolonged droughts, possibly triggering more episodes of dieback and, in turn, limiting the capacity of these marshes to adjust to other

Table 8.2. Chlorophyll *a* fluorescence measurements including effective quantum yield ($\Delta F/F_m'$) and maximum-quantum yield (F_v/F_m) for salt marsh halophytes under extreme salt- and temperature-stress conditions. Data show that when *Spartina patens* and *Juncus roemerianus* were grown in hypersaline conditions (salinity of 45 psu for 8 weeks) there were significant reductions in effective- and maximum-quantum yields. This salinity response was not observed in *S. alterniflora*. Similarly, heat shock (45°C for 1 h) resulted in declines in quantum yields for all three species when acclimated to salinity of 0 psu, but not for plants acclimated to 45 psu. Although hypersalinity resulted in decreased quantum yields, in general there were no further declines in effective- and maximum-quantum yields due to heat stress, suggesting that physiological modifications that support growth in highly saline conditions also provide some degree of photosystem II thermoprotection in some halophytic species.

Species	Temperature (°C)	Salinity (psu)	$\Delta F/F_m'$	% change	F_v/F_m	% change
S. alterniflora	Ambient	0	0.741 ± 0.017^A		0.807 ± 0.011^A	
	Ambient	45	0.733 ± 0.016^A		0.756 ± 0.066^A	
	45	0	0.237 ± 0.035^B	−68.1 ± 4.2	0.644 ± 0.024^B	−20.2 ± 2.9
	45	45	0.456 ± 0.054^{AB}	−37.8 ± 7.3*	0.733 ± 0.034^A	−3.0 ± 4.5*
S. patens	Ambient	0	0.750 ± 0.011^A		0.794 ± 0.007^A	
	Ambient	45	0.453 ± 0.018^B		0.597 ± 0.062^B	
	45	0	0.435 ± 0.076^B	−42.0 ± 10.1	0.612 ± 0.071^B	−23.0 ± 8.9
	45	45	0.405 ± 0.048^B	−10.5 ± 10.6	0.672 ± 0.027^B	12.5 ± 4.6*
J. roemerianus	Ambient	0	0.770 ± 0.015^A		0.831 ± 0.006^A	
	Ambient	45	0.562 ± 0.079^B		0.423 ± 0.153^B	
	45	0	0.658 ± 0.073^B	−14.5 ± 9.6	0.726 ± 0.034^B	−12.7 ± 4.4
	45	45	0.590 ± 0.053^B	5.1 ± 9.5	0.677 ± 0.051^B	29.1 ± 30.4

Data were statistically analysed using Kruskal–Wallis ANOVAs (quantum yields) and Mann–Whitney rank sum test (% change). Data are presented as means ± 1 SE, and statistical differences are indicated by letters (quantum yields) or asterisks (% change).

climate-related constraints such as sea-level rise (Hughes *et al.*, 2012).

8.7 Conclusions

Salt marshes are dynamic, highly productive ecosystems, controlled by a variety of biotic and abiotic influences. Abiotic factors such as soil salinity and waterlogging are important drivers in determining plant occurrence, abundance and distribution. Climate change-linked processes that alter or disrupt these drivers are likely to negatively impact coastal wetlands and promote a global net loss of these ecologically and economically important systems. Sea-level rise will, in some instances, decrease overall plant diversity by selecting for species that are more tolerant to sustained flooding and/or high soil salinities, or through the loss of mid- and high marsh species that are less competitive to changing conditions (e.g. displacement of *Spartina patens* in New England). Under more dramatic conditions, rising sea levels will promote the conversion of marshes into open water systems, especially in areas where soil accretion fails to keep pace with rising waters and where landward migration is impeded. Changes in macroclimate will also probably influence coastal wetlands. In regions where precipitation is expected to decline (or become seasonal), periods of soil and/or water hypersalinity may promote indirect die-offs (elevated stress, coupled with other biotic or abiotic agents) or direct die-offs. Nevertheless, the amount of coastal wetland loss remains uncertain due, in part, to the natural complexity of these systems. Regional differences in climate, geology, hydrology and topography, along with the degree of anthropogenic disturbance, coupled with a variety of biotic controls (e.g. physiological tolerance, competition, herbivory) will probably determine which marshes withstand the challenges of climate change and which marshes will perish.

References

Adam, P. (1990) *Saltmarsh Ecology*. Cambridge University Press, Cambridge, UK, pp. 461.

Adams, J.B. and Bate, G.C. (1999) Growth and photosynthetic performance of *Phragmites australis* in estuarine waters: A field and experimental evaluation. *Aquatic Botany* 64, 359–367.

Ahmad, I. and Wainwright, S.I. (1976) Ecotype differences in leaf surface properties of *Agrostis stolonifera* from salt marsh, spray zone and inland habitats. *New Phytologist* 76, 361–366.

Alber, M., Swenson, E.M., Adamowicz, S.C. and Mendelssohn, I.A. (2008) Salt marsh dieback: an overview of recent events in the US. *Estuarine, Coastal and Shelf Science* 80, 1–11.

Badger, K.S. and Ungar, I.A. (1989) The effects of salinity and temperature on the germination of the inland halophyte *Hordeum jubatum*. *Canadian Journal of Botany* 67, 1420–1425.

Baisakh, N., Subudhi, P.K. and Parami, N.P. (2006) cDNA-AFLP analysis reveals differential gene expression in response to salt stress in a halophyte *Spartina alterniflora* Loisel. *Plant Science* 170, 1141–1149.

Barbour, M.G. (1978) Salt spray as a microenvironmental factor in the distribution of beach plants at Point Reyes, California. *Oecologia* 32, 213–224.

Barbour, M.G. and Davis, C.B. (1970) Salt tolerance of five California salt marsh plants. *American Midland Naturalist* 84, 262–265.

Barhoumi, Z., Djebali, W., Smaoui, A., Chaïbi, W. and Abdelly, C. (2007) Contribution of NaCl excretion to salt resistance of *Aeluropus littoralis* (Willd) Parl. *Journal of Plant Physiology* 164, 842–850.

Bartlett, M.K., Scoffoni, C. and Sack, L. (2012) The determinants of leaf turgor loss point and prediction of drought tolerance of species and biomes: A global meta-analysis. *Ecology Letters* 15, 393–405.

Bertness, M.D., Gough, L. and Shumway, S.W. (1992) Salt tolerances and the distribution of fugitive salt marsh plants. *Ecology* 73, 1842–1851.

Bertness, M.D., Brisson, C.P., Bevil, M.C. and Crotty, S.M. (2014) Herbivory drives the spread of salt marsh die-off. *PloS One* 9. doi:10.1371/journal.pone.0092916.

Blum, M.D. and Roberts, H.H. (2009) Drowning of the Mississippi Delta due to insufficient sediment supply and global sea-level rise. *Nature Geoscience* 2, 488–491.

Bolaños, J.A. and Longstreth, D.J. (1984) Salinity effects on water potential components and bulk elastic modulus of *Alternanthera philoxeroides* (Mart.) Griseb. *Plant Physiology* 75, 281–284.

Brisson, C.P., Coverdale, T.C. and Bertness, M.D. (2014) Salt marsh die-off and recovery reveal disparity between the recovery of ecosystem structure and service provision. *Biological Conservation* 179, 1–5.

Burdick, D.M., Buchsbaum, R. and Holt, E. (2001) Variation in soil salinity associated with expansion of *Phragmites australis* in salt marshes. *Environmental and Experimental Botany* 46, 247–261.

Chapman, V.J. (1960) *Salt Marshes and Salt Deserts of the World*. Interscience Publishers Inc., New York.

Chapman, V.J. (1977) Introduction. In: Chapman, V.J. (ed.) *Ecosystem of the World: Vol. 1, Wet Coastal Ecosystems*. Elsevier, New York, pp. 1–29.

Cheplick, G.P. and Demetri, H. (1999) Impact of saltwater spray and sand deposition on the coastal annual *Triplasis purpurea* (Poaceae). *American Journal of Botany* 86, 703–710.

Clarke, L.D. and Hannon, N.J. (1970) The mangrove swamp and salt marsh communities of the Sydney district: III. Plant growth in relation to salinity and waterlogging. *The Journal of Ecology* 58, 351–369.

Clifford, S.C., Arndt, S.K., Corlett, J.E., Joshi, S., Sankhla, N. *et al.* (1998) The role of solute accumulation, osmotic adjustment and changes in cell wall elasticity in drought tolerance in *Ziziphus mauritiana* (Lamk.). *Journal of Experimental Botany* 49, 967–977.

Cooper, A. (1982) The effects of salinity and waterlogging on the growth and cation uptake of salt marsh plants. *New Phytologist* 90, 263–275.

Couvillion, B.R., Barras, J.A., Steyer, G.D., Sleavin, W., Fischer, M. *et al.* (2011) Land area change in coastal Louisiana from 1932 to 2010. US Geological Survey Scientific Investigations Map 3164, scale 1: 265,000. Reston, Virginia, USA, 12 pp.

Craft, C., Clough, J., Ehman, J., Joye, S., Park, R. *et al.* (2009) Forecasting the effects of accelerated sea-level rise on tidal marsh ecosystem services. *Frontiers in Ecology and the Environment* 7, 73–78.

Crosby, S.C., Sax, D.F., Palmer, M.E., Booth, H.S., Deegan, L.A. *et al.* (2016) Salt marsh persistence is threatened by predicted sea-level rise. *Estuarine, Coastal and Shelf Science* 181, 93–99.

Dainty, J., Vinters, H. and Tyree, M.T. (1974) A study of transcellular osmosis and the kinetics of swelling and shrinking in cells of *Chara coralline*. In: Zimmermann, U. and Dainty, J. (eds), *Membrane Transport in Plants*. Springer, Berlin, pp. 59–63.

Dassanayake, M. and Larkin, J.C. (2017) Making plants break a sweat: The structure, function, and evolution of plant salt glands. *Frontiers in Plant Science* 8: 406. doi:10.3389/fpls.2017.00406.

Davidson, N.C. (2014) How much wetland has the world lost? Long-term and recent trends in global wetland area. *Marine and Freshwater Research* 65, 934–941.

Di Bella, C.E., Jacobo, E., Golluscio, R.A. and Rodríguez, A.M. (2014) Effect of cattle grazing on soil salinity and vegetation composition along an elevation gradient in a temperate coastal salt marsh of Samborombón Bay (Argentina). *Wetlands Ecology and Management* 22, 1–13.

Donnelly, J.P. and Bertness, M.D. (2001) Rapid shoreward encroachment of salt marsh cordgrass in response to accelerated sea-level rise. *Proceedings of the National Academy of Sciences* 98, 14218–14223.

Flowers, T.J. and Colmer, T.D. (2015) Plant salt tolerance: Adaptations in halophytes. *Annals of Botany* 115, 327–331.

Flowers, T.J., Troke, P.F. and Yeo, A.R. (1977) The mechanism of salt tolerance in halophytes. *Annual Review of Plant Physiology* 28, 89–121.

Flowers, T.J., Galal, H.K. and Bromham, L. (2010) Evolution of halophytes: Multiple origins of salt tolerance in land plants. *Functional Plant Biology* 37, 604–612.

González-Alcaraz, M.N., Jiménez-Cárceles, F.J., Álvarez, Y. and Álvarez-Rogel, J. (2014) Gradients of soil salinity and moisture, and plant distribution, in a Mediterranean semiarid saline watershed: a model of soil–plant relationships for contributing to the management. *Catena* 115, 150–158.

Graves, J.S. and Gutknecht, J. (1976) Ion transport studies and determination of the cell wall elastic modulus in the marine alga *Halicystis parvula*. *Journal of General Physiology* 67, 579–597.

Griffiths, M.E. (2006) Salt spray and edaphic factors maintain dwarf stature and community composition in coastal sandplain heathlands. *Plant Ecology* 186, 69–86.

Griffiths, M.E. and Orians, C.M. (2003) Salt spray differentially affects water status, necrosis, and growth in coastal sandplain heathland species. *American Journal of Botany* 90, 1188–1196.

Grigore, M.N., Ivanescu, L. and Toma, C. (2014) *Halophytes: An Integrative Anatomical Study*. Springer, Cham, Switzerland, pp. 5–25.

Gulzar, S., Khan, M.A. and Ungar, I.A. (2003) Salt tolerance of a coastal salt marsh grass. *Communications in Soil Science and Plant Analysis* 34, 2595–2605.

Gustafsson, M.E. and Franzén, L.G. (1996) Dry deposition and concentration of marine aerosols in a coastal area, SW Sweden. *Atmospheric Environment* 30, 977–989.

Hartig, E.K., Gornitz, V., Kolker, A., Mushacke, F. and Fallon, D. (2002) Anthropogenic and climate-change impacts on salt marshes of Jamaica Bay, New York City. *Wetlands* 22, 71–89.

Hughes, A.L., Wilson, A.M. and Morris, J.T. (2012) Hydrologic variability in a salt marsh: Assessing the links between drought and acute marsh dieback. *Estuarine, Coastal and Shelf Science* 111, 95–106.

Hwang, Y.H. and Morris, J.T. (1994) Whole-plant gas exchange responses of *Spartina alterniflora* (Poaceae) to a range of constant and transient salinities. *American Journal of Botany* 81, 659–665.

IPCC (Intergovernmental Panel on Climate Change) (2013) Climate Change 2013: The Physical Science Basis. Contribution of Working Group I to the Fifth Assessment Report of the IPCC. Stocker, T.F., Qin, D., Plattner, G.-K., Tignor, M., Allen, S.K. *et al.* (eds). Cambridge University Press, Cambridge, UK, pp. 1535.

Iversen, J. (1936) *Biologische Pflanzentypen als Hilfsmittel in der Vegetationsforschung*. Ein Beitrag zur Ökologischen Characteriserung und Anordnung der Pflanzengesellschaften. *Meddelelser fra Skalling-Laboriet* 4, 1–224.

Kearney, M.S., Rogers, A.S., Townshend, J.R., Rizzo, E., Stutzer, D. *et al.* (2002) Landsat imagery shows decline of coastal marshes in Chesapeake and Delaware Bays. *EOS, Transactions American Geophysical Union* 83, 173–178.

Kearney, T.H. (1904) Are plants of sea beaches and dunes true halophytes? *Botanical Gazette* 37, 424–436.

Khan, M. and Ungar, I. (1997) Effects of thermoperiod on recovery of seed germination of halophytes from saline conditions. *American Journal of Botany* 84, 279.

Khan, M.A., Gul, B. and Weber, D.J. (2000) Germination responses of *Salicornia rubra* to temperature and salinity. *Journal of Arid Environments* 45, 207–214.

Kirst, G.O. (1989) Salinity tolerance of eukaryotic marine algae. *Annual Review of Plant Physiology and Molecular Biology* 40, 21–53.

Kirwan, M.L. and Megonigal, J.P. (2013) Tidal wetland stability in the face of human impacts and sea–level rise. *Nature* 504, 53. doi: 10.1038/nature12856.

Kirwan, M.L., Temmerman, S., Skeehan, E.E., Guntenspergen, G.R. and Fagherazzi, S. (2016) Overestimation of marsh vulnerability to sea level rise. *Nature Climate Change* 6, 253–260. doi:10.1038/nclimate2909.

Kobayashi, H. (2008) Ion secretion via salt glands in Poaceae. *Japanese Journal of Plant Science* 2, 1–8.

Kramer, P.J. and Boyer, J.S. (1995) *Water Relations of Plants and Soils*. Academic Press, San Diego, CA, USA.

Kuhn, N.L. and Zedler, J.B. (1997). Differential effects of salinity and soil saturation on native and exotic plants of a coastal salt marsh. *Estuaries* 20, 391–403.

Larcher, W. (2003) *Physiological Plant Ecology: Ecophysiology and Stress Physiology of Functional Groups*, 4th edition. Springer-Verlag, Berlin, pp. 514.

Mahajan, S. and Tuteja, N. (2005) Cold, salinity and drought stresses: An overview. *Archives of Biochemistry and Biophysics* 444, 139–158.

Mahall, B.E. and Park, R.B. (1976) The ecotone between *Spartina foliosa* Trin. and *Salicornia virginica* L. in salt marshes of Northern San Francisco Bay: II. soil water and salinity. *Journal of Ecology* 64, 793–809.

Marcum, K.B. and Murdoch, C.L. (1992) Salt tolerance of the coastal salt marsh grass, *Sporobolus virginicus* (L.) Kunth. *New Phytologist* 120, 281–288.

Marschner, H. (1995) *Mineral Nutrition of Higher Plants*. 2nd edn. Academic Press, San Diego, CA, USA.

Mitsch, W.J. and Gosselink, J.G. (2015) *Wetlands*. 5th edn. Wiley, NJ, USA.

Morris, J.T., Sundareshwar, P.V., Nietch, C.T., Kjerfve, B. and Cahoon, D. R. (2002) Responses of coastal wetlands to rising sea level. *Ecology* 83, 2869–2877.

Munns, R. (2002) Comparative physiology of salt and water stress. *Plant, Cell and Environment* 25, 239–250.

Murray, N.J., Clemens, R.S., Phinn, S.R., Possingham, H.P. and Fuller, R.A. (2014) Tracking the rapid loss of tidal wetlands in the Yellow Sea. *Frontiers in Ecology and the Environment* 12, 267–272.

Osland, M.J., Enwright, N., Day, R.H. and Doyle, T.W. (2013) Winter climate change and coastal wetland foundation species: Salt marshes vs. mangrove forests in the southeastern United States. *Global Change Biology* 19, 1482–1494.

Osland, M.J., Enwright, N.M., Day, R.H., Gabler, C.A., Stagg, C.L., *et al.* (2016) Beyond just sea-level rise: considering macroclimatic drivers within coastal wetland vulnerability assessments to climate change. *Global Change Biology* 22, 1–11.

Parida, A.K. and Das, A.B. (2005) Salt tolerance and salinity effects on plants: a review. *Ecotoxicology and Environmental Safety* 60, 324–349.

Partridge, T.R. and Wilson, J.B. (1987) Salt tolerance of salt marsh plants of Otago, New Zealand. *New Zealand Journal of Botany* 25, 559–566.

Phleger, C.F. (1971) Effect of salinity on growth of a salt marsh grass. *Ecology* 52, 908–911.

Raposa, K.B., Weber, R.L., Ekberg, M.C. and Ferguson, W. (2017) Vegetation dynamics in Rhode Island salt marshes during a period of accelerating sea level rise and extreme sea level events. *Estuaries and Coasts* 40, 640–650.

Redfield, A.C. (1972) Development of a New England salt marsh. *Ecological Monographs* 42, 201–237.

Reed, J.F. (1947) The relation of the *Spartinetum glabrae* near Beaufort, North Carolina, to certain edaphic factors. *American Midland Naturalist* 38, 605–614.

Richards, C.L., White, S.N., McGuire, M.A., Franks, S.J., Donovan, L.A., *et al.* (2010) Plasticity, not adaptation to salt level, explains variation along a salinity gradient in a salt marsh perennial. *Estuaries and Coasts* 33, 840–852.

Rogel, J.A., Ariza, F.A. and Silla, R.O. (2000) Soil salinity and moisture gradients and plant zonation in Mediterranean salt marshes of Southeast Spain. *Wetlands* 20, 357–372.

Rozema, J., Van Manen, Y., Vugts, H.F. and Leusink, A. (1983) Airborne and soilborne salinity and the distribution of coastal and inland species of the genus *Elytrigia*. *Plant Biology* 32, 447–456.

Rozema, J., Bijwaard, P., Prast, G. and Broekman, R. (1985) Ecophysiological adaptations of coastal halophytes from foredunes and salt marshes. *Plant Ecology* 62, 499–521.

Saintilan, N., Wilson, N.C., Rogers, K., Rajkaran, A. and Krauss, K.W. (2014) Mangrove expansion and salt marsh decline at mangrove poleward limits. *Global Change Biology* 20, 147–157.

Silvestri, S., Defina, A. and Marani, M. (2005) Tidal regime, salinity and salt marsh plant zonation. *Estuarine, Coastal and Shelf Science* 62, 119–130.

Spencer, T., Schuerch, M., Nicholls, R.J., Hinkel, J., Lincke, D., *et al.* (2016) Global coastal wetland change under sea-level rise and related stresses: The DIVA Wetland Change Model. *Global and Planetary Change* 139, 15–30.

Srivastava, D.S. and Jefferies, R.L. (1995) Mosaics of vegetation and soil salinity: A consequence of goose foraging in an arctic salt marsh. *Canadian Journal of Botany* 73, 75–83.

Sykes, M.T. and Wilson, J.B. (1988) An experimental investigation into the response of some New Zealand sand dune species to salt spray. *Annals of Botany* 62, 159–166.

Torio, D.D. and Chmura, G.L. (2013) Assessing coastal squeeze of tidal wetlands. *Journal of Coastal Research* 29, 1049–1061.

Touchette, B.W. (2006) Salt tolerance in a *Juncus roemerianus* brackish marsh: spatial variations in plant water relations. *Journal of Experimental Marine Biology and Ecology* 337, 1–12.

Touchette, B.W. (2007) Seagrass-salinity interactions: physiological mechanisms used by submersed marine angiosperms for a life at sea. *Journal of Experimental Marine Biology and Ecology* 350, 194–215.

Touchette, B.W., Rhodes, K.L., Smith, G.A. and Poole, M. (2009a) Salt spray induces osmotic adjustment and tissue rigidity in smooth cordgrass, *Spartina alterniflora* (Loisel.). *Estuaries and Coasts* 32, 917–925.

Touchette, B.W., Smith, G.A., Rhodes, K.L. and Poole, M. (2009b) Tolerance and avoidance: two contrasting physiological responses to salt stress in mature marsh halophytes *Juncus roemerianus* Scheele and *Spartina alterniflora* Loisel. *Journal of Experimental Marine Biology and Ecology* 380, 106–112.

Touchette, B.W., Poole, M.R., McCullough, M.D., Smith, G.A. and Adams, E.C. (2010) Coastal plant growth and CO_2 enrichment: can the productivity of black needle rush keep pace with sea level rise? *Shifting Shorelines: Adapting to the Future*, Proceedings of the 21st International Meeting. The Coastal Society, Williamsburg, Vancouver, USA, pp. 4.

Touchette, B.W., Adams, E.C. and Laimbeer, P. (2012) Age-specific responses to elevated salinity in the coastal marsh plant black needlerush (*Juncus roemerianus* Scheele) as determined through polyphasic chlorophyll *a* fluorescence transients (OJIP). *Marine Biology* 159, 2137–2147.

Touchette, B.W., Marcus, S.E. and Adams, E.C. (2014) Bulk elastic moduli and solute potentials in leaves of freshwater, coastal and marine hydrophytes. Are marine plants more rigid? *AoB Plants* 6. doi:10.1093/aobpla/plu014.

Tyerman, S.D. (1982) Water relations of seagrasses: stationary volumetric elastic modulus and osmotic pressure of leaf cells of *Halophila ovalis*, *Zostera capricorni*, and *Posidonia australis*. *Plant Physiology* 69, 957–965.

Vasquez, E.A., Glenn, E.P., Guntenspergen, G.R., Brown, J.J. and Nelson, S.G. (2006) Salt tolerance and osmotic adjustment of *Spartina alterniflora* (Poaceae) and the invasive M haplotype of *Phragmites australis* (Poaceae) along a salinity gradient. *American Journal of Botany* 93, 1784–1790.

Warren, R.S. and Niering, W.A. (1993) Vegetation change on a northeast tidal marsh: interaction of sea-level rise and marsh accretion. *Ecology* 74, 96–103.

Weston, N.B. (2014) Declining sediments and rising seas: an unfortunate convergence for tidal wetlands. *Estuaries and Coasts* 37, 1–23.

Wiegert, R.G. (1979) Ecological processes characteristic of coastal *Spartina* marshes of the Southeastern USA. In: Jefferies, R.L. and Davy, A.J. (eds) *Ecological Processes in Coastal Environments*. Blackwell Scientific Publications, Oxford, UK, pp. 467–490.

Wilson, J.B. and Sykes, M.T. (1999) Is zonation on coastal sand dunes determined primarily by sand burial or by salt spray? A test in New Zealand dunes. *Ecology Letters* 2, 233–236.

Yamori, W., Hikosaka, K. and Way, D.A. (2014) Temperature response of photosynthesis in C_3, C_4, and CAM plants: Temperature acclimation and temperature adaptation. *Photosynthesis Research* 119, 101–117.

Yuan, H.J., Ma, Q., Wu, G.Q., Wang, P., Hu, J. *et al.* (2015) ZxNHX controls Na^+ and K^+ homeostasis at the whole-plant level in *Zygophyllum xanthoxylum* through feedback regulation of the expression of genes involved in their transport. *Annals of Botany* 115, 495–507.

Yuan, F., Leng, B. and Wang, B. (2016) Progress in studying salt secretion from the salt glands in recretohalophytes: How do plants secrete salt? *Frontiers in Plant Science* 7, 977. doi:10.3389/fpls.2016.00977.

Zörb, C., Sümer, A., Sungur, A., Flowers, T.J. and Özcan, H. (2013) Ranking of 11 coastal halophytes from salt marshes in northwest Turkey according to their salt tolerance. *Turkish Journal of Botany* 37, 1125–1133.

Section II

Mechanisms of Adaptation

9 Ion Accumulation Pattern of Halophytes

Doongar R. Chaudhary*
Central Salt and Marine Chemicals Research Institute, Bhavnagar, India

Abstract
Halophytes grow near the seashore, and in intertidal zones, coastal salt marshes, inland salt lakes and saline deserts under high salinity. They have socio-ecological values through their primary production, nutrient cycling, provision of wildlife habitats and stabilization of shorelines. Halophytes have attracted special attention from scientists because of their remarkable ability to tolerate higher salinity. This chapter summarizes current knowledge about ion absorption and accumulation patterns in the halophytes. These plants have different mechanisms to withstand salinity, such as succulence, salt exclusion, compartmentalization, compatible solutes and hair bladders. Halophytes that maintain higher K^+/Na^+ and Ca^{+2}/Na^+ ratios in their tissues are more tolerant of salinity. The electrical conductivity (salt concentration) of soils correlates well with Na^+ concentrations in halophytes, and elevated soil salinity reduces halophyte species diversity.

Keywords: Halophytes; Ion homeostasis; Osmotic stress; Soil salinity

9.1 Introduction

Soil degradation is an important global issue for the 21st century because of its adverse impact on food grain production for the world's increasing population. Among soil degradation processes, soil salinization is one of the most severe problems worldwide, challenging agricultural production in arid and semi-arid regions. Hence, it necessary to increase the productivity of typical soils and less productive soils, including salinized soil. To sustain grain production and address its scarcity, saline soils have to be rehabilitated and managed judiciously for food grain to meet the demands of the continually growing population. In conditions of higher soil salinity, the growth and development of agricultural crop plants is hampered and unproductive. This has led to the demand for salt-tolerant crop plants that can grow in saline conditions (saline soil/water). Worldwide, efforts are being made to improve plant performance in saline soils through the development of salt-tolerant varieties of conventional crop plants through natural genetic variation, and of transgenic plants through the introduction of novel genes, but limited success has been achieved because of the multigenic nature of salt-tolerance mechanisms (Brown *et al.*, 2014). Another approach may be to domesticate halophytes that have their own inherent genetic makeup to tolerate high salinity, and which are also of economic importance (Glenn *et al.*, 1999).

* Corresponding author e-mail: drchaudhary@csmcri.res.in

There are numerous definitions of halophytes which reflect the scientific background of researchers (Grigore *et al.*, 2014). Halophytes are capable of growing in 0.5% sodium chloride concentration (Chapman, 1942). They are defined as plants that grow and complete their life cycle in high salt concentration (Waisel, 1972). Breckle (1995) defined halophytes as those plants that grow and complete their entire life cycle in saline habitats. Further, they have adaptations at tissue and cellular levels, and at subcellular and biochemical levels, for coping with salinity (Breckle, 1995).They have also been defined as plants that survive to reproduce in environments where the salt concentration is around 200 mM NaCl or more (Flowers *et al.*, 1986; Flowers and Colmer, 2008). A more holistic definition has also been given:

> salt-tolerant plants (halophytes, including salt marsh and mangrove plants) are highly evolved and specialized organisms with well-adapted morphological and physiological characteristics allowing them to survive in the soils possessing high salt concentrations
>
> (Khan and Duke, 2001)

Plants growing under field conditions are subjected to a variety of environmental stress conditions such as salinity, heat, drought, nutrient limitation, toxicity and frost. Plant species have distinctly different adaptations to cope with these stress conditions. Plants growing in saline environmental conditions have various mechanisms to handle salinity. Halophytes have salinity-tolerance mechanisms that include combinations of: (1) salt exclusion; (2) excretion; (3) succulence; (4) transport and compartmentalization; and (5) compatible solutes (Breckle, 1995; Popp, 1995). Sometimes these effects are distinct from each other; sometimes they overlap. The response of plants to soil salinity depends on the ionic concentration and composition of soil solution, as well as on the genotype of the plant. Responses of the plant to salinity stress have been studied using anatomical, physiological and molecular approaches. The ionic physiology of halophytes is totally different from that of glycophytes: halophytes have quantitatively higher Na^+ (sodium) absorption, uptake and transport to maintain water potential and the turgor pressure required for water absorption (Flowers *et al.*, 1986; Flowers and Colmer, 2008). However, plants tolerant to salinity must be capable of osmotic adjustment, which depends on absorption, uptake and accumulation of ions and/or synthesis of organic solutes (Breckle, 1995; Popp, 1995). Generally, in halophytes, the quantitative abundance of ions follows the order $Na^+ > K^+ > Mg^{2+} > Ca^{2+}$ and $Cl^- > SO_4^{2-}$ (Krüger and Peinemann, 1996). In fact, different ions are accumulated by halophytes in different quantities; this is due to either the rate of entry into the plant or differences in their utilization pattern. However, closely related halophyte species or closely related ecotypes vary greatly in their ion uptake capability.

9.2 Ion Absorption and Accumulation

9.2.1 Sodium absorption and accumulation

Soil salinity is caused by a variety of salts in combinations of different concentrations, such as NaCl, Na_2SO_4, $NaHCO_3$, Na_2CO_3, $MgSO_4$, $CaSO_4$, $CaCl_2$, $MgCl_2$ and KCl; but NaCl is the dominant salt. The most common effect of soil salinity is inhibition in growth and development of plants, and this correlates well with internal Na^+ concentrations. Halophytes provide excellent illustrations of salinity tolerance, and have the capability to withstand and grow at high salt concentrations (Figs 9.1 and 9.2). They have developed different mechanisms to adapt to salinity, including: (1) Na^+ exclusion from roots; (2) Na^+ accumulation in shoots; (3) shedding of specialized leaves (Na^+ secretion from the leaf); (4) compartmentation of salts in vacuoles; and (5) salt excretion via salt glands (Zhang *et al.*, 2010; Volkov, 2015). The most important approach to control Na^+ uptake is reduction of Na^+ influx into the roots (Zhang *et al.*, 2010). Dicotyledonous halophytes grow optimally at salt concentrations of 100–250 mM, and generally accumulate salt in shoot tissues (Table 9.1); the salt optimum for monocotyledonous species is lower (Flowers *et al.*, 1986; Flowers and Colmer, 2008). Few halophyte species (e.g. *Suaeda maritima*) grow without significant yield reduction under irrigation with seawater (Greenway and Munns, 1980). The movement of ions across the plasma membrane is governed by two components: the chemical and electrical potentials created by the ion activity and charge, respectively. Higher concentrations of Na^+ in the soil solution will establish a large Na^+ electrochemical potential gradient, which favours the passive transport of Na^+ from soil solution to the cytosol (Blumwald *et al.*, 2000). Under an external saline environment, chemical potential will be the main driving force and the ratio of Na^+ concentration across the plasma membrane might well exceed a five-fold increase (Amtmann and Sanders, 1998). Transport of Na^+ in plants occurs passively via ion-selective proteinaceous pores of ion channels. Another pathway is via proteinaceous transporters (transporting one ion or co-transporting/antiporting several charged ions in one or opposite directions) with slower transport rates (Volkov, 2015). Recently, several classes of Na^+ transporters that play crucial roles in Na^+ homeostasis during salinity stress have been identified (Yamaguchi *et al.*, 2013). Considerable attention has been focused on NHX-, SOS1- and Class I HKT-type transporters, which have the ability to transport Na^+ across cellular membranes and are involved in Na^+ sequestration (vacuole), Na^+ extrusion and Na^+ circulation in plants in saline environments (Yamaguchi *et al.*, 2013).

Halophytes usually exhibit higher preference for Na^+ over K^+, and rates of Na^+ absorption are also high (Waisel, 1972). Sodium tends to accumulate in higher concentrations in leaves and shoots, while its content in roots is low (Table 9.1). Na^+ is not uniformly distributed in the leaves, but tends to accumulate in higher

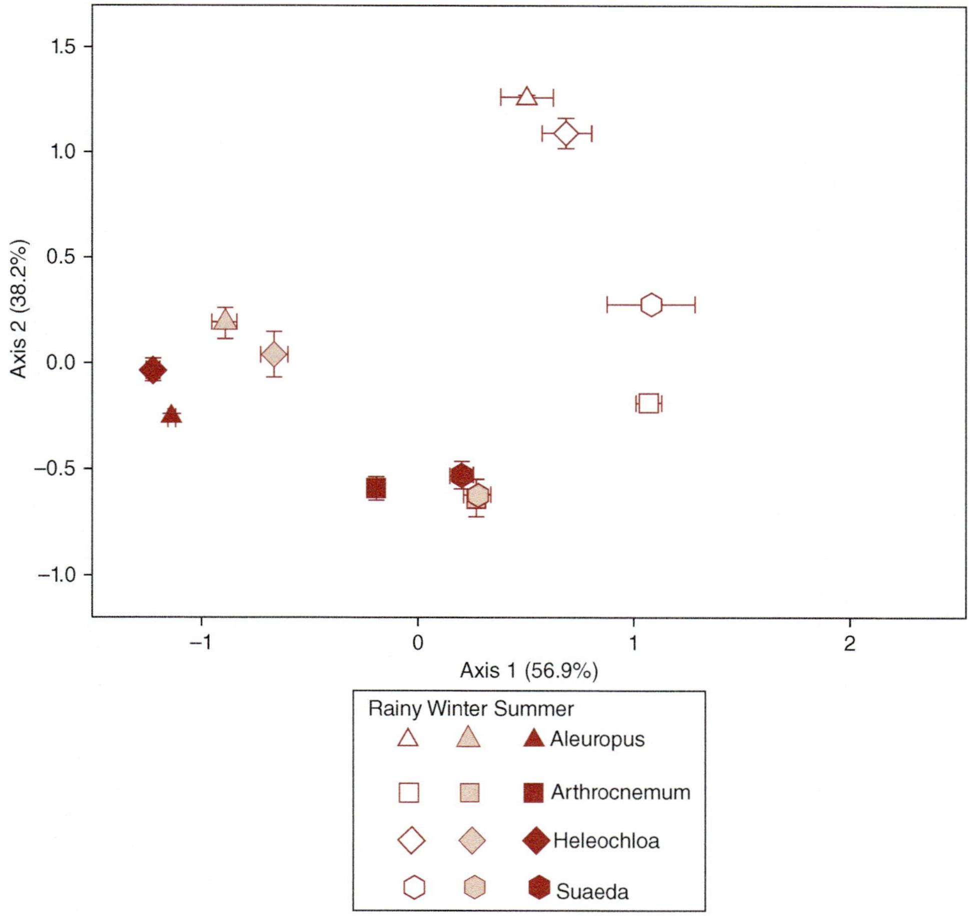

Fig. 9.1. Non-metric multidimensional scaling (NMS) representation of halophytes based on seasonal change in nutrient contents (unpublished data). [Aeluropus: *Aeluropus lagopoides*; Arthrocnemum: *Arthrocnemum indicum*; Heleochloa: *Heleochloa setulosa*; and Suaeda: *Suaeda nudiflora*].

Sesuvium portulacastrum *Suaeda* sp.

Population of *Salicornia brachiata* (young and mature stage)

Salicornia brachiata (young and mature stage)

Fig. 9.2. (a) and (b) Halophytes growing in the coastal intertidal soils of Gujarat, India.

concentrations in the epidermis, water storage tissues and phloem (in contrast to xylem) of *Bassia indica* (Pongrac *et al.*, 2013). At a cellular level, Na^+ was localized mainly in the cell symplast (probably in vacuoles) (Pongrac *et al.*, 2013). In the halophyte *Atriplex prostrata*, Na^+ was accumulated in the midrib parenchyma tissue surrounding the main leaf vascular bundle and in the mesophyll (Pongrac *et al.*, 2013). Therefore, both halophytes have different mechanisms for Na^+ sequestration. In *Spartina maritima*, Na^+ was localized in the epidermis and in parenchyma

Haloxylon salicornicum

Arthrocnemum indicum

Aeluropus lagopoides

Atriplex griffithii

Fig. 9.2. Continued.

tissues of the leaf (Pongrac *et al.*, 2013). Shoots of halophytes contain higher amounts of Na^+ compared to roots, because of the exclusive presence of the salt adaptation mechanisms in their shoots (Table 9.1); that is, they compartmentalize, forming salt crystals in vacuoles, excreting salt by salt glands or shedding salt trichomes (Breckle, 1990; Matinzadeh *et al.*, 2013). It has been shown that Na^+ contents in halophytes were about tenfold higher in dicotyledonous than in monocotyledonous species (Gil *et al.*, 2014). Generally, the salt tolerance in monocot halophytes is normally based on the exclusion of Na^+ and Cl^- ions for maintenance of lower cellular Na^+/K^+ ratios than in dicots (Rozema, 1991; Gil *et al.*, 2014). On the other hand, in dicot halophytes, salt-tolerance mechanisms appear to be primarily based on the efficient transportation of Na^+ and Cl^- ions and storage in the vacuoles (Rozema, 1991; Gil *et al.*, 2014). Monocot and dicot halophytes have shown different nutrient patterns during different seasons (Fig. 9.1). The Na^+ concentration depends on the halophyte species: there were minor seasonal changes in *Sarcocornia fruticosa* and *Inula crithmoides*; but in *Plantago crassifolia*, *Juncus maritimus* and *J. acutus* the seasonal changes correlated well with changes in electrical conductivity (Gil *et al.*, 2014). In an another study, Na^+ contents in the shoots of *S. persica* subsp. *rudshurensis* and *Climacoptera turcomanica* did not differ significantly during the growing season (Matinzadeh *et al.*, 2013). In other halophytes, the Na^+ concentration in shoots increased up to a certain period of growth and declined in a later period. This type of pattern was observed in *Halimocnemis pilifera*; this halophyte shed leaves which were replaced by smaller floral leaves during flowering (Matinzadeh *et al.*, 2013). This is a kind of self-regulatory mechanism by halophytes to cope with toxic concentrations of sodium during the reproduction phase of development. The Na^+ concentrations were

Table 9.1. Ionic concentrations (mmol kg^{-1}) in different halophyte species grown at high salinities (more than 200 mM NaCl range).

Halophyte	Plant part	Na^+	K^+	Ca^{2+}	Mg^{2+}	Cl^-	Reference
Haloxylon recurvum	Shoot	5285	27	60	60	4880	Khan *et al.* (2000b)
	Root	1362	23	28	13	866	
Atriplex griffithii	Leaf	6151	61	105	169	6590	Khan *et al.* (2000a)
	Stem	4114	12	53	102	1022	
	Root	2942	12	11	127	212	
Kochia prostrata	Shoot	3360	312	82	115	665	Karimi *et al.* (2005)
	Root	1208	74	39	21	296	
Suaeda salsa	Shoot	19315	320	12	56	5798	Liu *et al.* (2006)
	Root	2740	797	14	89	3192	
Limonium bicolor	Shoot	2170	355	12	78	2703	Liu *et al.* (2006)
	Root	1504	103	17	87	3334	
Sporobolus virginicus	Leaves	1401	404	126	150		Bell and O'Leary (2003)
	Root	704	95	51	44		
Sesuvium portulacastrum	Leaf	3291	378	1115			Venkatesalu *et al.* (1994)
	Stem	3035	286	272			
	Root	1943	120	89			
Armeria maritima	Leaves	1018	533			201	Kohl (1997)
Halopyrum mucronatum	Shoot	1685	340	12	46	1865	Khan *et al.* (1999)
	Root	786	400	20	50	506	
Atriplex barclayana	Leaves	5880	820	650		3670	Nerd and Pasternak (1992)
	Stem	2430	1100	630		2240	
Crithmum maritimum	Shoot	2993	224	756	46.2	2708	Amor *et al.* (2005)
	Root	1354	928	469	83	958	
Tecticornia indica	Shoot	7920	2000	300	90		Rabhi *et al.* (2010)
	Root	3100	2000	390	80		
Limoniastrum guyonianum	Leaves	6739	241	1099	609	2793	Zouhaier *et al.* (2015)
	Shoot	5304	317	460	108	1326	
	Root	2956	192	404	35	1044	

decreased during the growing season in *A. verrucifera* (Matinzadeh *et al.*, 2013) because plants in the genus *Atriplex* reduce salt toxicity by accumulating it in bladder-like trichomes and shedding them (Breckle *et al.*, 1990).

9.2.2 Potassium absorption and accumulation

Potassium is an essential nutrient for plant growth and is involved in numerous physiological functions. Plants usually accumulate 2–10% K^+ (dry weight basis), and this is achieved by a sophisticated K^+ transport system (potassium channels, different K-permeable non-selective cation channels and various transporters) (Marschner, 1995; Shabala and Pottosin, 2014). Under conditions of soil salinity, Na^+ concentrations increase dramatically, and at the same time K^+ content decreases. The ratio of K^+/Na^+ in the cytosol is an important determinant for salinity tolerance (Flowers and Colmer, 2008; Hauser and Horie, 2010; Volkov, 2015). Halophytes maintain a higher K^+/Na^+ ratio under saline conditions compared to glycophytes. Halophytes are known to rely heavily on the use of inorganic ions (mainly Na^+ and Cl^-) to maintain cell turgor pressure under hyperosmotic saline conditions, which is accomplished by the sequestration of these cytotoxic ions in the vacuoles of cells (Flowers and Colmer, 2008; Shabala and Pottosin, 2014). Simultaneously, an increase in the osmotic potential of the cytosol and cytosolic K^+ retention are also essential for this process (Shabala and Pottosin, 2014). Otherwise, plants have to invest in production of various organic osmolytes (proline, glycine, betaine, etc.) and consequently reduce their biomass (Shabala and Pottosin, 2014). Under salinity stress, the root plasma membrane is significantly depolarized (from typically −115 to −120 mV to as low as −40 to −50 mV).

This makes channel-mediated K^+ uptake impossible. At the same time, such depolarization also opens outward-rectifying depolarization-activated GORK channels, prompting massive K^+ efflux (Shabala and Pottosin, 2014; Wang *et al.*, 2017). Halophytes are known to have intrinsically more negative membrane potential compared with glycophyte species, even when grown under control conditions (Teakle *et al.*, 2013; Shabala *et al.*, 2016a). The same is true for genotypic variability within the same species (e.g. barley; Chen *et al.*, 2007); tolerant genotypes maintain more negative membrane potential. Finally, very recently the difference in membrane potential was shown to explain higher sensitivity of the root apex to NaCl (Shabala *et al.*, 2016b). However, sometimes K^+ absorption in saline soil also depends on the N source. If this is NO_3^- then K^+ uptake may be enhanced; and under NH_4^+ uptake of K^+ may reduce because of negative interaction between K^+ and NH_4^+ (Grattan and Grieve, 1999; Fageria *et al.*, 2011).

Similar to Na^+, K^+ is also accumulated in higher concentration in the epidermis, water storage tissues and phloem of *Bassia indica* (Pongrac *et al.*, 2013). Unlike Na^+, K^+ also localizes in the mesophyll and bundle sheath cells, indicating significant exclusion of Na from the photosynthetically active tissue of *Bassia indica* (Pongrac *et al.*, 2013). Potassium was predominantly found in the vascular bundle and epidermis of the *A. prostrata* leaf (Pongrac *et al.*, 2013). Seasonal effects on K^+ content in *P. crassifolia* have been found: the lowest K^+ levels were measured in the summer, when the highest Cl^- and Na^+ concentrations were detected; conversely, the highest K^+ values were observed in spring and corresponded to the lowest content of toxic ions (Na^+ and Cl^-) in the plants. Monocotyledonous halophytes, however, showed the reverse trend (Gil *et al.*, 2014). An increasing trend of K^+ content in the shoots of *S. persica* subsp. *rudshurensis* and *Climacoptera turcomanica* has been observed during the growing season (Matinzadeh *et al.*, 2013).

9.2.3 Calcium absorption and accumulation

Calcium is an essential element in the growth and development of plants. It has the ability to form intercellular linkages and plays an important role in maintaining the integrity, structure and stabilization of membranes and cell walls; it is also used as a second messenger in many signal transduction pathways (Hanson, 1984; Bush, 1995; Marschner, 1995). Ca^{2+} is phloem immobile, so once deposited in the shoot it is not readily redistributed. Ca^{2+} deficiency, ion uptake and osmoregulation in plants are imbalanced under situations of soil salinity (Yasar, 2007). The Ca^{2+} demand of the plant is higher under increased salt concentrations, but uptake of Ca^{2+} is reduced under saline conditions because of ionic interactions, precipitation and increase in ionic strength (Fageria *et al.*, 2011). Ca^{2+} alleviates the adverse effects of salinity on many crop plants as well as in halophytes (Colmer *et al.*, 1996; Tobe *et al.*, 2002; Amor *et al.* 2010) by altering ion selectivity absorption (Zhu, 2002), and reducing membrane leakage (Nedjimi and Daoud, 2009), improving nutrition under salt stress. Furthermore, as a divalent cation, Ca^{2+} can block the channel pore of non-selective cation channels (NSCC) (Demidchik and Tester, 2002). NSCC is considered to be a major pathway of Na^+ entry into the root, so such blockage will be beneficial in reducing the Na^+ load. Ca^{2+} can also block K^+ efflux via outward-rectifying channels (Shabala *et al.*, 2006). Taken together, that enables a better K^+/Na^+ ratio in the cytosol. An adequate supply of Ca^{2+} in soil stimulates the K^+/Na^+ ratio selectively by shifting the uptake in favour of K^+ at the expense of Na^+ (Grattan and Grieve, 1999). Dicotyledonous halophytes have been found to have higher Ca^{2+} concentrations than monocotyledonous halophytes (Gil *et al.*, 2014). Some halophytes have shown seasonal variability in Ca^{2+} concentrations, which increased in summer (Gil *et al.*, 2014). Generally, in halophytes the roots contain more Ca^{2+} than the shoots, to induce specific expression of aquaporin with an increased salt concentration (Munns and Tester, 2008).

9.2.4 Magnesium absorption and accumulation

Magnesium ions are cofactors required for the activity of different enzymes (involved in respiration, photosynthesis, and synthesis of DNA and RNA) and Mg^{2+} is also a structural part of the chlorophyll molecule (Marschner, 1995). Calcium is strongly competitive with Mg^{2+} for the binding sites on the root plasma membrane and Mg^{2+} appears to have less affinity as it is a highly hydrated ion (Marschner, 1995). Sometimes higher concentrations of Mg^{2+} are also toxic to halophytes. For example, Mg^{2+} was more toxic than Na^+ for seed germination and radicle survival of the halophyte *Kalidium capsicum*, and higher concentrations of Ca^{2+} were needed to alleviate its toxicity (Tobe *et al.*, 2002). Mg^{2+} was found to accumulate mostly in the bundle sheath cells of *Bassia indica* (Pongrac *et al.*, 2013).

Some experimental results have suggested that Mg^{2+} may play a similar role to that of Ca^{2+}, and protect cells from the toxic effects of NaCl (Grigore *et al.*, 2012). Progressive increments of Ca^{2+} and Mg^{2+} with increasing salinity (NaCl concentration) protected *Juncus acutus* from salt stress (Boscaiu *et al.*, 2011). These results indicate that Mg^{2+} can suppress or mitigate the deleterious effects of salt stress in *Plantago crassifolia* (Grigore *et al.*, 2012). Similar to Ca^{2+}, Mg^{2+} was shown to block K^+-efflux channels enabling its better retention in roots (Shabala *et al.*, 2003). Increased levels of $MgCl_2$ in the growth medium significantly increased K^+ secretion. The increase in K^+ secretion was far greater than that of Mg^{2+} secretion by salt glands, and decreased K^+ concentrations in leaves of Rhodes grass (*Chloris gayana*), perhaps owing to ion leakage from damaged cells (Kobayashi and Masaoka, 2008). Similar to Ca^{2+}, dicotyledonous halophytes also had higher Mg^{2+} concentration than monocotyledonous halophytes (Gil *et al.*, 2014).

9.2.5 Chloride absorption and accumulation

Chloride is an essential micronutrient for the growth and development of plants. It regulates enzyme activities in the cytoplasm; is an essential co-factor in photosynthesis; acts as a counter anion to stabilize membrane potential; and is involved in turgor and pH regulation (Marschner, 1995). Cl^- is the predominant anion in saline soils and its higher concentration in cytoplasm is metabolically toxic to plants (Teakle and Tyerman, 2010). Chloride is absorbed by the plant in the form of Cl^- from the soil solution through two pathways: symplastic (cytoplasmic; cell to cell through plasmodesmata; selective) and apoplastic (extracellular; through cell walls and water-filled spaces; non-selective) pathways. The symplastic pathway, however, is considered to dominate (Pitman, 1982; White and Broadley 2001). Levels of Cl^- are considered to be tenfold higher in dicotyledonous halophytes than in monocotyledonous halophytes (Gil *et al.*, 2014). Halophytes in general, and succulent halophytes in particular, have higher accumulations of Cl^- than of any other anions. Partitioning of Cl^- between different cell types in both roots and shoots contributes to salt tolerance in some species (Teakle and Tyerman, 2010). Higher accumulation of Cl^- occurred in the vascular bundle tissues of *B. indica* (Pongrac *et al.*, 2013). At cellular level, Cl^- was localized mainly in the cell symplast (vacuoles). In *A. prostrata*, Cl^- was accumulated in the midrib parenchyma tissue surrounding the main leaf vascular bundle and in the mesophyll (Pongrac *et al.*, 2013). Most Cl^- was evenly distributed, except for in the mesophyll of *Spartina maritima* leaves (Pongrac *et al.*, 2013). There is evidence that Cl^- especially accumulates in the epidermis, and reduces Cl^- toxicity in mesophyll cells, which are more important for leaf photosynthesis (Teakle and Tyerman, 2010). Another form of Cl^- compartmentation is the accumulation of Cl^- in salt glands or bladders, which are highly specialized cell structures in the leaves of halophytes; and accumulate Cl^- (along with Na^+) in salt glands on the leaf surfaces to lower internal leaf ion concentrations in *Leptochloa fusca* (Jeschke *et al.*, 1995; Teakle and Tyerman, 2010).

Terrestrial plants are divided into two major groups, halophytes and glycophytes, based on their sensitivity to salinity (Bose *et al.*, 2017). Greenway and Munns (1980) divided the halophytes into three categories on the basis of response to Cl^-: (1) species whose growth is stimulated (e.g. *Sueada maritima*, *Atriplex nummularia*); (2) species whose growth is little affected by 200 mM Cl^- (e.g. *Atriplex hastata*, *Spartina* spp., *Beta vulgaris*); and (3) species whose growth is reduced substantially by 100 mM Cl^-. The critical tissue Cl^- concentration for toxicity is about 4–7 mg g^{-1} and 15–50 mg g^{-1} dry weight for sensitive and tolerant plants, respectively (White and Broadley, 2001). Plants can withstand Cl^- toxicity by restricting Cl^- transport to the shoot (Bose *et al.*, 2017). There are substantial variations in the capacity of plants to accumulate Cl^-.

Halophytes generate turgor by accumulating high Cl^- content in plant tissues (340–475 mM), whereas glycophytes have lower concentrations (7–70 mM) (White and Broadley, 2001). Different parts of the plant also contain different concentrations of Cl^-. All halophyte leaves have similar concentrations of Cl^- (Jeschke *et al.*, 1995). Floral tissues, fruits and seeds have lower Cl^- concentrations than shoot tissues (Xu *et al.*, 2000; White and Broadley, 2001). Because of excretion by salt-secreting glands in *Leptochloa fusca* leaves and phloem export, Cl^- concentrations did not increase strongly with leaf age; however, Cl^- secretion increased strongly from younger to mature leaves (Klagges *et al.*, 1993). Similar to Na^+, Cl^- concentration depends on halophyte species, and minor seasonal changes occurred in *Sarcocornia fruticosa* and *Inula crithmoides*. In the cases of *Plantago crassifolia*, *Juncus maritimus* and *J. acutus* the seasonal changes correlated well with change in electrical conductivity (Gil *et al.*, 2004).

9.2.6 Sulfate absorption and accumulation

Sulfur (S) is an essential plant nutrient and a component of the amino acids cysteine and methionine, making it essential for protein synthesis (Marschner, 1995). Absorption, transport and metabolic reactions of sulfur in plants occur predominantly as an SO_4^{2-} ion that is rapidly assimilated into organic compounds. Usually, S concentrations can be higher in halophytes compared with glycophytes (Mayland and Robbins, 1994). However, SO_4^{2-} being a divalent ion, rates of its uptake are low (Waisel, 1972). Comparative studies have demonstrated that SO_4^{2-}-based solutions have a considerably stronger inhibitory effect on the germination and growth of *Prosopis strombulifera* (a legume halophyte) than Cl^-- based solutions at iso-osmotic concentrations (Sosa *et al.*, 2005; Reginato *et al.*, 2014). The different response of halophytes to the salts depends on the anion associated with sodium. Sulfate salinity treatment applied to *Prosopis strombulifera* induced structural changes in cells and tissues, resulting in changes in growth patterns compared to control plants or to plants grown in high chloride salinity (Reinoso *et al.*, 2005). Similar to Cl^-, S also accumulates in higher concentrations in vascular bundle tissues of *Bassia indica* leaves (Pongrac *et al.*, 2013). In halophytes, higher sulfate values were observed in leaves than in stems and roots (Salama and Ali, 2003). Among different halophytes (*Zygophyllum album*, *Suaeda pruinosa*, *Arthrocnemum glaucum*, *Halocnemum strobilaceum* and *Limoniastrum monopetalum*), highest sulfate values were recorded in *Zygophyllum album* leaves (17.2%-18.1%) (Salama and Ali, 2003). Sulfate concentrations in the laminae and sheaths of *Leptochloa fusca* leaves rose concurrently with leaf age, and content increased continuously from the youngest to the oldest leaves (Klagges *et al.*, 1993).

9.3 Ionic Ratios in Halophytes

The capacity of plants to maintain a low Na^+/K^+ ratio is likely to be one of the key determinants of salt tolerance. Although the K^+/Na^+ ratio could be a contributing factor for salt tolerance, it does not correlate with salt tolerance, suggesting that this ratio is not a good indicator of salt resistance (He and Cramer, 1993). Plant species that maintain higher cell K^+/Na^+ and Ca^{2+}/Na^+ ratios for cell integrity are more tolerant of high-salinity conditions (Maathuis and Amtmann, 1999). Calcium concentration in tissues correlated well with salinity tolerance (He and Cramer, 1992). The Na^+/K^+ ratios in tissues of shoots of halophytic grasses were commonly about 1.0 and are much higher for dicots (near to 10) (Flowers *et al.*, 1986). The ratios of Na^+/K^+ and Cl^-/K^+ in the dicotyledonous halophytes were significantly higher than in monocotyledonous halophytes under salinity conditions (Rozema, 1991). The shoots of dicotyledonous halophytes contain higher amounts of water with low water use efficiency compared to monocotyledonous halophytes, and this is linked with succulence, the swelling of leaves and stems, and the elastic properties of cell walls of dicotyledons which maintain the turgor pressure (Waisel, 1972; Rozema, 1991). The internal molecular ratio between $[K^+]$ and $[K^+ + Na^+]$ was found to be low in the shoots of dicots and higher in the monocots because dicots have higher ability to accumulate Na^+ (Tipirdamaz *et al.*, 2006).

suggesting specific traits that allow the dicots to absorb, accumulate and compartmentalize Na^+ to provide inexpensive osmotic particles for osmotic adjustment. Generally, the ratios of Na^+/K^+ in the roots of halophytes were lower than those of the leaves (Zörb *et al.*, 2013).

9.4 Soil–Halophyte Relationship

Under natural conditions, halophytes accurately reflect the chemical composition of sediments and the sediment solution. A soil–plant relationship exists between the plant community and sediment conditions. The electrical conductivity of soil significantly correlates with the Na^+ content of halophytes (Rathore *et al.*, 2016). There was higher correlation between soil electrical conductivity and leaf tissue sodium concentrations than for chloride (Zörb *et al.*, 2013). It was also observed that increased salinity reduces halophyte species diversity (González-Alcaraz *et al.*, 2014). Halophyte distribution has been used for predicting soil salinity, which helps in introducing salt-adapted species for fodder or forage. Sediment salinity determines the halophyte community composition and species dominance (Krüger and Peinemann, 1996), because of relative concentrations of Ca^{2+}, Na^+, Mg^{2+} K^+, Cl^- and SO_4^{2-} in the soil solution. The sediment solutions are considered important factors for the development of halophytes under saline environments (Asri and Ghorbanli, 1997). pH and electrical conductivity also play similar roles (Rathore *et al.*, 2016, 2017). Highly salt-tolerant species are likely to compete for resource availability (Maestre *et al.*, 2009). Studies have shown that there is a close relationship between plants and soil; even in a homogeneous topography, there are multifaceted varieties of microhabitats, which are related to edaphic variations and consequently affect the distribution patterns of halophyte communities. It was observed that the flooding period, total salinity, minimum Ca^{2+}/Na^+ ratio and the mean sodium adsorption ratio were main factors explaining halophyte zonation in a semi-arid Mediterranean salt marsh (Rogel *et al.*, 2001). The chenopod halophyte *Arthrocmemum macrostachyum* seems to be adapted to the most extreme habitats, since it supports maximum edaphic salinity levels and the most pronounced seasonal variations (Rogel *et al.*, 2001). In a study carried out on the soil salinity gradient and halophyte relationship, it was observed that leaf K^+, Ca^{2+} and Mg^{2+} of *Sarcobatus vermiculatus* did not match substrate availability; further, it was also observed that the increasing soil Na^+ did not interfere with K^+, Ca^{2+} and Mg^{2+} (Donovan *et al.*, 1997). Multivariate analysis showed that halophytes were closely related to soil moisture and salinity in a Mediterranean semi-arid saline watershed (González-Alcaraz *et al.*, 2014). These soils were dominated by *Lygeum spartum*, *Suaeda vera* and *Limonium* spp. in the driest (moisture $<\sim 20\%$) and least saline (EC $<\sim 30$ dS m^{-1}) soils, while *Phragmites australis*, *Sarcocornia fruticosa* and *Arthrocnemum macrostachyum* dominated in the most saline and wettest soils (González-Alcaraz *et al.*, 2014). Similarly, in semi-arid Mediterranean areas (Spain), salinity (sulfate ions), moisture and gravel content also accounted for variation in halophyte populations (Cañadas *et al.*, 2010).

Excretory species like *Aeluropus lagopoides* and *Sporobolus ioclados*, and *Haloxylon* spp., dominated the lower salinities while tussock grass *Lasiurus scindicus* was distributed mainly in lower and moderate salinities in the Cholistan desert (Naz *et al.*, 2013). Carbonate and calcium carbonate equivalents of the topsoil were positively correlated with establishment and growth of *Limonium iranicum*, but salinity and alkalinity were negatively correlated in Sejzy region, east of Isfahan, Iran (Chigani *et al.*, 2012). At the study site, growth of *Tamarix passerinoides* was positively correlated with soil organic matter, bicarbonate, total Ca^{2+} and Mg^{2+} contents, but negatively correlated with Na^+ content (Chigani *et al.*, 2012). The key edaphic factors that determined the halophyte zonation pattern were pH, Na^+ and Cl^- at Lake Tuz, central Anatolia, Turkey (Tug *et al.*, 2012). However, on coastal sand dunes on the west coast of central Italy, it was observed that the distribution and variation in abundance of plant species was mainly controlled by the salinity and soil texture gradients (from sandy to silt–clay) (Landi and Angiolini, 2015). Further, Landi and Angiolini (2015) observed that *Limbarda crithmoides* increased with the amount of sand; *Elymus repens* and *Phragmites australis* increased with decreasing salinity; *E. repens* also increased with the amount of silt–clay; *Sarcocornia fruticosa* increased with salinity and silt–clay; *Halimione portulacoides* increased with salinity and was less affected by soil texture; and *E. pycnanthus* increased with salinity and decreased with the amount of silt–clay. Halophytes grow in differentiation according to their salinity tolerance in natural saline habitats. Those with particularly succulent stems and leaves, such as *Suaeda prostrata* and *Salicornia fragilis*, have been found in clusters in higher salt concentrations, whereas those with salt glands (*Halocnemum portulacoides*, *Limonium graecum* and *Artemisia santonicum*) were more common in the lower salt concentrations (Zörb *et al.*, 2013).

9.5 Conclusions

Areas of soil salinity are increasing globally every day because of the irrigation with saline water, poor management of irrigation water and seawater ingress. Generally, excess salt concentrations in soil inhibit plant growth; however, halophytes can survive under adverse soil saline environmental conditions. These plants have different mechanisms and strategies (succulence, salt exclusion, excretion, compartmentalization, compatible solutes and broad leaves with hair bladders) to withstand salinity. The ionic patterns and relationships of halophytes are controlled by fixed genetic processes known as physiotypes (plant ecophysiology and systematics). Halophytes that belong to the family Chenopodiaceae are highly tolerant to saline conditions, and have the potential for use in reclamation of saline soils, as they have the ability to remove large amounts of salt. Generally, the salt tolerance in monocot halophytes is based on the exclusion of Na^+ and Cl^- ions by maintaining lower cellular Na^+/K^+ ratios (about 1), whereas in dicot halophytes the salt-tolerance mechanisms are primarily based on the efficient transport and storage of Na^+ and Cl^- ion vacuoles with higher Na^+/K^+ ratios (near to 10). Calcium is known to alleviate the harmful effects of salt in many plants through changing ion selectivity during absorption, and reduction in membrane leakage. Similarly, Mg^{2+} also plays a role in protection from the toxic effects of salt. Halophytes that maintain higher K^+/Na^+ and Ca^{2+}/Na^+ ratios in tissues for cell integrity are more tolerant of high-salinity conditions. The electrical conductivity of soil mostly correlates with the Na^+ content of halophytes. Increased soil salinity reduces halophyte species diversity.

References

Amor, N.B., Hamed, K.B., Debez, A., Grignon, C. and Abdelly, C. (2005) Physiological and antioxidant responses of the perennial halophyte *Crithmum maritimum* to salinity. *Plant Science* 168, 889–899.

Amor, N.B., Megdiche, W., Jimenez, A., Sevilla, F. and Abdelly, C. (2010) The effect of calcium on the antioxidant systems in the halophyte *Cakile maritima* under salt stress. *Acta Physiologiae Plantarum* 32, 453–461.

Amtmann, A. and Sanders, D. (1998) Mechanisms of Na^+ uptake by plant cells. *Advances in Botanical Research* 29, 75–112.

Asri, Y. and Ghorbanli, M. (1997) The halophilous vegetation of the Orumieh lake salt marshes, NW Iran. *Plant Ecology* 132, 155–170.

Bell, H.L. and O'Leary, J.W. (2003) Effects of salinity on growth and cation accumulation of *Sporobolus virginicus* (Poaceae). *American Journal of Botany* 90, 1416–1424.

Blumwald, E., Aharon, G.S. and Apse, M.P. (2000) Sodium transport in plant cells. *Biochimica et Biophysica Acta* 1465, 140–151.

Boscaiu, M., Ballesteros, G., Naranjo, M.A., Vicente, O. and Boira, H. (2011) Responses to salt stress in *Juncus acutus* and *J. maritimus* during seed germination and vegetative plant growth. *Plant Biosystems* 145, 770–777.

Bose, J., Munns, R., Shabala, S., Gilliham, M., Pogson, B. *et al.* (2017) Chloroplast function and ion regulation in plants growing on saline soils: Lessons from halophytes. *Journal of Experimental Botany* 68, 3129–3143.

Breckle, S.W. (1990) Salinity tolerance of different halophyte types. *Plant and Soil* 148, 167–175.

Breckle, S.W. (1995) How do halophytes overcome salinity? In: Khan, M.A. and Ungar, I.A. (eds) *Biology of Salt Tolerant Plants*. Department of Botany, University of Karachi, Karachi, Pakistan, pp. 199–213.

Breckle, S.W., Freitas, H. and Reimann, C. (1990) Sampling *Atriplex* bladders: a comparison of methods. *Plant, Cell and Environment* 13, 871–873.

Brown, J.J., Glenn, E.P. and Smith, S.E. (2014) Feasibility of halophyte domestication for high-salinity agriculture. In: Khan, M.A., Böer, B. Öztürk, M., Abdessalaam, T.Z.A., Clüsener-Godt, M. and Gul, B. (eds) *Sabkha Ecosystems: Volume IV: Cash Crop Halophyte and Biodiversity Conservation*. Springer, Dordrecht, Netherlands, pp. 73–80.

Bush, D.S. (1995) Calcium regulation in plant cells and its role in signalling. *Annual Review of Plant Physiology and Plant Molecular Biology* 46, 95–122.

Cañadas, E.M., Jiménez, M.N., Valle, F., Fernández-Ondoño, E., Martín-Peinado, F. *et al.* (2010) Soil-vegetation relationships in semi-arid Mediterranean old fields (SE Spain): Implications for management. *Journal of Arid Environment* 74, 1525–1533.

Chapman, V.J. (1942) The new perspective in the halophytes. *The Quarterly Review of Biology* 17(4), 291–311.

Chen, Z., Pottosin I.I., Cuin, T.A., Fuglsang, A.T., Tester, M. *et al.* (2007) Root plasma membrane transporters controlling K^+/Na^+ homeostasis in salt-stressed barley. *Plant Physiology* 145, 1714–1725.

Chigani, H.K., Khajeddin, S.J. and Karimzadeh, H.R. (2012) Soil-vegetation relationships of three arid land plant species and their use in rehabilitating degraded sites. *Land Degradation & Development* 23, 92–101.

Colmer, T.D., Fan, T.W.M., Highashi, R.M. and Lauchli, A. (1996) Interactive effects of Ca^{2+} and NaCl salinity on the ionic relations and proline accumulation in the primary root tip of *Sorghum bicolor*. *Physiologia Plantarum* 97, 421–424.

Demidchik, V. and Tester, M. (2002) Sodium fluxes through nonselective cation channels in the plasma membrane of protoplasts from Arabidopsis roots. *Plant Physiology* 128, 379–387.

Donovan, L.A., Richards, J.H. and Schaber, E.J. (1997) Nutrient relations of the halophytic shrub, *Sarcobatus vermiculatus*, along a soil salinity gradient. *Plant and Soil* 190, 105–117.

Fageria, N.K., Gheyi, H.R. and Moreira, A. (2011) Nutrient bioavailability in salt affected soils. *Journal of Plant Nutrition* 34, 945–962.

Flowers, T.J. and Colmer, T.D. (2008) Salinity tolerance in halophytes. *New Phytologist* 179, 945–963.

Flowers, T.J., Hajibagheri, M.A. and Clipson, N.J.W. (1986) Halophytes. *The Quarterly Review of Biology* 61, 313–337.

Gil, R., Bautista, R., Boscaiu, M., Lidon, A., Wankhade, S., Sanchez, H., Llinares, J. and Vicente, O. (2014) Responses of five Mediterranean halophytes to seasonal changes in environmental conditions. *AoB PLANTS* 6, plu049. doi: 10.1093/aobpla/plu049.

Glenn, E.P., Brown, J.J. and Blumwald, E. (1999) Salt tolerance and crop potential of halophytes. *Critical Reviews in Plant Sciences* 18, 227–255.

González-Alcaraz, M.N., Jiménez-Cárceles, F.J., Álvarez, Y. and Álvarez-Rogel, J. (2014) Gradients of soil salinity and moisture, and plant distribution, in a Mediterranean semiarid saline watershed: A model of soil-plant relationships for contributing to the management. *Catena* 115, 150–158.

Grattan, S.R. and Grieve, C.M. (1999) Salinity-mineral nutrient relations in horticulture crops. *Scientia Horticulturae* 78, 127–157.

Greenway, H. and Munns, R. (1980) Mechanisms of salt tolerance in non-halophytes. *Annual Review of Plant Physiology* 31, 149–190.

Grigore, M., Ivanescu, L. and Toma, C. (2014) Halophyte definitions and classifications. In: Grigore, M., Ivanescu, L. and Toma, C. (eds) *Halophytes: An Integrative Anatomical Study*. Springer International Publishing, Cham, Switzerland, pp. 5–25.

Grigore, M.N., Boscaiu, M., Llinares, J. and Vicente, O. (2012) Mitigation of salt stress-induced inhibition of *Plantago crassifolia* reproductive development by supplemental calcium or magnesium. *Notulae Botanicae Horti Agrobotanici Cluj-Napoca* 40, 58–66.

Hanson, J.B. (1984) The function of calcium in plant nutrition. In: Tinker, P.B. and Läuchli, A. (eds) *Advances in Plant Nutrition*. Praeger Scientific, New York, pp. 149–208.

Hauser, F. and Horie, T. (2010) A conserved salt tolerance mechanism mediated by HKT transporters: a mechanism for sodium exclusion and maintenance of high K^+/Na^+ ratio in leaves during salinity stress. *Plant, Cell and Environment* 33, 552–565.

He, T. and Cramer, G.R. (1992) Growth and mineral nutrition of six rapid-cycling Brassica species in response to seawater salinity. *Plant and Soil* 139, 285–294.

He, T. and Cramer, G.R. (1993) Salt tolerance of rapid-cycling Brassica species in relation to potassium/sodium ratio and selectivity at the whole plant and callus levels. *Journal of Plant Nutrition* 16, 1263–1277.

Jeschke, W.D., Klagges, S., Hilpert, A., Bhatti, A.S. and Sarwar, G. (1995) Partitioning and flow of ions and nutrients in salt treated plants of *Leptochloa fusca* L. Kunth. I. Cations and chloride. *New Phytologist* 130, 23–35.

Karimi, G., Ghorbanli, M., Heidari, H., Nejad, R.A.K. and Assareh, M.H. (2005) The effects of NaCl on growth, water relations, osmolytes and ion content in *Kochia prostrata*. *Biologia Plantarum* 49, 301–304.

Khan, M.A. and Duke, N.C. (2001) Halophytes - A resource for the future. *Wetlands Ecology and Management* 6, 455–456.

Khan, M.A., Ungar, I.A. and Showalter, A.M. (1999) Effects of salinity on growth, ion content, and osmotic relations in *Halopyrum mucronatum* (L.) Stapf. *Journal of Plant Nutrition* 22, 191–204.

Khan, M.A., Ungar, I.A. and Showalter, A.M. (2000a) Effects of salinity on growth, water relations and ion accumulation of the subtropical perennial halophyte *Atriplex griffithii* var. *stocksii*. *Annals of Botany* 85, 225–232.

Khan, M.A., Ungar, I.A. and Showalter, A.M. (2000b) Effects of sodium chloride treatments on growth and ion accumulation of the halophyte *Haloxylon recurvum*. *Communications in Soil Science and Plant Nutrition* 31, 2763–2774.

Klagges, S.E., Bhatti, A.S., Sarwar, G., Hilpert, A. and Jeschke, W.D. (1993) Ion distribution in relation to leaf age in *Leptochloa fusca* (L.) (Kallar grass). II. Anions. *New Phytologist* 125, 521–528.

Kobayashi, H. and Masaoka, Y. (2008) Salt secretion in Rhodes grass (*Chloris gayana* Kunth) under conditions of excess magnesium. *Soil Science and Plant Nutrition* 54, 393–399.

Kohl, K.I. (1997) The effect of NaCl on growth, dry matter allocation and ion uptake in salt marsh and inland populations of *Armeria maritima*. *New Phytologist* 135, 213–225.

Krüger, H.R. and Peinemann, N. (1996) Coastal plain halophytes and their relation to soil ionic composition. *Vegetatio* 122, 143–150.

Landi, M. and Angiolini, C. (2015) Soil-plant relationships in mediterranean salt marshes across dune-cultivated land gradient. *Journal of Coastal Research* 31, 588–594.

Liu, X., Duan, D., Li, W., Tadano, T. and Khan, M.A. (2006) A comparative study on responses of growth and solute composition in halophytes *Suaeda salsa* and *Limonium bicolor* to salinity. In: Khan, M.A. and Weber, D.J. (eds) *Ecophysiology of High Salinity Tolerant Plants*. Springer, Dordrecht, Netherlands, pp. 135–143.

Maathuis, F.J.M. and Amtmann, A. (1999) K^+ nutrition and Na^+ toxicity: the basis for cellular K^+/Na^+ ratios. *Annals of Botany* 84, 123–133.

Maestre, F.T., Callaway, R.M., Valladares, F. and Lortie, C.J. (2009) Refining the stress-gradient hypothesis for competition and facilitation in plant communities. *Journal of Ecology* 97, 199–205.

Marschner, H. (1995) *Mineral Nutrition of Higher Plants*. Academic Press, London.

Matinzadeh, Z., Breckle, S.W., Mirmassoumi, M. and Akhani, H. (2013) Ionic relationships in some halophytic Iranian Chenopodiaceae and their rhizospheres. *Plant and Soil* 372, 523–539.

Mayland, H.F. and Robbins, C.W. (1994) Sulfate uptake by salinity-tolerant plant species. *Communications in Soil Science and Plant Analysis* 25, 2523–2541.

Munns, R. and Tester, M. (2008) Mechanisms of salinity tolerance. *Annual Review of Plant Biology* 59, 651–681.

Naz, N., Hameed, M., Nawaz, T., Ahmad, M.S.A. and Ashraf, M. (2013) Soil-plant relationships in the arid saline desert of Cholistan. *Arid Land Research and Management* 27, 140–152.

Nedjimi, B. and Daoud, Y. (2009) Effects of calcium chloride on growth, membrane permeability and root hydraulic conductivity in two *Atriplex* species grown at high (sodium chloride) salinity. *Journal of Plant Nutrition* 32, 1818–1830.

Nerd, A. and Pasternak, D. (1992) Growth, ion accumulation, and nitrogen fractioning in *Atriplex barclayana* grown at various salinities. *Journal of Range Management* 45, 164–166.

Pitman, M.G. (1982) Transport across plant roots. *Quarterly Reviews of Biophysics* 15, 481–554.

Pongrac, P., Vogel-Mikuš, K., Regvar, M., Kaligaric, M., Vavpetic, P., *et al.* (2013) On the distribution and evaluation of Na, Mg and Cl in leaves of selected halophytes. *Nuclear Instruments and Methods in Physics Research Section B: Beam Interactions with Materials and Atoms* 306, 144–149.

Popp, M. (1995) Salt resistance in herbaceous halophytes and mangroves. In: Behnke, H.D., Luttge, U., Esser, K., Kadereit, J.W. and Runge, M. (eds) *Progress in Botany*, 56. Springer-Verlag, Berlin, Germany, pp. 416–429.

Rabhi, M., Hajji, S., Karray-Bouraoui, N., Giuntini, D., Castagna, A. *et al.* (2010) Nutrient uptake and management under saline conditions in the xerohalophyte: *Tecticornia indica* (Willd.) subsp. *indica*. *Acta Biologica Hungarica* 61, 486–497.

Rathore, A.P., Chaudhary, D.R. and Jha, B. (2016) Biomass production, nutrient cycling and carbon fixation by *Salicornia brachiata* Roxb.: A promising halophyte for coastal saline soil rehabilitation. *International Journal of Phytoremediation* 18, 808–818.

Rathore, A.P., Chaudhary, D.R. and Jha, B. (2017) Seasonal patterns of microbial community structure and enzyme activities in coastal saline soils of perennial halophytes. *Land Degradation & Development* 28, 1779–1790.

Reginato, M., Sosa, L., Llanes, A., Hampp, E., Vettorazzi, N. *et al.* (2014) Growth responses and ion accumulation in the halophytic legume *Prosopis strombulifera* are determined by Na_2SO_4 and NaCl. *Plant Biology* 16, 97–106.

Reinoso, H., Sosa, L., Reginato, M. and Luna, V. (2005) Histological alterations induced by sodium sulfate in the vegetative anatomy of *Prosopis strombulifera* (Lam.) Benth. *World Journal of Agricultural Sciences* 1, 109–119.

Rogel, J.A., Silla, R.O. and Ariza, F.A. (2001) Edaphic characterization and soil ionic composition influencing plant zonation in a semiarid Mediterranean salt marsh. *Geoderma* 99, 81–98.

Rozema, J. (1991) Growth, water and ion relationships of halophytic monocotyledonae and dicotyledonae; a unified concept. *Aquatic Botany* 39, 7–33.

Salama, H.M. and Ali, A.A. (2003) Variations in mineral ion composition of soil and some halophytes in the Mediterranean coastal north of Egypt. *Journal of Environmental Sciences* 26, 207–228.

Shabala, S., Shabala, L. and Volkenburgh, E.V. (2003) Effect of calcium on root development and root ion fluxes in salinised barley seedlings. *Functional Plant Biology* 30, 507–514.

Shabala, S., Demidchik, V., Shabala, L., Cuin, T.A., Smith, S.J. *et al.* (2006) Extracellular Ca^{2+} ameliorates NaCl-induced K^+ loss from Arabidopsis root and leaf cells by controlling plasma membrane K^+-permeable channels. *Plant Physiology* 141, 1653–1665.

Shabala, S. and Pottosin, I. (2014) Regulation of potassium transport in plants under hostile conditions: implications for abiotic and biotic stress tolerance. *Physiologia Plantarum* 151, 257–279.

Shabala, S., Bose, J., Fuglsang, A.T. and Pottosin, I. (2016a) On a quest for stress tolerance genes: membrane transporters in sensing and adapting to hostile soils. *Journal of Experimental Botany* 67, 1015–1031.

Shabala, L., Zhang, J., Pottosin, I., Bose, J., Zhu, M. *et al.* (2016b) Cell-type-specific H^+-ATPase activity in root tissues enables K^+ retention and mediates acclimation of barley (*Hordeum vulgare*) to salinity stress. *Plant Physiology* 172, 2445–2458.

Sosa, L., Llanes, A., Reinoso, H., Reginato, M. and Luna, V. (2005) Osmotic and specific ion effects on the germination of *Prosopis strombulifera*. *Annals of Botany* 96, 261–267.

Teakle, N.L. and Tyerman, S.D. (2010) Mechanisms of Cl^- transport contributing to salt tolerance. *Plant, Cell and Environment* 33, 566–589.

Teakle, N.L., Bazihizina, N., Shabala, S.N., Colmer, T.D., Barrett-Lennard, E.G. *et al.* (2013) Differential tolerance to combined salinity and O_2 deficiency in the halophytic grasses *Puccinellia ciliata* and *Thinopyrum ponticum*: The importance of K^+ retention in roots. *Environmental and Experimental Botany* 87, 69–78.

Tipirdamaz, R., Gagneul, D., Duhaze, C., Aınouche, A., Monnier, C. *et al.* (2006) Clustering of halophytes from an inland salt marsh in Turkey according to their ability to accumulate sodium and nitrogenous osmolytes. *Environmental and Experimental Botany* 57, 139–153.

Tobe, K., Li, X. and Omasa, K. (2002) Effect of sodium magnesium and calcium salts on seed germination and radicle survival of a halophyte, *Kalidium caspicum* (Chenopodiaceae). *Australian Journal of Botany* 50, 163–169.

Tug, G.N., Ketenoglu, O. and Bilgin, A. (2012) The relationships between plant zonation and edaphic factors in halophytic vegetation around Lake Tuz, Central Anatolia, Turkey. *Rendiconti Lincei* 23, 355–363.

Venkatesalu, V., Raj-Kumar, R. and Chellappan, K.P. (1994). Growth and mineral distribution of *Sesuvium portulacastrum* L., a salt marsh halophyte, under sodium chloride stress. *Communications in Soil Science and Plant Analysis* 25, 2797–2805.

Volkov, V. (2015). Salinity tolerance in plants. Quantitative approach to ion transport starting from halophytes and stepping to genetic and protein engineering for manipulating ion fluxes. *Frontiers in Plant Science* 6, 873. doi: 10.3389/fpls.2015.00873.

Waisel, Y. (1972) *Biology of Halophytes*. Academic Press, New York.

Wang, F., Chen, Z., Liu, X., Colmer, T.D., Shabala, L. *et al.* (2017) Revealing the roles of GORK channels and NADPH oxidase in acclimation to hypoxia in Arabidopsis. *Journal of Experimental Botany* 68, 3191–3204.

White, P.J. and Broadley, M.R. (2001) Chloride in soils and its uptake and movement within the plant: a review. *Annals of Botany* 88, 967–988.

Xu, G., Magen, H., Tarchitzky, J. and Kafkafi, V. (2000) Advances in chloride nutrition. *Advances in Agronomy* 68, 96–150.

Yamaguchi, T., Hamamoto, S. and Uozumi, N. (2013) Sodium transport system in plant cells. *Frontiers in Plant Science* 4, 410. doi:10.3389/fpls.2013.00410.

Yasar, F. (2007) Effects of salt stress on ion and lipid peroxidation content in green beans genotypes. *Asian Journal of Chemistry* 19, 1165–1169.

Zhang, J.L., Flowers, T.J. and Wang, S.M. (2010) Mechanisms of sodium uptake by roots of higher plants. *Plant and Soil* 326, 45–60.

Zhu, J.K. (2002) Salt and drought stress signal transduction in plants. *Annual Review of Plant Biology* 53, 247–273.

Zörb, C., Sümer, A., Sungur, A., Flowers, T.J. and Özcan, H. (2013) Ranking of 11 coastal halophytes from salt marshes in northwest Turkey according their salt tolerance. *Turkish Journal of Botany* 37, 1125–1133.

Zouhaier, B., Abdallah, A., Najla, T., Wahbi, D., Wided, C. *et al.* (2015) Scanning and transmission electron microscopy and X-ray analysis of leaf salt glands of *Limoniastrum guyonianum* Boiss. under NaCl salinity. *Micron* 78, 1–9.

10 Morpho-anatomical Traits of Halophytic Species

Dragana Rančić, Ilinka Pećinar, Svetlana Aćić and Zora Dajić Stevanović*
University of Belgrade, Belgrade, Serbia

Abstract

Some aspects of plant tolerance to salt stress are based on morphological and anatomical traits. Halophytes with different preferences for increased salinity conditions show differences in anatomical features, as a consequence of evolving adaptive responses to salt stress. In this chapter we discuss the general morphological and anatomical structure of plants of saline habitats, aiming to provide a detailed review of literature on typical and special morphological and anatomical characteristics of halophytes. Species in the group most tolerant to saline conditions develop two main strategies for avoiding the toxic effect of salt uptake: dilution of salts by increasing leaf or stem succulence (such as *Cakile maritima, Salsola soda, Suaeda maritima, Salicornia fruticosa*); and/or or by excreting excess salts from specialized secretory structures such as bladders and salt glands (e.g. *Halimione portulacoides, Limonium gmelinii*). In addition to succulent and salt-excreting halophytes, other species exhibit various xeromorphic characteristics, similar to plants of drought conditions (thick epidermis covered with cuticle and wax, abundant trichomes, small leaves, presence of bulliform cells, etc.). The latter species have usually developed various salt-exclusion adaptive mechanisms, and are a common flora of saline steppe in Europe (*Puccinellia distans, Camphorosma annua, Artemisia santonicum*, etc.).

Keywords: Abiotic stress; Adaptive responses; Salt bladders; Salt glands; Xeromorphism

10.1 Introduction

Different authors (Waisel, 1972; Grigore, 2008) have tried to organize salt-tolerant species into specific groups such as obligate and facultative halophytes; succulent and non-succulent halophytes; and salt-excreting and non-excreting halophytes. Each categorization of plant tolerance to salt stress is based upon particular criteria, such as morphological, anatomical and physiological traits. General differences in responses of plants to salinity are dependent on adaptive strategies amassed during evolution and natural selection, whereas salt tolerance relies upon mechanisms at all organizational levels of the plant (Dajić, 2006).

According to Grigore and Toma (2010a) halophytes are classified into three main groups, based on anatomical features and their ecological implications: extreme halophytes (euhalophytes), mesohalophytes and glycophytes. Euhalophytes are species well adapted to salinity, occurring almost exclusively in environments of high salinity, while mesohalophytes are species with intermediary

* Corresponding author e-mail: dajic@agrif.bg.ac.rs

anatomical adaptations between extreme halophytes and glycophytes, being able to grow in different habitats, both saline and non-saline. Glycophytes are species that are not tolerant to salinity, and which maintain low sodium levels in their aboveground tissues, especially in the leaves (Cheeseman, 2015); normally, these plants cannot grow in saline environments. It is considered that less than 1% of all flowering plants in the world are halophytes, and only about 0.2% are thought to be the most tolerant to salinity: that is, the extreme halophytes (Flowers *et al.*, 2010).

In this chapter the main anatomical features of different groups of halophytes are presented, with respect to their salt stress tolerance. All illustrations are our original micrographs, prepared in the Laboratory for Plant Anatomy of the Department of Applied Botany of Faculty of Agriculture, University of Belgrade.

10.2 General Morphological and Anatomical Adaptation of Plants to Saline Environments

Regulation of increasing tissue-salt concentrations in salt-tolerant plants induces specific changes at cell, tissue and organ levels through various mechanisms and special structural and physiological alternations (Koyro *et al.*, 2008). Some structural features that enable plants to overcome the high-salinity conditions of their habitats can be regarded as *xeromorphic* adaptations. These include morpho-anatomical features, which prevent water loss from the plant by development of thick epidermis and sclerenchyma; well-developed bulliform cells on the leaves, which serve for leaf rolling; and high density of protecting trichomes. These adaptations are also common in drought-tolerant plants, since increased soil salinity causes physiological drought (i.e. osmotic stress and inability to take up water regularly and control the water regime) (Fitter and Hay, 1989; Flowers *et al.*, 2010). Thus, maintaining the water balance is of crucial importance for survival in conditions of drought and salt stress. Apart from xerophytic features, commonly associated with salt-exclusion mechanisms (Dajić, 2006), the other adaptive strategies linked to regulation of the water content in the plant body under stress conditions are linked to increased capacity for moisture retention and development of succulence. In addition, special adaptive features for salt excretion have been evolved by a group of salt-tolerant plants, mainly referring to: glandular structures (bladder-like hairs and salt glands), which may be considered among the key adaptations in plant salinity tolerance (Hameed *et al.*, 2010, 2013).

10.2.1 Adaptation of upper vegetative plant parts (leaves and stem)

The presence of leaves arranged in a basal rosette has been found in several European halophyte species of Plantaginaceae (*Plantago schwarzenbergiana*, *P. crassifolia*, *P. coronopus*, *P. tenuiflora*); Asteraceae (*Taraxacum officinalis*, *T. bessarabicum*, *Scorzonera cana*, etc.); and the Brassicaceae family (e.g. *Lepidium cartilagineum*, *Rorippa kerneri*). The leaves are approximately parallel to the stem, probably as adaptation to uniform exposure to solar radiation, to make photosynthesis more efficient (Grigore *et al.*, 2012b). Grigore *et al.* (2014) reported about the phenomenon that the basal rosette may serve as a strategy to overcome transport of salts in upper-part plant organs through removal of salt-saturated rosette leaves. These strategies imply a relatively fast turnover of leaves and, as a consequence, the excess salts are removed from the plant when the leaf is shed (Koyro *et al.*, 2008).

Salt may affect leaf size, and it is reported that mangroves develop smaller leaves as salinity increases. Reductions in leaf size may be considered as water saving adaptations, since smaller leaves result in minimizing the need for evaporative cooling (Reef and Lovelock, 2015).

Halophytes quite often have leaves covered with a thick, waxy epidermal cuticle with protecting trichomes to decrease the impact of drought and ultraviolet light on leaf tissues (Koyro *et al.*, 2008) and therefore minimize the development of reactive oxygen species (ROS). Oxidative stress occurs as a consequence of hyperosmolarity caused by drought or salt stress conditions, resulting in the appearance of harmful reactive oxygen molecules such as hydroxyl radicals, hydrogen peroxide and superoxide anions (Xiong *et al.*, 2002). ROS scavenging in plants exposed to salt and drought stress is associated with inclusion of enzymatic pathways, particularly superoxide dismutase, glutathione peroxidase and catalase (Sekmen *et al.*, 2012) and the presence of osmoprotectant compounds such as mannitol, betaine and proline (Xiong *et al.*, 2002), as well as the synthesis of antioxidant compounds such as carotenoids, glutathione, ascorbic acid and polyphenols (Xiong and Zhu, 2002; Ksouri *et al.*, 2007).

The different morphological traits of the xeric leaf structure helps to prevent water loss caused by transpiration and thus limits the import of salts into the photosynthetic tissues (Koyro *et al.*, 2008; Hameed *et al.*, 2010).

The density and size of stomata are also critical for controlling water loss from the leaf surface due to transpiration. Under the conditions of salt stress, the density of the stomata is usually reduced (Omamt *et al.*, 2006; Maricle *et al.*, 2009; Akcin *et al.*, 2017); but, if stomata density is high, stomata are usually small and embedded in succulent and small-sized leaves. In most mangrove species, stomata are present only on the abaxial leaf surface and are usually positioned under the level of the epidermis (Reef and Lovelock, 2015), which correlates with other adaptations of xeromorphosis. When the stomata are not embedded, there are protective hairs above them (Grigore *et al.*, 2014).

The aerial parts of coastal halophytes are usually covered with various types of trichomes. These may have a protective function, mainly through their ability to reduce transpiration and heating, as well as to protect the leaf surface from droplets of seawater. These protective hairs have been observed on the stems and leaves of *Brassica* spp. (*B. hirsuta*, *B. hyssopifolia*, *B. sedoides*) and *Camphorosma annua* (Grigore *et al.*, 2014). The long protecting hairs are formed by a polygonal cell at the base (embedded between smaller epidermal cells), an intermediary rectangular cell and a very long terminal cell, with extremely thick walls and obtuse tip (Metcalfe and Chalk, 1972). The species *Trifolium fragiferum* is known for forming very similar protecting hairs on the leaves, but they are tricellular.

Some species avoid excessive transpiration by developing bulliform or motor cells. They are usually present in halophytes belonging to the families Poaceae (*Puccinellia distans*, *Agrostis stolonifera*, *Alopecurus arundinaceus*), Cyperaceae (*Bolboschoenus maritimus*, *Carex distans*, *C. vulpina*) and Juncaceae (*Juncus gerardii*, *J. trifidus*) (Grigore *et al.*, 2014). These thin-walled and water-containing cells are mostly involved in rolling the leaves in dry environmental conditions, to protect the leaf from dehydration and overheating. According to their function, bulliform cells could contribute to reducing the transpiration rates in water deficit conditions by ability for leaf rolling (found in *Puccinellia distans*, *Juncus gerardii*, *Bolboschoenus maritimus*, *Carex distans*). These structures are considered as xeromorphic features, but they also play an important role in the overcoming of physiological drought caused by salinity (Stevanović *et al.*, 2010).

It has been already reported that high salinity often causes marked anatomical alterations in the structure of the leaf, such as organization and distance among the vascular bundles, and diameter and number of xylem vessels (Al-Taisan and Gabr, 2017). At increased salinity levels, the diameter of the xylem vessels is reduced, as shown in *Salicornia freitagii* (Akcin *et al.*, 2017), which could be important in reducing the risk of xylem embolism development (Reef and Lovelock, 2015).

Some halophytes growing in sandy habitats, such as *Salsola kali*, *Eryngium maritimum*, *Echinophora spinosa* and *Carthamus lanatus* are characterized by slightly thorny leaves which undergo some lignification. In several species of the genus *Limonium* (Plumbaginaceae) there are mechanical idioblasts in the leaf mesophyll; these vary in their shape, size and orientation (Grigore *et al.*, 2014).

The photosynthetic pathway plays an important ecological role in habitats exposed to extreme environmental conditions such as aridity and salinity. C_4 plants show a phenomenon in which the light-dependent and dark phases of photosynthesis are physically separated and have specific structural patterns known as Kranz leaf structures. These are characterized by large bundle-sheath cells located in circles around the leaf veins; these cells are usually rich in chloroplasts. The light-dependent reactions (carbon assimilation) occur in the mesophyll leaf cells (outer layer of chlorenchyma), while the Calvin cycle occurs in the bundle-sheath cells (inner layer of chlorenchyma) and could be performed even if stomata are closed which increases photosynthesis efficiency compared to C_3 plants. It is well established that, among the dicot families, species from Chenopodiaceae have the largest number of C_4 species, and consequently the greatest diversity in leaf anatomy: C_3, C_4-Kranz anatomy and C_4 single cell type (Edwards and Voznesenskaya, 2011). The Kranz-type leaf arrangement in Chenopodiaceae includes mesophyll and pronounced bundle sheath-type cells forming continuous green rings around the periphery of the leaf (or stems in aphyllous species), with their central part mostly occupied by water-storing parenchyma and vascular tissues. Although this type of leaf anatomy also occurs in non-halophytic species, their presence may be regarded as a typical adaptation in halophytes (Grigore *et al.*, 2012a). Kranz anatomy of C_4 Chenopodiaceae plants includes six or seven types: Atriplicoid, Kochioid, Salsoloid, Salsinoid, Schoberioid and Kranz-Tecticornioid (Kadereit *et al.*, 2014) and Suaedioid (Edwards and Voznesenskaya, 2011), which differ in the presence or absence of a hypodermis, the position of the chlorenchyma, the position and amount of water-storage tissue and the position and orientation of vascular bundles. Some halophytes are able to shift from a C_3 to a C_4 pathway (Meinzer and Zhu, 1999).

It is reported that increased mucilage content could be connected to salt stress. In the halophyte species *Kosteletzkya virginica*, for example, the amount of mucilage in vegetative parts increased (especially in leaf epidermis) as a response to increased salt stress (Ghanem *et al.*, 2010). Also, multicellular mucilage glands may occur at the leaf surface in some halophytic members of the Plumbaginaceae family (genus *Limonium*) (Batanouny *et al.*, 1992). It has been reported that some succulent halophytes, such as *Salicornia brachiata* and *Suaeda maritima*, increase in mucilage content under short-term salt stress (Jaiswar and Kazi, 2016). Mucilage may provide plants with the ability to overcome physiological drought due to its hydrophilic properties, which can increase tissue capacity for storing larger amounts of water (Clarke *et al.*, 1979). Moreover, mucilage is frequently found in the xylem vessels of plants that root in high-salinity water, and may have a role in regulating water and ion transport (Ghanem *et al.*, 2010).

10.2.2 Ion excretion: salt bladders and salt glands

Apart from the strategy of succulent halophytes to dilute excess salts in photosynthetic tissue, another group of salt-tolerant species evolved the strategy of salt excretion by specialized structures of the aerial organs, and this plays an important role in removing salts from buds and young stems. Salt excretion is a mechanism typical for species that have developed special features, mostly localized at the leaf and epidermis, known as salt glands, salt hairs or salt bladders (Fahn, 1988). These structures are a key trait of salt-excreting plants, the so-called crynohalophytes. There are about 370 crynohalophyte species worldwide (Yuan *et al.*, 2016). Excess salt can be released through the leaf cuticle, by dissolving in the guttation fluid or by exclusion via special secretory structures (Stenlid, 1956). About half of all halophytes possess salt bladders or salt glands (Flowers *et al.*, 2008), which shift ions from leaf tissues to leaf surfaces, where saline solutions crystallize and can be blown away by wind or washed away by rain. These structural adaptations are found in many halophytic members of different plant families including Scrophulariaceae, Verbenaceae, Convolvulaceae, Plumbagenaceae, Tamaricaceae, Frankeniaceae, Primulaceae, Myrsinaceae, Acanthaceae, Sonneratiaceae and Poaceae (Yuan *et al.*, 2016), Amaranthaceae (including Chenopodiaceae) and Aizoaceae. They are very common in many halophytic genera such as *Cressa* (Convolvulaceae), *Frankenia* (Frankeniaceae), *Spartina*, *Chloris*, *Aeluropus* (Poaceae), *Atriplex* (Chenopodiaceae), *Limonium*, *Plumbago*, *Armeria* (Plumbaginaceae), *Glaux* (Primulaceae), *Tamarix* and *Reamuria* (Tamaricaceae), as well as in some mangrove species such as *Avicennia*, *Aegialitis*, *Aegiceras* and *Acanthus* (Waisel, 1972; Crawford, 1989; Popp, 1995).

Salt-excreting structures vary in structure, position, physiological mechanism and related ecological significance.

10.2.2.1 Salt bladders

Salt bladders are highly specialized structures involved in the accumulation and removal of excess salts. Salt bladders are epidermal trichomes, which consist of a large secretory or bladder-like cell on the top and a stalk consisting of one or sometimes a few cells (Dickison, 2000; Shabala *et al.*, 2014). Stalk cells transport ions from the leaf mesophyll to the bladder cell that accumulates salt. The external wall of the bladder and stalk cell is covered by a cuticle which is impermeable to both water and salt, while the inner primary walls are relatively thin (Thomson and Platt-Aloia, 1979). Salts are collected in the central vacuole of the bladder cell (Dickison, 2000; Grigore *et al.*, 2014). The bladder cells grow and often burst, releasing the salt contents onto the leaf surface. The presence of many vesicular trichomes is usually evident in younger organs, which are more sensitive to salt stress than mature plant parts; this emphasizes the existence of an efficient and specialized salt-eliminating system (Grigore *et al.*, 2014). The bladder salt type of trichome is typical in members of the family Aizoaceae and Amaranthaceae (including former Chenopodiaceae) (Dassanayake and Larkin, 2017; Kiani-Pouya *et al.*, 2017) such as *Atriplex tatarica*, *A. leucoclada*, *A. littoralis* and *A. glauca*, as well as *Halimione verrucifera* and *H. portulacoides* (Grigore *et al.*, 2014). Once the bladder cell collapses, crystallized salts remaining on the leaf surface provides a reflective coating that protects the leaves from strong sunlight; the sharp salt crystals may also serve as a mechanical deterrent to herbivores (Dickison, 2000).

10.2.2.2 Salt glands

Salt glands are more complex and consist of several specialized cells usually located in slight depressions of the leaf epidermis and covered by the cuticle. Salt glands occur in at least 65 members of 11 dicotyledon plant families (Yuan *et al.*, 2016) and in one family of monocotyledons, the Poaceae (Evert, 2006).

In the Poaceae, salt glands exist in more than 30 species belonging to the subfamily Chloridoideae (Céccoli *et al.*, 2015). Salt glands in the Poaceae are generally simple, usually consisting of two cells: a cap cell on the top, and a basal cell (e.g. the grass *Diplachne fusca*) (Semenova *et al.*, 2010; Céccoli *et al.*, 2015). A continuous cuticle covers the outer protruding portion of the gland and the adjoining epidermal cells, but the side walls of the basal cell are not cutinized. Neither cap cell nor basal cell contains a large central vacuole. The outer cap cell wall forms a collecting chamber, and salty water is eliminated though fine pores in the cuticle (Evert, 2007). The size of salt glands in the Poaceae ranges from 15 μm in *Distichlis* to 70 μm in *Buchlon* (Marcum, 1999), found on both leaf surfaces (Liphschitz and Waisel, 1974; Marcum, 1999). Salt glands may be sunken, subsunken, extending out of the epidermis or recumbent to the leaf surface in parallel rows on the top intercostal ridges (Liphschitz and Waisel, 1974; Marcum and Murdoch, 1994; Marcum *et al.*, 1998).

In dicotyledonous species the salt glands are structurally more complex, consisting of a basal cell, cutinized stalk cells and terminal secretory cells covered by a thin perforated cuticle that is separated from the apical cell wall (Evert, 2007). The number of cells may vary from 6 up to 40 in different genera (Fahn, 1990). Multicellular salt glands are characteristic for the species in the Aviceniaceae, Acanthaceae, Frankiaceae, Plumbaginaceae and Tamaricaceae (Mauseth, 1988; Thomson *et al.*, 1988; Marcum and Murdoch, 1994). For example, in *Frankenia grandifolia* the salt gland contains 8 cells, 2 collecting cells and 6 secretory cells (Balsamo and Thomson, 1993); in *Tamarix aphylla* the salt glands consist of 2 basal collecting cells and 6 outer secretory cells (Mauseth, 1988); in *Glaux maritima* the salt gland comprises 6–10 cells, 1 basal cell and several secretory cells (Rozema *et al.*, 1977); the glands of *Avicennia* sp. comprise 11–13 cells, 1 basal cell and several secretory cells (Shimony *et al.*, 1973; Rozema *et al.*, 1977); and in *Limonium* the salt gland is composed of 20 cells, 4 basal cells and a complex of 16 secretory cells (Batanouny *et al.*, 1992; Colombo, 2002). A symplastic continuity, due to the presence of plasmodesmata, exists between the basal and secretory cells and basal and mesophyll cells (Zeigler and Luttge, 1967). The secretory cells have heavily cutinized walls, except in the region of contact with the basal cells. The protoplasts of the transfer cells contain numerous small vesicles in which the salts accumulate. Fusion of the vesicles with the plasmalemma releases the salt into the wall, from which is it further moved through the inner non-cutinized layer towards the surface of the gland and, finally, exported through the small pores (Thompson *et al.*, 1988). In contrast to salt bladders, the salt glands eliminate salts outside of the cells; the ions that have previously accumulated in the salt gland are then excreted through salt gland pores (Yuan *et al.*, 2016). The position of the epidermal glands may be lateral (*Tamarix* sp.), or positioned in the epidermal excavations, such as in *Glaux maritima* (Rozema and Gude, 1981), as well as projected out of the abaxial surface of leaf-like trichomes in *Avicennia* (Thomson *et al.*, 1988).

In addition to the main two types of salt-excreting structures, the salt bladders and salt glands there is one further type, found in the drought- and/or salt-tolerant species of the genus *Mesembryanthemum* (Aizoaceae); this has an intermediary position between salt bladders and salt glands (Grigore and Toma, 2010b).

10.2.3 Adaptation of root

Depending on the plant species, as well as the level of salinity stress and the composition of the external solution, the root growth may be stimulated, inhibited or unaffected (Delane *et al.*, 1982; Waisel, 1985). Salinity usually induces changes in root morphology and anatomy, as a consequence of inhibition of cell division and cell expansion process during root development (Zidan *et al.*, 1990; Hameed *et al.*, 2010), but in some plants root growth is less affected by salinity (Maas and Nieman, 1978) or could be even provoked in some halophytic grasses (Marcum *et al.*, 1998). The roots of succulent euhalophytes such as *Cakile maritima* and *Salsola kali* are approximately 1.2–1.3 m long although the shoots usually only reach 20–30 cm, as a result of a xero-halophytic set of adaptations, including the ability to develop a deep root system to reach the water in maritime sand habitats (Grigore *et al.*, 2012b).

The root of a typical halophyte has well-developed cork and a lacunar primary cortex with sclereids and secretory cells, as well as well-developed mechanical and vascular tissue. The anatomical characteristic of roots of saline desert plants is a reduced cortex, which shortens the distance between epidermis and vascular bundle in the central part of the root (Dajić, 2006).

A common adaptive strategy of halophytes to survive in high-salinity conditions is salt exclusion (Dajić, 2006). It is considered that mangroves exclude 99% of salts through the roots (Tomlinson, 1986). Since the endodermis represents an important barrier that controls salt penetration into the plant body, and plays a crucial role in salt exclusion from the inner tissues (Ding *et al.*, 2010; Krishnamurthy *et al.*, 2014), suberization of the exodermis and endodermis is evident (Reef and Lovelock, 2015). It has been shown that the Casparian strip of the root endodermis of halophytes is usually much wider, as in *Artemisia santonicum*, *Juncus gerardii*, *Lactuca saligna* and *Salicornia* sp. (Schreiber *et al.*, 1999; Grigore and Toma, 2010c; Shabala and Mackay, 2011).

The root of *Arthrocnemum fruticosum*, as well as that of some other halophyte species, is rich in lignified tissue related to vessels and lignified parenchyma (Saadeddin and Doddema, 1986). The presence of the large quantities of lignification and many libriform fibres in the central cylinder of the root and stem (Grigore *et al.*, 2014) may be the result of successive cambia activity induced by excessive soil salinity (Bickenbach, 1932; Reef and Lovelock, 2015). In some Chenopodiaceae a higher amount of lignin has been found in the structure of a root than in a stem (Grigore *et al.*, 2014), because the stem of many halophytes is unaffected by successive cambia activity (Grigore and Toma, 2006, 2007). The appearance of successive cambia is also evident in the root of some *Atriplex* species (e.g. *A. tatarica*). According to Grigore and Toma (2014), the phenomenon of successive cambia has an ecological and adaptive value. Phloem and parenchyma cells in close proximity to xylem may be connected with storage of water and sugars near the xylem vessels, which could facilitate repair of embolisms (Reef and Lovelock, 2015). Moreover, the increased lignin content associated with this structure may confer cellular resistance to high osmotic pressure in order to exceed the osmotic pressure of the soil solution (Kozlowski, 1997). In old root of the desert halophytes *Zygophyllum* sp. and *Nitraria* sp. the secondary xylem vessels are embedded in lignified cells (fibres); this could be a mechanism that helps protect from embolism, since the presence of fibres around vessels contributes to cavitation resistance (Jacobsen *et al.*, 2005). A large amount of lignin in the cell walls may also be important for providing rigidity to the root.

Halophytes usually develop many shallow roots. If growing in muddy or loose sandy soil, like some mangrove species, they can develop many stilt or prop roots from the aerial branches of the stem, or a large number of adventitious root buttresses from the basal parts of tree trunks, which provide sufficient support to the plants. Mangrove halophytes (e.g. some *Rhizophora* spp.) have stilt roots developing from stem branches or underground roots, necessary for efficient anchorage in muddy or loose sandy soil (Hameed *et al.*, 2010).

Aerenchyma is well expressed, especially in the underground vegetative organs such as rhizomes and roots. The aerenchyma tissue is found in halophytes in relation to waterlogging or flooded saline habitats (Flowers and Colmer, 2008; Hameed *et al.*, 2010), as in *Juncus gerardii*, *Carex distans*, *C. vulpina*, *Lysimachia maritima*, *Triglochin maritima*, *Crithmum maritimum* and *Cymodocea nodosa*, where the plants are subjected to hypoxia or even anoxia and have to store oxygen in the root aerenchyma (Schulze *et al.*, 2005). The soil in coastal regions is low in oxygen and, to compensate for the lack of soil aeration, plants develop negatively geotropic breathing roots called pneumatophores. The cortex of pneumatophores is made of aerenchyma with large air chambers which are in contact with numerous lenticels on the root surface. These 'breathing pores' enable gaseous exchange in poorly aerated soil, and prominent aerenchyma enclosing large air cavities inside the roots help in the conduction of air down to the subterranean or submerged roots. It has been suggested that lenticels on these roots may have a significant role in the uptake of CO_2 (Reef and Lovelock, 2015), which is beneficial for photosynthesis. Such roots have been well described from many mangrove halophytes such as *Avicennia marina* (Ish-Shalom-Gordon and Dubinsky, 1992; Purnobasuki and Suzuki, 2005).

10.2.4 Leaf and stem succulence

In natural conditions, the salt content in the vegetative organs of halophytes increases with ageing, and might reach toxic levels. Halophytic species, therefore, tend to lower such unfavourable salt concentrations by increasing the water content in their tissues and/or by enhanced growth (Dajić, 2006). One of the most evident adaptive features of halophytes is the phenomenon of leaf and stem succulence, which represents a very efficient way of mitigating the adverse osmotic and toxic effects of ions through dilution (Waisel, 1972). Leaf succulence is very rare in monocots (Hameed *et al.*, 2010), but dicotyledonous plants commonly develop succulence in response to salinity (Flowers and Colmer, 2008). The halophytes *Salicornia bigelovii* and *Salsola kali* exhibit greater succulence when grown in highly saline conditions (Parks *et al.*, 2002); and stem thickness, and the length and width of water-storing tissue increased significantly in *Salicornia freitagii* when this plant was exposed to increased salinity (Akcin *et al.*, 2017). Salt-induced succulence was also reported for *Salsola kali* when this species was grown in a saline medium: plants exhibited leaf thickness about three times greater than the control plants (Rilke and Reimann, 1996).

Succulent halophytes are characterized by thick and fleshy stems and leaves, and sometimes may be more or less translucent (Grigore *et al.*, 2014). The thickness of the leaf is a consequence of enlargement of the mesophyll water-storage parenchyma with small intercellular spaces (Strogonov, 1973). Succulence is probably induced by accumulation of salts in the cytoplasm, which leads to an increase in the osmotic concentration of cell sap and makes these plants able to easily absorb salty water. Succulent halophytes accumulate large amounts of ions in their cells or tissues (Youssef, 2009). The contents of the salt ions in succulent halophytes can be more than one-half of the weight of the dry leaf (Eshel, 1985). The concentration of salt ions (mainly sodium ions) in the cell sap is much higher than in a soil solution, allowing further water uptake and dilution of salts accumulated in the vacuole, which occupies more than 70% of the water-storage parenchyma cell (Hajibagheri *et al.*, 1984). This kind of leaf structure provides more space for ions and increases the total plant water content (Hameed *et al.*, 2010). Leaf succulence is typical for range of European (eu)halophytes, such as *Suaeda maritima*, *Salsola soda*, *Bassia hirsuta*, *B. sedoides*, *P. crassifolia*, *P. maritima*, *P. tenuiflora*, *Spergularia salina*, *Crithmum maritimum* and *Cakile maritima* (Grigore *et al.*, 2014). In addition, some leafless halophytes are known for their high-salinity tolerance (mainly from the Salicornioideae), including *Salicornia europea*, *S. fruticosa* and *Halocnemum strobilaceum*. Succulence represents a very efficient way to enable an erect position in a species whose mechanical tissue is less developed (Grigore, 2008a; Grigore and Toma, 2010a). The stem of the succulent halophyte *Salicornia fruticosa* has a simple cortex and a one-layer epidermis which is thin-walled; the photosynthetic tissue has palisade and parenchymatous cells for storage of water (Fahn, 1990). The succulent species *Arthrocnemum macrostachyum* has a very thick stem cortex comprising an external assimilating area made of 2–3 layers of long palisade cells embedded in the water-storage parenchyma, and an internal thicker area of tissue consisting of very large water-storage cells (Grigore and Toma, 2014). Between the water-storage tissue and the epidermis in *Arthrocnemum* spp. a number of cells known as vessel-like tracheoid idioblasts (Saadeddin and Doddema, 1986) or brachysclereids (Grigore and Toma, 2014) are found. These tracheid-like cells have also been found in *Salicornia* and *Suaeda* species (Grigore and Toma, 2007; Polić *et al.*, 2009). These cells are positioned parallel with the palisade cells and quite apart from the vascular bundles. It has been suggested that tracheid-like cells are alive (and contain the nuclei, Hess *et al.*, 1975) and are involved in the uptake of dew by the stomata (Saadeddin and Doddema, 1986), and therefore could have a role in the secretion of excess of salt (Bercu and Bavaru 2005). Furthermore, they could have a role in water transport towards peripheral tissues (Keshavarzi and Zare, 2006).

It is known that succulence is an important trait in the regulation of temperature and water balance (Fitter and Hay, 1989), which is crucial for the survival of many halophytes of dry and/or highly saline habitats (Weiglin and Winter, 1991).

10.3 Review of Literature on Typical and Special Morphological and Anatomical Characteristics of Halophytes

Morphological and anatomical characteristics of different halophyte species based on available literature data are provided in Table 10.1.

10.4 Examples of Different Types of Halophytes Related to Morphological and Anatomical Characteristics

10.4.1 Mesohalophytes

10.4.1.1 Atriplex littoralis

Atriplex littoralis (Amaranthaceae) is a shrub species that grows along beaches and salt marshes. Its leaves are characterized by well-developed palisade tissue below both leaf sides (bilateral palisade tissue) and reduced spongy tissue (Figs 10.1.1 and 10.1.2), which could be considered as adaptations to increased insolation. On the leaf surface is a single-layered epidermis with stomata at the epidermal level (Fig. 10.1.2). Collenchyma and parenchyma cells are visible in the region of the midvein, surrounding the vascular bundle. (Fig. 10.1.1) These are particular anatomical adaptations to saline conditions.

10.4.1.2 Puccinellia distans

Puccinellia distans (Poaceae) is a perennial grass that grows in moist and moderately saline habitats native to Europe. The leaves of *P. distans* are characterized by large bulliform cells in the upper epidermis (Figs 10.1.3 and 10.1.4), enabling leaf movement (enrollment) which may protect the stomata and reduce transpiration. Bulliform cells of grasses also occur in nonhalophytic species and represent a typical xeromorphic adaptation (Fahn and Cutler, 1992).

10.4.2 Euhalophytes

10.4.2.1 Xeromorphic

Camphorosma monspeliaca (Amaranthaceae) is a perennial shrub halophyte which grows in dry saline areas of the Mediterranean. *C. monspeliaca* is a C_4 plant, and a leaf section reveals Kranz anatomy is noticeable. The leaf surface has a one-layered epidermis with some enlarged cells; these represent the basal cells of long, non-glandular trichomes. Under the epidermis is a layer of small isodiametric, hypodermal cells, and below is a layer of palisade cells interrupted abaxially in the zone of the midvein (Fig. 10.1.5). On the internal side of the palisade layer is a layer of isodiametric, internal chlorenchyma cells. In the central part of the leaf is a large vascular bundle surrounded by very large parenchymatous cells (the internal chlorenchyma). There are also numerous small vascular bundles located at the contact line between the internal chlorenchyma and central parenchyma cells (Fig. 10.1.6).

10.4.2.2 Succulent

Cakile maritima (Brassicaceae) is a frequent annual coastal halophyte distributed along sandy beaches and dunes in Europe, North Africa and

Table 10.1. Morphological and anatomical characteristics of different halophyte species.

Family	Species	Anatomy of plant organ	References
Acanthaceae	*Avicennia germinans*	Salt-excreting leaf	Balsamo and Thomson (1993)
	Avicennia marina	Salt-excreting leaf	Campbell and Thomson (1976); Salama *et al.* (1999); Naidoo (2016)
	Avicennia officinalis	Root	Kong and Li (2008)
Aizoaceae	*Mesembryanthemum crystallinum*	Salt-excreting leaf	Steudle *et al.* (1975); Agarie *et al.* (2007); Grigore *et al.* (2014)
	Mesembryanthemum nodiflorum	Salt-excreting leaf	Steudle *et al.* (1975); Agarie *et al.* (2007); Grigore *et al.* (2014)
	Aizoon canariense	Salt-excreting leaf	Grigore *et al.* (2014)
Apiaceae	*Crithmum maritimum*	Rhizome, aerial stem, leaf succulent xerohalophytic structure	Cornara *et al.* (2009); Grigore *et al.* (2012b, 2014); Grigore and Toma (2017)
Asteraceae	*Acanthus ebracteatus*	Salt-excreting leaf	Das (2002); Ong and Gong (2013)
	Acanthus ilicifolius	Salt-excreting leaf	Das (2002); Ong and Gong (2013)
	Artemisia santonicum	Root, stem, leaf xerohalophytic structure	Grigore and Toma (2006); Badea and Zamirache (2011); Grigore *et al.* (2012c, 2014)
	Aster pannonicus	Root, rhizome, stem, leaf mesohalophytic structure	Grigore and Toma (2006, 2010b); Geissler *et al.* (2009); Bercu *et al.* (2012)
	Baccharis spartioides	Stem, leaf xerohalophytic structure	Cuadra and Cambi (2014)
	Baccharis tenella	Stem, leaf xerohalophytic structure	Cuadra and Cambi (2014)
	Inula crithmoides	Root (adventitious), rhizome, aerial stem, leaf succulent mesohalophytic structure	Grigore *et al.* (2011a, b, 2014); Grigore and Toma (2017)
	Lactuca saligna	Root, stem, leaf mesohalophytic structure	Grigore and Toma (2006); Grigore *et al.* (2014); Abu-Serag (2015)
	Scorzonera cana	Root, stem, rhizome, leaf mesohalophytic structure	Makbul *et al.* (2011); Grigore *et al.* (2012c); Grigore and Toma (2014); Norouzi *et al.* (2016)
Brassicaceae	*Cakile arabica*	Stem, leaf succulent structure	Al-Taisan and Gabr (2017)
	Cakile maritima	Root, stem, C_4 leaf succulent structure	Ciccarelli *et al.* (2010, 2012b, 2014); Jianu *et al.* (2014)
	Lepidium cartilagineum	Root, rhizome, aerial stem, leaf mesohalophytic structure	Grigore and Toma (2008b); Grigore *et al.* (2014)
	Lepidium latifolium	Rhizome, aerial stem, leaf mesohalophytic structure	Grigore and Toma (2008b); Grigore *et al.* (2014)
	Lepidium perfoliatum	Root, rhizome, aerial stem, leaf mesohalophytic structure	Grigore and Toma (2008b); Grigore *et al.* (2014)

Caryophyllaceae	*marina*	Root, stem and leaf succulent structure	Akcin *et al*. (2014)
	Spergularia media	Root, stem and C_3 leaf succulent structure mesohalophytic structure	Grigore and Toma (2008a, 2014); Grigore *et al*. (2012c); Grigore and Toma (2017)
Chenopodiaceae	*Arthrocnemum macrostachyum*	Stem succulent structure	Kashavarzi and Zare (2006); Grigore *et al*. (2013, 2014); Grigore and Toma (2017)
	Atriplex amnicola	Salt-excreting leaf	Malcolm *et al*. (2003)
	Atriplex canescens	Salt-excreting leaf	Pan *et al*. (2016)
	Atriplex cristata	Root, stem, leaf	Fahn and Zimmermann (1982); Jáuregui *et al*. (2014)
	Atriplex farinosa	Salt-excreting leaf	Salama *et al*. (1999)
	Atriplex glauca	Root, stem, salt-excreting leaf	Grigore and Toma (2017)
	Atriplex halimus	Root, stem, C_4 leaf succulent structure	Fahn and Zimmermann (1982); Salama *et al*. (1999); Smail-Saadoun (2005); Boughalleb *et al*. (2009); Grigore *et al*. (2014); Mahi *et al*. (2015); Grigore and Toma (2017)
	Atriplex littoralis	Root, stem, leaf succulence euhalophyte	Grigore and Toma (2010a, 2017); Grigore *et al*. (2012b, 2014)
	Atriplex oestophora	Root, stem, leaf	Fahn and Zimmermann (1982); Jáuregui *et al*. (2014)
	Atriplex prostrata	Root, stem, leaf euhalophyte	Grigore and Toma (2010a, 2017); Grigore *et al*. (2012b, 2014); Oakley *et al*. (2014)
	Atriplex tatarica	Root, stem, leaf C_4 euhalophyte	Grigore and Toma (2007, 2010a, 2017); Grigore *et al*. (2012a, 2012b, 2014)
	Bassia hirsuta	Root, stem, C_4 kochoid leaf type succulent structure euhalophyte	Grigore and Toma (2007, 2017); Grigore *et al*. (2014); Safiallah *et al*. (2017)
	Bassia hyssopifolia	Stem, leaf C_4 kochoid type	Grigore *et al*. (2012b, 2014); Kadereit *et al*. (2014); Grigore and Toma (2017)
	Bassia littorea	Kranz anatomy structure (Kochioid subtype) leaf succulent structure	Kadereit *et al*. (2014)
	Bassia pilosa	Kranz anatomy structure (Kochioid subtype) leaf succulent structure	Grigore and Toma (2017); Safiallah *et al*. (2017)
	Bassia prostrata	Kranz anatomy structure (Kochioid subtype) leaf succulent structure	Freitag and Kadereit (2014); Kadereit *et al*. (2014); Grigore and Toma (2017); Safiallah *et al*. (2017)
	Bassia salsoloides	Kranz anatomy structure (Kochioid subtype) leaf succulent structure	Kadereit *et al*. (2014)
	Bassia scoparia	Kranz anatomy structure (Kochioid subtype) leaf succulent structure	Freitag and Kadereit (2014); Kadereit *et al*. (2014); Safiallah *et al*. (2017)

Continued

Table 10.1. Continued.

Family	Species	Anatomy of plant organ	References
	Bassia sedoides	Main and adventitious root, stem, leaf succulent structure	Grigore *et al.* (2014); Kadereit *et al.* (2014); Grigore and Toma (2017)
	Bassia stellaris	Kranz anatomy structure (Kochioid subtype) succulent structure	Kadereit *et al.* (2014); Grigore and Toma (2017); Safiallah *et al.* (2017)
	Bassia turkestanica	Kranz anatomy structure (Kochioid subtype) leaf succulent structure	Safiallah *et al.* (2017); Grigore and Toma (2017)
	Bienertia sinuspersici	Salt-excreting leaf	Akhani *et al.* (2005); Park *et al.* (2009)
	Camphorosma annua	Root, stem and succulent leaves C_4 salsoloid type xeromorphic euhalophyte	Jacobs (2001); Grigore and Toma (2007, 2010a, 2017); Grigore *et al.* (2012a, 2012b, 2014); Kadereit *et al.* (2014)
	Camphorosma monspeliaca	Salsoloid type of leaf xeromorphic euhalophyte	Grigore *et al.* (2012a); Grigore and Toma (2017)
	Chenopodium album	Root, stem, leaf mesohalophytic structure	Artschwager (1920); Costa and Tavares (2006); Tajadod and Moogouei (2012); Grigore *et al.* (2014); Grigore and Toma (2017)
	Halimione portulacoides	Stem, succulent C_3 leaf salt-excreting	Grigore *et al.* (2014); Grigore and Toma (2017)
	Halimione verrucifera	Root, stem, C_3 leaf salt-excreting	Grigore and Toma (2010a, 2017); Grigore *et al.* (2012c); Grigore *et al.* (2014)
	Halocnemum strobilaceum	Stem succulent structure euhalophyte	Keshavarzi and Zare (2006); Grigore *et al.* (2014); Grigore and Toma (2017)
	Petrosimonia oppositifolia	Root, stem, C_4 salsoloid subtype leaf succulent structure euhalophyte	Grigore and Toma (2007, 2010a, 2017); Grigore *et al.* (2012a, 2014)
	Petrosimonia triandra	Root, stem, C_4 salsoloid subtype leaf, succulent structure euhalophyte	Grigore and Toma (2007, 2010a, 2017); Grigore *et al.* (2012a, 2012b, 2014)
	Salicornia europaea	Root, stem, leaf C_3 succulent structure euhalophyte	Kashavarzi and Zare (2006); Grigore and Toma (2007, 2010a, 2014, 2017); Milic *et al.* (2011); Grigore *et al.* (2012c)
	Salicornia freitagii	Root, stem succulent structure	Akcin *et al.* (2017)
	Salicornia ramosissima	Root, stem succulent structure	Milic *et al.* (2011); Grigore *et al.* (2013, 2014); Grigore and Toma (2017)
	Salicornia virginica	Stem, leaf succulent structure	Seliskar (1981, 1983)
	Salsola kali	Root, stem, leaf salsoloid C_4 Kranz leaf subtype succulent structure	Patrignani *et al.* (1993); Bercu and Bavaru (2004); Grigore *et al.* (2012b, 2014); Grigore and Toma (2017)

	Salsola oppositifolia	Root, stem, leaf salsoloid type C_4 succulent structure	Grigore *et al*. (2014); Grigore and Toma (2017)
	Salsola soda	Stem, C_4 salsoloid leaf type succulent structure	Carraro *et al*. (1993); Milic *et al*. (2011, 2013)
	Sarcocornia fruticosa	Main and lateral root, stem, seed succulent structure	Shepherd *et al*. (2005); Smail-Saadoun (2005); Grigore *et al*. (2013, 2014); Grigore and Toma (2017)
	Suaeda aralocaspica	Salt-excreting leaf	Park *et al*. (2009)
	Suaeda maritima	Root, stem, leaf bassioid anatomic type C_3 succulent structure euhalophyte	Grigore and Toma (2007); Milic Polic *et al*. (2009); Grigore and Toma (2012b, 2014, 2017)
	Suaeda salsa	Root, stem succulent structure	Kong and Li (2008)
	Suaeda spicata	Root, stem, leaf succulent structure	Grigore *et al*. (2011b, 2014); Grigore and Toma (2017)
	Suaeda splendens	Root, stem, leaf succulent structure	Grigore *et al*. (2014); Grigore and Toma (2017)
	Suaeda vera	Stem, leaf succulent structure	Muhaidat *et al*. (2007); Abd Elbar and Abd El-Maboud (2013); Grigore *et al*. (2014)
Convolvulaceae	*Cressa cretica*	Stem	Weiglin and Winter (1988); Al-Adily *et al*. (2016)
Cyperaceae	*Bolboschoenus maritimus*	Rhizome, aerial stem, leaf xerohalophytic structure	Grigore *et al*. (2010, 2014); Rad and Hroudová (2013); Grigore and Toma (2017)
	Carex distans	Rhizome, aerial stem, leaf mesohalophytic structure	Zarrinkamar *et al*. (2002); Grigore *et al*. (2010, 2012b, 2014); Grigore and Toma (2017)
	Carex vulpina	Rhizome, aerial stem, leaf mesohalophytic structure	Grigore *et al*. (2010, 2014); Grigore and Toma (2017)
Fabaceae	*Alhagi maurorum*	Stem	Al-Adily *et al*. (2016)
	Lotus tenuis	Root, rhizome, aerial stem, leaf mesohalophytic structure	Grigore *et al*. (2014); Paz *et al*. (2014)
	Trifolium fragiferum	Root, rhizome, aerial stem, leaf and stipule mesohalophytic structure	Grigore *et al*. (2014)
	Prosopis strombulifera	Root, stem, hypocotyls, leaf, xerohalophytic structure	Reinoso *et al*. (2004, 2005); Reginato *et al*. (2016)
Frankeniaceae	*Frankenia grandiflora*	Salt-excreting leaf	Balsamo and Thomson (1993)
	Frankenia juniperoides	Salt-excreting leaf, stem	Cuadra and Cambi (2014)
	Frankenia laevis	Salt-excreting leaf, stem	Grigore *et al*. (2012b, 2014); Grigore and Toma (2017)
	Frankenia pulverulenta	Salt-excreting leaf, stem	Cuadra and Cambi (2014)
	Frankenia revoluta	Salt-excreting leaf, stem	Salama *et al*. (1999)
Iridaceae	*Iris halophila*	Rhizome, aerial stem, leaf, mesohalophytic structure	Grigore *et al*. (2014); Ivanescu *et al*. (2016)

Continued

Table 10.1. Continued.

Family	Species	Anatomy of plant organ	References
Juncaceae	*Juncus acutus*	Root, stem xerohalophytic structure	Al Hassan *et al.* (2015); Al-Adily *et al.* (2016)
	Juncus articulatus	Root, stem xerohalophytic structure	Al Hassan *et al.* (2015)
	Juncus gerardii	Rhizome, aerial stem, leaf sheath leaf, mesohalophytic structure	Grigore and Toma (2008a, 2008b, 2010a, 2010b, 2017); Grigore *et al.* (2012c); Grigore *et al.* (2014)
Juncaginaceae	*Triglochin maritima*	Salt-excreting leaf, root, aerial stem	Grigore *et al.* (2014)
Myoporaceae	*Myoporum bontioides*	Leaf salt gland	Das *et al.* (2014)
Oleaceae	*Phillyrea latifolia*	Leaf salt gland	Weiglin and Winter (1988); Gucci *et al.* (1997)
Polygonaceae	*Polygonum equisetiforme*	Stem, leaf mesohalophytic structure	Crivellaro and Schweingruber (2015); Grigore *et al.* (2014)
Plumbaginaceae	*Armeria canescens*	Salt excreting leaf	Scassellati *et al.*, 2016
	Limoniastrum guyonianum	Salt-excreting leaf	Ioannidou-Akoumianaki *et al.* (2015); Zouhaier *et al.* (2015)
	Limoniastrum monopetalum	Salt-excreting leaf	Batanouny and Sitta (1977); Zouhaiera *et al.* (2015)
	Limonium anfractum	Salt-excreting leaf	Zoric *et al.* (2013)
	Limonium axillare		Salama *et al.* (1999)
	Limonium bicolor	Salt-excreting leaf	Feng *et al.* (2015)
	Limonium furfuraceum	Rhizome, aerial stem, leaf salt-excreting	Grigore *et al.* (2014)
	Limonium girardianum	Root (adventitious), rhizome, aerial stem, leaf salt-excreting	Grigore *et al.* (2014)
	Limonium gmelinii	Salt-excreting stem, leaf	Grigore and Toma (2010b, 2017); Aydosova *et al.* (2012); Grigore *et al.* (2012b); Akhani *et al.* (2013); Zoric *et al.* (2013)
	Limonium monopetalum	Salt-excreting leaf	Salama *et al.* (1999)
	Limonium narborense	Salt-excreting leaf, stem	Grigore and Toma (2010b); Grigore *et al.* (2014)
	Limonium parvibracteatum	Salt-excreting leaf, root, rhizome, aerial stem	Grigore *et al.* (2014)
	Limonium pruinosum	Salt-excreting leaf	Salama *et al.* (1999)
	Limonium santapolense	Salt-excreting leaf, root, rhizome, aerial stem	Grigore *et al.* (2014)
	Limonium sinense	Salt-excreting stem, root	Kong and Li (2008)
Plantaginaceae	*Plantago coronopus*	Root, stem, leaf mesohalophytic structure	Grigore *et al.* (2012b, 2014)
	Plantago crassifolia	Root (adventitious), rhizome, leaf succulence mesohalophytic structure	Grigore *et al.* (2014); Grigore and Toma (2017)

	Plantago schwarzenbergiana	Root, rhizome, aerial stem, leaf mesohalophytic structure	Grigore and Toma (2008a, 2008b, 2014); Ianovici (2011)
	Plantago tenuiflora	Root, stem, leaf mesohalophytic structure	Grigore *et al.* (2014); Grigore and Toma (2017)
Poaceae	*Aeluropus lagopoides*	Salt-excreting leaf, stem	Salama *et al.* (1999); Al-Adily *et al.* (2016)
	Aeluropus littoralis	Salt-excreting leaf, aerial stem	Liu *et al.* (2006); Barhoumi *et al.* (2008); Grigore *et al.* (2014)
	Agrostis stolonifera	Rhizome, aerial stem, leaf xerohalophytic structure	Pammenter *et al.* (1986); Grigore *et al.* (2010, 2014)
	Alopecurus arundinaceus	Rhizome, aerial stem, leaf mesohalophytic structure	Grigore *et al.* (2012c); Grigore *et al.* (2010, 2014); Grigore and Toma (2017)
	Buchloe dactyloides	Salt-excreting leaf	Koyro *et al.* (2008)
	Chloris gayana	Salt-excreting leaf	Takao *et al.* (2013)
	Cynodon dactylon	Root, stem, leaf xerohalophytic structure	Oross and Thomson (1982); Amarasinghe and Watson (1988); Marcum (2006, 2008); Koyro *et al.* (2008); Hameed *et al.* (2010, 2013); Vandana (2013)
	Dactyloctenium aegyptium	Salt-excreting leaf	Liphschitz and Waisel (1974)
	Diplachne fusca	Salt-excreting leaf	Céccoli *et al.* (2015)
	Distichlis spicata	Salt-excreting leaf	Liphschitz and Waisel (1974); Oross and Thomson (1982); Marcum (2006); Koyro *et al.* (2008); Semenova *et al.* (2010); Céccoli *et al.* (2015)
	Eleusine indica	Salt-excreting leaf	Liphschitz and Waisel (1974)
	Leptochloa digitata	Salt-excreting leaf	Amarasinghe and Watson (1988)
	Leptochloa fusca	Salt-excreting leaf	Wieneke *et al.* (1987)
	Munroa argentina	Salt-excreting leaf	Céccoli *et al.* (2015)
	Odyssea paucinervis	Salt-excreting leaf	Somaru *et al.* (2002)
	Pappophorum philippianum	Salt-excreting leaf	Taleisnik and Anton (1988); Céccoli *et al.* (2015)
	Porteresia coarctata	Salt-excreting leaf	Flowers *et al.* (1990); Sengupta and Majumder (2010)
	Puccinellia distans ssp. *limosa*	Aerial stem, leaf xerohalophytic structure	Grigore and Toma (2010a, 2010b, 2017); Grigore *et al.* (2014)
	Sporobolus airoides	Salt-excreting leaf	Koyro *et al.* (2008); Weragodavidana (2016)
	Sporobolus cryptandrus	Salt-excreting leaf	Weragodavidana (2016)
	Sporobolus pungens	Root, aerial stem, rhizome salt-excreting xerohalophytic structure	Grigore *et al.* (2012b, 2014)
	Sporobolus virginicus	Salt-excreting leaf	Amarasinghe and Watson (1988); Naidoo and Naidoo (1998)
	Spartina foliosa	Salt-excreting leaf	Levering and Thomson (1971)

Continued

Table 10.1. Continued.

Family	Species	Anatomy of plant organ	References
	Zoysia matrella	Salt-excreting leaf	Rao (2011)
Primulaceae	*Lysimachia maritima*	Salt-excreting leaf, root, stem	Rozema *et al.* (1977), Grigore *et al.* (2014); Grigore and Toma (2017)
Tamaricaceae	*Tamarix androssowii*	Salt-excreting leaf	Campbell and Strong (1964)
	Tamarix amplexicaulis	Salt-excreting leaf	Salama *et al.* (1999)
	Tamarix aphylla	Salt-excreting root	Thomson *et al.* (1969); Salama *et al.* (1999); Bosabalidis (2010); Oladi *et al.* (2017)
	Tamarix minoa	Salt-excreting leaf	Villar *et al.* (2015)
	Tamarix nilotica	Salt-excreting leaf	Salama *et al.* (1999)
	Tamarix passerinoides	Salt-excreting leaf	Salama *et al.* (1999)
	Tamarix pentrandra	Salt-excreting leaf	Campbell and Strong (1964)
	Tamarix usneoides	Salt-excreting leaf	Wilson *et al.* (2016)

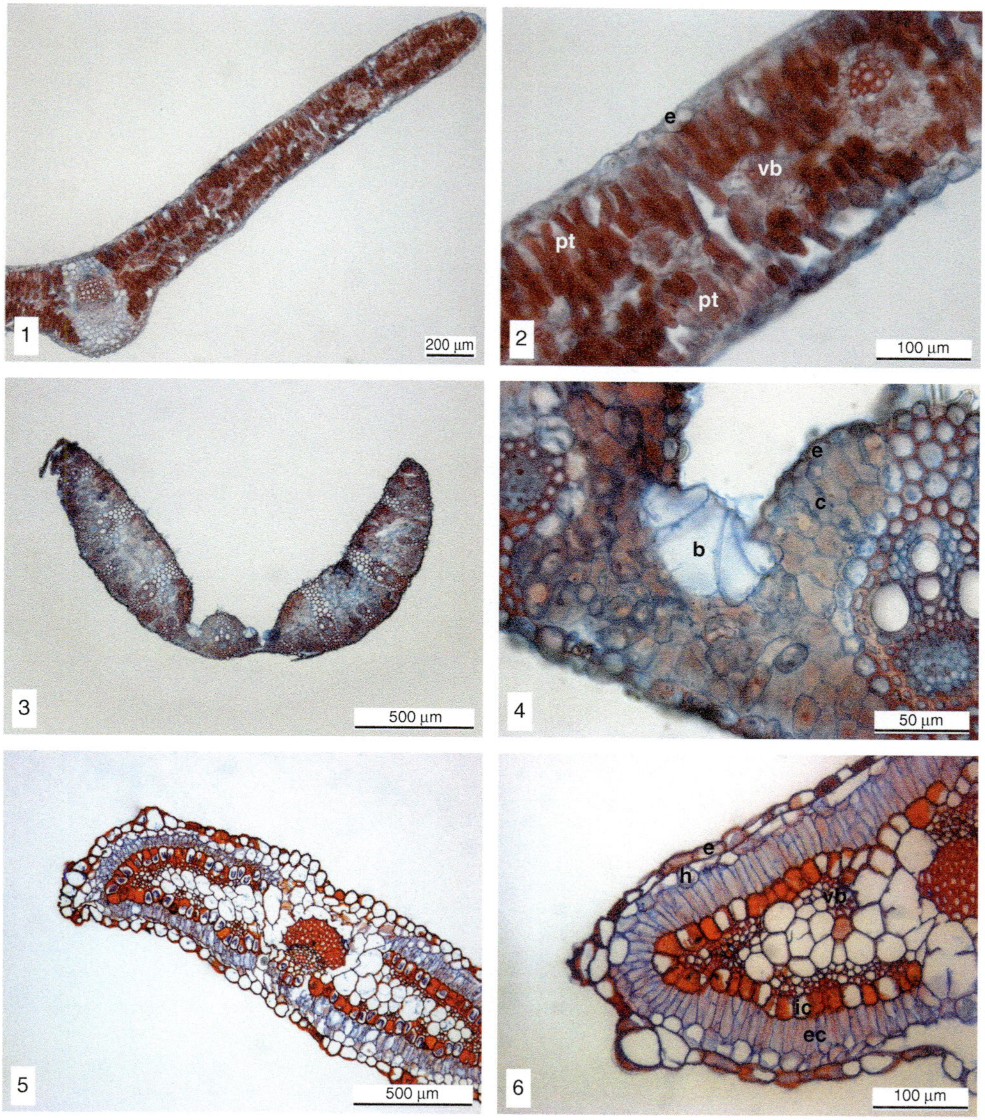

Fig. 10.1. Transverse section through the leaves of *Atriplex littoralis* (1 and 2), *Puccinellia distans* (3 and 4) and *Camphorosma monspeliaca* (5 and 6): e, epidermis; pt, palisade tissue; vb, vascular bundle; c, chlorenchyma; b, bulliform cell; h, hypodermis; ec, external chlorenchyma; ic, internal chlorenchyma.

western Asia. It belongs to the C_3 group of plants. Epidermal cells of *C. maritima* are covered with a cuticle. The mesophyll is composed of chlorenchyma and parenchyma tissue (water-storage tissue). The chlorenchyma is distributed in the outer part of the leaf, in the form of 3–4 layers of narrow cells of palisade tissue; the water-storage tissue is composed of 4–6 layers of large thin-walled, round to polygonal parenchyma cells (Figs 10.2.1 and 10.2.2). The vascular system is represented by several collateral bundles.

Salsola soda (Amaranthaceae) is an annual shrub native to the Mediterranean basin. Both the leaves and stem of *S. soda* are succulent. Transverse leaf sections are triangular (Fig. 10.2.3). The leaves have a single layer of epidermis covered with a thick cuticle. Below the epidermis is one layer of thin-walled cells (hypodermis),

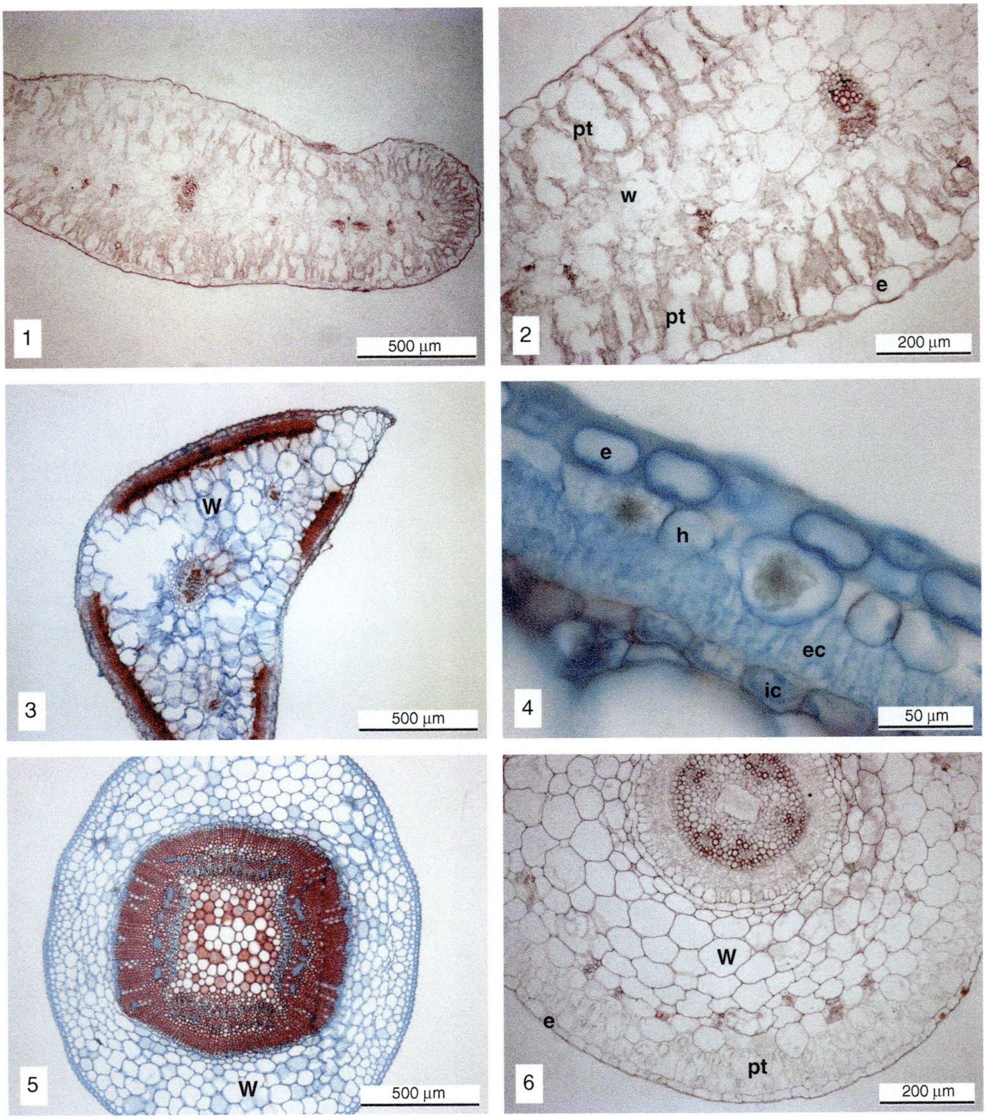

Fig. 10.2. Transverse section through the leaf of *Cakile maritima* (1 and 2), *Salsola soda* (3 and 4), transverse section through the stem of *Salsola soda* (5) and through the succulent shoot of *Salicornia fruticosa* (6): e, epidermis; pt, palisade tissue; w, water-storage tissue; h, hypodermis; ec, external chlorenchyma; ic, internal chlorenchyma.

which may contain crystals of calcium oxalate. In the central part of the leaf lies the main vascular bundle surrounded by thin-walled water-storage parenchyma tissue. *S. soda* is a C_4 plant with two types of photosynthetic tissue. Between the hypodermis and the water-storage tissue are two different layers of chlorenchyma: the outer is palisade-like, and the inner is bundle-sheath-like, and this layer is positioned close to the smaller peripherally located bundles (Fig. 10.2.4). The chlorenchyma layers are not continuous, and are interrupted in the region of the central vascular bundle as well as on the leaf edges (Fig. 10.2.3).

The stem is covered by epidermis on the surface, under which is distributed a very well-developed cortex made of water-storage parenchyma

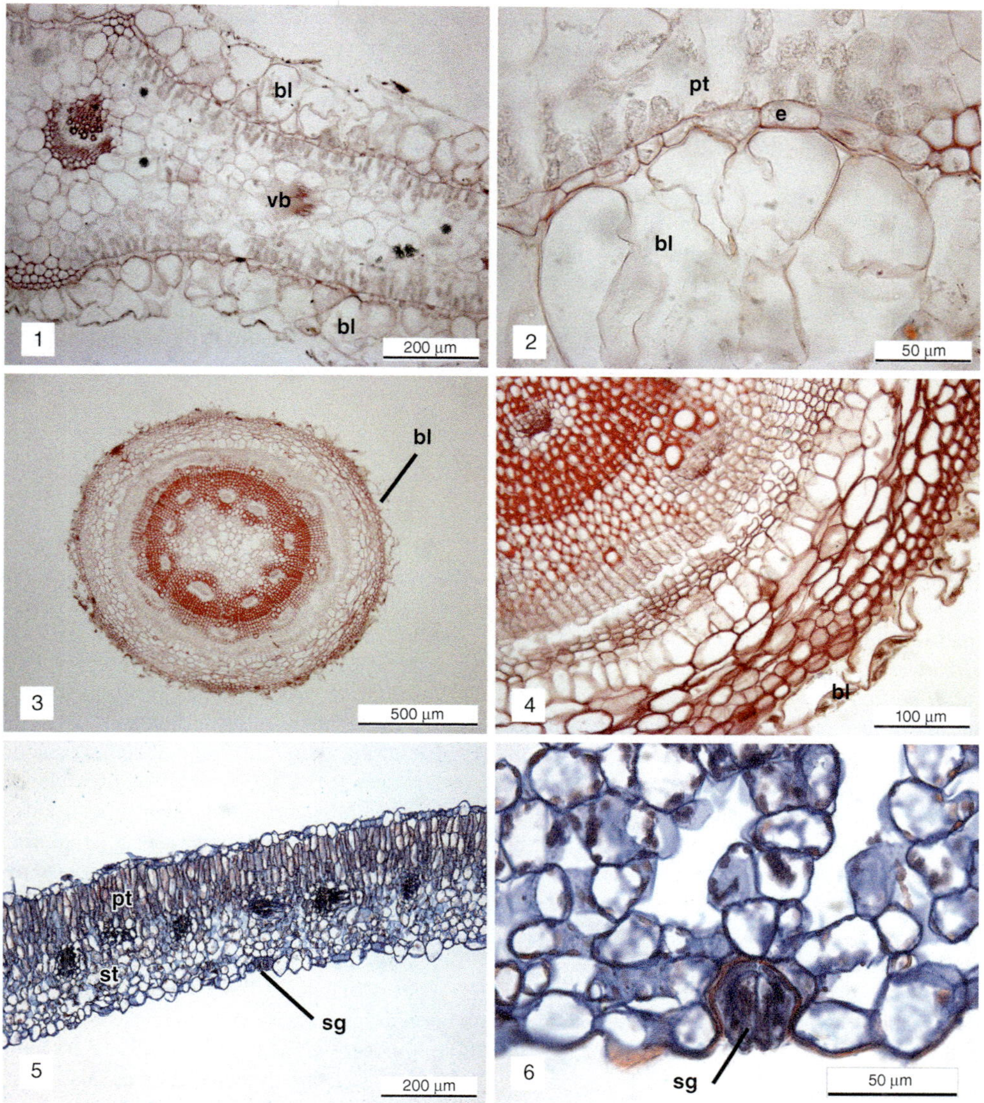

Fig. 10.3. Transverse section through the leaf of *Halimione portulacoides* (1 and 2); transverse section through the stem of *Halimione portulacoides* (3 and 4) and transverse section through the leaf of *Limonium gmelinii* (5 and 6): e, epidermis; pt, palisade tissue; bl, salt bladder; st, spongy tissue; vb, vascular bundle; sg, salt gland.

cells; the central cylinder with secondary elements of phloem and xylem forms a ring in the central part of the stem (Fig. 10.2.5).

Salicornia fruticosa (Amaranthaceae) is a succulent halophyte growing in the salt marshes and beaches of North America, Europe, South Africa and south Asia. The stems of *S. fruticosa* have reduced leaves, since the upper side of the leaf is fused with the stem (Fahn and Arzee, 1959). On the surface of these succulent shoots is a single-layered epidermis with cutinized outer cell walls. *S. fruticosa* is a C_3 species. Below the epidermis is palisade tissue made of 2–3 rows of elongated cells. In the palisade tissue the elongated elements similar to tracheids (tracheoid idioblasts) are found, with characteristic spiral thickenings. These are located among the palisade cells and are not directly connected to the epidermis and to vascular system of the plant. Beneath the palisade tissue is a well-developed parenchymatous cortex representing water-storage tissue. The last layer of the cortex is a thin-walled endodermis, and beneath the endodermis is a single layer of pericycle. In the central cylinder, the open collateral vascular bundles are arranged in a circle, and in the central part of the shoot there are pith parenchyma cells forming a small central cavity (Fig. 10.2.6).

10.4.3 Salt-excreting plants

10.4.3.1 Salt bladders: Halimione portulacoides

Halimione portulacoides is small shrub, found in salt marshes and coastal dunes, and widely distributed in temperate Eurasia and parts of Africa. *Halimione* is a genus of salt-excreting halophytes. In *H. portulacoides* the leaves are covered with salt trichomes (salt bladders) that form a dense layer on both leaf surfaces (Fig. 10.3.1), enabling elimination of excess salts from the plant. The salt bladder structures consist of a stalk cell and a large secretory cell (bladder cell) on the top. Below the single-layered epidermis is palisade tissue (Fig. 10.3.2), and in the central part of the leaf the vascular bundles are found, surrounded by water-storage tissue (Fig. 10.3.1).

On the surface of the stem is the epidermis, which also contains numerous salt bladders. Below the epidermis is the cortex, and below the cortex is a central cylinder showing the beginning of a secondary growth. A continuous ring of cambium is evident, forming secondary phloem and xylem, and in the middle of the stem is a large amount of parenchyma (Figs 10.3.3 and 10.3.4).

10.4.3.2 Salt glands: Limonium gmelinii

Limonium gmelinii is a perennial species that can grow in saline water meadows, as well as on dry slopes. It is distributed from central and south-east Europe, through central Asia to Siberia. The leaves of *L. gmelinii* exhibit a dorsiventral structure. The leaf mesophyll consists of palisade and spongy tissue in a ratio of about 1:1 (Fig. 10.3.5). Sclereid idioblasts of irregular shape may be observed in the mesophyll. The leaf surface of *L. gmelinii* is covered by a single-layered epidermis. Stomata occur on both leaf sides. Salt glands are also present on both adaxial and abaxial leaf surfaces (Fig. 10.3.6), determining the species as a crynohalophyte. At maturity, the salt glands are composed of 16 secretory cells arranged in four circles, overlying four sub-basal collecting cells (Dickison, 2000; Zoric *et al.*, 2013). Salts are deposited in the subcuticular cavity between the cell wall of secretory cells and the cuticle. A cuticle over the gland is perforated above the secretory cells to allow salt excretion.

10.5 Conclusions: Relations of Plant Structure and Soil Salinity

Halophytes with different preferences for salinity conditions also show differences in expression of anatomical adaptations to saline habitats.

According to Ding *et al.* (2010), core halophytes may be classified as excretive and succulent. Besides developing either well-developed water-storing tissues, and leaf and stem succulence or salt secretory structures (the salt trichomes and salt glands), their anatomical adaptations to saline conditions often include a foliar Kranz anatomy structure; widening of the Casparian band and well-developed root endodermis; and the phenomenon of successive cambia in the roots with increased lignin content (Grigore and Toma, 2010a). Mucilage cells may also be abundant in some species. Extreme halophytes (the most salt-tolerant species), which are strictly restricted to saline areas, exhibit the most expressed anatomical adaptations.

Mesohalophytes can grow in different environmental conditions, but usually not in extremely salinized habitats. They rely on other physiological mechanisms of salt tolerance (e.g. salt exclusion), rather than on succulence based upon vacuolar compartmentation as in euhalophtes. Mesohalophytes usually show some xerophytic features, which enable them to grow in conditions of increased salinity, including thick epidermis covered with cuticle and wax; abundant trichomes; smaller leaves; sunken stomata, small stomata and reduced number of stomata; a well-developed endodermis; changes in diameter of xylem vessels and the presence of bulliform cells, as well as developing water-storage tissue and increasing succulence in some cases. Therefore, this group could be placed between xerophytes and succulent halophytes (Grigore and Toma, 2010a).

Plant responses to salt and water stress have much in common and it is not always easy to distinguish between the morphological and anatomical adaptations serving to avoid conditions of water deficit, and those adaptations that enable survival in saline habitats. Both dry and saline habitats cause the effect of physiological drought, which can be overcome by similar morpho-anatomical, physiological and ecological solutions. In comparison to water deficit, salinity affects plant homeostasis not only by osmotic stress, but also by ion toxicity, which targets a range of physiological and metabolic processes. In summary, plant tolerance to salinity involves two main strategies, which can be observed at a morpho-anatomical level, for avoiding the toxic effect of salt uptake: (1) dilution of salts (increasing succulence); and (2) excretion of salts by specialized peripheral structures, the salt trichomes and salt glands. Halophytes that evolved such mechanisms are able to survive in the most extreme saline habitats and can successfully germinate, grow and complete their life cycle under these conditions.

References

Abd Elbar, O.H. and Abd El-Maboud, M.M. (2013) Anatomical and physiological responses of three species of *Suaeda* Forssk. ex Scop. under different habitat conditions. *Journal of Applied Sciences Research* 9(8), 5370–5379.

Abu-Serag, N.A. (2015) The stem anatomy of the stem abatomy of *Lactuca* L. (Asteraceae) and related genera in Iraq. *Journal of Environmental Sciences* 44, 731–739.

Agarie, S., Shimoda, T., Shimizu, Y., Baumann, K., Sunagawa, H., *et al*. (2007) Salt tolerance, salt accumulation, and ionic homeostasis in an epidermal bladder-cell-less mutant of the common ice plant *Mesembryanthemum crystallinum*. *Journal of Experimental Botany* 58, 1957–1967.

Akcin, A.T., Akcin, A. and Yalcin, E. (2014) Anatomical adaptations to salinity in *Spergularia marina* (Caryophyllaceae) from Turkey. *Proceedings of the National Academy of Sciences, India - Section B. Biological Sciences* 85, 625–634.

Akcin, A.T., Akcin, A. and Yalcin, E. (2017) Anatomical changes induced by salinity stress in *Salicornia freitagii* (Amaranthaceae). *Brazilian Journal of Botany* 40, 1013–1018.

Akhani, H., Barroca, J., Koteeva, N., Voznesenskaya, E., Franceschi, V., *et al*. (2005) *Bienertia sinuspersici* (Chenopodiaceae): A new species from southwest asia and discovery of a third terrestrial C_4 plant without Kranz anatomy. *Systematic Botany* 30, 290–301.

Akhani, H., Malekmohammadi, M., Mahdavi, P., Gharibiyan, A. and Chase, M.W. (2013) Phylogenetics of the Irano-Turanian taxa of *Limonium* (Plumbaginaceae) based on ITS nrDNA sequences and leaf anatomy provides evidence for species delimitation and relationships of lineages. *Botanical Journal of the Linnean Society* 171, 519–550.

Al Hassan, M., Gohari, G., Boscaiu, M., Vicente, O. and Grigore, M.N. (2015) Anatomical modifications in two *Juncus* species under salt stress conditions. *Notulae Botanicae Horti Agrobotanici* 43, 501–506.

Al-Adily, B.M.H., Al-Ani, H.A. and Al-Taee, M.M.S. (2016) Anatomical variation of some halophytes from different sites in Babylon Provinus. *Mesopotamia Environment Journal* 2, 19–26.

Al-Taisan, W.A. and Gabr, D.G. (2017) Comparative morphological and anatomical characters of *Cakile arabica* from different habitat in eastern region of Saudi Arabia. *Saudi Journal of Biological Sciences* 24, 226–233.

Amarasinghe, V. and Watson, L. (1988) Comparative ultrastructure of microhairs in grasses. *Botanical Journal of the Linnean Society* 98, 303–319.

Artschwager, E.F. (1920) On the anatomy of *Chenopodium album* L. *American Journal of Botany* 7, 252–260.

Aydosova, S.S., Zhussupova, G.E., Akhtayeva, N.Z., Gadetskaya, A.V. and Abilov, Z.A. (2012) Macro- and microscopy of upper parts from *Limonium gmelinii* genus plants. *International Journal of Biology and Chemistry* 4, 3–6.

Badea, L.M. and Zamfirache, M.M. (2011) Anatomical research of *Artemisia santonica* and *Artenisia scoparia* (Asteraceae). *Analele ştiinţifice ale Universităţii Al. I. Cuza" Iaşi Biologie vegetală* 57(2) (s. 2 a), 21–24.

Balsamo, R.A. and Thomson, W.W. (1993) Ultrastructural features associated with secretion in the salt glands of *Frankenia grandifolia* (Frankeniaceae) and *Avicennia germinans* (Avicenniaceae). *American Journal of Botany* 80, 1276–1283.

Barhoumi, Z., Djebali, W., Abdelly, C., Chaïbi, W. and Smaoui, A. (2008) Ultrastructure of *Aeluropus littoralis* leaf salt glands under NaCl stress. *Protoplasma* 233, 195–202.

Batanouny, K.H. and Sitta, A.Y.M. (1977) Eco-physiological studies on halophytes in arid and semi-arid zones. I-Autecology of the salt-excreting halophyte *Limoniastrum monopetalum*. *Acta Botanica Academiae Scientiarum Hungaricae* 23, 13–31.

Batanouny K.H., Hassan, A.H. and Fahmy, G.M. (1992) Eco-physiological studies on halophytes in arid and semi-arid zones. II. Eco-physiology of *Limonium delicatulum* (GIR.) KT. *Flora* 186, 105–116.

Bercu, R. and Bavaru, E. (2004) Anatomical aspects of *Salsola kali* subsp. *ruthenica* (Chenopodiaceae). *Phytologia Balcanica* 10, 227–232.

Bercu, R. and Bavaru, E. (2005) Contributii la cunoasterea anatomiei speciei *Salicornia europaea* L. (Chenopodiaceae). *Lucr. St. Univ. St. Agr. Med. Vet."Ion Ionescu de la Brad" Iasi Ser Hort.* 47, 625–630.

Bercu, R., Făgăraş, M. and Broască, L. (2012) Anatomical features of *Aster tripolium* L. (Asteraceae) to saline enironments. *Annals of the Romanian Society for Cell Biology* 17, 271–278.

Bickenbach, K. (1932) Zur anatomie und physiologie einiger strand und dünenpflanzen. Beitrage zum halophytenproblem. *Beiträge zur Biologie der Pflanzen* 19, 334–370.

Bosabalidis, A.M. (2010) Wall protuberance formation and function in secreting salt glands of *Tamarix aphylla* L. *Acta Botanica Croatica* 69, 229–235.

Boughalleb, F., Denden, M. and Tiba, B.B. (2009) Anatomical changes induced by increasing NaCl salinity in three fodder shrubs, *Nitraria retusa, Atriplex halimus* and *Medicago arborea*. *Acta Physiologia Plantarum* 31, 947–960.

Campbell, C.J. and Strong, J.E. (1964) Salt gland anatomy in *Tamarix penandra* (Tamaricaceae). *The Southwest Naturalist* 9, 232–238.

Campbell, N. and Thomson, W.W. (1976) The ultrastructure of *Frankenia* salt glands. *Annals of Botany* 40, 681–686.

Carraro, L., Patrignani, G., Iacumin, P. and Orsenigo, M. (1993) Leaf morphology and carbon isotope discrimination in members of the genus *Salsola*. II. Studies on *Salsola soda* L. *Caryologia* 46, 335–342.

Céccoli, G., Ramos, J., Pilatti, V., Dellaferrera, I., Tivano, J.C., *et al.* (2015) Salt glands in the Poaceae family and their relationship to salinity tolerance. *The Botanical Review* 81, 162–178.

Cheeseman, J.M. (2015) The evolution of halophytes, glycophytes and crops, and its implications for food security under saline conditions. *New Phytologist* 206, 557–570.

Ciccarelli, D., Balestri, M., Pagni, A.M. and Laura, F.M.C. (2010) Morpho-functional adaptations in *Cakile maritima* Scop. subsp. *maritima*: comparation of two different morphological types. *Caryologia* 63, 411–421.

Clarke, A.E., Andreson, R.L. and Stone, B.A. (1979) Form and function of arabinogalactans and arabinogalactan-proteins. *Phytochemistry* 18, 521–540.

Colombo, P. (2002) Morpho-anatomical and taxonomical remarks on *Limonium* (Plumbaginaceae) in Sicily. *Flora Mediterranea* 12, 389–412.

Cornara, L., D'Arrigo, C., Pioli, F., Borghesi, B., Bottino, C., *et al.* (2009). Micromorphological investigation on the leaves of the rock samphire (*Crithmum maritimum* L.): Occurrence of hesperidin and diosmin crystals. *Plant Biosystems* 143, 283–292.

Costa, M.V.L. and Tavares, E.S. (2006) Anatomia foliar de *Chenopodium ambrosioides* L. (Chenopodiaceae) – erva-de-Santa-Maria. *Revista Brasileira de Plantas Medicinais* 8, 63–71.

Crawford, R.M.M. (1989) *Studies in Plant Survival. Ecological Case Histories of Plant Adaptation to Adversity*. Blackwell, Oxford, UK.

Crivellaro, A. and Schweingruber, F.H. (2015) *Stem anatomical features of dicotyledons: Xylem, phloem, cortex and periderm characteristics for ecological and taxonomical analysis*. Kessel Publishing House, Remagen-Oberwinter, Germany.

Cuadra, P.V. and Cambi, V. (2014) Morphoanatomical functional traits in xerophytic species of a saline environment. *Phyton* 83, 389–396.

Dajic, Z. (2006) Salt stress - salinity and tolerance mechanisms in plants. In: Madhava Rao, K.V., Raghavendra, A.S. and Reddy, K.J. (eds) *Physiology and Molecular Biology of Stress Tolerance in Plants*. Springer, Dordrecht, Netherlands, pp. 41–99.

Das, S. (2002) On the ontogeny of stomata and glandular hairs in some Indian mangroves. *Acta Botanica Croatica* 61, 199–205.

Das, S., Chiou-Rong, S. and Yuen-Po Yang (2014) Leaf micromorphology and leaf glandular hair ontogeny of *Myoporum bontioides* A. Gray. *Feddes Repertorium* 124, 50–60.

Dassanayake, M. and Larkin, J.C. (2017) Making plants break a sweat: the structure, function, and evolution of plant salt glands. *Frontiers in Plant Science* 8, 724. doi: 10.3389/fpls.2017.00724.

Delane, R., Greenway, H., Munns, R. and Gibbs, J. (1982) Ion concentration and carbohydrate status of the elongating leaf tissue of *Hordeum vulgare* growing at high external NaCl. 1. Relationship between solute concentration and growth. *Journal of Experimental Botany* 33, 557–573.

Dickison, W.C. (2000) *Integrative Plant Anatomy*. Harcourt Academic Press, San Diego, CA, USA.

Ding F., Yang, J.C., Yuan, F. and Wang, B.S. (2010) Progress in mechanism of salt excretion in recretohalopytes. *Frontiers in Biology* 5, 164–170.

Edwards, G.E. and Voznesenskaya, E.V. (2011) C_4 photosynthesis: Kranz forms and single-cell C_4 in terrestrial plants. In: Raghavendra, A.S. and Sage, R.F. (eds) *C_4 Photosynthesis and Related CO_2 Concentrating Mechanisms*. Springer, Netherlands, pp. 29 –61.

Eshel, A. (1985) Effects of NaCl and KCl on growth and ionic composition of the halophytic C_4 succulent chenopods *Salsola kali, Suaeda monoica* and *Suaeda aegyptiaca*. *Australian Journal of Plant Physiology* 64, 308–315.

Evert, R.F. (2006) *Esau's Plant Anatomy: Meristems, Cells, and Tissues of the Plant Body: Their Structure, Function, and Development*. 3rd edn. John Wiley & Sons, Hoboken, NJ, USA.

Fahn, A. (1988) Secretory tissues in vascular plants. *New Phytologist* 108, 229–257.

Fahn, A. (1990) *Plant Anatomy*. 4th edn. Pergamon Press, Oxford, UK.

Fahn, A. and Arzee, T. (1959) Vascularization of the articulated Chenopodiaceae and the nature of their fleshy cortex. *American Journal of Botany* 46, 330–338.

Fahn, A. and Cutler, D.F. (1992) *Xerophytes*. Handbuch der Pflanzenanatomie XIII (3), Gebrüd er Borntra eger, Stuttgart, Berlin.

Fahn, A. and Zimmermann, M.H. (1982) Development of the successive cambia in *Atriplex halimus* (Chenopodiaceae). *Botanical Gazette* 143, 353–357.

Feng, Z., Deng, Y., Zhang, S., Liang, X., Yuan, F., *et al.* (2015) K^+ accumulation in the cytoplasm and nucleus of the salt gland cells of *Limonium bicolor* accompanies increased rates of salt secretion under NaCl treatment using NanoSIMS. *Plant Science* 238, 286–296.

Fitter, A.H. and Hay, R.K.M. (1989) *Environmental Physiology of Plants*. 2nd edn. Academic Press, London, pp. 157.

Flowers, T.J. and Colmer, T.D. (2008) Salinity tolerance in halophytes. *New Phytologist* 179, 945–963.

Flowers, T.J., Flowers, S.A., Hajibagheri, M.A. and Yeo, A.R. (1990) Salt tolerance in the halophytic wild rice, *Porteresia coarctata* Tateoka. *New Phytologist* 114, 675–684.

Flowers, T.J., Galal, H.K. and Bromham, L. (2010) Evolution of halophytes: Multiple origins of salt tolerance in land plants. *Functional Plant Biology* 37, 604–612.

Freitag, H. and Kadereit, G. (2014) C_3 and C_4 leaf anatomy types in Camphorosmeae (Camphorosmoideae, Chenopodiaceae). *Plant Systematics and Evolution* 300, 665–687.

Geissler, L.N., Hussin, S. and Koyro, H.W. (2009) Evaluated atmospheric CO_2 concentration ameliorates effects of NaCl salinity on photosynthesis and leaf structure of *Aster tripolium*. *Journal of Experimental Botany* 60, 137–151.

Ghanem, M.E., Han, R.M., Classen, B., Quetin-Leclerq, J., Mahy, G., *et al.* (2010) Mucilage and polysaccharides in the halophyte plant species *Kosteletzkya virginica*: Localization and composition in relation to salt stress. *Journal of Plant Physiology* 167, 382–392.

Grigore, M.N. (2008) *Halofitotaxonomia. Lista Plantelor de Sărătură din România*. Pim. Iași, 1–8.

Grigore, M.N. and Toma, C. (2006) Ecological anatomy elements related to *Asteraceae* halophytes species. *Studii și Comunicări* 21, 94–98.

Grigore, M.N. and Toma, C. (2007) Histo-anatomical strategies of Chenopodiaceae halophytes: Adaptive, ecological and evolutionary implications. *WSEAS Transactions on Biology and Biomedicine* 12, 204–218.

Grigore, M.N. and Toma, C. (2008a) Ecological anatomy investigations related to some halophyte species from Moldavia. *Romanian Journal of Biology –Plant Biology* 53, 23–30.

Grigore, M.N. and Toma, C. (2008b) A histo-anatomical study on some halophylous species of the *Lepidium* genus. *Studia Universitatis "Vasile Goldiș", Seria Științele Vieții [Life Sciences Series]* 18, 27–31.

Grigore, M.N. and Toma, C. (2010a) A proposal for a new halophytes classification, based on integrative anatyomy observations, Muzeul Olteniei Craiova. Oltenia. *Studii și comunicări* 26, 45–50.

Grigore, M.N. and Toma, C. (2010b) *Structuri Secretoare de Săruri la Halofite. O Abordare Integrativă* [*Salt-secreting structures of Halophytes. An integrative approach*]. Academiei Române, București, pp. 243–253.

Grigore, M.N. and Toma, C. (2010c) *Halofitele. Aspecte De Anatomie Ecologica*. Universitatii Alexandru Ioan Cuza, Iasi, Romania.

Grigore, N.M. and Toma, C. (2017) *Anatomical Adaptations of Halophytes. A Review of Classic Literature and Recent Findings*. Springer, Swizerland, pp. 548.

Grigore, M.N., Toma, C. and Boscaiu, M. (2010) Ecological implications of bulliform cells on halophytes, in salt and water stress natural conditions. *Analele ştiinţifice ale Universităţii "Al. I. Cuza". Biologie vegetală* 56(2), (2 a), 5–15.

Grigore, M.N., Toma, C. and Boscaiu, M. (2011a) Ecological notes on halophytes species from Mediterranean climate. *Lucr. Ş. t. (Horticultură), USAMV"Ion Ionescu de la Brad"Iasi* 54, 29–34.

Grigore, M.N., Boscaiu, M. and Vicente, O. (2011b) Ecological notes in Mediterranean halophytes. Towards an integrative approach. *Ecological Questions* 14, 11–14.

Grigore, M.N., Toma, C., Zamfirache, M.M., Boscaiu, M., Olteanu, Z., *et al.* (2012a) Ecological anatomy in halophytes with C_4 photosynthesis: Discussing adaptative features in endangered ecosystems. *Carpathian Journal of Earth and Environmental Sciences* 7(2), 13–21.

Grigore, M.N., Toma, C., Boscaiu, M., Zamfirache, M.M. and Vanescu, L. (2012b) Anatomical and ecological observations on psammo-halophytes species (Eastern part of Spain). *Seria Horticultura* 55, 19–24.

Grigore, M.N., Toma, C., Zamfirache, M.M. and Ivănescu, L. (2012c) A survey of anatomical adaptations in Romanian halophytes. Towards an ecological interpretation. *Fresenius Environmental Bulletin* 21, 11b, 3370–3375.

Grigore, M.N., Toma, C., Zamfirache, M.M., Ivanescu, L. and Daraban, I. (2013) Anatomical and ecological observations in succulent (articulated) halophytes from Chenopodiaceae. *Lucr. Şt. (Horticultură), USAMV „Ion Ionescu de la Brad"Iaşi* 56(1), 25–30.

Grigore, M.N., Ivanescu, L. and Toma, C. (2014) *Halophytes: An Integrative Anatomical Study*, Springer, Cham, Switzerland, pp. 548.

Gucci, R., Aronne, G., Lombardini, L. and Tattini, M. (1997) Salinity tolerance in *Phillyrea* species. *New Phytologist* 135, 227–234.

Hajibagheri, M.A., Hall, J.L. and Flowers, T.J. (1984) Stereological analysis of leaf cells of the halophyte *Suaeda maritima* (L.) Dum. *Journal of Experimental Botany* 35, 1547–1557.

Hameed, M., Ashraf, M., Naz, N. and Al-Qurainy, F. (2010) Anatomical adaptations of *Cynodon dactylon* (L.) Pers., from the salt range Pakistan, to salinity stress. I. Root and stem anatomy. *Pakistan Journal of Botany* 42, 279–289.

Hameed, M., Ashraf, M., Naz, N., Nawaz, T., Batool, R., *et al.* (2013) Anatomical adaptations of *Cynodon dactylon* (L.) Pers. from the salt range (Pakistan) to salinity stress. II leaf anatomy. *Pakistan Journal of Botany* 45, 133–142.

Hess, W.M., Hansen, D.J. and Weber, D.J. (1975) Light and electron microscopy localization of chloride ions in cells of *Salicornia pacifica* var. *utahensis*. *Canadian Journal of Botany* 53, 1176–1187.

Ianovici, N. (2011) Histoanatomical and ecophysiological studies on some halophytes from Romania - *Plantago schwarzenbergiana*. *Annals of West University of Timişoara, ser. Biology* 14, 53–64.

Ioannidou-Akoumianaki, A., Spentza, R.P. and Fasseas, C. (2015) *Limoniastrum monopetalum* (L.) Boiss, a candidate plant for use in urban and suburban areas with adverse conditions. An anatomical and histochemical study. *Bulletin of the University of Agricultural Sciences and Veterinary Medicine Cluj-Napoca Horticulture* 72, 438–440.

Ish-Shalom-Gordon, N. and Dubinsky, Z. (1992) Ultrastructure of the pneumatophores of the mangrove *Avicennia marina*. *South African Journal of Botany* 58, 358–362.

Ivanescu, L., Grigore, M.N. and Toma, C. (2016) Contributions to the knowledge of vegetative organs structure of *Iris halophila* Pall. *Analele Stiintifice ale Universitatii Alexandru Ioan Cuza din Iasi. Biologie Vegetala* 62 (2A),108–109.

Jacobs, S.W.L. (2001) Review of leaf anatomy and ultrastructure in the Chenopodiaceae (Caryophyllales). *Journal of the Torrey Botanical Society* 128, 236–253.

Jacobsen, A.L., Ewers, F.W., Pratt, R.B., Paddock, W.A. and Davis, S.D. (2005) Do xylem fibers affect vessel cavitation resistance? *Plant Physiology* 139, 546–556.

Jaiswar, S. and Kazi, M.A. (2016) Composition and localization of mucilage under short term NaCl stress in *Salicornia brachiata* Roxb. and *Suaeda maritime* (L.) Dumort. *Indian Journal of Natural Products and Resources* 7, 162–168.

Jáuregui, D., Castro, M., Ruiz-Zapata, T. and Lapp, M. (2014) Anatomía de los órganos vegetativos de dos especies de Atriplex (Chenopodiaceae) de Venezuela. *Revista de Biología Tropical* 62, 1625–1636.

Jianu, L.D., Bercu, R. and Popoviciu, R.D. (2014) Anatomical features of the endegered plant *Cakile maritime* Scop. subsp. *euxina* (Pobed.). NYÁR. *Annals of West University of Timişoara, ser. Biology* 17(2), 79–86.
Kadereit, G., Lauterbach, M., Pirie, M.D., Arafeh, R. and Freitag, H. (2014) When do different C_4 leaf anatomies indicate independent C_4 origins? Parallel evolution of C_4 leaf types in Camphorosmeae (Chenopodiaceae). *Journal of Experimental Botany* 65(13), 3499–3511.
Keshavarzi, M. and Zare, G. (2006) Anatomical study of Salicorniae Dumort. (Chenopodiacea Vent.) native to Iran. *International Journal of Botany* 2(3), 278–285.
Kiani-Pouya, A., Roessner, U., Jayasinghe, N.S., Lutz, A., Rupasinghe, T., *et al.* (2017) Epidermal bladder cells confer salinity stress tolerance in the halophyte quinoa and Atriplex species. *Plant, Cell and Environment* 40, 1900–1915.
Kong, L. and Li, F. (2008) Anomalous secondary structures in stem and root of two typical halophytes. *Nordic Journal of Botany* 26, 364–367.
Koyro, H.W., Geißler, N., Hussin, S., Debez, A. and Huchzermeyer, B. (2008) Strategies of halophytes to survive in a salty environment. In: Khan, N.A. and Singh, S. (eds) *Abiotic Stress and Plant Responses*. I.K. International Publishing House, New Delhi, pp. 83–104.
Kozlowsk, T.T. (1997) *Responses of Woody Plants to Fooding and Salinity*. Tree Physiology Monograph No. 1, Heron Publishing, Victoria, B.C., Canada, pp. 1–29.
Krishnamurthy, P., Jyothi-Prakash, P.A., Qin, L., He, J., Lin, Q. *et al.* (2014) Role of root hydrophobic barriers in salt exclusion of a mangrove plant *Avicennia officinalis*. *Plant, Cell and Environment* 37, 1656–1671.
Ksouri, R., Megdiche, W., Debez, A., Falleh, H., Grignon, C. *et al.* (2007) Salinity effects on polyphenol content and antioxidant activities in leaves of the halophyte *Cakile maritima*. *Plant Physiology and Biochemistry* 45, 244–249.
Levering, C.A. and Thomson, W.W. (1971) The ultrastructure of the salt gland of *Spartina foliosa*. *Planta* 97, 183–196.
Liphschitz, N. and Waisel, Y. (1974) Existence of salt glands in various genera of the Gramineae. *New Phytologist* 73, 507–513.
Liu, Z.H., Shi, L.R. and Zhao, K.F. (2006) The morphological structure of salt gland and salt secretion in *Aeluropus littoralis* var. *sinensis* Debeaux. *Journal of Plant Physiology and Molecular Biology* 32, 420–426.
Maas, E.V. and Nieman, R.H. (1978) Physiology of plant tolerance to salinity. In: Jung, G.A. (ed.) *Crop Tolerance to Suboptimal Land Conditions*. American Society of Agronomy, Madison, WI, USA, pp. 277–299.
Mahi Z., Dedaldechamp, F., Belkhodja, M. and Lemoine, R. (2015) Anatomical features of *Atriplex halimus* L. to saline environments. *International Journal of Scientific Research in Science and Technology* 1, 69–76.
Makbul S., Coskuncelebi, K., Türkmen, Z. and Beyazoglu, O. (2011) Comparison of foliar anatomy of *Scorzonera* L. (Asteraceae) taxa from north east Anatolia. *Pakistan Journal of Botany* 43, 135–155.
Malcolm, C.V., Lindley, V.A., O'Leary, J.W., Runciman, H.V. and BarrettLennard, E.G. (2003) Halophyte and glycophyte salt tolerance at germination and the establishment of halophyte shrubs in saline environments. *Plant Soil* 253, 171–185.
Marcum, K.B. (1999) Salinity tolerance mechanisms of grasses in the subfamily Chloridoideae. *Crop Science* 39, 1153–1160.
Marcum, K.B. (2006) Use of saline and non-potable water in the turfgrass industry: Constraints and developments. *Agricultural Water Management* 80, 132–146.
Marcum, K.B. (2008) Saline tolerance physiology in grasses. In: Khan, M.A. and Weber, D.J. (eds) *Ecophysiology of High Salinity Tolerant Plants*. Springer, Berlin, pp. 157–172.
Marcum, K.B. and Murdoch, C.L. (1994) Salinity tolerance mechanisms of six C_4 turfgrasses. *Journal American Society Horticultural Science* 119, 779–784.
Marcum, K.B., Anderson, S.J. and Engelke, M.C. (1998) Salt gland ion secretion: A salinity tolerance mechanism among five zoysiagrass species. *Crop Science* 38, 806–810.
Maricle, B.R., Koteyeva, N.K., Voznesenskaya, E.V., Thomasson, J.R. and Edwards, G.E. (2009) Diversity in leaf anatomy, and stomatal distribution and conductance, between salt marsh and freshwater species in the C_4 genus Spartina (Poaceae). *New Phytologist* 184, 216–233.
Mauseth, J.D. (1988) *Plant Anatomy*. Benjamin Cummings Publishing Co., Menlo Park, CA, USA, pp. 560.
Meinzer, F.C. and Zhu, J. (1999) Efficiency of C_4 photosynthesis in *Atriplex lentiformis* under salinity stress. *Functional Plant Biology* 26, 79–86.

Metcalfe, C.R. and Chalk, L. (1972) *Anatomy of the Dicotyledons. Leaves, Stem, and Wood in Relation to Taxonomy*. Oxford University Press, Oxford, UK.

Milic, D., Lukovic, J., Đan, M., Zoric, L., Obreht, D., *et al.* (2011) Identification of *Salicornia population*: Anatomical characterization and RAPD fingerprinting. *Archive of Biological Science* 63, 1087–1098.

Milic, D., Lukovic, J.Ž., Zoric, L. and Merkulov, L.S. (2013) Structural adaptation of *Salsola soda* L. (Chenopodiaceae) from inland and maritime saline area. *Journal for Natural Sciences* 125, 55–67.

Muhaidat, R., Sage, R.F. and Dengler, N.G., (2007) Diversity of Kranz anatomy and biochemistry in C_4 eudicots. *American Journal of Botany* 94, 362–381.

Naidoo, G. (2016) The mangroves of South Africa: An ecophysiological review. *South African Journal of Botany* 107, 101–113.

Naidoo, Y. and Naidoo, G. (1998) Salt glands of *Sporobolus virginicus*: morphology and ultrastructure. *South African Journal of Botany* 64, 198–204.

Norouzi, M., Ghahremaninejad, F., Maassoumi, A.A. and Safavi, S.R. (2016) Anatomical studies on *Scorzonera* (Asteraceae) species, subgenera *Podospermum* and *Pseudopodospermum* in Iran. *Modern Phytomorphology* 9, 51–68.

Oakley, J.C., Sultmanis, S., Stinson, C.R., Sage, T.L. and Sage, R.F. (2014) Comparative studies of C_3 and C_4 *Atriplex* hybrids in the genomics era: Physiological assessments. *Journal of Experimental Botany* 65, 3637–3647.

Oladi, R., Gorgij, R., Emami-Nasab, M. and Nasiriani, S. (2017) Wood anatomy and physical and chemical properties of fast growing Athel tamarisk (*Tamarix aphylla* L.). *Iranian Journal of Wood and Paper Industries* 7, 511–522.

Omamt, E., Hammes, P. and Robbertse, P. (2006) Differences in salinity tolerance for growth and water-use efficiency in some amaranth (*Amaranthus* spp.) genotypes. *New Zealand Journal of Crop and Horticultural Science* 34, 11–22.

Ong, J.E. and Gong, W.K. (2013) Structure, function and management of mangrove ecosystems. In: *International Society for Mangrove Ecosystems* (ISME). In: Chan, H.T. (ed.) (Yokohama: International Tropical Timber Organization), Okinawa, Japan, 1–71.

Oross, J.W. and Thomson, W.W. (1982) The ultrastructure of the salt glands of *Cynodon* and *Distichlis* (Poaceae). *American Journal of Botany* 69, 939–949.

Pammenter, N.W., Drennan, P.M. and Smith, V.R. (1986) Physiological and anatomical aspects of photosynthesis of two agrostis species at a sub-antarctic island. *New Phytologist* 102, 143–160.

Pan, Y., Guo, H., Wang, S., Zhao, B., Zhang, J., *et al.* (2016) The photosynthesis, Na^+/K^+ homeostasis and osmotic adjustment of *Atriplex canescens* in response to salinity. *Frontiers in Plant Science* 7, 848. doi:10.3389/fpls.2016.00848.

Park, J., Okita, T.W. and Edwards, G.E. (2009) Salt tolerant mechanisms in single-cell C_4 species *Bienertia sinuspersici* and *Suaeda aralocaspica* (Chenopodiaceae). *Plant Science* 176, 616–626.

Parks, G.E., Dietrich, M.A. and Schumaker, K.S. (2002) Increased vacuolar Na^+/H^+ exchange activity in *Salicornia bigelovii* Torr. in response to NaCl. *Journal of Experimental Botany* 53, 1055–1065.

Patrignani, G., Carrara, L., Lacumin, P. and Orsenigo, M. (1993) Leaf morphology and carbon isotope discrimination in members of the genus *Salsola*. I Studies on *Salsola kali* L. *Caryologia* 46, 283–291.

Paz, R.C., Reinoso, H., Espasandin, F.D., Antivilo, G.F.A., Sansberro, P.A., *et al.* (2014) Akaline, saline and mixed saline-alkaline stresses induce physiological and morpho-anatomical changes in *Lotus tenuis* shoots. *Plant Biology* (Stuttgart) 16, 1042–1049.

Perez Cuadra, V. and Cambi, V. (2014) Morphoanatomical functional traits in xerophytic species of a saline environment. *Phyton* 83, 389–396.

Polić, M.D., Luković, J., Zorić, L., Boža, P., Merkulov, L., *et al.* (2009) Morpho-anatomical differentiation of *Suaeda maritima* (L.) Dumort. 1827. (Chenopodiaceae) populations from inland and maritime saline area. *Central European Journal of Biology* 4, 117–129.

Popp, M. (1995) Salt resistance in herbaceous halophytes and mangroves. *Progress in Botany* 56, 416–429.

Purnobasuki, H. and Suzuki, M. (2005) Functional anatomy of air conducting network on the pneumatophores of a mangrove plant, *Avicennia marina* (Forsk.) Vierh. *Asian Journal of Plant Science* 4, 334–347.

Rad, A.M. and Hroudová, Z. (2013) Leaf and stem anatomy in Iranian *Bolboschoenus* species (Cyperaceae), as related to their taxonomy and ecology. *Iranian Journal of Botany* 19, 221–234.

Rao, S.S. (2011) Elucidation of mechanisms of salinity tolerance in *Zoysia matrella* cultivars: a study of structure and function of salt glands. PhD thesis. Texas A&M University, Bryan, USA, 1–146.

Reef, R. and Lovelock, C.E. (2015) Regulation of water balance in mangroves. *Annals of Botany* 115, 385–395.
Reginato, M., Travaglia, C., Reinoso, H., Garello, F. and Luna, V. (2016) Salt mixtures induce anatomical modifications in the halophyte *Prosopis strombulifera* (Fabaceae: Mimosoideae). *Flora* 218, 75–85.
Reinoso, H., Sosa, L., Ramirez, L. and Luna, V. (2004) Salt-induced changes in the vegetative anatomy of *Prosopis strombulifera* (Leguminosae). *Canadian Journal of Botany* 82, 618–628.
Reinoso, H., Sosa, L., Reginato, M. and Luna, V. (2005) Histological alterations induced by sodium sulfate in the vegetative anatomy of *Prosopis strombulifera* (Lam.) Benth . *World Journal of Agricultural Science* 1, 109–119.
Rilke, S. and Reimann, C. (1996) Morphological and ecophysiological differences between the subspecies of *Salsola kali* L. in Europe: Results of culture experiments. *Flora* 191, 363–376.
Rozema, J., Riphagen, I. and Sminia, T. (1977) A light and electron-microscopical study on the structure and function of the salt gland of *Glaux maritima*. *New Phytologist* 79, 665–671.
Rozema, J. and Gude, H. (1981) An ecophysiological study of the salt secretion of four halophytes. *New Phytologist* 89, 201–217.
Saadeddin, R.A. and Doddema, H. (1986) Anatomy of the 'extreme' halophyte *Arthrocnemum fruticosum* (L.) Moq. in relation to its physiology. *Annals of Botany* 57, 531–544.
Safiallah, S., Mahdi Hamdi, S.M., Grigore, M.N. and Jalili, S. (2017) Micromorphology and leaf ecological anatomy of *Bassia* halophyte species (Amaranthaceae) from Iran. *Acta Biologica Szegediensis* 61, 85–93.
Sage, R.F. (2016) A portrait of the C4 photosynthetic family on the 50th anniversary of its discovery: species number, evolutionary lineages, and Hall of Fame. *Journal of Experimental Botany* 67(14), 4039–4056.
Salama, F.M., El-Naggar, S.M. and Ramadan, T. (1999) Salt glands of some halophytes in Egypt. *Phyton (Horn Austria)* 39, 91–105.
Scassellati, E., Pasqua, G., Valletta, A. and Abbate, G. (2016) Salt glands of *Armeria canescens* (Host) Boiss: morphological and functional aspects. *Plant Biosystem* 150, 1134–1139.
Schreiber, L., Hartmann, K., Skrabs, M. and Zeier, J. (1999) Apoplastic barriers in roots: chemical composition of endodermal and hypodermal cell walls. *Journal of Experimental Botany* 50, 1267–1280.
Schulze, E.D., Beck, E. and Mçller-Hohenstein, K. (2005) *Plant Ecology*. Springer, New York.
Sekmen, A.H., Turkan, I., Tanyolac, Z.O., Ozfidan, C. and Dinc, A. (2012) Different antioxidant defense responses to salt stress during germination and vegetative stages of endemic halophyte *Gypsophila oblanceolata* Bark. *Environmental and Experimental Botany* 77, 63–76.
Seliskar, D.M. (1981) Morphological and anatomical responses of selected coastal salt marsh plants to soil moisture. PhD thesis. Oregon State University, Oregon, USA.
Seliskar, D.M. (1983) Root and rhizome distribution as an indicator of upper salt marsh wetland limits. *Hydrobiologia* 107, 231–236.
Semenova, G.A., Fomina, I.R. and Biel, K.Y. (2010) Structural features of the salt glands of the leaf of *Distichlis spicata* "Yensen 4a" (Poaceae). *Protoplasma* 240, 75–82.
Sengupta, S. and Majumder, A.L. (2010) *Porteresia coarctata* (Roxb.) Tateoka, a wild rice: a potential model for studying salt-stress biology in rice. *Plant, Cell and Environment* 33, 526–542.
Shabala, S. and Mackay, A. (2011) Ion transport in halophytes. *Advanced Botanical Research* 57, 151–199.
Shabala, S., Bose, J. and Hedrich, R. (2014) Salt bladders: Do they matter? *Trends in Plant Science* 19, 687–691.
Shepherd, K.A., Macfarlane, T.D. and Colmer, T.D. (2005) Morphology, anatomy and histochemistry of Salicornioideae (Chenopodiaceae) Fruits and Seeds. *Annals of Botany* 95, 917–933.
Shimony, C., Fahn, A. and Reinhold, L. (1973) Ultrastructure and ion gradients in the salt glands of *Avicennia marina* (Forssk.) Vierh. *New Phytologist* 72, 27–37.
Smail-Saadoun, N. (2005) Réponse adaptative de l'anatomie des Chénopodiacées du Sahara algérien à des conditions de vie d'aridité extreme. *Sécheresse* 16, 121–124.
Somaru, R., Naidoo, Y. and Naidoo, G. (2002) Morphology and ultrastructure of the leaf salt glands of *Odyssea paucinervis* (Stapf) (Poaceae). *Flora* 197, 67–75.
Stenlid, G. (1956) Salt losses and redistribution of salts in higher plants. *Encyclopedia of Plant Physiology* 4, 615–637.
Steudle, E., Lüttge, U. and Zimmermann, U. (1975) Water relations of the epidermal bladder cells of the halophytic species *Mesembryanthemum crystallinum*: Direct measurements of hydrostatic pressure and hydraulic conductivity. *Planta* 126, 229–246.

Stevanović, D.Z., Kresović, M., Pećinar, I., Aćić, S., Obratov-Petković, D. *et al.* (2010) Distribution of the halophytic grass *Puccinellia limosa* (Schur.) Holomb. on salt affected soils in Serbia in relation to its main adaptive responses to salinity. *Ekológia (Bratislava)* 29(3), 258–268.

Strogonov, B.P. (1973) *Metabolism rastenij r uslovijah zosolenija* (*Metabolism of plants under salinity conditions*). Moskva: Izdatelstvo Nauka. Stumpf.

Tajadod, G. and Moogouei, R. (2012) The study of anatomical changes under the influence of cesium uptake in *Chenopodium album* L. *Annals of Biological Research* 3, 1582–1592.

Takao, O., Kanjun, H., Taniguchia, M. and Miyake, H. (2013) Salt excretion from the salt glands in Rhodes grass (*Chloris gayana* Kunth) as evidenced by low-vacuum scanning electron microscopy. *Flora - Morphology, Distribution, Functional Ecology of Plants* 208, 52–57.

Taleisnik, E.L. and Anton, A.M. (1988) Salt glands in *Pappophorum* (Poaceae). *Annals of Botany* 62, 383–388.

Thomson, W.W. and Platt-Aloia, K. (1979) Ultrastructural transitions associated with the development of the bladder cells of the trichomes of *Atriplex*. *Cytobios* 25, 105–14.

Thomson, W.W., Berry, W.L. and Liu, L.L. (1969) Localization an secretion of salt by the salt glands of *Tamarix aphylla*. *Botany* 63, 310–317.

Thomson, W.W., Faraday, C.D. and Oross, J.W. (1988) Salt glands. In: Baker, D.A. and Hall, J.L. (eds) *Solute Transport in Plant Cells and Tissues*. Longman Scientific & Technical, Essex, UK, pp. 498–537.

Tomlinson, P.B. (1986) *The botany of mangroves*. Cambridge University Press, Cambridge, UK, p. 413.

Vandana, S. (2013) Anatomical and phytochemical study on durva (*Cynodon dactylon* Linn, Pers.) – an ayurvedic drug. *International Ayurvedic Medical Journal* 1, 1–7.

Villar, J.L., Turland, N.J., Juan, A., Gaskin, J.F., Alonso, M.A., *et al.* (2015) *Tamarix minoa* (Tamaricaceae), a new species from the island of Crete (Greece) based on morphological and plastid molecular sequence data. *Willdenowia* 45, 161–172.

Waisel, Y. (1972) *Biology of Halophytes*. Academic Press, New York, pp. 40–45.

Waisel, Y. (1985) The stimulating effects of NaCl on root growth of Rhodes grass (*Chloris gayana*). *Physiologia Plantarum* 64, 519–522.

Weiglin, C. and Winter, E. (1988) Studies on the ultrastructure and development of the glandular trichomes of *Cressa cretica* L. *Flora* 181, 19–27.

Weiglin, C. and Winter, E. (1991) Leaf structures of xerohalophytes from an East Jordanian salt pan. *Flora* 185, 405–424.

Weragodavidana, P.S. (2016) Salt gland excretion efficiency and salinity tolerance of *Sporobolus* species. MSc thesis. United Arab Emirates University College of Science Department of Biology, Abu Dhabi, United Arab Emirates.

Wieneke, J., Sarwar, G. and Roeb, M. (1987) Existence of salt glands on leaves of kallar grass (*Leptochloa fusca* L. Kunth). *Journal of Plant Nutrition* 10, 805–819.

Wilson, H., Mycock, D. and Weiersbye, I.M. (2016) The salt glands of *Tamarix usneoides* E. Mey. ex Bunge (South African salt cedar). *International Journal of Phytoremediation* 19, 587–595.

Xiong, L. and Zhu, J.K. (2002) Molecular and genetic aspects of plant responses to osmotic stress. *Plant, Cell and Environment* 25, 131–139.

Xiong, L., Schumaker, K.S. and Zhu, J.K. (2002) Cell signaling during cold, drought, and salt stress. *Plant Cell* 14, 165–183.

Youssef, A.M. (2009) Salt tolerance mechanisms in some halophytes from Saudi Arabia and Egypt. *Research Journal of Agriculture and Biological Sciences* 5, 191–206.

Yuan, F., Leng, B. and Wang, B. (2016) Progress in studying salt secretion from the salt glands in recretohalophytes: How do plants secrete salt? *Frontiers in Plant Science* 7, 977. doi.org/10.3389/fpls.2016.00977.

Zarrinkamar, F., Jalili, A., Hamzeh'ee, B., Asri, Y., Hodgson, J.G., *et al.* (2002) Foliar anatomy of some *Carex* species in Arasbaran. *Iranian Journal of Botany* 9, 261–270.

Zidan, I., Azaizeh, H. and Neumann, P.M. (1990) Does salinity reduce growth in maize root epidermal cells by inhibiting their capacity for cell wall acidification? *Plant Physiology* 93, 7–11.

Ziegler, H. and Lüttge, U. (1967) Die Salzdrüsen von *Limonium vulgare*. *Planta* 74, 1–17.

Zoric, L.N., Anačkov, G.T., Karanović, D.S. and Luković, J.Ž. (2013) Leaf structural adaptations of two *Limonium miller* (Plumbaginales, Plumbaginaceae) taxa. *Journal Natural Science, Matica Srpska Novi Sad* 125, 43–54.

Zouhaiera, B., Abdallah, A., Najlaa, T., Wahbi, D., Wided, C., *et al.* (2015) Scanning and transmission electron microscopy and X-ray analysis of leaf salt glands of *Limoniastrum guyonianum* Boiss. under NaCl salinity. *Micron* 78, 1–9.

11 ROS Signalling, and Antioxidant Defence Network in Halophytes

Ewa Surówka[1,*], Dariusz Latowski[2], Marta Libik-Konieczny[1] and Zbigniew Miszalski[1]
[1]Polish Academy of Sciences, Kraków, Poland;
[2]Jagiellonian University, Kraków, Poland

Abstract

Halophytes have evolved a range of adaptations to tolerate disturbances in reactive oxygen species (ROS) homeostasis caused by high salinity alone or in combination with other stresses. Singlet oxygen ($^1O_2$), superoxide anion ($O_2^{\cdot-}$), hydroxyl radical ($OH^{\cdot}$) and hydrogen peroxide (H_2O_2) are the major ROS generated during plant metabolism. The main sources of ROS are chloroplasts, mitochondria, peroxisomes, apoplasts, plasma membranes and cell walls. Regulation of ROS levels in these compartments plays a key role in response to salt stress. Many various defence strategies have been discovered in chloroplasts. These include stress-avoidance mechanisms (e.g. alteration in proteins involved in electron transport chains, changes in lipid composition, harmless dissipation of excess energy) and ROS-scavenging systems (enzymatic and non-enzymatic antioxidants, some osmoprotectants). The efficient antioxidant mechanisms in halophytes can prepare plants to cope with subsequent stress factors by inducing a signal to activate the stress response and suppress the oxidative burst within a short time after the stimulus action. The interrelationship between ROS scavengers leads to spatio-temporal modulation of the ROS signalling network. Although the ROS level is modulated by both: (1) specific mechanisms enabling delay/reduction in the oxidative burst; and (2) ROS-scavenging systems in each cell compartment, interactions between cell organelle and cytoplasm are critical in adjusting redox homeostasis, and the ROS signalling network plays a role in inducing salinity tolerance at the cellular and whole-plant levels.

Keywords: Antioxidant defence; Catalase; Glutathione; Hydrogen peroxide, Lipid peroxidation; Oxidative stress

11.1 Introduction: the Role of ROS in Adaptation to Salinity

Salt tolerance is a complex phenomenon and halophytes have evolved various mechanisms responsible for acclimation to cope with ion imbalance, and osmotic and oxidative stresses (excess of reactive oxygen species, ROS) (Flowers and Colmer, 2015). These mechanisms have been analyzed in whole organs or tissues as well as in subcellular compartments. According to Yeo (1998), compartmentation of defence pathways and antioxidant activities in subcellular compartments determine the functioning of salt-tolerance mechanisms, including those dependent on ROS. Thus, analyzing antioxidant defence at the subcellular level allows a better understanding of how halophytes respond to salinity.

The main sources of ROS production in plants are chloroplasts, mitochondria, peroxisomes/

* Corresponding author e-mail: ewa.surowka@gmail.com

glyoxysomes, apoplasts, plasma membranes and cell walls (Schmitt *et al.*, 2014; Hossain and Dietz, 2016). ROS are produced as by-products of metabolic processes as well as 'deliberately' generated by ROS-producing enzymatic complexes localized on the cell surface or in the apoplast. Among the apoplastic ROS sources, NADPH oxidases – termed respiratory burst oxidase homologues (RBOHs) (e.g. plasma membrane-bound NADPH oxidases, cell wall-bound NAD(P)H oxidase-peroxidase) – are the most prominent. However, other apoplastic ROS sources, such as amine oxidases (AOs; including copper-containing AOs EC 1.4.3.6 and flavin-containing polyamine oxidases, polyamine oxidases, PAOs, EC 1.5.3.11), oxalate oxidases (EC 1.2.3.4) belonging to the germin-like protein (GLP) family, class III peroxidases (e.g. plasmalemma-bound POX isoforms), plasmalemma-localized phylloquinone (vitamin K1) or non-enzymatic degradation of ascorbate (AsA) or lipoxygenases (LOX) (Schmitt *et al.*, 2014, Podgórska *et al.* 2017) also need to be mentioned.

ROS, particularly singlet oxygen ($^1O_2$), superoxide anion ($O_2^{\bullet -}$), hydrogen peroxide (H_2O_2) and hydroxyl radical ($OH^\bullet$) are involved in damaging processes; however, ROS can play a key regulatory role when maintained at the proper cellular concentrations and in the spatio-temporal accumulation pattern (Hossain and Dietz, 2016; Noctor *et al.*, 2018). H_2O_2 is the only ROS that is able to cross the cellular membrane, thus it can directly trigger chemical reactions and affect responsive targets such as metabolites or proteins, inside and outside the organelles of its origin. Other ROS species, such as $^1O_2$ and $O_2^{\bullet -}$, can also be involved in the induction of stress-response mechanisms (Schmitt *et al.*, 2014; Noctor *et al.*, 2018). The salt-tolerance mechanism is coordinated with the balance between ROS generation and scavenging, and is mainly controlled by antioxidant enzymes and molecular antioxidants as well as by some compatible solutes (Tables 11.1 and 11.2). Oxidants and ROS scavengers operate as a part of cellular stress signalling network that integrates information from many metabolic pathways to (1) maintain cellular redox balance; (2) orchestrate the readjustment of redox homeostasis and redox/ROS-dependent signalling pathways; (3) regulate differential expression of salt-responsive genes and transcription factors; and (4) activate the antioxidant/osmotic response network to modulate defence responses (Schmitt *et al.*, 2014; Hossain and Dietz, 2016; Noctor *et al.*, 2018).

11.2 Antioxidant Defence in Halophytes and Avoidance Mechanisms Enabling Adaptation to Salinity Stress at the Cellular Level

Comparative studies between halophytes and glycophytes, as well as different halophytic species, have demonstrated different mechanisms of ROS content regulation. Obligatory halophytes have efficient mechanisms of Na^+ exclusion from the cytosol that protect from ROS over-accumulation and may not require a high level of antioxidant activity (Bose *et al.*, 2014; Kumari *et al.*, 2015). Accordingly, Shabala and Munns (2017) concluded that the increased antioxidant activity in obligatory halophytes should be treated as a damage control mechanism rather than a trait of direct salinity tolerance. However, general comparative studies between halophyte and glycophyte species even revealed several-fold higher levels of antioxidants or stress-related metabolites in halophytic plants (Bose *et al.*, 2014; Mishra and Tanna, 2017). An elevated induction of antioxidants is commonly believed to be associated with increased salinity tolerance. Halophytes have a potential to differentially modulate antioxidant system capacity depending on the nature, duration and intensity of stress. The variation in the antioxidants' level reflects ecological requirements and depends on the halophytes' developmental stage as well as the type of their organs (leaves, shoots and roots) (e.g. Boestfleisch *et al.*, 2014; Bose *et al.*, 2014; Houmani *et al.*, 2016; Lee *et al.*, 2016; Hassan *et al.*, 2017). Hossain and Dietz (2016), however, have emphasized that changes in antioxidant patterns are often difficult to interpret, because although the up-regulation of antioxidant enzymes may represent the response to oxidative stress, it can also be considered a defence reaction, leading to ROS level reduction.

Among different antioxidant enzymes, superoxide dismutase (SOD) exhibits a particularly high level of activity in halophytes compared to glycophytes (Bose *et al.*, 2014). SODs scavenge superoxide radicals – the primary products of the univalent pathway of molecular oxygen (O_2) reduction and the precursors of most ROS, thus

Table 11.1. Enzymatic antioxidants, their localization in the cell and reaction catalysed.

Enzymatic antioxidant	Enzyme commission number	Reaction catalysed	Subcellular location
Superoxide dismutase (SOD)	1.15.1.1	$O_2^{\bullet-} + O_2^{\bullet-} + 2H^+ \rightarrow 2H_2O_2 + O_2$	Peroxisomes, mitochondria, cytosol, chloroplasts
Catalase (CAT)	1.11.1.6	$2H_2O_2 \rightarrow 2H_2O + O_2$	Peroxisomes, mitochondria
Ascorbate oxidase (APX)	1.11.1.11	H_2O_2 + AsC → $2H_2O$ + DHA	Peroxisomes, mitochondria, cytosol, chloroplasts
Monodehydroascorbate reductase (MDHAR)	1.6.5.4	2MDHA + NADH → 2AsC + NAD	Mitochondria, cytoplasm, chloroplasts
Dehydroascorbate reductase (DHAR)	1.8.5.1	DHA + 2GSH → AsC + GSSG	Mitochondria, cytoplasm, chloroplasts
Glutathione reductase (GR)	1.6.4.2	GSSG + NADPH + H^+ → 2GSH + $NADP^+$	Mitochondria, cytoplasm, chloroplasts
Glutathione peroxidase (GPx)	1.11.1.9	2GSH + H_2O_2 → GSSG + $2H_2O$	Mitochondria, cytoplasm, chloroplasts
Glutathione transferase (GST)	2.5.1.18	RX + GSH → HX + R-S-GSH	Cytosol, mitochondria
Guaiacol peroxidase (GPX)	1.11.1.7	2GSH + H_2O_2 → GSSG + $2H_2O$ 2GSH + ROOH → GSSG + ROH + H_2O	Mitochondria, cytoplasm, chloroplast, ER
Thioredoxin reductase (TrxR)	1.6.4.5	(TrxR): NADPH + H^+ + $TrxS_2$ → $NADP^+$ + $Trx(SH)_2$.	Cytosol, plastids, mitochondria
Thioredoxins (Trx)		TrxR-S_2 + NADPH + H^+ → TrxR-$(SH)_2$ + $NADP^+$ Trx-S_2 + TrxR-$(SH)_2$ → Trx-$(SH)_2$ + TrxR-S_2 Protein-S_2 + Trx-$(SH)_2$ → Protein-$(SH)_2$ + Trx-S_2	Cytosol, plastids, mitochondria
Thioredoxin peroxidase		Trx + H_2O_2 → Trx-$(SH)_2$ + H_2O	Cytosol, plastids, mitochondria
Ferredoxin-NADP+ oxidoreductase (FNR)	1.18.1.2	2Fd(red) + $NADP^+$ + H^+ → 2 Fd(ox) + NADPH	Chloroplasts, plastids
Peroxiredoxins (Prxs)	1.11.1.5	2R′-SH + ROOH → R′-S-S-R′ + H_2O + ROH	Cytosol, plastids, mitochondria
Glutaredoxins (Grxs)		RSSR′ + R″SH → RSSR″ + R′SH	Cytosol, plastids, mitochondria

they limit the induction of other antioxidants. In many halophytes, the increase in individual SOD activities is due to the rise in salt concentration and it correlates with the enhanced level of lipid peroxidation, measured by malondialdehyde (MDA) content (e.g. Ellouzi *et al.*, 2011; Ozgur *et al.*, 2013; Hameed *et al.*, 2015; Houmani *et al.*, 2016). Bose *et al.* (2014) postulated that a higher SOD level in halophytes is required for rapid induction of the H_2O_2 'signature' to initiate a cascade of adaptive responses at the genetic and physiological level, while other enzymatic antioxidants take part in modulating the basal level of H_2O_2. Thus, the balance between SODs and peroxide-scavenging enzymes in cells is considered to be crucial for suppressing toxic H_2O_2 levels and determining salinity tolerance. We suggest that, in some cases, SOD may reduce the superoxide signal from different cellular compartments, thereby playing a decisive role in defence signal induction.

Table 11.2. Non-enzymatic scavengers, their functions and cellular localization.

Non-enzymatic ROS scavengers	Function	Subcellular localization
Ascorbic acid (AsA, reducing agent)	Scavenges $O_2^{\bullet-}$ and H_2O_2 directly or through the Asada–Halliwell–Foyer pathway, component of a cellular redox buffer, recycles the lipid-soluble antioxidant α-tocopherol	Cytosol, chloroplasts, mitochondria, peroxisome, vacuole, apoplast
Glutathione (reduced, GSH)	Acts as a detoxifying co-substrate for enzymes such as peroxidases, glutathione reductase (GR) and glutathione transferase (GST), a component of the cellular redox buffer	Cytosol, chloroplasts, mitochondria, peroxisome, vacuole, apoplast
α-Tocopherol (TC)	Physically quenches $^1O_2$, reduces $OH^\bullet$, scavenges lipid peroxyl radicals ($LOO^\bullet$), acts as a hydrogen donor; a component of the cellular redox buffer	Predominates in chloroplasts (thylakoid membranes and plastoglobuli)
Carotenoids (Car)	Carotenoids: de-excitation of $^3Chl^*$, quenchers of excess energy from photosystem and light-harvesting antenna complexes (LHC) before reactive oxygen can be formed; Xanthophylls (oxygen derivatives of carotenoids, i.e. zeaxanthin): lipid peroxidation inhibitor (alone or in interaction with α-tocopherol), thermal energy dissipation by the xanthophyll cycle via the PsbS protein, plays a role as pre-emptive prevention of ROS formation	Chloroplasts and other non-green plastids
Proline (Pro)	Direct scavenger of $OH^\bullet$ or $^1O_2$ and indirect scavenger of $O_2^{\bullet-}$ and H_2O_2, osmoprotectant, involved in cellular energy and redox homeostasis (e.g. NADP+/NADPH and GSH/GSSG ratios), protects photochemical efficiency of PSII, acts as an electron shuttle to balance redox potentials between chloroplasts and mitochondria, precursor of other osmoprotectants, e.g. proline-betaine, shared with compatible solutes, e.g. glycine betaine, involved in polyamine metabolism, an alternative and important energy source during abiotic stress – it participates in other pathways, such as the tricarboxylic acid cycle, urea cycle, acetyl-glutamate pathway, pentose phosphate pathway and GSH biosynthesis	Mitochondria, cytosol, chloroplasts, vacuoles
Polyamines (PAs); agmatine (Agm), putrescine (Put), spermidine (Spd), spermine (Spm)	PAs catabolism gives rise to the production of H_2O_2, PAs anabolism is involved in reducing ROS (mainly $O_2^{\bullet-}$ and H_2O_2), Spm and Spd are physical quenchers of $^1O_2$, $OH^\bullet$, PAs promote the induction of antioxidants (e.g. catalase), activate signalling molecules in the ABA-regulated stress-response pathway and by H_2O_2 production	Cytoplasm, peroxisomes, apoplast
Flavonoids	Direct scavengers of H_2O_2, $^1O_2$ and $OH^\bullet$	Vacuole

Coordinated regulation of enzymatic and non-enzymatic components of the antioxidant system enables halophytes to protect themselves against oxidative damage. For example, glutathione reductase (GR) and glutathione peroxidase (GPX) play a key role in ROS elimination in *Salsola crassa* under short-term salt exposure (to 15 days), while longer exposure to salt stress (30 days) or high NaCl concentration (1000–1500 mM) promotes ROS detoxification through low molecular antioxidants such as AsA and glutathione (GSH) (Yildiztugay *et al.*, 2014). According to Lokhande *et al.* (2013), AsA efficiently protects *Sesuvium portulacastrum* tissues even at concentrations of NaCl as high as 1000 mM. Stimulation of AsA accumulation under salinity correlates with the rapid initial increase of H_2O_2 and lipid peroxidation (Rozema and Schatt, 2013) as well as the fast induction of GSH synthesis (Ellouzi *et al.*, 2017).

Furthermore, a strong correlation between salinity content and polyphenol accumulation, as well as high antioxidant potential, was found in both photosynthetic active tissues and roots of many halophytic species growing under salinity (e.g. Frary *et al.*, 2010; Ramos *et al.*, 2014; Reginato *et al.*, 2014; Grigore and Oprica, 2015; Quasim *et al.*, 2017). Polyphenols scavenge $O_2^{\cdot-}$, $OH^{\cdot}$, H_2O_2 and $^1O_2$ (Waśkiewicz *et al.*, 2013) and the activity of polyphenol oxidase is lower in halophytes than in glycophytes (Sharma *et al.*, 1983). According to Bose *et al.* (2014), non-enzymatic antioxidants are the only effective means of preventing harmful effects of hydroxyl radicals on cellular structures.

Srivastava *et al.* (2015) showed that NADPH oxidase (NOX) activity significantly decreased in a glycophyte (*Brassica juncea*) under salt stress, while in halophytes (*S. portulacastrum*) it was unaffected. These authors concluded that halophytes were not only able to activate more efficiently their antioxidant defence under stress, but that halophyte enzymes were also comparably more resistant and stable than their glycophytic counterparts. The authors also stated that the coordinated subtle regulation of antioxidant defence at the cellular level was a key determinant of salt tolerance in halophytes. Ellouzi *et al.* (2011) showed that a halophyte (*Cakile maritima*) accumulated a high H_2O_2 concentration (within the first hours of salinity), followed by a rapid decrease in its content and activation of efficient H_2O_2 scavenging mechanisms upon signalling completion. In contrast, *Arabidopsis thaliana* exhibited a continuous increase in H_2O_2 concentration during the entire period of salt treatment. Similar results were obtained in the experiments with *Thellungiella salsuginea* (also known as *Thellungiella halophila*, *Eutrema salsugineum*) (Wiciarz *et al.*, 2015; Pilarska *et al.*, 2016). In contrast, *Arabidopsis thaliana* exhibited continuous increase in H_2O_2 concentration during the entire period of salt treatment. Consistent with the above results, Kurusu *et al.* (2015) reported that the basal level of H_2O_2 in halophytic plants activated antioxidant enzymes in order to efficiently cope with ROS over-production under salinity stress. According to Kaur *et al.* (2016), the elevated H_2O_2 level considered as a key determinant for conferring salt-stress tolerance might have originated from an alternative route of photocatalytic chlorophyll activity. Hossain *et al.* (2017) showed that the unusual inhibition of ROS accumulation under severe salinity resulted from the involvement of specific *cis*-elements in the promoters of antioxidant genes in halophytes, such as sugar beet (*Beta vulgaris*) in adjusting redox and ROS networks, which are fundamental to developing high-salinity tolerance. Additionally, Ellouzi *et al.* (2017) showed that *C. maritima* plants produced H_2O_2 under combined stress conditions, which enabled them to activate the antioxidant system directly after stress action and to prepare antioxidant mechanisms for their rapid and efficient stimulation under subsequent stresses. The potential halophyte adaptation to different stresses is, at least partially, related to the phenomenon known as 'cross-tolerance' (e.g. Libik-Konieczny *et al.*, 2011; Gabara *et al.*, 2012; Meng *et al.*, 2016; Shamustakimova *et al.*, 2017).

The diversity of antioxidant defences in halophytes under salt stress is also manifested in a different pattern of osmoprotectants, which enable plants to scavenge ROS and modulate the cellular redox balance (Singh *et al.*, 2015; Hossain and Dietz, 2016). The synthesis of osmoprotectants is higher in halophytes than in glycophytes (Bose *et al.*, 2014). Compatible solutes have been described in halophytes from the main families, such as Amaranthaceae, Plumbaginaceae, Plantaginaceae, Aizoaceae, Poaceae and Brassicaceae (Slama *et al.*, 2015).

Aliphatic polyamines (PAs) have, among others, been shown to be involved in the regulation of ROS and antioxidant levels (Table 11.2) (Pottosin *et al.*, 2014; Gupta *et al.*, 2016) or the protection of the photosynthetic apparatus (Elso *et al.*, 2012). A different balance between individual PAs, such as agmatine (Agm), putrescine (Put), spermidine (Spd) and spermine (Spm) has been found in various halophytic species (e.g. Elso *et al.*, 2012; Minocha *et al.*, 2014, Nejad *et al.*, 2014; Surówka *et al.*, 2016).

Moreover, the accumulation of PAs or PA conjugates with a stronger antioxidant capacity than free polyamines is an organ-specific process in *Mesembryanthemum crystallinum* plants (Kuznetsov and Shevyakova, 2007), while in *Cymodocea nodosa* it depends on the plant developmental stage (Elso *et al.*, 2012). The response of plants to salinity involves crosstalk between polyamine transduction and oxidative and/or Ca^{2+}-dependent pathways (Bose *et al.*, 2014; Gupta *et al.*, 2016). A major interaction between PAs and ROS presumably occurs at the level of the plasma membrane where PAs (due to their strong positive charge) can effectively block cation channels (Minocha *et al.*, 2014). Kuznetsov *et al.* (2007) showed that the accumulation of cadaverine in *M. crystallinum* was NaCl- and ethylene-dependent under salt stress. Since polyamines and ethylene biosynthetic pathways are linked by the same precursor, such as S-adenosylmethionine (S-AdoMet), it can be suggested that the regulation of S-AdoMet or methionine pools determines metabolic pathways responsible for salt tolerance (Sauter *et al.*, 2013).

In addition, proline enables plants to have better 'stress-anticipatory preparedness' under salt stress through its involvement in many metabolic pathways (Table 11.2). These include protection of PSII photochemical efficiency, ROS scavenging and maintaining cellular redox homeostasis (Askari *et al.*, 2006; Monteoliva *et al.*, 2014; Rejeb *et al.*, 2014; Surówka *et al.*, 2016; Slama *et al.*, 2017). According to Askari *et al.* (2006), proline synthesis and catabolism are both essential to promote stress tolerance. The synthesis of proline via the competition with many metabolic processes for carbon skeletons leads to the avoidance of ROS production as well NADP and ADP recycling. Proline can be accumulated in different plant organs, together with anthocyanins and polyphenols, even at 1000 mM of NaCl (Shabala and Mackay, 2011). The accumulation of proline has been shown to be higher in halophytes than in glycophytes (Gong *et al.*, 2005; Yaish *et al.*, 2015). Accordingly, we postulate that the accumulation of proline, as a precursor of compounds involved in antioxidant defence, may function as a mechanism enabling rapid response to changes in environmental factors.

11.3 ROS Signalling Network

G proteins are considered as the first ROS-induced components of the stress-response signalling pathway (Ozgur *et al.*, 2013). Induction of GTP-binding protein due to salinity stress was recorded in the EST databases for the halophytic species *Suaeda salsa* and *T. salsuginea* (Zhang *et al.*, 2008). One of the best-characterized signalling pathways specific to salinity involves calcium as a secondary messenger in the intracellular signalling cascade. Salinity-induced elevation of Ca^{2+} concentration and calcium-dependent induction of NOX activity has been described in many studies (Bose *et al.*, 2011; Kurusu *et al.*, 2015). Induction of the adaptive response to salinity is caused by the subtle integration of different signalling pathways, such as the connection between the salt overly sensitive (SOS) pathway, ROS signalling and phytohormones (Verslues *et al.*, 2007; Mishra and Tanna, 2017). One of the components of the SOS pathway – kinase phosphorylating SOS1 (SOS2) – and H_2O_2 are able to control nucleoside triphosphate kinase 2 (NDPK2), which can stimulate key elements of phosphatidic acid signalling, inducing responses to oxidative stress proteins, MAPK3 and MAPK6 (Verslues *et al.*, 2007). Moreover, SOS2 interacts with catalase (CAT2, CAT3 isoforms), regulating H_2O_2 level in the cells. Although the described results were reported for *Arabidopsis*, similar data were also obtained in halophytes. The overexpression of a putative homologue of SOS2, in the *Hordeum brevisubulatum* HbCIPK2 gene, was associated with salt tolerance in transgenic *A. thaliana* deficient in SOS2-1 (Li *et al.*, 2012).

11.4 Response to Salinity at the Subcellular Level

ROS level can be controlled independently at each cell compartment since all these regions have specific antioxidants and mechanisms that maintain ROS homeostasis (Fig. 11.1). However, interactions between chloroplasts, mitochondria, peroxisomes and cytosol integrate functions not only of these subcellular compartments but also the cellular metabolism of the entire plant under

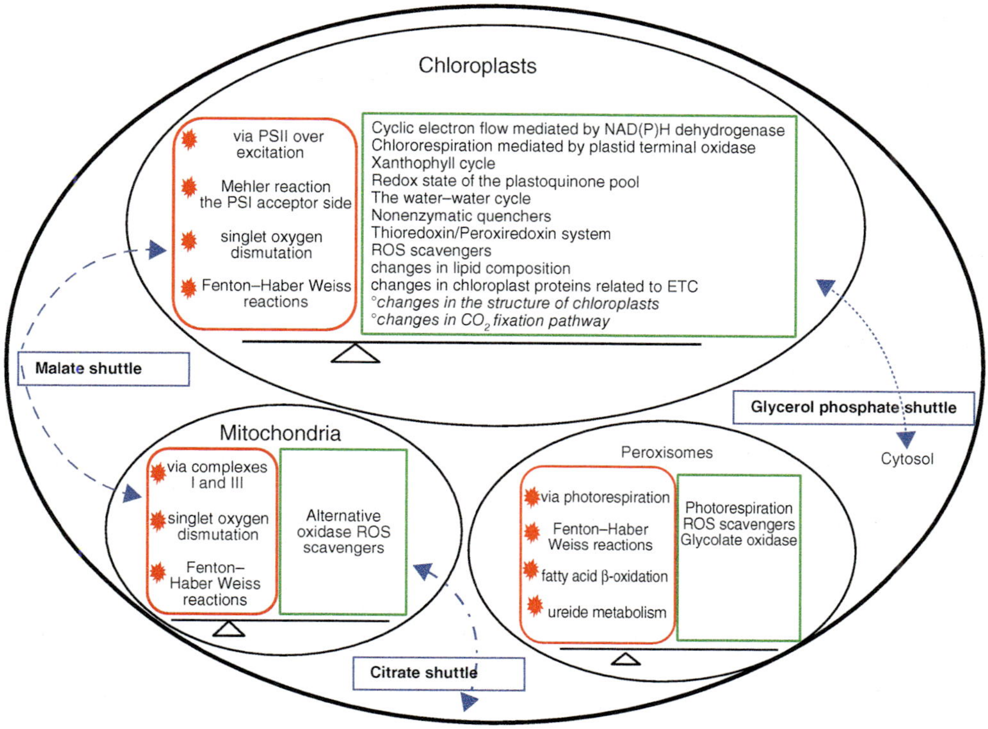

Fig. 11.1. Overview of ROS sources and defence strategies at the cellular level in halophytes. Each cellular compartment has specific ROS sources (marked in the figure as red stars, described in red squares). ROS generation in chloroplasts, mitochondria and peroxisomes is usually related to electron transport chains (ETC). Cell compartments developed different strategies either to avoid ROS generation or to scavenge the generated ROS (described in green squares). ROS homeostasis in the cell is also regulated by the transport of reducing equivalents via the activity of malate, citrate and glycerol phosphate shuttles (marked with blue dotted lines ending with arrows, described in blue boxes). Subtle coordination of the defence system makes it possible to maintain redox homeostasis in the cell and to induce salt tolerance in all organs of the halophytic plant.

different environmental conditions (Hossain and Dietz, 2016; Noctor *et al.*, 2018).

11.4.1 Chloroplasts

Chloroplasts are the most sensitive organelles to salt stress. They act as sensors of intra- and extracellular stimuli and communicate via redox signalling. Their redox signal depends on the functional and energetic state of the photosynthetic electron transport chain (PETC) and ROS generation (Hossain and Dietz, 2016; Noctor *et al.*, 2018). According to Bose *et al.* (2017), some processes – such as electron transport and photosynthetic efficiency – are influenced in halophytes by the coordinated traffic of ions (Na^+, Cl^-, K^+) across chloroplast membrane systems. ROS level in chloroplasts has a strong impact on the orchestration of acclimation processes (Noctor *et al.*, 2018).

11.4.1.1 Strategies for avoiding ROS generation

Several studies have demonstrated that NAD(P)H dehydrogenase (NDH complex) mediated cyclic electron flow (CEF) by preventing excessive reduction of the electron transport chain, thereby avoiding ROS production under salt stress (Ishikawa *et al.*, 2016). Significant up-regulation of CEF was found in soybean (Yang *et al.*, 2007) and a halotolerant cyanobacterium, *Aphanothece halophytica* under salinity stress (Hibino *et al.*, 1996). According to He *et al.* (2015), the enhanced NDH-dependent CEF supplies extra ATP that is used to sequester Na^+ in the vacuole, and thus prevents Na^+ over-accumulation and ROS generation in chloroplasts. The NDH complex, together with plastid terminal oxidase (PTOX), may play a regulatory role by poising the redox state of intersystem electron carriers in thylakoid domains, in which CEF is prevalent in PSI (Ishikawa *et al.*, 2016). Comparative studies on glycophytes (*A. thaliana*) and halophytes (*T. halophila*) showed that PTOX in halophytes was able to support up to 30% of total PSII electron transport flux, preventing over-reduction of PSII and mitigating the reduction of the plastoquinone (PQ) pool, thereby helping to reduce ROS production (especially $O_2^{\cdot-}$) and thus to protect PSII from photoinhibition (Johnson and Stepien, 2016). The key role of PTOX in protecting the photosynthetic apparatus against photoinhibition was also demonstrated in a halophyte, *Sarcocornia fruticosa* (Redondo-Gómez *et al.*, 2006). Several studies documented a positive correlation between the induction of PTOX and salinity levels (Nawrocki *et al.*, 2015; Uzilday *et al.*, 2015). Even in the control (without salt treatment), the protein content and activity of PTOX are higher in halophytes (*E. salsugineum*) than in glycophytes (*A. thaliana*) (Wiciarz *et al.*, 2015; Johnson and Stepien, 2016). Moreover, Wiciarz *et al.* (2015) showed that isolated thylakoids and chloroplasts of *T. halophila* expressing PTOX exhibited high levels of H_2O_2 production and concluded that PTOX, when coupled with an antioxidant system, could act as a safety valve under NaCl stress.

The redox state of the PQ pool and H_2O_2 generation are the main signal sources in chloroplasts under oxidative stress (Ślesak *et al.*, 2003). A comparative analysis revealed that the more salt-resistant *Hordeum maritimum* showed a higher PQ reduction both in control plants and after salinity treatment when compared to a glycophyte, *Hordeum vulgare* (Degl'Innocenti, 2009). *E. salsugineum* also retained highly reduced states of the PQ pool and P700 even under control conditions in comparison to *A. thaliana* plants. This phenomenon may form the basis for a high H_2O_2 formation under salt stress, which in turn activates defence mechanisms to prevent the production of more toxic forms of ROS (Wiciarz *et al.*, 2017). Similarly, Niewiadomska *et al.* (2011) showed a strongly reduced pool of electron acceptors from PSI in *M. crystallinum* after salinity treatment, while Ślesak *et al.* (2003) indicated that changes in the redox state of the PQ pool and accumulation of H_2O_2 in chloroplasts of *M. crystallinum* influenced nuclear gene expression, including ascorbate peroxidase (APX), through retrograde signalling and played a key role in C_3/CAM (crassulacean acid metabolism) transition regulation.

According to Asrar *et al.* (2017), protective strategies also enable plants to avoid the production of ROS in the proximity of the major functional chloroplast protein complexes, such as oxygen evolving complexes (OEC), PSI, PSII, APT-synthase and b6-f cytochrome, which are involved in light energy harvesting. Additionally, several subunits that form these proteins complexes, such as *PsbO, D1 (PsbA), PetD* or *AtpA* were shown to be altered or their synthesis changed under saline conditions along with the Rubisco protein. Moreover, the increase in the level of proteins, such as PsbB and PsaE, enables the optimal rate of photosynthetic electron transport and NADPH production and can be considered an important mechanism protecting against ROS production in halophytes (Sengupta and Majumder, 2010). Additionally, Askari *et al.* (2006) suggested that the efficient utilization of redox energy by the Calvin cycle can be considered as another mechanism that reduces the risk of ROS production in photosynthetic electron transport, and thus also improves plant salt resistance. In a halophyte, *Suaeda aegyptiaca*, an up-regulation of CO_2 assimilation rate was accompanied by an increase in D2 protein abundance, which is stabilized by PsbB and S-adenosyl-L-methionine (SAM) synthetase – synthesizing a methyl donor required for the biosynthesis of several osmoprotective and ROS-scavenging compounds.

Furthermore, light-stimulated Rubisco activase promotes the release of inhibitory ribulose 1,5-bisphosphate (RuBP) from the catalytic sites and activates Rubisco, and additionally prevents ROS production (Debez *et al.*, 2012; Wiciarz *et al.*, 2015). Increased ATP synthase levels are consistent with increased photosynthetic activity and the up-regulation of the Calvin cycle enzymes. Enhanced ATP synthesis in salt-treated plants may reflect the requirements of active transport mechanisms across the membranes (Debez *et al.*, 2012; Cheng *et al.*, 2015). Furthermore, ferredoxin-NADP reductase (FNR) participates in a variety of redox pathways and is involved in defence against oxidative damage (Valderrama *et al.*, 2006). Increased FNR in *Puccinellia tenuiflora* leaves enhanced the redox function despite the parallel photosynthesis photoinhibition (Yu *et al.*, 2011). Stimulation of FNR activity was also documented in *Tangut nitraria* (Cheng *et al.*, 2015). In addition, chlororespiration is another alternative pathway aimed at avoiding ROS generation during photosynthesis, which has been documented in *Thellungiella* sp. plants exposed to salinity (Johnson and Stepien, 2016). The water–water cycle has also been shown to prevent excess ROS accumulation in chloroplasts, and the induction of water–water cycle enzyme activities or their expressions has been documented in *Eutrema parvulum* (syn. *Thellungiella parvula*) (Uzilday *et al.*, 2015) and *C. maritima* (Amor *et al.*, 2006).

Halophytes are also capable of decreasing ROS production under salinity stress by: (1) changing the CO_2 fixation pathway from C_3 to CAM (e.g. *M. crystallinum*) or C_4 (e.g. *Portulaca oleracea*); or (2) a flexible photorespiration rate; that is, through a change of the carbon fixation pathway from C_3 to C_4 photosynthesis (e.g. *Atriplex lentiformis*) (Bose *et al.*, 2014). Furthermore, the halophyte *Thellungiella* sp. is able to accumulate more chloroplasts and starch grains than glycophytes under salinity (Wang *et al.*, 2013). Salt-induced changes in the chloroplast structure in *M. crystallinum* have also been documented (Niewiadomska *et al.*, 2011; Gabara *et al.*, 2012). Morphological alterations in chloroplasts play a crucial role in the adaptation to salt stress.

11.4.1.2 The antioxidant system in chloroplasts

Modulation of ROS levels is closely associated with redox regulatory enzymes (Table 11.1). Increased expression of peroxiredoxin (PrxR)/thioredoxin (Trx) defence system components were reported in leaves of *Reaumuria trigyna* (Dang *et al.*, 2013), *T. nitraria* (Cheng *et al.*, 2015) and *E. parvulum* (Uzilday *et al.*, 2015), while decreased expression was shown in *P. tenuiflora* (Yu *et al.*, 2011). The aforementioned authors stated that the efficient expression of thiol-containing redox regulatory enzymes in chloroplasts is responsible for the control of redox homeostasis in chloroplasts under salinity.

Carotenoids and their derivatives also protect chloroplasts against ROS generation (Table 11.2) (Demmig-Adams *et al.*, 2013). The basic level of carotenoids in halophytes is higher than in glycophytes, and their regulation is organ-specific (Yang *et al.*, 2009). Many halophytes, such as *S. portulacastrum*, accumulate large quantities of carotenoids with rising salt concentrations, which has been explained by the increase in the photosynthetic apparatus photoprotection (Slama *et al.*, 2017). It was postulated that the higher level of carotenoids observed in *E. salsugineum* leaves in control and salt-stressed plants was caused by the higher expression of the PTOX protein as a cofactor in carotenoid biosynthesis (Nawrocki *et al.*, 2015). In some halophytes (e.g. *Atriplex centralasiatica*) an efficient energy dissipation mechanism, such as the xanthophyll cycle, is involved in preventing excessive photosystem reduction and thus protecting them from photoinhibitory damage (Qiu *et al.*, 2003). It has been observed in *Aster tripolium* that salinity induces the increase in auroxanthin (violaxanthin analogue) level. These authors concluded that auroxanthin level and the process of efficient non-photochemical quenching might contribute to reducing the ROS load under salinity and prevent photosynthetic machinery over-excitation (Duarte *et al.*, 2017).

Detoxification of ROS species occurs via de-excitation or re-reduction by tocopherols or zeaxanthin, respectively, and possibly by other carotenoids (Table 11.2) (Demmig-Adams *et al.*, 2013).

α-Tocopherol, as a lipophilic antioxidant, plays an important role both in glycophytes and halophytes under salinity stress (Munné-Bosch and Alegre, 2002) and its level in both types of plants is regulated differently. The level of α-tocopherol was shown to decrease even by 50% in salt-treated *A. thaliana* plants, while it remained constant in a dicotyledonous halophyte, *C. maritima*; in contrast, its concentration increased in a monocotyledonous halophyte, *P. tenuiflora* (Ellouzi *et al.*, 2014). The increase in α-tocopherol with salinity and its involvement in ROS scavenging in *P. tenuiflora* was also confirmed by, for example, Yu *et al.* (2011).

Stimulation of antioxidants, such as chloroplastic SOD, thylakoid-bound APX or GR and stromal APX has been observed in the chloroplasts of halophytes during salinity conditions

(Table 11.1). Among them, chloroplastic SOD was shown to be activated prior to the activation of other ROS-detoxifying enzymes (Miszalski *et al.*, 1998; Ślesak *et al.*, 2003). Regulation of chloroplastic APXs is particularly important in maintaining redox signalling and redox homeostasis of AsA and GSH. It also plays a key role in the regulation of high photosynthetic performance (Ślesak *et al.*, 2003; Schmitt *et al.*, 2014). In addition, the induction of the ascorbate–glutathione cycle in *E. salsugineum* under salt stress allowed avoidance of harmful singlet oxygen production in PSII (Wiciarz *et al.*, 2017). Moreover, salt-inducible chloroplastic monodehydroascorbate reductase from the halophyte *Avicennia marina* was shown to improve tolerance to oxidative stress in transgenic plants (Kavitha *et al.*, 2010). In parallel, the ability of halophytes to intensively accumulate H_2O_2 in chloroplasts just after salinity conditions might be an important element of their preparedness for stress (Wiciarz *et al.*, 2015).

11.4.1.3 Lipids as antioxidants in the defence against oxidative stress

Halophytes and glycophytes also exhibit differences in the composition of lipids, such as glycolipids (monogalactosyldiacylglycerol, MGDG; digalactosyldiacylglycerol, DGDG; sulfoquinovosyldiacylglycerol, SQDG) and phospholipids (phosphatidylglycerol, PG). The MGDG/DGDG ratio and phospholipid contents decline with increasing halophility, while the glycolipids/phospholipids ratio increases (Rozentsvet *et al.*, 2017). Thylakoid glycerolipids are essential for the formation of the lipid bilayer and activity of photosynthetic complexes, while glycolipids are crucial for maintaining optimal photosynthesis efficiency (Kobayashi *et al.*, 2016). Salinity-induced changes in the MGDG/DGDG ratio and the level of SQDG and PG determine the structure, stability and function of the membrane and directly affect various photosynthesis-related processes (e.g. structural organization of PSI and PSII complexes, capacity of electron transport/transfer and state transition, status of the PQ pool), and in this manner they may control ROS levels (Sui *et al.*, 2010; Wang *et al.*, 2014; Kobayashi *et al.*, 2016, Rozentsvet *et al.*, 2017). Moreover, galactolipids and SQDG influence the activity of the b6-f cytochrome complex and CF0-CF1 ATPase (Minoda *et al.*, 2003; Kobayashi *et al.*, 2016). The SQDG level was found to be considerably higher in many salt-treated halophytes (e.g. *A. tripolium*, *S. portulacastrum*) when compared to glycophytes (Ramani *et al.*, 2006; Rozentsvet *et al.*, 2017). Lipids, especially those with polyunsaturated fatty acids (PUFA), efficiently protect both PSI and PSII from the damaging effects of salt stress (Sui *et al.*, 2010). A strong correlation between the increase of tolerance to salt stress and overexpression of rice MGDG synthase was observed in transgenic tobacco (Wang *et al.*, 2014). MGDG, containing significant amounts of PUFA and localized near photosystems, has been suggested to be an efficient scavenger of $^1O_2$ as well as hydroxyl radicals formed mainly in PSII. A new cyclic antioxidant mechanism has been proposed, in which lipid oxidation products such as MDA are introduced into plastid galactolipids, where they are again fragmented into MDA. This mechanism postulates a protective role for PUFA as ROS sinks (Schmid-Siegert *et al.*, 2016). The analysis of halophyte fatty acid profiles under salt stress showed that the most important compounds were represented mainly by linolenic acid (C18:3), linoleic acid (C18:2), oleic acid (C18:1) and palmitoleic acid (C16:1) (Duarte *et al.*, 2017). In most plants (which, according to the degree of halophilism, belong to various ecomorphs), the content of unsaturated fatty acids declined in the following series: C18:3 > C18:2 > C18:1 and the C18:2/C18:3 exchange ratio (Rozentsvet *et al.*, 2017). It is postulated that C18:3, belonging to omega-3-trienoic fatty acids, acts as a direct non-enzymatic scavenger of ROS under salt stress, while C18:1 and C16:1 improve the stabilization of light-harvesting complexes (LHC) of PSII and protect photosystems against oxidative stress. It was detected in *A. tripolium* that C16:1 was able to inhibit the energy dissipation mechanisms most probably caused by a higher level of auroxanthin, which was perceived as further evidence of LHC structure stabilization by C16:1 (Duarte *et al.*, 2017).

11.4.2 Mitochondrial ROS defence and salinity tolerance

Enhanced ROS generation, in parallel with increased MnSOD activity, was found in *M. crystallinum* mitochondria, as an early response to high salinity (Miszalski *et al.*, 1998). Jacoby *et al.* (2010), by comparing salt-tolerant and salt-sensitive wheat varieties, showed that only the salt-tolerant variety expressed a high level of the mitochondrial SOD isoform (MnSOD). According to these authors, the mitochondrial ROS defence enabled higher respiratory flux under salt stress, which is associated with alternative oxidase (AOX) induction, as shown previously in several halophytes. An increase of AOX expression level was described in *Populus euphratica* mitochondria, as a result of salt stress (Ottow *et al.*, 2005). In turn, a decrease in ROS production was observed in *A. thaliana* with a constitutive AOX overexpression (*Ataox1a*) (Smith *et al.*, 2009). On the other hand, the expression of AOX genes under stress conditions was shown to be associated with stress-dependent ROS production (Rogov *et al.*, 2014). The AOX pathway optimizes photosynthesis by regulating cellular ROS, malate or citrate valves and antioxidative systems (Dinakar *et al.*, 2016). Proline metabolism may play a protective role in over-reduction of mitochondrial electron transport and ROS generation, since it is catabolized in the mitochondria where it releases $FADH_2$ and NADH. The strong link between mitochondrial antioxidant defences and respiration allows the induction of salinity tolerance mechanisms (Jacoby *et al.*, 2010).

11.4.3 Peroxisomes and glyoxysomes in salinity tolerance

It has been reported that photorespiration and cyclic electron transport are two physiological processes that prevent oxidative damage, while optimizing photosynthesis (Voss *et al.*, 2013) in many halophytes, such as *S. fruticosa* (Redondo-Gómez *et al.*, 2006) or *P. tenuiflora* (Yu *et al.*, 2011). Simultaneously, glycolate oxidase and CAT levels increased in *P. tenuiflora* under salinity, enhancing glycolate oxidation and H_2O_2 scavenging (Yu *et al.*, 2011). The salt-induced increase in CAT activity – in parallel with significant H_2O_2 accumulation – in an aquatic macrophyte, *Najas graminea*, and a mangrove plant, *Suaeda maritima* (Mallik *et al.*, 2011), may suggest involvement of photorespiration in the salt-stress response. The salt-induced elevated expression of peroxisomal APX in *A. marina* halophyte (Kavitha *et al.*, 2008), might be due to H_2O_2 scavenging from peroxisomes by APX and the involvement of this enzyme in salt-induced defence signalling. In turn, the overexpression of the peroxisomal APX (SbpAPX) gene from *Salicornia brachiata* in peanut (*Arachis hypogaea*) plants conferred abiotic stress endurance and salt-stress tolerance (Singh *et al.*, 2014).

11.5 Subcellular Integration in Antioxidant Defence in Halophytes

Mechanisms such as malate and citrate valves integrate the metabolic activity of chloroplasts, mitochondria and peroxisomes, and limit the excessive reduction of photosynthetic electron transport chain components as well as prevent energy imbalances and ROS generation (Kumari *et al.*, 2015; Igamberdiev and Eprintsev, 2016). Gong *et al.* (2005) showed that citric acid, malic acid and succinic acid (along with proline) had constitutively higher levels in the salinity-resistant *T. halophila* than in a glycophyte, *A. thaliana*. Malate and citrate are accumulated in many halophytes under salinity conditions, for example in *M. crystallinum* (Miszalski *et al.*, 1998; Gawronska and Niewiadomska, 2015) and *Chloris virgata* (Wang *et al.*, 2016). The antioxidant nature of NADP-dehydrogenases results from the catalysis of the NADPH recycling reaction, which is essential for the functioning of the ascorbate–glutathione cycle (Valderrama *et al.*, 2006; Scheibe and Dietz, 2012; Igamberdiev and Eprintsev, 2016). Stimulation of enzymes involved in the malate valve, such as NADP-MDH and NAD-MDH, have been shown in salt-tolerant rice (Kumar *et al.*, 2000), and NADP-MDH in *M. crystallinum* under salinity (Gawronska *et al.*, 2013). The upregulation of NADP-MDH activity and the 'malate valve' caused by salinity are specific defence mechanisms of halophytic plants (Gawronska *et al.*, 2013). The citrate cycle also plays a role in antioxidative stress defence (Gawronska and Niewiadomska, 2015). Accumulation of citrate, isocitrate and activation of the cytosolic NADP-dependent isocitrate dehydrogenase were found in salt-treated *M. crystallinum* plants (Popova *et al.*, 2002; Gawronska and Niewiadomska, 2015). In *Leymus chinensis* plants, the activated citrate cycle was shown to be involved in the regulation of amino acid synthesis; for example, proline playing a role in plant adaptation to high salinity (Li *et al.*, 2017). Plants can regenerate NAD^+ from NADH by using DHAP/GAP-3-P-glycerate (glycerol phosphate) shuttle, thereby providing energy for the increased activity of H^+-ATPases during salt stress. Debez *et al.* (2012) showed in *C. maritima* that salt stress increased the abundance of proteins involved in carbohydrate metabolism and glycolysis, and suggested that energy transfer from chloroplasts to the cytosol was mediated by the glycerol phosphate valve. Interactions between organelles can improve cellular metabolic synchronization under stress conditions, thereby reducing ROS generation and contributing to salt-stress tolerance.

References

Al Hassan, M., Chaura, J., Donat-Torres, M.P., Boscaiu, M., and Vicente, O. (2017a) Antioxidant responses under salinity and drought in three closely related wild monocots with different ecological optima. *AoB Plants* 9(2), plx009. doi: 10.1093/aobpla/plx009.

Al Hassan, M., Estrelles, E., Soriano, P., López-Gresa, M.P., Bellés, J.M. *et al.* (2017b) Unravelling salt tolerance mechanisms in halophytes: a comparative study on four Mediterranean Limonium species with different geographic distribution patterns. *Frontiers in Plant Science* 8, 1438. http://doi.org/10.3389/fpls.2017.01438.

Amor, N.B., Jiménez, A., Megdiche, W., Lundqvist, M., Sevilla, F. *et al.* (2006) The response of antioxidant systems to NaCl stress in the halophyte *Cakile maritima*. *Physiologia Plantarum* 126, 446–457.

Askari, H., Edqvist, J., Hajheidari, M., Kafi, M. and Salekdeh, G.H. (2006) Effects of salinity levels on proteome of *Suaeda aegyptiaca* leaves. *Proteomics* 6(8), 2542–2554.

Asrar, H., Hussain, T., Hadi, S.M.S., Gul, B., Nielsen, B.L. *et al.* (2017) Salinity induced changes in light harvesting and carbon assimilating complexes of *Desmostachya bipinnata* (L.). *Environmental and Experimental Botany* 135, 86–95.

Boestfleisch, C., Wagenseil, N.B., Buhmann, A.K., Seal, C.E., Wade, E.M. *et al.* (2014) Manipulating the antioxidant capacity of halophytes to increase their cultural and economic value through saline cultivation. *AoB Plants* 6, plu046. doi: 10.1093/aobpla/plu046.

Bose, J., Pottosin, I.I., Shabala, S.S., Palmgren, M.G. and Shabala, S. (2011) Calcium efflux systems in stress signaling and adaptation in plants. *Frontiers in Plant Science* 2, 85. doi: 10.3389/fpls.2011.00085.

Bose, J., Rodrigo-Moreno, A. and Shabala, S. (2014) ROS homeostasis in halophytes in the context of salinity stress tolerance. *Journal of Experimental Botany* 65(5), 1241–1257.

Bose, J., Munns, R., Shabala, S., Gilliham, M., Pogson, B. *et al.* (2017) Chloroplast function and ion regulation in plants growing on saline soils: lessons from halophytes. *Journal of Experimental Botany* 68, 3129–3143.

Cheng, T., Chen, J., Zhang, J., Shi, S., Zhou, Y. *et al.* (2015) Physiological and proteomic analyses of leaves from the halophyte *Tangut nitraria* reveals diverse response pathways critical for high salinity tolerance. *Frontiers in Plant Science* 6, 30. doi: 10.3389/fpls.2015.00030.

Dang, Z.H., Zheng, L.L., Wang, J., Gao, Z., Wu, S.B. *et al.* (2013) Transcriptomic profiling of the salt-stress response in the wild recretohalophyte *Reaumuria trigyna*. *BMC Genomics* 14, 29. doi:10.1186/1471-2164-14-29.

Debez, A., Braun, H-P., Pich, A., Taamalli, W., Koyro, H-W. *et al.* (2012) Proteomic and physiological responses of the halophyte *Cakile maritima* to moderate salinity at the germinative and vegetative stages. *Journal of Proteomics* 75(18), 5667–5694.

Degl'Innocenti, E., Hafsi, C., Guidi, L. and Navari-Izzo, F. (2009) The effect of salinity on photosynthetic activity in potassium-deficient barley species. *Journal of Plant Physiology* 166(18), 1968–1981.

Demmig-Adams, B., Cohu, Ch.M., Amiard, V., van Zadelhoff, G. *et al.* (2013) Emerging trade-offs – impact of photoprotectants (PsbS, xanthophylls, and vitamin E) on oxylipins as regulators of development and defence. *New Phytologist* 197, 720–729.

Dinakar, C., Vishwakarma, A., Raghavendra, A.S. and Padmasree, K. (2016) Alternative oxidase pathway optimizes photosynthesis during osmotic and temperature stress by regulating cellular ROS, malate valve and antioxidative systems. *Frontiers in Plant Sciences* 7, 68. doi: 10.3389/fpls.2016.00068.

Duarte, B., Cabrita, M.T., Gameiro, C., Matos, A.R., Godinho, R. *et al.* (2017) Disentangling the photochemical salinity tolerance in *Aster tripolium* L.: Connecting biophysical traits with changes in fatty acid composition. *Plant Biology* 19(2), 239–248.

Ellouzi, H., Ben Hamed, K., Cela, J., Munné-Bosch, S. and Abdelly, C. (2011) Early effects of salt stress on the physiological and oxidative status of *Cakile maritima* (halophyte) and *Arabidopsis thaliana* (glycophyte). *Physiologia Plantarum* 142, 128–143.

Ellouzi, H., Ben Hamed, K., Hernández, I., Cela, J., Müller, M. *et al.* (2014) A comparative study of the early osmotic, ionic, redox and hormonal signaling response in leaves and roots of two halophytes and a glycophyte to salinity. *Planta* 240(6), 1299–1317.

Ellouzi, H., Sghayar, S. and Abdelly, C. (2017) H_2O_2 seed priming improves tolerance to salinity; drought and their combined effect more than mannitol in *Cakile maritima* when compared to *Eutrema salsugineum*. *Journal of Plant Physiology* 210, 38–50.

Elso, M.Z., García-Jiménez, P. and Robaina, R.R. (2012) Endogenous polyamine content and photosynthetic performance under hypo-osmotic conditions reveal *Cymodocea nodosa* as an obligate halophyte. *Aquatic Biology* 17(1), 7–17.

Flowers, T.J. and Colmer, T.D. (2015) Plant salt tolerance: Adaptations in halophytes. *Annals of Botany* 115(3), 327–331.

Frary, A., Göl, D., Keleş, D., Ökmen, B., Pınar, H. *et al.* (2010) Salt tolerance in *Solanum pennellii*: Antioxidant response and related QTL. *BMC Plant Biology* 10, 58. doi: 10.1186/1471-2229-10-58.

Gabara, B., Kuźniak-Gębarowska, E., Skłodowska, M., Surówka, E. and Miszalski, Z. (2012) Ultrastructural and metabolic modifications at the plant–pathogen interface in *Mesembryanthemum crystallinum* leaves infected by *Botrytis cinerea*. *Environmental and Experimental Botany* 77, 33–43.

Gawronska, K. and Niewiadomska, E. (2015) Participation of citric acid and isocitric acid in the diurnal cycle of carboxylation and decarboxylation in the common ice plant. *Acta Physiologia Plantarum* 37, 61. doi: 10.1007/s11738-015-1807-x.

Gawronska, K., Romanowska, E., Miszalski, Z. and Niewiadomska, E. (2013) Limitation of C_3–CAM shift in the common ice plant under high irradiance. *Journal of Plant Physiology* 170(2), 129–135.

Gong, Q., Li, P., Ma, S., Rupassara, I.S. and Bohnert, H.J. (2005) Salinity stress adaptation competence in the extremophile *Thellungiella halophila* in comparison with its relative *Arabidopsis thaliana*. *Plant Journal* 44(5), 826–839.

Grigore, M-N. and Oprica, L. (2015) Halophytes as possible source of antioxidant compounds, in a scenario based on threatened agriculture and food crisis. *Iranian Journal of Public Health* 44(8), 1153–1155.

Gupta, K., Sengupta, A., Chakraborty, M. and Gupta, B. (2016) Hydrogen peroxide and polyamines act as double edged swords in plant abiotic stress responses. *Frontiers in Plant Science* 7, 1343. doi: 10.3389/fpls.2016.01343.

Hameed, A., Gulzar, S., Aziz, I., Hussain, T., Gul, B. *et al.* (2015) Effects of salinity and ascorbic acid on growth, water status and antioxidant system in a perennial halophyte. *AoB Plants* 7, plv004. doi: 10.1093/aobpla/plv004.

He, Y., Fu, J., Yu, C., Wang, X., Jiang, Q. *et al.* (2015) Increasing cyclic electron flow is related to Na^+ sequestration into vacuoles for salt tolerance in soybean. *Journal of Experimental Botany* 66(21), 6877–6889.

Hibino, T., Lee, B.H., Rai, A.K., Ishikawa, H., Kojima, H. *et al.* (1996) Salt enhances photosystem I content and cyclic electron flow via NAD(P)H dehydrogenase in the halotolerant cyanobacterium *Aphanothece halophytica*. *Australian Journal of Plant Physiology* 23(3), 321–330.

Hossain, M.S. and Dietz, K.J. (2016) Tuning of redox regulatory mechanisms, reactive oxygen species and redox homeostasis under salinity stress. *Frontiers in Plant Science* 7, 548. doi: 10.3389/fpls.2016.00548.

Hossain, M.S., ElSayed, A.I., Moore, M. and Dietz, K.J. (2017) Redox and reactive oxygen species network in acclimation for salinity tolerance in sugar beet. *Journal of Experimental Botany* 68(5), 1283–1298.

Houmani, H., Rodríguez-Ruiz, M., Palma, J.M., Abdelly, C. and Corpas, F.J. (2016) Modulation of superoxide dismutase (SOD) isozymes by organ development and high long-term salinity in the halophyte *Cakile maritima*. *Protoplasma* 253(3), 885–894.

Igamberdiev, A.U. and Eprintsev, A.T. (2016) Organic acids: the pools of fixed carbon involved in redox regulation and energy balance in higher plants. *Frontiers in Plant Science* 7, 1042. doi: 10.3389/fpls.2016.01042.

Ishikawa, N., Takabayashi, A., Sato, F. and Endo, T. (2016) Accumulation of the components of cyclic electron flow around photosystem I in C4 plants, with respect to the requirements for ATP. *Photosynthesis Research* 129(3), 261–277.

Jacoby, R.P., Millar, A.H. and Taylor, N.L. (2010) Wheat mitochondrial proteomes provide new links between antioxidant defence and plant salinity tolerance. *Journal of Proteome Research* 9(12), 6595–6604.

Johnson, G.N. and Stepien, P. (2016) Plastid terminal oxidase as a route to improving plant stress tolerance: known knowns and known unknowns. *Plant and Cell Physiology* 57(7), 1387–1396.

Kaur, N., Dhawan, M., Sharma, I. and Pati, P.K. (2016) Interdependency of reactive oxygen species generating and scavenging system in salt sensitive and salt tolerant cultivars of rice. *BMC Plant Biology* 16, 131. doi: 10.1186/s12870-016-0824-2.

Kavitha, K., Venkataraman, G. and Parida, A. (2008) An oxidative and salinity stress induced peroxisomal ascorbate peroxidase from *Avicennia marina*: Molecular and functional characterization. *Plant Physiology and Biochemistry* 46(8), 794–804.

Kavitha, K., George, S., Venkataraman, G. and Parida, A. (2010) A salt-inducible chloroplastic monodehydroascorbate reductase from halophyte *Avicennia marina* confers salt stress tolerance on transgenic plants. *Biochimie* 92, 1321–1329.

Kobayashi, K., Endo, K. and Wada, H. (2016) Roles of lipids in photosynthesis. In: Nakamura, Y. and Li-Beisson, Y. (eds) *Lipids in Plant and Algae Development*. Springer, Cham, Switzerland, pp. 21–49.

Kumar, R.G., Shah, K. and Dubey, R.S. (2000) Salinity induced behavioural changes in malate dehydrogenase and glutamate dehydrogenase activities in rice seedlings of differing salt tolerance. *Plant Science* 156(1), 23–34.

Kumari, A., Das, P., Parida, A.K. and Agarwal, P.K. (2015) Proteomics, metabolomics, and ionomics perspectives of salinity tolerance in halophytes. *Frontiers in Plant Science* 6, 537. doi: 10.3389/fpls.2015.00537.

Kurusu, T., Kuchitsu, K. and Tada, Y. (2015) Plant signaling networks involving Ca^{2+} and Rboh/Nox-mediated ROS production under salinity stress. *Frontiers in Plant Science* 6, 427. doi: 10.3389/fpls.2015.00427.

Kuznetsov, V.V. and Shevyakova, N.I. (2007) Polyamines and stress tolerance of plants. *Plant Stress* 1, 50–71.

Kuznetsov, V.V., Shorina, M., Aronova, E., Stetsenko, L., Rakitin, V. *et al.* (2007) NaCl- and ethylene-dependent cadaverine accumulation and its possible protective role in the adaptation of the common ice plant to salt stress. *Plant Science* 172, 363–370.

Lee, S.J., Jeong, E.M., Ki, A.Y., Oh, K-S., Kwon, J. *et al.* (2016) Oxidative defense metabolites induced by salinity stress in roots of *Salicornia herbacea*. *Journal of Plant Physiology* 206, 133–142.

Li, J., Cui, G., Hu, G., Wang, M., Zhang, P. *et al.* (2017) Proteome dynamics and physiological responses to short-term salt stress in *Leymus chinensis* leaves. *PLoS ONE*. 12, e0183615. doi: 10.1371/journal.pone.0183615.

Li, R., Zhang, J., Wu, G., Wang, H., Chen, Y. *et al.* (2012) HbCIPK2, a novel CBL-interacting protein kinase from halophyte *Hordeum brevisubulatum*, confers salt and osmotic stress tolerance. *Plant, Cell and Environment* 35(9), 1582–1600.

Libik-Konieczny, M., Surówka, E., Kuźniak, E., Nosek, M. and Miszalski, Z. (2011) Effects of *Botrytis cinerea* and *Pseudomonas syringae* infection on the antioxidant profile of *Mesembryanthemum crystallinum* C_3/CAM intermediate plant. *Journal of Plant Physiology* 168, 1052–1059.

Lokhande, V.H., Gor, B.K., Desai, N.S., Nikam, T.D. and Suprasanna, P. (2013) *Sesuvium portulacastrum*, a plant for drought, salt stress, sand fixation, food and phytoremediation. A review. *Agronomy for Sustainable Development* 33(2), 329–348.

Mallik, S., Nayak, M., Sahu, B.B., Panigrahi, A.K. and Shaw, B.P. (2011) Response of antioxidant enzymes to high NaCl concentration in different salt-tolerant plants. *Biologia Plantarum* 55, 191–195.

Meng, X, Zhao, Q., Jin, Y., Yu, J., Yin, Z. *et al.* (2016) Chilling-responsive mechanisms in halophyte *Puccinellia tenuiflora* seedlings revealed from proteomics analysis. *Journal of Proteomics* 143, 365–381.

Minocha, R., Majumdar, R. and Minocha, S.C. (2014) Polyamines and abiotic stress in plants: a complex relationship. *Frontiers in Plant Science* 5, 175. doi: 10.3389/fpls.2014.00175.

Minoda, A., Sonoike, K., Okada, K., Sato, N. and Tsuzuki, M. (2003) Decrease in the efficiency of the electron donation to tyrosine Z of photosystem II in an SQDG-deficient mutant of Chlamydomonas. *FEBS Letters* 553, 109–112.

Mishra, A. and Tanna, B. (2017) Halophytes: Potential resources for salt stress tolerance genes and promoters. *Frontiers in Plant Science* 8, 829. doi: 10.3389/fpls.2017.00829.

Miszalski, Z., Ś lesak, I., Niewiadomska, E., Bączek-Kwinta, R., Lüttge, U. *et al.* (1998) Subcellular localization and stress response of superoxide dismutase isoforms from leaves in the C_3-CAM intermediate halophyte *Mesembryanthemum crystallinum* L. *Plant, Cell and Environment* 21, 169–179.

Monteoliva, M.I., Rizzi, Y.S., Cecchini, N.M., Hajirezaei, M.R. and Alvarez, M.E. (2014) Context of action of proline dehydrogenase (ProDH) in the hypersensitive response of Arabidopsis. *BMC Plant Biology* 14, 21. doi.org/10.1186/1471-2229-14-21.

Munné-Bosch, S. and Alegre, L. (2002) The function of tocopherols and tocotrienols in plants. *Critical Reviews in Plant Sciences* 21, 31–57.

Nawrocki, W.J., Tourasse, N.J., Taly, A., Rappaport, F. and Wollman, F.A. (2015) The plastid terminal oxidase: Its elusive function points to multiple contributions to plastid physiology. *Annual Review of Plant Biology* 66, 49–74.

Nejad, E.S., Askari, H. and Gholami, M. (2014) Evaluation of cellular responses of the halophyte *Aeluropus littoralis* to salinity stress through cell growth, sugars and polyamines content. *Plant Knowledge Journal* 3, 1–7.

Niewiadomska, E., Bilger, W., Gruca, M., Mulisch, M., Miszalski, Z. *et al.* (2011) CAM-related changes in chloroplastic metabolism of *Mesembryanthemum crystallinum* L. *Planta* 233(2), 275–285.

Noctor, G., Reichheld, J.-P. and Foyer, Ch.H. (2018) ROS-related redox regulation and signaling in plants. *Seminars in Cell and Developmental Biology* doi: 10.1016/j.semcdb.2017.07.013.

Ottow, E.A., Brinker, M., Teichmann, T., Fritz, E., Kaiser, W. *et al.* (2005) *Populus euphratica* displays apoplastic sodium accumulation, osmotic adjustment by decreases in calcium and soluble carbohydrates, and develops leaf succulence under salt stress. *Plant Physiology* 139(4), 1762–1772.

Ozgur, R., Uzilday, B., Sekmen, A.H. and Turkan, I. (2013) Reactive oxygen species regulation and antioxidant defence in halophytes. *Functional Plant Biology* 40, 832–847.

Pilarska, M., Wiciarz, M., Jajić, I., Kozieradzka-Kiszkurno, M., Dobrev, P. *et al.* (2016) A different pattern of production and scavenging of reactive oxygen species in halophytic *Eutrema salsugineum* (*Thellungiella salsuginea*) plants in comparison to *Arabidopsis thaliana* and its relation to salt stress signaling. *Frontiers in Plant Science* 7, 1179. doi: 10.3389/fpls.2016.01179.

Podgórska, A., Burian, M. and Szal, B. (2017) Extra-cellular but extra-ordinarily important for cells: apoplastic reactive oxygen species metabolism. *Frontiers in Plant Science* 8, 1353. doi:10.3389/fpls.2017.01353.

Popova, O.V., Ismailov, S.F., Popova, T.N., Dietz, K-J. and Golldack, D. (2002) Salt-induced expression of NADP-dependent isocitrate dehydrogenase and ferredoxin-dependent glutamate synthase in *Mesembryanthemum crystallinum*. *Planta* 215(6), 906–913.

Pottosin, I., Velarde-Buendía, A.M., Bose, J., Zepeda-Jazo, I., Shabala, S. *et al.* (2014) Cross-talk between reactive oxygen species and polyamines in regulation of ion transport across the plasma membrane: implications for plant adaptive responses. *Journal of Experimental Botany* 65(5), 1271–1283.

Qasim, M., Abideen, Z., Adnan, M.Y., Gulzar, S., Gul, B. *et al.* (2017) Antioxidant properties, phenolic composition, bioactive compounds and nutritive value of medicinal halophytes commonly used as herbal teas. *South African Journal of Botany* 110, 240–250.

Qiu, N., Lu, Q. and Lu, C. (2003) Photosynthesis, photosystem II efficiency and the xanthophyll cycle in the salt-adapted halophyte *Atriplex centralasiatica*. *New Phytologist* 159(2), 479–486.

Ramani, B., Reeck, T., Debez, A., Stelzer, R., Huchzermeyer, B., *et al.* (2006) *Aster tripolium* L. and *Sesuvium portulacastrum* L.: two halophytes, two strategies to survive in saline habitats. *Plant Physiology and Biochemistry* 44(5–6), 395–408.

Ramos, P.A.B., Santos, S.A.O., Guerra, Â.R., Guerreiro, O., Freire, C.S.R. *et al.* (2014) Phenolic composition and antioxidant activity of different morphological parts of *Cynara cardunculus* L. var. *altilis* (DC). *Industrial Crops and Products* 61, 460–471.

Redondo-Gómez, S., Wharmby, C., Castillo, J.M., Mateos-Naranjo, E., Luque, C.J. *et al.* (2006) Growth and photosynthetic responses to salinity in an extreme halophyte, *Sarcocornia fruticosa*. *Physiologia Plantarum* 128(1), 116–124.

Reginato, M.A., Castagna, A., Furlán, A., Castro, S., Ranieri, A. *et al.* (2014) Physiological responses of a halophytic shrub to salt stress by Na_2SO_4 and NaCl: oxidative damage and the role of polyphenols in antioxidant protection. *AoB Plants* 6, plu042. doi: 10.1093/aobpla/plu042.

Rejeb, B.K., Abdelly, C. and Savouré, A. (2014) How reactive oxygen species and proline face stress together. *Plant Physiology and Biochemistry* 80, 278–284.

Rogov, A.G., Sukhanova, E.I., Uralskaya, L.A., Aliverdieva, D.A. and Zvyagilskaya, R.A. (2014) Alternative oxidase: Distribution, induction, properties, structure, regulation, and functions. *Biochemistry* 79(13), 1615–1634.

Rozema, J. and Schat, H. (2013) Salt tolerance of halophytes, research questions reviewed in the perspective of saline agriculture. *Environmental and Experimental Botany* 92, 83–95.

Rozentsvet, O.A., Nesterov, V.N. and Bogdanova, E.S. (2017) Structural, physiological, and biochemical aspects of salinity tolerance of halophytes. *Russian Journal of Plant Physiology* 64(4), 464–477.

Sauter, M., Moffatt, B., Saecho, M.C., Hell, R. and Wirtz, M. (2013) Methionine salvage and S-adenosylmethionine: essential links between sulfur, ethylene and polyamine biosynthesis. *Biochemical Journal* 451(2), 145–154.

Scheibe, R. and Dietz, K.J. (2012) Reduction-oxidation network for flexible adjustment of cellular metabolism in photoautotrophic cells. *Plant, Cell and Environment* 35, 202–216.

Schmid-Siegert, E., Stepushenko, O., Glauser, G. and Farmer, E.E. (2016) Membranes as structural antioxidants - recycling of malondialdehyde to its source in oxidation-sensitive chloroplast fatty acids. *The Journal of Biological Chemistry* 291(25), 13005–13013.

Schmitt, F.J., Renger, G., Friedrich, T., Kreslavski, V.D., Zharmukhamedov, S.K. *et al.* (2014) Reactive oxygen species: re-evaluation of generation, monitoring and role in stress-signaling in phototrophic organisms. *Biochimica et Biophysica Acta* 1837(6), 835–848.

Sengupta, S. and Majumder, A.L. (2010) *Porteresia coarctata* (Roxb.) Tateoka, a wild rice: a potential model for studying salt-stress biology in rice. *Plant, Cell and Environment* 33(4), 526–542.

Shabala, S. and MacKay, A. (2011) Ion transport in halophytes. In: Turkan, I. (ed.) *Plant Responses to Drought and* Salinity Stress *: Developments in a Post-genomic Era*. Advances in Botanical Research, Academic Press Ltd/Elsevier Science Ltd, London, pp. 151–199.

Shabala, S. and Munns, R. (2017) Salinity stress: Physiological constraints and adaptive mechanisms. In: Shabala, S. (ed.) *Plant Stress Physiology*. CABI, Wallingford, UK, pp. 24–63.

Shamustakimova, A.O., Leonova, T.G., Taranov, V.V., de Boer, A.H. and Babakov, A.V. (2017) Cold stress increases salt tolerance of the extremophytes *Eutrema salsugineum* (*Thellungiella salsuginea*) and *Eutrema* (*Thellungiella*) *botschantzevii*. *Journal of Plant Physiology* 208, 128–138.

Sharma, S., Bal, A. and Joshi, Y. (1983) Polyphenol oxidase activity in glycophytes and alkali halophytes under salt stress. *Current Agriculture* 7, 71–74.

Singh, M., Kumar, J., Singh, S., Singh, V.P. and Prasad, S.M. (2015) Roles of osmoprotectants in improving salinity and drought tolerance in plants: A review. *Reviews in Environmental Science and Bio/Technology* 14, 407–426.

Singh, N., Mishra A. and Jha, B. (2014) Over-expression of the peroxisomal ascorbate peroxidase (*SbpAPX*) gene cloned from halophyte *Salicornia brachiata* confers salt and drought stress tolerance in transgenic tobacco. *Marine Biotechnology* 16(3), 321–332.

Slama, I., Abdelly, C., Bouchereau, A., Flowers, T. and Savouré, A. (2015) Diversity, distribution and roles of osmoprotective compounds accumulated in halophytes under abiotic stress. *Annals of Botany* 115(3), 433–447.

Slama, I., M'Rabet, R., Ksouri, R., Talbi, O., Debez, A. *et al.* (2017) Effects of salt treatment on growth, lipid membrane peroxidation, polyphenol content, and antioxidant activities in leaves of *Sesuvium portulacastrum* L. *Arid Land Research and Management* 31(4), 404–417.

Ślesak, I., Karpinska, B., Surówka, E., Miszalski, Z. and Karpinski, S. (2003) Redox changes in the chloroplast and hydrogen peroxide are essential for regulation of C_3-CAM transition and photooxidative stress responses in the facultative CAM plant *Mesembryanthemum crystallinum* L. *Plant and Cell Physiology* 44, 573–581.

Smith, C.A., Melino, V.J., Sweetman, C. and Soole, K.L. (2009) Manipulation of alternative oxidase can influence salt tolerance in *Arabidopsis thaliana*. *Physiologia Plantarum* 137(4), 459–472.

Srivastava, A.K., Srivastava, S., Lokhande, V.H., D'souza, S.F. and Suprasanna, P. (2015) Salt stress reveals differential antioxidant and energetics responses in glycophyte (*Brassica juncea* L.) and halophyte (*Sesuvium portulacastrum* L.). *Frontiers in Environmental Science* 3, 19. doi: 10.3389/fenvs.2015.00019.

Sui, N., Li, M., Li, K., Song, J. and Wang, B.-S. (2010) Increase in unsaturated fatty acids in membrane lipids of *Suaeda salsa* L. enhances protection of photosystem II under high salinity. *Photosynthetica* 48(4), 623–629.

Surówka, E., Dziurka, M., Kocurek, M., Goraj, S., Rapacz, M. *et al.* (2016) Effects of exogenously applied hydrogen peroxide on antioxidant and osmoprotectant profiles and the C_3-CAM shift in the halophyte *Mesembryanthemum crystallinum* L. *Journal of Plant Physiology* 200, 102–110.

Uzilday, B., Ozgur, R., Sekmen, A.H., Yildiztugay, E. and Turkan, I. (2015) Changes in the alternative electron sinks and antioxidant defence in chloroplasts of the extreme halophyte *Eutrema parvulum* (*Thellungiella parvula*) under salinity. *Annals of Botany* 115(3), 449–463.

Valderrama, R., Corpas, F.J., Carreras, A., Gómez-Rodríguez, M.V., Chaki, M. *et al.* (2006) The dehydrogenase-mediated recycling of NADPH is a key antioxidant system against salt-induced oxidative stress in olive plants. *Plant, Cell and Environment* 29(7), 1449–1459.

Verslues, P.E., Batelli, G., Grillo, S., Agius, F., Kim, Y-S. *et al.* (2007) Interaction of SOS2 with nucleoside diphosphate kinase 2 and catalases reveals a point of connection between salt stress and H_2O_2 signaling in *Arabidopsis thaliana*. *Molecular and Cellular Biology* 27(22), 7771–7780.

Voss, I., Sunil, B., Scheibe, R. and Raghavendra, A.S. (2013) Emerging concept for the role of photorespiration as an important part of abiotic stress response. *Plant Biology* 15(4), 713–722.

Wang, H., Bai, B., Bai, Z., Shi, L., Ye, J. *et al.* (2016) Enzymatic regulation of organic acid metabolism in an alkali-tolerant halophyte *Chloris virgata* during response to salt and alkali stresses. *African Journal of Biotechnology* 15(40), 2243–2250.

Wang, S., Uddin, M.I., Tanaka, K., Yin. L., Shi, Z. *et al.* (2014) Maintenance of chloroplast structure and function by overexpression of the rice *MONOGALACTOSYLDIACYLGLYCEROL SYNTHASE* gene leads to enhanced salt tolerance in Tobacco. *Plant Physiology* 165(3), 1144–1155.

Wang, X., Chang, L., Wang, B., Wang, D., Li, P. *et al.* (2013) Comparative proteomics of *Thellungiella halophila* leaves from plants subjected to salinity reveals the importance of chloroplastic starch and soluble sugars in halophyte salt tolerance. *Molecular and Cellular Proteomics* 12(8), 2174–2195.

Waśkiewicz A., Muzolf-Panek, M. and Goliński, P. (2013) Phenolic content changes in plants under salt stress. In: Ahmad, P., Azooz, M.M. and Prasad, M.N.V. (eds) *Ecophysiology and Responses of Plants under Salt Stress*. Springer, New York, pp. 283–324.

Wiciarz, M., Gubernator, B., Kruk, J. and Niewiadomska, E. (2015) Enhanced chloroplastic generation of H_2O_2 in stress-resistant *Thellungiella salsuginea* in comparison to *Arabidopsis thaliana*. *Physiologia Plantarum* 153(3), 467–476.

Wiciarz, M., Niewiadomska, E. and Kruk, J. (2017) Effects of salt stress on low molecular antioxidants and redox state of plastoquinone and P700 in *Arabidopsis thaliana* (glycophyte) and *Eutrema salsugineum* (halophyte). *Photosynthetica* 56(3), 811–819. doi:10.1007/s11099-017-0733.

Yaish, M.W. (2015) Proline accumulation is a general response to abiotic stress in the date palm tree (*Phoenix dactylifera* L.). *Genetics and Molecular Research* 14(3), 9943–9950.

Yang, Y., Yan, C-Q., Cao, B-H., Xu, H-X., Chen, J-P. *et al.* (2007) Some photosynthetic responses to salinity resistance are transferred into the somatic hybrid descendants from the wild soybean *Glycine cyrtoloba* ACC547. *Physiologia Plantarum* 129(3), 658–669.

Yang, C.-W., Zhang, M.-L., Liu, J., Shi, D.-C. and Wang, D.-L. (2009) Effects of buffer capacity on growth, photosynthesis, and solute accumulation of a glycophyte (wheat) and a halophyte (*Chloris virgata*). *Photosynthetica* 47(1), 55–60.

Yeo, A. (1998) Molecular biology of salt tolerance in the context of whole-plant physiology. *Journal of Experimental Botany* 49, 915–929.

Yildiztugay, E., Ozfidan-Konakci, C. and Kucukoduk, M. (2014) The role of antioxidant responses on the tolerance range of extreme halophyte *Salsola crassa* grown under toxic salt concentrations. *Ecotoxicology and Environmental Safety* 110, 21–30.

Yu, J., Chen, S., Zhao, Q., Wang, T., Yang, C. *et al.* (2011) Physiological and proteomic analysis of salinity tolerance in *Puccinellia tenuiflora*. *Journal of Proteome Research* 10(9), 3852–3870.

Zhang, Y., Lai, J., Sun, S., Li, Y., Liu, Y. *et al.* (2008) Comparison analysis of transcripts from the halophyte *Thellungiella halophila*. *Journal of Integrative Plant Biology* 50(10), 1327–1335.

12 Antioxidant Defence in Halophytes under High Salinity

Neeraj Kumar[1], Shubham Lamba[1,*], Ashwani Kumar[2], Pratima Kumar[3], Anita Mann[2], Sarita Devi[1], Pooja[4], Anita Kumari[1] and Babita Rani[1]

[1]CCS Haryana Agricultural University, Hisar, India; [2]Central Soil Salinity Research Institute, Karnal, India; [3]Prannath Parnami Institute of Management and Technology, Hisar, India; [4]Sugarcane Breeding Institute, Karnal, India

Abstract

Understanding the various mechanisms of salt-stress tolerance is important for crop improvement, especially when approximately 20% of agricultural land and 50% of crop land in the world is under salt stress. Stomatal closure due to salt stress reduces the CO_2: O_2 ratio inside the leaf tissues and inhibits CO_2 fixation. This inhibition leads to over-reduction of the photosynthetic electron transport chain and causes the generation of reactive oxygen species (ROS) such as singlet oxygen ($^1O_2$), superoxide anion ($O_2^{\bullet-}$), hydrogen peroxide (H_2O_2) and hydroxyl radical ($OH^{\bullet}$). These ROS are responsible for various stress-induced damages to macromolecules and ultimately to cellular structure. Hence, to avoid excessive ROS accumulation during stress and maintain the correct levels of ROS for signalling, plants possess a complex antioxidant defence system including non-enzymatic antioxidants such as ascorbic acid, glutathione, tocopherols and carotenoids; and enzymatic antioxidants such as superoxide dismutase (SOD), catalase (CAT), peroxidase (POX) and ascorbate peroxidase (APX). Although ROS production and defence mechanisms are common in glycophytes and halophytes, detoxification strategies may vary in response to salinity with regard to total antioxidant activity and type of iso-enzymes expressed. In this chapter we describe salinity, the classification of saline soils, genetic variability within halophytes, sources and overproduction of ROS under high-salinity conditions, the antioxidant defence system, enzymatic antioxidant system and non-enzymatic antioxidant system.

Keywords: Antioxidants; Halophytes; Reactive oxygen species; Redox regulatory enzymes; Salinity

12.1 Introduction

Worldwide, around 1.2 billion ha land is affected by the problem of salinity and sodicity (FAO, 2007). About 1128 million (M) ha of salt-affected soils are in the Middle East (189 M ha) followed by Australia (169 M ha) and North Africa (144 M ha). A major part of the total global area (~85%) is only slightly to moderately affected by high salt concentrations while the remaining 15% suffers from severe to extreme limitations for crop cultivation (Wicke *et al.*, 2011). The latest data provided by the Central Soil Salinity Research Institute (CSSRI), Karnal, India (www.cssri.org) indicates that the total salt-affected area in India is 6.73 Mha, out of which 3.79 Mha is affected with sodicity and the remaining 2.95 Mha is affected with salinity problems. Salt-affected soils

* Corresponding author e-mail: shubham.hau@gmail.com

exist across the length and breadth of India, covering 16 states/union territories. The states of Gujarat (2.23 M ha), Uttar Pradesh (1.37 M ha), Maharashtra (0.61 M ha), West Bengal (0.44 M ha) and Rajasthan (0.38 M ha) together account for almost 75% of saline and sodic soils in the country. In Haryana state, an area of 0.232 M ha is affected with alkalinity and 0.255 M ha with salinity and waterlogging (Government of Haryana, 2018).

In most salt-affected environments, poor-quality (saline and sodic) soils and groundwater are prevalent. The states of Rajasthan, Haryana and Punjab, lying in the north-western arid part of the country, suffer greatly from the problem of water of marginal quality (Singh, 2009). Salinity affects plant growth in several ways through osmotic and ionic effects (Munns, 2005). Soluble salts in soil decrease the water potential, requiring osmotic adjustment for the plant to absorb water. As a response to this phenomenon, plants can accumulate ions and other solutes to decrease their osmotic potential and hence their water potential below that of the surrounding soil. In this respect, responses by plants to drought and salinity are similar; however, the physical conditions of the environment are different. Water is limited in soil under drought; but, in saline soils, although water is present it is less available for plants to use.

The ionic effects of salinity result from their accumulation in plant tissues and especially in leaves. Transpiration is the driving force that moves salts from the roots to shoots and, as a consequence, this results in accumulation of ions in the leaf cells. The toxic effects of these ions are seen first in older leaves. As younger leaves expand, accumulated salts are diluted and this delays the occurrence of toxic effects of these ions (Munns, 2005). However, inhibition of growth due to an osmotic effect removes this advantage for younger leaves. Increased levels of salinity negatively influence germination, plant growth and reproducibility, physiological processes (photosynthesis, respiration and transpiration), membrane properties, nutrient balance, enzymatic and metabolic activities, cellular homeostasis and hormone regulation; and consequently lead to the production of reactive oxygen species (ROS). In cases of severe stress, it leads to plant death (Mahajan *et al.*, 2006; Hasanuzzaman *et al.*, 2012).

12.2 Halophytes

Plants capable of growing in highly saline environment are commonly known as halophytes. Highly saline environments may include mangrove swamps, marshes and sloughs, and seashores. The word halophyte is formed from two Greek words: *halas*, meaning salt and *phyton*, meaning plant. The halophytes are defined on the basis of different criteria. According to Schimper (1903), halophytes are plants that are capable of normal growth in saline soil as well as in normal soils. Stocker (1928) defined halophytes as plants that can thrive in salt concentrations over 0.5% throughout their life cycle. Dansereau (1957) defined halophytes as plants exclusively growing in saline soils. Munns and Greenway (1980) defined halophytes as a flora that is native of saline soils where osmotic potential is not less than 3.3 bars.

12.2.1 Classification of halophytes

Halophytes can be classified based on their ecological aspects as: (i) obligate halophytes; (ii) facultative halophytes; and (iii) habitat-indifferent halophytes (Sengbusch, 2003). Obligate halophytes are specific to saline habitats and thrive well only in saline environments (e.g. Chenopodiaceae family). Facultative halophytes, on the other hand, perform well in salt-free or in low-salt conditions. Habitat-indifferent halophytes are those that are not necessarily native to saline habitats but are capable of living in them. Being exotic to saline habitats they compete with plants sensitive to salt and under extreme conditions are capable of salt tolerance. There are diversified species of halophytes suited for growing in different saline regions throughout the world, such as coastal saline soil, soils of mangrove forests, wet land, marshy land, arid and semiarid regions, and agricultural fields. In 1989, Aronson enumerated 1550 salt-tolerant species on the basis of high salt tolerance (>80 mM NaCl ≈ 7.8 dSm^{-1} EC). Most Poaceae, Cyperaceae and Brassicaceae species, as well as a large number of dicotyledons like *Aster tripolium*, *Glaux maritima* and *Plantago maritima* belong to this group. These species constitute approximately 1% of the world's flora and are adapted to live in soil containing a high concentration of salt, and benefiting from it. They thus represent an ideal model to understand complex physiological and genetic mechanisms of salinity stress tolerance. Halophytes can effectively improve saline soil as they are well adjusted in the salt environment because of their diversified adaptation mechanisms, which include ion compartmentalization, osmotic adjustment, succulence, ion transport and uptake, antioxidant systems, maintenance of redox status and salt inclusion or excretion (Lokhande and Suprasanna, 2012). In recent years, plants having the ability to remove salts from contaminated soils have been identified and studied by many researchers (Rabhi *et al.*, 2008; Ashraf *et al.*, 2010).

12.3 Antioxidant Defence System

12.3.1 ROS generation in plants

An unavoidable consequence of aerobic metabolism is the production of ROS. ROS are partially reduced or excited forms of atmospheric oxygen. Activation of O_2 occurs by two different mechanisms: (1) by the transfer of one, two or three electrons to oxygen to produce, respectively, a superoxide radical ($O_2^{\bullet-}$), hydrogen peroxide (H_2O_2), or hydroxy radical ($HO^{\bullet}$); or (2) by the excitation of oxygen to produce singlet oxygen ($^1O_2$) (Fig. 12.1). The first step during oxygen reduction is the formation of superoxide ($O_2^{\bullet-}$) that is quickly converted to hydrogen peroxide (H_2O_2) by the action of superoxide dismutases (SODs). In the presence of transition metals, especially Fe: $O_2^{\bullet-}$ and H_2O_2, form highly toxic hydroxyl radicals ($OH^{\bullet}$) by the Haber–Weiss reaction. The final step, which involves the oxidation of Fe^{+2} by H_2O_2, is referred to as Fenton's reaction. Another non-radical ROS that can be produced in plant cells is singlet oxygen ($^1O_2$) by excitation of oxygen (Sharma *et al.*, 2012; Rodrigo-Moreno *et al.*, 2013). H_2O_2 is the most stable of the ROS. While the half-lives of $^1O_2$, $O_2^{\bullet-}$ and $OH\cdot$ are <1 μs, the half-life of H_2O_2 is in minutes (Pitzschke *et al.*, 2006). Increased ROS production has been reported in response to a variety of exogenous

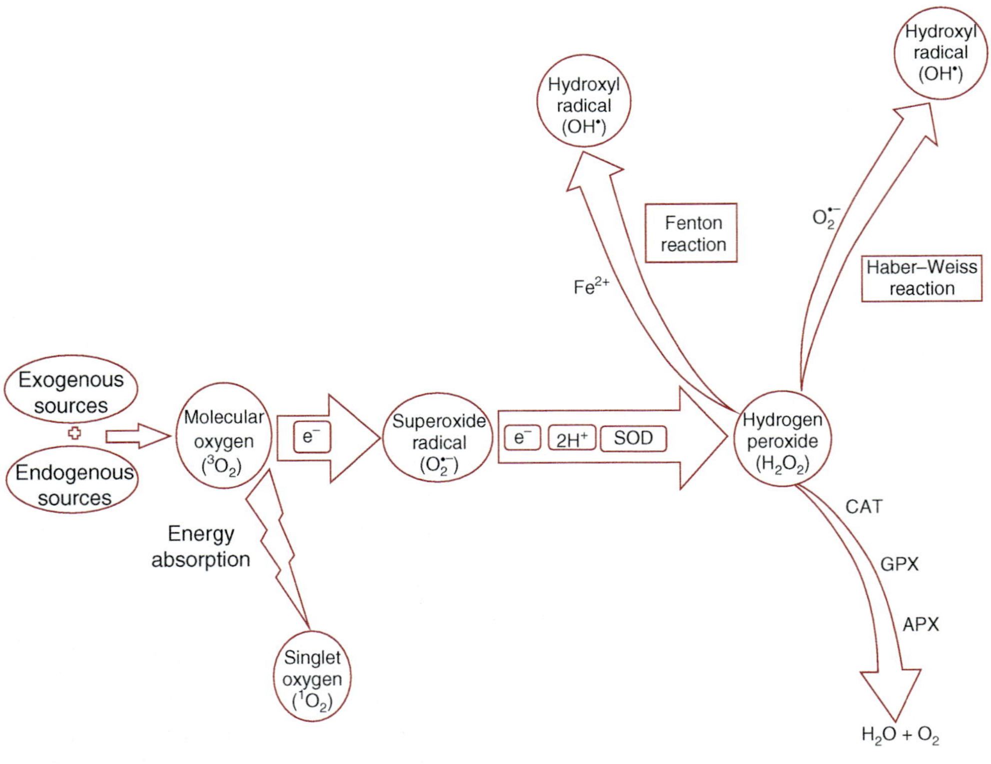

Fig. 12.1. Generation and detoxification of ROS.

abiotic stresses including drought, salinity, flooding, heat, cold, ozone, heavy metals, air pollution, nutrient deprivation, excess light and UV radiation (Mittler, 2002; Miller *et al.*, 2008).

12.3.2 Sites of ROS production in different cell organelles

In plants, ROS are routinely generated during normal plant metabolic processes and numerous subcellular compartments/organelles always contribute to ROS production (Table 12.1). Chloroplasts, mitochondria and peroxisomes are considered to be the main ROS producers during abiotic stresses (Mittler *et al.*, 2004; Asada, 2006). In addition, ROS can be produced in the apoplastic space by plant NADPH oxidases, cell wall-associated peroxidases (POXs) and oxalate oxidases (Kawano, 2003).

In chloroplasts, various forms of ROS production such as $O_2^{\bullet-}$ and H_2O_2 are generated during the electron transport chain (ETC) in photosynthetic photo system -I (PSI) and photosynthetic photo system -II (PS II) under illuminated conditions (Asada, 2006). The $O_2^{\bullet-}$ is formed at the antenna pigments and by the Mehler reaction at PSI, is then converted to H_2O_2 by SOD and then further reduced by the action of ascorbate peroxidases (APXs) in the water–water cycle (Asada, 2006). In addition, singlet oxygen ($^1O_2$) is generated in PSII by the electron transfer of the excited triplet state of chlorophyll to molecular oxygen ($^3O_2$), mainly under high light intensities (Laloi *et al.*, 2004).

Table 12.1. Sources of ROS generation in various cell organelles.

Cell organelle	Source of ROS generation
Chloroplast	ROS production at PS-I & II during photosynthetic electron transport
Peroxisome	Photorespiration, fatty acid β oxidation, xanthine oxidase, flavin oxidase
Mitochondria	During respiratory electron transport chain at complexes I, II & III
Cytoplasm	Quinone oxidase of plasma membrane
Apoplast	NADPH oxidase associated with plasma membrane, amine oxidase, oxalate oxidase, peroxidase associated with cell wall
Plasma membrane	Electron transporting oxidoreductases, quinone oxidase, NADPH oxidase
Endoplasmic reticulum	NADPH-dependent electron transport including Cyt P_{450}
Cell wall	Diamine oxidase, cell wall-associated peroxidase

Peroxisomes are small, usually spherical subcellular organelles with a single membrane and high metabolic plasticity. They are probably the major site of intracellular H_2O_2 during different processes, including the photorespiration glycolate oxidase reaction, an enzymatic reaction of flavin oxidases, fatty acid β-oxidation and disproportionation of $O_2^{\bullet-}$ (Baker and Graham, 2002; del Río *et al.*, 2002; Foyer and Noctor, 2003). Like chloroplasts and mitochondria, peroxisomes also produce $O_2^{\bullet-}$ at two sites, i.e. the organelle matrix and the integral peroxisomal membrane (Corpas *et al.*, 2001).

Plant mitochondria, being the 'energy factory' of the plant cell, also produce ROS at several sites. The two main sites are complexes I and III in the mitochondrial electron transport chain, where $O_2^{\bullet-}$ is generated and rapidly catalysed into H_2O_2 (Raha and Robinson, 2000; Moller, 2001; Sweetlove and Foyer, 2004). The levels of mitochondrial ROS in leaves are considerably lower than in peroxisomes or chloroplasts, especially under light conditions, and are estimated to be 30–100 times lower than in chloroplasts (Foyer and Noctor, 2003). However, in non-green tissues, mitochondria would be the main source of ROS. Indeed, in roots, mitochondria are considered a major source of ROS production. Several enzymes present in the mitochondrial matrix can produce ROS. Some of these (e.g. aconitase) produce ROS directly, whereas others (e.g. 1-galactonic-γ-lactone dehydrogenase (GAL)) are able to feed electron to ETC (Sharma *et al.*, 2012).

In the endoplasmic reticulum, NADPH-dependent electron transport involving Cyt P_{450} produces $O_2^{\bullet-}$ (Mittler, 2002). The organic substrate RH reacts first with Cyt P_{450} and then is reduced by a flavoprotein to form a radical intermediate (Cyt P_{450}R−). Triplet oxygen can readily react with this radical intermediate as each has one unpaired electron. This oxygenated complex (Cyt P_{450}-ROO−) may be reduced by cytochrome b, or occasionally the complexes may decompose releasing $O_2^{\bullet-}$. ROS are also produced in the apoplastic space by NADPH oxidases, cell wall-associated peroxidases that generate $O_2^{\bullet-}$ by oxidizing NADPH and transferring the electron to oxygen (O_2) (Sagi and Fluhr, 2006). This apoplastic ROS generation has important roles in the hypersensitive response to pathogen attack and also in developmental and growth processes, as well as in controlling programmed cell death (PCD) (Foreman *et al.*, 2003; Gechev and Hille, 2005; Gapper and Dolan, 2006; Sagi and Fluhr, 2006). There is a contribution to apoplastic ROS generation from the amine- and germin-like oxidases, and by pH-dependent oxalate peroxidases (Bolwell and Wojtaszek, 1997).

NADPH-dependent oxido reductases have been proposed to play a key role and lead to the generation of ROS at the plasma membrane. NADPH oxidase catalyses the transfer of electrons from cytoplasmic NADPH to O_2 to form $O_2^{\bullet-}$, followed by dismutation to H_2O_2 either spontaneously or by SOD activity (Apel and Hirt, 2004).

Cell walls are also regarded as sites of active metabolism and ROS production. The role of cell-wall-associated peroxidase in H_2O_2 generation has been shown. Diamine oxidases are also involved in the production of activated oxygen in the cell wall using diamine or polyamines (putrescine, spermidine, cadaverine, etc.) to reduce a quinone that auto-oxidizes to form peroxides (Spiteller, 2003; Higuchi, 2006). The LOX reaction, which catalyses the hydro-peroxidation of PUFA, also involved in the response against various stresses and a possible source of ROS such as $O_2^{\bullet-}$, H_2O_2, $OH^{\bullet}$ and $^1O_2$. Generation of $OH^{\bullet}$ was observed *in vivo* as well as *in vitro* within the cell wall of several plant species (Spiteller, 2003).

12.3.3 Role of ROS production and detoxification

ROS production occurs in all organisms as a by-product of metabolism under oxygen exposure. ROS are both beneficial and harmful for plants: their role depends upon their concentration. At low concentrations ROS act as signaling molecules, whereas their high concentration was found to be deteriorative to plant cells. These detrimental effects of ROS are a result of their ability to cause lipid peroxidation in cellular membranes, DNA damage, protein denaturation, carbohydrate oxidation, pigment breakdown and impairment of enzymatic activity (Scandalios, 1993; Noctor and Foyer, 1998). This dual action of ROS has been discussed in detail with specific instances of tolerance to environmental stress conditions with low concentrations of ROS (Dat *et al.*, 2000; Sharma *et al.*, 2012).

Plants have evolved complex antioxidant defence mechanisms, consisting of enzymatic and non-enzymatic pathways, to eliminate excessive ROS production while maintaining an optimum level of ROS for signalling, and halophytes are no exception to this. The enzymatic component consists of enzymes such as SOD, catalase (CAT), peroxidase (POX), ascorbate peroxidase (APX) and those of ascorbate–glutathione cycle (monodehydroascorbate reductase (MDHAR), dehydroascorbate reductase (DHAR) and glutathione reductase (GR)). As cells do not have mechanisms to detoxify $OH^{\bullet}$ enzymatically, plants can minimize the $OH^{\bullet}$-induced damage to cell membranes by preventing its formation, or by detoxifying it by low-molecular weight non-enzymatic antioxidant molecules such as tocopherols, carotenoids, ascorbate, glutathione and polyphenols (Sharma *et al*, 2012).

A number of reviews have concentrated on the link between salt stress and antioxidative pathways in plants (Bohnert and Jensen, 1996; Dat *et al.*, 2000; Van Breusegem *et al.*, 2001; Arora *et al.*, 2002; Borsani *et al.*, 2003). In these cases, enhancement of antioxidant functions has been suggested to be the more radical approach for increasing stress tolerance capacity because: (1) water stress imposed by increased salinity disrupts cellular homeostasis and chloroplast functions (Bohnert and Jensen, 1996); and (2) antioxidants have been implicated in stress acclimation (Dat *et al.*, 2000).

12.4 Enzymatic Component

12.4.1 SOD

SODs dismutase $O_2^{\bullet-}$ into H_2O_2 and have been considered to act as the 'first line of defence' against oxidative stress in plants (Alscher *et al.*, 2002). SODs are categorized into three main groups on the basis of the metal cofactor (Alscher *et al.*, 2002):

1. Cu/Zn SOD: These enzymes have copper and zinc as their cofactors and in plants are localized mainly in the cytosol and chloroplasts. Cytoplasmic Cu/Zn SODs are homodimeric and chloroplastic Cu/Zn SODs are homotetrameric in plants.
2. Mn SOD: This has manganese as its cofactor and is localized in mitochondria and the peroxisomes.
3. Fe SOD: In plants, these are found predominantly in chloroplasts; they are absent in animals.

As a general rule, up-regulation of Fe SOD and Mn SOD correlated well with the salt tolerance of several halophytes (Jithesh *et al.*, 2006), while changes in Cu/Zn SOD activity showed no clear trends in halophytes. However, overexpression of cytosolic halophytic Cu/Zn SOD of *Avicennia marina* in glycophyte indicates improved oxidative and salt tolerance, implying that Cu/Zn SOD also plays a significant role in stress tolerance (Prashanth *et al.*, 2008). Also, overexpression of Fe SOD in tobacco and of Mn SOD in rice conferred tolerance against salt and oxidative stresses (Van Camp *et al.*, 1996; Tanaka *et al.*, 1999). For example, in *H. marinum*, Mn SOD4 and Cu/Zn SOD7 were only detected under stress conditions (Seckin *et al.*, 2010), whereas in *Suaeda salsa* no new SOD isoenzymes were triggered by stress (Qiu-Fang *et al.*, 2005).

12.4.2 Catalase (CAT)

The main role of CAT is in catalysing the decomposition of H_2O_2 into water and oxygen. Catalases (CAT) are haem-containing tetrameric enzymes involved in the removal of H_2O_2 (Guan and Scandalios, 1993). Plant catalases are involved in photorespiratory functions and scavenging of H_2O_2 during β-oxidation of fatty acids in germinating seeds (Willekens *et al.*, 1995) and also during salt stress and other abiotic stress conditions. The multifarious functions of plant catalases require that plants evolve different isozymes of this important enzyme for the various roles it plays in plant systems. Plants have therefore evolved multiple CAT isozymes, which are encoded by members of a small gene family: two isozymes in castor bean (Ota *et al.*, 1992), and three in tobacco *Nicotiana tabacum* (Havir and McHale, 1987), *Nicotiana plumbaginifolia* (Willekens *et al.*, 1995), maize (Guan and Scandalios, 1993) and *Arabidopsis* (Frugoli *et al.*, 1996). The turnover rate of CAT is very high; every second, one unit of CAT protein complex can decompose millions of molecules of H_2O_2. Like SODs, CATs also have different isoforms localized mainly in peroxisomes, but a mitochondrial localization has also been noted in some species (Jithesh *et al.*, 2006). In a range of halophytes, increasing the salt concentration has yielded all the possible scenarios (increase, decrease and no change) in CAT expression/activity (Jithesh *et al.*, 2006).

12.4.3 Ascorbate peroxidase (APX)

Ascorbate peroxidases (APX) are antioxidants performing the same general function as catalases. However, unlike catalase, they catalyse the removal of H_2O_2 by using ascorbate as a reductant. APX play an important role in the regulation of the intracellular level of H_2O_2 in higher plants (Van Breusegem *et al.*, 2001; Shigeoka *et al.*, 2002). APX is a family of isozymes with different characteristics. APX is localized in all four major cellular compartments: stromal (sAPX) and thylakoid-membrane-bound APX in chloroplasts (tAPX), micro body (including glyoxysomes and peroxisomes) membrane-bound APX (mAPX), cytosolic APX (cAPX) and lastly a mitochondrial-membrane-bound form (mitAPX) (Shigeoka *et al.*, 2002). The APX isozymes use ascorbate as a reductant and are located in the stroma of chloroplasts, mitochondria, cytosol and the membrane of peroxisomes and chloroplasts. As a general rule, up-regulation of APX expression/activity increases the stress tolerance of both glycophytes and halophytes (Shigeoka *et al.*, 2002; Jithesh *et al.*, 2006).

12.4.4 Peroxidase (POX)

Plants contain abundant amounts of peroxidases (POX) that are also involved in H_2O_2 scavenging. They are associated with the cell wall where they generate phenoxy compounds from cinnamic acids. POX is also involved in other plant defence mechanisms, including responses to insects (Salin, 1987; Bi and Felton, 1995), and in a coordinated response known as the oxidative burst (Kawano, 2003).

Redox regulatory enzymes such as monodehydroascorbate reductase (MDAR; EC 1.6.5.4), GR, glutathione peroxidase (GPX), glutathione *S*-transferases (GST), thiol peroxidase type II peroxiredoxin (Prx) and APX also perform antioxidant functions in glycophytes and halophytes (Mittler 2002; Ozgur *et al.*, 2013). Several studies showed that halophytes have higher constitutive antioxidant defence activity as compared with glycophytes. For instance, the halophyte *Hordeum marinum* had higher constitutive levels of CAT, POX, APX and GR than *H. vulgare* (Seckin *et al.*, 2010). In addition to this, the SOD gene of *Thellungiella salsuginea* was expressed at much higher levels under unstressed conditions than in *Arabidopsis* under salinity (Taji *et al.*, 2004). Ellouzi *et al.* (2011) also showed that the POX, SOD and CAT activities of *Cakile maritima* were higher than those of *A. thaliana.*

12.5 Non-enzymatic Component

In addition to antioxidative enzymes, non-enzymatic antioxidants also play a key role in scavenging free radicals in plants during salinity stress. The non-enzymatic component comprises molecules such as ascorbic acid, glutathione, tocopherols, carotenoids and phenolic compounds that can scavenge ROS.

12.5.1 Ascorbic acid (AA)

AA or Vitamin C is one of the most powerful low molecular weight antioxidants and is present in the human diet and most plant cell types, organelles and the apoplast (Horemans *et al.*, 2000; Smirnoff, 2000). Under physiological conditions, AA exists mostly in the reduced form (90% of the ascorbate pool) in chloroplasts (Smirnoff, 2000). AA functions as a reductant, and its ability to donate electrons in a wide range of enzymatic and non-enzymatic reactions makes it a main ROS-detoxifying compound, which helps in minimizing the damage caused by oxidative stress. AA can directly scavenge $O_2^{\cdot-}$, OH• and $_1O^2$, and can reduce H_2O_2 to H_2O via the APX reaction (Noctor and Foyer, 1998). In chloroplasts, AA acts as a cofactor of violaxanthin de-epoxidase, thereby dissipating excess excitation energy (Smirnoff, 2000). AA regenerates TOC from the toco peroxyl radical (TOC•), which provides membrane protection (Horemans *et al.*, 2000; Smirnoff, 2000). Thus, elevated levels of endogenous AA in plants are necessary to offset oxidative stress in addition to regulating another plant metabolic process (Smirnoff, 2000). Oxidation of AA occurs in two sequential steps. First, MDHA is produced and, second, if MDHA is not rapidly re-reduced to ascorbate, it disproportionates into AA and DHA.

12.5.2 Glutathione

Glutathione (GSH, γ-glutamyl-cysteinyl-glycine) is a tripeptide, which is among the crucial low-molecular weight nonprotein thiols that plays an important role in intracellular defence against ROS-induced oxidative damage (May *et al.*, 1998). It has been detected in virtually all cell compartments such as cytosol, chloroplasts, endoplasmic reticulum, vacuoles and mitochondria (Noctor and Foyer, 1998). GSH is synthesized in the cytosol and chloroplasts of plant cells by compartment-specific isoforms of γ-glutamyl-cysteinyl synthetase (γ-ECS) and glutathione synthetase (GS). The balance between the GSH and glutathione disulfide (GSSG) is a central component in maintaining the cellular redox state. In combination with its oxidized form (GSSG), GSH maintain redox equilibrium in the cellular compartments is of great biological importance because it allows the fine-tuning of the cellular redox environment under normal conditions and, upon the onset of stress, provides the basis for GSH stress signalling (Wang *et al.*, 2008). Owing to its reducing power, GSH plays an important role in diverse biological processes including growth/division, regulation of sulfate transport, signal transduction, conjugation of metabolites, enzymatic regulation, synthesis of proteins and nucleic acids, synthesis of phytochelatins for metal chelation, detoxification of xenobiotics and the expression of the stress-responsive genes. A central nucleophilic cysteine residue is responsible for the high reductive potential of GSH, which scavenges cytotoxic H_2O_2 and reacts non-enzymatically with another ROS, such as $^1O_2$, $O_2^{\bullet-}$ or $OH^{\bullet}$ (Noctor and Foyer, 1998; Wang *et al.*, 2008); it can, therefore, function directly as a free-radical scavenger. GSH can protect macromolecules (i.e. proteins, lipids, DNA) either by the formation of adducts directly with reactive electrophiles (glutathionylation) or by acting as a proton donor in the presence of ROS or organic free radicals, yielding GSSG. The central role of GSH in the antioxidative defence is due to its ability to regenerate another powerful water-soluble antioxidant, AA, via the AsA–GSH cycle (Noctor and Foyer, 1998; Halliwell, 2006). The concentration of cellular GSH has a major effect on its antioxidant function and it varies considerably under abiotic stresses.

An important role for ascorbate as a major non-enzymatic means of scavenging superoxide radicals in mangroves during oxidative stress conditions was, therefore, suggested (Cheeseman *et al.*, 1997). In addition, ascorbate is the reductant necessary for the functioning of ascorbate peroxidase. Glutathione is required for conversion of dehydroascorbate to AA, in the Halliwell–Asada pathway (Bowler *et al.*, 1992).

Parida *et al.* (2004) studied the effect of salt stress on concentrations of AA and glutathione in the halophyte *Bruguiera parviflora*. A reduction in AA and glutathione levels was observed during salt stress in *B. parviflora*. However, this reduction was accompanied by an increase in the activities of APX and GR. This suggested that a reduction in the levels of glutathione and AA was due to consumption of these antioxidants in the ascorbate–glutathione oxidative pathway (Parida *et al.*, 2004).

12.5.3 Tocopherols

Tocopherols (α, β, γ and δ) represent a group of lipophilic antioxidants synthesized by all plants. Tocopherols are an essential component of biological membranes and are involved in scavenging of oxygen containing free radicals, lipid peroxyl radicals and singlet oxygen radicals. (Kiffin *et al.*, 2006). The relative antioxidant activity of the tocopherol isomers *in vivo* is $\alpha > \beta > \gamma > \delta$, which is due to the methylation pattern and a number of methyl groups attached to the phenolic ring of the polar head structure. Tocopherols have poor ability to donate electrons and thereby transfer single hydrogen atoms, making them efficient antioxidants (Arora *et al.*, 2002). In higher plants, tocopherols are known to protect lipids and other membrane components by physically quenching and chemically reacting with O_2 in chloroplasts, thus protecting the structure and function of PSII (Igamberdiev *et al.*, 2004). They prevent the chain propagation step in lipid autoxidation. A fully substituted benzoquinone ring and fully reduced phytol chain of tocopherol act as antioxidants in redox interactions with singlet $^1O_2$ (Igamberdiev *et al.*, 2004). $^1O_2$ oxygen quenching by tocopherols is highly efficient, and it is estimated that a single α-tocopherol molecule can neutralize up to 220 $^1O_2$ molecules *in vitro* before being degraded. Regeneration of the oxidized tocopherol back to its reduced form can be achieved by AsA, GSH or coenzyme Q. In *Suaeda maritima* seeds, the concentration of α-tocopherols was strictly bound with seed viability as protection from oxidative damage (Seal *et al.*, 2010). Yu *et al.* (2011) also reported that the α-tocopherol content of *Puccinellia tenuiora* was increased by salinity as a non-enzymatic antioxidant. These reports indicate that ability to regenerate oxidized tocopherol is important under salinity.

12.5.4 Carotenoids

Carotenoids are lipophilic organic compounds located in the plastids of both photosynthetic and non-photosynthetic plant tissues. They are referred to as antenna molecules as they absorb light in the region from 450 nm to 570 nm of the visible spectrum and pass the captured energy on to the chlorophyll (Chl). Carotenoids perform a multitude of functions in plant metabolism, including a role in oxidative stress tolerance. In chloroplasts, the carotenoids act as accessory pigments in light harvesting; and play an important role in detoxification of various forms of activated oxygen and triplet chlorophyll ($^3Chl^*$) which are produced as a consequence of photosynthetic complexes excitation by light. Carotenoids can exist both in ground state as well as in excited state after the absorption of light energy. In terms of their antioxidant properties, they can protect photosystems in one of four ways: (1) by reacting with LP products to terminate chain reactions; (2) by scavenging $^1O_2$ and dissipating the energy as heat; (3) by reacting with $^3Chl^*$ or excited chlorophyll (Chl*) molecules to prevent the formation of $^1O_2$; or (4) by dissipating excess excitation energy through the xanthophyll cycle. The main protective role of β-carotene in photosynthetic tissue may be accomplished via direct quenching of $^3Chl^*$, which prevents $^1O_2$ generation and thereby inhibits oxidative damage (Collins, 2001). During quenching of $^3Chl^*$, energy is transferred from Chl to CAR, which subsequently dissipates the energy in a non-radiative form (i.e. heat). Thus, CARs act as a competitive inhibitor of $^1O_2$ formation and this is aided by their proximity to chlorophyll in the light-harvesting complex. Another form of CAR, zeaxanthin, has been implicated in the dissipation of thermal energy, but the precise mechanism underlying this dissipation has not been resolved. Zeaxanthin appears to facilitate the conversion of $^3Chl^*$ to $^1Chl^*$ more efficiently than does β-carotene (Mortensen *et al.*, 2001). Carotenoids contain a chain of isoprene residues bearing numerous conjugated double bonds, which allows easy energy uptake from excited molecules and dissipation of excess energy as heat. The carotenoid content of the halophyte *Plantago coronopus* was enhanced by increasing salt treatment (Koyro, 2006). However, the carotenoid content of *Salicornia persica* and *S. europaea* decreased over the range of 100–600 mM NaCl treatments that could be related to photoinhibition or ROS formation (Aghaleh *et al.*, 2009).

12.5.5 Polyphenols

Polyphenols are another important diverse group of non-enzymatic antioxidants (flavonoids, tannins, anthocyanins and lignin) that possess antioxidant properties (Smirnoff, 2005). They are abundant in plant tissues and have strong antioxidant capacities depending on the donation of electrons to hydroxyl groups. Polyphenols contain an aromatic ring with OH or OCH_3 substituents, which together contribute to their biological activity, including antioxidant action. These can chelate transition metal ions, directly scavenge molecular species of active oxygen and inhibit lipid peroxidation by trapping the lipid alkoxyl radical. They also modify lipid packing order and decrease fluidity of the membranes. These changes could strictly hinder diffusion of free radicals and restrict peroxidative reactions. There are many studies on polyphenol contents of halophytes in the literature (Kalita and Saikia, 2004; Meot-Duros and Magne, 2009; Trabelsi *et al.*, 2010; Gómez-Caravaca *et al.*, 2012). Ksouri *et al.* (2009) reported relatively high polyphenol content and antioxidant defence capacity in the medicinal halophyte *Tamarix gallica*. Furthermore, the polyphenol content of halophytes can vary in the same genus: *Mesembryanthemum edule* showed greater total phenolic content compared with *M. crystallinum* and *M. nodiflorum* (Falleh *et al.*, 2009).

Although most studies concentrate on enzymatic scavenging systems, non-enzymatic antioxidants are also vital because they can scavenge ROS that cannot be detoxified by enzymatic systems, such as $^1O_2$ and $OH^•$. Previously, it has been shown that these ROS were major causes of lipid peroxidation (Triantaphylides *et al.*, 2008). Once the enzymatic detoxification system is overwhelmed, non-enzymatic antioxidants such as tocopherols, carotenoids and polyphenols are needed to prevent oxidative damage. As mentioned above, these molecules are accumulated to higher concentrations in unstressed conditions in halophytes as compared with glycophytes, which is compatible with the higher enzymatic antioxidant activity of these plants. Therefore, accumulations of these molecules are clearly advantageous to prevent damage caused by salinity-induced oxidative stress.

12.6 Conclusion

Salt stress is a major element among the many environmental stresses faced by plants since it is associated with the productive properties of soils, the very environment in which plants grow and survive. Moreover, among various soil problems around the globe, soil salinity problems are most severe and widely spread. Studying those plants showing salt tolerance, the halophytes, is therefore of significance.

Salt stress leads to overproduction of ROS. Halophytes possess some of the major antioxidant enzymes and low molecular weight antioxidant molecules involved in detoxification mechanisms, such as SOD, CAT, APX, POX, AA, glutathione, tocopherols, carotenoids and phenolic compounds, which are related to salt tolerance. This chapter suggests that both antioxidant enzymes and antioxidants itself possess protective properties against ROS, thus preventing lipid peroxidation during salt stress. This infers that the enzymatic and non-enzymatic antioxidants are integral components of the adaptive defence mechanism of halophytes against salt stress, and explains why most halophyte studies are concerned with these enzymatic and non-enzymatic antioxidants.

In totality it can be inferred that induction of enzymatic and non-enzymatic antioxidants is part of the integrated strategy adopted by plants against adverse stress conditions such as salt stress. This strategy is more pronounced in the halophytes and serves as a biochemical marker for studying salt-tolerance traits in plants.

References

Aghaleh, M., Nikham, V., Ebrahimzadeh, H. and Razavi, K. (2009) Salt stress effects on growth, pigments, proteins and lipid peroxidation in *Salicornia persica* and *S. europaea*. *Biologia Plantarum* 53, 243–248.

Alscher, R.G., Erturk, N. and Heath, L.S. (2002) Role of superoxide dismutases (SODs) in controlling oxidative stress in plants. *Journal of Experimental Botany* 53, 1331–1341.

Apel, K. and Hirt, H. (2004) Reactive oxygen species: Metabolism, oxidative stress, and signal transduction. *Annual Review of Plant Biology* 55, 373–399.

Aronson, J.A. (1989) *HALOPH a Database of Salt Tolerant Plants of the World Office Arid Land Studies*. University of Arizona, Tucson, AZ, USA.

Arora, A., Sairam, R.K. and Srivastava, G.C. (2002) Oxidative stress and antioxidative systems in plants. *Current Science* 82, 1227–1238.

Asada, K. (2006) Production and scavenging of reactive oxygen species in chloroplasts and their functions. *Plant Physiology* 141, 391–396.

Ashraf, M.Y., Ashraf, M. and Mahmood, K., Akhter, J., Hussain, F. *et al.* (2010) Phytoremediation of saline soils for sustainable agricultural productivity. In: Ashraf, M., Ozturk, M. and Ahmad, M.S.A. (eds) *Plant Adaptation and Phytoremediation*. Springer, Berlin, Germany, pp. 335–355.

Baker, A. and Graham, I. (2002) *Plant Peroxisomes: Biochemistry, Cell Biology and Biotechnological Applications*. Kluwer Academic Publishers, Dordrecht, The Netherlands.

Bi, J.L. and Felton, G.W. (1995) Foliar oxidative stress and insect herbivory: Primary compounds, secondary metabolites, and reactive oxygen species as components of induced resistance. *Journal of Chemical Ecology* 21, 1511–1530.

Bohnert, H.J. and Jensen, R.G. (1996) Strategies for engineering water-stress tolerance in plants *Trends in Biotechnology* 14, 89–97.

Bolwell, G.P. and Wojtaszek, P. (1997) Mechanisms for the generation of reactive oxygen species in plant defence—a broad perspective. *Physiological and Molecular Plant Pathology* 51, 347–366.

Borsani, O., Valpuesta, V. and Botella, M.A. (2003) Developing salt tolerant plants in a new century: a molecular biology approach. *Plant Cell, Tissue and Organ Culture* 73, 101–115.

Bowler, C., Van Montagu, M. and Inze, D. (1992) Superoxide dismutase and stress tolerance. *Annual Review of Plant Physiology and Plant Molecular Biology* 43, 83–116.

Cheeseman, J., Herendeen, L., Cheeseman, A. and Clough, B.F. (1997) Photosynthesis and photoprotection in mangroves under field conditions. *Plant, Cell and Environment* 20, 579–588.

Collins, A. (2001) Carotenoids and genomic stability. Production and action of active oxygen species in photosynthetic tissues. In: Foyer, C.H. and Mullineaux, P.M. (eds) *Causes of Photooxidative Stress and Amelioration of Defense Systems in Plants*. CRC Press, Boca Raton, FL, USA, pp 77–104.

Corpas, F.J., Barroso, J.B. and del Río, L.A. (2001) Peroxisomes as a source of reactive oxygen species and nitric oxide signal molecules in plant cells. *Trends in Plant Sciences* 6, 145–150.

Dansereau, P. (1957) *Biogeography: An Ecological Perspective*, Ronald Press, New York.

Dat, J.F., Vandenabeele, E., Vranova, E., Van Montagu, M., Inzé, D. *et al.* (2000) Dual action of the active oxygen species during plant stress responses. *Cellular and Molecular Life Sciences* 57, 779–795.

del Río, L.A., Corpas, F.J., Sandalio, L.M., Palma, J.M., Gómez, M. and Barroso, J.B. (2002) Reactive oxygen species, antioxidant systems and nitric oxide in peroxisomes. *Journal of Experimental Botany* 53, 1255–1272.

Ellouzi, H., Ben Hamed, K., Cela, J., Munné-Bosch, S. and Abdelly, C. (2011) Early effects of salt stress on the physiological and oxidative status of *Cakile maritima* (halophyte) and *Arabidopsis thaliana* (glycophyte). *Physiologia Plantarum* 142, 128–143.

Falleh, H., Ksouri, R., Oueslati, S., Sylvain, G., Christian, M. *et al.* (2009) Interspecific variability of antioxidant activities and phenolic composition in *Mesembryanthemum* genus. *Food and Chemical Toxicology* 47, 2308–2313.

FAO (Food and Agriculture Organization of the United Nations) (2007) Statistical Database. FAO Bulletin of Statistics and the FAO Production, Trade and Fertilizer Year Book, Rome.

Foreman, J., Demidchik, V., Bothwell, J.H., Mylona, P., Miedema, H., *et al.* (2003) Reactive oxygen species produced by NADPH oxidase regulate plant cell growth. *Nature* 422, 442–446.

Foyer, C.H. and Noctor, G. (2003) Redox sensing and signalling associated with reactive oxygen in chloroplasts, peroxisomes and mitochondria. *Physiologia Plantarum* 119, 355–364.

Frugoli, J.A., Zhong, H.H., Nuccio, M.L., McCourt, P., McPeek, M.A., *et al.* (1996) Catalase is encoded by a multigene family in *Arabidopsis thaliana* (L.) Heynh. *Plant Physiology* 112, 327–336.

Gapper, C. and Dolan, L. (2006) Control of plant development by reactive oxygen species. *Plant Physiology* 141, 341–345.

Gechev, T.S. and Hille, J. (2005) Hydrogen peroxide as a signal controlling plant programmed cell death. *Journal of Cell Biology* 168, 17–20.

Gómez-Caravaca, A.M., Iafelice, G., Lavini, A., Pulvento, C., Caboni, M.F., *et al.* (2012) Phenolic compounds and saponins in quinoa samples (*Chenopodium quinoa* Willd.) grown under different saline and non-saline irrigation regimens. *Journal of Agricultural Food Chemistry* 60, 4620–4627.

Government of Haryana (2018) About Haryana. Available at: https://haryana.gov.in/index.html (accessed 14 January 2019).

Guan, L. and Scandalios, J.G. (1993) Characterization of the catalase antioxidant defense gene Cat1 of maize, and its developmentally regulated expression in transgenic tobacco. *Plant Journal* 3, 527–536.

Halliwell, B. (2006) Oxidative stress and neurodegeneration: Where are we now? *Journal of Neurochemistry* 97, 1634–1658.

Hasanuzzaman, M., Hossain, M.A., da Silva, J.A.T. and Fujita, M. (2012) Plant responses and tolerance to abiotic oxidative stress: antioxidant defense is a key factor. In: Bandi, V., Shanker, A.K., Shanker, C. and Mandapaka, M. (eds) *Crop Stress and its Management: Perspectives and Strategies*. Springer, Berlin, pp. 261–316.

Havir, E.A. and McHale, N.A. (1987) Biochemical and developmental characterization of multiple forms of catalase in tobacco leaves. *Plant Physiology* 84, 450–455.

Higuchi, T. (2006) Look back over the studies of lignin biochemistry. *Journal of Wood Science* 52, 2–8.

Horemans, N., Foyer, C.H., Potters, G. and Asard, H. (2000) Ascorbate function and associated transport systems in plants. *Plant Physiology and Biochemistry* 38, 531–540.

Igamberdiev, A.U., Seregelyes, C., Manach, N. and Hill, R.D. (2004) NADH-dependent metabolism of nitric oxide in alfalfa root cultures expressing barley haemoglobin. *Planta* 219, 95–102.

Jithesh, M.N., Prashanth, S.R., Sivaprakash, K.R., and Parida, A.K. (2006) Antioxidative response mechanisms in halophytes: Their role in stress defence. *Journal of Genetics* 85, 237–254.

Kalita, D. and Saikia, C.N. (2004) Chemical constituents and energy content of some latex bearing plants. *Bioresource Technology* 92, 219–227.

Kawano, T. (2003) Roles of the reactive oxygen species-generating peroxidase reactions in plant defence and growth induction. *Plant Cell Reports* 21, 829–837.

Kiffin, R., Bandyopadhyay, U. and Cuervo, A.M. (2006) Oxidative stress and autophagy. *Antioxidants & Redox Signaling* 8, 152–162.

Koyro, H.W. (2006) Effect of salinity on growth, photosynthesis, water relations and solute composition of the potential cash crop halophyte. *Plantago coronopus* (L.). *Environmental and Experimental Botany* 56, 136–146.

Ksouri, R., Falleh, H., Megdiche, W., Trabelsi, N., Mhamdi, B. *et al.* (2009) Antioxidant and antimicrobial activities of the edible medicinal halophyte *Tamarix gallica* L. and related polyphenolic constituents. *Food and Chemical Toxicology* 47, 2083–2091.

Laloi, C., Apel, K. and Danon, A. (2004) Reactive oxygen signalling: The latest news. *Current Opinion in Plant Biology* 7, 323–328.

Mahajan, S., Sopory, S.K. and Tuteja, N. (2006) Cloning and characterization of CBL-CIPK signalling components from a legume (*Pisum sativum*). *FEBS J* 273, 907–925.

May, M.J., Vernoux, T., Leaver, C., Montagu, M.V. and Inze, D. (1998) Glutathione homeostasis in plants: Implications for environmental sensing and plant development. *Journal of Experimental Botany* 49, 649–667.

Meot-Duros, L. and Magne, C. (2009) Antioxidant activity and phenol content of *Crithmum maritimum* L. leaves. *Plant Physiology and Biochemistry* 47, 37–41.

Miller, G., Shulaev, V. and Mittler, R. (2008) Reactive oxygen signalling and abiotic stress. *Physiologia Plantarum* 133, 481–489.

Mittler, R. (2002) Oxidative stress, antioxidants and stress tolerance. *Trends in Plant Sciences* 7, 405–410.

Moller, I.M. (2001) Plant mitochondria and oxidative stress: Electron transport, NADPH turnover, and metabolism of reactive oxygen species. *Annual Review of Plant Physiology* and *Plant Molecular Biology* 52, 561–591.

Mortensen, A., Skibsted, L.H. and Truscott, T.G. (2001) The interaction of dietary carotenoids with radical species. *Archives of Biochemistry and Biophysics* 385, 13–19.

Munns, R. (2005) Genes and salt tolerance: Bringing them together. *New Phytologist* 167, 645–663.

Noctor, G. and Foyer, C.H. (1998) Ascorbate and glutathione: keeping active oxygen under control. *Annual Review of Plant Physiology and Plant Molecular Biology* 49, 249–279.

Ota, Y., Ario, T., Hayashi, K., Nakagawa, T., Hattori, T., *et al.* (1992) Tissue-specific isoforms of catalase subunits in castor bean seedlings. *Plant and Cell Physiology* 33, 225–232.

Ozgur, R., Uzilday, B., Sekmen, A.H. and Turkan, I. (2013) Reactive oxygen species regulation and antioxidant defence in halophytes. *Functional Plant Biology* 40, 832–847.

Parida, A.K., Das, A.B. and Mohanty, P. (2004) Defense potentials to NaCl in a mangrove, *Bruguiera parviflora*: Differential changes of isoforms of some antioxidative enzymes. *Journal of Plant Physiology* 161, 531–542.

Pitzschke, A., Forzani, C. and Hirt, H. (2006) Reactive oxygen species signalling in plants. *Antioxidants & Redox Signaling* 8, 57–64.

Prashanth, S.R., Sadhasivam, V. and Parida, A. (2008) Over expression of cytosolic copper/zinc superoxide dismutase from a mangrove plant *Avicennia marina* in indica rice var Pusa Basmati-1 confers abiotic stress tolerance. *Transgenic Research* 17, 281–291.

Qiu-Fang, Z., Yuan, L.Y., Hong, P.C., Cong-Ming, L. and Bao-Shan, W. (2005) NaCl enhances thylakoid-bound SOD activity in the leaves of C_3 halophyte *Suaeda salsa* L. *Plant Science* 168, 423–430.

Raha, S. and Robinson, B.H. (2000) Mitochondria, oxygen free radicals, disease and ageing. *Trends in Biochemical Sciences* 25, 502–508.

Sagi, M. and Fluhr, R. (2006) Production of reactive oxygen species by plant NADPH oxidases. *Plant Physiology* 141, 336–340.

Salin, M.L. (1987) Toxic oxygen species and protective systems of the chloroplast. *Physiologia Plantarum* 72, 681–689.

Scandalios, J.G. (1993) Oxygen stress and superoxide dismutases. *Plant Physiology* 101, 7–12.

Schimper, A.F.W. (1903) *Plant Geography Upon a Physiological Basis*. Clarendon Press, Oxford, UK.

Seal, C.E., Zammit, R., Scott, P., Flowers, T.J. and Kranner, I. (2010) Glutathione half-cell reduction potential and α tocopherol as viability markers during the prolonged storage of *Suaeda maritima* seeds. *Seed Science and Research* 20, 47–53.

Seckin, B., Turkan, I., Sekmen, A.H. and Ozfidan, C. (2010) The role of antioxidant defense systems at differential salt tolerance of *Hordeum marinum* Huds. (Sea barleygrass) and *Hordeum vulgare* L. (cultivated barley). *Environmental and Experimental Botany* 69, 76–85.

Sengbusch, P. von (2003) Halophytes. *Botanik Online* University of Hamburg.

Sharma, P., Jha, A.B., Dubey, R.S. and Pessarakli, M. (2012) Reactive oxygen species, oxidative damage, and antioxidative defense mechanism in plants under stressful conditions. *Journal of Botany* 2012, 217037. doi:10.1155/2012/217037.

Shigeoka, S., Ishikawa, T., Tamoi, M., Miyagawa, Y., Takeda, T., *et al.* (2002) Regulation and function of ascorbate peroxidase isoenzymes. *Journal of Experimental Botany* 53, 1305–1319.

Singh, G. (2009) Salinity-related desertification and management strategies: The Indian experience. *Land Degradation and Development* 20, 367–385.

Smirnoff, N. (2000) Ascorbic acid: Metabolism and functions of a multi-facetted molecule. *Current Opinion in Plant Biology* 3, 229–235.

Spiteller, G. (2003) The relationship between changes in the cell wall, lipid peroxidation, proliferation, senescence and cell death. *Physiologia Plantarum* 119, 5–18.

Stocker, O. (1928) Das Halophytenproblem. In: Frisch, K.V., Goldschmidt, R., Ruhland, W. and Winterstein, H. (eds) *Ergebnisse der Biologie*. Springer, Berlin, pp. 266–353.

Sweetlove, L.J. and Foyer, C.H. (2004) Roles for reactive oxygen species and antioxidants in plant mitochondria. In: Day, D.A., Millar, A.H. and Whelan, J. (eds) *Plant Mitochondria: from Genome to Function, Advances in Photosynthesis and Respiration*. Kluwer Academic Press, Dordrecht, The Netherlands, pp. 307–320.

Taji, T., Seki, M., Satou, M. Sakurai, T., Kobayashi, M., *et al.* (2004) Comparative genomics in salt tolerance between *Arabidopsis* and *Arabidopsis*-related halophyte salt cress using *Arabidopsis* microarray. *Plant Physiology* 135, 1697–1709.

Tanaka, Y., Hibino, T., Hayashi, Y., Tanaka, A., Kishitani, S., *et al.* (1999) Salt tolerance of transgenic rice overexpressing yeast mitochondrial Mn-SOD in chloroplasts. *Plant Science* 148, 131–138.

Trabelsi, N., Megdiche, W., Ksouri, R., Falleh, H., Oueslati, S., *et al.* (2010) Solvent effects on phenolic contents and biological activities of the halophyte *Limoniastrum monopetalum* leaves. *Food Science and Technology* 43, 632–639.

Triantaphylides, C., Krischke, M., Hoeberichts, F.A., Ksas, B., Gresser, G., *et al.* (2008) Singlet oxygen is the major reactive oxygen species involved in photooxidative damage to plants. *Plant Physiology* 148, 960–968.

Van Breusegem, F., Vranova, E., Dat, J.F. and Inzé, D. (2001) The role of active oxygen species in plant signal transduction. *Plant Science* 161, 405–414.

Van Camp, W., Capiau, K., Van Montagu, M., Inze, D. and Slooten, L. (1996) Enhancement of oxidative stress tolerance in transgenic tobacco plants overproducing Fe-superoxide dismutase in chloroplasts. *Plant Physiology* 112, 1703–1714.

Wang, J.P., Li, Y.L. and Zhang, J.G. (2008) Effect of high-temperature and excessive-light stress on APX activity in apple peel. *Acta Agriculturae Boreali-Sinica* 23, 144–147.

Wicke, B., Smeets, E., Dornburg, V., Vashev, B., Gaiser, T., *et al.* (2011) The global technical and economic potential of bioenergy from salt-affected soils. *Energy & Environmental Science* 4, 2669–2681.

Willekens, H., Inze, D., Van Montagu, M. and van Camp, W. (1995) Catalase in plants. *Molecular Breeding* 1, 207–228.

Yu, J., Chen, S., Zhao, Q., Wang, T., Yang, C., *et al.* (2011) Physiological and proteomic analysis of salinity tolerance in *Puccinellia tenuiora*. *Journal of Proteome Research* 10, 3852–3870.

13 Soil Chemical Composition Modifies the Morpho-physiological Responses of *Prosopis strombulifera*, a Halophyte Native to South America

Mariana Reginato¶, Analía Llanes¶ and Virginia Luna*
National University of Río Cuarto, Río Cuarto-Córdoba, Argentina

Abstract

In some salty areas of South America, and especially in central Argentina, high levels of sodium sulfate (Na_2SO_4) are found together with sodium chloride (NaCl). Plant species show differences in their susceptibility to growth in the presence of these salts. Some studies showed that Na_2SO_4 may inhibit the growth of species such as wheat, sugarcane, beet, tomato, wild potato and barley more than NaCl. However, studies focusing on how sodium sulfate can affect the biochemical and physiological processes of plants are very scarce. This chapter provides an overview of the tolerance/non-tolerance mechanisms of the halophyte *Prosopis strombulifera*, with a special emphasis on the effects of Na_2SO_4 on growth parameters, ion accumulation, production of secondary metabolites, antioxidant system and hormonal regulation, showing that the presence of the SO_4^{2-} anion in the culture medium was determinant in the toxicity observed in *P. strombulifera* plants treated with Na_2SO_4. It is proposed that, as SO_4^{2-} assimilation may be limited by the high concentration in the culture medium, the sulfur that has not been metabolized to cysteine would be in excess; it may be binding to cytochrome b559 of PSII, blocking its activity partially or completely, and thus inhibiting photosynthesis. Carbon metabolism and partitioning of Na_2SO_4 treated plants are also affected, and energy resources should be diverted to synthesis of secondary metabolites such as condensed tannins and lignin, and polyphenol precursors, to cope with the high oxidative stress. As a consequence, there is a strong inhibition in the growth of Na_2SO_4 treated plants, leading to chlorosis, necrosis and foliar abscission.

Keywords: Halophytes; Ion compartmentation; NaCl; Osmolytes; Oxidative stress; Phytohormones; Plant water use; Salt tolerance

13.1 Introduction

Abiotic stress pressure is likely to increase in the near future due to anthropogenic contributions and climate change, exacerbating the detrimental effects on plant development and crop yields. Climate change may bring about rising sea levels and increasing drought, both of which will contribute to increasing salinization in many regions worldwide; this will mainly affect crops, which cannot withstand significant salinization (Flowers and Muscolo, 2015). The negative effects of soil salinity on plant growth and development are generally attributed to the toxicity of

* Corresponding author e-mail: vluna@exa.unrc.edu.ar
¶ These authors contributed equally to this chapter

Na^+ ions, which at high cytosolic concentrations can disrupt the homeostasis of K^+ ions and negatively affect a large group of enzymes; however, in some species, the Cl^- ion may be more toxic than Na^+. Actually, most studies on salt tolerance have been conducted by using NaCl, which has been proposed to be generally the most abundant salt in saline soils (Reich *et al.*, 2017).

Differences in the susceptibility of plant species to NaCl and Na_2SO_4 have been reported. Some pioneering work showed that Na_2SO_4 may be more toxic than NaCl in species such as wheat (Datta *et al.*, 1995), sugarcane (Joshi and Naik, 1980), beet, tomato, wild potato (Bilski *et al.*, 1988) and barley (Huang and Redmann, 1995). Recently, Reich *et al.* (2017) demonstrated the higher toxicity of Na_2SO_4 over NaCl in *Brassica rapa* growing in hydroponics. In these plants, the stronger decrease in calcium, magnesium and manganese caused by sulfate salts may contribute to their increased toxicity compared to chloride salts. However, there are few references related to how sulfate can affect biochemical and physiological processes in plants growing in soils where this salt is abundant. For that reason, there has been an increasing interest in comparing the effects of sulfate with those of NaCl in order to better understand plant responses to the two most abundant salts in salinized soils of several countries (Manivannan *et al.*, 2008). Research on halophytes also has a distinct place in aiding our understanding of salt tolerance in plants, knowledge that is likely to be of importance as climate change and population growth combine to challenge our ability to feed the world's increasing human population (Flowers and Muscolo, 2015). Thus, studies on plants that grow and develop naturally in salty soils can be used to deepen our knowledge of the mechanisms of salinity tolerance that operate in model plants like *Arabidopsis thaliana* and *Thellungiella halophila*, or to identify novel ones.

In Argentina, numerous legumes have successfully adapted to saline environments, and are evolutionarily differentiated in their strategy of salt tolerance. These species are of great environmental importance as they can improve the fertility of saline soils by their ability to associate with N_2-fixing bacteria. Among them, the genus *Prosopis* includes important components of salty areas including sub-shrubs, shrubs and trees that behave as opportunists regarding water use, maintaining pulses of growth in times of water availability. They have exceptional physiological properties such as high photosynthetic rates, nitrogen fixation even at temperatures of 45°C and growth in alkaline soils with pH up to 9 (Felker *et al.*, 2008).

The species *Prosopis strombulifera* (Burkart, 1976) is distributed in salinized areas in Peru; northern and central areas of Argentina. It is a woody, spiny shrub with a subcoriaceous yellow legume containing 5–12 seeds (Fig. 13.1). It is very abundant in the saline areas of Córdoba and San Luis provinces (Argentina), where high levels of sodium sulfate (Na_2SO_4) were found at concentrations similar to those of sodium chloride (NaCl), and even higher in some soil samples (Sosa *et al.*, 2005). Comparative studies of the effects of different Cl^- and SO_4^{2-} salts performed at the germination stage demonstrated that solutions containing SO_4^{2-} anions were considerably

Fig. 13.1. *Prosopis strombulifera* plants in their natural habitat (Salinas del Bebedero, San Luis), detailing the inflorescence and fruit.

more inhibitory than Cl^- solutions at iso-osmotic concentrations (Llanes *et al.*, 2005; Sosa *et al.*, 2005). These results suggested that this species was an interesting model to study the mechanisms of salt tolerance in halophytic plants. An understanding of the contribution of sulfate salt to the toxicity of soil salinity is important in understanding plant responses under conditions in which not only NaCl, but additionally sulfate salts, are also present.

This chapter provides an overview of the different tolerance/non-tolerance mechanisms of the halophyte *P. strombulifera*, with special emphasis on the effects of Na_2SO_4 on growth parameters, ion accumulation, secondary metabolite production, antioxidant system response and hormonal regulation.

13.2 Growth Responses and Anatomical Modifications Under Na_2SO_4 Treatment

Soil salinity is a threat to agricultural production and a better understanding of the physiological basis underlying salt stress will be essential to improve the salt tolerance of crop plants (Flowers and Yeo, 1995; Chinnusamy *et al.*, 2005; Ashraf *et al.*, 2008). The effects of salt stress are the result of complex interactions between morphological, physiological and biochemical processes, resulting in growth inhibition as a general response. However, these responses vary according to the type of salt and concentration, length of exposure, the species or cultivar and the growth stage.

Results obtained with *P. strombulifera* plants hydroponically cultivated for 48 days with increasing concentrations of NaCl and Na_2SO_4 were very interesting and contrasting, showing that plant responses depended not only on the osmotic potential generated in the culture medium, but mainly on the chemical composition of the salt in the medium. Plants cultivated with Na_2SO_4 showed strong growth inhibition from the beginning of salinization ($\Psi o = -1$ MPa, 29 days), with evident symptoms of senescence such as chlorosis, necrosis and leaf abscission, and changes in plant morphology and anatomy. In contrast, the response to NaCl was of the halophytic type, with a slight stimulation of growth up to a concentration of 450 mM ($\Psi o = -1.71$ MPa, 40 days) without toxicity symptoms, and tolerance up to a concentration of 700 mM ($\Psi o = -2.6$ MPa, 48 days) without toxicity symptoms (Reginato *et al.*, 2014a). When a mixture of both salts was added to the culture medium to imitate their respective concentrations in the soil, the growth data obtained confirmed the existence of specific toxicity phenomena caused by the SO_4^{2-} anion. Indeed, they were partially alleviated in the salt mixture, probably due to dilution and ionic antagonism and/or mutual competition between Cl^- and SO_4^{2-}anions at the membrane level (Reginato *et al.*, 2014a).

Highly tolerant species such as halophytes possess specific structural adaptations which include specialized organs for salt excretion (salt glands), succulence, increased cuticle thickness, changes in the diameter and number of xylem vessels, early development of Caspary bands and lignification. These changes may be adaptations to saline stress or, in some cases, symptoms of damage and disorders of normal developmental processes (Reinoso *et al.*, 2004, 2005; Hameed *et al.*, 2009; Flowers and Colmer, 2015).

Anatomical studies performed of NaCl-treated *P. strombulifera* seedlings showed absence of salt glands in the leaves, their tissues develop great vacuolization and the root system undergoes precocious lignification and/or suberization of endodermal cells, with Casparian strips found much closer to the root tip than in glycophytes (Reinoso *et al.*, 2004, 2005; Reginato *et al.*, 2016). These plants can therefore filter soil solutions more efficiently to prevent passage of excess ions to the xylem (Reinoso *et al.*, 2004). Particularly under Na_2SO_4, seedlings presented specific morphological and anatomical modifications, which included a significant decrease in foliar area that reached 60% at high salinity ($\Psi o = -2.6$ MPa; 530 mM), a marked reduction in leaflet thickness and a significant increase in stomatal density in both abaxial and adaxial leaf sides in relation to control plants (Reginato *et al.*, 2013). These xerophytic features are related to a need to regulate stomatal conductance to balance photosynthesis versus transpiration rate, which is higher than in NaCl-treated plants. In the primary root, a greater number of cortical cell layers and early development of the endodermis were observed. These features make it possible to efficiently filter the soil solution and prevent the passage of excess ions to the xylem, as well as to increase water retention in the roots. Roots and hypocotyls showed a reduction in cambial activity and reduction in the formation of secondary xylem and phloem. The stems developed 1–3 layers of highly vacuolized cortical parenchyma cells, which has been observed in many halophytes and would be an adaptive mechanism to regulate the ionic concentration. These anatomical changes were accompanied by a significant increase in the synthesis and accumulation of tannins and lignins in all organs as salinity increased (Reinoso *et al.*, 2005). When the seedlings were cultured in the presence of both salts, a very interesting response was obtained. The ionic interactions also determined specific anatomical responses that differed in their degree or intensity from those observed in the monosaline cultures, showing a strong correlation with the chemical nature of the ions present in the medium (Reginato *et al.*, 2016).

13.3 Effects of Na_2SO_4 on Ion Compartmentation

Under salinity, and in addition to the reduction in soil water potential, antagonistic or synergistic effects on the absorption of ions occur: ions in high concentration in the external solution (e.g. Na^+, Cl^-) are incorporated at high speed, causing excessive accumulation in the tissue. In turn, they can inhibit the incorporation of other ions by the root (e.g. K^+, Ca^{2+}) and their transport to the stem via the xylem, eventually provoking deficiency symptoms in the upper tissues.

Under salt stress, plants may accumulate large amounts of Na^+. In *P. strombulifera*, Na_2SO_4 treatment caused significant accumulation of Na^+ in the tissues, both in roots and leaves, although it was higher in leaves treated with NaCl (21 g/100 g DW in $\Psi o = -2.6$ MPa). Generally, leaves are more vulnerable to Na^+ than roots, accumulating larger amounts of this ion due to unidirectional transport from the root by the transpiration process. Some halophytes may accumulate unusual amounts of Na^+ in the aerial part, where it can be used as an osmoticum and cell turgor regulator (even by the stomata cells) to regulate stomatal opening in saline conditions (Tester and Davenport, 2003). However, the ability of plants to tolerate saline stress generally depends on their ability to maintain adequate K^+ nutrition for osmoregulation, as well as to ensure a normal metabolism (Adem *et al.*, 2014). In *P. strombulifera* seedlings, Na_2SO_4 treatment caused a progressive decrease in K^+ content as salinity increased. It has been proposed that, owing to its physicochemical structure being similar to that of Na^+, competition occurs for the K^+ transporters and results in K^+ deficiency (Apse and Blumwald, 2002). This competition is the cause of Na^+ toxicity, since key biochemical processes of the cell (for which K^+ is essential) can be inhibited. In addition, Na^+ can displace Ca^{2+} from the surface of the plasma membrane, inducing membrane depolarization and causing leakage of cytosolic K^+ out of the cell (Apse and Blumwald, 2002). Calcium is an essential macronutrient and component of cell walls, also fulfilling a key role as a versatile messenger, mediating a variety of developmental, hormonal, and biotic and abiotic stress signals (Reddy and Reddy, 2004). Na_2SO_4 salinization decreased the Ca^{2+} content in *P. strombulifera* seedlings in both leaves and roots, similar to that reported for *Brassica napa* (Reich *et al.*, 2017). This large difference in Na^+ concentration in relation to that of Ca^{2+} within cells could be due to the presence of non-selective cation channels in the membrane, allowing the movement of monovalent or divalent cations indistinctly. This constitutes the primary entrance pathway of Ca^{2+} to cells, so in a medium with high Na^+ concentrations the entry of Na^+ over that of Ca^{2+} would be favoured (Demidchik, 2012). In our studies, it was observed that Ca^{2+} concentration was not affected by NaCl treatments, but it was strongly affected by the presence of the SO_4^{2-}anion (Reginato *et al.*, 2014a), reaffirming the hypothesis that the latter is more toxic.

Comparatively, anion accumulation was much lower than that of both cations (Reginato *et al.*, 2014a). Na_2SO_4 treatment caused a significant accumulation of SO_4^{2-} in tissues, mainly in leaves (6.7 g/100g DW with 530 mM Na_2SO_4), much higher than Cl^- and less than Na^+, but sufficient to provoke metabolic disorders at the cellular level, which were manifested as chlorosis, necrosis and finally leaf abscission. The deleterious effects of SO_4^{2-} may be the consequence of several metabolic reactions that will be considered later in this chapter.

Cl^- tolerance may vary among species, and some halophytes require this anion to maintain stomatal activity by performing osmotic adjustment with equivalent amounts of Na^+ and Cl^- in vacuoles (Kefu *et al.*, 2003). *P. strombulifera* does not allow the entry of large amounts of Cl^-, by restricting its incorporation (Reginato *et al.*, 2014a), a strategy that has been correlated with salinity tolerance in many species.

In summary, *P. strombulifera* plants that showed growth stimulation and higher salinity tolerance (treated with NaCl) were those accumulating larger amounts of Na^+ in leaves (21 g/100 g of DW in $\Psi o = -2.6$ MPa) and a lower proportion in roots. Those that showed strong growth inhibition and toxicity symptoms (treated with Na_2SO_4) had a different Na^+ accumulation dynamic, with higher levels in roots than in leaves. These results suggest that the presence of the SO_4^{2-} anion in the solution caused impairment in Na^+ transport to the shoot, which would result in a failure in the mechanism of osmoregulation by ion accumulation (Reginato *et al.*, 2014a, Devinar, 2015).

13.4 Effects of Na_2SO_4 on Hydric Relations

Halophytic plants are characterized by their ability to adjust water potential in the tissues through the accumulation of ions and compatible solutes for maintenance of water flow from the soil to the roots. Leaves of *P. strombulifera* seedlings grown in high concentrations of NaCl ($\Psi o = -2.6$ MPa) showed the ability to maintain a relative water content greater than controls to alleviate the impact of excessive Na^+ in the tissues. Anatomical adaptations such as the radial arrangement of the chlorenchyma around the conductive tissues, the modification of number and size of stomata and an efficient control of the transpiration rate allow this species to display more efficient use of water when NaCl is present in the medium (Reinoso *et al.*, 2004; Reginato *et al.*, 2013). In contrast, Na_2SO_4-treated plants ($\Psi o = -2.6$ MPa) showed reduced osmotic potential, extremely negative water potential and low relative water content. These responses reflect failure of an efficient osmotic adjustment. In addition, these plants exhibited a high transpiration rate, which caused the water imbalance (Llanes *et al.*, 2013, 2014) (Fig. 13.2). Therefore, when the sulfate anion is present in the medium, the capacities for ion compartmentalization and osmotic adjustment are reduced, resulting in water imbalance and symptoms of toxicity.

13.5 Effects of Na_2SO_4 in Compatible Solutes Production

As mentioned above, plants need to regulate their water potential to allow the entry of water and achieve the necessary turgidity for growth under saline environments. This can be accomplished through incorporating ions as inorganic solutes; but, given their toxicity in the cytoplasm, they must be compartmentalized in the vacuole. Thus, the cytoplasm must accumulate the so-called osmolytes (organic, low-molecular-weight compounds) to counter the osmotic potential (commonly polyhydroxylic compounds such as sugars and polyalcohols, amino acids such as proline and quaternary ammonium compounds). A common feature of osmolytes is that they can accumulate in high concentrations without causing disruption in intracellular biochemistry, minimally affecting the pH and charge balance of the cytosol or organelle lumen. Their synthesis is often performed by divergence of basic metabolites towards unique biochemical reactions, stress being one of the main causes of this metabolic divergence (Koyro *et al.*, 2013).

Both Na_2SO_4- and NaCl-treated *P. strombulifera* plants showed high proline accumulation, especially in leaves, which could be performing other functions besides osmotic adjustment such as the activation of peroxidases, the main enzymes involved in lignin synthesis (Llanes *et al.*, 2013). In this sense, Na_2SO_4-treated plants showed a remarkable increase in fibres and lignified tissues, which give them xeromorphic traits (Reinoso *et al.*, 2005). Ghars *et al.* (2007) demonstrated that proline accumulation *per se* does not confer salt tolerance in *Thellungiella halophila*. Similarly, in *P. strombulifera* the proline content increased significantly at high salinity regardless of the salt composition of the medium, suggesting that proline synthesis is an indicator of the osmotic stress intensity rather than a tolerance indicator. This species did not accumulate glycine betaine under any treatment.

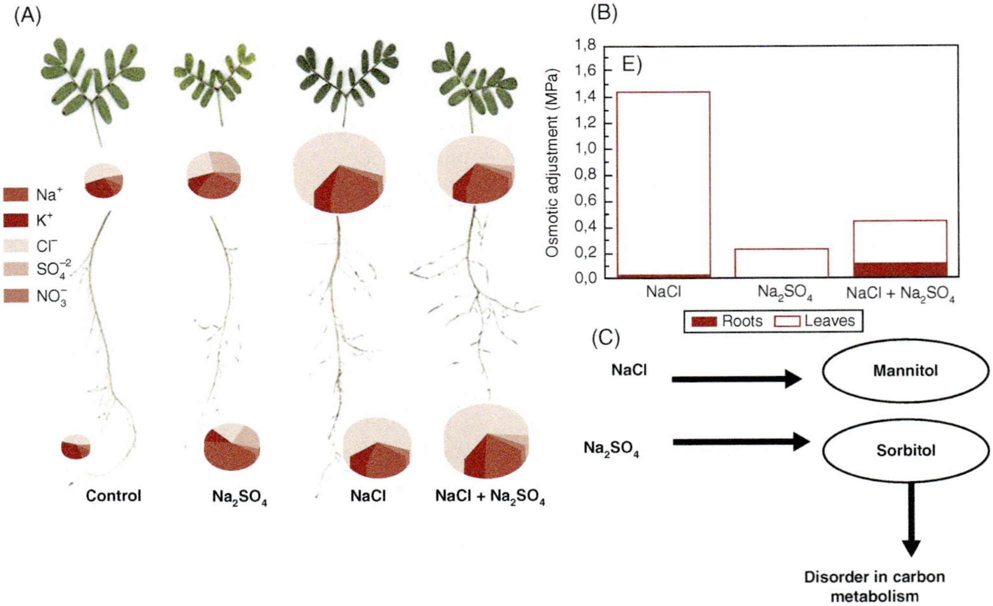

Fig. 13.2. Effect of NaCl, Na_2SO_4 and their iso-osmotic mixture on the ion content (A), osmotic adjustment (B) and polyalcohol accumulation (C) in *Prosopis strombulifera* plants.

Carbohydrates level is also affected by stress. In *P. strombulifera*, total soluble carbohydrates and reducing sugar (glucose and fructose) contents increased concomitantly with the increase in salt concentration in the medium, mainly in the roots of sulfate-treated plants, showing an alteration in carbon partitioning (Llanes *et al.*, 2013). In these plant roots, reducing sugars may contribute to the synthesis of polyalcohols and lignin, since they showed early vascular tissue differentiation with highly lignified elements (Reinoso *et al.*, 2005).

The synthesis of polyalcohols was also affected by the concentration and type of salt in the medium. Mannitol and sorbitol content in roots was not affected by salt treatments in this species. In leaves, mannitol accumulated only in NaCl-treated plants at moderate and high salinity, whereas leaves of plants treated with Na_2SO_4 or the iso-osmotic salt mixture mainly accumulated sorbitol. Although the functions of mannitol are not well understood in halophytes, it has been proposed that it would increase salt tolerance through protective mechanisms such as stabilization of macromolecules and stimulation of synthesis of antioxidant compounds (Noiraud *et al.*, 2001). In relation to sorbitol and in agreement with our results, Sheveleva *et al.* (1998) observed that transgenic tobacco lines with high sorbitol production showed lower chlorophyll and sucrose content, leaf necrotic spots and reduced growth rate. The main carbohydrate in response to all salt treatments in our experiments was pinitol, accumulating in both leaves and roots. Considering that inositol and its derivatives (such as pinitol and ononitol) have properties as compatible solutes and serve as carbon reservoirs for plants, its accumulation would represent an advantage for the metabolic and osmotic requirement necessary for growth under conditions of saline stress. However, as *P. strombulifera* accumulates large amounts of sodium and pinitol in leaves under salinity, the synthesis of this polyalcohol seems to be regulated by the salt concentration in the medium, both in the presence of sulfates and chlorides. In summary, plants treated with NaCl showed a high capacity of Na^+ compartmentalization in leaves, balanced by a high concentration of pinitol and mannitol in their cytoplasm. However, Na_2SO_4-treated plants failed in osmoregulation, although the level of proline and pinitol in leaves was similar to that determined in NaCl-treated plants. The distinctive feature was a large accumulation of sorbitol. According to previous comments in relation to the physiological role of sorbitol in plants, we suggest that, in this species, its presence could be considered an indication of a disorder in carbon metabolism (Llanes *et al.*, 2013).

13.6 Effects of Na_2SO_4 on the Metabolomic Profile

Plant responses to salinity involve changes in the activity of genes and proteins, which lead to modifications in plant metabolism. Metabolome changes in *P. strombulifera* plants grown hydroponically in iso-osmotic solutions of NaCl and Na_2SO_4, and their combination at $\Psi o = -1, -1.8$ and -2.6 MPa, were studied by means of GC/MS and UPLC/ESI-QqTOF-MS metabolite profiling. Analysis of the primary and secondary metabolomic profile showed a differential response to salt treatments according to salt type and concentration, organ and age of plants (Llanes *et al.*, 2016). Amino acid accumulation was observed in Na_2SO_4-treated plants, and this is associated with cellular damage possibly caused by an excessive increase in respiration. An abrupt increase of proline content was confirmed. The roots of these plants showed the highest number of altered compounds, with large accumulation of carbohydrates. The regulation of metabolites derived from glycolysis/gluconeogenesis would suggest that metabolic flux through these pathways is essential for the production of important intermediates for salinity-tolerance acquisition. However, the high concentration of some carbohydrates found in Na_2SO_4-treated plants is related to the detrimental effect of this anion in the culture medium. For example, high levels of a sorbitol derivative, maltitol (4-O-α-glucopyranosyl-D-sorbitol) were observed. Its role in plants is unknown but, in the fungus *Aspergillus niger* F-08, it was reported to induce (up to eightfold) the activity of a glucoamylase (RSDG) involved in starch digestion (Sun *et al.*, 2008). In plants, the induction of glucoamylases under stress conditions may be reflecting the need to obtain soluble sugars for osmotic adjustment or for signalling. Maltotritol, a polyalcohol synthesized from maltose, accumulated under NaCl and bisaline treatments but not under Na_2SO_4 treatment. This polyalcohol can inhibit α-glucosidase and maltose metabolism in *Streptococcus mutans* (Miyasawa, 2005). It could be thought that inhibition of glucosidases, maltose metabolism and excessive acid production under NaCl treatment could work as part of a fine regulation of sugar metabolism and signalling to improve tolerance to this salt. In Na_2SO_4-treated plants, the large increase in fructofuranosyl and cellobiose residues, especially in roots, is consistent with reduced synthesis and transport of sucrose. Therefore, carbohydrates constitute an interconnected and self-regulated network to maintain metabolic balance according to demand, to face adverse conditions such as salinity. Finally, the stimulation of metabolites related to protection against oxidative stress – such as acids related to ascorbic acid metabolism, the phenolic compound catechin and the diamine cadaverine (unusual PA in plants) were all detected exclusively in leaves of sulfate-treated plants (Reginato *et al.*, 2012). All these compounds are metabolic responses associated with the sequestering of free radicals to maintain cellular integrity.

13.7 Effects of Na_2SO_4 on Endogenous Phytohormones

It is widely known that phytohormones regulate a large number of responses by adjusting plant growth and development to adverse environmental conditions (Balderas-Hernandez *et al.*, 2013). Although plant responses to salinity depend on several factors, phytohormones are among the main endogenous substances involved in the modulation of physiological responses that eventually lead to adaptation to an unfavourable environment (Sharma *et al.*, 2013). Thus, understanding the role of each phytohormone and its signal cross-linkings provides valuable information on the physiological and molecular mechanisms that confer adaptation to salty soils.

Abscisic acid (ABA) is a phytohormone that plays a fundamental role in the regulation of plant responses to several abiotic stresses, and is called the 'stress hormone'. These responses include: changes in stomatal conductance (Wilkinson and Davies, 2010), growth regulation and osmolyte accumulation (Alves and Setter, 2004) and specific gene expression (Raghavendra *et al.*, 2010). Regulation of these processes is mediated by changes in ABA *de novo* synthesis and metabolism (hydroxylation and conjugation). In our studies, the endogenous levels of ABA in *P. strombulifera* were different according to the type of salt and its concentration, the organ analyzed and the plant age. Low levels of ABA were observed in roots of NaCl-treated plants at $\Psi o = -1.9$ MPa (500 mM), suggesting that this salt concentration does not imply a stress for this species, in concordance with optimum growth. In the presence of Na_2SO_4, the leaves showed high ABA content, which was higher at the highest concentration, coinciding with toxicity symptoms. In all salt treatments, roots and leaves mainly accumulated the glucose conjugate (ABA-GE), which reached high levels at high salt concentrations ($\Psi o = -2.6$ MPa) mainly with Na_2SO_4. ABA-GE was synthesized in roots and transported to leaves where the highest levels of free ABA and ABA-glycosylase activity were recorded (Llanes *et al.*, 2014). Therefore, the presence of Na_2SO_4 triggers ABA accumulation as a result of dehydration, due to the failure in osmotic adjustment and the extremely negative water potential of these plants. However, in previous studies we have proposed that the presence of the SO_4^- anion blocks ABA activity, thus inhibiting stomatal closure and so increasing transpiration (Llanes *et al.*, 2014). In contrast, ABA levels in NaCl-treated plants were slightly modified by the salt concentration, which shows the high adaptive capacity of this species under NaCl (Devinar *et al.*, 2013).

It has been proposed that phytohormones known as 'growth promoters' such as gibberellins, auxins and cytokines could have a role under stress, but their roles in halophytes are scarcely known. Thus, we tried to find out the relationships between these plant growth promoters and the physiological responses described above in salinized *P. strombulifera* plants. A much higher accumulation of gibberellins GA_4 and GA_1 in relation to GA_3 was observed. Although GA_1 and GA_3 are the most studied active gibberellins, it is known that GA_4 is also present in several species and would be the most important GA in *Arabidopsis* sp. and in some members of the genus *Cucurbita*. In leaves and roots of *P. strombulifera*, a large accumulation of GA_4 and GA_1 was observed at 500 and 700 mM NaCl ($\Psi o =-1.9$ MPa and -2.6 MPa), coinciding with low levels of ABA and ABA-GE and optimal plant growth. Conversely, leaves of Na_2SO_4-treated plants at $\Psi o =-2.6$ MPa showed very low levels of GA_4 and GA_1 in coincidence with high levels of ABA and ABA-GE and strong growth inhibition. Plants treated with the salt mixture showed intermediate levels of both gibberellins, demonstrating that the ionic effect of SO_4^- was also affecting GAs biosynthesis; ionic interactions and/or dilution in the salt mixture alleviated SO_4^- toxicity (Llanes *et al.*, 2015). On the other hand, inactivation of active GAs is essential for plants to regulate their levels and maintain their homeostasis. The most prevalent mechanism of GA inactivation is the enzymatic 2β-hydroxylation by gibberellin 2-oxidases. These, for example, inactivate GA_4 and GA_1, producing GA_{34} and GA_8, respectively (Hedden and Thomas, 2012). In concordance with the high levels of GA_4 found in NaCl-treated plants, lower levels of its metabolite GA_{34} were observed. Conversely, the low GA_4 levels in Na_2SO_4-treated plants were correlated with higher GA_{34} content, suggesting that the inactivation of GA_4 is a sign of SO_4^- toxicity. The low levels of GA_8 detected in all treatments indicates that, for some reason, there was less oxidation of GA_1 in these experiments (Llanes *et al.*, 2015).

In relation to auxins (IAA) our studies show its role in the induction of lateral roots by ionic stress caused in *P. strombulifera* roots by the presence of sulfate in the culture medium. The initiation of lateral roots plays an important role in plant development, determining the root system architecture, providing physical support and ensuring water absorption and nutrient incorporation. IAA increase in the central cylinder of the root triggers the signal for lateral root morphogenesis (Dubrovsky *et al.*, 2008). In *P. strombulifera*, high levels of IAA in roots of Na_2SO_4-treated plants was concomitant with a remarkable increase in lateral root formation (Reinoso *et al.*, 2005). However, in the presence of NaCl, plant roots showed low auxin levels without lateral root formation or any other histological changes, as compared to control plants (Reinoso *et al.*, 2004).

In summary, plants with low levels of ABA/GA and ABA/AIA, and a high Z level may be exhibiting higher salt tolerance than those with other levels. The high levels of ABA and inactive GAs would be a sign of salt toxicity. These hormonal patterns generated by both sodium salts indicate that primary signalling pathways are modified by the chemical composition of the salts present in the medium, as can be seen in Fig. 13.3.

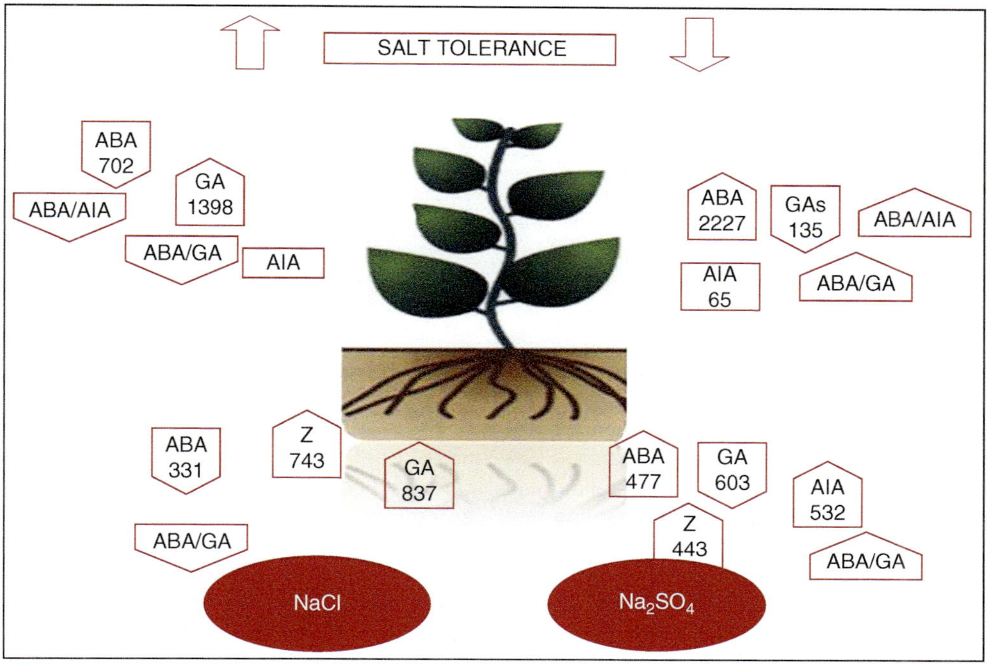

Fig. 13.3. Hormonal profile in leaves and roots of *Prosopis strombulifera* with differential tolerance to NaCl (left) and Na_2SO_4 (right). The arrows indicate the relative increase (up arrow), decrease (down arrow) or no change (rectangle) in each phytohormone at −2.6 MPa. The numbers inside the arrows indicate the absolute concentration of each hormone.

Understanding the interaction between hormonal changes and stress tolerance will allow identification of key steps in signalling pathways leading to plant adaptive responses. However, the chain of events leading to these metabolic differences caused by the presence of chlorides and sulfates in the medium is far from being understood, and adds complexity to the already known ionic factors involved in plant responses to salinity.

13.8 Oxidative Damage and Antioxidant System Response under Na_2SO_4

Reactive oxygen species (ROS) represent partially reduced or excited forms of atmospheric oxygen such as superoxide anion ($O_2^{\bullet -}$), radical hydroxyl ($OH^\bullet$) singlet oxygen ($^1O_2$) and hydrogen peroxide (H_2O_2). ROS are continuously formed in different compartments of the cell as an unavoidable product of aerobic metabolism, but are also formed as a result of activation of specific ROS-producing enzymes that mediate signal transduction processes (Mittler, 2002; König *et al.*, 2012; Foyer and Noctor, 2013). The major ROS-producing compartments in plants are the chloroplast ($O_2^{\bullet -}$, H_2O_2 and $^1O_2$ as a product of photosynthesis); the mitochondrion ($O_2^{\bullet -}$ and H_2O_2 as a product of respiration); the peroxisome (H_2O_2 as a product of photorespiration); and the apoplast, which mainly produces $O_2^{\bullet -}$ (and subsequently H_2O_2) as key signalling molecules generated by NADPH oxidases termed respiratory burst oxidase homologue (RBOH), peroxidases and superoxide dismutases (SODs; Mittler *et al.*, 2011; Vaahtera *et al.*, 2014; Mignolet-Spruyt *et al.*, 2016). Oxidative stress is a ubiquitous phenomenon, and plants not only use ROS as signals against stress, but also in normal growth and development (Apel and Hirt, 2004). The generation of ROS in plants is triggered by different kinds of environmental stresses, such as high light, high or low temperature, salinity, drought, nutrient deficiency and pathogen attack. This generalized response ('oxidative stress') occurs when the balance between production and removal of ROS is disturbed (Demidchick, 2012). These radicals can react with different cellular components and cause their oxidation and consequently oxidative damage at the cellular level (Fridovich, 1978; Halliwell and Gutteridge, 2015).

In roots and leaves of *P. strombulifera* plants growing in Na_2SO_4, increases in both H_2O_2 and $O_2{\bullet}^-$ content were found, in relation to control and NaCl-treated plants. They also had lower total soluble protein content, a significant increase in malondialdehyde (MDA) concentration and a generalized metabolic disorder causing growth inhibition (Llanes *et al.*, 2013; Reginato *et al.*, 2014a). MDA is considered an important oxidative stress indicator resulting from membrane lipid peroxidation, organelle disintegration and oxidation and loss of function of membrane proteins, affecting almost all cell metabolic activities, often irreversibly (Halliwell and Gutteridge, 2007). As a first line of defence, ROS generation is usually avoided through dissipative mechanisms, where most of the excess energy is dissipated as heat. When these mechanisms are overcome by the intensity of stress, the second line of defence comes into action: avoiding ROS accumulation through chemical and enzymatic destruction by antioxidants. Although there is considerable debate in the literature about the activity of antioxidant enzymes in plants exposed to stressing conditions, some authors argue that in halophytic species these enzymes increase their activity as the concentration of salts to which plants are exposed increases (Bose *et al.*, 2014). Modifications in the specific activity of catalase (CAT) and superoxide dismutase (SOD) enzymes were observed in both roots and leaves in *P. strombulifera* Na_2SO_4-treated plants. CAT plays an important role in early detoxification of H_2O_2 in roots, whereas SOD has a predominant role in leaves, reflecting its important role in detoxification of $O_2{\bullet}^-$ anions. Unlike in NaCl treatment, increasing concentrations of Na_2SO_4 stimulated production of tannin and lignin in roots, stems and leaves (Reinoso *et al.*, 2004, 2005; Reginato *et al.*, 2016) and of a large quantity of flavonoid-type polyphenolic compounds, particularly flavan-3-ols (Reginato *et al.*, 2014b) (Fig. 13.4). These polyphenolic compounds are involved in various physiological processes and constitute an important part of the plant's non-enzymatic antioxidant defence system owing to the stability of its phenolic radicals (Simic and Jovanovich, 1994). This significant accumulation of flavonoids with strong antioxidant activity indicates their role in cellular protection, mainly in the detoxification of ROS that cannot be attacked enzymatically as in the case of singlet oxygen and hydroxyl radicals (Zhan and Zhao, 2003). These compounds, in addition to their redox properties, possess metal ion chelating properties, so their synthesis under salinity could be stimulated to sequester excess Na^+ and thus avoid their permanence in the cytoplasm when the vacuoles have been saturated. It has recently been proposed that, in quinoa plants salinized with 400 mM of NaCl, the flavonoid rutin would act in the cytoplasm of the cells by activating Na^+ pumping outwards, sequestering hydroxyl radicals and thus preventing the loss of K^+ through the efflux pathway activated by ROS (Ismail *et al.*, 2016). That is, flavonoid accumulation would also have a direct correlation with ionic homeostasis in salinity conditions.

13.9 Conclusion

In plants, sulfate is absorbed by the root and constitutes the main source of sulfur for plant growth. Sulfate is reduced in chloroplasts and plastids for later assimilation into cysteine, which is the precursor for most of the sulfur organic compounds in plants. Its root uptake, distribution in the plant and reduction in the chloroplast/plastids are processes finely regulated by the sulfur demand for growth (Hawkesford and De Kok, 2006). Several of the intermediates in the sulfate reduction pathway are toxic, although it has been mentioned that it could have toxic effects *per se* on plant metabolism. In recent years we demonstrated that the presence of the SO_4^{2-} anion in the culture medium was the determinant of the toxicity observed in *P. strombulifera* plants treated with Na_2SO_4. This conclusion was reached by determinations of relative water content, osmotic potential, osmotic adjustment capacity, transpiration rate, oxidative damage, ion content, soluble carbohydrates, organic acids, polyalcohols, proteins, total and specific amino acids such as proline, phytohormone levels, chlorophyll fluorescence, photosynthetic electron flow, respiratory rate, polyphenol synthesis and metabolomic

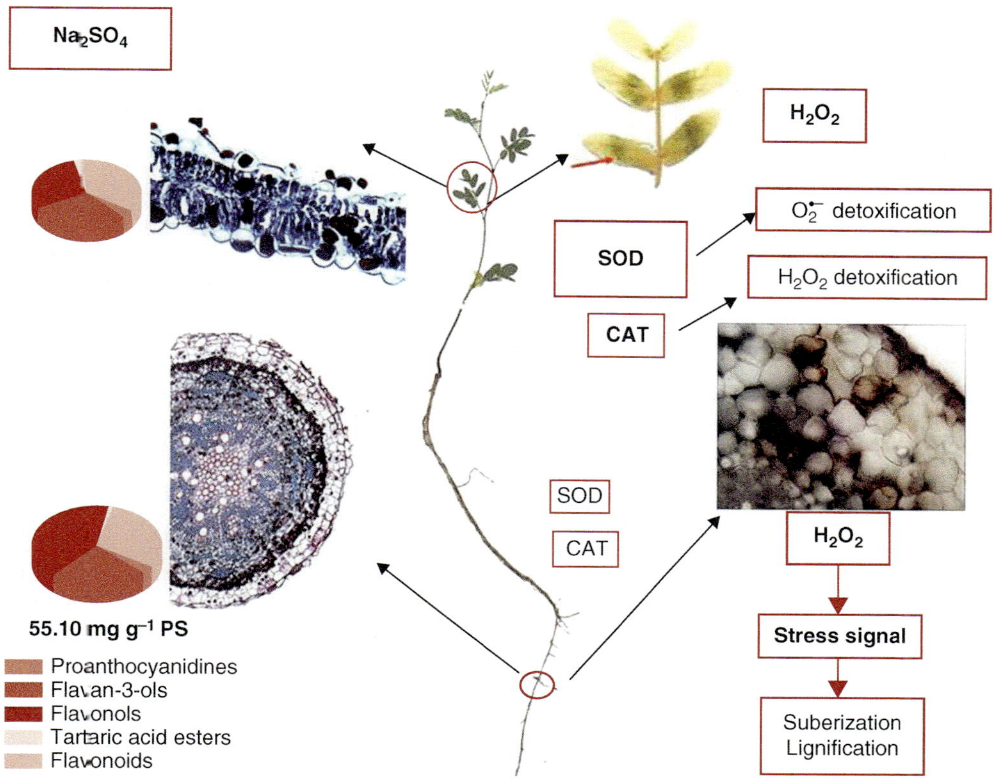

Fig. 13.4. Na_2SO_4 effects on H_2O_2 accumulation, catalase (CAT) and superoxide dismutase (SOD) activity and polyphenol synthesis in *Prosopis strombulifera* plants ($\Psi o = -2.6$ MPa, 48 days of culture). Area of boxes for each enzyme is approximately proportional to their activity. Left: pictures (cross section) and pie charts show polyphenol accumulation for leaves and roots. Right: H_2O_2 localization in leaf (above) and root tissues (cross section).

analysis (Devinar *et al.*, 2013, Llanes *et al.*, 2013, 2014, 2015, 2016; Reginato *et al.*, 2014a, 2014b, 2016). Based on these results, a possible explanation for the high SO_4^{2-} toxicity is proposed. Once in the stem, SO_4^{2-} is transported to the chloroplasts for its reduction. S_2^- is generated during the assimilation process and the reactions from sulfate to cysteine take place in the chloroplasts. The S_2^- formed can bind to the cytochromes involved in the respiratory chain, affecting this process as well, as it can bind to other complexes possessing the haemo group. Recently, Kobayashi *et al.* (2015) reported that sulfite (SO_3^{2-}), an intermediate compound in the SO_4^{2-} assimilation pathway, could repress 90% photosynthesis in the cyanobacteria *Synechococcus elongatus* PCC 7942. These authors demonstrated an inactivation of PSII with consequent growth inhibition. Since the SH_2 generated in the SO_4^{2-} pathway can be attached to iron-bearing haemo groups, as mentioned above, it may be expected that it could also attach to the haemo group of cytochrome b559, which is an intrinsic component of PSII in all photosynthetic aerobic organisms. In our experiments, since the assimilation of SO_4^{2-} is limited because of its high concentration in the culture medium, we propose that excess SH_2 could be binding to cytochrome b559 of PSII, blocking its activity and causing photosynthesis inhibition, as shown by the observed values of photosynthetic electron transfer rate (ETR) and photochemical efficiency of PSII (Fv/Fm) (Devinar, 2015). This hypothesis would explain the toxic effects of SO_4^{2-}, since Na_2SO_4-treated plants show growth inhibition, chlorosis and a strong water imbalance due to excessive ion incorporation by alterations in the signal coordination of the ion channels. In addition, carbon metabolism and partition are seriously affected, and energy resources must be diverted to synthesis of secondary metabolites such as condensed tannins, lignin precursors and polyphenols to cope with the great oxidative stress. Biochemical and molecular studies are ongoing in our laboratory to elucidate the causes of these differential plant responses to chlorides and sulfates, and to be able to provide tools for greater success in programmes for salinity tolerance improvement in plants.

References

Adem, G.D., Roy, S., Zhou, M., Bowman, J. and Shabala, S. (2014) Evaluating contribution of ionic, osmotic and oxidative stress components towards salinity tolerance in barley. *BMC Plant Biology* 14, 113.

Alves, A. and Setter, T. (2004) Response of cassava leaf area expansion to water deficit: Cell proliferation, cell expansion and delayed development. *Annals of Botany* 94, 605–613.

Apel, K. and Hirt, H. (2004) Reactive oxygen species: Metabolism, oxidative stress, and signal transduction. *Annual Reviews in Plant Biology* 55, 373–399.

Apse, M.P. and Blumwald, E. (2002) Engineering salt tolerance in plants. *Current Opinion in Biotechnology* 13, 146–150.

Ashraf, M., Athar, H.R., Harris, P.J.C. and Kwon, T.R. (2008) Some prospective strategies for improving crop salt tolerance. *Advances in Agronomy* 97, 45–110.

Balderas-Hernández, V.E., Alvarado-Rodríguez, M. and Fraire-Velázquez, S. (2013) Conserved versatile master regulators in signaling pathways in response to stress in plants. *AoB Plants* 5, plt033. doi:10.1093/aobpla/plt033.

Bilski, J., Nelson, D. and Conlon, R. (1988) Response of six wild potato species to chloride and sulfate salinity. *American Potato Journal* 65, 605–612.

Bose, J., Shabala, L., Pottosin, I., Zeng, F., Velarde-Buendía, A. *et al.* (2014) Kinetics of xylem loading, membrane potential maintenance, and sensitivity of K^+-permeable channels to reactive oxygen species: Physiological traits that differentiate salinity tolerance between pea and barley. *Plant, Cell and Environment* 37, 589–600.

Burkart, A. (1976) A monograph of the genus *Prosopis* (Leguminosae subfam. Mimosoideae). Catalogue of the recognized species of *Prosopis*. *Journal of the Arnold Arboretum* 57, 450–525.

Chinnusamy, V., Jagendorf, A. and Zhu, J.K. (2005) Understanding and improving salt tolerance in plants. *Crop Science* 45, 437–448.

Datta, K.S., Kumar, A., Varma, S.K. and Angrish, R. (1995) Differentiation of chloride and sulphate salinity on the basis of ionic distribution in genetically diverse cultivars of wheat. *Journal of Plant Nutrition* 18, 2199–2212.

Demidchik, V. (2012) Reactive oxygen species y oxidative stress in plants. In: Shabala, S. (ed.) *Plant Stress Physiology*. CABI, Wallingford, UK, pp. 24–58.

Devinar, G. (2015) Papel del ácido abscísico, salicílico, etileno y óxido nítrico en la homeostasis.iónica y tolerancia a salinidad de Prosopis strombulifera. PhD thesis. Universidad Nacional de Río Cuarto, Argentina.

Devinar, G., Llanes, A., Masciarelli, O. and Luna, V. (2013) Abscisic acid and salicylic acid levels induced by different relative humidity and salinity conditions in the halophyte *Prosopis strombulifera*. *Plant Growth Regulation* 70, 247–256.

Dubrovsky, J., Sauer, M., Napsucialy-Mendivil, S., Ivanchenko, M., Friml, J., *et al.* (2008) Auxin acts as a local morphogenetic trigger to specify lateral root founder cells. *Proceedings of the National Academy of Sciences* 105, 8790–8794.

Felker, P., Ewens, M., Velarde, M. and Medina, D. (2008) Initial evaluation of *Prosopis alba* Griseb clones selected for growth at seawater salinities. *Arid Land Research and Management* 22, 334–345.

Flowers, T.J. and Colmer, T.D. (2015) Plant salt tolerance: Adaptations in halophytes. *Annals of Botany* 115, 327–331.

Flowers, T.J. and Muscolo, A. (2015) Introduction to the special issue: Halophytes in a changing world. *AoB Plants* 7, plv020. doi: 10.1093/aobpla/plv020.

Flowers, T.J. and Yeo, A.R. (1995) Breeding for salinity resistance in crop plants – where next. *Australian Journal of Plant Physiology* 22, 875–884.

Foyer, C.H. and Noctor, G. (2013) Redox signaling in plants. *Antioxidants and Redox Signaling* 18, 2087–2090.

Fridovich, I. (1978) The biology of oxygen radicals. *Science* 201, 875–880.

Ghars, M., Parre, E., Debez, A., Bordenave, M., Richard, L., *et al.* (2007) Comparative salt tolerance analysis between *Arabidopsis thaliana* and *Thellungiella halophila*, with special emphasis on K^+/Na^+ selectivity and proline accumulation. *Journal of Plant Physiology* 165, 588–599.

Halliwell, B. and Gutteridge, J. (2007) *Free Radicals in Biology and Medicine*. Clarendon Press/Oxford University Press, Oxford, UK.

Halliwell, B. and Gutteridge, J.M.C. (2015) *Free Radicals in Biology and Medicine*. Oxford University Press, New York.

Hameed, M., Ashraf, M. and Naz, N. (2009) Anatomical adaptations to salinity in cogon grass [*Imperata cylindrica* (L.) Raeuschel] from the Salt Range, Pakistan. *Plant and Soil* 322, 229–238.

Hawkesford, M.J. and De Kok, L.J. (2006) Managing sulphur metabolism in plants. *Plant, Cell and Environment* 29, 382–395.

Hedden, P. and Thomas, G.S. (2012) Gibberellin biosynthesis and its regulation. *Biochemistry Journal* 444, 11–25.

Huang, J. and Redmann, R. (1995) Salt tolerance of *Hordeum* and *Brassica* species during germination and early seedling growth. *Canadian Journal of Plant Science* 75, 815–819.

Ismail, H., Maksimovic, J., Maksimovic, V., Shabala, L., Živanovic, V. *et al.* (2016) Rutin, a flavonoid with antioxidant activity, improves plant salinity tolerance by regulating K^+ retention and Na^+ exclusion from leaf mesophyll in quinoa and broad beans. *Functional Plant Biology* 43, 75–86.

Joshi, G. and Naik, R. (1980) Response of sugarcane to different types of salt stress. *Plant and Soil* 56, 255–263.

Kefu, Z., Hai, F., San, Z. and Jie, S. (2003) Study on the salt and drought tolerance of *Suaeda salsa* and *Kalanchoe claigremontiana* under iso-osmotic salt and water stress. *Plant Science* 165, 837–844.

Kobayashi, K., Fujii, S. and Sato, M. (2015) Specific role of phosphatydilglycerol and functional overlaps with other thylacoid lipids in *Arabidopsis* chloroplast biogenesis. *Plant Cell Reports* 34, 361–642.

König, J., Muthuramalingam, M. and Dietz, K.J. (2012) Mechanisms and dynamics in the thiol/disulfide redox regulatory network: Transmitters, sensors and targets. *Current Opinion in Plant Biology* 15, 261–268.

Koyro, H., Geissler, N., Seenivasan, R. and Huchzermeyer, B. (2013) Plant stress physiology: Physiological and biochemical strategies allowing plants/crops to thrive under ionic stress. In: Pessarakli, M. (ed.) *Handbook of Plant and Crop Physiology*. CRC Press, Boca Raton, Florida, USA.

Llanes, A., Reinoso, H. and Luna, V. (2005) Germination and early growth of *Prosopis strombulifera* seedlings in different saline solutions. *World Journal of Agricultural Sciences* 1(2), 120–128.

Llanes, A., Bertazza, G., Palacio, G. and Luna, V. (2013) Different sodium salts cause different solute accumulation in the halophyte *Prosopis strombulifera*. *Plant Biology* 15, 118–125.

Llanes, A., Masciarelli, O., Ordoñez, R., Isla, M. and Luna, V. (2014) Differential growth responses to sodium salts involve different ABA catabolism and transport in the halophyte *Prosopis strombulifera*. *Biologia Plantarum* 58, 80–88.

Llanes, A., Masciarelli, O. and Luna, V. (2015) Growth responses to sulfate and chloride are related to different phytohormone profiles in the halophyte *Prosopis strombulifera*. *Emirates Journal of Food Agriculture* 26, 1097–1113.

Llanes, A., Arbona, V., Gómez-Cadenas, A. and Luna, V. (2016) Metabolomic profiling of the halophyte *Prosopis strombulifera* shows sodium salt-specific response. *Plant Physiology and Biochemistry* 108, 145–157.

Manivannan, P., Jaleel, C., Kishorekumar, A., Sankar, B., Somasundaram, R., *et al.* (2008) Protection of *Vigna unguiculata* (L.) Walp. plants from salt stress by paclobutrazol. *Colloids and Surfaces B: Biointerfaces* 61, 315–318.

Mignolet-Spruyt, L., Xu, E., Idänheimo, N., Hoeberichts, F.A., Mühlenbock, P., *et al.* (2016) Spreading the news: Subcellular and organellar reactive oxygen species production and signalling. *Journal of Experimental Botany* 67, 3831–3844.

Mittler, R. (2002) Oxidative stress, antioxidants and stress tolerance. *Trends in Plant Science* 7, 405–410.

Mittler, R., Vanderauwera, S., Suzuki, N., Miller, G., Tognetti, V.B. *et al.* (2011) ROS signaling: The new wave? *Trends in Plant Science* 16, 300–309.

Miyazawa, M., Yagi, N. and Taguchi, K. (2005) Inhibitory compounds of α-glucosidase activity from *Arctium lappa* L. *Journal of Oleo Science* 54, 589–594.

Noiraud, N., Maurousset, L. and Lemoine, R. (2001) Transport of polyols in higher plants. *Plant Physiology and Biochemistry* 39, 717–728.

Raghavendra, A., Gonugunta, V., Christmann, A. and Grill, E. (2010) ABA perception and signalling. *Trends in Plant Science* 15, 395–401.

Reddy, V. and Reddy, A. (2004) Proteomics of calcium-signaling components in plants. *Phytochemistry* 65, 1745–1776.

Reginato, M., Abdala, G., Miersch, O., Ruiz, O., Moschetti, E., *et al.* (2012) Changes in the levels of jasmonates and free polyamines induced by Na_2SO_4 and NaCl in roots and leaves of the halophyte *Prosopis strombulifera*. *Biologia* 67, 689–697.

Reginato, M., Reinoso, H., Llanes, A. and Luna, V. (2013) Stomatal abundance and distribution in *Prosopis strombulifera* plants growing under different iso-osmotic salt treatments. *American Journal of Plant Sciences* 4, 80–90.

Reginato, M., Sosa, L., Llanes, A., Hampp, E., Vettorazzi, N. *et al.* (2014a) Na_2SO_4 and NaCl determine different growth responses and ion accumulation in the halophytic legume *Prosopis strombulifera*. *Plant Biology* 16, 97–106.

Reginato, M., Castagna, A., Furlán, A., Castro, S., Ranieri, A., *et al.* (2014b) Analysis of the oxidative damage in the halophyte *Prosopis strombulifera* salinized with NaCl and Na_2SO_4. Role of polyphenols as antioxidant protection. *AoB Plants* 6. doi: 10.1093/aobpla/plu042, plu042.

Reginato, M., Travaglia, C., Reinoso, H., Garello, F. and Luna, V. (2016) Anatomical modifications in the halophyte *Prosopis strombulifera* caused by salt mixture. *Flora* 218, 75–85.

Reich, M., Aghajanzadeh, T., Helm, J., Parmar, S., Hawkesford, M., *et al.* (2017) Chloride and sulfate salinity differently affect biomass, mineral nutrient composition and expression of sulphate transport and assimilation genes in *Brassica rapa*. *Plant and Soil* 411, 319–332.

Reinoso, H., Sosa, L., Ramirez, L. and Luna, V. (2004) Salt-induced changes in the vegetative anatomy of *Prosopis strombulifera* (Leguminosae). *Canadian Journal of Botany* 82, 618–628.

Reinoso, H., Sosa, L., Reginato, M. and Luna, V. (2005) Histological alterations induced by sodium sulfate in the vegetative anatomy of *Prosopis strombulifera* (Lam.) Benth. *World Journal of Agricultural Science* 1, 109–119.

Sharma, R., De Vleesschauwer, D., Sharma, M. and Ronald, P. (2013) Recent advances in dissecting stress–regulatory crosstalk in rice. *Molecular Plant* 6, 250–260.

Shelevela, E., Marquez, S., Chmara, W., Zegeer, A., Jensen, R. *et al.* (1998) Sorbitol-6 phosphate dehydrogenase expression in transgenic tobacco: high amounts of sorbitol lead to necrotic lesions. *Plant Physiology* 117, 831–839.

Simic, M. and Jovanovic, S. (1994) Inactivation of oxygen radicals by dietary phenolic compounds in anticarcinogenesis. *Food Phytochemicals for Cancer Prevention II*. American Chemical Society, Washington, DC, pp. 20–39.

Sosa, L., Llanes, A., Reinoso, H., Reginato, M. and Luna, V. (2005) Osmotic y specific ion effects on the germination of *Prosopis strombulifera*. *Annals of Botany* 96, 261–267.

Sun, A.Y., Wang, Q., Simonyi, A. and Sun, G.Y. (2008) Botanical phenolics and brain health. *Neuromolecular Medicine* 10, 259–274.

Tester, M. and Davenport, R. (2003) Na^+ tolerance and Na^+ transport in higher plants. *Annals of Botany* 91, 503–527.

Vaahtera, L., Brosché, M., Wrzaczek, M. and Kangasjärvi, J. (2014) Specificity in ROS signaling and transcript signatures. *Antioxidants & Redox Signaling* 21, 1422–1441.

Wilkinson, S. and Davies, W. (2010) Drought, ozone, ABA and ethylene: New insights from cell to plant to community. *Plant, Cell and Environment* 33, 510–525.

Zhan, X. and Zhao, X. (2003) Mechanism of adsorption from aqueous solutions using an adsorbent synthesized from natural condensed tannin. *Water Research* 37, 3905–3912.

14 Elimination of Salt by Recretion: Salt Glands and Gland-supported Bladders in Recretohalophytes

Ulrich Lüttge*
Technical University of Darmstadt, Darmstadt, Germany

Abstract

Elimination of mineral ions and salts by salt glands and salt hairs is called recretion, which has given the name recretohalophytes to plants using this mechanism for tolerance of salinity. There are about 370 vascular plant species of recretohalophytes comprising about 15% of all halophytes and 0.15% of angiosperms. Their evolutional origin is polyphyletic. Anatomical complexity shows trends from hydathodes to salt glands. There is different complexity between simple epidermal bladders and trichome-like salt hairs concentrating salt in terminal bladder cells. Transport pathways comprise cell-to-cell apoplastic and symplastic transport in salt gland and salt hair complexes, membrane transport by ion transporters and H^+-ATPases (eccrine recretion) and possibly vesicle transport (granulocrine recretion). Salt-gland cell cytology is characterized by a very dense cytoplasm enriched in mitochondria, but lacking functional chloroplasts. The plasma-membrane surface is often inflated by a labyrinth of cell wall protuberances, typical of transfer cells. Materials recreted are preferentially Na^+ and Cl^-. However, many other ions and also some organic solutes may be found. Isolation of salt glands embedded in the epidermis has so far not been achieved, and salt hairs with their glandular stalk cells (which are easier to isolate) have not been used for identifying recretion-specific ion transporters. Nevertheless, transcriptomics and proteomics have been applied to compare plants with differing recretory activity, including specific recretion mutants, to identify transporters putatively involved in recretion. Recretion is under genetic control. Although the number of recretohalophytes among angiosperm halophytes is relatively low (ca 15%), individual recretohalophyte species can play an important role in saline habitats. Ecophysiologically the efficiency of recretion depends on the site characteristics to which plants are adapted. Water loss with the recreted fluid is an ecophysiological challenge. Recretion does not reduce habitat salinity, because recreted salt is recycled to the environment. Recretohalophytes provide multiple services to anthropogenic management of salinity-affected ecosystems under increasingly severe environmental challenges now and in the future of global changes. There are many promising applications of recretohalophytes in saline agriculture. They serve as pasture and fodder plants and have entered human diets. Due to salt recycling they do not appear to be applicable to phytoremediation of salinity but they may clear metals from soils, and may serve as sources of bio-energy and biofuel. Recretohalophyte species can be used in the restoration ecology of deteriorated land. Recretohalophytes provide gene pools for gene engineering and synthetic biology in managing future challenges.

Keywords: Ecophysiology; Guttation; Hydathodes; Membranes; Membrane ATPases; Sodium chloride; Transfer cells; Transporters; Water relations.

* Corresponding author e-mail: luettge@bio.tu-darmstadt.de

14.1 Introduction

Elimination of solutes by plants is of fundamental importance in their physiology and their interactions with the environment. There is a large variety of elimination processes (Lüttge and Pitman, 1976), mediated by various types of transport mechanisms. These can be performed by any kind of cells without particular specialization. A prominent example is the exudation of roots determining the relations of the rhizosphere. Beyond that there is specialization for elimination, which we see in the individual cells and cell complexes of glands. There are many different types of glands in plants, which are usually defined and distinguished by the major solute contained at particularly high concentration in eliminated fluid. The glands typically eliminate solutes to the outside of the plants, for example nectaries eliminating sugars (Lüttge and Schnepf, 1976), the glands of carnivorous plants eliminating polysaccharide mucilage functioning as glue for the capture of prey as well as digestive enzymes (i.e. two rather different types of compounds) (Lüttge and Schnepf, 1976) and salt glands eliminating inorganic salts (Hill and Hill, 1976). In addition, there is also internal elimination, where so-called transfer cells are involved when elimination of solutes from the cytoplasm of cells remains internal. Examples include the gland-like companion cells of sieve tubes in the phloem, vascular parenchyma cells in the xylem and the cells of resin ducts. Internal elimination is also the sequestering of solutes from the cytoplasm into the large vacuoles of epidermal bladder cells, trichomes and gland hairs, although in this case the solutes are mostly destined to be released eventually to the outside. A wealth of examples is seen among gland hairs and epidermal epithelia eliminating secondary plant products such as various terpenes, resins, essential oils, flavonoid aglycones, fatty oils, quinones, fat and waxes (Lüttge and Schnepf, 1976).

Applying the terminology of secretion, excretion and recretion allows a systematic view of this diversity (Lüttge and Pitman, 1976). There are three ways of doing this:

1. A teleological or ecological distinction is derived from animal physiology, using excretion for end products of metabolism serving no further purposes, and secretion for the elimination of substances functioning in relation to the environment.

2. In a structural definition excretion is elimination out of the plant and secretion into compartments within the plant (Haberlandt, 1918).

3. A dynamic physiological and metabolic definition considers secretion as the elimination of substances originating from assimilative metabolism after the resorption of precursors, and excretion as the elimination of substances that have been subject to dissimilative metabolism after resorption and assimilation. Among the three this is the only one that then also adds a further term, namely recretion for the elimination of substances that are exported non-metabolized (i.e. as they were absorbed) (Frey-Wyssling, 1935, 1972).

Recretion is predominantly, if not exclusively, the elimination of mineral ions and salts; hence it is the appropriate term for the elimination of material by salt glands and salt bladders. It is inconsequent that one finds in the literature publications that accordingly name halophytes using salt glands or salt bladders for adaptation to salinity stress recretohalophytes, but which nevertheless at the same time speak of secretion of salt. There is plasticity, so the distinctions made by the three definitions are not strictly exclusive (Lüttge and Pitman, 1976). For instance, in the case of mineral ions we cannot really say if phosphate, nitrate or sulfate are not in fact involved in metabolic reactions before elimination. Nevertheless, it appears very useful to use 'recretion' for the massive elimination of NaCl by recretohalophytes adapted to environmental salinity, and this is the definition adopted in this chapter.

For those who like diversity it is worth mentioning at the end of a consideration of systematic principles that there are even salt glands that do not eliminate salt but take it up. On the abaxial surface of floating leaves of water lilies (*Nymphaea*) are glands known as hydropotes (literally translated, 'water drinkers'), which absorb mineral nutrients (Lüttge, 1964; Lüttge and Krapf, 1969; Lüttge *et al.*, 1971).

14.2 Polyphyletic Evolution of Salt Glands

There are about 370 known vascular plant species of recretohalophytes (Lüttge, 1975; Breckle, 2002; Yuan *et al.*, 2016a). Their families and the phylogenetic distance of the respective orders are shown in Fig. 14.1. It is evident that evolution of salt glands occurred several times and the trait of salt glands is polyphyletic.

14.3 Structural Aspects

14.3.1 Increased anatomical complexity

14.3.1.1 From hydathodes to salt glands

Guttation is the elimination of water via hydathodes, mostly on the leaves of plants. However, the fluid of guttation is never pure water, because for the transport of water an osmotic driving force is required. The guttated fluid always contains some osmotically active material, often mineral ions but also organic solutes (Singh, 2014). We may envisage that salt glands evolved from hydathodes with increasing structural complexity (Fig. 14.2). Hydathodes are an extremely heterogeneous group of structures both anatomically and functionally (Perrin, 1972). According to Haberlandt (1918) we can distinguish passive and active hydathodes. The former are simple openings in the surface or pores formed by non-functional stomata (Fig. 14.2 left). The guttation fluid is pressed out due to the root pressure, for which the osmotic driving force is generated in the roots. By exchange with cells along the route, the composition of the fluid eventually eliminated can be modified (Dieffenbach *et al.*, 1980). However, a special tissue or epithem between the xylem vessel endings and the periphery may also be involved in recreting salts for establishing an osmotic potential gradient driving the guttation (Fig. 14.2 middle), and then we have active hydathodes (Haberlandt, 1918; Ziegenspeck, 1949; Ivanoff, 1963). Hydathodes not only eliminate salt but may also reabsorb ions from the xylem stream, thus contributing to retrieval before recretion, where metabolically important phosphate can be re-translocated while chloride is recreted (Wilson *et al.*, 1991; Nagai *et al.*, 2013).

As the salts recreted increase in concentration there is a gradual change from hydathodes to salt glands with increasing structural and functional complexity (Fig. 14.2 right) (Lüttge, 1975). Salt glands of dicotyledonous plants are epidermal structures sunken into the leaf surface. Considering possible anatomical evolutionary trends it is intriguing to note that comparative transcriptome analyses during developmental stages of leaves reveal a strong homology between the 16-cell complex of *Limonium* salt glands and epidermal trichomes of other species not recreting concentrated salt (Yuan *et al.*, 2015a).

14.3.1.2 Epidermal bladders versus salt hairs

The annual halophyte *Mesembryanthemum crystallinum* L., which has become a model plant for studies of stress physiology and molecular biology

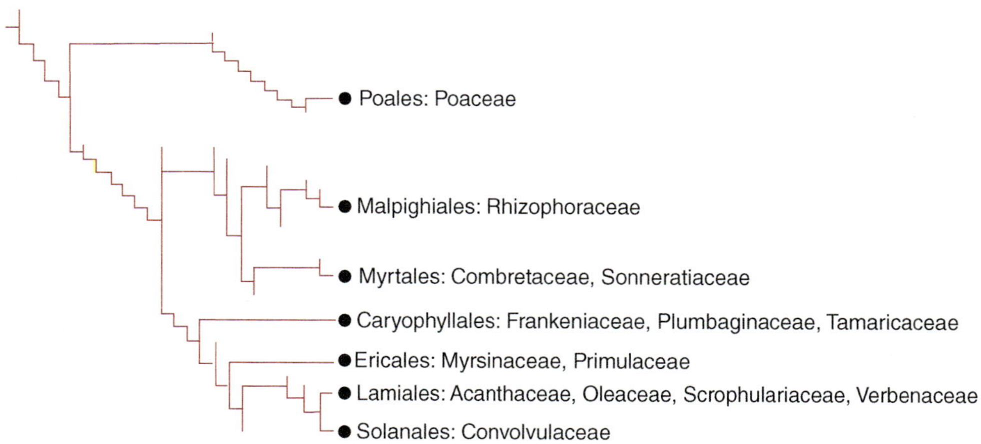

Fig. 14.1. Families and respective orders of plants having species with salt glands after lists of Lüttge (1975), Breckle (2002) and Yuan *et al.* (2016a; for Oleaceae see Gucci *et al.*, 1997). Phylogenetic distances according to Stevens (from Kadereit *et al.*, 2014).

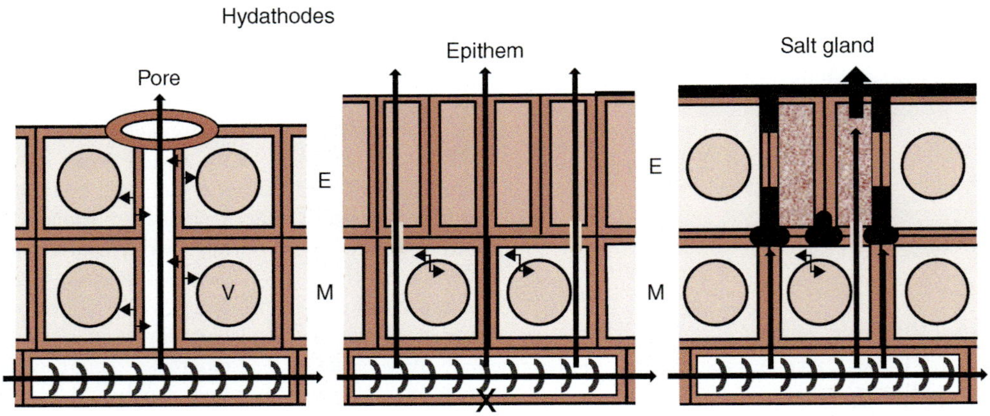

Fig. 14.2. Scheme of hydathodes and salt glands delineating increasing complexity with a putative evolutionary trend from hydathodes with simple pores and with epithems to salt-gland cell complexes. E, epidermis; M, mesophyll; V, vacuole; X, vessel of xylem vein endings. Arrows mark the pathways of transport as described in the text. Dark black labels in the right-hand diagram indicate the cuticular envelope of salt glands by cell wall incrustations and adcrustations.

under salinity, has very large epidermal bladder cells of a volume of up to 2 mm^3 (Fig. 14.3 left). An early study demonstrated that under stress of salinity in the root medium these bladder cells do not accumulate salt above the levels found in the metabolically active mesophyll tissue, even at NaCl concentrations building up to 0.6 M in both tissues (Lüttge *et al.*, 1978). This was corroborated by later analyses where salt concentrations were sometimes a little larger and sometimes even smaller in the bladder cells than in the leaf tissue (Table 14.1). Therefore, these bladders cannot be considered as salt hairs with gland-like functions of active salt transport. Haberlandt (1918) has described these bladder cells covering the entire plants of *M. crystallinum*, especially under salinity stress (Lüttge, 2002) as forming an external water storage tissue. This is supported by measurements of water relation parameters. The hydraulic conductivity (Lp) of the membranes between the bladder cells and the underlying mesophyll is comparatively large and the cell wall elastic modulus (ε) of the bladder cell wall is pressure dependent; that is, positively correlated with pressure and volume (Steudle *et al.*, 1975, 1977). The inorganic salt, with other solutes including sugars, sugar alcohols such as pinitol, proline and other protein and non-protein amino acids and organic acids in the bladders, will serve osmotic regulation and as driving force for water storage (Adams *et al.*, 1992; Barkla and Vera-Estrella, 2015). Barkla and Vera-Estrella (2015) and Oh *et al.* (2015) have used vacuum aspiration with a fine-gauge needle to obtain the contents of the bladder cells selectively, which allowed them to perform bladder cell-specific transcriptomics and metabolomics to characterize salinity-responsive metabolites and genes including those of ion transporters of membranes.

With their huge volume, the bladders contribute to the sequestering of salt without concentrating it. Studies of mutants lacking the epidermal bladders have demonstrated that they are significantly involved in salinity tolerance of *M. crystallinum* (Agarie *et al.*, 2007). Unlike for hydathodes and salt glands, there is no gradual change from such epidermal bladders to salt hairs, whose structure is fundamentally different. The bladders of *M. crystallinum* are not real hairs or trichomes but, more simply, just inflated epidermal cells. Conversely, the salt hairs – such as those of halophytic Chenopodiaceae of the genus *Atriplex* – emerge from the epidermis. Their essential feature is that the terminal salt bladder has a gland-like stalk cell (Fig. 14.3 right) which, by active transport, considerably concentrates the salt (Table 14.1) and which is missing in *M. crystallinum*.

14.3.2 Transport pathways

Salt glands are multicellular complexes of up to 40 or even more cells in the epidermis, mostly in the leaves of plants (e.g. *Limonium* or the mangrove

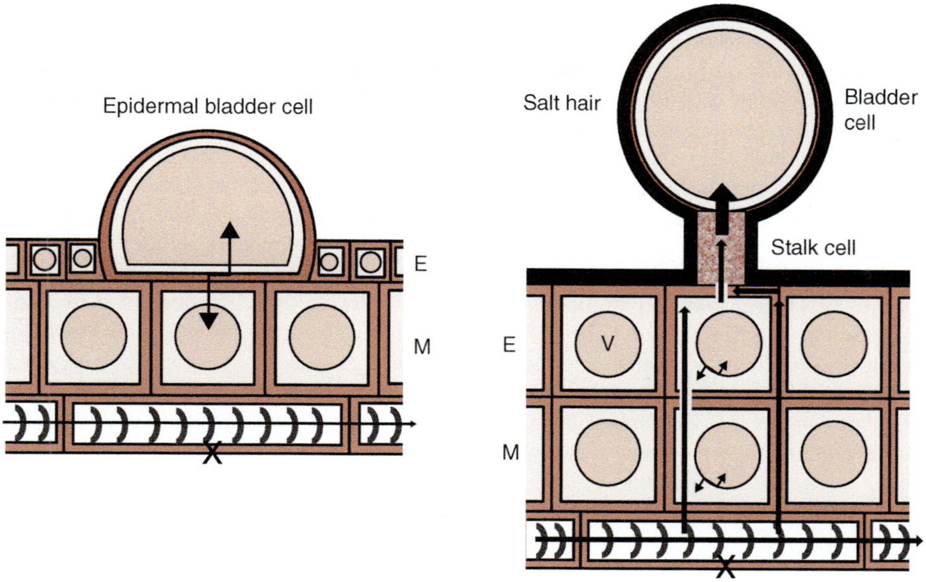

Fig. 14.3. Schemes comparing epidermal bladder cells (left, example of *Mesembryanthemum crystallinum*) and salt hair bladders, where the essential feature is a gland-like stalk cell (right, example of Chenopodiaceae). Other details as in Fig. 14.2.

Table 14.1. Comparison of salt levels in epidermal bladders of *Mesembryanthemum crystallinum* and in bladder cells of salt hairs over photosynthetically active leaf tissue.

Species	Concentration in the root medium	Ratio of ion levels bladder/leaf	Reference
Mesembryanthemum crystallinum L.	200 mM NaCl	Na^+ 1.2	Barkla *et al.* (2002)
Mesembryanthemum crystallinum L.	400 mM NaCl	Na^+ 0.77 Cl^- 0.92	Agarie *et al.* (2007)
Atriplex spongiosa F. Muell.	250 mM NaCl	Na^+ 3.6 Cl^- 5.5	Osmond *et al.* (1969)
Atriplex nummularia Lindl.	250 mM NaCl	Cl^- 3	Osmond *et al.* (1969)
Atriplex halimus L.	100 mM NaCl + 100 mM KCl	Electrical conductivity 60	Mozafar and Goodin (1970)
Atriplex hortensis L.	1 mM Na^+ + 1 mM K^+	Na^+ 13.9 (young leaves) Na^+ 3.0 (mature leaves)	Jeschke and Stelter (1983)
Atriplex amnicola Paul G. Wilson	200 mM NaCl 400 mM NaCl	Na^+ 4.5 Na^+ 3.3 (young leaves)	Aslam *et al.* (1986)
Atriplex gmelinii C. A. Meyer	250 mM NaCl	Na^+ 3.9	Tsutsumi *et al.* (2015)
Porteresia coarctata Tateoka	200 mM NaCl	Na^+ 2.0 Cl^- 1.6	Flowers *et al.* (1990)
Halimione portulacoides (L.) Aellen	500 mM NaCl	Na^+ 2.1 Cl^- 2.9	Baumeister and Kloos (1974)

Aegiceras) (Yuan *et al.*, 2016a). In addition to the recretory gland cells there are other types of cells, such as stalk cells, basal cells and collecting cells. In grasses (Poaceae), such as *Spartina*, *Aeluropus* and *Chloris*, the complex is only two-celled, with one cell sited above the other, and the salt glands come closer to salt hairs (Lüttge, 1975; Hill and Hill, 1976).

The transpiration stream delivers salt to the mesophyll via vessel endings close to the salt glands, where basal cells may serve as collecting cells. The distribution of salt glands on leaves along veins and in relation to each other and to mesophyll cells modulates gland activity (Leng *et al.*, 2018). Transport from the collecting cells can be both apoplastic and symplastic (Ziegler and Lüttge, 1967; Faraday *et al.*, 1986). However, apoplastic transport is blocked at the recretory gland cells by cell wall incrustations – that is, lignification, cutinization and suberization – preventing movement of water and solutes within the free space of the cell wall at strategic points. This ensures that a cytoplasmic phase is involved before recretion. Moreover, salt glands as well as salt hairs (Mozafar and Goodin, 1970) are encased in cuticular envelopes also involving adcrustation of cell walls with cutin and waxes, preventing transport across them (Figs 14.2 right, 14.3 right and 14.4; Lüttge, 1975; Hill and Hill, 1976; Fitzgerald and Allaway, 1991). Membrane transport of the ions across the plasma membrane into the living cytoplasm is, therefore, enforced before they can be recreted (Drennan *et al.*, 1987). This assures molecular and metabolic control of the recretion process. Symplastic transport via the salt glands of *Limonium vulgare* Mill. (Ziegler and Lüttge, 1967) and via the gland-like stalk cells with their very dense cytoplasm into the peripheral layer of cytoplasm in the large vacuolated bladder cells of *Atriplex spongiosa* F. Muell. (Osmond *et al.*, 1969) has been visualized by micro-autoradiography using labelled $^{36}Cl^-$. The symplastic transport across cell boundaries from cell to cell in salt-gland complexes is mediated by plasmodesmata (Ziegler and Lüttge, 1967; Faraday *et al.*, 1986; Sager and Lee, 2014).

Recretion from the gland cell cytoplasm can be eccrine mediated by transport molecules such as ion channels and energized by ATPases in the plasma membrane (Section 14.5.2). It may also be granulocrine via the extrusion of vesicles (Ding *et al.*, 2010), in which membrane transport must be involved loading the vesicles (Fig. 14.4). Ultrastructural studies of membrane trafficking have supported the consideration of granulocrine recretion (Balsamo and Thomson, 1993). On the other hand, on the basis of electrochemical measurements and thermodynamic considerations, Hill and Hill (1976) take recretion by extrusion of vesicles as unlikely. Conversely, recent transcriptome analyses show that genes of vesicle-mediated transport are up-regulated under salinity stress in *Limonium bicolor* (Bunge) O. Kuntze (Yuan *et al.*, 2015a, 2016b; see Sections 14.5.2 and 14.5.3). A special view has been developed by Vassilyev and Stepanova (1990). In salt glands of *Limonium platyphyllum* Linzs., these authors observed that the tonoplast of the cell vacuoles establishes close contact with the plasma membrane, similar to the tight junctions of membranes in animal cells, and that this mediates recretion without extrusion of vesicles being involved.

In many cases, such as in *Statice gmelinii* Willd., *Spartina townsendii* J. et J. Groves, *Tamarix*, *Avicennia*, *Distichlis*, *Ceratostigma* and *Limonium* there are pores in the cuticle (for a list of species and references see Feng *et al.*, 2014) above the gland through which the recreted fluid can be eliminated. The eliminated fluid may collect between the cuticle and the cell wall on top of the glands or salt hair bladder cells under the osmotic driving force of the salt recreted. In a rhythmic process, as hydrostatic pressure builds up, the pores in the cuticle open and the recretion is released, leading to relaxation of the pressure and again closing of the pores in a valve-like fashion. The molecular composition of the cuticle can also serve particular functions. Yuan *et al.* (2013) have shown that the cuticle encapsulating salt glands of *L. bicolor* and the mangrove *Aegialitis rotundifolia* Roxb. emits blue autofluorescence under UV-excitation of 330–380 nm, and they identified this to be due to ferulic acid (Deng *et al.*, 2015). They obtained mutants of *L. bicolor* where autofluorescence was increased (*fii*) and decreased (*fid*) and compared them with the wild type (WT). Na^+ concentration in the recretion did not differ among them. However, the volume of the recreted fluid was higher in *fii* and lower in *fid* than in WT. Ferulic acid thus increased recretory salt elimination. The authors speculated that this is related to hydraulic pressure relationships, as ferulic acid is known to decrease cuticle extensibility and provides rigidity.

14.3.3 Gland cytology

The recretory gland cells, as well as the stalk cells of the bladder cells of salt hairs, are characterized by very dense cytoplasm (Lüttge, 1975;

Hill and Hill, 1976). Numerous mitochondria can ensure the massive supply of the energy required for the recretion process. The plastids present are non-functional chloroplasts without well-developed thylakoid lamellae and grana. Where photosynthetically active radiation energizes recretion, this must be due to coupling with the underlying photosynthetically active mesophyll (see Section 14.5.1). The gland cells do not have large central vacuoles but often there are very many small vesicles, suggesting granulocrine recretion (Section 14.3.2; Figs 14.4 and 14.5).

A conspicuous and regular feature of recretory salt-gland cells is that the outer cell wall forms many protuberances towards the cell interior (Lüttge, 1975; Hill and Hill, 1976; Faraday and Thomson, 1986a,b; Faraday *et al.*, 1986). This is typical for so-called 'transfer cells' also

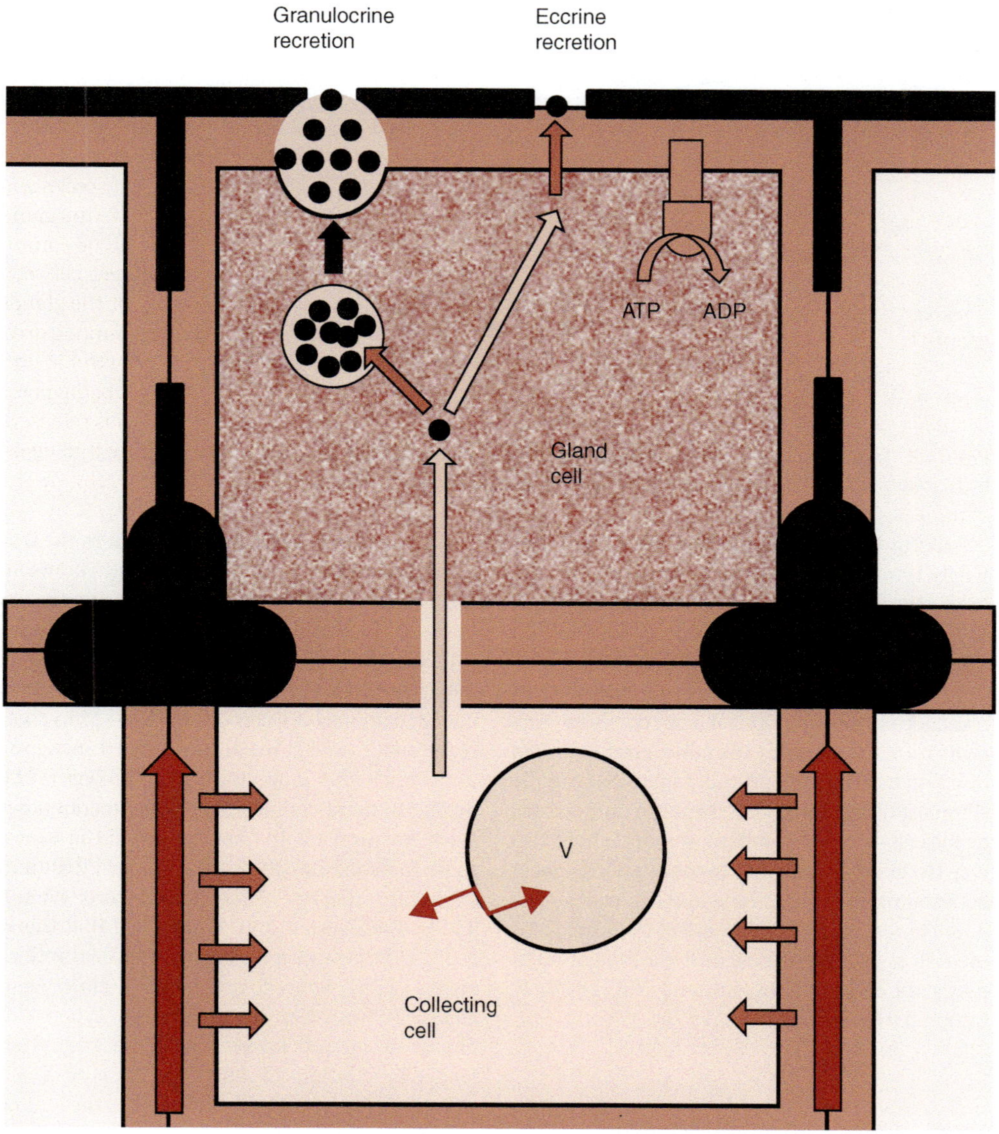

Fig. 14.4. Scheme of the pathways of salt recretion by salt glands with apoplastic transport, membrane transport and symplastic transport, granulocrine and eccrine recretion, as described in the text. Other details as in Fig. 14.2.

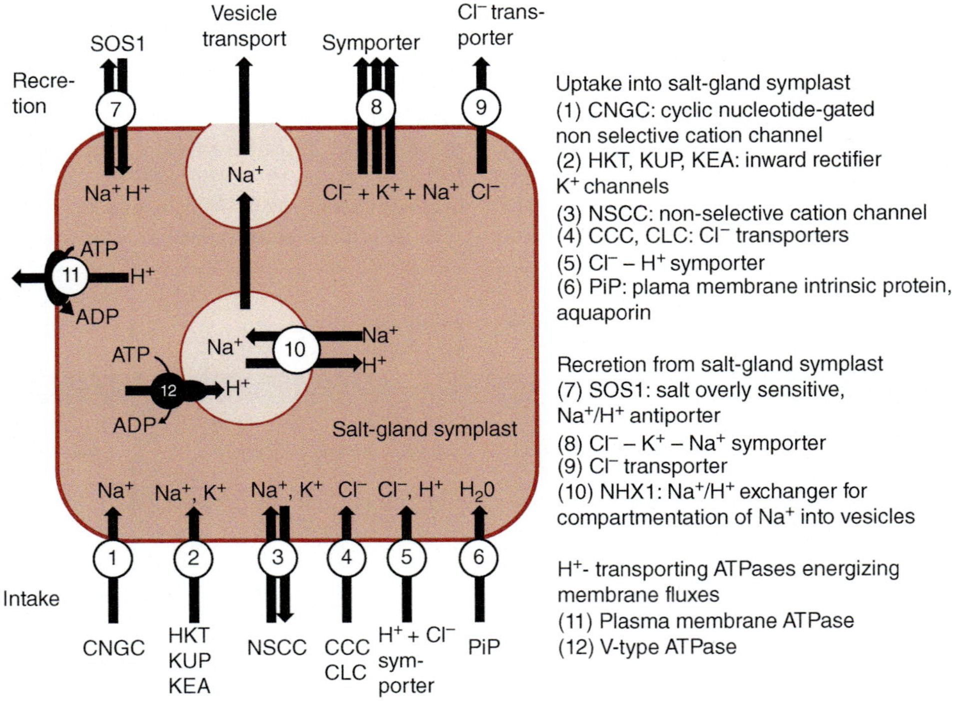

Fig. 14.5. Molecular identification of transporter molecules putatively involved in recretion of salt glands by high-throughput RNA sequencing for transcriptome analyses (after Yuan *et al*., 2016a, b).

occurring at other critical locations within plants, where massive 'transfer' of solutes via membrane transport is required (Gunning and Pate, 1969; Pate and Gunning, 1972; see also Section 14.1). This provides an increased area of the plasma membrane for the accommodation of transporters and pumps for ion recretion (Section 14.5.2). Standing osmotic gradients can build up between the protuberances outside the symplast of the gland cells. Ions are pumped out of the cells and water follows osmotically. In this way the exudate is formed in the coupling spaces between plasma membrane and cell wall (Oross and Thomson, 1982, 1984) before moving by mass flow in the apoplast and via the cuticular pores out of the gland (Hill and Hill, 1973a,b; Lüttge, 1975; Hill and Hill, 1976).

14.4 Materials Recreted

The habitats where recretohalophytes typically grow are characterized by strong salinity of NaCl. Thus NaCl is the dominant salt in the recreted fluid (Berry, 1970; Faraday and Thomson, 1986c; Sobrado and Greaves, 2000; Suárez and Medina, 2008; Ma *et al.*, 2011; Yuan *et al.*, 2016a). Various recretohalophytes growing in seawater (0.48 M Na^+, 0.54 M Cl^-) showed NaCl concentrations between 0.2 and 1.0 M in their recretion (Table 5.1 in Hill and Hill, 1976). However, other ions may also be found. Depending on the mineral composition of the medium besides Na^+ and Cl^- in *Tamarix aphylla* L. Karst. these included Br^-, K^+, Li^+, Mg^{2+}, Ca^{2+}, Zn^{2+} and even Cd^{2+} (Berry, 1970; Pollack and Waisel, 1970; Baumeister and Kloos, 1974; Rozema *et al.*, 1981; Hagemeyer and Waisel, 1988; Gucci *et al.*, 1997; Sobrado and Greaves, 2000; Ding *et al.*, 2010; Yuan *et al.*, 2016a). When salt glands of excised leaves were artificially challenged by solutions of different salts, such as KI, KCl, NaCl, Na_2SO_4, $MgCl_2$ and $MgSO_4$, the ions of the challenging solution were dominantly recreted (Faraday and Thomson, 1986c). Organic compounds, such as amino acids, have also been found (Pollak and Waisel, 1970).

14.5 Transport Mechanisms

14.5.1 Electrochemical gradients, active transport and supply of energy

Transport of salt from the leaf lamina into the recretion of salt glands and salt hairs (Table 14.1) occurs against large concentration gradients. Thermodynamic studies demonstrate that electrochemical gradients are built up (Lüttge, 1975); thus active transport must be involved in at least one step of the overall recretion process. Discussions of the various transport pathways provided by the architecture of the entire leaf-gland system (Section 14.3.2, Figs 14.2–14.4) have prompted extensive considerations about the point at which this may occur: into the cytoplasm by loading the symplast, or out of the symplast by recretion (Lüttge, 1975). Active transport of ions is always membrane transport. The plasma membranes of the collecting cells (loading) or of the gland cells (eccrine recretion); vesicle membranes of the gland cells (granulocrine recretion); or tonoplasts (recretion in salt bladders) could be involved. It is possible that more than one participates, but mostly it is considered that the active process should reside in the glands.

The dense cytoplasm of salt-gland cells is often found to be packed with mitochondria, and respiration is likely to produce much of the energy needed for recretion (Ziegler and Lüttge, 1966; Oross and Thomson, 1984). However, studies using various metabolic inhibitors, anaerobiosis, light and darkness have shown that it is not a specific pathway of energy metabolism, but any mechanism in the network that can supply ATP which is providing the metabolic driving force for recretion by salt glands and salt hairs (Lüttge, 1975). It is intriguing that light is energizing recretion via the activity of photosynthesis, but gland cells of *Limonium* – as well as the stalk and bladder cells of the salt hairs of *Atriplex* – do not possess photosynthetically active chloroplasts (Section 14.3.3; Osmond *et al.*, 1969; Hill and Hill, 1976; Lüttge, 1978). When *A. spongiosa* was accumulating NaCl several-fold in its bladder cells during growth at 0.25 M NaCl (Table 14.1), photosynthetic CO_2 fixation was 10.63 and 0.17 μmol g_{FW}^{-1} h^{-1}, respectively, while light stimulation of Cl^- uptake was 1.9× and 4.7×, respectively, in the photosynthetically active mesophyll cells and in the salt hairs (Osmond *et al.*, 1969; Lüttge, 1978). In the absence of chloroplasts in the gland hairs, this strong light stimulation of Cl^- accumulation may either be due to the active transport step being localized in the green mesophyll rather than in the salt hairs, or to transport of energy-rich substrates from the photosynthetically active mesophyll to the salt hairs driving the active recretion there. The latter is quite likely. Studies of electrical coupling between salt hairs and lamina have shown that there is good symplastic contact between them (Lüttge and Pallaghy, 1969; Osmond *et al.*, 1969; Pallaghy and Lüttge, 1970; Lüttge, 1978).

14.5.2 Transporter molecules and H^+-ATPases

The development of molecular biology now allows us to study the dynamics of transcription and translation by analysing the dependence of entire transcriptomes and proteomes on conditions, environment and time. This has been applied to halophytes, and among them recretohalophytes, to identify genes encoding transporter molecules and membrane ATPases. However, it is mostly done on bulk leaf preparations so that we do not know if the studied molecules are specific for the recretory glands or just constituent of the general equipment of the halophytes. Image-providing localization techniques producing staining, fluorescence or luminescence (e.g. by using reporter genes such as for ß-glucuronidase, ß-galactosidase, aequorin and luciferase) should be used for identifying gland-specific molecular mechanisms. Transformation of *L. bicolor* with the trans-gene reporter of ß-glucuronidase is promising (Yuan *et al.*, 2014). Another approach is using mutagenesis. With ^{60}Co gamma irradiation, two mutants of *L. biolor* have been obtained, which show differences in salt-gland development and the rate of salt recretion per individual salt gland (Yuan *et al.*, 2015b). Methods of enzymolysis and microdissection should be developed for the isolation of salt glands (Yuan *et al.*, 2016a). Older studies have, in fact, already demonstrated that it is not too difficult to isolate *Atriplex* bladder hairs together with their gland-like stalk cells (Osmond *et al.*, 1969). Even a small Californian desert rodent, *Dipodomys microps* Merriam, was observed to do it using its broadened chisel-shaped lower incisors to shave off the salt hair layers from leaves of *Atriplex confertifolia* (Torr & Frém.) S. Watson where the less salt-loaded leaf lamina proved to be tastier (Kenagy, 1972, 1973).

Transporter molecules involved in salinity tolerance of halophytes are listed by Volkov (2015). To circumvent the limits still imposed by the problem of not being able to extract single salt glands from leaves of recretohalophytes, a promising approach was to compare: (1) salt-stressed leaves and control leaves; and (2) mutants of *L. bicolor* with low and high recretion rates with the WT. This allowed correlation of increased rates of recretion with highly transcribed genes up-regulated under salinity. Yuan *et al.* (2016b) have used high-throughput RNA sequencing for transcriptome analyses and the identification of transporter genes highly transcribed under salinity stress. This has already allowed molecular identification of the transporter molecules most likely involved in salt glands (Yuan *et al.*, 2016a, b). A summary of the results is given in Fig. 14.5. It is noted that the distribution of the various transporters is taken to be asymmetric in the gland cell complex to ensure intake and recretion transport across it, into and from the symplast, respectively. It remains pure speculation so far that a similar set of transporters also operates in salt hairs of the Chenopodiaceae (Shabala *et al.*, 2014), although by contrast to the salt glands which are embedded in the epidermis, the peripheral salt hairs could so easily be isolated for direct molecular analysis.

14.5.3 Genetic control

As early as the beginning of the 1970s, long before the era of molecular biology, it had already been shown that chloride recretion by the salt glands of *Limonium* is induced by the substrate chloride (Hill, 1970; Hill and Hill, 1973c; reviewed by Lüttge, 1975, and by Hill and Hill, 1976). Hill and Hill in these studies distinguished three phases after the addition of NaCl to leaf discs: a lag phase of recretion which they called induction phase; an exponential rise of recretory activity named assembly phase; and a final recretory phase. They used a pharmacological approach applying inhibitors of transcription and translation, such as actinomycin, fluorouracil, puromycin and 5-fluorophenylalanine, to demonstrate that the substrate induction was under the control of gene activation, while the phases following induction were insensitive to the inhibitors (Fig. 14.6). In a microsomal membrane fraction isolated from *Limonium* leaves, Hill and Hill (1973c) found a Cl^--stimulated ATPase whose activity was induced by Cl^- just as gland activity in Fig. 14.6. Such induction was specific

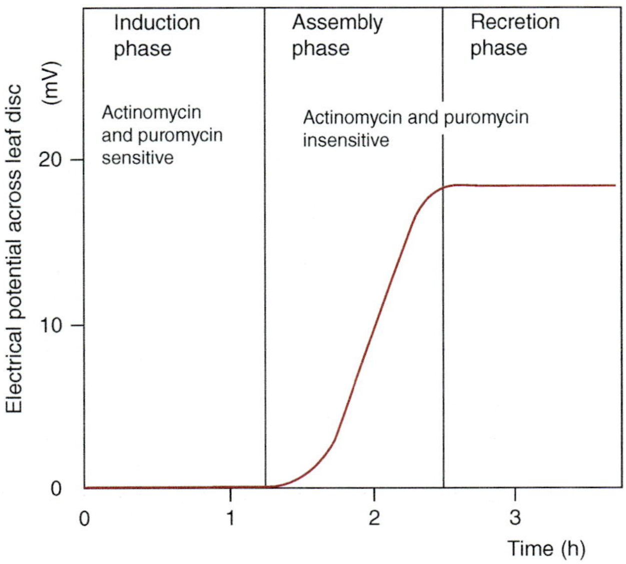

Fig. 14.6. Induction of recretory gland activity by leaves of *Limonium* after treatment with 100 mM NaCl at time zero. Recretory activity is given by the electrical potential difference across recreting leaf discs. The induction phase is sensitive to transcription and translation inhibitors, whereas the following phases are insensitive (after Hill, 1970; Hill and Hill, 1973b).

and only obtained with Cl^- and not with SO_4^{2-} and was also suppressed in the presence of the inhibitor puromycin. This may be part of the recretory machinery; thus, recretion is clearly under genetic control.

In the development of emergent leaves of the recretohalophyte *L. bicolor*, salt glands are the first differentiated epidermal structures. They differentiate before stomata so they are ready to recrete from expanding leaves before the transpirational pull by evaporation of water via the stomata delivers too much salt (Yuan *et al.*, 2015a). Yuan *et al.* (2015a) have used high-throughput RNA sequencing, annotation and identification of genes at five stages of development of *L. bicolor* leaves to reveal the dynamics of gene expression documenting up-and down-regulation of genes under salinity stress. The transcriptomic survey showed the involvement of 26 genes in salt-gland differentiation. Among these were genes encoding proteins for structural features important in the pathways of salt transport in the glands, such as those for the organization of plasmodesmata (i.e. plasmodesmata callose-binding protein, PDCB, and plasmodesmata-located protein, PDLP) and for the cuticular envelope (i.e. for displaying autofluorescence). There were also genes for vesicle-mediated transport (i.e. ER vesicle transporter proteins (LB111423 and LB103860) and of the vesicle-associated membrane protein family). Genes for specific transporter molecules have already been reviewed in Section 14.5.2.

14.6 Ecophysiology

This book demonstrates that the ecology of halophytes is a major field, given the broad global distribution of hygrohalophytes in coastal salt marshes, mangroves and inland salt lakes, and of xerohalophytes in vast arid regions (see also Breckle, 2002). Whole-plant adaptations to salinity stress comprise three major mechanisms: salt exclusion, accumulation and recretion. Recretohalophytes are just a small sub-cohort of halophytes. The majority of halophytes do not rely on recretion by salt glands or salt hairs for adapting to high external salinity. While halophytes comprise about 1% of the world's flora (Flowers and Colmer, 2008) and about 250,000 angiosperms are known to science, there are about 2500 species of halophytes. The 370 known recretohalophytes (Section 14.2) make up about 15% of these, or 0.15% of all angiosperms. Thus, relatively speaking, recretion is a rare adaptation. Nevertheless, individual recretohalophyte species may dominate habitats and can play important roles in saline ecosystems.

In some species more than one of the major adaptations may be involved, such as both exclusion and recretion in *Limonium* (Alcarón *et al.*, 1999). Site salinity and the salt load within the root zones are not only determined by sea tides and/or rain events (Bradley and Morris, 1991), but the three major plant adaptations also contribute to them. Exclusion leads to salinization of the root medium. Salt accumulated is returned to the environment with litter by shedding of old leaves full of salt. Salt recreted will fall from leaves as salty 'rain'. During dry and hot days, gland recretion may form salt crystals on the leaves of mangroves which dissolve again hygroscopically at high relative air humidity, especially during the morning hours in the tropics, when they drip down like rain (Lüttge, 2008).

The efficiency of recretion depends on the site characteristics to which plants are adapted (Table 14.2). Recretion selectively eliminates

Table 14.2. Efficiency of recretion in percent of Na^+ or salt absorbed.

Species	Recretion efficiency (%)	Site	Reference
Spartina anglica C. E. Hubbard	60	Lower part of salt marshes, flooded daily with seawater	Rozema *et al.* (1981)
Limonium vulgare Mill.	33		
Avicennia marina (Forssk.) Vierh.	40	Mangrove	Waisel *et al.* (1986)
Glaux maritima L.	20	Less saline coastal sites	Rozema *et al.* (1981)
Armeria maritima (Mill.) Willd.	4		

NaCl (Section 14.4). It maintains homeostatic cytoplasmic levels of Cl^- (Jeschke *et al.*, 1995). Recretion of NaCl also keeps cytoplasmic K^+/Na^+ ratios high. Among its other functions, K^+ is particularly required for charge balance, serving as a counter ion for fixed negative charges of proteins and nucleic acids. Conversely, Na^+ is toxic. Favourable K^+/Na^+ ratios in the cytoplasm and nucleus are necessary for performing gene transcriptions and metabolic functions of enzymes (Breckle, 2002; Feng *et al.*, 2015). This does not only apply to the regulation of metabolic competence of whole organs, but is also specifically essential for the metabolically dependent operation of the salt glands themselves. Feng *et al.* (2015) have used secondary ion mass spectrometry (SIMS) for imaging at the nano-scale (nanoSIMS) after preparation of samples by high-pressure freezing followed by freeze substitution to preserve fine structure and localization of elements at the ultrastructural level. They found that by stressing plants of *Limonium bicolor* L. with 200 mM NaCl salinity, the K^+ signal in the cytoplasm and in the nucleus of salt-gland cells was 16.5- and 21.9-fold, respectively, the signal observed in non-stressed plants. The Na^+ signal also increased under NaCl salinity, but the increase was not much larger than for K^+: an increase in the cytoplasm and nucleus of 24.5- and 26.1-fold respectively. Thus the special accumulation of K^+ in these compartments kept the K^+/Na^+ ratios under control. Jeschke and Stelter (1983) and Aslam *et al.* (1986) have compared young and old *Atriplex* leaves and found that Na^+ sequestration in the bladder cell vacuoles is particularly high in relatively young leaves (Table 14.1), keeping K^+/Na^+ ratios sufficiently high in these leaves to sustain their growth. The recretion fluid of salt glands may be alkaline and can bind CO_2, by which it may support photosynthesis (Waisel, 1991).

The water loss with the recreted fluid is a challenge for water relationships (Ramadan, 1998). Water can be reabsorbed via aquaporins. In the mangrove *Avicennia officinalis* L. two aquaporin genes were preferentially expressed in the salt-gland cells and induced in response to increasing salt concentration (Tan *et al.*, 2013). In another mangrove, *Laguncularia racemosa* (L.) Gaertn., the rate of recretion was found to increase exponentially with increasing medium salinity, and at the same time leaf conductance for water vapour decreased hyperbolically as a trade-off between fluid recretion and transpirational water loss (Sobrado, 2002, 2004).

Salt recretion is highly effective as compared to salt exclusion. Measurements by Atkinson *et al.* (1967) showed that the mature leaves of the recreto-mangrove *Aegialitis annulata* R. Br. even had relatively lower salt concentrations (356 mM Na^+ and 386 mM Cl^-) than the salt-excluder mangrove *Rhizophora mucronata* Lam. (461 mM Na^+ and 522 mM Cl^-), although it allowed a much higher concentration in its xylem sap (85–122 mM and 17 mM Cl^-, respectively).

14.7 Conclusion and Outlook: Potential Uses of Recretohalophytes and Their Ecosystem Services Under Environmental Challenges

As this chapter demonstrates, recretion is a very important mechanism in the salt tolerance of higher plants. How can recretohalophytes serve in the anthropogenic management of salinity-affected ecosystems under increasingly severe environmental challenges now and in the future?

Potential uses of recretohalophytes in agriculture on salinized soils appear to have obtained little specific attention. However, recretohalophytes successfully mingle with non-recreting halophytes in many promising applications of saline halophyte agriculture (Panta *et al.*, 2014). The thermophilic C_4 grass *Leptochloa fusca* (L.) Kunth. has been considered suitable for pasture reclamation (Gorham, 1987). Other grasses with salt glands, including species of *Spartina* as well as species of *Atriplex* with their salt hairs, are used in pastures and as fodder (Panta *et al.*, 2014). Chenopodiaceae with salt hairs (*Atriplex*, *Chenopodium*) and even mangroves with salt glands have been incorporated in human diets in different parts of the world, and *Chenopodium glaucum* L. is a source of oilseeds (Panta *et al.*, 2014). A wild species of rice, *Porteresia coarctata* Tateoka (formerly *Oryza coarctata* Roxb.) has unicellular salt hairs with large vacuoles concentrating NaCl under salinity stress (Table 14.1) (Flowers *et al.*, 1990). The hairs are in deep folds or furrows in the leaves of these grasses. There are no gland-like stalk cells, but underlying epidermal cells, as depicted in Fig. 6b of Flowers *et al.* (1990) may drive the recretion.

Porteresia coarctata can tolerate up to 20% seawater, but is slow growing. One may envisage trying to engineer its genes for the formation of salt hairs into the genome of *Oryza sativa* L. to increase the salt tolerance of rice, but the trade-off of slow growth may be discouraging (Flowers *et al.*, 1990). A globally important recretohalophyte with steadily increasing rates of export from the producing countries is the traditional Andean crop plant of the Incas in South America, *Chenopodium quinoa* Willd. It is a salt-tolerant facultative halophyte with salt hairs, increasingly cherished for its high caloric and qualitative nutritional value as a cereal substitute (Adolf *et al.*, 2013).

The salt hair-bearing Chenopodiaceae could provide a basis for molecular engineering of non-halophyte Chenopodiaceae crops such as spinach (*Spinacea oleracea* L.) and beet (*Beta vulgaris* L.) (Shabala *et al.*, 2014), where the latter notably has a halophyte background originating from *Beta maritima* (L.) Arcang. Generally, transcriptome analyses of plants subjected to salinity, such as those performed with *L. bicolor* by the team of Bao-Shan Wang, which has now developed into a leading group in recretohalophyte research (Sections 14.5.2 and 14.5.3), may contribute the identification of salt-tolerance genes suitable for bioengineering in crops (Yuan *et al.*, 2015a; Yuan *et al.*, 2016b).

By contrast to non-recreting halophytes, recretohalophtyes do not offer promising applications in phytoremediation of saline sites because the recreted salt is recycled to the environment (Section 14.6). They may, however, be used in phytoremediation of metal pollution of soils; examples of these include *Atriplex halimus* L. and *Tamarix smyrnensis* Bunge (Manousaki and Kalogerakis, 2011; Panta *et al.*, 2014). *Tamarix chinensis* Lour. has been evaluated as an energy plant for biofuel (Liu *et al.*, 2012; Panta *et al.*, 2014). Ecosystem services can be important in other respects, such as for the reclamation of salinized land (Flowers and Colmer, 2008; Song and Wang, 2015). There are recretohalophytes among both hygrohalophytes (mangroves, *Limonium* and some grasses) and xerohalophytes in arid sites (species of *Atriplex*). In restoration ecology they could be used for the rehabilitation of land that, owing to anthropogenic exploitation including agriculture, is increasingly lost through salinization.

References

Adams, P., Thomas, J.C., Vernon, D.M., Bohnert, H.J. and Jensen, R.G. (1992) Distinct cellular and organismic responses to salt stress. *Plant Cell Physiology* 33, 1215–1223.

Adolf, V.I., Jacobsen, S.E. and Shabala, S. (2013) Salt tolerance mechanisms in quinoa (*Chenopodium quinoa* Willd.). *Environmental and Experimental Botany* 92, 43–54.

Agarie, S., Shimoda, T., Shimizu, Y., Baumann, K., Sunagawa, H. *et al.* (2007) Salt tolerance, salt accumulation, and ionic homeostasis in an epidermal bladder-cell-less mutant of the common ice plant *Mesembryanthemum crystallinum*. *Journal of Experimental Botany* 58, 1957–1967.

Alcarón, J.J., Morales, M.A., Torrecillas, A. and Sánchez-Blaner, M.J. (1999) Growth, water relations and accumulation of organic and inorganic solutes in the halophyte *Limonium latifolium* cv. *Avignon* and its interspecific hybrid *Limonium caspia* x *Limonium latifolium* cv. *Beltlaard* during salt stress. *Journal of Plant Physiology* 154, 795–801.

Aslam, Z., Jeschke, W.D., Barrett-Lennard, E.G., Stetter, T.L., Watkin, E., *et al.* (1986) Effects of external NaCl on the growth of *Atriplex amnicola* and the ion relations and carbohydrate status of leaves. *Plant, Cell and Environment* 9, 571–580.

Atkinson, M.R., Findlay, G.P., Hope, A.B., Pitman, M.G., Saddle, H.D.W. *et al.* (1967) Salt regulation in the mangroves *Rhizophora mucronata* Lam. and *Aegialitis annulata* R.Br. *Australian Journal of Biological Science* 20, 589–599.

Balsamo, R.A. and Thomson, W.W. (1993) Ultrastructural features associated with secretion in the salt glands of *Frankenia grandifolia* (Frankeniaceae) and *Avicennia germinans* (Avicenniaceae). *American Journal of Botany* 80, 1276–1283.

Barkla, B.J. and Vera-Estrella, R. (2015) Single cell-type comparative metabolomics of epidemal bladder cells from the halophyte *Mesembryanthemum crystallinum*. *Frontiers in Plant Science* 6, 435. doi: 10.3389/fpls.2015.00435.

Barkla, B.J., Vera-Estrella, R., Camacho-Emiterio, J. and Pantoja, O. (2002) Na^+/H^+ exchange in the halophyte *Mesembryanthemum crystallinum* is associated with cellular sites of Na^+ storage. *Functional Plant Biology* 29, 1017–1024.

Baumeister, W. and Kloos, G. (1974) Über die Salzsekretion bei *Halimione portulacoides* (L.) Aellen. *Flora* 163, 310–326.

Berry, W.L. (1970) Characteristics of salts secreted by *Tamarix aphylla*. *American Journal of Botany* 57, 1226–1230.

Bradley, P.M. and Morris, J.T. (1991) Relative importance of ion exclusion, secretion and accumulation in *Spartina alternifolia* Loisel. *Journal of Experimental Botany* 42, 1525–1532.

Breckle, S.W. (2002) Salinity, halophytes and salt affected natural ecosystems. In: Läuchli, A. and Lüttge, U. (eds) *Salinity: Environment – Plants – Molecules*. Springer, Dordrecht, Netherlands, pp. 53–77.

Deng, Y., Feng, Z., Yuan, F., Guo, J., Suo, S., *et al*. (2015) Identification and functional analysis of the autofluorescent substance in *Limonium bicolor* salt glands. *Plant Physiology and Biochemistry* 97, 20–27.

Dieffenbach, H., Kramer, D. and Lüttge, U. (1980) Release of guttation fluid from passive hydathodes of intact barley plants. I. Structural and cytological aspects. *Annals of Botany* 45, 397–401.

Ding, F., Yang, J.C., Yuan, F. and Wang, B.S. (2010) Progress in mechanism of salt excretion in recretohalophytes. *Frontiers of Biology* 5, 164–170.

Drennan, P.M., Berjak, P., Lawton, J.R. and Pammenter, N.W. (1987) Ultrastructure of the salt glands of the mangrove, *Avicennia marina* (Forssk.) Vierh., as indicated by the use of selective membrane staining. *Planta* 172, 176–183.

Faraday, C.D. and Thomson, W.W. (1986a) Structural aspects of the salt glands of the Plumbaginaceae. *Journal of Experimental Botany* 37, 461–470.

Faraday, C.D. and Thomson, W.W. (1986b) Morphometric analysis of *Limonium* salt glands in relation to ion efflux. *Journal of Experimental Botany* 37, 471–481.

Faraday, C.D. and Thomson, W.W. (1986c) Functional aspects of the salt glands of the Plumbaginaceae. *Journal of Experimental Botany* 37, 1129–1135.

Faraday, C.D., Quinton, P.M. and Thomson, W.W. (1986) Ion fluxes across the transfusion zone of secreting *Limonium* salt glands determined from secretion rates, transfusion zone areas and plasmodesmatal frequencies. *Journal of Experimental Botany* 37, 482–494.

Feng, Z., Sun, Q., Deng, Y., Sun, S., Zhang, J., *et al*. (2014) Study on pathway and characteristics of ion secretion of salt glands of *Limonium bicolor*. *Acta Physiologia Plantarum* 36, 2729–2741.

Feng, Z.T., Deng, Y.Q., Zhang, S.C., Liang, X., Yuan, F., *et al*. (2015) K^+ accumulation in the cytoplasm and nucleus of the salt gland cells of *Limonium bicolor* accompanies increased rates of salt secretion under NaCl treatment using NanoSIMS. *Plant Science* 238, 286–296.

Fitzgerald, M.A. and Allaway, W.G. (1991) Apoplastic and symplastic pathways in the leaf of the grey mangrove *Avicennia marina* (Forssk.) Vierh. *New Phytologist* 119, 217–226.

Flowers, T.J. and Colmer, T.D. (2008) Salinity tolerance in halophytes. *New Phytologist* 179, 945–963.

Flowers, T.J., Flowers, S.A., Hajibagheri, M.A. and Yeo, A.R. (1990) Salt tolerance in the halophytic wild rice, *Porteresia coarcata* Tateoka. *New Phytologist* 114, 675–684.

Frey-Wyssling, A. (1935) *Die Stoffausscheidungen der Höheren Pflanzen*. Springer, Berlin.

Frey-Wyssling, A. (1972) Elimination processes in higher plants. *Saussurea* 3, 79–90.

Gorham, J. (1987) Photosynthesis, transpiration and salt fluxes through leaves of *Leptochloa fusca* L. Kunth. *Plant, Cell and Environment* 10, 191–196.

Gucci, R., Aronne, G., Lombardini, L. and Tattini, M. (1997) Salinity tolerance in *Phillyrea* species. *New Phytologist* 135, 227–234.

Gunning, B.E.S. and Pate, J.S. (1969) "Transfer cells". Plant cells with wall ingrowths, specialized in relation to short distance transport of solutes - their occurrence, structure, and development. *Protoplasma* 68, 107–133.

Haberlandt, G. (1918) *Physiologische Pflanzenanatomie*. W. Engelmann, Leipzig, Germany.

Hagemeyer, J. and Waisel, Y. (1988) Excretion of ions (Cd^{2+}, Li^+, Na^+, and Cl^-) by *Tamarix aphylla*. *Physiologia Plantarum* 7, 541–546.

Hill, A.E. (1970) Ion and water transport in *Limonium*. IV. Delay effects in the transport process. *Biochimica et Biophysica Acta* 196, 73–79.

Hill, A.E. and Hill, B.S. (1973a) The electrogenic chloride pump of the *Limonium* salt gland. *Journal of Membrane Biology* 12, 129–144.

Hill, A.E. and Hill, B.S. (1973b) The *Limonium* salt gland: A biophysical and structural study. *International Reviews of Cytology* 35, 299–320.

Hill, B.S. and Hill, A.E. (1973c) ATP-driven chloride pumping and ATPase activity in the *Limonium* salt gland. *Journal of Membrane Biology* 12, 145–158.

Hill, A.E. and Hill, B.S. (1976) Mineral ions. In: Lüttge, U. and Pitman, M.G. (eds) *Transport in Plants II, Part B Tissues and Organs. Encyclopedia of Plant Physiology*. Springer, Berlin, pp. 225–243.

Ivanoff, S.S. (1963) Guttation injuries of plants. *Botanical Review* 29, 202–229.

Jeschke, W.D. and Stelter, W. (1983) Ionic relations of garden orache, *Atriplex hortensis* L.: Growth and ion distribution at moderate salinity and the function of bladder hairs. *Journal of Experimental Botany* 34, 795–810.

Jeschke, W.D., Klagges, S., Hilpert, A., Bhatti, A.S. and Sarwar, G. (1995) Partitioning and flows of ions and nutrients in salt-treated plants of *Leptochloa fusca* L. Kunth. *New Phytologist* 130, 23–35.

Kadereit, J.W., Körner, C., Kost, B. and Sonnewald, U. (2014) *Strasburger. Lehrbuch der Pflanzenwissenschaften.* 37th edn. Springer-Spektrum, Berlin.

Kenagy, G.J. (1972) Saltbush leaves: Excision of hypersaline tissue by a kangaroo rat. *Science* 178, 1094–1096.

Kenagy, G.J. (1973) Adaptation for leaf eating in the Great Basin kangaroo rat, *Dipodomys microps*. *Oecologia* 12, 383–412.

Leng, B.Y., Yuan, F., Dong, X.X., Wang, J. and Wang, B.S. (2018) Distribution pattern and salt secretion rate of salt glands in recretohalophytes. *South African Journal of Botany* 115, 74–80.

Liu, X.Z., Wang, C.Z., Su, Q. and Li, C.K. (2012) The potential resource of halophytes for developing bio-energy in China coastal zone. *Herald Journal of Agriculture and Food Science Research* 1, 44–51.

Lüttge, U. (1964) Mikroautoradiographische Untersuchungen über die Funktion der Hydropoten von *Nymphaea*. *Protoplasma* 59, 157–162.

Lüttge, U. (1975) Salt glands. In: Baker, D.A. and Hall, J.L. (eds) *Ion Transport in Plant Cells and Tissues*. North Holland Publishing Company, Amsterdam; Oxford, UK, pp. 335–376.

Lüttge, U. (1978) Transport und Integration bei der Evolution der Pflanzen. *Biologie in unserer Zeit* 8, 129–139.

Lüttge, U. (2002) Performance of plants with C_4-carboxylation modes of photosynthesis under salinity. In: Läuchli, A. and Lüttge, U. (eds) *Salinity: Environment – Plants – Molecules*. Springer, Dordrecht, Netherlands, pp. 341–360.

Lüttge, U. (2008) *Pysiological Ecology of Tropical Plants*. 2nd edn. Springer, Berlin.

Lüttge, U. and Krapf, G. (1969) Die Ultrastruktur der *Nymphaea*-Hydropoten in Zusammenhang mit ihrer Funktion als Salz-transportierende Drüsen. *Cytobiologie* 1, 121–131.

Lüttge, U. and Pallaghy, C.K. (1969) Light triggered transient changes of membrane potentials in green cells in relation to photosynthetic electron transport. *Zeitschrift für Pflanzenphysiologie* 61, 58–67.

Lüttge, U. and Pitman, M.G. (1976) Elimination processes by glands. General introduction. In: Lüttge, U. and Pitman, M.G. (eds) *Transport in Plants II, Part B Tissues and Organs. Encyclopedia of Plant Physiology*. Springer, Berlin, pp. 222–224.

Lüttge, U. and Schnepf, E. (1976) Elimination processes by glands. Organic substances. In: Lüttge, U. and Pitman, M.G. (eds) *Transport in Plants II, Part B Tissues and Organs. Encyclopedia of Plant Physiology*. Springer, Berlin, pp. 244–277.

Lüttge, U., Pallaghy, C.K. and von Willert, K. (1971) Microautoradiographic investigations of sulfate uptake by glands and epidermal cells of water lily (*Nymphaea*) leaves with special reference to the effect of poly-L-lysine. *Journal of Membrane Biology* 4, 395–407.

Lüttge, U., Fischer, E. and Steudle, E. (1978) Membrane potentials and salt distribution in epidermal bladders and photosynthetic tissue of *Mesembryanthemum crystallinum* L. *Plant, Cell and Environment* 1, 121–129.

Ma, H.Y., Tian, C.Y., Feng, G. and Yuan, J.F. (2011) Ability of multicellular salt glands in *Tamarix* species to secrete Na^+ and K^+ selectively. *Science China Life Science* 54, 282–289.

Manousaki, E. and Kalogerakis, N. (2011) Halophytes present new opportunities in phytoremediation of heavy metals and saline soils. *Industrial and Engineering Chemistry Research* 50, 656–660.

Mozafar, A. and Goodin, J.R. (1970) Vesiculated hairs: A mechanism for salt tolerance in *Atriplex halimus* L. *Plant Physiology* 45, 62–65.

Nagai, M., Ohnishi, M., Uehara, T., Yamagami, M., Miura, E. *et al.* (2013) Ion gradients in xylem exudate and guttation fluid related to tissue ion levels along primary leaves in barley. *Plant, Cell and Environment* 36, 1826–1837.

Oh, D.-H., Barkla, B.J., Vera-Estrella, R., Pantoja, O., Lee, S. *et al.* (2015) Cell type-specific response to salinity – the epidermal bladder cell transcriptome of *Mesembryanthemum crystallinum. New Phytologist* 207, 627–644.
Oross, J.W. and Thomson, W.W. (1982) The ultrastructure of the salt glands of *Cynodon* and *Distichlis* (Poaceae). *American Journal of Botany* 69, 939–949.
Oross, J.W. and Thomson, W.W. (1984) The ultrastructure of *Cynodon* salt glands: secreting and non secreting. *European Journal of Cell Biology* 34, 287–291.
Osmond, C.B., Lüttge, U., West, K.R., Pallaghy, C.K. and Shacher-Hill, B. (1969) Ion absorption in *Atriplex* leaf tissue. II. Secretion of ions to epidermal bladders. *Australian Journal of Biological Sciences* 22, 797–814.
Pallaghy, C.K. and Lüttge, U. (1970) Light-induced H^+-ion fluxes and bioelectric phenomena in mesophyll cells of *Atriplex spongiosa. Zeitschrift für Pflanzenphysiologie* 62, 417–425.
Panta, S., Flowers, T., Lane, P., Doyle, R., Haros, G. *et al.* (2014) Halophyte agriculture: Success stories. *Environmental and Experimental Botany* 107, 71– 83.
Pate, J.S. and Gunning, B.E.S. (1972) Transfer cells. *Annual Reviews of Plant Physiology* 23, 173–196.
Perrin, A. (1972) Contribution á l'étude de l'organisation et du fonctionnement des hydathodes : Recherches anatomiques, ultrastructurales et physiologiques. PhD thesis, Université de Lyon, Lyon, France.
Pollak, G. and Waisel, Y. (1970) Salt secretion in *Aeluropus littoralis* (Willd.) Parl. *Annals of Botany* 34, 879–888.
Ramadan, T. (1998) Ecophysiology of salt excretion in the xero-halophyte *Reaumuria hirtella. New Phytologist* 139, 273–281.
Rozema, J., Gude, H. and Pollak, G. (1981) An ecophysiological study of the salt secretion of four halophytes. *New Phytologist* 89, 201–217.
Sager, R. and Lee, J.Y. (2014) Plasmodesmata in integrated signalling: insights from development and environmental signals and stresses. *Journal of Experimental Botany* 65, 6337–6358.
Shabala, S., Bose, J. and Hedrich, R. (2014) Salt bladders: Do they matter? *Trends in Plant Science* 19, 687–691.
Singh, S. (2014) Guttation: Quantification, microbiology and implications for phytopathology. *Progress in Botany* 75, 187–214.
Sobrado, M.A. (2002) Effect of drought on leaf gland secretion of the mangrove *Avicennia germinans* L. *Trees: Structure and Function* 16, 1–4.
Sobrado, M.A. (2004) Influence of external salinity on the osmolality of xylem sap, leaf tissue and leaf gland secretion of the mangrove *Laguncularia racemosa* (L.) Gaertn. *Trees: Structure and Function* 18, 422–427.
Sobrado, M.A. and Greaves, E.D. (2000) Leaf secretion composition of the mangrove species *Avicennia germinans* (L.) in relation to salinity: a case study by using total-reflection X-ray fluorescence analysis. *Plant Science* 159, 1–5.
Song, J. and Wang, B.S. (2015) Using euhalophytes to understand salt tolerance and to develop saline agriculture: *Suadea salsa* as a promising model. *Annals of Botany* 115, 541–553.
Steudle, E., Zimmermann, U. and Lüttge, U. (1975) Water relations of the epidermal bladder cells of the halophytic species *Mesembryanthemum crystallinum*: Direct measurements of hydrostatic pressure and hydraulic conductivity. *Planta* 126, 229–246.
Steudle, E., Zimmermann, U. and Lüttge, U. (1977) Effect of turgor pressure and cell size on the wall elasticity of plant cells. *Plant Physiology* 59, 285–289.
Suárez, N. and Medina, E. (2008) Salinity effects on leaf ion composition and salt secretion rate in *Avicennia germinans* (L.). *Brazilian Journal of Plant Physiology* 20, 131–140.
Tan, W.K., Lin, Q., Lim, T.M., Kumar, P. and Loh, C.S. (2013) Dynamic secretion changes in the salt glands of the mangrove tree species *Avicennia officinalis* in response to a changing saline environment. *Plant, Cell and Environment* 36, 1410–1422.
Tsutsumi, K., Yamada, N., Cha-um, S., Tanaka, Y. and Takabe, T. (2015) Differential accumulation of glycinebetaine and choline monooxygenase in bladder hairs and lamina leaves of *Atriplex gmelini* under high salinity. *Journal of Plant Physiology* 176, 101–107.
Vassilyev, A.E. and Stepanova, A.A. (1990) The ultrastructure of ion-secreting and non-secreting salt glands of *Limonium platyphyllum. Journal of Experimental Botany* 41, 41–46.
Volkov, V. (2015) Salinity tolerance in plants. Quantitative approach to ion transport starting from halophytes and stepping to genetic and protein engineering for manipulating ion fluxes. *Frontiers in Plant Science* 6, 873. doi: 10.3389/fpls.2015.00873.

Waisel, Y. (1991) The glands of *Tamarix aphylla*: a system for salt recretion or for carbon concentration. *Physiologia Plantarum* 83, 506–510.

Waisel, Y., Eshel, A. and Agami, M. (1986) Salt balance of leaves of the mangrove *Avicennia marina*. *Physiologia Plantarum* 67, 67–72.

Wilson, T.P., Canny, M.J. and McCully, M.E. (1991) Leaf teeth, transpiration and the retrieval of apoplastic solutes in balsam poplar. *Physiologia Plantarum* 83, 225–232.

Yuan, F., Chen, M., Leng, B.Y. and Wang, B.S. (2013) An efficient autofluoresence method for screening *Limonium bicolor* mutants for abnormal salt gland density and salt secretion. *South African Journal of Botany* 88, 110–117.

Yuan, F., Chen, M., Yang, J., Leng, B. and Wang, B. (2014) A system for the transformation and regeneration of the recretohalophyte *Limonium bicolor*. *In Vitro Cellular and Developmental Biology of Plants* 50, 610–617.

Yuan, F., Amy Lyu, M.J., Leng, B.Y., Zheng, G.Y., Feng, Z.T. *et al.* (2015a) Comparative transcriptome analysis of developmental stages of the *Limonium bicolor* leaf generates insights into salt gland differentiation. *Plant, Cell and Environment* 38, 1637–1657.

Yuan, F., Chen, M., Yang, J., Song, J. and Wang, B. (2015b) The optimal dosage of ^{60}Co gamma irradiation for obtaining salt gland mutants of exo-recretohalophyte *Limonium bicolor* (Bunge) O. Kuntze. *Pakistan Journal of Botany* 47, 71–76.

Yuan, F., Leng, B. and Wang, B. (2016a) Progress in studying salt secretion from the salt glands in recretohalophytes: How do plants secrete salt? *Frontiers in Plant Science* 7. doi.org/10.3389/fpls.2016.00977.

Yuan, F., Amy Lyu, M.J., Leng, B.Y., Zhu, X.G. and Wang, B.S. (2016b) The transcriptome of NaCl-treated *Limonium bicolor* leaves reveals the genes controlling salt secretion of salt gland. *Plant Molecular Biology* 91, 241–256.

Ziegenspeck, H. (1949) Zur Physiologie der Hydathoden. *Phyton* 1, 302–318.

Ziegler, H. and Lüttge, U. (1966) Die Salzdrüsen von *Limonium vulgare*. I. Mitteilung Die Feinstruktur. *Planta* 70, 193–206.

Ziegler, H. and Lüttge, U. (1967) Die Salzdrüsen von *Limonium vulgare*. II. Mitteilung Die Lokalisierung des Chlorids. *Planta* 74, 1–17.

15 Synergic Effects of Rhizobacteria: Increasing Use of Halophytes in a Changing World

Jennifer Mesa-Marín, Enrique Mateos-Naranjo, Ignacio D. Rodríguez-Llorente, Eloísa Pajuelo and Susana Redondo-Gómez*
University of Seville, Seville, Spain

Abstract

Halophytes have special features that enable them to adapt to saline environments. They are also associated with some microbiomes, which play an important role in their survival and special adaptation to adverse situations. Among them, bacteria intimately associated with plant tissues may interact with and affect their growth. Particularly, some bacteria from the rhizosphere, known as rhizobacteria, can facilitate the growth of plants or their adaptation to different stresses using a wide range of different mechanisms. Recently, studies have shown that bacterial treatments in halophytes improved their physiologic parameters and inner adaptive capacities, as heavy metal or salt phytoaccumulation, as well as crop adaptation. We are witnessing the emergence of a nascent natural resource that may preserve native ecosystems, and at the same time need to address current challenges in an ever-changing world subjected to climate change.

Keywords: Climate change; Plant–microbe interactions; Phytoremediation; Rhizobacteria; Salt stress

15.1 Introduction

Our world is changing, and there is some growing evidence that halophytes are plant species with a high number of potential uses to face changing environments, due to their special features and adaptive mechanisms. Despite this, there is a natural resource closely associated to these species that has never been addressed: their microbiome.

Bacteria intimately associated with plant tissues may interact with and affect their growth. Particularly, some bacteria from the rhizosphere, known as rhizobacteria, can facilitate the growth of plants or their adaptation to different stresses using a wide range of different mechanisms (Glick, 2012). However, there are very few studies in this regard using halophytes.

Recently, this lack of knowledge has been exploited, and several approaches have been studied in this line. The studies above-mentioned throughout this chapter have shown that bacterial treatments in halophytes may improve their physiology and inner adaptive capacities, as heavy metal and salt bioremediation, or crop adaptation.

This strategy opens up a wide opportunity for study and underscores the critical importance of halophytes. We are witnessing the emergence of a nascent natural resource to preserve

* Corresponding author e-mail: susana@us.es

native ecosystems while facing current challenges like pollution, salinity, anthropogenic intervention or biotechnological demands in a more and more changing world subjected to climate change.

In this chapter, we look deeper into the bacterial rhizospheric and endophytic world. Their mechanisms of beneficial interaction with the plant will be briefly described; the contribution of these effects on halophytes to different fields of action will be revealed; and the current status of the rhizosphere–halophyte strategy as a promising and environmentally friendly biotool is discussed.

15.2 Soil Bacteria and Their Interaction With Plants

The soil is replete with microscopic life forms such as bacteria, fungi, protozoa and algae. Bacteria represent by far the largest group of these microorganisms (about 95%). It is known that soil can host a large number of bacteria, between 10^8 and 10^9 cells/g of soil, and that only 1% is cultivable in the laboratory (Glick, 2012). A variety of strategies have to be followed to study them in a laboratory, according to the purpose in hand. For example, using crop-dependent techniques, the bacteria can be isolated from the soil and reproduced in culture media; or, using culture-independent techniques, their whole genomes or part of them may be studied using polymerase chain reaction (PCR), sequencing and/or genetic fingerprinting techniques (Ma *et al.*, 2016a).

15.2.1 Bacteria in the rhizosphere

The bacterial population is not evenly distributed throughout the soil. Its concentration is usually higher around roots because of the presence of root exudates such as sugars, amino acids, organic acids and other nutrients, which promote bacterial growth on the root surface and its surroundings (Glick, 2012). This area is defined as the rhizosphere.

The advantages from bacteria growing in the rhizosphere (known as rhizobacteria) are widely known. The term 'plant growth-promoting rhizobacteria' (PGPR) refers to those soil bacteria that inhabit the rhizosphere of plants and which have a positive effect on plant growth, while themselves benefiting from root exudates (Kloepper and Schroth, 1978). Research into endophytes (bacteria living inside plant tissues) and their role in plant growth (referred to as plant growth-promoting bacterial endophytes, PGPBE) has also been carried out (Santoyo *et al.*, 2016). We now speak more generally of 'plant growth-promoting bacteria' (PGPB) (Glick, 2012). These plant–microbe interactions may occur in the phyllosphere (the aerial parts of the plant), endosphere (related to the internal transport system) and the rhizosphere (Bhattacharyya and Jha, 2012). Sometimes there is no general parameter for differentiating PGPR from PGPBE, as most rhizobacteria also colonize the root surface (rhizoplane) and even penetrate into the root cortex, reaching the endosphere (Dimkpa *et al.*, 2009).

Microbial communities constitute a complex and dynamic structure in soils. In this sense, bacterial communities in brackish or saline land ecosystems remain understudied in comparison to upland ecosystems. A few studies about microbiome community composition in salt marshes suggest that it is distinct to the host species, and evidence points to environmental factors such as oxygen, organic matter content, pH, sulfur chemistry, pollutants and nutrient availability as ecological drivers of microbiome community structure (Hong *et al.*, 2015; Cleary *et al.*, 2016; Mesa *et al.*, 2016; Rietl *et al.*, 2016). However, some studies challenge this view, showing that microbiome composition does not always vary significantly between different halophyte plant species (Rietl *et al.*, 2016).

15.2.2 PGPB

Some rhizospheric and endophytic bacteria have several mechanisms to promote plant growth. We may classify them into direct and indirect mechanisms.

15.2.2.1 Direct mechanisms of plant growth promotion

1. Facilitation of nutrient acquisition: fixation of atmospheric nitrogen (Olivares, 2008), phosphate solubilization (Rodríguez and Fraga, 1999) and siderophore production for iron uptake (Neilands, 1995).

2. Production of plant growth-modulating phytohormones: cytokinins (CKs), gibberellins (GA), salicylic acid (SA), abscisic acid (ABA), brassinosteroids (BRs) and jasmonates (JA), plant regulators involved in the cell cycle and several developmental processes (Werner *et al.*, 2001; Davière and Achard, 2013; Fahad *et al.*, 2015). Also, production of auxins for root development (the most common is indole-3-acetic acid, IAA) (Duca *et al.*, 2014) and the enzyme 1-aminocyclopropane-1-carboxylate deaminase (ACCD) that cleaves ethylene, is crucial in the stress response (Gamalero and Glick, 2015).

3. Production of antibiotics, bacteriocins, metabolic by-products and lytic enzymes to defend against plant pathogens (Beneduzi *et al.*, 2012; Subramanian and Smith, 2015).

15.2.2.2 Indirect mechanisms of plant growth promotion

1. Biofilm production: improves bacterial adhesion to the surface of the root tissue, and intercellular communication. This facilitates competition against pathogens, nutrient uptake and adaptation to changing environmental conditions (Seneviratne *et al.*, 2010), in addition to the immobilization and compartmentalization of toxic elements (Das *et al.*, 2012).

2. Induction of defence: the presence of microorganisms may prime the whole plant body for enhanced defence against a broad range of pathogens and insect herbivores (Pieterse *et al.*, 2014).

3. Competition: rhizospheric and endophytic bacteria, more abundant than pathogens, colonize the surface and interior of plant tissues, using available spaces and nutrients, making it difficult for pathogens to grow (Glick, 2012).

15.3 Bacterial Strategies

PGPB used to promote growth and ameliorate stress should be ubiquitous in nature, common to soils around the world, and pose no threat to humans, wildlife or the environment. The PGPB used should be classified as Biosafety Level 1 (the safest possible designation) (Gerhardt *et al.*, 2017).

This ecological strategy may be applied by biostimulation of microbial populations, bacterial bioaugmentation or co-inoculation of seeds:

- Biostimulation: addition of electron acceptors, electron donors or nutrients to stimulate naturally occurring microbial populations, to enhance the activity of indigenous microbial degraders or to promote co-metabolism (Pajuelo *et al.*, 2014).
- Bioaugmentation: enrichment of the native beneficial microbial populations present in the rhizosphere of the plants. This is achieved by inoculation with exogenous microorganisms with good plant growth-promoting properties or pollutant-degradation abilities. However, inoculation with microorganisms previously isolated from the same rhizosphere is preferable, because the adaptation and competitiveness of inocula are guaranteed (Pajuelo *et al.*, 2014). In fact, the term indigenous microorganism (IMO)-based technology can be found in the literature (Kumar and Gopal, 2015).
- Co-inoculation of seeds (Fahad *et al.*, 2015): seeds are normally surface disinfected and submerged in a suspension of the selected PGPR for a predetermined period of time.

For each of the strategies proposed, one bacterial strain or a mixture of strains, known as a consortium, may be used. Different PGPB may have different modes of action for promoting plant growth that could be additive or synergistic in a microbial mix, further increasing the odds of success in the field (Kuiper *et al.*, 2004). Sometimes, however, it is preferable to use a single strain of PGPB to avoid antagonistic effects (S. Qin *et al.*, 2016; Y. Qin *et al.*, 2016).

15.4 PGPB–Halophyte Biotechnology

In contrast to human microbiome research, plant microbiome research is in its infancy (Bai *et al.*, 2015). We are still in the discovery phase, characterizing the microbial community in model plants, crop plants and freshwater or upland terrestrial ecosystems (Berg *et al.*, 2014), while the ecology of a diversity of halophyte microbiomes around the world is just beginning to be explored.

Climate changes have multifaceted effects, threatening the sustainability and productivity of agricultural systems. Among them, induced soil salinization is a constant threat to agriculture and ecology worldwide, explaining why halophytes are gaining attention: they are naturally adapted to survive and maintain productivity in saline water (Glenn *et al.*, 1999). Modern agriculture faces other abiotic stresses, such as drought, waterlogging, heavy metal and organic pollutants, chilling and heat, all of which are major factors affecting crop yields. Under these unfavourable environmental conditions, PGPB play an important role in augmenting the stress tolerance in plants (Fahad *et al.*, 2015). To date, this potential has been underexploited in halophytes.

Very few studies of the microbiome of halophytes have been carried out. In some cases, bacteria have been isolated from the rhizosphere and screened for plant growth-promoting properties; for example, in the species *Sonchus arvensis*, *Solanum surrattense*, *Lactuca dissecta* and *Chrysopogon aucheri* from Khewra Salt Range and Attock, Pakistan (Naz and Bano, 2009). The halophyte *Sesuvium verrucosum* has been grown under extreme soil salinity in Jeddah, Saudi Arabia (El-Awady *et al.*, 2015), and *Salsola grandis* has been grown in Ankara, Turkey (Kataoka *et al.*, 2017). Cultivable bacteria have been also studied in the highly polluted Tinto and Odiel salt marshes in Spain, where the halophytes *Spartina maritima*, *S. densiflora* and *Arthrocnemum macrostachyum* grow naturally (Fig. 15.1) (Mesa *et al.*, 2015a, b; Navarro-Torre *et al.*, 2016; Paredes-Páliz *et al.*, 2016a). Endophytes have been analysed from halophytes such as *Salicornia brachiata*, *Sphaeranthus indicus*, *Cressa cretica* and *Suaeda nudiflora*, growing in coastal Gujarat, India (Arora *et al.*, 2014). In all cases, sodium-resistant bacterial strains with plant growth-promoting properties have been isolated, belonging to different genera such as *Rhizobium*, *Arthrobacter*, *Bacillus*, *Enterobacter*, *Vibrio* and *Escherichia*.

Other molecular techniques have been employed to study metabolic profiles of the halophyte rhizospheric microbiome, such as the case of *Salicornia europaea* in Poland, to determine potential bacterial mechanisms responsible for the high tolerance of halophytes to saline conditions (Szymańska *et al.*, 2014), or diazotroph diversity in the rhizosphere of *Spartina densiflora* (Lovell *et al.*, 2000).

Although halophytes have been poorly explored from a microbiological point of view, they are now being found to be a source of novel bacterial species from different genera, for example *Marinomonas* and *Vibrio* (Lucena *et al.*, 2016, 2017), among many others.

15.4.1 PGPB: halophytes and ecological restoration

Coastal wetlands are of great ecological, environmental and socioeconomic importance because they are among the most biologically productive ecosystems on the planet (Day *et al.*, 2013). They are being lost at an alarming rate, however, because of natural and anthropogenic factors including land use change, eutrophication, sea level rise, subsidence, extreme weather events and shoreline erosion. It has been estimated that approximately half of all salt marshes have been lost worldwide (Silliman *et al.*, 2015). Halophytic plants predominate in these scenarios and play a determining role in the structure and dynamics of the marshes. Some halophytes, though, are suffering sudden vegetation dieback (SVD), a phenomenon that causes rapid death followed by no or slow recovery of the salt marsh flora. In some cases, the implication of the microbiome has been studied (Elmer *et al.*, 2017). Through their beneficial effects on root systems, PGPB may improve halophyte capacity to recolonize degraded shores and stabilize soils. (Mesa *et al.*, 2016). It should be emphasized that care should be exerted to avoid the use of invasive halophyte species, because this may lead to biodiversity alteration, coast remodelling and economic losses (Kittelson and Boyd, 1997; Ayres *et al.*, 2008). In the same way, using indigenous PGPB with halophytes may rehabilitate and restore arid lands where desertification processes are taking place (Bashan and De-Bashan, 2010).

Plant-associated microbial communities also provide critical ecosystem services in salt marshes by mediating biogeochemical cycles (Kostka *et al.*, 2002). For example, in *Spartina*-dominated

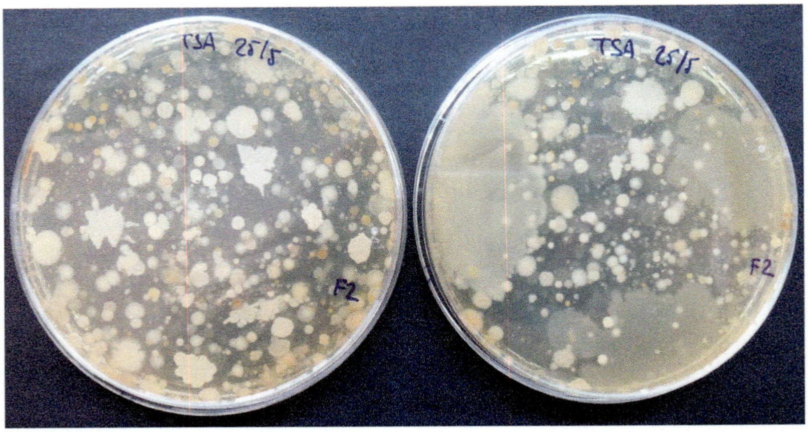

Fig. 15.1. Cultivable bacteria isolated from the rhizosphere of *Spartina maritima* in the Tinto salt marsh, Spain (from Mesa *et al.*, 2015).

marshes in Georgia, sediments are rich in iron minerals and bathed in sulfate-rich seawater. Sediment anoxia and sulfide accumulation, in particular, limit nitrogen uptake and growth by *Spartina* (Mendelssohn and Morris, 2002). Through catalysis of sulfide and iron oxidation, microbial processes consume toxic sulfur and regulate redox poise in the root zone to benefit plants (Weiss *et al.*, 2003; Thomas *et al.*, 2014).

15.4.2 PGPB: halophytes and pollution

Mechanisms of adaptation that allow halophytes to survive high salt concentrations may be not exclusive to sodium and chloride, and may confer tolerance to other toxic ions. It has been shown recently that a number of these halophytes do indeed have the ability to accumulate heavy metals or to tolerate high levels of toxic pollutants in the environment (Redondo-Gómez, 2013; Paredes-Páliz *et al.*, 2016a, b; Thijs *et al.*, 2017).

These abilities make some halophytes excellent candidates for phytoremediation of heavy metals and other xenobiotics in polluted soils. This strategy is now preferred to traditional remediation for inorganic and organic pollutants, which required high technical complexity and, therefore, economic cost (Glass, 1999).

For successful phytoremediation, substantial plant biomass production is essential. However, the presence of toxins inevitably leads to plant stress and deterioration (Duarte *et al.*, 2013). In search of strategies to implement plant bioremediation capacity, some authors have proposed genetic improvements to species such as *Spartina alterniflora* (Czakó *et al.*, 2006). The environmentally friendly method that is gaining the attention of researchers, however, is the association with metal-resistant PGPB inoculants to enhance plant adaptation and ameliorate their phytoremediative capacity (Ma *et al.*, 2016a).

Bacteria promote halophyte phytoremediation capacity by acting on the plant and/or the pollutant by various mechanisms, depending on the organic or inorganic nature of the toxin. They may reduce the plant stress response to the pollutant through the production of 1-aminocyclopropane-1-carboxylate deaminase (ACCD); or by modulating the activity of reactive oxygen species (ROS) enzymes such as superoxide anions (O^{2-}), hydrogen peroxide (H_2O_2), hydroxyl radicals ($HO^{\bullet-}$) or singlet oxygen, reducing oxidative damage in the cells (Mittler *et al.*, 2004). Bacteria may also produce detoxification of pollutants by different chemical pathways (e.g. precipitation, biosorption, oxidation or methylation), increasing their bioavailability for phytoextractive plants (releasing chelating agents such as siderophores, biosurfactants or organic acids), or immobilizing and decreasing their bioavailability (ion complexation, biofilm production) in the case of phytostabilizing plants. Finally, rhizobacteria and endophytes possess a variety of catabolic enzymes that can degrade and detoxify the xenobiotics. The toxic effects of pollutants on the plants can be mitigated by soil microbes, leading to a cheap, energy-efficient phytoremediation method. This suggests that PGPB are potential tools for sustainable agriculture in the future (Fahad *et al.*, 2015; Ma *et al.*, 2016b; Arslan *et al.*, 2017; Ashraf *et al.*, 2017; Thijs *et al.*, 2017).

Several species of the genus *Spartina* are able not only to tolerate but also to accumulate organic and inorganic contaminants in their tissues (Mackova *et al.*, 2006; Redondo-Gómez *et al.*, 2011). This, coupled with their key role as engineers of coastal ecosystems and their widespread distribution in wetlands around the world, make them valuable biotools for the phytoremediation of contaminated soils (Mackova *et al.*, 2006). In particular, species such as *S. densiflora* and *S. maritima* have demonstrated a great capacity for bioaccumulation of heavy metals in roots and leaves, and are known as hyperaccumulating plants (Redondo-Gómez, 2013). These are plants that can accumulate contaminants in their tissues at concentrations 100 times higher than other species grown under the same conditions, without severe damage to their growth and their physiological processes (Thakur *et al.*, 2016). Researchers looking for strategies to restore one of the most polluted estuaries in the word, the Tinto estuary, Spain, used PGPR-mediated inoculation and demonstrated good results in greenhouse conditions. Inoculation on both *Spartina* species with bacterial consortia isolated from their rhizospheres improved their root growth and photosynthetic parameters, and also incremented root accumulation of heavy metals. In the case of *S. maritima* this was up to 50% for copper (Fig. 15.2) (Mateos-Naranjo *et al.*, 2015; Mesa *et al.*, 2015a, b; Paredes-Páliz *et al.*, 2016b; Paredes-Páliz *et al.*, 2017). Similarly, *Arthrocnemum macrostachyum* improved its metal accumulation capacity after bioaugmentation treatments with native rhizobacteria (Navarro-Torre *et al.*, 2017b).

With respect to organic toxins, microbe-assisted phytoremediation is also being studied as an alternative to hydrocarbon remediation in oil-contaminated soils, using halophytes such as *Juncus acutus* or *Halocnemum strobilaceum* (Al-Mailem *et al.*, 2010; Zhao *et al.*, 2013; Oliveira *et al.*, 2015; Syranidou *et al.*, 2017).

Fig. 15.2. In a greenhouse experiment, *Spartina densiflora* developed a higher root biomass when a native rhizobacterial consortium bioaugmentation treatment was applied (left) than did the non-inoculated control (right) (from Mateos-Naranjo *et al.*, 2015).

15.4.3 PGPB: halophytes and salinity

Salinity is a serious and growing problem for agriculture worldwide. Plants growing in salt-impacted soils are prone to the combined detrimental effects of suboptimal soil conditions (e.g. low organic content and poor soil structure), water stress and nutritional deficiencies, which result in substandard plant growth (Gerhardt *et al.*, 2017). The amount of salt-affected land represents at least 20% of the world's cultivated area and this percentage continues to increase as a direct consequence of irrigation (Glick, 2014).

In this scenario halophytes have been regarded both as a rich source of potential new crops and as a biotool for phytoremediation strategies (Glenn *et al.*, 1999; Qadir *et al.*, 2006). There is considerable diversity in halophytes; they accumulate high salt concentrations in their tissues; have been tested as vegetable, forage and oilseed crops in agronomic field trials;

and have high yield potential when irrigated with seawater (Glenn *et al.*, 1999). However, plant tolerance is also connected with complex ecological processes within the rhizosphere and phyllosphere. Thus, the halophyte microbiome may be significantly implicated in their well-being and salinity tolerance (Ruppel *et al.*, 2013).

On one hand, PGPB can enhance the uptake of nutrient elements in plants. Controlling Na^+ homeostasis is critically important to maintain normal plant growth during salt stress, and many PGPB have been shown to improve Na^+ exclusion and K^+ uptake in plants, thereby increasing the K^+/Na^+ ratio. This is achieved, for example, by promoting biofilm formation on root surfaces, thus restricting Na^+ influx into roots; or via tissue-specific regulation of HKT1, a plasma membrane Na^+ uniporter (Zhang *et al.*, 2008; Gerhardt *et al.*, 2017).

On the other hand, some PGPB improve plant salt tolerance by altering the endogenous hormone status (Vurukonda *et al.*, 2016). For instance, when subjected to salinity, drought and pathogen stresses, plants are known to produce excessive ethylene, which severely retards root development. ACC deaminase-producing bacteria reduce ethylene levels, overcoming salt-induced growth inhibition (Y. Qin *et al.*, 2016). Considerable attention has thus been paid to the isolation of ACC deaminase-producing microbes for their use in direct plant growth promotion in saline environments. Another important bacterial feature is indole-3-acetic acid (IAA) production, increasing the number of root hairs under stress. These hairs contribute substantially to the surface area of roots, where most of the ion uptake (including Na^+) occurs. Moreover, ABA-producing PGPR and ABA-metabolizing PGPB may modulate plant ABA status, causing variable plant responses to salinity stress, provided that ABA plays a crucial role in signalling shoot-to-root flows, which may mitigate plant adaptation to water deficits (Gerhardt *et al.*, 2017). Like pollutants, high salinity leads to an increased production of ROS, and PGPB may regulate salt-induced oxidative stress, as discussed above. Another bacterial capacity is related to root hydraulic conductance. For salinity accommodation, plants must decrease their cell water potential to continue to take up water from salinized environments. Water uptake capacity depends on root hydraulic conductance, determined by the activity of aquaporins. Rhizobacteria inoculation has shown to up-regulate aquaporin gene expression (Gond *et al.*, 2015; S. Qin *et al.*, 2016; Vurukonda *et al.*, 2016). Thus, PGPB-treated plants have a greater capacity for Na^+ uptake from the soil than untreated plants, which generally corresponds to decreases in both soil salinity and sodicity.

The rhizosphere of halophytes provides excellent niches where salt-tolerant rhizobacteria, with the ability to promote plant growth and salinity tolerance, can be found (Sáenz-Mata *et al.*, 2016). Salt-tolerant PGPB have begun to emerge as an important alternative to help restore salt-affected soils. Treatments with native rhizobacteria and endophytes have widely shown better salt uptake capacity, as well as ameliorative effects in the physiological parameters of halophytes such as *Puccinellia* (Niu *et al.*, 2016), *Salicornia* (Mapelli *et al.*, 2013; Rueda-Puente *et al.*, 2013) or *Arthrocnemum* (Navarro-Torre *et al.*, 2017b).

Although PGPB alleviate salt stress in halophytes to some extent, another outstanding application has been studied for protection against salt stress in non-host glycophytes of agricultural interest (Sáenz-Mata *et al.*, 2016). Plant growth-promoting halotolerant bacteria have been isolated from halophytes to improve plant growth and salt tolerance in common crops such as summer barley (Suarez *et al.*, 2015). The rhizosphere of the American halophyte *Prosopis strombulifera* has also been proved to host microorganisms capable of improving salt tolerance in crops (Reginato *et al.*, 2012). PGPR from *Atriplex leucoclada*, *Haloxylon salicornicum*, *Lespedeza bicolor*, *Suaeda fruticosa* and *Salicornia virginica* from Pakistan enhanced maize shoots and root length and dry mass, the accumulation of osmolytes and the antioxidant enzyme activity when inoculated in saline soils (Ullah and Bano, 2015). Another example can be found in halotolerant bacteria from the roots of *Arthrocnemum indicum*, which successfully colonized the roots of peanut seedlings, maintaining ion homeostasis, accumulating less ROS and showing enhanced growth compared to non-inoculated seedlings (Sharma *et al.*, 2016). Also, ACC deaminase-producing bacteria isolated from halophytes promoted sugar beet growth under conditions of saline stress (Zhou *et al.*, 2017).

From a genetic point of view, the halophyte microbiome has been found to be enriched in genes contributing to salt stress acclimatization, nutrient solubilization and competitive root colonization (Yuan *et al.*, 2016). On the whole, it is feasible to think that halophytes may be linked to a specialized and adapted microbiome to confront soil salinity, and this may be exploited in the design of bacterial bioinoculants for salt phytoremediation or to sustain crop production in salty soils.

15.4.4 PGPB: halophytes and enzymes of interest

Halophytes have other potential applications where associated PGPB may play a crucial role. As a result of adaptation to their environment, many microorganisms have evolved unique properties of considerable biotechnological and, therefore, commercial significance (Wang and Dai, 2010; Glick, 2012; Brader *et al.*, 2014; Nair and Padmavathy, 2014). These studies have also concluded that the rhizobacteria of halophytes have potential for exploitation in bioprospecting work. For example, rhizobacteria isolated from the halophyte *Suaeda maritima* from Indian solar salterns were found to be excellent producers of industrially important hydrolytic enzymes such as protease, lipase and amylase (Jayaprakashvel *et al.*, 2015).

15.5 Conclusion and Future Perspectives

The potential of the halophyte rhizosphere as reservoir of beneficial microbes has been widely demonstrated. Halophyte–microbe-assisted techniques have undeniable advantages and have demonstrated considerable promise in both the laboratory and greenhouse. However, these techniques have general limitations, as they are significantly affected by environmental factors and the particular features of each ecosystem. These include the properties of the soil itself; or the bioavailability of the components and plant root depth in the case of phytoremediation purposes. Despite the fact that they are inexpensive, sustainable and environmentally friendly techniques, they are usually slow (Pilon-Smits, 2005; Oliveira *et al.*, 2015). Many authors also agree that several aspects of this field of research require more attention and need to be addressed:

1. In-depth understanding of the biological processes associated with halophyte–bacteria interaction has not been fully described. Thus, halophyte–bacteria–soil interaction at the molecular level requires further research. Here, the 'omic' sciences play an important role in further linking microbiome structure with ecosystem function and biogeochemical cycles.
2. Research must now specifically address the scaling up from laboratory to the field for practical implementation of this approach *in situ*. It is essential to deal with technical aspects such as large-scale cultivation of microorganisms, transport, application over large areas, determining microbial colonization, the duration of experiments and their monitoring over time. In the case of halophytes, different hosting ecosystems may present technical difficulties. For example, many halophytes grow in estuaries and coastal salt marshes. In these cases, optimization of inoculation should be carried out to minimize the dilution or wash effect of the tide (Fig. 15.3). These areas are barely accessible: muddy soil predominates, making inoculation arduous (Fig. 15.4). Well-established field application could drive commercialization, with the appropriate linkages between scientific organizations and industries (Bhattacharyya and Jha, 2012).

Finally, it is important that the general public is educated about the importance of soil bacteria and the use of the halophyte microbiome (Pilon-Smits, 2005; Thijs *et al.*, 2017).

Other currently attractive ideas of research may be considered for study. For example, the application of genetically engineered PGPR to target particular needs (Bhattacharyya and Jha, 2012), or applying PGPR in other current fields of halophyte research such as animal feed (Attia-Ismail, 2015), biofuel (Abideen *et al.*, 2011) or human health (Patel, 2016; Barreira *et al.*, 2017).

In general, a better understanding of soil microbiota and smart manipulation of the

Fig. 15.3. The halophyte *Spartina maritima* in the Odiel salt marsh, Spain is covered twice daily by the tide (red arrow). In these cases, applying bacterial strategies could be a challenge.

Fig. 15.4. The halophyte *Spartina maritima* in the Tinto salt marsh, Spain, when the tide recedes. The muddy soil complicates accessibility to the area.

halophyte microbiome may create a reliable future route for contributing to crop adaptation and for taking advantage of all the potential that halophytes provide to face the environmental consequences of climate change.

Acknowledgements

We thank the Spanish Economy, Industry and Competitiveness Ministry for its support (project CGL2016-75550-R AEI/FEDER, UE).

References

Abideen, Z., Ansari, R. and Khan, M.A. (2011) Halophytes: Potential source of ligno-cellulosic biomass for ethanol production. *Biomass and Bioenergy* 35, 1818–1822. doi:10.1016/j.biombioe.2011.01.023.

Al-Mailem, D.M., Sorkhoh, N.A., Marafie, M., Al-Awadhi, H., Eliyas, M. *et al.* (2010) Oil phytoremediation potential of hypersaline coasts of the Arabian Gulf using rhizosphere technology. *Bioresource Technology* 101, 5786–5792. doi:10.1016/j.biortech.2010.02.082.

Arora, S., Patel, P.N., Vanza, M.J. and Rao, G.G. (2014) Isolation and characterization of endophytic bacteria colonizing halophyte and other salt tolerant plant species from coastal Gujarat. *African Journal of Microbiology Research* 8, 1779–1788. doi:10.5897/AJMR2013.5557.

Arslan, M., Imran, A., Khan, Q.M. and Afzal, M. (2017) Plant–bacteria partnerships for the remediation of persistent organic pollutants. *Environmental Science and Pollution Research* 24, 4322–4336. doi:10.1007/s11356-015-4935-3.

Ashraf, M.A., Hussain, I., Rasheed, R., Iqbal, M., Riaz, M. *et al.* (2017) Advances in microbe-assisted reclamation of heavy metal contaminated soils over the last decade: A review. *Journal of Environmental Management* 198, 132–143. doi:10.1016/j.jenvman.2017.04.060.

Attia-Ismail, S. (2015) Nutritional and feed value of halophytes and salt tolerant plants. In: El Shaer, H.M. and Squires, V.R. (eds) *Halophytic and Salt-Tolerant Feedstuffs: Impacts on Nutrition, Physiology and Reproduction of Livestock*. CRC Press, Boca Raton, Florida, pp. 106–126. doi:10.1201/b19862-10.

Ayres, D.R., Grotkopp, E., Zaremba, K., Sloop, C.M., Blum, M.J. *et al.* (2008) Hybridization between invasive *Spartina densiflora* (Poaceae) and native *S. foliosa* in San Francisco Bay, California, USA. *American Journal of Botany* 95, 713–719. doi:10.3732/ajb.2007358.

Bai, Y., Müller, D.B., Srinivas, G., Garrido-Oter, R., Potthoff, E. *et al.* (2015) Functional overlap of the *Arabidopsis* leaf and root microbiota. *Nature* 528, 364–369. doi:10.1038/nature16192.

Barreira, L., Resek, E., Rodrigues, M.J., Rocha, M.I., Pereira, H. *et al.* (2017) Halophytes: Gourmet food with nutritional health benefits? *Journal of Food Composition and Analysis* 59, 35–42. doi:10.1016/j.jfca.2017.02.003.

Bashan, Y. and De-Bashan, L.E. (2010) Microbial populations of arid lands and their potential for restoration of deserts. In: Dion, P. (ed.) *Soil Biology and Agriculture in the Tropics*. Springer, Berlin, Germany, pp. 109–137. doi:10.1007/978-3-642-05076-3_6.

Beneduzi, A., Ambrosini, A. and Passaglia, L.M.P. (2012) Plant growth-promoting rhizobacteria (PGPR): Their potential as antagonists and biocontrol agents. *Genetics and Molecular Biology* 35, 1044–1051.

Berg, G., Grube, M., Schloter, M. and Smalla, K. (2014) Unraveling the plant microbiome: looking back and future perspectives. *Frontiers in Microbiology* 5, 148. doi:10.3389/fmicb.2014.00148.

Bhattacharyya, P.N. and Jha, D.K. (2012) Plant growth-promoting rhizobacteria (PGPR): emergence in agriculture. *World Journal of Microbiology and Biotechnology* 28, 1327–1350. doi:10.1007/s11274-011-0979-9.

Brader, G., Compant, S., Mitter, B., Trognitz, F. and Sessitsch, A. (2014) Metabolic potential of endophytic bacteria. *Current Opinion in Biotechnology* 27, 30–37. doi:10.1016/j.copbio.2013.09.012.

Cleary, D.F.R., Polónia, A.R.M., Sousa, A.I., Lillebø, A.I., Queiroga, H. and Gomes, N.C.M. (2016) Temporal dynamics of sediment bacterial communities in monospecific stands of *Juncus maritimus* and *Spartina maritima*. *Plant Biology* 18, 824–834. doi:10.1111/plb.12459.

Czakó, M., Feng, X., He, Y., Liang, D. and Márton, L. (2006) Transgenic *Spartina alterniflora* for phytoremediation. *Environmental Geochemistry and Health* 28, 103–110. doi:10.1007/s10653-005-9019-8.

Das, N., Basak, L.V.G., Salam, J.A. and Abigail, E.A. (2012) Application of biofilms on remediation of pollutants – an overview. *Journal of Microbiology and Biotechnology Research* 2, 783–790.

Davière, J.M. and Achard, P. (2013) Gibberellin signaling in plants. *Development* 140, 1147–51. doi:10.1242/dev.087650.

Day, J.W., Kemp, W.M., Yáñez-Arancibia, A. and Crump, B.C. (2013) *Estuarine Ecology*. Wiley, Weinheim, Germany.

Dimkpa, C., Weinand, T. and Asch, F. (2009) Plant-rhizobacteria interactions alleviate abiotic stress conditions. *Plant, Cell and Environment* 32, 1682–1694. doi:10.1111/j.1365-3040.2009.02028.x.

Duarte, B., Santos, D. and Caçador, I. (2013) Halophyte anti-oxidant feedback seasonality in two salt marshes with different degrees of metal contamination: Search for an efficient biomarker. *Functional Plant Biology* 40, 922–930. doi:10.1071/FP12315.

Duca, D., Lorv, J., Patten, C.L., Rose, D. and Glick, B.R. (2014) Indole-3-acetic acid in plant–microbe interactions. *Antonie Van Leeuwenhoek* 106, 85–125. doi:10.1007/s10482-013-0095-y.

El-Awady, M.A.M., Hassan, M.M. and Al-Sodany, Y.M. (2015) Isolation and characterization of salt tolerant endophytic and rhizospheric Plant Growth Promoting Bacteria (PGPB) associated with the halophyte plant (*Sesuvium verrucosum*) grown in KSA. *International Journal of Applied Sciences and Biotechnology* 33, 552–560. doi:10.3126/ijasbt.v3i3.13440.

Elmer, W.H., Thiel, P. and Steven, B. (2017) Response of sediment bacterial communities to sudden vegetation dieback in a coastal wetland. *Phytobiomes* 1, 5–13. doi:10.1094/PBIOMES-09-16-0006-R.

Fahad, S., Hussain, S., Bano, A., Saud, S., Hassan, S. *et al.* (2015) Potential role of phytohormones and plant growth-promoting rhizobacteria in abiotic stresses: Consequences for changing environment. *Environmental Science and Pollution Research* 22, 4907–4921. doi:10.1007/s11356-014-3754-2.

Gamalero, E. and Glick, B.R. (2015) Bacterial modulation of plant ethylene levels. *Plant Physiology* 169, 13–22. doi:10.1104/pp.15.00284.

Gerhardt, K.E., Macneill, G.J., Gerwing, P.D. and Greenberg, B.M. (2017) Phytoremediation of salt-impacted soils and use of Plant Growth-Promoting Rhizobacteria (PGPR) to enhance phytoremediation. In: Ansari, A., Gill, S., Gill, R., Lanza, G.R., Newman, L. (eds) *Phytoremediation*. Springer, Cham, Switzerland.

Glass, D.J. (1999) U.S. and International Markets for Phytoremediation, 1999–2000. Needham, MA, USA.

Glenn, E.P., Brown, J.J. and Blumwald, E. (1999) Salt tolerance and crop potential of halophytes. *Critical Reviews in Plant Sciences* 18, 227–255. doi:10.1080/07352689991309207.

Glick, B.R. (2012) Plant growth-promoting bacteria: Mechanisms and applications. *Scientifica* 2012. doi:10.6064/2012/963401.

Glick, B.R. (2014) Bacteria with ACC deaminase can promote plant growth and help to feed the world. *Microbiological Research* 169, 30–39. doi:10.1016/j.micres.2013.09.009.

Gond, S.K., Torres, M.S., Bergen, M.S., Helsel, Z. and White, J.F. (2015) Induction of salt tolerance and up-regulation of aquaporin genes in tropical corn by rhizobacterium *Pantoea agglomerans*. *Letters in Applied Microbiology* 60, 392–399. doi:10.1111/lam.12385.

Hong, Y., Liao, D., Hu, A., Wang, H., Chen, J. *et al.* (2015) Diversity of endophytic and rhizoplane bacterial communities associated with exotic *Spartina alterniflora* and native mangrove using Illumina amplicon sequencing. *Canadian Journal of Microbiology* 61, 723–733. doi:10.1139/cjm-2015-0079.

Jayaprakashvel, M., Rajan, A. and Hussain, A.J. (2015) Bioprospecting of halotolerant bacteria from the rhizosphere of a halophyte for the production of industrial enzymes. *International Journal of Pure & Applied Bioscience* 3, 193–201.

Kataoka, R., Güneri, E., Turgay, O.C., Yaprak, A.E., Sevilir, B. *et al.* (2017) Sodium-resistant plant growth-promoting rhizobacteria isolated from a halophyte, *Salsola grandis*, in saline-alkaline soils of Turkey. *Eurasian Journal of Soil Science* 6, 216–225. doi:10.18393/ejss.289460.

Kittelson, P.M. and Boyd, M.J. (1997) Mechanisms of expansion for an introduced species of cordgrass, *Spartina densiflora*, in Humboldt Bay, California. *Estuaries* 20, 770. doi:10.2307/1352250.

Kloepper, J.W. and Schroth, M.N. (1978) Plant growth promoting rhizobacteria on radishes. In: *Proceedings of the 4th International Conference in Plant Pathogenic Bacteria.* Station de Pathologie Vegetale et Phytobacteriologie, Angers, France.

Kostka, J.E., Gribsholt, B., Petrie, E., Dalton, D., Skelton, H. *et al.* (2002) The rates and pathways of carbon oxidation in bioturbated saltmarsh sediments. *Limnology and Oceanography* 47, 230–240. doi:10.4319/lo.2002.47.1.0230.

Kuiper, I., Lagendijk, E.L., Bloemberg, G.V. and Lugtenberg, B.J.J. (2004) Rhizoremediation: a beneficial plant-microbe interaction. *Molecular Plant-Microbe Interactions* 17, 6–15. doi:10.1094/MPMI.2004.17.1.6.

Kumar, B.L. and Gopal, D.V.R.S. (2015) Effective role of indigenous microorganisms for sustainable environment. *3 Biotech* 5(6), 867–876. doi:10.1007/s13205-015-0293-6.

Lovell, C.R., Piceno, Y.M., Quattro, J.M. and Bagwell, C.E. (2000) Molecular analysis of diazotroph diversity in the rhizosphere of the smooth cordgrass, *Spartina alterniflora*. *Applied and Environmental Microbiology* 66, 3814–3822. doi:10.1128/AEM.66.9.3814-3822.2000.

Lucena, T., Mesa, J., Rodriguez-Llorente, I.D., Pajuelo, E., Caviedes, M.Á. *et al.* (2016) *Marinomonas spartinae* sp. nov., a novel species with plant-beneficial properties. *International Journal of Systematic and Evolutionary Microbiology* 66, 1686–1691. doi:10.1099/ijsem.0.000929.

Lucena, T., Arahal, D.R., Ruvira, M.A., Navarro-Torre, S., Mesa, J. *et al.* (2017) *Vibrio palustris* sp. nov. and *Vibrio spartinae* sp. nov., two novel members of the Gazogenes clade, isolated from salt-marsh plants (*Arthrocnemum macrostachyum* and *Spartina maritima*). *International Journal of Systematic and Evolutionary Microbiology* 67, 3506–3512. doi:10.1099/ijsem.0.002155.

Ma, Y., Oliveira, R.S., Freitas, H. and Zhang, C. (2016a) Biochemical and molecular mechanisms of plant-microbe-metal interactions: relevance for phytoremediation. *Frontiers in Plant Science* 7, 918. doi:10.3389/fpls.2016.00918.

Ma, Y., Rajkumar, M., Zhang, C. and Freitas, H. (2016b) Beneficial role of bacterial endophytes in heavy metal phytoremediation. *Journal of Environmental Management* 174, 14–25. doi:10.1016/j.jenvman.2016.02.047.

Mackova, M., Dowling, D. and Macek, T. (2006) *Phytoremediation and Rhizoremediation, Focus on Biotechnology*. Springer, Amsterdam, Netherlands.

Mapelli, F., Marasco, R., Rolli, E., Barbato, M., Cherif, H. *et al.* (2013) Potential for plant growth promotion of rhizobacteria associated with *Salicornia* growing in Tunisian hypersaline soils. *BioMed Research International* 2013, 248078. doi:10.1155/2013/248078.

Mateos-Naranjo, E., Mesa, J., Pajuelo, E., Perez-Martin, A., Caviedes, M.A. and Rodríguez-Llorente, I.D. (2015) Deciphering the role of plant growth-promoting rhizobacteria in the tolerance of the invasive cordgrass *Spartina densiflora* to physicochemical properties of salt-marsh soils. *Plant and Soil* 394, 45–55. doi:10.1007/s11104-015-2504-7.

Mendelssohn, I.A. and Morris, J.T. (2002) Eco-physiological controls on the productivity of *Spartina alterniflora* Loisel. In: Weinstein, M.P. and Kreeger, D.A. (eds) *Concepts and Controversies in Tidal Marsh Ecology*. Springer, Dordrecht, Netherlands, pp. 59–80. doi:10.1007/0-306-47534-0_5.

Mesa, J., Mateos-Naranjo, E., Caviedes, M.A., Redondo-Gómez, S., Pajuelo, E. *et al.* (2015a) Scouting contaminated estuaries: Heavy metal resistant and plant growth promoting rhizobacteria in the native metal rhizoaccumulator *Spartina maritima*. *Marine Pollution Bulletin* 90, 150–159. doi:10.1016/j.marpolbul.2014.11.002.

Mesa, J., Rodríguez-Llorente, I.D., Pajuelo, E., Piedras, J.M.B., Caviedes, M.A. *et al.* (2015b) Moving closer towards restoration of contaminated estuaries: bioaugmentation with autochthonous rhizobacteria improves metal rhizoaccumulation in native *Spartina maritima*. *Journal of Hazardous Materials* 300, 263–271. doi:10.1016/j.jhazmat.2015.07.006.

Mesa, J., Mateos-Naranjo, E., Pajuelo, E., Caviedes, M.A. and Rodríguez-Llorente, I.D. (2016) Heavy metal pollution structures soil bacterial community dynamics in SW Spain polluted salt marshes. *Water, Air, & Soil Pollution* 227, 466. doi:10.1007/s11270-016-3176-5.

Mittler, R., Vanderauwera, S., Gollery, M. and Van Breusegem, F. (2004) Reactive oxygen gene network of plants. *Trends in Plant Science* 9, 490–498. doi:10.1016/j.tplants.2004.08.009.

Nair, D.N. and Padmavathy, S. (2014) Impact of endophytic microorganisms on plants, environment and humans. *Scientific World Journal* 2014, 250693. doi:10.1155/2014/250693.

Navarro-Torre, S., Mateos-Naranjo, E., Caviedes, M.A., Pajuelo, E. and Rodríguez-Llorente, I.D. (2016) Isolation of plant-growth-promoting and metal-resistant cultivable bacteria from *Arthrocnemum macrostachyum* in the Odiel marshes with potential use in phytoremediation. *Marine Pollution Bulletin* 110, 133–142. doi:10.1016/j.marpolbul.2016.06.070.

Navarro-Torre, S., Barcia-Piedras, J.M., Caviedes, M.A., Pajuelo, E., Redondo-Gómez, S. *et al.* (2017a) Bioaugmentation with bacteria selected from the microbiome enhances *Arthrocnemum macrostachyum* metal accumulation and tolerance. *Marine Pollution Bulletin* 117, 340–347. doi:10.1016/j.marpolbul.2017.02.008.

Navarro-Torre, S., Barcia-Piedras, J.M., Mateos-Naranjo, E., Redondo-Gómez, S., Camacho, M. *et al.* (2017b) Assessing the role of endophytic bacteria in the halophyte *Arthrocnemum macrostachyum* salt tolerance. *Plant Biology* 19, 249–256. doi:10.1111/plb.12521.

Naz, I. and Bano, A. (2009) Morphological, biochemical and molecular characterization of rhizobia from halophytes of Khewra Salt Range and Attock. *Pakistan Journal of Botany* 41, 3159–3168.

Neilands, J.B. (1995) Siderophores : structure and function of microbial iron transport compounds. *Journal of Biological Chemistry* 270, 26723–26726. doi:10.1074/jbc.270.45.26723.

Niu, S.Q., Li, H.-R., Paré, P.W., Aziz, M., Wang, S.M. *et al.* (2016) Induced growth promotion and higher salt tolerance in the halophyte grass *Puccinellia tenuiflora* by beneficial rhizobacteria. *Plant and Soil* 407, 217–230. doi:10.1007/s11104-015-2767-z.

Olivares, J. (2008) Fijación biológica del nitrógeno. http://www.eez.csic.es/~olivares/ciencia/fijacion/ (accessed 15 January 2018).

Oliveira, V., Gomes, N., Almeida, A., Silva, A.M.S., Silva, H. *et al.* (2015) Microbe-assisted phytoremediation of hydrocarbons in estuarine environments. *Microbial Ecology* 69, 1–12. doi:10.1007/s00248-014-0455-9.

Pajuelo, E., Rodríguez-Llorente, I.D., Lafuente, A.I., Pérez-Palacios, P., Doukkali, B. *et al.* (2014) Engineering the rhizosphere for the purpose of bioremediation: an overview. *CAB Reviews* 9. doi:10.1079/PAVSNNR20140001.

Paredes-Páliz, K.I., Caviedes, M.A., Doukkali, B., Mateos-Naranjo, E., Rodríguez-Llorente, I.D. *et al.* (2016a) Screening beneficial rhizobacteria from *Spartina maritima* for phytoremediation of metal polluted salt marshes: comparison of gram-positive and gram-negative strains. *Environmental Science and Pollution Research* 23, 19825–19837. doi: 10.1007/s11356-016-7184-1.

Paredes-Páliz, K.I., Pajuelo, E., Doukkali, B., Ángel, M., Rodríguez-Llorente, I.D. *et al.* (2016b) Bacterial inoculants for enhanced seed germination of *Spartina densiflora*: Implications for restoration of metal polluted areas. *Marine Pollution Bulletin* 110, 6–10. doi:10.1016/j.marpolbul.2016.06.036.

Paredes-Páliz, K.I., Mateos-Naranjo, E., Doukkali, B., Caviedes, M.A., Redondo-Gómez, S. *et al.* (2017) Modulation of *Spartina densiflora* plant growth and metal accumulation upon selective inoculation treatments: A comparison of gram negative and gram positive rhizobacteria. *Marine Pollution Bulletin* 125, 77–85. doi:10.1016/j.marpolbul.2017.07.072.

Patel, S. (2016) *Salicornia*: evaluating the halophytic extremophile as a food and a pharmaceutical candidate. *3 Biotech* 6, 104. doi:10.1007/s13205-016-0418-6.

Pieterse, C.M.J., Zamioudis, C., Berendsen, R.L., Weller, D.M., Van Wees, S.C.M. *et al.* (2014) Induced systemic resistance by beneficial microbes. *Annual Review of Phytopathology* 52, 347–375. doi:10.1146/annurev-phyto-082712-102340.

Pilon-Smits, E. (2005) Phytoremediation. *Annual Review of Plant Biology* 56, 15–39. doi:10.1146/annurev.arplant.56.032604.144214.

Qadir, M., Noble, A.D., Oster, J.D., Schubert, S. and Ghafoor, A. (2006) Driving forces for sodium removal during phytoremediation of calcareous sodic and saline-sodic soils: a review. *Soil Use and Management* 21, 173–180. doi:10.1111/j.1475-2743.2005.tb00122.x.

Qin, S., Zhou, W., Li, Z. and Lyu, D. (2016) Effects of rhizobacteria on the respiration and growth of *Cerasus sachalinensis* Kom. seedlings. *Spanish Journal of Agricultural Research* 14. doi: 10.5424/sjar/2016142-6848.

Qin, Y., Druzhinina, I.S., Pan, X. and Yuan, Z. (2016) Microbially mediated plant salt tolerance and microbiome-based solutions for saline agriculture. *Biotechnology Advances* 34(7), 1245–1259. doi:10.1016/j.biotechadv.2016.08.005.

Redondo-Gómez, S. (2013) Bioaccumulation of heavy metals in *Spartina*. *Functional Plant Biology* 40(9), 913. doi: 10.1071/FP12271.

Redondo-Gómez, S., Andrades-Moreno, L., Parra, R., Valera-Burgos, J., Real, M. *et al.* (2011) *Spartina densiflora* demonstrates high tolerance to phenanthrene in soil and reduces it concentration. *Marine Pollution Bulletin* 62, 1800–1808. doi:10.1016/j.marpolbul.2011.05.018.

Reginato, M., Sgroy, V., Llanes, A., Cassán, F. and Luna, V. (2012) The American halophyte *Prosopis strombulifera*, a new potential source to confer salt tolerance to crops, In: Ashraf, M., Öztürk, M., Ahmad, M. and Aksoy, A. (eds) *Crop Production for Agricultural Improvement*. Springer, Dordrecht, Netherlands.

Rietl, A.J., Overlander, M.E., Nyman, A.J. and Jackson, C.R. (2016) Microbial community composition and extracellular enzyme activities associated with *Juncus roemerianus* and *Spartina alterniflora* vegetated sediments in Louisiana saltmarshes. *Microbial Ecology* 71, 290–303. doi:10.1007/s00248-015-0651-2.

Rodríguez, H. and Fraga, R. (1999) Phosphate solubilizing bacteria and their role in plant growth promotion. *Biotechnology Advances* 17, 319–339. doi:10.1016/S0734-9750(99)00014-2.

Rueda-Puente, E.O., Prabhaharan, R., Murillo-Amador, B., Ruiz-Espinoza, F., Puente, M. *et al.* (2013) Ameliorative effects of salt resistance on physiological parameters in the halophyte *Salicornia bigelovii*

Torr. with plant growth-promoting rhizobacteria. *African Journal of Biotechnology* 12, 5278–5284. doi:10.5897/AJB12.2750.

Ruppel, S., Franken, P. and Witzel, K. (2013) Properties of the halophyte microbiome and their implications for plant salt tolerance. *Functional Plant Biology* 40, 940–951. doi:10.1071/FP12355.

Sáenz-Mata, J., Palacio-Rodríguez, R., Sánchez-Galván, H. and Balagurusamy, N. (2016) Plant growth promoting rhizobacteria associated to halophytes: potential applications in agriculture. In: Khan, M., Boër, B., Özturk, M., Clüsener-Godt, M., Gul, B. and Breckle, S.W. (eds) *Sabkha Ecosystems. Tasks for Vegetation Science*. Springer, Cham, Switzerland.

Santoyo, G., Moreno-Hagelsieb, G., del Carmen Orozco-Mosqueda, M. and Glick, B.R. (2016) Plant growth-promoting bacterial endophytes. *Microbiological Research* 183, 92–99. doi:10.1016/j.micres.2015.11.008.

Seneviratne, G., Weerasekara, M.L.M.A.W., Seneviratne, K.A.C.N., Zavahir, J.S., Kecskés, M.L. *et al.* (2010) Importance of biofilm formation in plant growth promoting rhizobacterial action. In: Maheshwari, D.K. (ed.) *Plant Growth and Health Promoting Bacteria.* Springer, Berlin.

Sharma, S., Kulkarni, J. and Jha, B. (2016) Halotolerant rhizobacteria promote growth and enhance salinity tolerance in Peanut. *Frontiers in Microbiology* 7, 1600. doi:10.3389/fmicb.2016.01600.

Silliman, B.R., Schrack, E., He, Q., Cope, R., Santoni, *et al.* (2015) Facilitation shifts paradigms and can amplify coastal restoration efforts. *Proceedings of the National Academy of Sciences of the United States of America* 112, 14295– 14300. doi:10.1073/pnas.1515297112.

Suarez, C., Cardinale, M., Ratering, S., Steffens, D., Jung, S. *et al.* (2015) Plant growth-promoting effects of *Hartmannibacter diazotrophicus* on summer barley (*Hordeum vulgare* L.) under salt stress. *Applied Soil Ecology* 95, 23–30. doi:10.1016/j.apsoil.2015.04.017.

Subramanian, S. and Smith, D.L. (2015) Bacteriocins from the rhizosphere microbiome - from an agriculture perspective. *Frontiers in Plant Science* 6, 909. doi:10.3389/fpls.2015.00909.

Syranidou, E., Christofilopoulos, S., Politi, M., Weyens, N., Venieri, D. *et al.* (2017) Bisphenol-A removal by the halophyte *Juncus acutus* in a phytoremediation pilot: characterization and potential role of the endophytic community. *Journal of Hazardous Materials* 323, 350–358. doi:10.1016/j.jhazmat.2016.05.034.

Szymańska, S., Piernik, A., Baum, C., Złoch, M. and Hrynkiewicz, K. (2014) Metabolic profiles of microorganisms associated with the halophyte *Salicornia europaea* in soils with different levels of salinity. *Écoscience* 21, 114–122. doi:10.2980/21-2-3705.

Thakur, S., Singh, L., Wahid, Z.A., Siddiqui, M.F., Atnaw, S.M. *et al.* (2016) Plant-driven removal of heavy metals from soil: uptake, translocation, tolerance mechanism, challenges, and future perspectives. *Environmental Monitoring and Assessment* 188. doi:10.1007/s10661-016-5211-9.

Thijs, S., Sillen, W., Weyens, N. and Vangronsveld, J. (2017) Phytoremediation: state-of-the-art and a key role for the plant microbiome in future trends and research prospects. *International Journal of Phytoremediation* 19, 23–38. doi:10.1080/15226514.2016.1216076.

Thomas, F., Giblin, A.E., Cardon, Z.G. and Sievert, S.M. (2014) Rhizosphere heterogeneity shapes abundance and activity of sulfur-oxidizing bacteria in vegetated salt marsh sediments. *Frontiers in Microbiology* 5, 309. doi:10.3389/fmicb.2014.00309.

Ullah, S. and Bano, A. (2015) Isolation of plant-growth-promoting rhizobacteria from rhizospheric soil of halophytes and their impact on maize (*Zea mays* L.) under induced soil salinity. *Canadian Journal of Microbiology* 61, 307–313. doi:10.1139/cjm-2014-0668.

Vurukonda, S.S.K.P., Vardharajula, S., Shrivastava, M. and SkZ, A. (2016) Enhancement of drought stress tolerance in crops by plant growth promoting rhizobacteria. *Microbiological Research* 184, 13–24. doi:10.1016/j.micres.2015.12.003.

Wang, Y. and Dai, C.C. (2010) Endophytes: a potential resource for biosynthesis, biotransformation, and biodegradation. *Annals of Microbiology* 61, 207–215. doi:10.1007/s13213-010-0120-6.

Weiss, J.V., Emerson, D., Backer, S.M. and Megonigal, J.P. (2003) Enumeration of Fe(II)-oxidizing and Fe(III)-reducing bacteria in the root zone of wetland plants: Implications for a rhizosphere iron cycle. *Biogeochemistry* 64, 77–96. doi:10.1023/A:1024953027726.

Werner, T., Motyka, V., Strnad, M. and Schmülling, T. (2001) Regulation of plant growth by cytokinin. *Proceedings of the National Academy of Sciences of the United States of America* 98, 10487– 10492. doi:10.1073/pnas.171304098.

Yuan, Z., Druzhinina, I.S., Labbé, J., Redman, R., Qin, Y. *et al.* (2016) Specialized microbiome of a halophyte and its role in helping non-host plants to withstand salinity. *Nature Scientific Reports* 6, 32467. doi:10.1038/srep32467.

Zhang, H., Kim, M.S., Sun, Y., Dowd, S.E., Shi, H. *et al.* (2008) Soil bacteria confer plant salt tolerance by tissue-specific regulation of the sodium transporter HKT1. *Molecular Plant-Microbe Interactions* 21, 737–744. doi:10.1094/MPMI-21-6-0737.

Zhao, S., Zhou, N., Wang, L. and Tian, C.Y. (2013) Halophyte-endophyte coupling: a promising bioremediation system for oil-contaminated soil in northwest China. *Environmental Science & Technology* 47, 11938–11939. doi:10.1021/es404219j.

Zhou, N., Zhao, S. and Tian, C.Y. (2017) Effect of halotolerant rhizobacteria isolated from halophytes on the growth of sugar beet (*Beta vulgaris* L.) under salt stress. *FEMS Microbiology Letters* 364, 574–581. doi:10.1093/femsle/fnx091.

16 Arsenic Tolerance Mechanisms in Halophytes: the Case of *Tamarix gallica*

Dhouha Belhadj Sghaier[1,*], Sílvia Pedro[2], Bernardo Duarte[2], Isabel Caçador[2] and Noomene Sleimi[1]

[1]*University of Carthage, Bizerte, Tunisia;* [2]*University of Lisbon, Lisbon, Portugal*

Abstract

Toxic compounds in the ecosphere are the consequence of environmental pollution, and have a disruptive influence in the environment. They affect ecosystems, entering food chains and ultimately touching human health. Metal consumption has increased by 300% in the last 50 years and the anthropogenic release of metallic elements such as Pb, Hg, As, Cd, Al and Cr has increased since the beginning of the industrial era. Different strategies have been identified to overcome metallic stress. Knowledge of plant stress responses and adaptations at physiological, biochemical and cellular levels is a priority in understanding the impact of these constraints on plant biodiversity. These adaptations have evolved naturally in halophytes as responses to their colonization of saline ecosystems, and therefore make halophytes good model plants. In this chapter we discuss the biophysical mechanisms underlying energy capture and transduction in halophytes and their relation to pigment profile alteration, compartmentation and subcellular localization, to devise sustainable strategies for environmental or ecosystem management and safety.

Keywords: Abiotic stress; Halophytes; Metalloids; Phytoremediation; Reactive oxygen species

16.1 Introduction

Increasing industrial and agricultural development induces the accumulation of trace metals elements, and therefore soil contamination, which becomes a serious environmental problem (Naz *et al.*, 2015). They have deleterious effects on plants, and induce directly or indirectly a potential health risk to animals and also to humans, by integrating into the food chain (Miransari, 2011).

We are treating Arsenic (As) and its toxicity as an element which can be detected in all environmental compartments. In soil, the inorganic forms of As, oxidized arsenate (As^{5+}) and reduced arsenite (As^{3+}), are generally more toxic than the organic form (Mateos-Naranjo *et al.*, 2012). Inorganic As has been known to cause growth inhibition and membrane disintegration (Moreno-Jiménez *et al.*, 2012). In soil, the As-inorganic form is generally more harmful than the As-organic form (Mateos-Naranjo *et al.*, 2012). Arsenic toxicity is induced by oxidative stress through generation of reactive oxygen species (ROS) and the inhibition of antioxidant defences (Mateos-Naranjo *et al.*, 2012).

Phytoremediation is considered to be one of the most important cleaning solutions, using

* Corresponding author e-mail: dhouhasghaier@hotmail.fr

plants to remove trace metals elements from the soil and concentrating them in the harvested parts (Pulford and Watson, 2003). In fact, the selection of suitable plants for restoration is one of the key factors in the phytoremediation of sites contaminated by trace metals elements (Duarte *et al.*, 2016).

The tolerance of plants to environmental constraints, and more particularly to pollution by trace metals elements, is a promising criterion to be taken into account in the treatment of polluted areas (Singh *et al.*, 2016). Understanding the mechanisms of trace metals element toxicity and the tolerant species that can be used in the depollution processes is important. Thus, a study of the different responses of plants to metallic pollution can allow us to select an effective phytorehabilitation technique (Krämer, 2010).

In recent decades, several studies have been published that identify plants tolerant to trace metals elements (Reeves and Baker, 2000; Reboreda and Caçador, 2007). Halophyte species living in saline coastal environments and in land deserts are known for their ability to survive in extreme environmental conditions (Lokhande and Suprasanna, 2012; Flowers and Colmer, 2015). Moreover, halophytes have adapted naturally to the environmental stresses, including the trace metal elements (Anjum *et al.*, 2014). Wang *et al.* (2013) reported that the high resistance of halophytes to trace metals elements is strongly associated with their salt-tolerance characteristics. Given the large amount of biomass produced by some halophytic plant species and their deep-rooting systems, perennial halophytes could be putative candidates for phytoremediation (Hasanuzzaman *et al.*, 2014; Sruthi *et al.*, 2016).

Among the halophyte flora, the genus *Tamarix* is considered a relatively long-living plant resistant to a wide range of environmental conditions, and it can tolerate abiotic stresses such as salt, high temperature and drought stresses (Saïdana *et al.*, 2008). In addition, owing to the presence of salts glands or trichomes, there may be special interest in *Tamarix*. In Tunisia, for instance, *Tamarix gallica* occurs in coastal, desert regions and also in some saline depressions, which are usually used as sites for the accumulation of industrial and urban effluents contaminated by metals (Ghnaya *et al.*, 2007). To survive in these sites, and depending on their tolerance capacity, the plants may have the mechanisms to regulate internal and cell-wall metal concentrations that determine their survival (Sousa *et al.*, 2008). Understanding the mechanisms involved in As tolerance and accumulation will enable better use of this species in phytoremediation.

16.2 Accumulation of Metals

After exposure to different concentrations of As, *T. gallica* accumulated 92 and 94 μg/g^{-1} As in shoots and roots, respectively (Fig. 16.1). The As uptake showed an increase with elevation of the As concentrations in the medium. Reeves and Baker (2000) and Van der Ent *et al.* (2013) defined a plant hyperaccumulator as one that accumulates at least 1000 μg g^{-1} of Co, Cu, Cr, Pb, As or Ni, or 10,000 μg g^{-1} of Mn or Zn in the dry matter.

Taken together, *T. gallica* could not be considered as an accumulator, despite the translocation factor (TF) from root to shoot being almost 1. The high As TF demonstrated that *T. gallica* behaved as an As-tolerant plant. The presence of salt glands may play a role in the high TF value (Sruthi *et al.*, 2016), and this TF value may also be used as an indicator of As phytoextraction potential (Campos *et al.*, 2014). Li *et al.* (2016) reported that accumulator plants store high concentrations of toxic metals in the shoots. In tolerant plants, the root system may be considered the dominant absorption path for metals, with only small amounts transported to the aerial parts (Caçador *et al.*, 2009). In fact, the limitation of translocation can be explained by: (1) blocking trace metals element transport at the epidermal cells (Solis-Dominguez *et al.*, 2007) and immobilizing them in the compartment by complexing with organic acids such as malate, citrate and oxalate (Lutts *et al.*, 2004); or (2) complexation by phytochelatins or metallothioneins followed by sequestration in the vacuolar system (Sousa *et al.*, 2008).

Knowing *T. gallica* as a salt cedar plant, their glandular structures are not specific to Na^+, as described by Sookbirsingh *et al.* (2010), and have the ability to uptake high concentrations of ions and concentrate them in their shoots to be excreted at a later stage. Manousaki

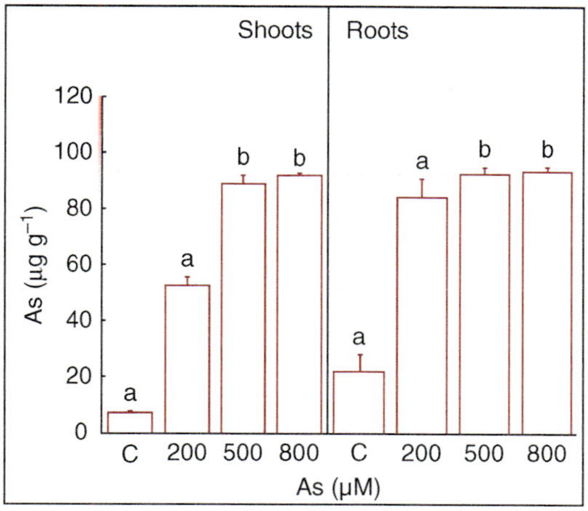

Fig. 16.1. Arsenic (As) uptake in *Tamarix gallica*. Statistical significance at $p < 0.05$ between control and treated plants. Bars marked with the same letter are not significantly different at $p = 0.05$. The concept of this figure is based on Sghaier *et al.* (2015).

and Kalogerakis (2011) reported that other toxic metals (e.g. Cd, Zn, Pb and Cu) are accumulated and excreted by salt glands or trichomes on the surface of the leaves. According to Redondo-Gómez (2013) the *Spartina* genus has been shown to excrete the trace metals elements in salt crystals. Similarly, other studies on *T. smyrnensis* and *T. aphylla* showed that their salt glands accumulate and excrete Cd and Pb on the leaf surface, suggesting that they use a salt excretion mechanism to excrete excess metals on their leaf surface as a detoxification mechanism (Lefèvre *et al.*, 2009).

16.3 Morphological Adaptations

Exposure to trace metals elements can cause physiological disorders, the intensity depending on the level of accumulation of the metallic ion in the tissue and on the plant species (Chibuike and Obiora, 2014). In fact, plants growing on polluted soils have developed the ability to accumulate high amounts of metals in their tissues without deposition toxicity symptoms (Baker *et al.*, 1991). Under As stress, *T. gallica* demonstrated a decline in biomass (dry shoots and roots) (Fig. 16.2). This variation was related to the As level. In fact, plants treated with higher As concentrations presented an important loss in dry weight compared to the control (Fig. 16.2). *Spartina densiflora* treated with arsenic has shown a drop in growth parameters (Mateos-Naranjo *et al.*, 2012). The same result was observed in *Atriplex atacamensis* after exposure to As (Vromman *et al.*, 2017).

Nevertheless, *T. gallica* did not show clear phytotoxicity symptoms at leaf concentrations at As stress, and nor were significant alterations observed in chlorophyll *a* (Chl *a*) and chlorophyll *b* (Chl *b*) contents ($p > 0.05$). In the same way, the levels of total chlorophylls and carotenoids did not demonstrate any variation in their concentrations, independently of the As concentration exposure (Fig. 16.2).

Exposure to different concentrations of trace metals elements generally affected plant growth and changed various aspects of plant physiology (Guala *et al.*, 2010). Chlorophyll quantification can be used to determine the physiological state of a plant. Previous studies have found that pigment contents were the main target of the metal toxicity (Adrees *et al.*, 2015) and had direct or indirect consequences on chlorophyll content (Duarte *et al.*, 2013; Santos *et al.*, 2015) through metabolic disturbances (Santos *et al.*, 2014) or inhibition of chlorophyll biosynthesis enzymes (Padmaja *et al.*, 1990). Indeed, the biosynthesis of chlorophyll implies NADPH protochlorophyllide oxidoreductase (POR) enzyme in the formation of Chl *a* from the protochlorophyllide (Mysliwa-Kurdziel *et al.*, 2002). This essential process for the formation of chlorophyll has been shown to be very sensitive to the effects of trace metals elements. Cr and Cd may inhibit POR activity and thus decrease the accumulation of Chl *a* in the cell (Mysliwa-Kurdziel and Strzalka, 2002). The substitution of Mg by Zn,

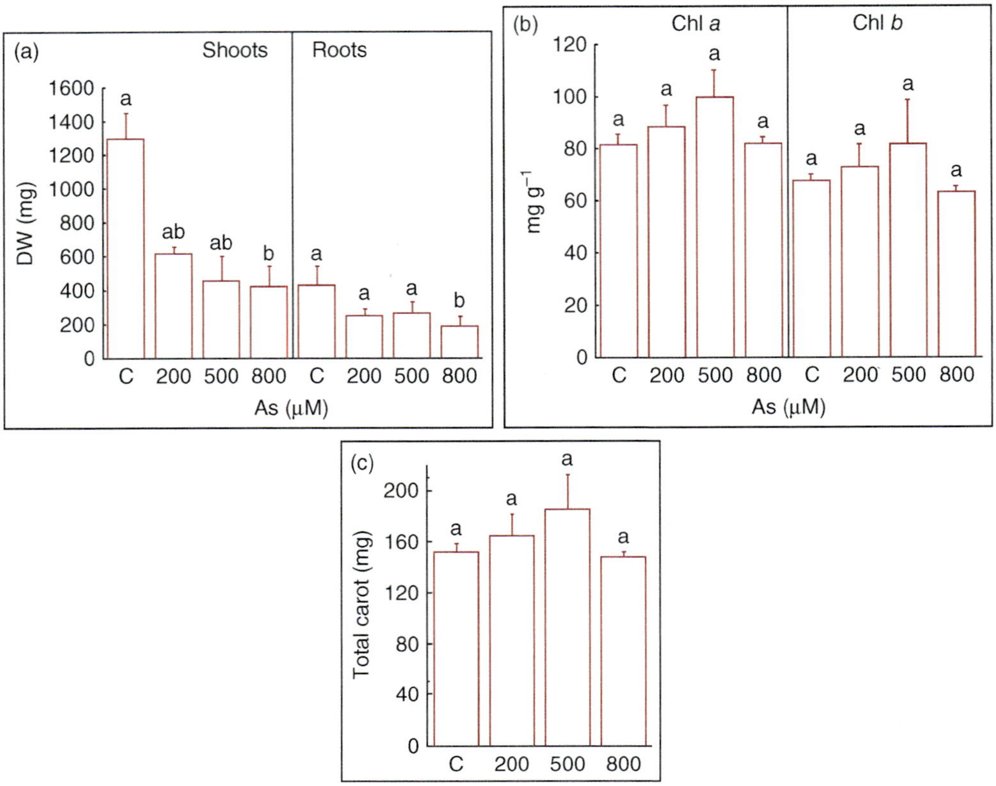

Fig. 16.2. Dry weight of shoots and roots, chlorophyll *a* and *b* and total carotenoid (carot) content of *Tamarix gallica* after exposure to different concentrations of arsenic (As). The concept of this figure is based on Sghaier *et al.* (2015).

Cu and Cd induces damage in the chlorophyll synthesis system, allowing the accumulation of degradation products such as phaeophythins (Santos *et al.*, 2015). *Spartina densiflora* exposed to a high level of As toxicity showed degradation and an alteration in chlorophyll content (Mateos-Naranjo *et al.*, 2012). Further, rising of metal content of leaves might induce damage to chloroplast functioning (Vitorello *et al.*, 2005). For example, a decrease in β-carotene or xanthophyll content can induce a perturbation in photosynthetic activity (Duarte *et al.*, 2014). In contrast, other studies demonstrated the capacity of plants to support metal toxicity; for example, *Atriplex atacamensis* presented a similar trend compared to control plants (Vromman *et al.*, 2017), and *Erica andevalensis* (Márquez-García and Córdoba, 2010) and *Borreria verticillata* (Campos *et al.*, 2014) did not present visual symptoms of toxicity.

16.4 Biophysical Constraints

The photosynthetic apparatus, and especially photosystem II (PSII), is known to be affected by various environmental stresses (Force *et al.*, 2003). The energy models in PSII may also be used as tools to understand environmental adaptation (Endo *et al.*, 2014). In fact, the photochemical efficiency of the PSII value (Fv/Fm) can be used as an indicator for plant tolerance to environmental factors (Sekmen *et al.*, 2007). Before any irreversible structural damage occurs, changes in PSII photochemical efficiency appear (Force *et al.*, 2003). The decline in PSII quantum yields is probably owing to a drop in basal fluorescence, F_0, due to changes in the thylakoid structure on the PSII electron donor sites. This affects electron transport and/or leads to the destruction of chlorophyll pigment (Duarte *et al.*, 2014).

In fact, it is well established that excess accumulation of trace metals elements leads to pigment destruction, reduction in ferrodoxin NADP+ oxidoreductase activity and arrests plastoquinone synthesis, as well as perturbations during the reception of energy or to the transduction (and as a result distribution) in maximum PSII photochemistry (Duarte *et al.*, 2014; Anjum *et al.*, 2016).

Arsenic exposure did not lead to a significant alteration in PSII efficiencies, as reported above for the photosynthetic pigments. *T. gallica* exposed to various concentrations of As showed that there were no changes in F_0, Fm and Fv/Fm, even at high concentrations (Fig. 16.3). Moreover, this can indicate that there were no perturbations in energy transfer from the LHCII, thereby maintaining the chlorophyll content (Gusman *et al.*, 2013). These results pointed out the maintenance of PSII function and the photosynthetic activity which could appear as an important strategy for *T. gallica* developing in contaminated soil. On the other hand, the rapid transient curves, and the absorbed, trapped and dissipated energy fluxes did not reveal any changes (Fig. 16.3).

Similarly, no different behaviour was observed in the Kautsky curves derived from the OJIP transient analysis (Fig. 16.4). This is surprising

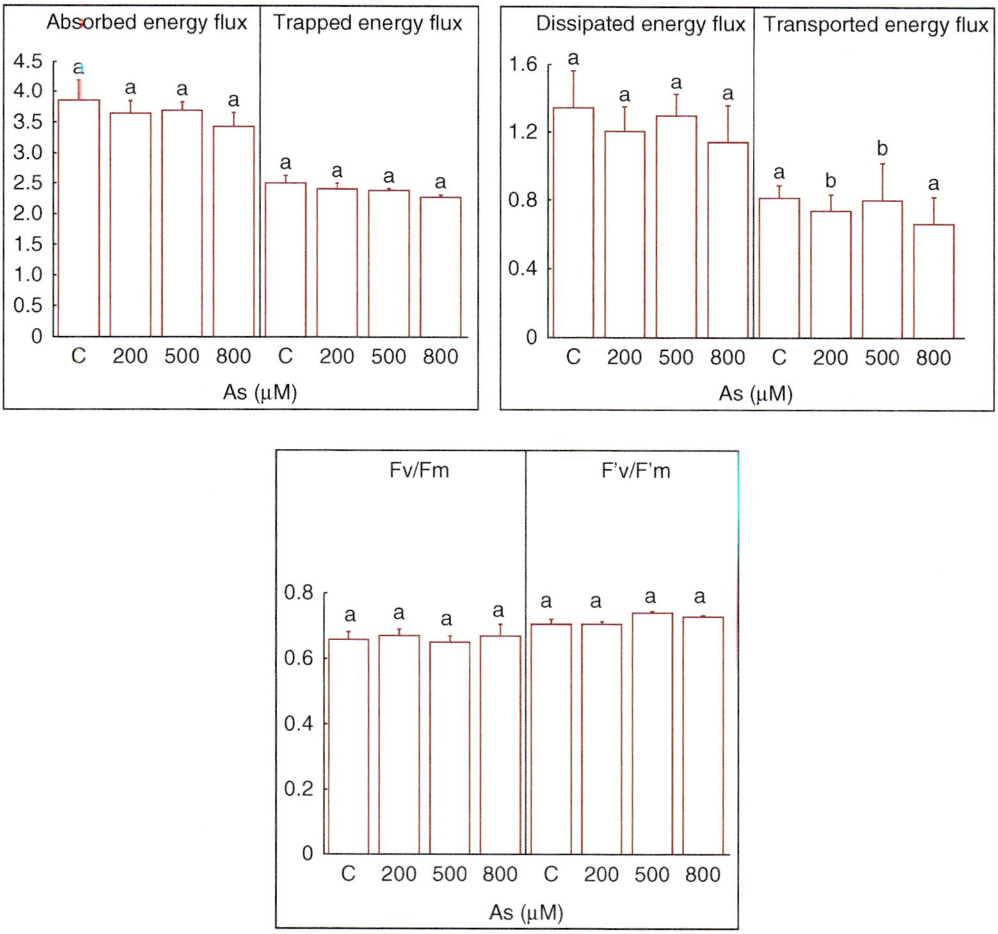

Fig. 16.3. Energy fluxes and quantum yield in dark and light state Fv/Fm and F'v/F'm in *Tamarix gallica* under As stress. Statistical significance at $p < 0.05$ between control and treated plants. Bars marked with the same letter are not significantly different at $p = 0.05$. The concept of this figure is based on Sghaier *et al.* (2015).

since it is commonly considered that plants exposed to ion toxicities have their photosynthesis activity limited as a result (Megdiche *et al.*, 2008). Thus limitations in photosynthesis may result in exposure to excess energy, with inevitable consequences for PSII if the dissipation mechanisms are insufficient (Lu *et al.*, 2003; Caçador and Duarte, 2015).

In the presence of trace metals elements, the transport of electrons via PSII and photosystem I (PSI) and the CO_2 fixation process are reduced. Under these conditions, to protect the photosynthetic apparatus, a rise in the dissipation of energy in the form of heat by the stimulation of the xanthophyll cycle will be established (Hendrickson *et al.*, 2003; Heber *et al.*, 2006). Further, taking into account the xanthophyll concentration, another evident signal of environmental stress appears: xanthophyll cycle functioning, as illustrated by the de-epoxidation state (DES) index (Fig. 16.4) (Demmig-Adams and Adams, 1996). The non-photochemical dissipation of energy regulated by the xanthophyll cycle allows the synthesis of zeaxanthin (Anjum *et al.*, 2016).

Violaxanthin was reversibly converted to zeaxanthin through the intermediate pigment antheraxanthin (Demmig-Adams and Adams, 1996). This regulating role of the xanthophyll cycle depends on the transthylacoidal proton gradient, which is closely linked to the oxidation–reduction state of PSII–PSI electron transport. Zeaxanthin may also be found in a free form in thylakoid membranes, where it acts as an antioxidant, protecting membranes against ROS-induced lipid peroxidation (Moronsito *et al.*, 2003). For example, when *Jungermannia exsertifolia* subsp. *cordifolia* was treated with Cd, the ratio of zeaxanthin to violaxanthin and antheroxanthin was enhanced (Otero *et al.*, 2006).

Another possible way to counteract excessive energy accumulation is through heat

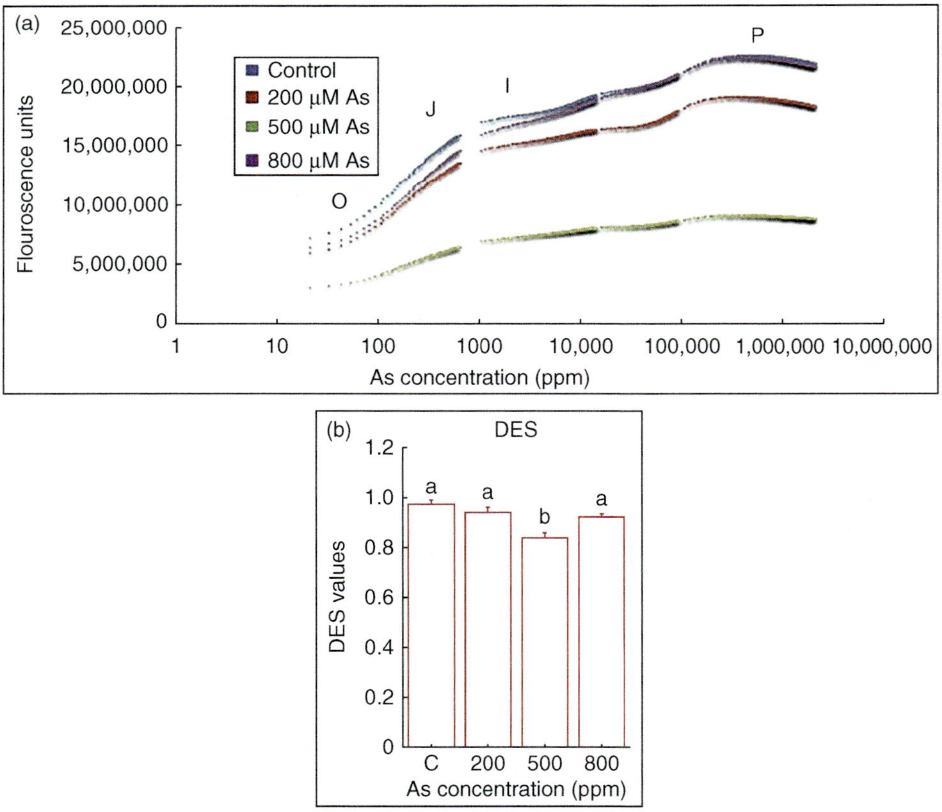

Fig. 16.4. (a) Kautsky curves in dark-adapted leaves; and (b) De-epoxidation state values of *Tamarix gallica* after exposure to different concentrations of arsenic (from Sghaier *et al.* 2015).

dissipation (Duarte *et al.*, 2013). Stressed plants dissipate more energy in order to overcome the accumulation of the minimizing power of the excessive ions and inhibit the photo-destruction of the photosynthetic apparatus (Duarte *et al.*, 2014). The dissipation of this excessive energy in PSII as heat, or in non-damaging chemical reactions, represents an adaptive mechanism to deal with stress, rather than merely a negative consequence of it (Lu *et al.*, 2003). Konrad *et al.* (2005) suggested that it is an internal protection mechanism for the photosynthetic apparatus. Indeed, in *T. gallica* treated with high concentrations of As, different chlorophyll fluorescence parameters were not affected. There was only an increase in the energy dissipation fluxes, suggesting a suitable mechanism adopted by *T. gallica* to cope with an excess of trace metals elements (Sghaier *et al.*, 2015).

16.5 Compartmentation and Subcellular Localization

The exposure of plants to trace metals elements induced an alteration in cellular mechanisms (Choppala *et al.*, 2014) and gene expression (Chafai and Koyama, 2011; Hossain *et al.*, 2012). The mechanisms used to alleviate metal toxicity involved chelation by ligands and/or sequestration into the vacuole or the cell wall. Subcellular compartmentalization of metals was considered to be a potential key element in the removal of trace metals elements from the active metabolic sites (Sousa *et al.*, 2008), restricting them to a limited area to reduce their toxicity.

In fact, *T. gallica* accumulated As in the leaves preferentially in the polysaccharide fraction of the cell wall (more than 70%); a small amount was accumulated in the proteic fraction and also in the intracellular central part of cells. In the roots, the metal ion was distributed at similar concentrations in the different polysaccharide, aqueous and protein fractions.

It has been shown that a large proportion of As had been linked to cell wall components in the leaves, which seem to function as the first barrier protecting the plant from toxicity (Sghaier *et al.*, 2016). It is possible that As forms a strong bond to cell wall components such as cellulose, pectin and lignin (Koch *et al.*, 1999). The cell wall is rich in pectic and histidyl groups and plays a key role in the immobilization of metal ions, which alleviates metal toxicity to plant cells. It provides a large number of metal-binding sites (DalCorso *et al.*, 2013). In the leaves of *Halimione portulacoides*, more than 50% of metals were accumulated by cell wall polymers (Sousa *et al.*, 2008). Reboredo (2012) supported the view that the preferential Zn binding sites were carbohydrates of the cell walls (cellulose, hemicellulose and pectins). The trapping of As in the cell walls could be considered a very important mechanism responsible for the detoxification of this ion (Sghaier *et al.*, 2016). In addition, the cell wall has been shown to represent a central storage site to other metals like Cd (Carrier *et al.*, 2003) and it is the first barrier to block cell penetration (Xiong *et al.*, 2010). Thus, Cd deposition in the cell wall is considered a crucial mechanism of Cd tolerance in this plant species. In *Lactuca sativa* leaves, 64% of the total Cd was bound to cell walls (Ramos *et al.*, 2002) and a similar proportion of Cd associated with the cell wall fraction was reported in *Lupinus albus* (Zornoza *et al.*, 2002).

The compartmental pattern appears to be closely related to the species studied and the metals in question. Besides the cell wall, the vacuole appeared to be another preferential site for Ni, Cd and AS (Psaras *et al.*, 2000; Küpper *et al.*, 2002).

Several studies have revealed that As is located in the intracellular, central part of cells. Since most of this area is occupied by the vacuole, this suggests that As is predominantly contained in vacuoles (Lombi *et al.*, 2002). The protein fraction has also been shown to be of great importance in the compartmentalization of As in the roots. Transport and storage in the vacuole requires increasing levels of sulfur-rich peptides, including phytochelatins and organic acids (di Toppi and Gabbrielli, 1999). The inactivation of toxic metal ions by the synthesis of phytochelatins, followed by the formation of metal–phytochelatin complexes, is a general homeostasis mechanism in plants (Zenk, 1996). Depending on the acidic conditions that confer the stability of these complexes, As– phytochelatin complexes could be stored in vacuoles (Meharg and Hartley-Whitaker, 2002). GSH may also detoxify toxic metals by forming gluthation (GSH)–HM complexes and sequestering in vacuoles, which might be excluded from the apoplast

(Krämer, 2010). Metallothiones are other low-molecular-weight chelating protein molecules enriched in cysteine, responsible for complex formation with toxic metals (MT–HM complexes) (Anjum *et al.*, 2015).

16.6 Conclusions

This study may provide a new insight into the arsenic remediation by *T. gallica* as a plant that could become a potential candidate for phytoextraction. Phytoexcretion has great potential for the removal of some trace metals elements from contaminated soils. Halophytes are expected to be able to cope with multiple stress factors. Numerous physiological targets directly affected by trace metals elements are also affected by salinity, and it may be hypothesized that plants naturally resistant to salinity may exhibit promising adaptive properties to sustain the external presence or even accumulation of trace metals elements in their tissues. Data concerning halophytes' tolerance mechanisms to the trace metals elements were well established, however, fewer studies had been done using *T. gallica*. Furthermore, *T. gallica* shows different responses to As compared to other species, which are probably correlated with plant partitioning. The high capacity to sequester As and its efficient system for maintaining ROS at low levels allows the accumulation of As.

References

Adrees, M., Ali, S., Rizwan, M., Zia-ur-Rehman, M., Ibrahim, M., *et al.* (2015) Mechanisms of silicon-mediated alleviation of heavy metal toxicity in plants: A review. *Ecotoxicology and Environmental Safety* 119, 186–197.

Anjum, N.A., Gill, S.S., Gill, R., Hasanuzzaman, M., Duarte, A.C., *et al.* (2014) Metal/metalloid stress tolerance in plants: role of ascorbate, its redox couple, and associated enzymes. *Protoplasma* 25, 1265–1283.

Anjum, N.A., Hasanuzzaman, M., Hossain, M.A., Thangavel, P., Roychoudhury, A., *et al.* (2015) Jacks of metal/metalloid chelation trade in plants - an overview. *Frontiers in Plant Sciences* 6, 192. doi: 10.3389/fpls.2015.00192.

Anjum, N.A., Sharma, P., Gill, S.S., Hasanuzzaman, M., Khan, E.A., *et al.* (2016) Catalase and ascorbate peroxidase—representative H_2O_2-detoxifying heme enzymes in plants. *Environmental Science and Pollution Research* 23, 19002–19029.

Baker, N.R. (1991) A possible role for photosystem II in environmental perturbations of photosynthesis. *Physiologia Plantarum* 81, 563–570.

Caçador, I. and Duarte, B. (2015) Chromium phyto-transformation in salt marshes: the role of halophytes. In: Ansari, A., Gill, S., Gill, R., Lanza, G. and Newman, L. (eds) *Phytoremediation*. Springer, Cham, Switzerland, pp. 211–217. Available at https://doi.org/10.1007/978-3-319-10969-5_18.

Caçador, I., Caetano, M., Duarte, B. and Vale, C. (2009) Stock and losses of trace metals from salt marsh plants. *Marine Environmental Research* 67, 75–82.

Campos, N.V., Loureiro, M.E. and Azevedo, A.A. (2014) Differences in phosphorus translocation contributes to differential arsenic tolerance between plants of *Borreria verticillata* (Rubiaceae) from mine and non-mine sites. *Environmental Science and Pollution Research* 21, 5586–5596.

Carrier, P., Baryla, A. and Havaux, M. (2003) Cadmium distribution and microlocalization in oilseed rape (*Brassica napus*) after long term growth on cadmium contaminated soil. *Planta* 216, 939–950.

Chaffai, R. and Koyama, H. (2011) Heavy metal tolerance in *Arabidopsis thaliana*. *Advances in Botanical Research* 60, 1– 49.

Chibuike, G.U. and Obiora, S.C. (2014) Heavy metal polluted soils: Effect on plants and bioremediation methods. *Applied and Environmental Soil Science* 2014, 1–12.

Choppala, G., Saifullah, Bolan, N., Bibi, S., Iqbal, M., *et al.* (2014) Cellular mechanisms in higher plants governing tolerance to cadmium toxicity. *Critical Reviews in Plant Sciences* 33, 374–391.

DalCorso, G., Manara, A. and Furini, A. (2013) An overview of heavy metal challenge in plants: from roots to shoots. *Metallomics* 5, 1117–1132.

Demmig-Adams, B. and Adams, W.W. (1996) The role of xanthophyll cycle carotenoids in the protection of photosynthesis. *Trends in Plant Sciences* 1, 21–26.

di Toppi, S.L. and Gabbrielli, R. (1999) Response to cadmium in higher plants. *Environmental and Experimental Botany* 41, 105–130.

Duarte, B., Santos, D., Marquez, J.C. and Caçador, I. (2013) Ecophysiological adaptations of two halophytes to salt stress: photosynthesis, PS II photochemistry and anti-oxidant feedback: implications for resilience in climate change. *Plant Physiology and Biochemistry* 67, 178–188.

Duarte, B., Sleimi, N. and Caçador, I. (2014) Biophysical and biochemical constraints imposed by salt stress: learning from halophytes. *Frontiers in Plant Science* 5. doi: 10.3389/fpls.2014.00746.

Duarte, B., Marques, J.C. and Caçador, I. (2016) Ecophysiological responses of native and invasive *Spartina* species to extreme temperature events in Mediterranean marshes. *Biological Invasions* 18, 2189–2205.

Endo, T., Uebayashi, N., Ishida, S., Ikeuchi, M. and Sato, F. (2014) Light energy allocation at PSII under field light conditions: how much energy is lost in NPQ-associated dissipation? *Plant Physiology and Biochemistry* 81, 115–120.

Flowers, T.J. and Colmer, T.D. (2015) Plant salt tolerance: adaptations in halophytes. *Annals of Botany* 115, 327–331.

Force, L., Critchley, C. and Van Rensen, J.J.S. (2003) New fluorescence parameters for monitoring photosynthesis in plants. *Photosynthesis Research* 78, 17–33.

Ghnaya, T., Slama, I., Messedi, D., Grignon, C., Ghorbel, M.H. *et al.* (2007) Cd-induced growth reduction in the halophyte *Sesuvium portulacastrum* is significantly improved by NaCl. *Journal of Plant Research* 120, 309–316.

Guala, S.D., Vega, F.A. and Covelo, E.F. (2010) The dynamics of heavy metals in plant-soil interactions. *Ecological Modelling* 221, 1148–1152.

Gusman, G.S., Oliveira, J.A., Farnese, F.S. and Cambraia, J. (2013) Arsenate and arsenite: the toxic effects on photosynthesis and growth of lettuce plants. *Acta Physiologiae Plantarum* 35, 1201–1209.

Hasanuzzaman, M., Nahar, K., Alam, M.M., Bhowmik, P.C., Hossain, M.A., *et al.* (2014) Potential use of halophytes to remediate saline soils. *BioMed Research International* 1–12. doi: 10.1155/2014/589341.

Heber, U., Bilger, W. and Shuvalov, V.A. (2006) Thermal energy dissipation in reaction centres and in the antenna of photosystem II protects desiccated poikilohydric mosses against photo-oxidation. *Journal of Experimental Botany* 57, 2993–3006.

Hendrickson, L., Bail, M.C., Osmond, C.B., Furbank, R.T. and Chow, W.S. (2003) Assessment of photoprotection mechanisms of grapevines at low temperature. *Functional Plant Biology* 30, 631–642.

Hossain, M.A., Piyatida, P., Teixeira da Silva, J.A. and Fujita, M. (2012) Molecular mechanism of heavy metal toxicity and tolerance in plants, central role of glutathione in detoxification of reactive oxygen species and methylglyoxal and in heavy metal chelation. *Journal of Botany*, 872875. doi: 10.1155/2012/872875.

Koch, I., Feldmann, J., Wang, L.X., Andrewes, P., Reimer, K.J., *et al.* (1999) Arsenic in the meager creek hot springs environment, British Columbia, Canada. *Science of the Total Environment* 236, 101–117.

Konrad M.L.F., Da Silva, J.A.B., Furlani, P.R. and Machado, E.C. (2005) Gas exchange and chlorophyll fluorescence in six coffee cultivars under aluminum stress. *Bragantia* 64, 339–347.

Krämer, U. (2010) Metal hyperaccumulation in plants. *Annual Review of Plant Biology* 61, 517–534.

Küpper, H., Lombi, E., Zhao, F., Wieshammer, G. and McGrath, S.P. (2002) Cellular compartmentation of nickel in the hyperaccumulators *Alyssum lesbiacum*, *Alyssum bertolonii* and *Thlaspi goesingense*. *Journal of Experimental Botany* 52, 2291–2300.

Lefèvre, I., Marchal, G., Meerts, P., Corréal, E. and Lutts, S. (2009) Chloride salinity reduces cadmium accumulation by the Mediterranean halophyte species *Atriplex halimus* L. *Environmental and Experimental Botany* 65, 142–152.

Li, N., Wang, J. and Song, W.Y. (2016) Arsenic uptake and translocation in plants. *Plant Cell Physiology* 57, 4–13.

Lokhande, V.H. and Suprasanna, P. (2012) Prospects of halophytes in understanding and managing abiotic stress tolerance. In: Ahmad, P. and Prasad, M.N.V. (eds) *Environmental Adaptations and Stress Tolerance of Plants in the Era of Climate Change*. Springer, New York, pp. 29–56.

Lombi, E., Zhao, F.J., Fuhrmann, M., Ma, L.Q. and McGrath, S.P. (2002) Arsenic distribution and speciation in the fronds of the hyperaccumulator *Pteris vittata*. *New Phytologist* 156, 195–203.

Lu, C., Qiu, N., Lu, Q. and Marcelis, L.F.M. (2003) Photoinhibition and the xanthophyll cycle are not enhanced in the salt-acclimated halophyte *Artemisia anethifolia*. *Physiologia Plantarum* 118, 532–537.

Lutts, S., Lefèvre, I., Delpérée, C., Kivits, S., Dechamps, C., *et al.* (2004) Heavy metal accumulation by the halophyte species Mediterranean saltbush. *Journal of Environmental Quality* 33, 1271–1279.

Manousaki, E. and Kalogerakis, N. (2011) Halophytes present new opportunities in phytoremediation of heavy metals and saline soils. *Industrial & Engineering Chemistry Research* 50, 656–660.

Márquez-García, B. and Córdoba, F. (2010) Antioxidative system in wild populations of *Erica andevalensis*. *Environmental and Experimental Botany* 68, 58–65.

Mateos-Naranjo, E., Andrades-Moreno, L. and Redondo-Gómez, S. (2012) Tolerance and accumulation of arsenic in the cordgrass *Spartina densiflora* Brongn. *Bioresource Technology* 104, 187–194.

Megdiche, W., Hessini, K., Gharbi, F., Jaleel, C.A., Ksouri, R., *et al.* (2008) Photosynthesis and photosystem-2 efficiency of two salt-adapted halophytic seashore *Cakile maritima* ecotypes. *Photosynthetica* 46, 410–419.

Meharg, A.A. and Hartley-Whitaker, J. (2002) Arsenic uptake and metabolism in arsenic resistant and nonresistant plant species. *New Phytologist* 154, 29–44.

Miransari, M. (2011) Hyperaccumulators, arbuscular mycorrhizal fungi and stress of heavy metals. *Biotechnology Advances* 29, 645–653.

Moreno-Jiménez, E., Esteban, E. and Peñalosa, J.M. (2012) The fate of arsenic in the soil-plant system. *Reviews of Environmental Contamination and Toxicology* 215, 1–37.

Moronsito, T., Caffari, S., Dall'Osto, L. and Bassi, R. (2003) Mechanistic aspect of the xanthophyll dynamics in higher plant thylakoids. *Physiologia Plantarum* 119, 347–354.

Mysliwa-Kurdziel, B., Prasad, M.N.V. and Strzalka, K. (2002) Heavy metal influence on the light phase of photosynthesis. In: Prasad, M.N.V. and Strzalka, K. (eds) *Physiology and Biochemistry of Metal Toxicity and Tolerance in Plants*. Springer, Dordrecht, Netherlands, pp 229–255. doi.org/10.1007/978-94-017-2660-3_9.

Naz, A., Khan, S., Muhammad, S., Khalid, S., Alam, S., *et al.* (2015) Toxicity and bioaccumulation of heavy metals in spinach (*Spinacia oleracea*) grown in a controlled environment. *International Journal of Environmental Research and Public Health* 12, 7400–7416.

Otero, S., Nunez-Olivera, E., Martinez-Abaigar, J., Tomas, R., Arroniz-Crespo, M., *et al.* (2006). Effects of cadmium and enhanced UV radiation on the physiology and the concentration of UV-absorbing compounds of the aquatic liverwort *Jungermannia exsertifolia* subsp. *cordifolia*. *Photochemical and Photobiological Science* 5, 760–769.

Padmaja, K., Prasad, D.D.K. and Prasad, A.R.K. (1990) Inhibition of chlorophyll synthesis in *Phaseolus vulgaris* L. seedlings by cadmium acetate. *Photosynthetica* 24, 399–405.

Psaras, G.K., Constantinidis, T.H., Cotsopoulos, B. and Manetas, Y. (2000) Relative abundance of nickel in the leaf epidermis of eight hyperaccumulators, evidence that the metal is excluded from both guard cells and trichomes. *Annals of Botany* 86, 73–78.

Pulford, L.D. and Watson, C. (2003) Phytoremediation of heavy metal-contaminated land by trees—a review. *International Journal of Environment* 29, 529–540.

Ramos, I., Esteban, E., Lucena, J.J. and Gàrate, A. (2002) Cadmium uptake and subcellular distribution in plants of *Lactuca* sp. Cd-Mn interaction. *Plant Sciences* 162, 761–767.

Reboredo, R. (2012) Zinc compartmentation in *Halimione portulacoides* (L.) Aellen and some effects on leaf ultrastructure. *Environmental Science and Pollution Research* 19, 2644–2657.

Reboreda, R. and Caçador, I. (2007) Halophyte vegetation influences in salt marsh retention capacity for heavy metals. *Environmental Pollution* 146, 147–154.

Redondo-Gómez, S. (2013) Bioaccumulation of heavy metals in *Spartina*. *Functional Plant Biology* 40, 913–921.

Reeves, R.D. and Baker, A.J. (2000) Metal-accumulating plants. In: Raskin, I. and Ensley, B.D. (eds) *Phytoremediation of Toxic Metals: Using Plants to Clean up the Environment*. Wiley, New York, pp. 193–229.

Saïdana, D., Mahjoub, M.A., Boussaada, O., Chria, J., Chéraif, I., *et al.* (2007) Chemical composition and antimicrobial activity of volatile compounds of *Tamarix boveana* (Tamaricaceae). *Microbiological Research* 163, 445–455.

Santos, D., Duarte, B. and Caçador, I. (2014) Unveiling Zn hyperaccumulation in *Juncus acutus*: implications on the electronic energy fluxes and on oxidative stress with emphasis on non-functional Zn-chlorophylls. *Journal of Photochemistry and Photobiology B: Biology* 140, 228–239.

Santos, D., Duarte, B. and Caçador, I. (2015) Biochemical and photochemical feedbacks of acute Cd toxicity in *Juncus acutus* seedlings: the role of non-functional. *Estuarine, Coastal and Shelf Science* 167, 228–239.

Sekmen, A.H., Turkan, I. and Takio, S. (2007) Differential responses of antioxidative enzymes and lipid peroxidation to salt stress in salt-tolerant *Plantago maritima* and salt-sensitive *Plantago media*. *Physiologia Plantarum* 131, 399–411.

Sghaier, D.B., Duarte, B., Bankaji, I., Caçador, I. and Sleimi, N. (2015) Growth, chlorophyll fluorescence and mineral nutrition in the halophyte *Tamarix gallica* cultivated in combined stress conditions: arsenic and NaCl. *Journal of Photochemistry and Photobiology B: Biology* 149, 204–214.

Sghaier, D.B., Pedro, S., Diniz, M.S., Duarte, B., Caçador, I., *et al.* (2016) Tissue localization and distribution of As and Al in the halophyte *Tamarix gallica* under controlled conditions. *Frontiers in Marine Science* 3, 274. doi: 10.3389/fmars.2016.00274.

Singh, S., Parihar, P., Singh, R., Singh, V.P. and Prasad, S.M. (2016) Heavy metal tolerance in plants: role of transcriptomics, proteomics, metabolomics, and ionomics. *Frontiers in Plant Science* 6, 1143. doi: 10.3389/fpls.2015.01143.

Solís-Domínguez, F.A., González-Chávez, M.C., Carrillo-González, R. and Rodríguez-Vázquez, R. (2007) Accumulation and localization of cadmium in *Echinochloa polystachya* grown within a hydroponic system. *Journal of Hazardous Materials* 141, 630–636.

Sookbirsingh, R., Castillo, K., Gill, T. and Chianelli, R. (2010) Salt separation processes in the saltcedar *Tamarix ramosissima* (Ledeb.). *Communications in Soil Science and Plant Analysis* 41, 1271–1281.

Sousa, A.I., Caçador, I., Lillebø, A.I. and Pardal, M.A. (2008) Heavy metal accumulation in *Halimione portulacoides*, Intra- and extra-cellular metal binding. *Chemosphere* 70, 850–857.

Sruthi, P., Majeed, A. and Jos, S. (2016) Heavy metal detoxification mechanisms in halophytes: an overview. *Wetlands Ecology and Management* 25(2), 129–148.

Van Der Ent, A., Baker, A.J.M., Reeves, R.D., Pollard, A.J. and Schat, H. (2013) Hyperaccumulators of metal and metalloid trace elements: Facts and fiction. *Plant and Soil* 362, 319–334.

Vitorello, V.A., Capaldi, F.R.C. and Stefanuto, V.A. (2005) Recent advances in aluminum toxicity and resistance in higher plants. *Brazilian Journal of Plant Physiology* 17, 129–143.

Vromman, D., Martínez, J.P. and Lutts, S. (2017) Phosphorus deficiency modifies As translocation in the halophyte plant species *Atriplex atacamensis*. *Ecotoxicology and Environmental Safety* 139, 344–351.

Wang, Y., Chantreau, M., Sibout, R. and Hawkins, S. (2013) Plant cell wall lignifications and monolignol metabolism. *Frontiers in Plant Science* 4, 220. doi : 10.3389/fpls.2013.00220.

Xiong, J., Fu, G., Tao, L. and Zhu, C. (2010) Roles of nitric oxide in alleviating heavy metal toxicity in plants. *Archives of Biochemistry and Biophysics* 497, 13–20.

Zenk, M.H. (1996) Heavy metal detoxification in higher plants. *Gene* 179, 21–30.

Zornoza, P., Vazquez, S., Esteban, E., Fernandez-Pascual, M. and Carpena, R. (2002) Cadmium-stress in nodulated white lupin, strategies to avoid toxicity. *Plant Physiology and Biochemistry* 40, 1003–1009.

17 Thylakoid Electron Transfer in *Salicornia veneta* under Different Salinity Levels: a Fluorescence-based Study

Romina Cannata and Roberto Barbato*
University of Eastern Piemont, Alessandria, Italy

Abstract

The halophyte *Arthrocnemum macrostachyum* is able to survive in extreme environments where salt concentrations may rise to 1M NaCl. Here we report the results of a study in which the photochemical properties of both photosystem (PS) PSII and PSI have been investigated in individuals living in their natural environment (Sacca di Bellocchio, Prato Zangheri, Ravenna, Italy) by using Pulse amplitude modulation and fast fluorescence techniques. Results from quenching analysis (for both photosystems) and fluorescence decay in the sub-millisecond range (for PSII) are reported and discussed in terms of adaptation mechanism(s) of these plants to this extreme environment.

Keywords: Halophytes; Photosystems; Plant adaptation; Salinity

17.1 Introduction

Halophytes are a heterogeneous group of plants highly adapted to salty environments and often require high salt concentrations for optimal growth. Selective absorption of K^+ over Na^+ and compartmentalization and/or secretion of most of the absorbed salts (Flowers and Colmer, 2008) represent their survival strategy. Absorbed salts are then accumulated in the vacuole where a strong negative water potential is developed. Dehydration of the cytoplasm is prevented by the synthesis of osmocompatible solutes which, besides osmotic counterbalance, exert a protective role against the chaotropic effect of highly damaging cations. The way in which these halophytes protect their photosynthetic apparatus, instead, is not known, even though the increase in the antioxidant system (Rangani *et al.*, 2016), the active extrusion from chloroplasts of accumulated Na^+ (Mueller *et al.*, 2014) and the rearrangement of the thylakoid membrane organization (see Duarte *et al.*, 2014 and Jayakumar *et al.*, 2017 for recent reviews) may have important roles. Recently, it has been shown that in extreme halophytes such as *Salicornia veneta* and *Arthrocnemum macrostachyum* external salt concentrations affected the polypeptide composition of the thylakoid membrane, with particular reference to the extrinsic polypeptides of photosystem II (PSII), the so-called PsbO, PsbP and PsbQ proteins (Suorsa and Aro, 2007). As an example, it was found that in *S. veneta* plants living in a naturally saline environment, the PsbQ protein was totally absent and only 30% of the PsbP polypeptide (when compared with the spinach glycophyte)

* Corresponding author e-mail: roberto.barbato@uniupo.it

was detected (Pagliano *et al.*, 2009). The conclusion was that in such growing conditions PsbQ was largely dispensable, whereas even a limited amount of PsbP was sufficient for normal growth. A further study with *A. macrostachyum* grown in different salt conditions showed that the stoichiometry of extrinsic PSII polypeptides was affected by external salt concentrations in a complex way, suggesting that function besides structure of the photosystem may be affected by this stress (Trotta *et al.*, 2012). Moreover, in all these cases, vesciculation of the thylakoids was observed (Pagliano *et al.*, 2009; Trotta *et al.*, 2012; Zhou *et al.*, 2015) as well. With these observations in mind, we decided to undertake a functional study of the photochemical properties of PSII, based on pulse-modulated fluorimetry and chlorophyll (chl) transient analysis, using *S. veneta* plants collected in their natural environment and subsequently treated for 3 weeks with different amounts of NaCl in a greenhouse.

17.2 Methodology

17.2.1 Plant material and growth conditions

S. veneta plants were collected in Sacca di Bellocchio (Ravenna, Italy) and transferred to a growth chamber in hydroponic conditions. For this experiment Hoagland solution containing the desired amount of NaCl (0%, 0.5%, 1%, 3% and 6%) was used. Treatment was carried out for the following 3 weeks, and plants were then collected and analysed.

17.2.2 Fluorescence analysis

Pulse-modulated fluorescence was measured by using a Dual Pulse Amplitude Modulated (PAM)-100 fluorometer (Walz, Effeltrich, Germany). Light curves were obtained by irradiating plants with light of increasing intensity (11, 18, 27, 58, 100, 131, 221, 344, 536 and 830 μmol m^{-2} s^{-1}) for 1 min. At the end of each step, a saturating pulse (3000 μmol m^{-2} s^{-1} for 1.0 s) was fired, allowing the determination of different PSI and PSII yield parameters. Induction/recovery curves were recorded using different actinic light intensities of 22, 166, 430 and 1029 μmol m^{-2} s^{-1}. The initial values of chlorophyll fluorescence, F_0 and $F_{0'}$, were determined using light with an intensity of 6 μmol m^{-2} s^{-1}. Different yield parameters for both PSI and PSII were calculated according to Klughammer and Schreiber (2008). Fluorescence induction (i.e. the transient from F_0 to F_M) was measured by using a photosynthesis efficiency analyzer (PEA) Handy fluorometer (Hansatech Instruments Ltd, Norfolk, UK), and data were analysed according to Stirbet and Govindjie (2011).

17.2.3 Thylakoid isolation and gel electrophoresis

Thylakoids were isolated as described by Trotta *et al.* (2012). Chlorophyll concentration was measured according to Arnon (1949). SDS-PAGE and coomassie staining were carried out as described by Barbato *et al.* (1992).

17.3 Results and Discussion

In Fig. 17.1 a coomassie-stained gel containing thylakoids from *S. veneta* treated with different

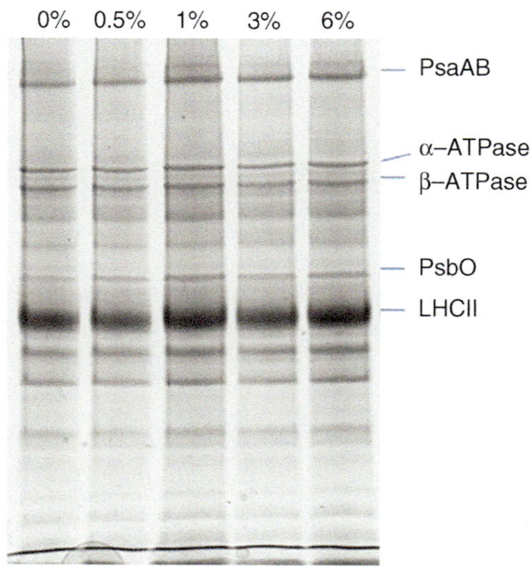

Fig. 17.1. Coomassie-stained SDS gel containing thylakoids isolated from *Salicornia veneta* treated with the indicated amounts of NaCl. Each gel lane contained 8 μg chlorophyll.

salt levels is shown. As already reported by Trotta *et al.* (2012) with *A. macrostachyum*, several differences could be detected by this method, the most important of which is a decreased amount of the PsbO protein in the sample with no NaCl). However, when immunoblotting analysis was conducted in parallel, a complex effect on PSII polypeptide composition has been reported a number of times. These included the modification of the PsbP protein level in both salt-sensitive and salt-tolerant rice cultivars (Wang *et al.*, 2009); the absence (Trotta *et al.*, 2012) or down-regulation of the PsbQ subunit (Pagliano *et al.*, 2009; Trotta *et al.*, 2012); increased expression of PsbO in *S. europea* (Zhou *et al.*, 2015); and general modulation in the stoichiometry of other PSII proteins such as CP47, CP43, D1 and D2 (Trotta *et al.* 2012; Zhou *et al.*, 2015). In another study, at variance, no effect was detected in the expression profile of both *psbp* and *psbq* genes after exposure to 3% NaCl (Momonoki *et al.*, 2009).

Figure 17.2 shows a plot reporting the 1-*q*P parameter as a function of light intensity for plants grown in different salt conditions. This parameter is proportional to the amount of PSII centres with a reduced Q_A primary acceptor (Q_A^-) and is interpreted as an 'excitonic pressure' (see Maxwell and Johnson, 2000). In this figure, it may be observed that, in low salt conditions and for light intensity up to about 400 µmol m^{-2} s^{-1}, this parameter is higher than in other samples. This suggests that in this salt condition (already at relatively low light intensity), most PSII centres had a reduced primary acceptor, thus increasing the risk of photoinhibition (Vass *et al.*, 1992). In parallel, F_V/F_M (which is equivalent to the quantum yield Y(II) in the dark; see Fig. 17.3a), is lower in 0% salt than in any other condition (0.65 in 0% sample versus about 0.80 in other samples). Similar results have also been reported by de Melo *et al.* (2017) with *Atriplex nummularia*. At variance, at 0.5, 1, 3 and 6% NaCl, this parameter is not significantly affected (Fig. 17.3a), even though the lowest values are observed for the 6% NaCl sample. Apart from Y(II), other PSII yield parameters were calculated: Y(NO) (the amount of absorbed light degraded to health in a not regulated way) (Fig. 17.3b), Y(NPQ) (the thermal dissipation under control of the PsbS/xanthophyll system) (Fig. 17.3c) and ETR(II) (the relative electron transport rate around PSII) (Fig. 17.3d). In the 0% NaCl sample, a higher value of Y(NPQ) was observed at any light intensity as compared to samples grown at 0.5, 1, 3 and 6% NaCl, according to an increased 'excitonic pressure'. An increase in NPQ as a result of salt treatment

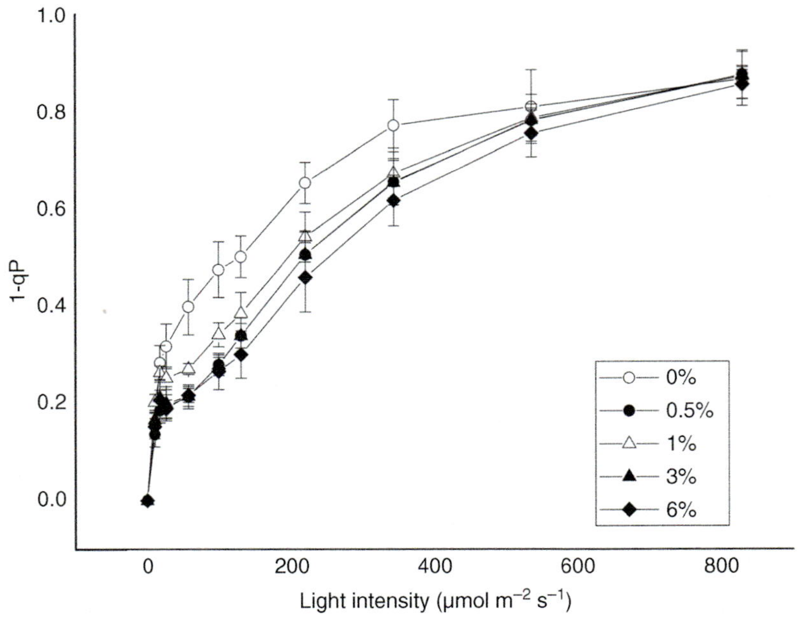

Fig. 17.2. (1-qP) parameter obtained from *Salicornia veneta* plants treated with the indicated amounts of salt.

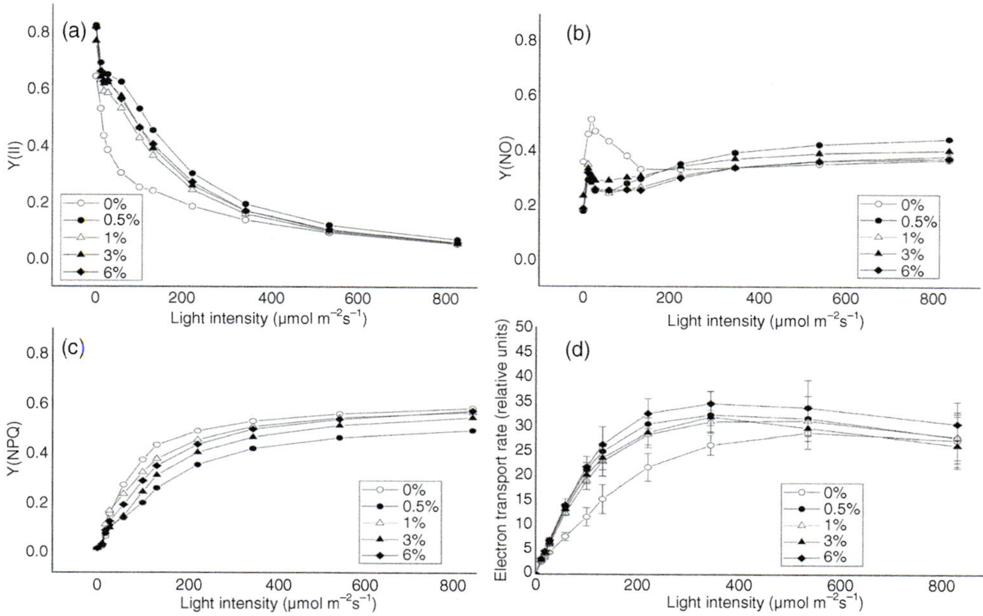

Fig. 17.3. Photosystem II yields parameters: (a) Y(II); (b) Y(NO); (c) Y(NPQ) and (d) PSII relative electron transfer rates; *Salicornia veneta* plants were treated with the indicated amounts of salt.

was also observed in the halophyte *Aster tripolium* by Duarte *et al.* (2017), as well as by Qiu *et al.* (2003) with salt-adapted *Atriplex centralasiatica*; in the latter case, the increase in NPQ was linked to an increase in zeaxanthin content. The Y(NO) parameter showed, in the 0% sample, a spike at low light which could indicate some perturbation in the dark to light transition (see below). As a consequence of these modifications in use of light for the 0% sample, a decrease in the relative electron transfer around PSII, that is ETR(II), was observed (Fig. 17.3d). The development of NPQ (measured as Y(NPQ)) was then analysed at different actinic light intensities: 22, 166, 430 and 1029 μmol m^{-2} s^{-1}, using the induction/recovery technique. In the 0% sample a relatively strong NPQ was already observed at the very low actinic light of 22 μmol m^{-2} s^{-1}, much higher than in any other sample (Fig. 17.4a). In this sample, a full NPQ had already developed at 166 μmol m^{-2} s^{-1}, whereas in other samples a full NPQ was observed at higher light intensities. In the 6% sample, the maximum value observed was lower than in the 0.5, 1 and 3% samples. It should also be noted that, in the 0% NaCl sample, the decay of NPQ was slower than in other samples and, in our 12-minute window of time, NPQ did not completely relax. The nature of this long-living quencher, whether or not it inactivated PSII centres (Vass *et al.*, 1992), remains to be established.

A further analysis was performed by taking into account the fluorescence transient measured by PEA. The unnormalized fluorescence rise from F_0 to F_M is reported in Fig. 17.5a, where the lowest value for F_M was observed in the 0% sample and the highest in the 0.5% sample. Double normalization at F_0 and F_M (Fig. 17.5b) showed some important differences in the shape of transients. These were analysed semiquantitatively by calculating ΔV_{OP} (Fig. 17.5c), using the 3% sample as control (i.e. the situation most similar to natural conditions). This figure shows that a somewhat negative band appeared at 2 ms in the 6 and 0% samples. This is interpreted as evidence for accumulation of centres with the primary quinone acceptor in a reduced form; that is, Q_A^- (Stirbet and Govindjie, 2011), confirming data obtained by PAM fluorimetry reported in Fig. 17.2. In the 0% a second very deep band appeared at 20 ms, which indicated that electron transfer beyond intersystem electron transfer to PSI acceptors is highly retarded. This finding suggests that in this condition PSI, as well as PSII, is affected by salt.

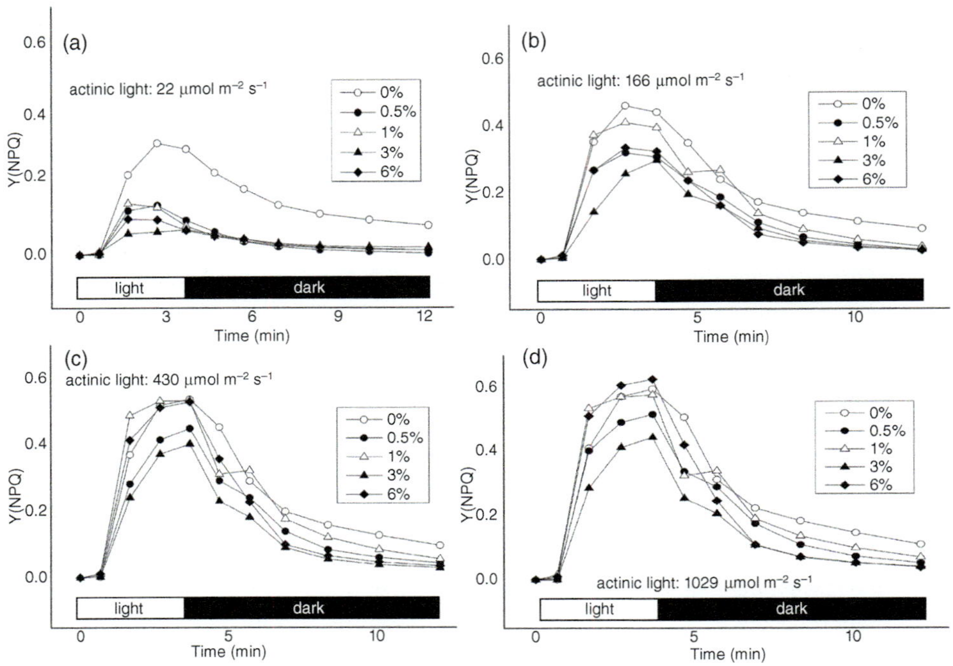

Fig. 17.4. NPQ induction (light) and decay (dark) at different actinic light intensities: (a) 22 μmol m^{-2} s^{-1}; (b) 166 μmol m^{-2} s^{-1}; (c) 430 μmol m^{-2} s^{-1}; (d) 1029 μmol m^{-2} s^{-1}; *Salicornia veneta* plants were treated with the indicated amounts of salt.

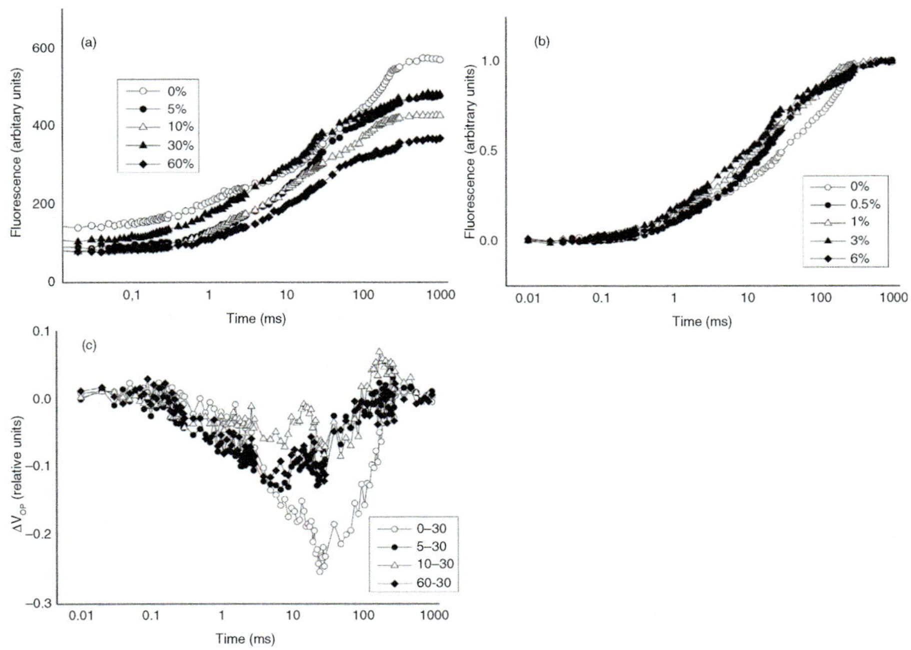

Fig. 17.5. Chlorophyll fluorescence transients in *Salicornia veneta* plants treated with the indicated amounts of salt: (a) original unnormalized transients; (b) double normalized transients expressed as $F=(F-F_0)/(F_M-F_0)$; (c) ΔV_{OP} was calculated by subtracting from the 0%, 0.5%, 1% and 6% transients the 3% NaCl transient, which was used as a control.

Thus, it seemed that under more extreme conditions (i.e. 0% and 6% NaCl), perturbation in both PSI and PSII occurred. Accumulation of centres with a reduced primary quinone acceptor (i.e. Q_A^-) may make plants prone to acceptor-side photoinhibition.

17.4 Conclusion

The photosynthetic apparatus of *S. veneta* is strongly affected by external salt concentrations. When salt concentrations are higher or lower than optimal, an effect on both photosystems are observed. In particular, PSII showed a very high sensibility to low salt conditions, as accumulation of centres with a reduced primary Q_A acceptor was observed. This condition might bring about double reduction of Q_A promoting acceptor-side type photoinhibition (Vass *et al.*, 1992). PSI was also affected by this condition, indicating that damage to both photosystems occurred under low salt conditions. The observed damages could be responsible for the reorganization of photosynthetic apparatus described previously with the similar halophyte *Arthrocnemum macrostachyum* (Trotta *et al.*, 2012). Thus, a decrease in salinity linked to climate change may represent a potential risk factor for halophytes.

References

Arnon, D.I. (1949) Copper enzymes in isolated chloroplast polyphenoloxidase in *Beta vulgaris*. *Plant Physiology* 24, 1–14.

Barbato, R., Friso, G., Rigoni, F., Dalla Vecchia, F. and Giacometti, G.M. (1992) Structural changes and lateral redistribution of photosystem II during donor side photoinhibition of thylakoids. *Journal of Cell Biology* 119, 325–335.

de Melo, H.F., de Souza, E.R. and Cunha, J.C. (2017) Fluorescence of chlorophyll a and photosynthetic pigments in *Atriplex nummularia* under abiotic stress. *Rivista Brasileira de Engegneria Agricola e Ambiental* 21, 232–237.

Duarte, B., Sleimi, N. and Caçador, I. (2014) Biophysical and biochemical constraints imposed by salt stress: learning from halophytes. *Frontiers in Plant Sciences* 5, 746. doi:10.3389/fpls.2014.00746.

Duarte, B., Cabrita, T., Gameiro, C., Matos, A.R., Godonho, R., *et al.* (2017) Disentangling the photochemical salinity tolerance in *Aster tripolium* L.: Connecting biophysical traits with changes in fatty acid composition. *Plant Biology* 19, 239–248.

Flowers, T.J. and Colmer, T.D. (2008) Salinity tolerance in halophytes. *New Phytologist* 179, 945–963.

Jayakumar, B., Munns, R., Shabala, S., Gilliham, M., *et al.* (2017) Chloroplast function and ion regulation in plants growing on saline soil: Lessons from halophytes. *Journal of Experimental Botany* 68, 3129–3143.

Klughammer, C. and Schreiber, U. (2008) Complementary PS II quantum yields calculated from simple fluorescence parameters measured by PAM fluorometry and the Saturation Pulse method. *PAM Application Notes* 1, 27–35.

Maxwell, G. and Johnson, G. (2000) Chlorophyll fluorescence – A practical guide. *Journal of Experimental Botany* 51, 659–668.

Momonoki, Y., Yamamoto, K. and Ogurim, S. (2009) Molecular cloning of oxygen-evolving enhancer genes induced by salt treatment in a halophyte, *Salicornia europea* L. *Plant Production Science* 12, 193–198.

Mueller, M., Kunz, H.-H., Schroeder, J.I., Kemp, G., *et al.* (2014) Decreased capacity for sodium export out of *Arabidopsis* chloroplasts impairs salt tolerance, photosynthesis and plant performances. *Plant Journal* 78, 646–658.

Pagliano, C., La Rocca, N., Andreucci, F., Deak, S., Vass, I., *et al.* (2009) The extreme halophyte *Salicornia veneta* is depleted of the extrinsic PsbQ and PsbP proteins of the oxygen-evolving complex without loss of functional activity. *Annals of Botany* 103, 505–515.

Qiu, N., Lu, Q. and Lu, C. (2003) Photosynthesis, photosystem II efficiency and the zanthophyll cycle in the salt-adapted halophyte *Atriplex centralasiatica*. *New Phytologist* 159, 479–486.

Rangani, J., Parida, A.K., Panda, A. and Kumari, A. (2016) Coordinated changes in antioxidant enzymes protect the photosynthetic machinery from salinity induced oxidative damage and confer salt tolerance in an extreme halophyte *Salvadora persica* L. *Frontiers in Plant Sciences* 7, 50. doi: 10.3389/fpls.2016.00050.

Stirbet, A. and Govindjee (2011) On the relation between the Kautsky effect (chlorophyll a fluorescence induction) and Photosystem II: Basics and applications of the OJIP fluorescence transient. *Journal of Photochemistry and Photobiology B: Biology* 104, 236–257.

Suorsa, M. and Aro, E.M. (2007) Expression, assembly, and auxillary functions of photosystem II oxygen-evolving proteins in higher plants. *Photosynthesis Research* 93, 89–100.

Trotta, A., Redondo-Gomez, S., Pagliano, C., Figueroa Clemente, E., Rascio, N. *et al.* (2012) Chloroplast ultrastructure and thylakoid polypeptide composition are affected by different salt concentrations in the halophytic plant *Arthrocnemum macrostachyum*. *Journal of Plant Physiology* 169, 111–116.

Vass, I., Styring, S., Hundal, T., Koivuniemi, A., Aro, E.M., *et al.* (1992) Reversible and irreversible intermediate during photoinhibition of Photosystem II: Stable reduced Q_A species promote chlorophyll triplet formation. *Proceedings of National Academy of Sciences U.S.A.* 89, 14908–1412.

Wang, R.T.L., Hua, C., Zhou, F. and Zhou, Q.C. (2009) Effects of NaCl stress on photochemical activity and thylakoid membrane polypeptide composition of a salt-tolerant and a salt-sensitive rice cultivar. *Photosynthetica* 47, 125–127.

Zhou, F., Hua, C., Nianwei, Q., Zheng, C. and Wang, R. (2015) Promotion of growth and upregulation of thylakoid membrane proteins in the halophyte *Salicornia bigelowii* Torr. under saline conditions. *Acta Physiologiae Plantarum* 37, 20.

Section III

Practical Uses

18 Introgression of Halophytic Salt Stress-responsive Genes for Developing Stress Tolerance in Crop Plants

Rajesh Kumar Jha, Jaykumar Patel, Avinash Mishra* and Bhavanath Jha
CSIR-Central Salt and Marine Chemicals Research Institute, Bhavnagar, India; and Academy of Scientific and Innovative Research, Ghaziabad, India

Abstract

The world's population is increasing daily, with corresponding demands for sustainable food production, but about 800 million ha of land is affected by salt. Salinization is gradually increasing for several reasons, including scanty rainfall, poor irrigation practices, salt ingression and natural calamities. Salinity is considered a major abiotic stress that adversely affects the growth and productivity of crop plants. Commonly, crop plants are salt sensitive (glycophytes) and so cannot grow in the salt-affected areas. Some plants have natural ability to grow in the high saline areas and are known as halophytes. Halophytes require salt to complete their life cycle and are thus considered potential sources for salt-responsive genes and promoters. The salt-tolerance mechanism is a very complex process which is coordinated from stress perception to signal transduction, and thus provides stress endurance. Several potential salinity-stress responsive and tolerance genes have been isolated from halophytes, functionally characterized and explored for developing transgenic crop plants for sustainable agriculture in the salt-affected areas. About one-quarter of the entire *Arabidopsis* genome responds to salt stress, and so the search continues for promising stress-responsive genes that can modulate physiological traits and metabolic pathways without imposing yield penalties. This chapter focuses on the examination of halophytes for salt-responsive genes, their functional validation and further utilization to engineer crop plants.

Keywords: Abiotic stress; Halophyte; Salinity; Stress tolerance; Transgenic plants

18.1 Introduction

Salinity negatively impacts the growth, development, yield and seed quality of crops and other plants. Similarly, healthy soil and its microbial community, and a large area of fertile land are transformed into barren land by the salinization process. The process detrimentally affects plant uptake of nutrients, vegetative growth and microbiological activity in the soil (Grattan and Grieve, 1998; Shrivastava and Kumar, 2015). According to FAO reports, salinity is gradually increasing; about 800 million ha of land is affected by soil salinity, which thus limits crop productivity worldwide (FAO, 2008). Halophytes are salt-tolerant plants that can thrive and complete their life cycle in saline conditions. Salt tolerance is usually measured as the percentage biomass gain by the plant in the saline environment, compared to control conditions over a definite

* Corresponding author e-mail: avinash@csmcri.res.in

 Halophytes and Climate Change: Adaptive Mechanisms and Potential Uses (eds M. Hasanuzzaman, S. Shabala and M. Fujita) period (Munns, 2002). The growth of some halophytic plant species, especially from the Amaranthaceae, is stimulated by higher salinity (150–300 mM NaCl) (Rozema and Schat, 2013).

The mechanism of stress tolerance is hard to explore, because of the complex genetic make-up and multigene origin of stress endurance. The stress-tolerance molecular mechanism of halophytic plant species has been investigated, and the halophytes quinoa (*Chenopodium quinoa*) and *Thellungiella* emerge as models for the study of the abiotic stress-tolerance mechanism (Wang *et al.*, 2004; Amtmann, 2009; Adolf *et al.*, 2013; Yasui *et al.*, 2016; Jarvis *et al.*, 2017; Mishra and Tanna, 2017; Zou *et al.*, 2017). Quinoa, an allotetraploid crop, shows great environmental adaptability and contains remarkable nutritional properties (Adolf *et al.*, 2013; Kiani-Pouya *et al.*, 2017; Sun *et al.*, 2017; Zou *et al.*, 2017). Halophytes are considered as potential resources for salt-tolerance genes and promoters. Introgression of their salt stress-tolerance genes to other crop species by employing genetic engineering is vital in gaining knowledge of the salt-tolerance mechanism and sustaining crop productivity (Mishra and Tanna, 2017).

Several promising candidate genes have been isolated from halophytes, functionally characterized by transgene approaches and explored to allow molecular, physiological and biochemical intervention in crop plants. The search continues to identify and explore the function of additional and novel promising stress-responsive genes (Mishra and Tanna, 2017). *Salicornia brachiata* is considered an extreme succulent halophyte, and the source of some key novel and salt-responsive genes and promoters (Tiwari *et al.*, 2014; Udawat *et al.*, 2014; Singh *et al.*, 2016; Tiwari *et al.*, 2016; Udawat *et al.*, 2017), which were transferred to castor, cumin, *Jatropha* and groundnut crops. In this chapter, we discuss some important salt stress-tolerance or responsive genes from halophytic species and their further use to improve stress tolerance in crop plants.

18.2 Salinity-stress Tolerance Mechanisms: an Overview

Salinity adversely affects plant growth and productivity by different mechanisms (Munns and Tester, 2008). It has been shown that high salinity leads to programmed cell death in higher plants, which could be considered as a salt adaptation mechanism (Katsuhara, 1997; Huh *et al.*, 2002). Programmed cell death occurs in a coordinated series of events and plays a key role to maintain cellular homeostasis, regulate development or metamorphosis, or control ageing in multicellular organisms, including plants. Under saline conditions: (1) Na^+ enters the cell through a non-selective cation channel (NSCC); (2) causes significant membrane depolarization; (3) this results in massive K^+ leakage from the cell through depolarization-activated TEA^+-sensitive KOR channels (Shabala *et al.*, 2006; Shabala and Cuin, 2008); (4) at the same time, salinity induces elevation in cytosolic Ca^{2+} (Tracey *et al.*, 2008), which leads to a dramatic increase in reactive oxygen species (ROS) level; (5) this results in activation of NADPH oxidase via a positive feedback mechanism (Lecourieux *et al.*, 2002); (6) this causes an additional K^+ efflux via ROS-activated NSCC channels (Demidchik *et al.*, 2003); (7) this leads to a decrease in the cytosolic K^+ pool and may activate caspase-like proteases; and (8) finally, programmed cell death occurs (Shabala, 2009).

Broadly, the mechanism of salinity tolerance is divided into three categories: salt avoidance, compartmentalization and exclusion. During the initial exposure to salinity stress, plants show salt avoidance and reduce the Na^+ influx. The osmotic stress induces decrease in cell expansion in root tips and infantile leaves, followed by stomatal closure. Plant tissues accumulate salts by compartmentalizing at the cellular and intracellular level; this decreases accumulation of toxic concentrations within the cytoplasm, particularly in the mesophyll cells of the leaf. Na^+ exclusion through roots ensures that Na^+ does not accumulate to a lethal concentration within the leaves. A failure in Na^+ exclusion revealed the toxic effect after days or weeks, depending on the species, and thus caused premature death of older leaves (Munns and Tester, 2008).

Different approaches have been reported to combat salinity-stress conditions, including ion homeostasis, signal transduction, transcriptional regulation of gene expression, accumulation of compatible osmolytes and antioxidant regulation (Qiu *et al.*, 2004; James *et al.*, 2011; Hayat *et al.*, 2012; Das and Roychoudhury, 2014; Patel *et al.*, 2016; Singh *et al.*, 2016). Ion homeostasis is an essential process in normal plant growth and is maintained by ion channels and transporters. In halophytes, high salt concentrations within their cytoplasm is either conveyed to the vacuole or organelle-like chloroplast (Bose *et al.*, 2017) or sequestered in old tissues through ion channels and transporters (James *et al.*, 2011). Several signalling pathways control the expression and activities of ion transporters to sustain a low cytoplasmic concentration of Na^+ under salt stress conditions. Salt stress tolerance is mostly governed by the salt overly sensitive (SOS) signalling pathway, which comprises SOS3, SOS2 and SOS1 genes (Qiu *et al.*, 2004). Molecular interactions and complementation analysis showed that SOS3 is essential for the activation of SOS2, which regulates the transcription of SOS1 (Shi *et al.*, 2000). The Ca^{2+} sensitive SOS3–SOS2 protein kinase pathway intermediates Na^+ ion regulation of the expression and activities of Na^+ transporters (Halfter *et al.*, 2000). Hyperosmolarity is anticipated to induce the production of ABA, which in turn up-regulates the transcription of NHX1 and other ion-transporter genes (Bartels and Sunkar, 2005; Flowers *et al.*, 2014).

The plasma membrane H^+ATPase (P-H^+ATPases) and vacuolar ATPases (V-ATPases), coupled with vacuolar H^+-PPase (V-H^+PPases), act at the plasma membrane and tonoplast, respectively. The tonoplast H^+ATPase and H^+PPase transport H^+ into the vacuole and generate proton motive force across the tonoplast, which operates Na^+/H^+ antiporters (Blumwald *et al.*, 2000; Hasegawa *et al.*, 2000). The lower Na^+/K^+ ratio is essential for maintaining cellular metabolism. Salt stress increases the transcript level of the root K^+ transporter gene KC and the expression of KMT1 (an AKT/KAT family member) along with various HAK/KUP- (high-affinity K^+/ K^+ uptake) transporter-type genes, while it decreases the expression of MKT1 (another AKT/KAT family member) (Su *et al.*, 2002; Gupta *et al.*, 2008).

Plants respond to external stimuli by commencing a signalling cascade that triggers the expression of several candidate genes. Signalling pathways contain a network of protein–protein communications and signalling molecules such as ROS, Ca^{2+}, AMP, IP3 and DAG (Pical *et al.*, 1999; Snedden and Fromm, 2001; Apel and Hirt, 2004). The mitogen-activated protein kinase (MAPK) cascade is a ubiquitous signalling module of eukaryotes, including plants, composed of three functionally linked protein kinases (Robinson and Cobb, 1997). MAP kinase (MAPK) activation requires the phosphorylation of conserved threonine and tyrosine residues of the Thr, Glu, Tyr (TEY) activation loop by a specific MAPK kinase (MAPKK), which is activated by MAPKK kinase (MAPKKK) via phosphorylation of conserved threonine and serine residues (Treisman, 1996). SNF-1-like kinases are involved in osmotic stress signalling sense of the ATP/AMP ratio and thus regulates fluxes between anabolism and catabolism through transcription of genes that encode enzymes associated with carbohydrate metabolism (Halford and Hardie, 1998).

Two major groups of phosphatases, including phosphoprotein phosphatases (PPases) and phosphotyrosine phosphatases (PTPases), act on the protein substrate and counteract a protein kinase that modulates the phosphoregulatory mechanisms (Cohen, 1989). The plasma membrane is a key component of the signalling cascade as it receives and transmits environmental signals. Osmotic stress frequently leads to alteration of membrane fluidity and thus causes variations in the phospholipids, mediating osmotic stress signals in plants. In the context of osmotic stress, inositol 1,4,5-triphosphate (IP3), diacylglycerol (DAG) and phosphatidic acid (PA) are key phospholipid signalling molecules (Pical *et al.*, 1999). Phospholipase C (PLC) cleaves the phospholipid phosphatidylinositol 4,5-bisphosphate (PIP2) into the soluble IP3 and the membrane-bound DAG (DeWald *et al.*, 2001). The role of Ca^{2+} signalling in response to osmotic and ionic stress is well documented (Bartels and Sunkar, 2005). At present, three main classes of Ca^{2+} sensors have been characterized: calmodulin, CBLs (calcineurin B-like proteins) and CDPKs (calcium-dependent protein kinase) (Yang and Poovaiah, 2003).

Salinity stress modifies the expression of several genes at a transcriptional, post-transcriptional or translational level and thus regulation of gene expression plays a significant role in stress response and tolerance. Various protein–protein and protein–DNA interactions often regulate the rate of transcription by activation/repression of a promoter under given environmental conditions (Tiwari *et al.*, 2014, 2016, 2019). Different cis-elements, including ABA-responsive element (ABRE), dehydration responsive element (DRE), basic region leucine zipper (bZIP) proteins, homeodomain–leucine zipper (HD-ZIP) proteins, Zn–finger proteins, AP2/ERF-type transcription factors, Myb-like proteins and Myc-like proteins are found to be involved in regulating the transcription of corresponding genes during salinity stress (Deinlein *et al.*, 2014; Flowers and Colmer, 2015). All these transcription factors, with other interrelating associated molecules, constitute a tortuous network which has been extensively studied but not completely understood so far. Many have been used to increase salinity-stress tolerance in different crops such as rice, wheat, soybean and maize (Nakashima *et al.*, 2014; Singh and Laxmi, 2015).

Compatible osmolytes are non-toxic molecules that do not affect normal metabolism and accumulate mainly in the cytoplasm at high concentrations under osmotic stress conditions. Several physiological studies have proposed that, under stress conditions: (1) nonstructural carbohydrates such as sucrose, hexoses, trehalose and sugar alcohols; (2) cyclic polyols like D-pinitol and D-ononitol; (3) amino acids including proline, cysteine, arginine and methionine; and (4) glycine betaine are accumulated to a variable degree in different plant species to counter the adverse effects of stress conditions (Vernon and Bohnert, 1992; Streeter *et al.*, 2001; Rontein *et al.*, 2002).

Similarly, protective proteins are also involved in stress tolerance. Proteins such as late embryogenesis-abundant (LEA), aquaporins, heat shock proteins (HSPs), proteases, proteinase inhibitors and some polyamines play a major role in abiotic stress tolerance (Ingram and Bartels, 1996; Sarda *et al.*, 1999; Sun *et al.*, 2001; Chattopadhayay *et al.*, 2002). During embryogenesis, LEA proteins accumulate; this increases the ABA level and thus provides desiccation tolerance (Ingram, 1996). Aquaporins are members of a large superfamily of membrane-spanning intrinsic proteins (MIPs) and increase the water permeability of the membrane by forming water-specific pores (Sarda *et al.*, 1999). *In vitro* experiments have proposed that cytosolic HSPs work as molecular chaperones, preventing thermal aggregation of substrate proteins and assisting subsequent reactivation (Sun *et al.*, 2001). The polyamines are polycationic small aliphatic amines that have been associated with the various physiological processes of plants including growth and development (Chattopadhayay *et al.*, 2002). Polyamines probably exert their protective function by scavenging ROS, which occur as a result of osmotic stress (Tiburcio *et al.*, 1994). However, catabolism of polyamines by amine oxidases leads to the elevation of H_2O_2 and other ROS (Pottosin *et al.*, 2014).

Various environmental stresses, including salinity stress, lead to overproduction of ROS, causing oxidative damage and ultimately cell death. The ROS contain both free radicals (superoxide, hydroxyl, perhydroxy and alkoxy) and non-radical forms (H_2O_2; hydrogen peroxide and $^1O_2$; singlet oxygen). Several enzymatic and non-enzymatic antioxidants present in plant systems scavenge ROS and thereby protect plant cells from oxidative damage (Das and Roychoudhury, 2014). Plants have very efficient enzymatic (superoxide dismutase, SOD; catalase, CAT; ascorbate peroxidase, APX; glutathione reductase, GR; monodehydroascorbate reductase, MDHAR; dehydroascorbate reductase, DHAR; glutathione peroxidase, GPX; guaiacol peroxidase, GOPX; and glutathione *S*- transferase, GST) and non-enzymatic (ascorbic acid, glutathione, phenolic compounds, alkaloids, carotenoids, non-protein amino acids and α-tocopherols) antioxidant defence systems to combat stress conditions (Hasanuzzaman *et al.*, 2012; Sharma *et al.*, 2012; Das and Roychoudhury, 2014).

18.3 Halophytes as a Resource for Abiotic Stress-tolerance Genes

Halophytes are reservoirs of abiotic stress-tolerance genes and are extensively studied in various model species including *Arabidopsis*, *Nicotiana* and some agronomically important crop species (Roy *et al.*, 2014; Shabala *et al.*, 2014). Overexpression of genes from the halophytic reservoir regulates different tolerance pathways including ion homeostasis, antioxidative metabolism, organic solute, osmolyte synthesis and essential signalling machinery, and are therefore being examined for their salt stress tolerance (Ashraf and Akram, 2009). The role of halophytes as a resource for salt stress-responsive genes is well documented. Developing transgenic crop plants for better stress endurance might be achieved either by: (1) overexpressing the endogenous gene; or (2) introgression of promising genes from other species involved in the process of salinity tolerance through genetic engineering (Mishra and Tanna, 2017). Salt-responsive halophyte genes are identified and isolated through various genomic and transcriptomic procedures, and their functional validation is through transgenic approaches. The overexpression of different halophytic genes under the control of a non-specific constitutive 35SCaMV promoter has been reported to enhance abiotic stress tolerance in a glycophytic host plant (Mishra and Tanna, 2017). Several abiotic stress-responsive genes of halophytic origin have been used to improve stress tolerance in crop plants (Table 18.1).

A number of key genes, including Na^+/H^+ antiporters, potassium transporters, vacuolar pyrophosphates, antioxidant-encoding genes, ROS scavengers and genes from different downstream signalling pathways, have been isolated and characterized. Recently some novel abiotic stress-responsive genes have been reported from the extreme halophyte *S. brachiata* to enhance salt tolerance in transgenic tobacco plants (Singh *et al.*, 2016; Udawat *et al.*, 2017). Some halophytic homologues from *Suaeda* spp., *Atriplex* spp., *Thellungiella* spp., *Salsuginea* spp., *Aeluropus* spp. and Amaranthaceae (*Salicornia* spp.) are potential

Table 18.1. Halophytic salt stress responsive genes reported to enhance stress tolerance in glycophytic hosts (adopted from Mishra and Tanna, 2017).

Genes	Description	Sources	Host plant
AhDREB	Transcriptional regulator	*Atriplex hortensis*	*Nicotina tabacum*
AhProT	Proline transporter	*Atriplex hortensis*	*Arabidopsis thaliana*
AlNHX	Vacuolar Na^+/H^+ antiporter	*Aeluropus littoralis*	*Nicotina tabacum*
AlSAP	Transcriptional regulator	*Aeluropus littoralis*	*Nicotina tabacum*
AmDHAR	Monohydroascorboreductase	*Avicennia marina*	*Nicotina tabacum*
HcNHX1	Vacuolar Na^+/H^+ antiporter	*Halostachys caspica*	*Arabidopsis thaliana*
SaNHX1	Vacuolar Na^+/H^+ antiporter	*Spartina anglica*	*Oryza sativa*
SbASR1	Transcription factor	*Salicornia brachiata*	*Nicotiana tabacum*
SbGSTU	Tau class glutathione transferase	*Salicornia brachiata*	*Nicotiana tabacum*
SbMT-2	Metal responsive gene	*Salicornia brachiata*	*Nicotiana tabacum*
SbNHX1	Vacuolar Na^+/H^+ antiporter	*Salicornia brachiata*	*Nicotiana tabacum*
SbNHX1	Vacuolar Na^+/H^+ antiporter	*Salicornia brachiata*	*Ricinus communis*
SbNHX1	Vacuolar Na^+/H^+ antiporter	*Salicornia brachiata*	*Jatropha curcas*
SbNHX1	Vacuolar Na^+/H^+ antiporter	*Salicornia brachiata*	*Cuminum cyminum*
SbpAPX	Peroximal ascorbate peroxidase	*Salicornia brachiata*	*Arachis hypogaea*
SbpAPX	Peroximal ascorbate peroxidase	*Salicornia brachiata*	*Nicotiana tabacum*
SbSDR	Salt and drought responsive gene	*Salicornia brachiata*	*Nicotiana tabacum*
SbSI-1	Salt inducible protein	*Salicornia brachiata*	*Nicotiana tabacum*
SbSOS1	Membrane Na^+/H^+ antiporter	*Salicornia brachiata*	*Nicotiana tabacum*
SbSRP	Salt-responsive protein	*Salicornia brachiata*	*Nicotiana tabacum*
SbUSP	Universal stress protein	*Salicornia brachiata*	*Nicotiana tabacum*
ScVP	Vacuolar proton pumping pyrophosphatase	*Suaeda corniculata*	*Arabidopsis thaliana*
SeCMO	Choline monooxigenase	*Salicornia europaea*	*Nicotiana tabacum*
SlBADH	Betaine aldehyde dehydrogenase	*Suaeda hootungensis*	*Nicotiana tabacum*
SlCMO	Choline monooxigenase	*Suaeda liatungensis*	*Nicotiana tabacum*
SsPP	Vacuolar proton pumping pyrophosphatase	*Suaeda salsa*	*Arabidopsis thaliana*
ThCYPI	Stress signalling	*Thellungiella halophila*	*Nicotiana tabacum*
ThZFL	Transcriptional regulator	*Tamarix hispida*	*Nicotiana tabacum*
TsVP	Vacuolar proton pumping pyrophosphatase	*Thellungiella halophila*	*Gossypium hirsutum*
TsVP	Vacuolar proton pumping pyrophosphatase	*Thellungiella halophila*	*Nicotiana tabacum*

sources of salt-tolerance genes (Mishra and Tanna, 2017). In general, most halophytic cells contain more than 500 mM NaCl to maintain osmotic balance; however, the extreme halophytes *Tecticornia* spp. contain approximately 2000 mM Na^+ for positive osmotic adjustment (Flowers and Colmer, 2008).

The extreme halophyte *S. brachiata* is able to thrive and complete its life cycle in conditions of high salinity because of its unique characteristics, such as oligosaccharides, metabolites and sulfur-rich proteins (Jha *et al.*, 2011; Joshi *et al.*, 2012; Mishra *et al.*, 2013, 2015) and is considered a potential source of stress-responsive genes and promoters (Jha *et al.*, 2011; Chaturvedi *et al.*, 2012; Singh *et al.*, 2014a). *S. brachiata* is also a reservoir of many salt-responsive genes that have been used to develop transgenic crops including *Jatropha*, cumin, castor and groundnut (Joshi *et al.*, 2013; Singh *et al.*, 2014b; Patel *et al.*, 2015; Tiwari *et al.*, 2015a; Jha *et al.*, 2016; Pandey *et al.*, 2016) by employing different methods of genetic transformation (Singh *et al.*, 2010; Joshi *et al.*, 2011; Tiwari *et al.*, 2015b). Since the salt-tolerance mechanism is a complex and coordinated phenomenon and acquiring extensive knowledge requires considerable research, the search continues to identify promising stress-responsive genes that can intervene in plant physiology.

18.4 Functional Validation of Salt-responsive Genes

Salt stress tolerance is a complex mechanism at both physiological and molecular levels, and is influenced by different environmental factors. To gain knowledge of the molecular defence mechanism, a prerequisite will be the identification and functional validation of promising salt-tolerance genes through transgenics, and their further use in engineering crop plants. The overexpression of halophytic transporter-encoding genes (NHX, SOS, HKT, etc.) showed enhanced tolerance in the range of 400 mM NaCl under the control of a constitutive CaMV35S promoter (Sreeshan *et al.*, 2014; Volkov, 2015). The halophytic antiporter genes *AgNHX1* (*Atriplex gmelinii*), *SaNHX1* (*Spartina anglica*) and *SsNHX1* (*Suaeda salsa*) showed higher salinity tolerance in the range of 300–400 mM NaCl (Ohta *et al.*, 2002; Zhao *et al.*, 2006; Lan *et al.*, 2011). Similarly, *NHX1* genes from other halophytic homologues such as *S. brachiata* (*SbNHX1*) improved salt tolerance in terms of plant growth up to 100 mM in *Jatropha*, cumin and castor plants (Joshi *et al.*, 2013; Patel *et al.*, 2015; Pandey *et al.*, 2016). The overexpression of *SbpAPX*, *SbUSP*, *SbSDR*, *SbSRP* and *SbGSTU* genes attributed better salinity tolerance up to 300 mM NaCl in transgenic tobacco plants as compared to wild-type counterparts (Jha *et al.*, 2011; Singh *et al.*, 2014a, 2016; Udawat *et al.*, 2016, 2017). Transgenic groundnut plants overexpressing some salt-tolerance genes (*SbpAPX*, *SbNHX1* and *SbASR1*) showed higher salt tolerance up to 250 mM NaCl as compared to control wild-type plants (Singh *et al.*, 2014b; Tiwari *et al.*, 2015a). It was proved experimentally that numerous salt-responsive genes from halophytes and their glycophytic counterparts had been identified and functionally validated in the crop plants by using genetic engineering approaches under controlled containment conditions (Roy *et al.*, 2014). However, it is essential to evaluate salt-responsive genes, not only in greenhouse conditions, but also in the field. It has been asserted that plants showing salinity tolerance in greenhouse conditions might not have the same response in the field environment (Ashraf *et al.*, 2008).

18.5 Salt-tolerant Crops in Brief

Developing salt-tolerant GM crop plants is a big challenge, but breakthroughs made with the advent of genetic engineering provide a platform for understanding the mechanism of salt tolerance (Bartels and Dinakar, 2013). Conventional breeding is a conservative tool for manipulating the plant genome of the natural species and identifying the genetic variability to salinity among different cultivars of a crop. In recent years, plant breeders have widely exploited the genetic variation of crops at different breeding levels to introduce a salt-tolerant line or cultivar by employing conventional breeding programmes, and these have been intensively tested under field conditions (Ashraf *et al.*, 2008; Ashraf and Akram, 2009). Some agronomically important cultivars of *Medicago sativa* (alfalfa AZ-germ Salt II, AZ-90NDC-ST) have been tested successfully under field conditions (Dobrenz *et al.*, 1989;

Johnson *et al.*, 1991). Similarly, two wheat cultivars, S24 and KRL1-4, were evaluated in a natural saline environment and were reported to be salt-tolerant lines (Ashraf and O'Leary, 1996). The effectiveness of conventional breeding associated with molecular marker-assisted selection of salt-tolerant traits has been observed in durum wheat lines by using the *TmHKT1;5-A* gene, which showed 25% higher yield than control plants cultivated in the saline environment (Shabala, 2013). An allelic variant of the *HKT* gene also showed a salinity response in agronomically important crops such as rice and barley (Huang *et al.*, 2008). Although considerable conventional breeding has been carried out among different cultivars to produce a potentially salt-tolerant crop, success has been limited owing to the low magnitude of genetic variation in gene pools and reproductive barriers in most crop species (Ashraf *et al.*, 2008). In these circumstances, genetic engineering is an alternative approach to engineer salt-tolerant transgenic crop plants and potentially improve the yield. A review by Tester and Davenport (2003) noted that salt tolerance might be achieved by employing genetic modifications that amplify the process to reduce toxic Na^+ in the leaves generated during transpiration, limiting uptake of salt ions from the soil by exploring water conducting pipes (xylem) to enhance salinity endurance in plants. Plant biologists have focused on identifying a number of genes involved in the ion transport pathway, compatible solute biosynthesis, antioxidant synthesis and regulation of various transcriptions (Ashraf *et al.*, 2008, Ashraf and Akram, 2009).

The overexpression of a novel isoform of the Na^+/H^+ antiporter gene (*PgNHX1*) conferred high salinity tolerance in transgenic rice plants and modulated their life cycle in the presence of 150 mM NaCl (Verma *et al.*, 2007). Transgenic tomato plants were able to grow and flower in the presence of 200 mM salt conditions by overexpressing the vacuolar Na^+/H^+ antiporter gene and accumulating salt in leaf foliage without displaying it in fruits (Hong-Xia and Blumwald, 2001). Furthermore, it may be concluded that salt-tolerant transgenic crop plants provide an excellent platform for sustainable agriculture and worldwide food production.

18.6 Conclusion and Prospects

To ensure food security we need to develop crop varieties that are resilient to abiotic stress, able to cope with environmental constraints and grow without attracting a yield penalty. A basic understanding of the mechanisms underlying the function of stress-tolerance genes is necessary to develop transgenic crop plants. Halophytes are a reservoir of stress-responsive genes, many of which have been characterized through transgenic approaches. We need to explore the function of the regulatory genes and master switches to increase our understanding of the salinity-tolerance mechanism. Genetic engineering has valuable tools, but the exploration of promising candidate transgenes is impeded by a fundamental lack of knowledge of the mechanism of salt tolerance in halophytes. Identification, introgression and validation of promising salt-responsive genes and promoters from different halophytes need to be explored. Comprehensive genetic studies have shown that the abiotic stress-tolerance trait is multigenic in nature, and so it is essential to transfer multiple key genes to develop abiotic stress-tolerant crop plants. However, few studies have been reported in which more than one gene has been incorporated into the same plant to confer abiotic stress tolerance.

Acknowledgements

CSIR-CSMCRI Communication No.: PRIS-144/2017. CSIR-Young Scientist (YSP-02/2016-17) and SERB-DST (EMR/2016/000538) projects are acknowledged with thanks.

References

Adolf, V.I., Jacobsen, S.E. and Shabala, S. (2013) Salt tolerance mechanisms in quinoa (*Chenopodium quinoa* Willd.). *Environmental and Experimental Botany* 92, 43–54. doi: 10.1016/j.envexpbot.2012.07.004.

Amtmann, A. (2009) Learning from evolution: *Thellungiella* generates new knowledge on essential and critical components of abiotic stress tolerance in plants. *Molecular Plant* 2, 3–12. doi: 10.1093/mp/ssn094.

Apel, K. and Hirt, H. (2004) Reactive oxygen species: metabolism, oxidative stress, and signal transduction. *Annual Review of Plant Biology* 55, 373–399. doi: 10.1146/annurev.arplant.55.031903.141701.

Ashraf, M. and Akram, N.A. (2009) Improving salinity tolerance of plants through conventional breeding and genetic engineering: an analytical comparison. *Biotechnology Advances* 27(6), 744–752. doi: 10.1016/j.biotechadv.2009.05.026.

Ashraf, M. and O'Leary, J.W. (1996) Responses of some newly developed salt-tolerant genotypes of spring wheat to salt stress: I. yield components and ion distribution. *Journal of Agronomy and Crop Science* 176(2), 91–101. doi:10.1111/j.1439-037X.1996.tb00451.x.

Ashraf, M., Athar, H.R., Harris, P.J.C. and Kwon, T.R. (2008) Some prospective strategies for improving crop salt tolerance. *Advances in Agronomy* 97, 45–110. doi: 10.1016/S0065-2113(07)00002-8.

Bartels, D. and Dinakar, C. (2013) Balancing salinity stress responses in halophytes and non-halophytes: a comparison between *Thellungiella* and *Arabidopsis thaliana*. *Functional Plant Biology* 40(9), 819–831. doi: 10.1071/FP12299.

Bartels, D. and Sunkar, R. (2005) Drought and salt tolerance in plants. *Critical Reviews in Plant Sciences* 24(1), 23–58. doi: 10.1080/07352680590910410.

Blumwald, E., Aharon, G.S. and Apse, M.P. (2000) Sodium transport in plant cells. *Biochimica et Biophysica Acta (BBA)-Biomembranes* 1465, 140–151. doi: 10.1016/S0005-2736(00)00135-8.

Bose, J., Munns, R., Shabala, S., Gilliham, M., Pogson, B. and Tyerman, S.D. (2017) Chloroplast function and ion regulation in plants growing on saline soils: lessons from halophytes. *Journal of Experimental Botany* 68(12), 3129–3143. doi: 10.1093/jxb/erx142.

Chattopadhayay, M.K., Tiwari, B.S., Chattopadhyay, G., Bose, A., Sengupta, D.N. and Ghosh, B. (2002) Protective role of exogenous polyamines on salinity-stressed rice (*Oryza sativa*) plants. *Physiologia Plantarum* 116(2), 192–199. doi: 10.1034/j.1399–3054.2002.1160208.x.

Chaturvedi, A.K., Mishra, A., Tiwari, V. and Jha, B. (2012) Cloning and transcript analysis of type 2 metallothionein gene (*SbMT-2*) from extreme halophyte *Salicornia brachiata* and its heterologous expression in *E. coli*. *Gene* 499(2), 280–287. doi: 10.1016/j.gene.2012.03.001.

Cohen, P. (1989) The structure and regulation of protein phosphatases. *Annual Review of Biochemistry* 58(1), 453–508. doi: 10.1146/annurev.bi.58.070189.002321.

Das, K. and Roychoudhury, A. (2014) Reactive Oxygen Species (ROS) and response of antioxidants as ROS-scavengers during environmental stress in plants. *Frontiers in Environmental Science* 2, 53. doi: 10.3389/fenvs.2014.00053.

Deinlein, U., Stephan, A.B., Horie, T., Luo, W., Xu, G. and Schroeder, J.I. (2014) Plant salt-tolerance mechanisms. *Trends in Plant Science* 19(6), 371–379. doi: 10.1016/j.tplants.2014.02.001.

Demidchik, V., Shabala, S.N., Coutts, K.B., Tester, M.A. and Davies, J.M. (2003) Free oxygen radicals regulate plasma membrane Ca^{2+} and K^{+} permeable channels in plant root cells. *Journal of Cell Science* 116(1), 81–88. doi: 10.1242/jcs.00201.

DeWald, D.B., Torabinejad, J., Jones, C.A., Shope, J.C., Cangelosi, A.R. *et al.* (2001) Rapid accumulation of phosphatidylinositol 4, 5-bisphosphate and inositol 1, 4, 5-trisphosphate correlates with calcium mobilization in salt-stressed *Arabidopsis*. *Plant Physiology* 126(2), 759–769. doi: 10.1104/pp.126.2.759.

Dobrenz, A.K., Robinson, D.L., Smith, S.E. and Poteet, D.C. (1989) Registration of AZ-GERM SALT-II nondormant *alfalfa* germplasm. *Crop Science* 29(2), 493–493. doi: 10.2135/cropsci1989.0011183X002900020066x.

FAO (2008) *Land and Plant Nutrition Management Service*. FAO, Rome. Available at: https://www.eldis.org/organisation/A3599, accessed 10 November 2017.

Flowers, T.J. and Colmer, T.D. (2008) Salinity tolerance in halophytes. *New Phytologist* 179, 945–963. doi: 10.1111/j.1469-8137.2008.02531.x.

Flowers, T.J. and Colmer, T.D. (2015) Plant salt tolerance: adaptations in halophytes. *Annals of Botany* 115(3), 327–331. doi: 10.1093/aob/mcu267.

Flowers, T.J., Munns, R. and Colmer, T.D. (2014) Sodium chloride toxicity and the cellular basis of salt tolerance in halophytes. *Annals of Botany* 115, 419–431. doi: 10.1093/aob/mcu217.

Grattan, S.R. and Grieve, C.M. (1998) Salinity–mineral nutrient relations in horticultural crops. *Scientia Horticulturae* 78, 127–157. doi: 10.1016/S0304-4238(98)00192-7.

Gupta, M., Qiu, X., Wang, L., Xie, W., Zhang, C. *et al.* (2008) KT/HAK/KUP potassium transporters gene family and their whole-life cycle expression profile in rice (*Oryza sativa*). *Molecular Genetics and Genomics* 280, 437. doi: 10.1007/s00438-008-0377-7.

Halford, N.G. and Hardie, D.G. (1998) SNF1-related protein kinases: Global regulators of carbon metabolism in plants? *Plant Molecular Biology* 37, 735–748. doi: 10.1023/A:1006024231305.

Halfter, U., Ishitani, M. and Zhu, J.K. (2000) The *Arabidopsis* SOS2 protein kinase physically interacts with and is activated by the calcium-binding protein SOS3. *Proceedings of the National Academy of Sciences USA* 97(7), 3735–3740. doi: 10.1073%2Fpnas.040577697.

Hasanuzzaman, M., Hossain, M.A., da Silva, J.A.T. and Fujita, M. (2012) Plant response and tolerance to abiotic oxidative stress: antioxidant defense is a key factor. In: Venkateswarlu, B., Shanker, A., Shanker, C., Maheswari, M. (eds) *Crop Stress and its Management: Perspectives and Strategies*. Springer, Dordrecht, Netherlands, pp. 261–315. doi: 10.1007/978-94-007-2220-0_8.

Hasegawa, P.M., Bressan, R.A., Zhu, J.K. and Bohnert, H.J. (2000) Plant cellular and molecular responses to high salinity. *Annual Review of Plant Biology* 51, 463–499. doi: 10.1146/annurev.arplant.51.1.463.

Hayat, S., Hayat, Q., Alyemeni, M.N., Wani, A.S., Pichtel, J. *et al.* (2012) Role of proline under changing environments: a review. *Plant Signaling & Behavior* 7, 1456–1466. doi: 10.4161/psb.21949.

Hong-Xia, Z. and Blumwald, E. (2001) Transgenic salt-tolerant tomato plants accumulate salt in foliage but not in fruit. *Nature Biotechnology* 19, 765–768. doi: 10.1038/90824.

Huang, S., Spielmeyer, W., Lagudah, E.S. and Munns, R. (2008) Comparative mapping of *HKT* genes in wheat, barley, and rice, key determinants of Na^+ transport, and salt tolerance. *Journal of Experimental Botany* 59, 927–937. doi: 10.1093/jxb/ern033.

Huh, G.H., Damsz, B., Matsumoto, T.K., Reddy, M.P., Rus, A.M. *et al.* (2002) Salt causes ion disequilibrium-induced programmed cell death in yeast and plants. *The Plant Journal* 29, 649–659. doi: 10.1046/j.0960-7412.2001.01247.x.

Ingram, J. and Bartels, D. (1996) The molecular basis of dehydration tolerance in plants. *Annual Review of Plant Biology* 47, 377–403. doi: 10.1146/annurev.arplant.47.1.377.

James, R.A., Blake, C., Byrt, C.S. and Munns, R. (2011) Major genes for Na^+ exclusion, *Nax1* and *Nax2* (wheat *HKT1;4* and *HKT1;5*), decrease Na^+ accumulation in bread wheat leaves under saline and waterlogged conditions. *Journal of Experimental Botany* 62, 2939–2947. doi: 10.1093/jxb/err003.

Jarvis, D.E., Ho, Y.S., Lightfoot, D.J., Schmöckel, S.M., Li, B., *et al.* (2017) The genome of *Chenopodium quinoa*. *Nature* 542, 307–312. doi:10.1038/nature21370.

Jha, B., Sharma, A. and Mishra, A. (2011) Expression of *SbGSTU* (tau class glutathione S-transferase) gene isolated from *Salicornia brachiata* in tobacco for salt tolerance. *Molecular Biology Reports* 38, 4823–4832. doi: 10.1007/s11033-010-0625-x.

Jha, B., Mishra, A. and Chaturvedi, A.K. (2016) Engineering stress tolerance in peanut (*Arachis hypogaea* L.). In: Watson, R. and Preedy, V.R. (eds) *Genetically Modified Organisms (GMO) Foods: Production, Regulation and Public Health*. Elsevier, Philadelphia, PA, USA, pp. 305–311. doi: 10.1016/B978-0-12-802259-7.00027-0.

Johnson, D.W., Smith, S.E. and Dobrenz, A.K. (1991) Registration of AZ-90NDC-ST nondormant *alfalfa* germplasm with improved forage yield in saline environments. *Crop Science* 31, 1098–1099. doi: 10.2135/cropsci1991.0011183X003100040076x.

Joshi, M., Mishra, A. and Jha, B. (2011) Efficient genetic transformation of *Jatropha curcas* L. by microprojectile bombardment using embryo axes. *Industrial Crops and Products* 33, 67–77. doi: http://dx.doi.org/10.1016/j.indcrop.2010.09.002l.

Joshi, M., Mishra, A. and Jha, B. (2012) NaCl plays a key role for in vitro micropropagation of *Salicornia brachiata*, an extreme halophyte. *Industrial Crops and Products* 35, 313–316. doi: 10.1016/j.indcrop.2011.06.024.

Joshi, M., Jha, A., Mishra, A. and Jha, B. (2013) Developing transgenic Jatropha using the *SbNHX1* gene from an extreme halophyte for cultivation in saline wasteland. *PLoS One* 8(8), e71136. doi: 10.1371/journal.pone.0071136.

Katsuhara, M. (1997) Apoptosis-like cell death in barley roots under salt stress. *Plant and Cell Physiology* 38, 1091–1093. doi: 10.1093/oxfordjournals.pcp.a029277.

Kiani-Pouya, A., Roessner, U., Jayasinghe, N.S., Lutz, A., Rupasinghe, T. *et al.* (2017) Epidermal bladder cells confer salinity stress tolerance in the halophyte quinoa and *Atriplex* species. *Plant, Cell & Environment* 40, 1900–1915. doi: 10.1111/pce.12995.

Lan, T., Duan, Y., Wang, B., Zhou, Y. and Wu, W. (2011) Molecular cloning and functional characterization of a Na^+/H^+ antiporter gene from halophyte *Spartina anglica*. *Turkish Journal of Agriculture and Forestry* 35, 535–543. doi: 10.3906/tar-1003-2.

Lecourieux, D., Mazars, C., Pauly, N., Ranjeva, R. and Pugin, A. (2002) Analysis and effects of cytosolic free calcium increases in response to elicitors in *Nicotiana plumbaginifolia* cells. *The Plant Cell* 14, 2627–2641. doi: 10.1105/tpc.005579.

Mishra, A. and Tanna, B. (2017) Halophytes: Potential resources for salt stress tolerance genes and promoters. *Frontiers in Plant Science* 8, 829. doi: 10.3389/fpls.2017.00829.

Mishra, A., Joshi, M. and Jha, B. (2013) Oligosaccharide mass profiling of nutritionally important *Salicornia brachiata*, an extreme halophyte. *Carbohydrate Polymers* 92, 1942–1945. doi: 10.1016/j.carbpol.2012.11.055.

Mishra, A., Patel, M.K. and Jha, B. (2015) Non-targeted metabolomics and scavenging activity of reactive oxygen species reveal the potential of *Salicornia brachiata* as a functional food. *Journal of Functional Foods* 13, 21–31. doi: 10.3389%2Ffpls.2016.00431.

Munns, R. (2002) Comparative physiology of salt and water stress. *Plant, Cell and Environment* 25, 239–250. doi: 10.1046/j.0016-8025.2001.00808.x.

Munns, R. and Tester, M. (2008) Mechanisms of salinity tolerance. *Annual Review of Plant Biology* 59, 651–681. doi: 10.1146/annurev.arplant.59.032607.092911.

Nakashima, K., Yamaguchi-Shinozaki, K. and Shinozaki, K. (2014) The transcriptional regulatory network in the drought response and its crosstalk in abiotic stress responses including drought, cold, and heat. *Frontiers in Plant Science* 5, 170. doi: 10.3389%2Ffpls.2014.00170.

Ohta, M., Hayashi, Y., Nakashima, A., Hamada, A., Tanaka, A. *et al.* (2002) Introduction of a Na^+/H^+ antiporter gene from *Atriplex gmelini* confers salt tolerance to rice. *FEBS Letters* 532, 279–282. doi: 10.1016/S0014-5793(02)03679-7.

Pandey, S., Patel, M.K., Mishra, A. and Jha, B. (2016) *In planta* transformed cumin (*Cuminum cyminum* L.) plants, overexpressing the *SbNHX1* gene showed enhanced salt endurance. *PloS One* 11, e0159349. doi: 10.1371/journal.pone.0159349.

Patel, M.K., Joshi, M., Mishra, A. and Jha, B. (2015) Ectopic expression of *SbNHX1* gene in transgenic castor (*Ricinus communis* L.) enhances salt stress by modulating physiological process. *Plant Cell, Tissue and Organ Culture* 122, 477–490. doi: 10.1007/s11240-015-0785-4.

Patel, M.K., Mishra, A. and Jha, B. (2016) Untargeted metabolomics of halophytes. In: Kim, S. (ed.) *Marine Omics: Principles and Applications*. CRC Press, Boca Raton, FL, USA, pp. 309–325. doi: 10.1201/9781315372303-18.

Pical, C., Westergren, T., Dove, S.K., Larsson, C. and Sommarin, M. (1999) Salinity and hyperosmotic stress induce rapid increases in phosphatidylinositol 4, 5-bisphosphate, diacylglycerol pyrophosphate, and phosphatidylcholine in *Arabidopsis thaliana* cells. *Journal of Biological Chemistry* 274, 38232–38240. doi: 10.1074/jbc.274.53.38232.

Pottosin, I., Velarde-Buendía, A.M., Bose, J., Zepeda-Jazo, I., Shabala, S. *et al.* (2014) Cross-talk between reactive oxygen species and polyamines in regulation of ion transport across the plasma membrane: Implications for plant adaptive responses. *Journal of Experimental Botany* 65, 1271–1283. doi: 10.1093/jxb/ert423.

Qiu, Q.S., Guo, Y., Quintero, F.J., Pardo, J.M., Schumaker, K.S. *et al.* (2004) Regulation of vacuolar Na^+/H^+ exchange in *Arabidopsis thaliana* by the Salt-Overly-Sensitive (SOS) pathway. *Journal of Biological Chemistry* 279, 207–215. doi: 10.1074/jbc.M307982200.

Robinson, M.J. and Cobb, M.H. (1997) Mitogen-activated protein kinase pathways. *Current Opinion in Cell Biology* 9, 180–186. doi: 10.1016/S0955-0674(97)80061-0.

Rontein, D., Basset, G. and Hanson, A.D. (2002) Metabolic engineering of osmoprotectant accumulation in plants. *Metabolic Engineering* 4, 49–56. doi: 10.1006/mben.2001.0208.

Roy, S.J., Negrão, S. and Tester, M. (2014) Salt resistant crop plants. *Current Opinion in Biotechnology* 26, 115–124.

Rozema, J. and Schat, H. (2013) Salt tolerance of halophytes, research questions reviewed in the perspective of saline agriculture. *Environmental and Experimental Botany* 92, 83–95. doi: 10.1016/j.envexpbot.2012.08.004.

Sarda, X., Tousch, D., Ferrare, K., Cellier, F., Alcon, C. *et al.* (1999) Characterization of closely related *δ-TIP* genes encoding aquaporins which are differentially expressed in sunflower roots upon water deprivation through exposure to air. *Plant Molecular Biology* 40, 179–191. doi: 10.1023/A:1026488605778.

Shabala, S. (2009) Salinity and programmed cell death: unravelling mechanisms for ion specific signalling. *Journal of Experimental Botany* 60, 709–712. doi: 10.1093/jxb/erp013.

Shabala, S. (2013) Learning from halophytes: physiological basis and strategies to improve abiotic stress tolerance in crops. *Annals of Botany* 112, 1209–1221. doi: 10.1093/aob/mct205.

Shabala, S. and Cuin, T.A. (2008) Potassium transport and plant salt tolerance. *Physiologia Plantarum* 133, 651–669. doi: 10.1111/j.1399-3054.2007.01008.x.

Shabala, S., Demidchik, V., Shabala, L., Cuin, T.A., Smith, S.J. *et al.* (2006) Extracellular Ca^{2+} ameliorates NaCl induced K^+ loss from *Arabidopsis* root and leaf cells by controlling plasma membrane K^+ permeable channels. *Plant Physiology* 141, 1653–1665. doi: 10.1104/pp.106.082388.

Shabala, S., Bose, J. and Hedrich, R. (2014) Salt bladders: Do they matter? *Trends in Plant Science* 19, 687–691. doi: 10.1016/j.tplants.2014.09.001.

Sharma, P., Jha, A.B., Dubey, R.S. and Pessarakli, M. (2012) Reactive oxygen species, oxidative damage, and antioxidative defense mechanism in plants under stressful conditions. *Journal of Botany* 2012, Article ID 217037, 26 pp. doi: 10.1155/2012/217037.

Shi, H., Ishitani, M., Kim, C. and Zhu, J.K. (2000) The *Arabidopsis thaliana* salt tolerance gene SOS1 encodes a putative Na^+/H^+ antiporter. *Proceedings of the National Academy of Sciences USA* 97, 6896–6901. doi: 10.1073/pnas.120170197.

Shrivastava, P. and Kumar, R. (2015) Soil salinity: a serious environmental issue and plant growth promoting bacteria as one of the tools for its alleviation. *Saudi Journal of Biological Sciences* 22, 123–131. doi: 10.1016/j.sjbs.2014.12.001.

Singh, D. and Laxmi, A. (2015) Transcriptional regulation of drought response: a tortuous network of transcriptional factors. *Frontiers in Plant Science* 6, 895. doi: 10.3389/fpls.2015.00895.

Singh, N., Mishra, A., Joshi, M. and Jha, B. (2010) Microprojectile bombardment mediated genetic transformation of embryo axis and plant regeneration in cumin (*Cuminum cyminum* L.). *Plant Cell, Tissue and Organ Culture* 103. doi: 10.1007/s11240-010-9746-0.

Singh, N., Mishra, A. and Jha, B. (2014a) Over-expression of the peroxisomal ascorbate peroxidase (*SbpAPX*) gene cloned from halophyte *Salicornia brachiata* confers salt and drought stress tolerance in transgenic tobacco. *Marine Biotechnology* 16, 321–332. doi: 10.1007/s10126-013-9548-6.

Singh, N., Mishra, A. and Jha, B. (2014b) Ectopic over-expression of peroxisomal ascorbate peroxidase (*SbpAPX*) gene confers salt stress tolerance in transgenic peanut (*Arachis hypogaea*). *Gene* 547, 119–125. doi: 10.1016/j.gene.2014.06.037.

Singh, V.K., Mishra, A., Haque, I. and Jha, B. (2016) A novel transcription factor-like gene *SbSDR1* acts as a molecular switch and confers salt and osmotic endurance to transgenic tobacco. *Scientific Reports* 6, 31686. doi: 10.1038/srep31686.

Snedden, W.A. and Fromm, H. (2001) Calmodulin as a versatile calcium signal transducer in plants. *New Phytologist* 151, 35–66. doi: 10.1046/j.1469-8137.2001.00154.x.

Sreeshan, A., Meera, S.P. and Augustine, A. (2014) A review on transporters in salt tolerant mangroves. *Trees* 28, 957–960. doi: https://doi.org/10.1007/s00468-014-1034-x.

Streeter, J.G., Lohnes, D.G. and Fioritto, R.J. (2001) Patterns of pinitol accumulation in soybean plants and relationships to drought tolerance. *Plant, Cell and Environment* 24, 429–438. doi: 10.1046/j.1365-3040.2001.00690.x.

Su, H., Golldack, D., Zhao, C. and Bohnert, H.J. (2002) The expression of HAK-type K^+ transporters is regulated in response to salinity stress in common ice plant. *Plant Physiology* 129, 1482–1493. doi: 10.1104%2Fpp.001149.

Sun, W., Bernard, C., Van De Cotte, B., Van Montagu, M. and Verbruggen, N. (2001) *At-HSP17.6A*, encoding a small heat-shock protein in *Arabidopsis*, can enhance osmotolerance upon overexpression. *The Plant Journal* 27, 407–415. doi: 10.1046/j.1365-313X.2001.01107.x.

Sun, Y., Lindberg, S., Shabala, L., Morgan, S., Shabala, S. *et al.* (2017) A comparative analysis of cytosolic Na^+ changes under salinity between halophyte quinoa (*Chenopodium quinoa*) and glycophyte pea (*Pisum sativum*). *Environmental and Experimental Botany* 141, 154–160. doi: 10.1016/j.envexpbot.2017.07.003.

Tester, M. and Davenport, R. (2003) Na^+ tolerance and Na^+ transport in higher plants. *Annals of Botany* 91, 503–527. doi: 10.1093%2Faob%2Fmcg058.

Tiburcio, A.F., Besford, R.T., Capell, T., Borrell, A., Testillano, P.S. *et al.* (1994) Mechanisms of polyamine action during senescence responses induced by osmotic stress. *Journal of Experimental Botany* 45, 1789–1800. doi: 10.1093/jxb/45.12.1789.

Tiwari, V., Chaturvedi, A.K., Mishra, A. and Jha, B. (2014) The transcriptional regulatory mechanism of the peroxisomal ascorbate peroxidase (*pAPX*) gene cloned from an extreme halophyte, *Salicornia brachiata*. *Plant and Cell Physiology* 55, 1774–1471. doi: 10.1093/pcp/pct172.

Tiwari, V., Chaturvedi, A.K., Mishra, A. and Jha, B. (2015a) Introgression of the *SbASR-1* gene cloned from a halophyte *Salicornia brachiata* enhances salinity and drought endurance in transgenic groundnut (*Arachis hypogaea*) and acts as a transcription factor. *PLoS One* 10, e0131567. doi: 10.1371/journal.pone.0131567.

Tiwari, V., Chaturvedi, A.K., Mishra, A. and Jha, B. (2015b) An efficient method of *Agrobacterium* mediated genetic transformation and regeneration in local Indian cultivar of Groundnut (*Arachis hypogaea*) using grafting. *Applied Biochemistry and Biotechnology* 175, 436–453. doi: 10.1007/s12010-014-1286-3.

Tiwari, V., Patel, M.K., Chaturvedi, A.K., Mishra, A. and Jha, B. (2016) Functional characterization of the tau class glutathione-S-transferases gene (*SbGSTU*) promoter of *Salicornia brachiata* under salinity and osmotic stress. *PLoS One* 11, e0148494. doi: 10.1371/journal.pone.0148494.

Tiwari, V., Patel, M.K., Chaturvedi, A.K., Mishra, A. and Jha, B. (2019) Cloning and functional characterization of the Na^+/H^+ antiporter (*NHX1*) gene promoter from an extreme halophyte *Salicornia brachiata*. *Gene* 683, 233–242.

Tracy, F.E., Gilliham, M., Dodd, A.N., Webb, A.A. and Tester, M. (2008) NaCl induced changes in cytosolic free Ca^{2+} in *Arabidopsis thaliana* are heterogeneous and modified by external ionic composition. *Plant, Cell and Environment* 31, 1063–1073. doi: 10.1111/j.1365-3040.2008.01817.x.

Treisman, R. (1996) Regulation of transcription by MAP kinase cascades. *Current Opinion in Cell Biology* 8, 205–215. doi: 10.1016/S0955-0674(96)80067-6.

Udawat, P., Mishra, A., and Jha, B. (2014) Heterologous expression of an uncharacterized universal stress protein gene (*SbUSP*) from the extreme halophyte, *Salicornia brachiata*, which confers salt and osmotic tolerance to *E. coli*. *Gene* 536, 163–170. doi: 10.1016/j.gene.2013.11.020.

Udawat, P., Jha, R.K., Sinha, D., Mishra, A. and Jha, B. (2016) Overexpression of a cytosolic abiotic stress responsive universal stress protein (*SbUSP*) mitigates salt and osmotic stress in transgenic tobacco plants. *Frontiers in Plant Science* 7, 518. doi: 10.3389/fpls.2016.00518.

Udawat, P., Jha, R.K., Mishra, A. and Jha, B. (2017) Overexpression of a plasma membrane-localized *SbSRP*-like protein enhances salinity and osmotic stress tolerance in transgenic tobacco. *Frontiers in Plant Science* 8, 829. doi: 10.3389/fpls.2017.00582.

Verma, D., Singla-Pareek, S.L., Rajagopal, D., Reddy, M.K. and Sopory, S.K. (2007) Functional validation of a novel isoform of Na^+/H^+ antiporter from *Pennisetum glaucum* for enhancing salinity tolerance in rice. *Journal of Biosciences* 32. doi: 10.1007/s12038-007-0061-9.

Vernon, D.M. and Bohnert, H.J. (1992) A novel methyl transferase induced by osmotic stress in the facultative halophyte *Mesembryanthemum crystallinum*. *The EMBO Journal* 11, 2077. doi: 10.1002/j.1460-2075.1992.tb05266.x.

Volkov, V. (2015) Salinity tolerance in plants. Quantitative approach to ion transport starting from halophytes and stepping to genetic and protein engineering for manipulating ion fluxes. *Frontiers in Plant Science* 6, 873. doi: 10.3389/fpls.2015.00873.

Wang, Z.L., Li, P.H., Fredricksen, M., Gong, Z.Z., Kim, C.S. *et al.* (2004) Expressed sequence tags from *Thellungiella halophila*, a new model to study plant salt-tolerance. *Plant Science* 166, 609–616. doi: 10.1016/j.plantsci.2003.10.030.

Yang, T. and Poovaiah, B.W. (2003) Calcium/calmodulin-mediated signal network in plants. *Trends in Plant Science* 8, 505–512. doi: 10.1016/j.tplants.2003.09.004.

Yasui, Y., Hirakawa, H., Oikawa, T., Toyoshima, M., Matsuzaki, C., *et al.* (2016) Draft genome sequence of an inbred line of *Chenopodium quinoa*, an allotetraploid crop with great environmental adaptability and outstanding nutritional properties. *DNA Research* 23, 535–546. doi: 10.1093/dnares/dsw037.

Zhao, F.Y., Zhang, X.J., Li, P.H., Zhao, Y.X. and Zhang, H. (2006) Co-expression of the *Suaeda salsa SsNHX1* and *Arabidopsis AVP1* confer greater salt tolerance to transgenic rice than the single *SsNHX1*. *Molecular Breeding* 17, 341–353. doi: 10.1007/s11032-006-9005-6.

Zou, C., Chen, A., Xiao, L., Muller, H.M., Ache, P. *et al.* (2017) A high-quality genome assembly of quinoa provides insights into the molecular basis of salt bladder-based salinity tolerance and the exceptional nutritional value. *Cell Research* 27, 1327–1340. doi: 10.1038/cr.2017.124.

19 Halophytes: Potential Resources of Coastal Ecosystems and their Economic, Ecological and Bioprospecting Significance

Asish Kumar Parida*, Asha Kumari, Jaykumar Rangani and Monika Patel
CSIR-Central Salt and Marine Chemicals Research Institute (CSIR-CSMCRI), Bhavnagar, India

Abstract

Halophytes are a special group of plants that grow and flourish in highly saline environments. This chapter summarizes current knowledge exploring the enormous potential of halophytes for industrial applications in the form of nutraceuticals, essential oils, biofuels, alcohol, latex, cosmetics, fibres, etc., and the bioactive molecules rendering these properties. We critically analyse recent literature addressing halophytes as a potential source of valuable metabolites having nutraceutical and pharmaceutical value, and their potential for remediating nutrient-rich effluents from coastal aquaculture. The growing body of evidence discussed in this chapter supports the perception that halophytes can be incorporated easily into saltwater-based agriculture as a source of high-value products. Bioprospecting the biomedical compounds isolated from halophytes promises to help in the sustainable commercialization of the identified bioactive compounds. Cultivation of halophytes is both economically and ecologically beneficial: it encourages the extension of technologies to improve a country's economy by providing employment opportunities, and also helps to protect coastal wastelands and promotes ecosystem restoration. We discuss the ecological, economic, nutraceutical and therapeutic potential of halophytes, focusing on the primary and secondary bioactive compounds or metabolites applicable to nutraceuticals, pharmaceuticals or cosmetic applications; and also explore the potential of halophytes for reclamation of salt-affected and heavy metal-contaminated lands.

Keywords: Antioxidants; Bioactives; Biofilter; Biofuels; Bioprospection; Flavonoids; Glycolipids; Halophytes; Hyperaccumulator; Nutraceuticals; Phospholipids; Phytodesalinization; Phytoremediation; PUFA; Triglycerides

19.1 Introduction

Halophytes are defined as vascular plants which naturally inhabit saline environments and have the potential to tolerate high levels of salt in soil or water. 'They occupy niches from the marine to the arid, from salt deserts to salt marshes, and this range of habitats is reflected in a variety of recognized "physiotypes"' (Arora and Rao, 2017). However, these plants are more dominant in saline coastal and inland soils of arid and semiarid climates, and constitute about 1% of the world's flora (Flowers and Colmer, 2008). Halophytic plant species also play indispensable roles in habitat protection by maintaining ecological stability and restoration of salt-affected lands for agricultural development.

* Corresponding author e-mail: asishparida@csmcri.org

There is growing interest in the use of halophytes for reclamation of salt-affected and heavy metal-contaminated areas. Exploring the use of halophytic plants for phytostabilization or phytoextraction of salt and heavily contaminated soils is considered an economical and eco-friendly approach. Halophytes are also a rich source of proteins, oils and fats which are suitable for human and animal consumption. They are enriched with antioxidants that can be used in food and pharmaceutical industries. 'Halophytic plants provide options for livestock feeding in both arid and saline landscapes. These plants are variable in both biomass production and nutritive value; they are characterized by slow growth, low digestibility (therefore low metabolizable energy) and high content of nutritional and anti-nutritional factors' (Arora and Rao, 2017). They have received particular attention in the past few years, not just as model species in salt-tolerance research, but as potential forage, fibre and biomass crops as well as platforms for developing crop systems that use saline water and/or ameliorate salinized soils (Abdelly *et al.*, 2006; Fedoroff *et al.*, 2010; Van Oosten and Maggio, 2015). Many plant families, such as Poaceae, Fabaceae, Asteraceae and Chenopodiaceae, possess halophytic species. Among these, the family Chenopodiaceae contains the highest number of salt-tolerant species. Halophytes are categorized as: (i) facultative halophytes; and (ii) obligate halophytes (Wang *et al.*, 2011). Facultative halophytes are those that can grow in soil devoid of salt, making salt water an ecological requirement for their growth. However, for some halophytes a particular concentration of salt is essential to their survival, and they are referred to as obligate halophytes (Figs 19.1a and 19.1b). 'Halophytes have developed unique adaptation mechanisms to tolerate salinity by adjusting their internal water relations through ion compartmentation in the cell vacuoles, accumulation of compatible organic solutes, succulence and developing salt-secreting glands or bladders (Shabala, 2013)'. Some can grow in desert areas, and these are known as xerohalophytes (Fig. 19.1b D). Some examples of xerohalophytes are *Haloxylon salicornicum*, *Kali komarovii*, *K. paulsenii*, *K. tragus*, *Atriplex halimus* and *Zygophyllum fabago* (Parraga-Aguado *et al.*, 2016; Vromman *et al.*, 2016; Burundukova *et al.*, 2017). Halophytes have long been considered as the reservoirs of potentially valuable and recoverable bioactive compounds, rendering them potential sources of non-conventional food as well as raw material for the development of new products for health and industrial applications. They have already been used as vegetables and medicines; and in ornamental landscaping, environmental protection, wildlife support and various industrial applications, etc., owing to their special physiological characteristics and biochemical composition (Flowers and Colmer, 2008; Hameed and Khan, 2011; Ladeiro, 2012; Cassaniti *et al.*, 2013; Koyro *et al.*, 2013; Panta *et al.*, 2014). Modern awareness of the need for a healthier diet may promote additional markets for halophytes with their high nutritional potential, evident in the rapidly growing consumption of products from some halophytic plants like quinoa (Panta *et al.*, 2014). Halophytes have attracted recent interest because of their high content of bioactive compounds such as polyphenols, polyunsaturated fatty acids, vitamins, carotenoids, essential oils, glycosides and polysaccharides (Ksouri *et al.*, 2012b; Flowers and Muscolo, 2015). These bioactive compounds have been reported to have antimicrobial, anticancerous, anti-inflammatory and antioxidant properties, suggesting possible roles in the prevention of various diseases like cancer, cardiovascular disorders and atherosclerosis (Rodrigues *et al.*, 2014). Antioxidants are an important group of medicinal compounds that inhibit the detrimental changes in foods due to oxidation, suggesting their utility as food additives. Recently, antioxidants of natural origin have attracted interest due to the safety concerns of researchers (Joseph *et al.*, 2013). In comparison to glycophytes, halophytes have been reported to contain greater levels of many stress-related metabolites, contributing to their higher antioxidant properties (Panta *et al.*, 2014; Flowers and Muscolo, 2015; Kumari and Parida, 2018). Consequently, the antioxidant potential of halophytes has made them a novel and natural source of antioxidants, with numerous health benefits, to replace synthetic antioxidants in the agri-food, pharmaceutical and cosmetic industries.

Phytoremediation is the use of plants for efficient removal or reduction of pollutants in the environment (Liang *et al.*, 2016). Plants have the ability to clean up, transfer, stabilize and destroy contaminants from soil and groundwater. Halophytes can restore wetlands and revegetate them. Various halophytic species, such as *Suaeda*

Fig. 19.1a Halophytic species with economic, ecological and bioprospecting significance: (A) *Salicornia brachiata*; (B) *Arthrocnemum indicum;* (C) *Suaeda nudiflora*; (D) *Suaeda monoica*; (E) *Suaeda maritima*; (F) *Salvadora persica*; (G) *Atriplex stocksii*; (H) *Atriplex nummularia*.

maritima, *Sesuvium portulacastrum*, *Arthrocnemum indicum*, *Suaeda fruticosa*, *Tamarix aphylla*, *A. nummularia* and *A. halimus* have been documented as accumulating high amounts of salt in their tissues, thereby reducing salt levels in the saline soils (Gairola *et al.*, 2015). It has been reported that *Sarcocornia pacifica* has been used as a coastal filter for the restoration of salt marshes (Nelson and Zavaleta, 2012). Halophytes have the capability to tolerate other environmental

Fig. 19.1b Halophytic species with economic, ecological and bioprospecting significance: (A) *Juncus rigidus*; (B) *Aeluropus lagopoides*; (C) *Sesuvium portulacastrum*; (D) *Haloxylon salicornicum*; (E) *Heliotropium curassavicum*; (F) *Cressa cretica*; (G) *Heleochloa setulosa*; (H) *Thespesia populnea*.

stresses like drought and heavy metals (Panda *et al.*, 2017). It has been suggested that salinity and heavy metal tolerance in halophytes relies on the same physiological mechanisms, and halophytes are capable of accumulating heavy metals in root and shoot tissues (Panda *et al.*, 2017). They can, therefore, be used as potential candidates for phytoextraction of heavy metals in salinity-affected soils (Manousaki and Kalogerakis, 2011a,b; Van Oosten and Maggio, 2015; Liang *et al.*, 2016). In recent years, many studies have been carried out demonstrating the exploitation of halophytes for phytoremediation of heavy metal-contaminated saline soils (Manousaki *et al.*, 2014; Lutts and Lefèvre, 2015; Santos *et al.*, 2015; Santos Pedro *et al.*, 2015; Christofilopoulos *et al.*, 2016). Furthermore, *Scirpus* species have been used in the formation of wetlands for wastewater treatment, as this species has been found capable of eliminating nutrients and pathogens efficiently from effluent (Bissonnette and Skousen, 2001). It has been reported that *S. robustus* demonstrates excellent efficiency for removal of selenium (Se) from contaminated water, demonstrating its potential for selenium phytoremediation of wetlands (Lokhande and Suprasanna, 2012). Although there have been many reports highlighting the medicinal and industrial applications of halophytes (Ksouri *et al.*, 2012b; Panta *et al.*, 2014; Buhmann *et al.*, 2015; Lutts and Lefevre, 2015; Sharma *et al.*, 2016), reports focusing on the bioprospecting of halophytes and highlighting their economic and ecological potential remain scarce. In this chapter, the multifaceted applications of halophytes, ranging from production of animal feed to gourmet vegetables, and from ornamentals to biofuels, are emphasized.

19.2 Bioprospection Potential of Halophytes

Bioprospection is defined as the systematic and sustainable exploration and/or research into indigenous knowledge for the development of new sources of economically valuable metabolites, genes, and macro- and microorganisms from natural resources. In brief, it uses different means for the commercialization of available biodiversity (Marina and Sci, 2013). The bioprospection process involves four main steps:

1. Sample collection;
2. Extraction and identification of compounds;
3. Isolation, characterization and screening for bioactivity; and
4. Development and commercialization of the product (Fig. 19.2).

Bioprospection of natural resources can be applied to commercial activities in different industrial sectors including food and nutraceuticals, pharmaceuticals, crop protection, cosmetics

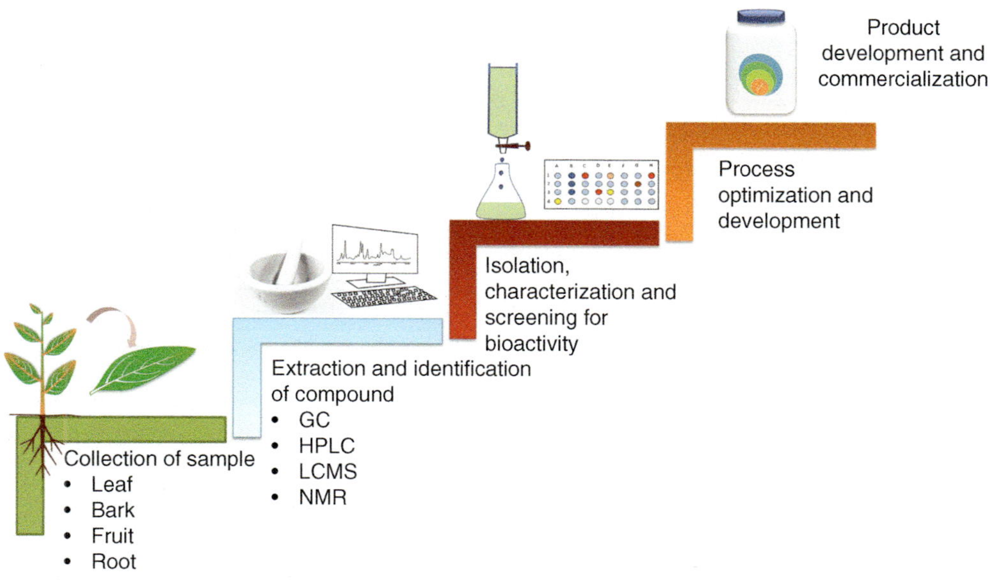

Fig. 19.2. Bioprospecting potential of halophytes: schematic overview.

and personal care (Marina and Sci, 2013). Halophytes, having a pool of valuable bioactive substances and being a source of polyunsaturated fatty acids (PUFAs), have the potential for being exported as a natural, non-conventional food resource, as well as for being used in the cosmetic and pharmaceutical industries (Ksouri *et al.*, 2012a). Although the majority of the identified bioactive compounds from halophytes are fatty acids (FAs) and phenolic compounds (Rodrigues *et al.*, 2014), some derivatives of lipids and glycosides also hold the potential to display antioxidant, anti-cancerous, antiviral, anti-inflammatory and antimicrobial properties (Plouguerné *et al.*, 2014). 'For example, mangrove plants are a rich source of flavonoids, triterpenes, alkaloids and saponins, as well as FAs, tannins and quinones. Extracts from different mangrove plants have been reported to possess diverse medicinal properties such as antibacterial, anticancer, antiulcer and antifungal activities' (Abdel-Aziz *et al.*, 2016). Falleh *et al.* (2011) reported bioactive phenolics in *Mesembryanthemum edule* (ice plant) by LC/ESI-MS/MS, demonstrating that this edible halophyte is a valuable source of propelargonidins and procyanidins. These polyphenol compounds, with several -OH groups, have been documented to show cardioprotective, anti-cancer, antibacterial and antiviral activities (Ksouri *et al.*, 2012a; Plouguerné *et al.*, 2014; Rodrigues *et al.*, 2014). Proanthocyanidins exclusively composed of catechin units are called procyanidins, whereas propelargonidins and prodelphinidins contain afzelechin and gallocatechin, respectively, and are usually mixed with procyanidins (Ksouri *et al.*, 2012a). Condensed tannins, which are oligomers and polymers of flavan-3-ol, are considered as dietary antioxidants. Therefore, due to strong antioxidant activity and phenolic abundance, *M. edule* could be used as a functional food, to prevent or moderate oxidative stress-related diseases (Falleh *et al.*, 2011; Ksouri *et al.*, 2012a). For instance, it has been reported that *Rhizophora stylosa* leaves comprise 17 kinds of amino acids of which seven are essential amino acids (Abdel-Aziz *et al.*, 2016). Table 19.1 describes the phytochemical constituents of various halophytes and their useful biological activities. The following section describes the biologically active components conferring antioxidant, anti-cancerous and other properties to halophytes, with reference to their importance in traditional medicines and industrial applications.

19.2.1 Phenolic compounds (flavonoids and phenolic acids)

Phenolic compounds are ubiquitously present in plants and are an important part of the human diet owing to their antioxidant properties. In plants, these compounds act as pigments, antioxidants and flavour precursors. In the human diet they have been related to maintenance of blood sugar levels, and anti-inflammatory and anti-ageing activity, etc. (Pandey and Rizvi, 2009). However, they mostly act as powerful antioxidants. The high antioxidant properties of phenolic compounds depend on the structure and nature of substitutions on the aromatic rings. 'The main difference between bioactive phenolic compounds that can act as nutraceuticals and other phenolic compounds without noticeable bioactivity is their metabolic origin. The first are derived from two biosynthetic routes: the shikimic acid and/or poly-acetate routes' (Bernal *et al.*, 2011). Halophytes are known to be equipped with powerful antioxidant systems that include polyphenols, responsible for their ability to quench toxic reactive oxygen species (ROS). Thus, halophytes have been reported as a novel source of phenolic molecules (Ksouri *et al.*, 2009; Falleh *et al.*, 2011; Jdey *et al.*, 2017).

The antioxidant and antimicrobial activities of *T. gallica* have been investigated by Ksouri *et al.* (2009), who reported syringic acid and isoquercitin, as well as catechin as the major phenolics present in *T. gallica* (Ksouri *et al.*, 2009). Furthermore, 17 phenolic acids such as p-coumaric, syringic, o-coumaric, 2,5-dihydroxybenzoic, vanillic, protocatechuic, rosmarinic, transhydroxycinnamic, ferulic and salicylic acids, and six flavonoids (naringin, flavone, luteolin, quercetin-3 rhamnosid, quercetin dihydrate and catechin hydrate) have been recognized in the medicinal halophyte *Retama raetam*, contributing to the antioxidant properties of the plant (Mariem *et al.*, 2014). *M. edule* has been reported to contain many phenolic compounds including rutin, hyperoside, ferulic acid, betacyanin and flavonol conjugates. These compounds have been reported to have high antioxidant, cardioprotective,

Table 19.1. Biologically active compounds reported in various halophyte species and their ethnopharmacological activities.

Species	Compounds	Ethnopharmacological activities	References
Phenolics			
Apocynum venetum	Plumbocatechin A, 8-O-methylretusin and kaempferol	Anti-hypertensive, cardiotonic, hepatoprotective	Kong *et al.* (2014)
Artemisia scoparia	Caffeic acid, hydroxyacetophenone and chlorogenic acid	Detoxicating liver, antidepressant and anti-inflammation, and anti-tumour action	Yao and Chen (2007)
Asparagus officinalis	Transresveratrol, flavonoids, ferulic acid, oligosaccharides, amino acid derivatives and steroidal saponins	Treatment of joint pains	Fukushi *et al.* (2000)
Atriplex halimus	Flavonols	Antioxidative	Benhammou *et al.* (2009)
Atriplex hortensis	Flavonol glycoside	Antioxidative	Bylka (2004)
Caesalpinia crista	Diterpenes, flavonoids	Antimalarial	Ksouri *et al.* (2012a)
Cakile maritima	2-methyl butyl glucosinolate, triterpenes, alkaloids and sulfur glycosides	Preventive against cardiovascular diseases, anticancer, antifungal	Radwan *et al.* (2008)
Crambe maritima	Sulfur glycosides, sterols, coumarins and triterpenes	Antioxidant	Radwan *et al.* (2008)
Limoniastrum guyonianum	Gallocatechin, epigallocatechin, epigallocatechin-3-O-gallate	Antioxidant	Trabelsi *et al.* (2012)
Limonium avei	m-coumaric acid, naringin and quercetin	Antimicrobial	Nostro *et al.* (2012)
Mesembryanthemum crystallinum	Flavonol, betacyanins	Antioxidative	Bouftira *et al.* (2007)
Mesembryanthemum edule	Propelargonidins, ferulic acid, rutin, betacyanin	Antifungal, antibacterial, anti-tumour properties	Falleh *et al.* (2011)
Nitraria tangutorum	Pelargonidin-3-O-diglucoside and cyanidin-3-O-diglucoside	Antioxidative, natural colourant or beneficial food	Zheng *et al.* (2011)
Plantago major	Baicalein, baicalin, luteolin, aucubin, caffeic acid, chlorogenic acid, ferulic acid, p-coumaric acid, oleanolic acid, ursolic acid	Antioxidant, anti-inflammatory and antiviral	Samuelsen (2000)
Retama raetam	Resorcinol, coumarin, gallic, ferulic, catechin hydrate, naringin, luteolin, quercetin dihydrate and flavone	Antioxidant, antibacterial	Mariem *et al.* (2014)
Salicornia herbacea	Quercetin 3-O-glucoside and tungtungmadic acid	Immunomodulatory and anti-inflammatory	Rhee *et al.* (2009)
Tamarix gallica	Syringic acid, gallic acid, vanillic, isoquercitin, catechin and apigenin	Antimicrobial, anti-carcinogenic, anti-inflammatory, antidiarrheic	Ksouri *et al.* (2009)
Teurium chamaedrys	β-caryophyllene and humulene, iridoid glycosides, diterpenes, flavonoids, phenylpropanes	Anti-inflammatory, hypolipidemic, hypoglycemic	Adams *et al.* (2009)
Zygophyllum album	Flavonoids	Antioxidant	Ksouri *et al.* (2013)

Continued

Table 19.1. Continued.

Species	Compounds	Ethnopharmacological activities	References
Polysaccharides and glycosides			
Althaea officinalis	Glucuronoxylans	Anti-inflammatory and lipoxygenase inhibition potency	Ksouri *et al.* (2012b)
Cakile maritima	4-pentyl glucosinolate, glucotropaeolin and ethyl glucosinolate	Antioxidant	Radwan *et al.* (2008)
Cynara cardunculus	Cynarin, silymarin, syringic, quercetin, epicatechin and transcinnamic acids	Acts as an antioxidant, anti-diabetic and hepatoprotective	Ksouri *et al.* (2012a)
Kosteletzkya virginica	Mucilage	Diluent, gelling agent and thickeners in oral liquids	Edmond *et al.* (2010)
Plantago lanceolata	Glucuronoxylans	Anti-inflammatory	Ksouri *et al.* (2012b)
Plantago major	Glucuronoxylans	Lipoxygenase inhibition potency	Ksouri *et al.* (2012b)
Salicornia herbacea	Isorhamnetin 3-O-glucoside and tungtungmadic acid	Treatment of obesity, constipation and diabetes	Ksouri *et al.* (2012a)
Salsola kali	Salsolidin, isochinolin and salsolin	Antioxidant, antibacterial	Ksouri *et al.* (2012a)
Urginea maritima	Glucoscillarene A, scillarene A, scilliphaeoside, flavonoids	Used for kidney disorder	Adams *et al.* (2009)
Lipids and alkaloids			
Artemisia judaica	Camphor and 2,6-dimethyl phenol	Used as flavouring agent	El-Massry *et al.* (2002)
Capparis decidua	Alkaloid, oleic acid	Anti-diabetic, anti-hyperlipidaemic	Sharma *et al.* (2010)
Crithmum maritimum	Saponins, linoleic acid, oleic acid, linolenic acid and essential oil	Anti-scorbutic, purgative, antimicrobial	Nabet *et al.* (2017)
Descurainia sophia	Polyunsaturated fatty acid, linolenic acid, arachidonic acid	Pre-substance for prostaglandin production	Yajun *et al.* (2003)
Desmostachya bipinnata	Camphene, isobornyl acetate, tricyclene, trans-2,6gamma-irone, Caryophyllene diepoxide, β-eudesmol, eseroline and calarene	Antibacterial	Kumar *et al.* (2010)
Eryngium maritimum	Saponins	Diuretic, cystotonic, antimicrobial and anti-inflammatory	Ksouri *et al.* (2012a)
Nitraria tangutorum	Spiropiperidines and 13-earbolines	Antispasmodic, antineuropathic and antiarrhythmic agent	Wang *et al.* (2007)
Nitraria retusa	Spiropiperidines, quinolizidines and quinazolines	Antispasmodic, antineuropathic, antiarrhythmic	Wang *et al.* (2007)
Salicornia bigelovii	Linoleic acid, β-carotene and omega-3 polyunsaturated fatty acid	Edible oil	Stuchlík and Žák (2002); Ventura *et al.* (2011)
Salsola kali	Salsolidin and salsolin	Sleep-inducing and anti-hypertonia	Sokolowska-Krzaczek *et al.* (2009)

Continued

Table 19.1. Continued.

Species	Compounds	Ethnopharmacological activities	References
Sesuvium portulacastrum	α humulene, tridecane α-terpinene	Antibacterial, antioxidative and remedy for fever and scurvy	Magwa *et al.* (2006)
Tamarix boveana	Hexadecanoic acid docosane, germacrene D, fenchyl acetate, benzyl benzoate	Antibacterial, antioxidant	Saïdana *et al.* (2008)
Suaeda fruticosa	Eugenol	Hypoglycaemic and antibacterial	Hameed and Khan (2011)
Teucrium abutiloides	Volatile oil, steroids, flavonoids, tannins	Used as flavouring agent and have anti-inflammatory and antirheumatic properties carminative and flavouring agents	Gaspar *et al.* (1996)
Teucrium chamaedrys	Sitosterol, clerosterol	Antioxidant and antimicrobial	Hachicha *et al.* (2009)
Teucrium polium	Volatile oil, steroids, flavonoids, tannins	Used as flavouring agent, anti-inflammatory and antirheumatic	Gaspar *et al.* (1996)
Spergularia marina	Phenyl lipid alkaloid	Antioxidant, anti-inflammatory, antimicrobial and anti-adipogenic	Cho *et al.* (2016)

anticancer, antibacterial and antiviral activities (Falleh *et al.*, 2011). The high radical scavenging properties of *Salicornia herbacea*, *A. indicum* and *S. brachiata* have been attributed to the presence of phenolic compounds like ferulic acid, caffeic acid, procatechuic acid, isohamnetin, quercetin and their derivative compounds (Joseph *et al.*, 2013). Flavonoids comprise the most studied polyphenol group. Researchers have reported 11 flavonoids from the leaves of the halophyte *Apocynum venetum*, including plumbocatechin A, 8-O-methylretusin and kaempferol 3-O-(6''-O-acetyl)-β-D-galactopyranoside; they have also demonstrated the antimicrobial activity of the leaf extract (Kong *et al.*, 2014). The antimicrobial and antiulcer properties of the mangrove *Excoecaria agallocha* have been attributed to the presence of the flavonoids 2′,4′,6′,4-tetramethoxychalcone and 3,5,7,3,5-pentahydroxy-2R,3R flavanonol 3-O-α-L-rhamnopyranoside (Mondal *et al.*, 2016). The potent antioxidant activity of *A. halimus* has been explained by the presence of flavonols, which were found to be the most abundant flavonoids in food (Benhammou *et al.*, 2009). Bylka (2004) has reported acetylated flavonol diglycoside from the aerial part of *A. littoralis*. Cyanidin-3-O-(trans-p-coumaroyl)-diglucoside and pelargonidin-3-O-(p-coumaroyl)-diglucoside have been reported as the main anthocyanins present in the purple and red fruits of *Nitraria tangutorum*. The presence of antocyanin pigments in the fruits of this plant is responsible for pigmentation of fruits and higher antioxidant activities, signifying it as a functional food and medicinal material (Zheng *et al.*, 2011). Flavonoid sulfates have also been identified in land halophytes such as *S. maritima*, *Nypa frutican*, *Limonium vulgare*, *Armeria maritima*, *Halophila ovalis* and species of *Frankenia*, *Tamarix and Atriplex*. Sea grasses such as *Zannichellia*, *Thalassia* and *Zostera* have also been recognized as containing flavonoid sulfates (Abdel-Aziz *et al.*, 2016). Samuelsen (2000) has investigated the biologically active compounds present in the halophyte *Plantago major* and found vanillic acid, baicalein, baicalin, luteolin, aucubin, caffeic acid, chlorogenic acid, ferulic acid, p-coumaric acid, oleanolic acid and ursolic acid to be the major constituents conferring high antioxidative properties to the plant. The putative therapeutic effects of the medicinal halophyte Z. *album* have been attributed to the existence of flavonoids, triterpenes and sterols (Ksouri *et al.*, 2013). Quercetin 3-O-rutinoside and isorhamnetin 3-O-rutinoside have been isolated from *Z. album*; however, *Z. simplex* has been reported to contain quercetin 3-O-glucoside and flavonol 3,7-di-O-glycosides (Hussein *et al.*, 2011).

The pharmacologically active ingredients of *S. herbacea* have been reported to be tungtungmadic acid (3-caffeoyl-4-dihydrocaffeoylquinic acid), quercetin 3-O-glucoside and isorhamnetin 3-O-glucoside, showing antioxidative, anti-inflammatory and immunomodulatory activities, etc., and a source of favourable chemicals for the development of new drugs (Rhee *et al.*, 2009; Ksouri *et al.*, 2012a). Some flavanone and chromone derivatives have been found to alleviate serious manifestations of septicaemia (Tuan *et al.*, 2015). Adams *et al.* (2009) have reported that the halophyte *Teurium chamaedrys* contained essential oil with β-caryophyllene and humulene, iridoid glycosides, diterpenes, flavonoids and phenylpropanes, contributing to the antimicrobial activity of *T. chamaedrys* leaf extract (Ksouri *et al.*, 2012a). Two halophytes, *Thellungiella salsuginea* and *Cakile maritima*, have been documented to be good sources of antioxidants owing to their high phenolic and glucosinolate content (Ksouri *et al.*, 2009). Radwan *et al.* (2008) have reported that *C. maritima* contained ascorbic acid, flavonoids, coumarins, alkaloids, triterpenes, sterols and sulfur glycosides. Young leaves of halophytes commonly consumed as ulam ('ulam' is an alternative term for 'condiment' in South-east Asian countries, particularly Malaysia, Thailand and Indonesia) include *Barringtonia racemosa*, *Centella asiatica*, *Cosmos caudatus*, *Murraya koenigii*, *Oenanthe javanica*, *Persicaria hydropiper* and *Piper sarmentosum*. Major chemical constituents reported in ulam herbs are flavonoids and phenolic acids (Chan *et al.*, 2017). These compounds are known to possess various health benefits including antioxidant, antimicrobial, anti-proliferative, anti-allergic, anti-inflammatory, anti-hypertensive and vasodilating properties (Pietta *et al.*, 2003; Yao *et al.*, 2004). These halophytes have shown great potential for use in the food and pharmaceutical industries.

19.2.2 Bioactive lipids from halophytes

Lipids are a group of natural compounds that includes FAs, waxes, sterols, tocopherols, mono-, di- and triacylglycerols (TGs), phospholipids (PLs) and glycolipids (GLs) (Maciel *et al.*, 2016). Halophytes are the source of large number of lipids and their presence raises the commercial value and biotechnological potential of these plants for medical, cosmetic, pharmaceutical, food and nutraceutical applications. The potential of identified bioactive lipids from halophytes has already been established with respect to their antioxidative, anti-proliferative, anti-inflammatory and antimicrobial activities (de Jesus Cortés-Sánchez *et al.*, 2013; Rodrigues *et al.*, 2014). In the following section we will describe the bioactive lipid classes identified in halophytes and their possible role in different industries.

19.2.2.1 FAs

Halophyte seeds are a good source of fats and fat-soluble vitamins that are important in human diet (Ksouri *et al.*, 2012a). Fatty acid profiling of oils from the seeds of various halophyte species have demonstrated the presence of high amounts of polyunsaturated fatty acids indicating their higher nutritional value as compared to traditionally used vegetable oils (Stuchlík and Žák, 2002; Ksouri *et al.*, 2012b). Several halophytes, like *S. fruticosa*, *A. macrostachyum*, *Kochia scoparia*, *A. indicum*, *A. rosea*, *Allenrolfea occidentialis*, *Halogeton glomeratus*, *H. stocksii*, *A. heterosperma*, *S. torreyana*, *Alhaji maurorum*, *Halopyrum mucronatum*, *C. cretica*, *Sarcobatus vermiculatus*, *S. brachiata*, *S. monoica*, *S. maritima* and *Descurainia sophia* have been reported to contain sufficient quantities of good-quality, edible oil of 70%–80% unsaturation (Qasim *et al.*, 2011). Seeds of the halophyte *S. bigelovii* have been found to contain about 30% oil having a pleasant, nutty flavour and high linoleic acid content comparable with levels in sunflower oil (Stuchlík and Žák, 2002). For this reason, this halophyte has been domesticated as a new edible oilseed crop. The oil content of the halophyte *D. sophia* has been reported to be significantly high, with 76.65% polyunsaturated fatty acid, 53.7% linolenic acid and 5.35% arachidonic acid (Yajun *et al.*, 2003). *Crithmum maritimum* seed oil has been found to be high in monounsaturated fatty acid (81%) and *C. maritima* seed oil has been reported as being rich in erucic acid (Ksouri *et al.*, 2012b). In contrast, *Z. album* seed oil is reported as having high levels of linoleic acid (Zarrouk *et al.*, 2003). The halophyte *Portulaca oleracea* has been found to be a rich source of omega-3 FAs, which help in strengthening the immune system and preventing heart attacks (Simopoulos, 2016). Lipid profiling of the halophyte *Eryngium maritimum* seeds has revealed high oil content rich in oleic and linoleic acid (Lajnef *et al.*, 2017). *A. tripolium* has been described as a potential cash crop for human consumption and nutraceuticals, owing to the presence of omega-3 and omega-6 FAs (linolenic acid and linoleic acid), palmitic acid and oleic acid (Duarte *et al.*, 2017). The high nutrition value of *Cichorium spinosum* has been described as being due to its high PUFA/saturated fatty acids (SFA) ratio (Petropoulos *et al.*, 2017). α-Linolenic acid, linoleic acid and palmitic acid have been reported as the most abundant FAs in the leaves of *C. spinosum* (Petropoulos *et al.*, 2017). Linoleic, oleic and palmitic acid have been described as the main component of the fruit and seed oil of *Lycium ruthenicum* (Wang *et al.*, 2017). A higher index of mono- and PUFAs have been reported in *S. salsa* seed oil, indicating high nutrition value for human consumption. *S. salsa* seed is reported as containing high amounts of linoleic, oleic and palmitic acid (Li *et al.*, 2017). *Artemisia lerchiana*, *A. pauciflora* and *A. santonica* have also been described as being rich sources of ω-3 and ω-6 FAs, indicating their potential for the food and nutraceutical industries (Bogdanova *et al.*, 2017).

19.2.2.2 Essential oil composition, phytosterol and tocopherol contents of halophytes

Halophytes represent a remarkable source of essential oils; these have been reported to have antioxidant and antibiotic activities which can be used in pharmaceutical industries. *C. maritimum* is important as its essential oils show considerable antioxidant and antibiotic activities. It has been reported that all parts of *C. maritimum* are rich in essential oils (Nabet *et al.*, 2017). Monoterpenes, phenylpropanoids and phenols have been reported as the main components of essential oils obtained from the aerial parts of the plant. The roots also contain monoterpene hydrocarbons, terpinolene (the major constituent of essential oil) and γ-terpinene. The essential oil extracted from the root of *C. maritimum* has been used in the cosmetics and food industries as a perfume and flavouring agent, respectively (Senatore *et al.*, 2000). The aerial parts of *D. bipinnata* have yielded 0.42% v/w of the yellow essential oil. The key components of the essential oil are monoterpene and sesquiterpene compounds including β-eudesmol, tricyclene, trans-2,6-gamma Irone, isobornyl acetate, calarene, eseroline and caryophyllene diepoxide (Kumar *et al.*, 2010). The essential oils from the fruits of *L. barbarum* and *L. ruthenicum* have been found to contain hexadecanoic acid, linoleic acid, myristic acid, β-elemene, ethyl hexadecanoate, ethyl linoleate, ethyl hexadecanoate, heptacosane and nonacosane (Altintas *et al.*, 2006). Investigation of the antioxidative activity of the volatile oil of *A. judaica* has shown the presence of 2,6-dimethyl phenol and camphor. The characteristic flavour of the oil has been attributed to presence of trans-ethyl cinnamate and it has been used in the food industry as a flavouring agent (El-Massry *et al.*, 2002). Furthermore, the essential oil from the seeds of *Spartium junceum* has been characterized to contain palmitic acid, palmitoleic acid, stearic acid, oleic acid, linoleic acid and linolenic acid, and it is rich in α-tocopherol, making it suitable for making alkyd resins, soaps and shampoos (Cerchiara *et al.*, 2010). The most interesting lipophilic constituents of halophytes have been phytosterols and tocopherols, as they have numerous beneficial properties for human consumption. Plant sterols, or phytosterols, can be classified based on their structure or biosynthesis: 4-desmethyl sterols (with no substituent on carbon-4), 4α-monomethyl sterols (with one methyl group at carbon-4) and 4,4-dimethyl sterols (with two methyl groups at carbon-4 (Maciel *et al.*, 2016)). Once incorporated in the membrane bilayer, sterols regulate its fluidity and permeability. The sterol/phospholipid (ST/PL) ratio can be used to indicate plant sensitivity to salt, with higher values helping to maintain structural integrity while decreasing the permeability of the membrane bilayer (Maciel *et al.*, 2016). Phytosterols play an important role in lowering serum cholesterol levels, whereas tocopherols have been known to prevent chronic diseases and certain types of cancer (Marangoni and Poli, 2010; Genser *et al.*, 2012). C28 and C29 sterols have a key role in nutraceutical industries as they are precursor molecules for cortisone and vitamin D2. Phytosterol, especially β-sitosterol, has also been shown to lower total and LDL cholesterol levels in humans by inhibiting cholesterol absorption from the intestine (Maciel *et al.*, 2016). Various steroidal compounds like β–sitosterol, β–sitostenone and (24R)-24-ethylcholesta-4, 22-dien-3-one have been reported from the extract of stem and twig of the medicinal mangrove *E. agallocha* (Mondal *et al.*, 2016). *Teucrium* seed oil has been a rich source of phytosterol. Many species of *Teucrium* have been investigated, and the different sterols in their aerial parts isolated. Sitosterol and clerosterol have been isolated from *T. chamaedrys*, whereas clerosterol, poriferasterol and (24S)-24-ethylcholesta-5, 22, 25-trien-3β-ol have been documented in *T. abutiloides* and *T. betonicum* (Gaspar *et al.*, 1996). Similarly, four sterols (clerosterol, sitosterol, stigmasterol and campesterol) have been isolated from the seed oil of *T. polium*, *T. nabli* and *T. alopecurus*, respectively (Hachicha *et al.*, 2009). *Teucrium* seed oil has also been documented as a good source of tocopherols containing α-isomer in higher amounts as compared to β-tocopherol, suggesting the importance of this halophyte to the pharmaceutical industries (Hachicha *et al.*, 2009). Several phytosterols (e.g. β-sitosterol, epicolestanol, Δ7-fucosterol, cholesterol, campasterol, stigmasterol, Δ7-campesterol, fucosterol, Δ7-sitosterol and Δ7-avenasterol) have been isolated from the seed oil of the medicinal halophyte *E. maritimum*, signifying the plant's high antioxidative and nutraceutical properties (Lajnef *et al.*, 2017).

19.2.3 Polysaccharides, glycosides and their derivative compounds

Polysaccharides have vital roles in living beings. These molecules act like structural components and are important for energy storage. Other beneficial effects include their prebiotic, antioxidant and anti-inflammatory activities (Bernal *et al.*, 2011). The role of polysaccharides isolated from *S. herbacea* have been investigated for their ability to activate macrophages, and it has been demonstrated that the isolated polysaccharides induce significant activation of nuclear factor-κB/Rel (Patel, 2016). *Kosteletzkya virginica* has been found to contain mucilage, a polysaccharide commonly found in rhizomes, roots and seed endosperms, where it may act primarily as a source of energy. The hydroxyl group of core polysaccharides has high water-binding capacity, which allows mucilage to hydrate and thus store large amounts of reserved water; this may offer plants the ability to resist physiological drought (Edmond *et al.*, 2010). Polysaccharides isolated from the fruit of *L. barbarum* have been reported to have antioxidant, immunomodulatory, antitumour, neuroprotective, radioprotective and anti-diabetic activity (Zhou *et al.*, 2016). Anti-cancerous, anti-diabetic, antibacterial, antineoplastic and anti-obese effects of glycosides have been reported (Erkucuk *et al.*, 2009). Adams *et al.* (2009) have revealed the presence of cardiac glycosides such as glucoscillarene A, scillarene A, scilliphaeoside, flavonoids and polysaccharides in *Urginea maritima*. Radwan *et al.* (2008) have reported four glucosinolates: glucotropaeolin, 2-methyl butyl glucosinolate, ethyl glucosinolate and 4-pentyl glucosinolate in the halophyte *C. maritima*. Leaf extract of the halophyte *Heritiera fomes* has been confirmed to contain saponins, glycosides and alkaloids. Furthermore, β-sitosterol, stigmast-4-en-3-one and stigmasterol have also been reported in the chloroform extract of *H. fomes* (Mahmud *et al.*, 2014). β-Sitosterol glycosides have been isolated from the halophyte *Spergularia marina* (Cho *et al.*, 2016). Saponins, which are amphipathic glycosides, have been also studied in halophytes. Herlt *et al.* (2002) isolated two major saponins from *B. asiatica* seeds. Other steroidal saponins including protodioscin, terrestrinins A and B, their sulfate derivatives, β-sitosterol-D-glucoside and tribulosin have been documented from the halophyte *Tribulus terrestris* (Ksouri *et al.*, 2012a). Thus, the polysaccharides and glycosides reported in various halophytes demonstrate their potential use in the pharmaceutical industry.

19.2.4 Alkaloids

Alkaloids represent an important group of natural products that exhibit various biological and pharmacological activities for treatment of numerous diseases (Stöckigt *et al.*, 2002). Alkaloids such as quinolizidines, spiropiperidines and quinazolines have been reported from different members of the genus *Nitraria* (Wang *et al.*, 2007). Two alkaloids, salsolin and salsolidin, with potential for treatment of hypertonia, have been identified from *S. kali* (Sokolowska-Krzaczek *et al.*, 2009). Similarly, an alkaloid-rich fraction of *Capparis decidua* has been demonstrated to have a positive effect in treatment of diabetes in mice (Sharma *et al.*, 2010). Nortropane alkaloids have been isolated from *L. chinense* root bark

(Wang *et al.*, 2017). Pyrrole alkaloids and pyrrolidine alkaloids have also been reported in the fruit and in the root bark, respectively, of *L. chinense* (Qian *et al.*, 2017). Six imidazole alkaloids have been detected in leaves of *L. cestroides*, and Na-[(*E*)-cinnamoyl]histamine has been isolated from *L. barbarum* leaves (Li *et al.*, 2014). Tropane alkaloids such as atropine, hyoscyamine and scopolamine have been isolated from *L. europaeum and L. halimifolium* (Qian *et al.*, 2017). Di-caffeoyl spermidine derivatives and lycibarbar spermidines have been documented in *L. barbarum* fruit and showed activity against Alzheimer's disease (Zhou *et al.*, 2016). N,N-bis-dhc-spermine and some spermidine derivatives, having significant antioxidant activity, have been identified from the fruit of *L. ruthenicum* (Wu *et al.*, 2016). Terpenoids (including monoterpenes, sesquiterpenes and diterpenoids) have been identified in the fruits of *L. chinense*, *L. barbarum* and *L. halimifolium*, establishing the valuable medicinal, functional and nutritional significance and industrial usefulness of these halophytes (Qian *et al.*, 2017). Phytochemical screening of the bark and leaf extract of *Lumnitzera racemosa* and *A. marina* has revealed the presence of alkaloids and terpenoids as the main constituents responsible for the antibacterial activity of mangroves (Abdel-Aziz *et al.*, 2016). A piperidine alkaloid, 2,4-dimethoxy-3-ψ,ψ-dimethylallyl-trans-cinnamoylpiperidide, has been documented from the mangrove *E. agallocha* (Mondal *et al.*, 2016). A phenyl alkaloid has been isolated from the halophyte *S. marina*, justifying the health benefits of using this halophyte as a vegetable (Cho *et al.*, 2016). A triterpenoid ester, triterpenes and steroids have been identified from the mangroves *Acrostichum aureum* and *R. apiculata*, justifying the use of the extracts of these halophytes in folk medicines (Abdel-Aziz *et al.*, 2016). Recently, five new triterpenoid paracaseolins have been isolated from the aerial parts of the mangrove *Sonneratia paracaseolaris* having potent anti-cancerous and antiviral activities (Gong *et al.*, 2017).

19.3 Ecological Importance of Halophytes

Anthropogenic development, pollution and increased human interference has put the coastal ecosystem in danger (Murray *et al.*, 2015). Natural calamities like fire, flood, storms and hurricanes also have negative impacts on salt marsh ecosystems (Parida and Jha, 2010). Throughout the world, increases in areas of salt marsh due to anthropogenic development have been a major problem. Halophytes, however, which can survive extreme conditions, can be used as a tool to maintain ecosystems in such environments. They possess many special anatomical, physiological and morphological features that help them to cope with the saline environment and stabilize the ecosystem by maintaining soil structure (Kathiresan and Bingham, 2001; Parida *et al.*, 2004; Parida and Jha, 2010). Their special adaptive features have led to them being explored for various ecological purposes such as phytoremediation, biofiltration, nutrient cycling and as a nursery for juvenile organisms (Fig. 19.3). Halophytic species found along the coastline have been important barriers for protection against coastal erosion by strong waves and currents (Parida and Jha, 2010). Some species of halophytes have been found to remove the salt from soil by excreting or accumulating it in various anatomical structures (Hasanuzzaman *et al.*, 2014), making the soil environment suitable for other organisms.

Halophytes increase the organic matter in soil, maintain moderate soil pH, lower bulk density and make nutrients available for the other components of saline ecosystems (Ravindran *et al.*, 2007; Nsanganwimana *et al.*, 2014; Christofilopoulos *et al.*, 2016; Liang *et al.*, 2016). Primary production and decomposition are major processes that influence carbon storage and other ecosystem services (Hooper *et al.*, 2012). Halophytes are the main primary producers in saline areas and provide a carbon source to other biotic components of the ecosystem. For example, mangroves are the most carbon-rich forest found in the tropics and fix 1100–5400 g m^{-2} carbon per year (Donato *et al.*, 2011). The higher primary productivity of mangroves allows a significant reduction of CO_2 from the atmosphere and its incorporation into biomass. Thus, mangroves have been reported to be major contributors to the carbon and nitrogen cycles in tropical coastal ecosystems. Nutrient enrichment in the coastal and estuarine parts of some areas has been a major problem due to run off from agriculture and urban land, and has led to changes in biogeochemical cycles, harmful phytoplankton

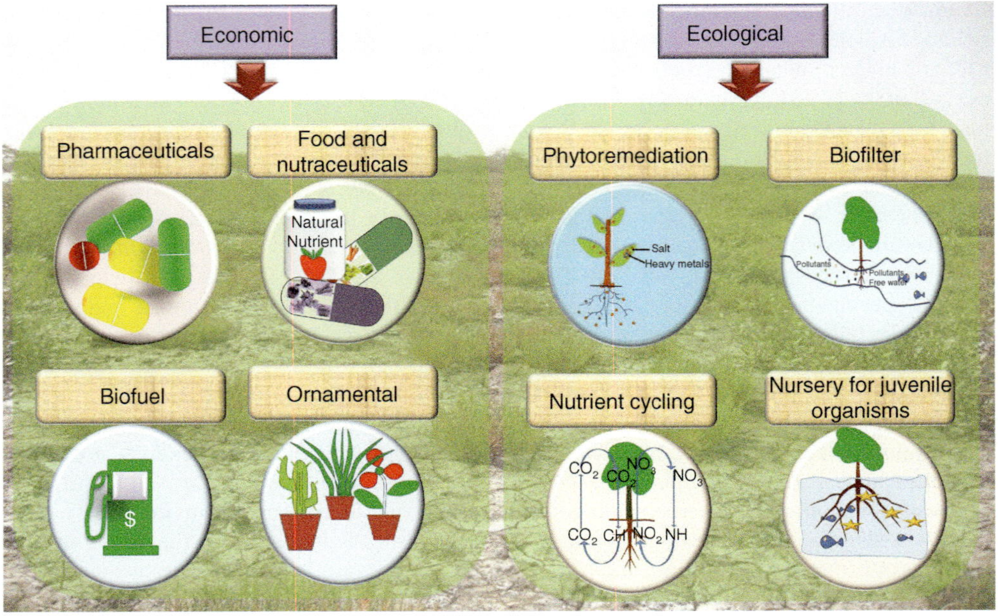

Fig. 19.3 Ecological and economic potential of halophytes: schematic representation.

blooms, changes in the food web and creation of anoxic regions in the oceans (Nelson and Zavaleta, 2012). Coastal salt marsh plants have been reported to serve as a robust nitrogen trap and coastal filter for the protection of the coastal area from eutrophication and its consequences (Parida and Jha, 2010). Mangroves serve as nurseries for juveniles of many organisms such as reef fishes, crabs and oysters, and provide protection against predators until they have grown (Mumby *et al.*, 2004). The salt marsh plants provide habitats for breeding, feeding and shelter to various species of insects, birds, mammals, reptiles, fish, amphibians, snails and crabs. These organisms also use the stems of plants as refuges from rising water levels, and as nesting material (Kathiresan and Bingham, 2001; Parida and Jha, 2010). Mangroves and their associated organisms are collectively known as a mangrove community. A mangrove community helps to maintain water quality by trapping and stabilizing sediments and contaminants from runoff; trapping nutrients; and providing a buffer between the marine environment and nearby terrestrial areas. Mangroves also help to stop erosion by creating a natural breakwater through the absorption of the tidal energy and storm-driven wind and wave action (Abdel-Aziz *et al.*, 2016). Table 19.2 lists different halophyte species used for various economic and ecological purposes.

19.3.1 Desalinization and phytoremediation potential of halophytes

Phytodesalination aims to improve crop production using halophytes or salt-tolerant plants, and has shown promising results in different types of remediation programmes, such as agricultural soil restoration, road runoff and wastewater desalination (Rabhi *et al.*, 2009, 2010; Shelef *et al.*, 2013; Jesus *et al.*, 2015; Suaire *et al.*, 2016). Depletion in the freshwater level due to the increased demand for drinking and irrigation has caused an irreversible spread of salinity in groundwater as well as on the soil surface (Rabhi *et al.*, 2009). Saline soil affects plant growth by inhibiting water and air movement, plant water-holding capacity and root penetration, creating an osmotic imbalance; this leads to disturbance in normal plant physiological functions such as photosynthesis, respiration and transpiration, leading to reduction in growth (Kumari *et al.*, 2015). Hydraulic, physical, chemical and biological methods

Table 19.2. Economic and ecological importance of various halophyte species.

Halophyte species	Family	Habit	References
Medicinal			
Alhagi maurorum	Papilionaceae	Shrub	Qasim *et al*. (2017)
Arthrocnemum indicum	Amaranthaceae	Herb	Qasim *et al*. (2011)
Chenopodium album	Amaranthaceae	Herb	Qureshi *et al*. (2010)
Cressa cretica	Convolvulaceae	Herb	Sunita *et al*. (2011)
Crithmum maritimum	Apiaceae	Herb	Ksouri *et al*. (2012a)
Cymbopogon jwarancusa	Poaceae	Grass	Qasim *et al*. (2011)
Cynara cardunculus	Asteraceae	Shrub	Falleh *et al*. (2008)
Cyperus rotundus	Cyeraceae	Grass	Qureshi *et al*. (2010)
Desmostachya bipinnata	Poaceae	Grass	Qureshi *et al*. (2010)
Enicostema hyssopifolium	Gentianaceae	Tree	Qasim *et al*. (2017)
Eucalyptus camaldulensis	Myrtaceae	Shrub	El-Massry *et al*. (2002)
Haloxylon stocksii	Amaranthaceae	Shrub	Ahmed *et al*. (2015)
Indigofera cordifolia	Papilionaceae	Herb	Qureshi and Bhatti (2008)
Indigofera oblongifolia	Papilionaceae	Shrub	Qureshi and Bhatti (2008)
Launaea resedifolia	Asteraceae	Herb	Qasim *et al*. (2011)
Mesembryanthemum edule	Aizoaceae	Herb	Ksouri *et al*. (2012a)
Parkinsonia aculeata	Caesalpiniaceae	Tree	Qasim *et al*. (2017)
Prosopis juliflora	Mimosaceae	Shrub to tree	Ahmed *et al*. (2015)
Salicornia brachiata	Amaranthaceae	Herb	Stanley (2008)
Salicornia herbacea	Amaranthaceae	Herb	Kong and Seo (2012)
Salsola baryosma	Amaranthaceae	Shrub	Taha and Alsayed (2000)
Salvadora persica	Salvadoraceae	Shrub to tree	Kumari and Parida (2016)
Sesuvium portulacastrum	Aizoaceae	Herb	Chandrasekaran *et al*. (2009)
Sesuvium verrucosum	Aizoaceae	Herb	Taha and Alsayed (2000)
Solanum surattense	Solanaceae	Herb	Qureshi and Bhatti (2008)
Suaeda maritima	Amaranthaceae	Herb	Chandrasekaran *et al*. (2009)
Suaeda monoica	Amaranthaceae	Herb	Chandrasekaran *et al*. (2009)
Tamarix gallica	Tamaricaceae	Shrub or small tree	Ksouri *et al*. (2009)
Thespesia populneoides	Malvaceae	Tree	Qasim *et al*. (2011)
Trianthema portulacastrum	Aizoaceae	Herb	Qasim *et al*. (2017)
Withania somnifera	Solanaceae	Shrub	Qasim *et al*. (2017)
Zygophyllum album	Zygophyllaceae	Shrub	Ksouri *et al*. (2013)
Zygophyllum gaetulum	Zygophyllaceae	Shrub	Bourgou *et al*. (2017)
Zygophyllum quatarense	Zygophyllaceae	Shrub	Taha and Alsayed (2000)
Zygophyllum simplex	Zygophyllaceae	Shrub	Hussein *et al*. (2011)
Food and nutraceuticals			
Alhagi maurorum	Fabaceae	Shrub	Weber *et al*. (2007)
Arthrocnemum indicum	Amaranthaceae	Herb	Weber *et al*. (2007)
Arthrocnemum macrostachyum	Amaranthaceae	Shrub	Custódio *et al*. (2012)
Batis maritima	Bataceae	Shrub	Marcone (2003)
Chenopodium quinoa	Amaranthaceae	Shrub	Varli and Şanlıer (2016)
Cressa cretica	Convolvulaceae	Herb	Weber *et al*. (2007)
Distichlis palmeri	Poaceae	Grass	Pearlstein *et al*. (2012)
Halopyrum mucronatum	Poaceae	Grass	Weber *et al*. (2007)
Haloxylon stocksii	Amaranthaceae	Shrub	Weber *et al*. (2007)
Kosteletzkya virginica	Malvaceae	Shrub	Panta *et al*. (2014)
Lepidium latifolium	Brassicaceae	Herb	Kaur *et al*. (2013)
Plantago coronopus	Plantaginaceae	Herb	Flowers and Muscolo (2015)
Salicornia bigelovii	Amaranthaceae	Herb	Ventura *et al*. (2011)
Salicornia europaea	Amaranthaceae	Herb	Flowers and Muscolo (2015)

Continued

Table 19.2. Continued.

Halophyte species	Family	Habit	References
Salsola soda	Amaranthaceae	Succulent shrub	Ventura *et al.* (2015)
Salvadora persica	Salvadoraceae	Shrub to tree	Kumari *et al.* (2017)
Salvia fruticosa	Lamiaceae	Herb	Weber *et al.* (2007)
Suaeda aralocaspica	Amaranthaceae	Herb	Wang *et al.* (2012b)
Tamarix gallica	Tamaricaceae	Shrub	Ksouri *et al.* (2009)
Tripolium pannonicum	Asteraceae	Shrub	Flowers and Muscolo (2015)
Fodder crops			
Arthrocnemum indicum	Amaranthaceae	Herb	Davidson and Galloway (1993)
Atriplex amnicola	Amaranthaceae	Shrub	Liang *et al.* (2016)
Atriplex halimus	Amaranthaceae	Shrub	Zarrouk *et al.* (2003)
Atriplex lentiformis	Amaranthaceae	Shrub	Glenn *et al.* (2009)
Atriplex leucoclada	Amaranthaceae	Shrub	Amouei (2013)
Atriplex nummularia	Amaranthaceae	Shrub	Ravindran *et al.* (2007)
Bienertia cycloptera	Amaranthaceae	Herb	Davidson and Galloway (1993)
Distichlis spicata	Poaceae	Grass	Bustan *et al.* (2005)
Haloxylon salicornicum	Amaranthaceae	Shrub	El Shaer (2010)
Juncus acutus	Juncaceae	Herb	El Shaer (2010)
Juncus gerardii	Juncaceae	Grass	Czyż *et al.* (2013)
Kochia eriophora	Amaranthaceae	Shrub	El Shaer (2010)
Limonium pruinosum	Plumbaginaceae	Shrub	El Shaer (2010)
Nitraria retusa	Nitrariaceae	Shrub	El Shaer (2010)
Panicum turgidum	Poaceae	Grass	Abideen *et al.* (2013)
Plantago maritima	Plantaginaceae	Herb	Czyż *et al.* (2013)
Salsola drummondii	Amaranthaceae	Herb	Davidson and Galloway (1993)
Sesbania sesban	Fabaceae	Tree	Oosting *et al.* (2011)
Suaeda fruticosa	Amaranthaceae	Shrub	Hameed and Khan (2011)
Triglochin palustris	Juncaginaceae	Grass	Czyż *et al.* (2013)
Triglochin maritimum	Juncaginaceae	Grass	Czyż *et al.* (2013)
Zygophyllum simplex	Zygophyllaceae	Herb	Hussein *et al.* (2011)
Edible oil			
Alhaji maurorum	Fabaceae	Shrub	Weber *et al.* (2007)
Allenrolfea occidentialis	Amaranthaceae	Shrub	Weber *et al.* (2001)
Arthrocnemum indicum	Amaranthaceae	Herb	Weber *et al.* (2007)
Arthrocnemum macrostachyum	Amaranthaceae	Shrub	Weber *et al.* (2007)
Atriplex heterosperma	Amaranthaceae	Herb	Weber *et al.* (2001)
Atriplex rosea	Amaranthaceae	Herb	Weber *et al.* (2001)
Cressa cretica	Convolvulaceae	Herb	Weber *et al.* (2007)
Descurainia sophia	Brassicaceae	Herb	Weber *et al.* (2007)
Halogeton glomeratus	Amaranthaceae	Herb	Weber *et al.* (2001)
Halopyrum mucronatum	Poaceae	Grass	Hameed and Khan (2011)
Haloxylon stocksii	Amaranthaceae	Shrub	Hameed and Khan (2011)
Kochia scoparia	Amaranthaceae	Herb	Weber *et al.* (2001)
Sarcobatus vermiculatus	Sarcobataceae	Shrub	Weber *et al.* (2001)
Suaeda fruticosa	Amaranthaceae	Shrub	Hameed and Khan (2011)
Suaeda torreyana	Amaranthaceae	Herb	Weber *et al.* (2001)
Biofuel			
Aeluropus lagopoides	Poaceae	Grass	Abideen *et al.* (2012)
Alhagi maurorum	Papilionaceae	Shrub	Abideen *et al.* (2012)
Arthrocnemum macrostachyum	Amaranthaceae	Shrub	Abideen *et al.* (2012)
Cressa cretica	Convolvulaceae	Herb	Abideen *et al.* (2012)
Crithmum maritimum	Apiaceae	Shrub	Abideen *et al.* (2014)
Descurainaia sophia	Brassicaceae	Herb	Abideen *et al.* (2011)

Continued

Table 19.2. Continued.

Halophyte species	Family	Habit	References
Desmostachya bipinnata	Poaceae	Grass	Hameed and Khan (2011)
Euphorbia tirucalli	Euphorbiaceae	Shrub or tree	Abideen *et al.* (2014)
Haloxylon stocksii	Amaranthaceae	Shrub	Abideen *et al.* (2012)
Kosteletzkya pentacarpos	Malvaceae	Shrub	Moser *et al.* (2013)
Kosteletzkya virginica	Malvaceae	Shrub	Abideen *et al.* (2014)
Miscanthus sinensis	Poaceae	Grass	Abideen *et al.* (2011)
Panicum turgidum	Poaceae	Grass	Hameed and Khan (2011)
Phragmites australis	Poaceae	Grass	Abideen *et al.* (2011)
Phragmites karka	Poaceae	Grass	Abideen *et al.* (2012)
Ricinus communis	Euphorbiaceae	Shrub	Abideen *et al.* (2014)
Salicornia bigelovii	Amaranthaceae	Herb	Eganathan *et al.* (2006)
Spartina alterniflora	Poaceae	Grass	Abideen *et al.* (2011)
Suaeda aralocaspica	Amaranthaceae	Shrub	Wang *et al.* (2012a)
Suaeda fruticosa	Amaranthaceae	Shrub	Abideen *et al.* (2014)
Tamarix indica	Tamaricaceae	Shrub or tree	Abideen *et al.* (2011)
Typha domingensis	Typhaceae	Herb	Hameed and Khan (2011)
Urochondra setulosa	Poaceae	Grass	Abideen *et al.* (2011)
Biofilter			
Aster tripolium	Asteraceae	Herb	de Lange and Paulissen (2016)
Atriplex barclayana	Amaranthaceae	Herb	Brown and Glenn (1999)
Bolboschoenus maritimus	Cyperaceae	Shrub	de Lange and Paulissen (2016)
Ficinia nodosa	Cyperaceae	Grass	Szota *et al.* (2015)
Glyceria maxima	Poaceae	Grass	de Lange *et al.* (2013)
Juncus kraussii	Juncaceae	Grass	Szota *et al.* (2015)
Phragmites australis	Poaceae	Grass	Hegedűs *et al.* (2010)
Salicornia bigelovii	Chenopodiaceae	Herb	Brown and Glenn (1999)
Salicornia europaea	Chenopodiaceae	Herb	Webb *et al.* (2012)
Salicornia persica	Chenopodiaceae	Herb	Shpigel *et al.* (2013)
Scirpus lacustris	Cyperaceae	Shrub	Hegedűs *et al.* (2010)
Spartina anglica	Poaceae	Grass	de Lange and Paulissen (2016)
Suaeda esteroa	Amaranthaceae	Shrub	Brown and Glenn (1999)
Tripolium pannonicum	Asteraceae	Shrub	Buhmann *et al.* (2015)
Typha angustifolia	Typhaceae	Herb	Hegedűs *et al.* (2010)
Phytoremediatiation and desalinization			
Atriplex halimus	Amaranthaceae	Shrub	Nedjimi and Daoud (2009)
Atriplex lentiformis	Amaranthaceae	Shrub	Glenn *et al.* (2009)
Atriplex nummularia	Amaranthaceae	Shrub	Jordan *et al.* (2002)
Clerodendron inerme	Lamiaceae	Tree	Jlassi *et al.* (2013)
Excoecaria agallocha	Euphorbiaceae	Tree	Jlassi *et al.* (2013)
Halimione portulacoides	Amaranthaceae	Shrub	Lokhande and Suprasanna (2012)
Heliotropium curassavicum	Boraginaceae	Shrub	Jlassi *et al.* (2013)
Ipomoea pes-caprae	Convolvulaceae	Shrub	Jlassi *et al.* (2013)
Juncus rigidus	Juncaceae	Grass	Jlassi *et al.* (2013)
Juncus acutus	Juncaceae	Herb	Jlassi *et al.* (2013)
Mesembryanthemum crystallinum	Aizoaceae	Herb	Abdelly *et al.* (2006)
Sesbania sesban	Fabaceae	Tree	Dan *et al.* (2011)
Sesuvium portulacastrum	Aizoaceae	Herb	Ghnaya *et al.* (2007)
Suaeda maritima	Amaranthaceae	Herb	Rabhi *et al.* (2009)
Sulla carnosa	Fabaceae	Herb	Jlassi *et al.* (2013)
Tamarix smyrnensis	Tamaricaceae	Tree	Manousaki *et al.* (2009)

have been used to reclaim saline soil (Hasanuzzaman *et al.*, 2014). Of these, biological methods appear more promising than cost-effective chemical amendments (Rabhi *et al.*, 2010; Hasanuzzaman *et al.*, 2014). Halophytes have provided a favourable microhabitat for improved growth of glycophytes by desalinization and improvement of the fertility of the rhizosphere (Rabhi *et al.*, 2010). Several authors have suggested the use of salt-accumulating species such as *S. maritima*, *S. portulacastrum*, *S. fruticosa*, *S. salsa*, *S. calceoliformis*, *Kalidium folium*, *Sesuvium portulacastrum*, *A. indicum*, *L. monopetalum*, *A. nummularia* and *A. prostrata* for desalinization of salt-affected areas, as these plants accumulate high concentrations of salt in their aerial tissues (Rabhi *et al.*, 2009; Lokhande *et al.*, 2013; de Souza *et al.*, 2014; Hasanuzzaman *et al.*, 2014; Manousaki *et al.*, 2014; Muchate *et al.*, 2016; Silva *et al.*, 2016). *S. fruticosa* has been found to remove more than 1088.6 kg of salt through its aerial parts each year (Manousaki and Kalogerakis, 2011b). Several studies on the use of halophytes in soil desalination have been reported in the literature. The obligate halophyte *S. portulacastrum*, for example, proved successful in desalinating soils, with the potential to remove up to 1 t Na^+ ha^{-1} through sequestration in roots and shoots (Rabhi *et al.*, 2009, 2010; Van Oosten and Maggio, 2015). *S. maritima* and *S. portulacastrum* have also been shown to bioaccumulate salts in their tissues and reduce soil salinity (Ravindran *et al.*, 2007; Rabhi *et al.*, 2009; Van Oosten and Maggio, 2015). Similarly, the halophyte *Sulla carnosa* showed desalinization capacity up to 0.3 metric tons Na^+ ha^{-1}), and showed potential as a forage crop (Jlassi *et al.*, 2013). Deep-rooted halophytes like *A. lentiformis* have been used to absorb salt from mildly saline effluent from wastewater treatment facilities, with the intent to protect aquifers in urban environments (Glenn *et al.*, 2009; Van Oosten and Maggio, 2015). Along with tolerance to sodium and chloride, halophytes can also tolerate other toxic ions, to some extent employing the same adaptation mechanism (Flowers and Colmer, 2008). Therefore, the enormous potential halophytes present to plant biology and agriculture is not just limited to salt tolerance. As momentum builds within the research community studying halophytes and salt tolerance, understanding how these species interact with other abiotic stresses will become more important (Wang *et al.*, 2014; Flowers and Colmer, 2015; Van Oosten and Maggio, 2015). Phytodesalination using *Echinochloa stagnina* has proved to be a cost-effective, eco-friendly method to restore saline clay soil (Ado *et al.*, 2016). *Spartina maritima* demonstrated significantly enhanced heavy metal uptake when grown in conjunction with low-weight organic acids. Application of citric acid to this species increased Zn uptake by 85% and acetic acid increased Ni concentration in roots by 139%. Moreover, chromium uptake was dramatically increased by application of low molecular weight organic acids, with acetic acid (1032%) and malic acid (770%) having the largest effects Van Oosten and Maggio, 2015; (Duarte *et al.*, 2017). Metal excretion has also been reported for other *Spartina* species, including *S. foliosa* and *S. anglica* (Rozema *et al.*, 2013). *Panicum turgidum*, a perennial halophytic grass, has been shown to be useful as an alternative cattle feed when irrigated with brackish water on saline soils. It has been reported that a *Panicum–Suaeda* intercropping system helped maintain soil salinity at low levels and that *P. turgidum* can produce more than 50 tons ha^{-1} biomass annually as fodder for cattle (Khan *et al.*, 2009). Mangrove leaves have been reported to contain high levels of salt, minerals, amino acids, vitamins and iodine, making them suitable as high-quality fodder (Abdel-Aziz *et al.*, 2016).

Phytoremediation is a low-cost, environmentally friendly and effective method for the decontamination of soils polluted by heavy metals (Ado *et al.*, 2016). Compared with glycophytes, halophytes have some additional advantages for phytoremediation in such soils, such as higher tolerance to heavy metals and greater amounts of heavy metal uptake (Taamalli *et al.*, 2014; Liang *et al.*, 2016). Moreover, salinity can increase heavy metal mobility in soils and promote the translocation of heavy metals from roots to shoots (Acosta *et al.*, 2011). The high resistance of halophytes to heavy metals is strongly linked to their salt-tolerance characteristics (Wang *et al.*, 2014). It has been proposed that heavy metal-tolerant plants and halophytes share a number of traits in common (Van Oosten and Maggio, 2015). Therefore, it has been suggested that halophytes are naturally better adapted to environmental stresses, including heavy metals, than glycophytes (Manousaki and Kalogerakis, 2011b; Hasanuzzaman *et al.*, 2014; Wang *et al.*, 2014). Some halophytes, such as *A. nummularia* (Silva *et al.*, 2016), *S. pectinata* (Korzeniowska and Stanislawska-Glubiak, 2015), *S. portulacastrum* (Ayyappan *et al.*, 2015),

T. africana (Santos *et al.*, 2015) and *Limoniastrum monopetalum* (Manousaki *et al.*, 2014) have proved to be well suited for phytoextraction or phytostabilization of heavy metal-contaminated saline soils. *A. halimus* typically accumulates low concentrations of Pb and Cd, but has high biomass, making it a viable candidate for use in phytoextraction in arid, saline soils (Manousaki *et al.*, 2009). *Bolboschoenus maritimus* has been reported for its use in phytoextraction of Cd-affected saline soil (Santos Pedro *et al.*, 2015). *S. maritima* is not a metal hyperaccumulator, but can accumulate reasonable levels of Cu and Zn, and tolerate both moist or dry saline soils (Milic *et al.*, 2012).

19.3.2 Use of halophytes as biofilters

Halophytes have the ability to grow in salt-affected habitats, which makes them useful in recycling the nutrient effluents from saline aquaculture (Buhmann and Papenbrock, 2013). Biofiltration systems are stormwater control measures specifically designed to reduce concentrations of pollutants discharged into urban waterways, including nitrogen, phosphorus and heavy metals (Fletcher *et al.*, 2014; Szota *et al.*, 2015). The majority of pollutants associated with particulates, including phosphorus and heavy metals, are physically removed from stormwater by the filter media, while dissolved forms can be removed with varying success through sorption and precipitation (Glaister *et al.*, 2014; Szota *et al.*, 2015). Vegetation can significantly increase the removal of dissolved nitrogen and phosphorus and, to a lesser extent, heavy metals (Szota *et al.*, 2015). Generally, biofilters include natural and artificially constructed wetlands for the treatment of municipal wastewater, industrial wastewater (e.g. from petrochemicals, textiles, paper, food and mining), agricultural wastewater (e.g. dairy and fish farm effluent) and runoff waters from highways, airports, landfill and after storms (Buhmann and Papenbrock, 2013). Aquaculture effluents have been found to contain various substances that are harmful to the environment as they contain high levels of nitrogen and phosphorus in organic and inorganic forms that are toxic to marine fishes and organisms in the surrounding environment (Boxman *et al.*, 2017). Toxicity occurs because unfiltered wastewater from the aquaculture effluent contains many toxic organic substances; heavy metals; and ions such as nitrate and phosphate, which cause hypertrophication of ecosystems. Nitrate has been assumed to be the limiting nutrient in most aquatic ecosystems, whereas in marine aquatic ecosystems phosphate is thought to be the limiting nutrient (Sundareshwar *et al.*, 2003; Buhmann and Papenbrock, 2013). Aquaculture effluent often contains high amounts of organic solids in settled, suspended and dissolved forms, apart from the dissolved inorganic nutrients (Buhmann and Papenbrock, 2013). The presence of organic solids and inorganic compounds results in high biological and chemical oxygen demand (Cao *et al.*, 2007). Use of glycophytes as biofilters for industrial and municipal wastewater treatment is well established (Kadlec and Wallace, 2009), but less focus has been given to the use of halophytes in saline wetlands as filters for saline aquaculture effluent. Now, the rising importance of seafood-derived protein in the human diet has increased demand for the use of halophytes in effluent treatment (Donato *et al.*, 2011; Ladeiro, 2012; Varli and Şanlıer, 2016; Filho *et al.*, 2017). In addition to taking up the nutrients, halophytes also contribute to the filtering effect by trapping suspended solids in the root and shoot system through slowing down the water flow and decreasing the biological oxygen demand due to microbial decomposition of the organic material. They also enhance bacterial growth by providing a large surface area and source of carbohydrate for bacterial consumption (Buhmann and Papenbrock, 2013; Hasanuzzaman *et al.*, 2014; Santos *et al.*, 2015). Several studies have investigated the use of halophytes as biofilters for different nutrient-rich saline effluents (Wu *et al.*, 2008; Buhmann and Papenbrock, 2013; Buhmann *et al.*, 2015). Studies on the application of halophytes as biofilters for aquaculture process water have been summarized by Buhmann and Papenbrock (2013). Díaz *et al.* (2013) have demonstrated good performance by various halophyte species (*Distichlis spicata*, *S. bigelovii*, *S. gracilis*, *Bassia hyssopifolia*, *A. occidentalis* and *A. lentiformis*) in a saline drainage water reuse system. In addition to the promising application of halophytes as biofilters and valuable co-products, it is highly beneficial to foster the use of saline water for agricultural production because freshwater is a declining resource in many regions of the world (Rozema and Schat, 2012; Buhmann *et al.*, 2015). Shpigel *et al.* (2013) have demonstrated construction of wetlands with *S. persica* for effective removal of nitrogen, phosphorus and total suspended solids from commercial and semi-recirculated aquacultural systems. Furthermore, three species of halophytes, *A. tripolium*, *B. maritimus* and *S. anglica* have been described as able to store nutrients and efficiently remove contaminants for the purification of contaminated water from aquaculture, saline agriculture or freshwater-based agriculture. It has been reported that, of the above species, *B. maritimus* had maximum uptake of phosphorus and nitrogen in shoots and *S. anglica* had highest uptake of phosphorus and nitrogen in roots (de Lange and Paulissen, 2016). Several salt-tolerant species such as *T. pannonicum*, *P. coronopus*, *S. persica*, *A. halimus*, *A. portolacoides*, *L. latifolium* and *S. dolicostachya* have been used as biofilters for effluent treatment in open marine aquaculture systems; in nutrient-rich saline municipal, agricultural and industrial wastewater; and in the growth and breeding of salt-tolerant crop plants, showing valuable secondary metabolites with market potential (Shpigel *et al.*, 2013; Buhmann *et al.*, 2015). 'Factors that influence the successful cultivation of salt-tolerant species, such as salinity, nutrient concentration, light conditions and availability of micronutrients, can also influence the performance of the respective halophyte biofilter' (Buhmann and Papenbrock, 2013).

19.4 Economic Prominence of Halophytes

Salt tolerance is a complex phenomenon, hence improving the salt tolerance of conventional crops and developing salt-tolerant crop varieties have been challenging. Using halophytes as non-conventional crop plants for food, forage, fuel and medicines is, therefore, an alternative solution to developing salt-tolerant traditional crops (Flowers and Colmer, 2015). Cultivation of halophytes on salt-affected wastelands by using saline water would reduce consumption of agricultural water and help to improve saline wastelands (Koyro *et al.*, 2011). Several studies have described the potential of halophytes as vegetables or as raw material for fodder, and to provide secondary metabolites that can be used for pharmaceuticals, functional foods, nutraceuticals and technical implementations (Ksouri *et al.*, 2012a; Buhmann and Papenbrock, 2013) (Fig. 19.3). A challenge for the combination of nutrient-rich saline wastewater purification by halophytes with the production of a valuable co-product is the limited knowledge about the cultivation of halophytes (Buhmann *et al.*, 2015). Different aspects of the cultivation of the edible halophyte genera *Salicornia* and *Sarcocornia* have been described (Ventura *et al.*, 2011, 2015; Rozema *et al.*, 2013; Ventura and Sagi, 2013). Other halophyte species, such as *T. pannonicum*, *P. coronopus* and *C. maritimum*, also have potential as saline vegetable crops, and initial studies into cultivation have been performed (Koyro, 2006; Koyro *et al.*, 2011; Ventura and Sagi, 2013). However, further research is required into cultivating wild halophyte species and improving yield. Besides this, halophytes have been reported to have huge potential for phytoremediation and desalinization of salt-affected areas (Hasanuzzaman *et al.*, 2014; Panta *et al.*, 2014). The unique ability of halophytes to thrive in saline environments, and their positive impact on the environment, also makes them potential candidates for the production of biofuels. The following sections describe various candidate halophyte species with economic potential.

19.4.1 Halophytes and nutraceuticals

The difference between foods and nutraceuticals depends upon the way in which they are consumed: functional food is consumed as regular food, and nutraceuticals are consumed as pills, tablets and capsules (Espin *et al.*, 2007; Buhmann *et al.*, 2013). Halophytic species have been used as an alternative to conventional food crops in regions with prevalent saline soil and saline groundwater for irrigation (Rozema *et al.*, 2013; Panta *et al.*, 2014; Ahmed *et al.*, 2015). So far, the potential of many halophytic species as crop plants have been evaluated and reviewed (Weber *et al.*, 2007; Koyro *et al.*, 2011; Rozema *et al.*, 2013; Ventura and Sagi, 2013; Panta *et al.*, 2014; Ahmed *et al.*, 2015; Ventura *et al.*, 2015; Patel, 2016). For example, *Salicornia* spp. have served in several European countries as a vegetable (Ventura *et al.*, 2011; Ventura and Sagi, 2013; Buhmann and Papenbrock, 2013). Many halophyte species, such as *P. coronopus*, *C. maritimum*, *S. soda*, *Crambe maritima*, *Beta maritima*, *A. tripolium* and *Salicornia* spp. have been documented as containing sugars, amino acids, quaternary ammonium compounds, polyols, antioxidants and essential oils that have the potential for nutritional and medicinal applications (Ventura *et al.*, 2015). In recent years, interest in halophyte agriculture has increased because of the unavailability of good quality land and water for cultivation (Rozema and Schat, 2012). Many halophytic species (e.g. *D. palmeri* and *C. quinoa*) have been used as staple food crops in many parts of the world (Pearlstein *et al.*, 2012; Filho *et al.*, 2017). *D. palmeri* has been reported to have a higher fibre content (8.4%) than wheat (2.6%) and is used for making biscuits and bread. This species is capable of producing 1.25 metric tons ha^{-1} of grain with high nutritional value, even in flooded and saline conditions (Pearlstein *et al.*, 2012). Similarly, quinoa has also been regarded as an exotic food. This species has a significant lysine, iron, magnesium and vitamin E content (Adolf *et al.*, 2012). The leaves of *A. triangularis* have been often used as regular salad crops in many parts of the world. This species has a fresh weight (FW) yield of 21.2 tonnes ha^{-1} (Panta *et al.*, 2014). The leaves of *S. bigelovii* are an alternative source of β-carotene and omega-3 polyunsaturated fatty acid to seaweed and spinach, and are used for human consumption. Quantities of β-carotene in this species range from 4.7 mg (Ventura *et al.*, 2015) to 15.9 mg 100 g^{-1} FW (Lu *et al.*, 2010), which is considerably higher than those found in seaweeds (4.0 mg 100 g^{-1} FW) and spinach (Isabelle *et al.*, 2010). Similarly, the seeds of *K. virginica* and *S. fruticosa* have been an important source of protein and FAs (Weber *et al.*, 2007; Panta *et al.*, 2014). The fatty acid profile of some halophytic species holds great promise for production of commercial high polyunsaturated vegetable oil equivalent to conventional oil seed crops such as canola and sunflower (Ksouri *et al.*, 2012a). Several mangrove species have also been used as food in many parts of the world; species such as *A. marina* and *A. germinans* have been used as salad (Ahmed *et al.*, 2015). Halophytic plants contain high amount of beneficial mono- and PUFAs which can be used as nutraceuticals (Kumar *et al.*, 2012; Varli and Şanlıer, 2016). The oil quantity and fatty acid composition of various halophytic species such as *A. indicum*, *Alhagi maurorum*, *C. cretica*, *H. mucronatum* and *H. stocksii* have been compared, and they all contain high amounts of linoleic acid (Weber *et al.*, 2007; Buhmann *et al.*, 2013). Seeds of *S. aralocaspica* and *A. indicum*, and the seed and leaves of *C. maritimum*, have been found to be rich in linoleic and oleic acids (Atia *et al.*, 2010; Custódio *et al.*, 2012; Wang *et al.*, 2012b). Furthermore, *C. maritimum* has been used as part of the human diet. The whole plant has a spicy taste, with a certain saltiness, justifying its use as an ingredient in salads (Jallali *et al.*, 2014). Their fresh young leaves and fermented products have been used as salad with yogurt, especially in Mediterranean regions (Nguir *et al.*, 2011). The antioxidative properties of halophytic plant extracts have raised the nutritional value of such species because the antioxidants provide a protective and health-promoting effect against ROS-generated diseases (Bouayed and Bohn, 2010). Various phenolic compounds, such as syringic acid, catechin and rutin hydrate, from the extract of different halophytic species, have contributed to their antioxidative properties (Oueslati *et al.*, 2012). Procyanidins, propelargonidins and quercitin are flavonoids with the highest antioxidative properties, and have been found in the extract of *M. edule*. In this species, the aerial parts (such as leaves and stem) have higher antioxidative activity than lower parts of the same plant, and act in an organ-specific manner (Falleh *et al.*, 2011). This high antioxidative property of aerial parts is due to the presence of numerous phenolic compounds in the stem of *M. edule* which is much higher than in many glycophytic medicinal plants (Ksouri *et al.*, 2012b). Similarly, in the halophyte species *M. crystallinum*, 2,6-bis (1,1-dimethylethyl)-4-methylphenol was found to be the main component contributing to the antioxidative property in a methanolic extract (Bouftira *et al.*, 2007). *T. gallica* has been reported to possess antiradical activity, mainly due to the presence of phenolic compounds (Ksouri *et al.*, 2009). An organ (leaf and flower) extract of this species had a high ferrous ion content with strong chelating abilities, as found in many vegetables, and was able to act as a powerful pro-oxidant (Ksouri *et al.*, 2009). *Cynara cardunculus* has been reported to have therapeutic potential due to presence of phenolic compounds such as gallic acid, sinapic acid, chlorogenic acid, quercetin and catechin (Falleh *et al.*, 2008). Halophyte seeds have been prospectively used as a source of edible oil. Seeds of several species (e.g. *A. indicum*, *Alhagi maurorum*,

C. cretica, *H. mucronatum*, *H. stocksii*, *S. aegyptiaca*, *Halocnemum strobilaceum* and *S. fruticosa*) have been analysed for their resources as edible oils. Of these, *S. fruticosa* and *S. aegyptiaca* have been documented as sources of edible oil for human consumption (Weber *et al*., 2007). *S. bigelovii* is a tidal marsh plant whose seed has been reported as yielding good-quality oil, comparable to conventional oil seed (Anwar *et al*., 2002).

19.4.2 Halophytes as biofuel crops and in industry

Because of rising demand for fuel, and the scarcity of available fossil fuels, a sustainable alternative is production of biofuels from plant sources. However, to avoid feed and fuel crops important for human consumption being used as biofuel feedstock, halophytes may be used instead (Koyro *et al*., 2011). The use of biofuels may also be a solution to reduce greenhouse gas emissions, and so diminish the effect of global warming produced by fossil fuel use (Sharma *et al*., 2016). Some halophytic species (e.g. *S. virginica*, *S. bigelovii*, *S. euphoraea*, *Rhizophora mangle*, *A. germinans* and sea grass) have been studied by the National Aeronautics and Space Administration (NASA) to optimize use of biomass produced by these halophytes as a viable alternative energy resource (Bomani *et al*., 2011). The advantages of using halophytes for production of biofuel are that they do not require fertile agricultural land and water for irrigation, and also that they will be less harmful to the environment than traditional fuels. Some researchers have suggested the potential candidature of halophytic grasses such as *P. turgidum*, *Phragmites karka*, *D. bipinnata*, *Typha domingensis* and *H. mucronatum* for bioethanol production (Abideen *et al*., 2011, 2012, 2014; Ventura *et al*., 2015). These species have been reported to have high growth rates for the production of good-quality ligno-cellulosic biomass, comprising cellulose 26%–37%, hemi-cellulose 24%–38% and lignin < 10% for ethanol production (Abideen *et al*., 2011). They are considered to be efficient producers of bioethanol owing to their massive biomass and low lignin content. Ligno-cellulosic biomass from halophytes is considered to be a potential source in biodiesel production (Koyro *et al*., 2011). In recent years the non-edible seed oil of *Calophyllum inophyllum* and *Pongamia pinnata* has been recognized as a potential source of biodiesel (Sangwan *et al*., 2010; Mardhiah *et al*., 2017; Marriboina *et al*., 2017). Studies show that many species of halophytes (e.g. *S. aralocaspica*, *S. bigelovii*, *C. maritimum*, *Ricinus communis*, *Euphorbia tirucalli*, *Kosteletzkya virginica* and *Descurainia sophia*) contain high concentrations of oil, more than 20% of the total seed dry weight (Abideen *et al*., 2012). For economic expansion of agricultural areas, oil-yielding crop plants are very important. It has been reported that the seed oil of *Salicornia* spp. contains 10-undecanoic acid and has high ester and saponification values, and has potential applications in protective coatings, plastics, plasticizers, dispersants, pharmaceuticals, cosmetics, detergents, soaps, textiles, surfactants, lubricant additives, organic pesticides, urethane derivatives and a variety of synthetic intermediates as stabilizers in plastic formulations (Eganathan *et al*., 2006). It also contained high amounts of linoleic acid, which make it suitable for paint and varnishes (Balat, 2011). Fatty acid methyl esters from the seeds of many halophyte species (e.g. *P. oleracea*, *E. maritimum*, *Conium maculatum*, *Prangos uechtritzii*, *Nigella damascena* and *Tribulus terrestris*) have been reported suitable for the production of biodiesel (Guil-Guerrero *et al*., 2017). Recently, a succulent halophyte, *Salicornia sinus-persica*, has been assessed for its potential as a feedstock for bioethanol production (Alassali *et al*., 2016).

19.4.3 Medicinal importance of halophytes

Halophytes have immense potential for medicinal applications and a great number of halophytes are still used in the treatment of various diseases, microbial infections and ageing processes, as part of traditional or folk medicines (Shafi *et al*., 2001; Adams *et al*., 2009; Ksouri *et al*., 2009, 2012a; Abeysinghe, 2010; Sher *et al*., 2010; Kumar *et al*., 2012; Mariem *et al*., 2014). Increasing attention is currently being shown to the potential of halophytes as medicines, nutraceuticals and functional foods, because of their high bioactive antioxidant secondary molecule content and big economic interests (Ksouri *et al*., 2012a). Halophytes are known to have powerful antioxidant defence systems that can tolerate unfavourable environmental conditions such as high salinity. Their enzymatic or non-enzymatic components can resist and reduce internally generated ROS owing to the production of polyphenols and other antioxidant compounds (Ksouri *et al.*, 2007). Some secondary metabolites, including terpenoids (carotenoids, diterpenes, sesquiterpenes and essential oil) and phenolic compounds have been identified so far in halophytes (Boestfleisch *et al.*, 2014). These antioxidant components may delay the oxidation of cellular molecules and repair tissue injuries that lead to the prevention of cell death (Pandey and Rizvi, 2009). Ethnobotanical studies on halophytes, therefore, have been essential in appraising their potential in medicine as well as their economic use, and to clarify their traditional use as curatives (Liebezeit, 2008; Wang *et al.*, 2012a). Numerous researchers have documented the traditional uses of medicinal halophytes in many rural and tribal areas of the world (Adams *et al.*, 2009; Ksouri *et al.*, 2009, 2012a; Ahmed *et al.*, 2015; Patel, 2016; Kumari *et al.*, 2017; Qasim *et al.*, 2017). Many of the secondary metabolites synthesized during adaptation to the saline environment are a major source of potential medicines against various health-related conditions and chronic diseases. Whole plant, or plant parts, have been used as a cure for various illnesses, indicating the presence of beneficial chemicals (Boestfleisch and Papenbrock, 2017). *S. herbacea* has been used in the dietary, pharmaceutical and cosmetic industries as a novel source of natural antioxidant owing to its strong biological activities derived from high contents of useful polyphenols, minerals and vitamins (Kang *et al.*, 2011). Isorhamnetin 3-O-β-D glucopyranoside, a component of glucopyranoside isolated from *S. herbacea*, appreciably suppressed adipogenic differentiation by blocking lipid accumulation; it may be effective in preventing obesity or obesity-related diseases (Kong and Seo, 2012). Several studies have indicated that many halophytic plant extracts have antimicrobial activity against some human pathogenic bacteria and fungi (e.g. *B. subtilis*, *E. coli*, *P. aeruginosa*, *A. fumigatus* and *C. albicans*) (Samiullah *et al.*, 2011; Nostro *et al.*, 2012; Sunita *et al.*, 2012; Ben Mansour *et al.*, 2016). The efficacy of antimicrobial leaf and stem extracts of *S. monoica* against various microorganisms has been established and it was reported that the leaves exhibited substantially higher antimicrobial activity than the shoot extract against *B. subtilis*, *B. megaterium*, *Lactobacillus acidophilus*, *E. coli*, *Enterobacter aerogenes*, *E. cloacae*, *Klebsiella pneumoniae*, *C. albicans*, *Mucor racemosus*, *Rhizoctonia solani*, *R. stolonifer* and *S. cerevisiae* (Lakshmi and Rao, 2013). The edible halophyte *C. maritimum* (also known as sea fennel) has been found to contain high amounts of vitamin C, iodine, carotenoids and phenolic compounds; owing to these properties it is widely used in medicine (Jallali *et al.*, 2014). Further, falcarindiol – obtained from the chloroform extract of *C. maritimum* leaves – has been reported to act as a strong growth inhibitor of *M. luteus* and *B. cereus* (Meot-Duros and Magné, 2009). Bioactive products like chlorogenic acid, epigallocatechin, vanillic acid and quercetin-3-galactoside have been found to be responsible for the antioxidative, antimicrobial and antiviral properties of *C. maritimum* (Meot-Duros and Magné, 2009; Jallali *et al.*, 2014). Extracts of various mangroves (e.g. *A. ilicifolius*, *A. marina*, *E. agallocha*, *R. apiculata*, *A. officinalis* and *B. sexangula*) have been documented as possessing diverse medicinal properties (Agoramoorthy *et al.*, 2007; Abeysinghe, 2010). Anticancer, antiulcer, antireverse transcriptase and antihistamine-release activities have been documented for the mangrove *E. agallocha* (Mondal *et al.*, 2016). The antibacterial and antifungal activity of *E. agallocha* fatty acid methyl ester extracts have been well studied (Agoramoorthy *et al.*, 2007). Antioxidant and antifilarial activity of the leaf extract of *E. agallocha* has been described (Mondal *et al.*, 2016). Another mangrove, *Acanthus ilicifolius*, has been reported to possess antiulcer, antifungal, osteoblastic, antileishmanial, anti-inflammatory, antidiabetic, anticancer and antioxidant activities (Abdel-Aziz *et al.*, 2016). Some researchers have carried out parallel studies on antioxidative and antimicrobial activity of halophytes such as *S. europaea*, *S. herbacea* (Yu *et al.*, 2012), *T. gallica* (Ksouri *et al.*, 2009) and *C. cardunculus* (Falleh *et al.*, 2008) against different microbes and diseases. Antioxidative and antimicrobial activity have been testified as being due to the presence of different substances, or to secondary substances in the same group but with different specificities (Boestfleisch *et al.*, 2014). The antimicrobial and anti-inflammatory effect of *L. monopetalum* leaf extract has been described as having high antioxidative activity but low antimicrobial activity

(Trabelsi *et al.*, 2012); whereas shoot extract of *S. brachiata* has been found effective against many Gram negative bacteria such as *Bacillus subtilis*, *B. pumilus*, *Micrococcus luteus* and *S. aureus* (Manikandan *et al.*, 2009). Many halophytic species have been used as purgatives, carminatives and anti-helminthics in the treatment of digestive disorders like stomach pain, vomiting and diarrhoea. For example, *H. stocksii* and *P. oleracea* have been used for the treatment of stomach disorders, ulcers, vomiting and gastric troubles (Agoramoorthy *et al.*, 2007; Qureshi *et al.*, 2010). *C. decidua* is popularly used as a carminative, and *Z. simplex* has been commonly used to cure helminthiases (Hammiche and Maiza, 2006; Ksouri *et al.*, 2009, 2012a; Hameed and Khan, 2011). Similarly, *Salvadora persica*, a highly tolerant facultative halophyte (Rangani *et al.*, 2016) has been used as a natural remedy to cure several diseases and conditions, such as asthma, bronchitis, scurvy, worm infestation and haemorrhoids, and in cases of poisoning (Kumar *et al.*, 2012). It has been reported that the fruit of *S. persica* has aphrodisiac, carminative, emollient and purgative properties. Moreover, the fruit is also used in the treatment of urinary and biliary calculi and in arthritis (Kumari *et al.*, 2017). Flowers of *S. persica* have been useful in gastric problems, constipation and sexual dysfunction (Kumar *et al.*, 2012). *Salvadora oleoides* has been used as a purgative (Shafi *et al.*, 2001). Many halophytic species have been useful in treating skin-related conditions such as infection, burning, acne and eruptions, and also for wound treatment (Ksouri *et al.*, 2012a; Kumari and Parida, 2016). *Fagonia indica* and *Heliotropium strigosum* have frequently been used in the treatment of wounds (Shafi *et al.*, 2001). Leaves of *S. monoica* have been used as an antiseptic ointment for skin eruptions (Agoramoorthy *et al.*, 2007). *H. curassavicum* (Qureshi *et al.*, 2010), *Ziziphus nummularia* (Ahmed *et al.*, 2015) and *Aerva javanica* (Qasim *et al.*, 2017) have been used for healing boils, skin infections and skin diseases respectively. *Xylocarpus moluccensis* has been used to treat gastrointestinal disorders, malaria and elephantiasis (Zhang and Xu, 2017). Extracts of different parts of *X. granatum* have been reported for their traditional use in treating fever, including one caused by malaria; inflammation; dysentery; diarrhoea; cholera; and other abdominal problems (Simlai and Roy, 2013). Halophytic species have also been useful in the treatment of respiratory conditions like coughs, asthma, bronchitis and sore throat (Ksouri *et al.*, 2012a). For example, *Solanum surattense* (Qureshi and Bhatti, 2008), *C. decidua* and *A. javanica* (Zhang and Xu, 2017) have been used to relieve cough and chest pain. *Solanum surattense*, *Achyranthes aspera* and *Cleome brachycarpa* have been used to treat joint pains (Hammiche and Maiza, 2006). Extracts of *Peganum harmala* and *Ipomoea pescaprae* have been useful in relieving muscle, back and ear pain (Hammiche and Maiza, 2006; Agoramoorthy *et al.*, 2007). *C. cretica* has various properties and is anti-diabetic, antiviral, antibacterial, antitubercular, expectorant, antibilious, antiasthmatic, stomachic and aphrodisiac; it has also been used to treat constipation, leprosy and urinary discharges (Khare *et al.*, 2013). *C. decandra* has been used for the treatment of ulcers, hepatitis and haemorrhage. The medicinal properties of this halophyte have been attributed to the presence of diterpenoids, triterpenoids and lignins (Boopathy *et al.*, 2011). One of the most important fields of application of halophyte-derived metabolites is the treatment of cancer. Extracts of various halophytic species (e.g. *S. bigelovii*, *S. herbacea*, *S. europaea* and *Glehnia littoralis*) have been found to have antiproliferative and toxicity effects on human colon cancer cells (Um *et al.*, 2010; Kang *et al.*, 2011; Altay *et al.*, 2017; Kim, 2017). Mangrove plants have been reported as having antibacterial, antiviral and antioxidant activities (Abdel-Aziz *et al.*, 2016). Many mangrove plants (*Aegiceras majus*, *E. agallocha*, *Avicennia africana*, *Kandelia rheedii* and *R. mangle*) have been widely used as for the treatment of various diseases like leprosy, tuberculosis, malaria, ulcers and skin diseases (Das *et al.*, 2014). *Sonneatia caseolaris* has been documented for its use in traditional medicine for the treatment of piles and haemorrhages. Antibacterial activity by the fresh leaves of *A. marina* and *Rhizophora stylosa* has been reported against some pathogens, such as *E. coli* and *S. aureus* (Abdel-Aziz *et al.*, 2016). Mangrove species such as *S. griffithii* have been reported to show antihyperglycemic activity (Tiwari *et al.*, 2008). Black tea extract from the mangrove plant *Ceriops decandra* has been proved to possess a potential oral anticancer effect (Abdel-Aziz *et al.*, 2016). The rootstock and leaves of *Desmostachya bipinnata* have been reported in Indian traditional medicine systems as curing skin and blood diseases, jaundice, biliousness, thirst, vaginal discharge, asthma, stranguary, vesicle calculi, urinary calculi and bladder diseases (Kumar *et al.*, 2010). The essential oil obtained from *D. bipinnata* has shown strong antimicrobial activity against *S. epidermidis*, *S. aureus*, *E. coli* and *P. aeruginosa*, and may have future applications in the food and pharmaceutical industries (Kumar *et al.*, 2010).

19.4.4 Halophytes as ornamentals

Success in the floriculture industry is largely based on the availability of new and attractive ornamental plant varieties, there being a constant demand for novelties (Cassaniti *et al.*, 2013; Ventura *et al.*, 2015). 'The use of halophytic plants for landscaping and ornamental purposes seems a very appropriate, economic use of saline soil, as many halophytes produce attractive flowers' (Ventura *et al.*, 2015). Cassaniti and Romano (2011) based their study on the eHALOPH database (https://www.sussex.ac.uk/affiliates/halophytes/) and investigated halophytes native to the Mediterranean region, listing 13 families with about 42 species having ornamental potential. 'Besides having value as a vegetable crop, *A. tripolium* produces attractive lilac flowers with central yellow stamens during its second year of growth' (Ventura *et al.*, 2015). *C. maritimum* seems to have the potential to become a multipurpose halophytic cash crop, as the aromatic, succulent leaves are appreciated as a salty vegetable (fresh or pickled); and the umbrella-like, delicate inflorescences may be used for ornamental purposes. *C. maritimum* grows in rocky coastal environments in the Mediterranean region, where it is often subjected to sea spray (Ben Hamed *et al.*, 2007; Ventura *et al.*, 2015). Because of their easy cultivation, even at seawater salinity, '*Salicornia* and *Sarcocornia* have been the subject of numerous research trials which demonstrated their multipurpose applications, including biodiesel, oil, bioremediation, forage, vegetable, probiotic and ornamental uses' (Ventura *et al.*, 2015). *M. crystallinum* is a well-known halophyte owing to its switching mechanism from C_3 to the CAM photosynthetic mechanism (Bouftira *et al.*, 2007). It can be grown as an ornamental groundcover. Many halophytes are useful ornamentals, including *A. tripolium*, *L. monopetalum*, *Batis maritima*, *T. nilotica*, *T. amnicola*, *Cistanche fistulosum*, *A. halimus*, *S. portulacastrum* and *Noronhia emarginata* (Ladeiro, 2012).

19.5 Conclusions and Future Perspectives

Halophytes are the natural flora of saline environments and constitute a diverse group of plants with the potential to revolutionize the pharmaceutical, nutraceutical, cosmetic, food and fuel industries. These plants have remarkable potential to fulfil human requirements related to food, fuel, fodder and medicines. They possess significant nutritional value and may serve as a potentially good source of non-conventional food and fodder for human and animal consumption. Many species of halophytes are used to treat diseases such as rheumatism, smallpox, hepatitis, leprosy, asthma and ulcers; and act as purgatives. However, many have not yet received much attention as the source of bioactive compounds owing to lack of commercial popularity. 'Since halophytes offer a wide array of bioactive molecules of therapeutic value, there is an urgent need to discover new compounds with diverse chemical structures and novel mechanisms of action' (Abdel-Aziz *et al.*, 2016). The benefits of cultivating halophytes meet the three 'E' criteria: ecology, economy and equity. Their ecological benefits include use of saline wastelands and the protection and development of coastal areas. Their economic importance includes production and marketing of their isolated, medicinally important bioactive compounds and poverty alleviation by providing employment opportunities. Equity benefits include collaboration between industry and institutions, facilitating technology extension and development of standard biotechniques. The high medicinal values of halophytes have been attributed to their considerable antioxidant properties and the presence of pharmacologically active compounds, as discussed in this chapter. Researchers believe that bioprospection of economically important halophytes will help in the identification of the most promising bioactive compounds for their use in the nutraceutical and pharmaceutical industries. Current concerns about food safety, and raising interest in health-promoting substances among consumers and industries, will allow this knowledge to lead to a new generation of non-conventional foods which may interface between drugs and food. Halophytes and their derived products may create a strong base for non-conventional food sources and safe alternatives to commercial drugs.

Halophytes have also been used to improve saline wastelands, phytoremediate areas polluted by heavy metals, and protect coastal land and biodiversity. Vegetative bioremediation or bioreclamation of salt- and heavy metal-affected soils have been proved to be economic solutions for developing countries, since chemical additives are becoming increasingly expensive and heavy metal deposition destroys soil fertility, making it unsuitable for agricultural crops. Several authors have proved that the hyperaccumulating ability of halophytes depends on the capacity of their aboveground biomass (Rabhi *et al.*, 2009, Hasanuzzaman *et al.*, 2014; Suaire *et al.*, 2016). This ability could be used to reduce the salt burden from the rhizosphere of plants in arid and semi-arid regions, where physiochemical removal methods are too expensive.

Environmentally safe and clean techniques to address the salinity and heavy metal toxicity include the introduction of salt- (ion) or heavy metal-removing species to control salinity and remove toxic metals and maintain the sustainability of the agricultural fields. 'Large-scale decontamination of the soils and underground water using phytoremediation techniques requires plants with high salt uptake rates, large biomass and tolerance to a wide array of environmental conditions and constraints. After reclamation, these cultivated halophytes can be used as animal fodder or for making organic compost' (Hasanuzzaman *et al.*, 2014). For example, the mangrove environment nutrient system mainly comprises mangrove leaves; as they are a good source of vitamins, amino acids, minerals and iodine, they will be useful for the nourishment, growth and development of marine organisms and of livestock. Mangrove foliage also plays a crucial role in the formation of detritus, which is exploited by marine and estuarine detritivorous organisms to produce silage; this is fed upon by marine animals. The use of mangrove leaves has led to the development of saline agriculture to support new sources of food, fodder, forage, medicine and important plant-based chemicals for human consumption (Ahmed *et al.*, 2015). This may also reduce the burden on crop plants, which are facing productivity problems due to abiotic stresses. Apart from accumulating salt from the saline habitat, many halophytes are able to remediate toxic metals (Hasanuzzaman *et al.*, 2014). Halophytes often adapt well to metal-affected habitats, compared to glycophytic plants, which makes them good candidates as eco-friendly and sustainable solutions for cleaning up contaminated coastal environments (Arora and Roa, 2017). Research into halophyte phytoremediation is anticipated to receive increased attention in future. In view of the potentially enormous contribution they could make, these plants need to be cultivated and conserved to promote sustainable development and improve the socio-economic status of poor people. We recommend the encouragement of investment to develop halophytes as a sustainable resource for the future.

Acknowledgements

The financial support from Science and Engineering Research Board (SERB) by the grant of SERB (SB/SO/PS-14/2014), DST, Government of India, New Delhi to AKP is duly acknowledged. The authors also acknowledge the financial support of Council of Scientific and Industrial Research (CSIR), New Delhi, India. AK and MP are thankful to DST, Government of India, New Delhi for providing Inspire Fellowships. This manuscript has been assigned CSIR-CSMCRI communication No.12/2018.

References

Abdel-Aziz, S.M., Mouafi, F.E., Moustafa, Y.A. and Abdelwahed, N.A.M. (2016) Medicinal importance of mangrove plants. In: Garg, N., Abdel-Aziz, S.M. and Aeron, A. (eds) *Microbes in Food and Health.* Springer, Cham, Switzerland, pp. 77–96.

Abdelly, C., Barhoumi, Z., Ghnaya, T., Debez, A., Hamed, K.B., *et al.* (2006) Potential utilisation of halophytes for the rehabilitation and valorisation of salt-affected areas in Tunisia. In: Öztürk, M.,

Waisel, Y., Khan, M.A. and Görk, G. (eds) *Biosaline Agriculture and Salinity Tolerance in Plants*. Springer, Basel, Switzerland, pp. 161–170.

Abeysinghe, P. (2010) Antibacterial activity of some medicinal mangroves against antibiotic resistant pathogenica bacteria. *Indian Journal of Pharmaceutical Sciences* 72, 167–172.

Abideen, Z., Ansari, R. and Khan, M.A. (2011) Halophytes: Potential source of ligno-cellulosic biomass for ethanol production. *Biomass and Bioenergy* 35, 1818–1822.

Abideen, Z., Ansari, R., Gul, B. and Khan, M.A. (2012) The place of halophytes in Pakistan's biofuel industry. *Biofuels* 3, 211–220.

Abideen, Z., Hameed, A., Koyro, H.W., Gul, B., Ansari, R. *et al.* (2014) Sustainable biofuel production from non-food sources-An overview. *Emirates Journal of Food and Agriculture* 26, 1057–1066.

Acosta, J.A., Jansen, B., Kalbitz, K., Faz, A. and Martínez-Martínez, S. (2011) Salinity increases mobility of heavy metals in soils. *Chemosphere* 85, 1318–1324.

Adams, M., Berset, C., Kessler, M. and Hamburger, M. (2009) Medicinal herbs for the treatment of rheumatic disorders-A survey of European herbals from the 16th and 17th century. *Journal of Ethnopharmacology* 121, 343–359.

Ado, N.M., Guero, Y., Michot, D., Soubeiga, B., Kiesse, T.S. *et al.* (2016) Phytodesalinization of irrigated saline Vertisols in the Niger Valley by *Echinochloa stagnina*. *Agricultural Water Managment* 177, 229–240.

Agoramoorthy, G., Chandrasekaran, M., Venkatesalu, V. and Hsu, M.J. (2007) Antibacterial and antifungal activities of fatty acid methyl esters of the blind-your-eye mangrove from India. *Brazilian Journal of Microbiology* 38, 739–742.

Ahmed, M.Z., Gul, B., Öztürk, M.A. and Khan, M.A. (2015) *Halophytes for Food Security in Dry Lands*. 5th edn. Academic Press, Cambridge, MA, USA.

Alassali, A., Cybulska, I., Galvan, A.R. and Thomsen, M.H. (2016) Wet fractionation of the succulent halophyte *Salicornia sinus-persica*, with the aim of low input (water saving) biorefining into bioethanol. *Applied Microbiology and Biotechnology* 101, 1769–1779.

Altay, A., Celep, G.S., Yaprak, A.E., Baskose, I. and Bozoglu, F. (2017) Glassworts as possible anticancer agents against human colorectal adenocarcinoma cells with their nutritive, antioxidant and phytochemical profiles. *Chemistry and Biodiversity* 14, 12–23.

Altintas, A., Kosar, M., Kirimer, N. Baser, K.H.C. and Demirci, B. (2006) Composition of the essential oils of *Lycium barbarum* and *L. ruthenicum* fruits. *Chemistry of Natural Compounds* 42, 24–25.

Amouei, A. (2013) Effect of saline soil levels stresses on agronomic parameters and fodder value of the halophyte *Atriplex leucoclada* L. (Chenopodiaceae). *African Journal of Agricultural Research* 8, 3007–3012.

Anwar, F., Bhanger, M.I., Nasir, M.K.A. and Ismail, S. (2002) Analytical characterization of *Salicornia bigelovii* seed oil cultivated in Pakistan. *Journal of Agriculture and Food Chemistry* 50, 4210–4214.

Atia, A. Debez, A., Barhoumi, Z. and Abdelly, C. (2010) Localization and composition of seed oils of *Crithmum maritimum* L. (Apiaceae). *African Journal of Biotechnology* 9, 6482–6485.

Ayyappan, D., Sathiyaraj, G. and Ravindran, K.C. (2015) Phytoextraction of heavy metals by *Sesuvium portulacastrum:* a salt marsh halophyte from tannery effluent. *International Journal of Phytoremediation* 18, 453–459.

Balat, M. (2011) Potential alternatives to edible oils for biodiesel production - A review of current work. *Energy Conversion* and *Management* 52, 1479–1492.

Ben Hamed, K., Castagna, A., Salem, E., Ranieri, A.M. and Abdelly, C. (2007) Sea fennel (*Crithmum maritimum* L.) under salinity conditions: a comparison of leaf and root antioxidant responses. *Plant Growth Regulation* 53, 185–194.

Benhammou, N., Atik, F. and Kadifkova, T. (2009) Antioxidant activity of methanolic extracts and some bioactive compounds of *Atriplex halimus*. *Comptes Rendus Chimie* 12, 1259–1266.

Ben Mansour, R., Ksouri, W.M., Cluzet, S., Krisa, S., Richard, T. *et al.* (2016) Assessment of antioxidant activity and neuroprotective capacity on PC12 cell line of *Frankenia thymifolia* and related phenolic LC-MS/MS identification. *Evidence-Based Complementary* and *Alternative Medicine* 2016, 1–8.

Bernal, J., Mendiola, J.A., Ibánez, E. and Cifuentes, A. (2011) Advanced analysis of nutraceuticals. *Journal of Pharmaceutical and Biomedical Analysis* 55, 758–774.

Bissonnette, G. and Skousen, J. (2001) Treatment of domestic wastewater by three plant species in constructed wetlands. *Water Air and Soil Pollution* 128, 283–295.

Boestfleisch, C. and Papenbrock, J. (2017) Changes in secondary metabolites in the halophytic putative crop species *Crithmum maritimum* L., *Triglochin maritima* L. and *Halimione portulacoides* (L.) Aellen as reaction to mild salinity. *PLoS One* 12. doi: 10.1371/journal.pone.0176303.

Boestfleisch, C., Wagenseil, N., Buhmann, A.K., Seal, C.E., Wade, E.M., *et al.* (2014) Manipulating the antioxidant capacity of halophytes to increase their cultural and economic value through saline cultivation. *AoB Plants* 6. doi: 10.1093/aobpla/plu046.

Bogdanova, E.S., Zakhozhiy, I.G., Tabalenkova, G.N. and Rozentsvet, V.A. (2017) Bioaccumulation of nutrients, lipids, pigments, antioxidants and essential oils in plants of genus *Artemisia* growing in lake Elton region (South-East of the European part of Russia). *Journal of Medicinal Plants Studies* 5, 270–275.

Bomani, B.M.M., Hendricks, R.C., Elbuluk, M., Okon, M., Lee, E. *et al.* (2011) *NASA's GreenLab Research Facility: A Guide for a Self-Sustainable Renewable Energy Ecosystem*. Glenn Research Center Cleveland, OH, USA. Nasa/Tp 217208.

Boopathy, N.S., Kathiresan, K. and Jeon, Y.J. (2011) Effect of mangrove black tea extract from *Ceriops decandra* (Griff.) on hematology and biochemical changes in dimethyl benz [a] anthracene-induced hamster buccal pouch carcinogenesis. *Environmental Toxicology* and *Pharmacology* 32, 193–200.

Bouayed, J. and Bohn, T. (2010) Exogenous antioxidants—double-edged swords in cellular redox state: health beneficial effects at physiologic doses versus deleterious effects at high doses. *Oxidative Medicine* and *Cellular Longevity* 3, 228–237.

Bouftira, I., Abdelly, C. and Sfar, S. (2007) Identification of a naturally occurring 2, 6-bis (1.1- dimethylethyl)-4-methylphenol from purple leaves of the halophyte plant *Mesembryanthemum crystallinum*. *African Journal of Biotechnology* 6, 1136–1139.

Bourgou, S., Megdiche, W. and Ksouri, R. (2017) The Halophytic genus *Zygophyllum* and *Nitraria* from North Africa : a phytochemical and pharmacological overview. In: Neffati, M., Najjaa, H. and Máthé, Á. (eds) *Medicinal and Aromatic Plants of the World - Africa*. Springer, Dordrecht, The Netherlands, pp. 345–356.

Boxman, S.E., Nystrom, M., Capodice, J.C., Ergas, S.J., Main, K.L. *et al.* (2017) Effect of support medium, hydraulic loading rate and plant density on water quality and growth of halophytes in marine aquaponic systems. *Aquaculture Research* 48, 2463–2477.

Brown, J.J. and Glenn, E.P. (1999) Reuse of highly saline aquaculture effluent to irrigate a potential forage halophyte, *Suaeda esteroa. Aquacultural Engineering* 20, 91–111.

Buhmann, A. and Papenbrock, J. (2013) Biofiltering of aquaculture effluents by halophytic plants: Basic principles, current uses and future perspectives. *Environmental and Experimental Botany* 92, 122–133.

Buhmann, A.K., Waller, U., Wecker, B. and Papenbrock, J. (2015) Optimization of culturing conditions and selection of species for the use of halophytes as biofilter for nutrient-rich saline water. *Agricultural Water Management* 149, 102–114.

Burundukova, O.L., Shuyskaya, E.V., Rakhmankulova, Z.F., Burkovskaya, E.V., Chubar, E.V., *et al.* (2017) *Kali komarovii* (Amaranthaceae) is a xero-halophyte with facultative NADP-ME subtype of C_4 photosynthesis. *Flora - Morphology, Distribution, Functional Ecology* of *Plants* 227, 25–35.

Bustan, A., Pasternak, D., Pirogova, I., Durikov, M., Devries, T.T. *et al.* (2005) Evaluation of saltgrass as a fodder crop for livestock. *Journal of the Science of Food and Agriculture* 85, 2077e2084.

Bylka, W. (2004) A new acylated flavonol diglycoside from *Atriplex littoralis*. *Acta Physiologiae Plantarum* 26, 393–394.

Cao, L., Wang, W., Yang, Y., Yang, C., Yuan, Z., *et al.* (2007) Environmental impact of aquaculture and countermeasures to aquaculture pollution in China. *Environmental Science* and *Pollution Research* 14, 452–462.

Cassaniti, C. and Romano, D. (2011) The use of halophytes for Mediterranean landscaping. Proceedings of the European COST Action FA901. *European Journal of Plant Science and Biotechnology* 5, 58–63.

Cassaniti, C., Romano, D., Hop, M.E.C.M. and Flowers, T.J. (2013) Growing floricultural crops with brackish water. *Environmental and Experimental Botany* 92, 165–175.

Cerchiara, T., Chidichimo, G., Ragusa, M.I., Belsito, E.L., Liguori, A. *et al.* (2010) Characterization and utilization of Spanish Broom (*Spartium junceum* L.) seed oil. *Industrial Crops and Products* 31, 423–426.

Chan, E.W.C., Baba, S., Chan, H.T., Kainuma, M., Inoue, T. *et al.* (2017) Ulam herbs: A review on the medicinal properties of *Anacardium occidentale* and *Barringtonia racemosa*. *Journal of Applied Pharmaceutical Science* 7(2), 241–247.

Chandrasekaran, M., Kannathasan, K., Venkatesalu, V. and Prabhakar, K. (2009) Antibacterial activity of some salt marsh halophytes and mangrove plants against methicillin resistant *Staphylococcus aureus*. *World Journal of Microbiology and Biotechnology* 25, 155–160.

Cho, J.Y., Kim, M.S., Lee, Y.G., Jeong, H.Y., Lee, H.J., *et al.* (2016) A phenyl lipid alkaloid and flavone C-diglucosides from *Spergularia marina*. *Food Science and Biotechnology* 25, 63–69.

Christofilopoulos, S., Syranidou, E. and Gkavrou, G. (2016) The role of halophyte *Juncus acutus* L. in the remediation of mixed contamination in a hydroponic greenhouse experiment. *Journal of Chemical Technology and Biotechnology* 91, 1665–1674.

Custódio, L., Ferreira, A.C., Pereira, H., Silvestre, L., Vizetto-Duarte, C. *et al.* (2012) The marine halophytes *Carpobrotus edulis* L. and *Arthrocnemum macrostachyum* L. are potential sources of nutritionally important PUFAs and metabolites with antioxidant, metal chelating and anticholinesterase inhibitory activities. *Botanica Marina* 55, 281–288.

Czyż, H., Kitczak, T. and Bury, M. (2013) The characteristics of coastal grassland in West Pomerania. *Plant Diversity* and *Evolution* 130, 229–237.

Dan, T.H., Quang, L.N., Chiem, N.H. and Brix, H. (2011) Treatment of high-strength wastewater in tropical constructed wetlands planted with *Sesbania sesban*: Horizontal subsurface flow versus vertical downflow. *Ecological Engineering* 37, 711–720.

Das, S.K., Samantaray, D. and Thatoi, H. (2014) Ethnomedicinal, antimicrobial and antidiarrhoeal studies on the mangrove plants of the genus *Xylocarpus*: A mini review. *Journal of Bioanalysis and Biomedicine* 12, 1–7.

Davidson, N. and Galloway, R. (eds) (1993) *Productive Use of Saline Land*. ACIAR Proceedings No. 42, ACIAR, Canberra, pp. 124.

de Jesus Cortes-Sanchez, A., Hernández-Sánchez, H. and Jaramillo-Flores, M.E. (2013) Biological activity of glycolipids produced by microorganisms: new trends and possible therapeutic alternatives. *Microbiological Research* 168(1), 22–32.

de Lange, H.J. and Paulissen, M.P.C.P. (2016) Efficiency of three halophyte species in removing nutrients from saline water: A pilot study. *Wetlands Ecology and Management* 24, 587–596.

de Lange, H.J., Paulissen, M.P.C.P. and Slim, P.A. (2013) "Halophyte filters": the potential of constructed wetlands for application in saline aquaculture. *International Journal of Phytoremediation* 15, 352–364.

de Souza, E.R., dos Santos Freire, M.B.G., de Melo, D.V.M. and de Assunção Montenegro, A.A. (2014) Management of *Atriplex nummularia* Lindl. in a salt affected soil in a semi arid region of Brazil. *International Journal of Phytoremediation* 16, 73–85.

Díaz, F.J., Benes, S.E. and Grattan, S.R. (2013) Field performance of halophytic species under irrigation with saline drainage water in the San Joaquin Valley of California. *Agricultural Water Management* 118, 59–69.

Donato, D.C., Kauffman, J.B., Murdiyarso, D., Kurnianto, S., Stidham, M. *et al.* (2011) Mangroves among the most carbon-rich forests in the tropics. *Nature Geoscience* 4, 293–297.

Duarte, B., Cabrita, M.T., Gameiro, C., Matos, A.R., Godinho, R., *et al.* (2017) Disentangling the photochemical salinity tolerance in *Aster tripolium* L.: connecting biophysical traits with changes in fatty acid composition. *Plant Biology* 19, 239–248.

Edmond, M., Han, R., Classen, B., Quetin-Leclerq, J., Mahy, G., *et al.* (2010) Mucilage and polysaccharides in the halophyte plant species *Kosteletzkya virginica* : Localization and composition in relation to salt stress. *Journal of Plant Physiology* 167, 382–392.

Eganathan, P., Subramanian, H.S., Latha, R. and Rao, C.S. (2006) Oil analysis in seeds of *Salicornia brachiata*. *Industrial Crops Products* 23, 177–179.

El-Massry, K.F., El-Ghorab, A.H. and Farouk, A. (2002) Antioxidant activity and volatile components of Egyptian *Artemisia judaica* L. *Food Chemistry* 79, 331–336.

El Shaer, H.M. (2010) Halophytes and salt-tolerant plants as potential forage for ruminants in the Near East region. *Small Ruminant Research* 91, 3–12.

Erdemoglu, N. and Kusmenoglu, S. (2003) Fatty acid composition of *Zygophyllum fabago* seeds. *Chemistry of Natural Compounds* 39, 595–596.

Erkucuk, A., Akgun, I.H. and Yesil-Celiktas, O. (2009) Supercritical CO_2 extraction of glycosides from *Stevia rebaudiana* leaves: Identification and optimization. *Journal of Supercritical Fluids* 51, 29–35.

Espin, J.C., Garcı, T. and Toma, F.A. (2007) Nutraceuticals: facts and fiction. *Phytochemistry* 68, 2986–3008.

Falleh, H., Ksouri, R., Chaieb, K., Karray-Bouraoui, N., Trabelsi, N., *et al.* (2008) Phenolic composition of *Cynara cardunculus* L. organs, and their biological activities. *Comptes Rendus Biologies* 331, 372–379.

Falleh, H., Ksouri, R., Medini, F., Guyot, S., Abdelly, C., *et al.* (2011) Antioxidant activity and phenolic composition of the medicinal and edible halophyte *Mesembryanthemum edule* L. *Industrial Crops Products* 34, 1066–1071.

Fedoroff, N.V., Battisti, D.S., Beachy, R.N., Cooper, P.J., Fischhoff, D.A., *et al.* (2010) Radically rethinking agriculture for the 21st century. *Science* 327, 833–834.

Filho, A.M.M., Pirozi, M.R., Borges, J.T.D.S., Pinheiro Sant'Ana, H.M., Chaves, J.B.P. *et al.* (2017) *Quinoa*: Nutritional, functional, and antinutritional aspects. *Critical Reviews in Food Science and Nutrition* 57, 1618–1630.

Fletcher, T.D., Shuster, W., Hunt, W.F., Ashley, R., Butler, D., *et al.* (2014) The evolution and application of terminology surrounding urban drainage. *Urban Water Journal* 32, 1–18.

Flowers, T.J. and Colmer, T.D. (2008) Salinity tolerance in halophytes. *New Phytologist* 179, 945–963.

Flowers, T.J. and Colmer, T.D. (2015) Plant salt tolerance: Adaptations in halophytes. *Annals of Botany* 115, 327–331.

Flowers, T.J. and Muscolo, A. (2015) Halophytes in a changing world. *AoB Plants* 7. doi: 10.1093/aobpla/plv020.

Fukushi, E., Onodera, S., Yamamori, A., Shiomi, N. and Kawabata, J. (2000) NMR analysis of tri- and tetrasaccharides from asparagus. *Magnetic Resonance in Chemistry* 38, 1005–1011.

Gairola, S., Bhatt, A. and El-Keblawy, A. (2015) A perspective on potential use of halophytes for reclamation of salt-affected lands. *Wulfenia Journal* 22, 88–97.

Gaspar, H., Palma, F.M.S.B., La, M.C.D.E. and Rodriguez, B. (1996) Sterols from *Teucrium abutiloides* and *T. betonicum*. *Phytochemistry* 43, 613–615.

Genser, B., Backer, G., De Bruckert, E., Bruckert, E., Carmena, R., *et al.* (2012) Plant sterols and cardiovascular disease : a systematic review and meta-analysis. *European Heart Journal* 33, 444–451.

Ghnaya, T., Slama, I., Messedi, D., Grignon, C., Ghorbel, M.H. *et al.* (2007) Effects of Cd^{2+} on K^+, Ca^{2+} and N uptake in two halophytes *Sesuvium portulacastrum* and *Mesembryanthemum crystallinum*: Consequences on growth. *Chemosphere* 67, 72–79.

Glaister, B.J., Fletcher, T.D., Cook, P.L.M. and Hatt, B.E. (2014) Co-optimisation of phosphorus and nitrogen removal in stormwater biofilters: the role of filter media, vegetation and saturated zone. *Water Science and Technology* 69, 1961–1969.

Glenn, E.P., Mckeon, C., Gerhart, V., Nagler, P.L., Jordan, F. *et al.* (2009) Deficit irrigation of a landscape halophyte for reuse of saline waste water in a desert city. *Landscape and Urban Planning* 89, 57–64.

Gong, K.K., Li, P.L., Qiao, D., Zhang, X.W., Chu, M.J., *et al.* (2017) Cytotoxic and antiviral triterpenoids from the mangrove plant *Sonneratia paracaseolaris*. *Molecules* 22, 1319–1330.

Guil-Guerrero, J.L., Guil-Laynez, J.L. and Guil-Laynez, Á. (2017) Bioprospecting for seed oils from wild plants in the Mediterranean Basin for biodiesel production. *Journal of Cleaner Production* 159, 180–193.

Hachicha, S.F., Barrek, S., Skanji, T., Zarrouk, H. and Ghrabi, Z.G. (2009) Fatty acid, tocopherol, and sterol content of three *Teucrium* species from Tunisia. *Chemistry of Natural Compounds* 45, 261–264.

Hameed, A. and Khan, M.A. (2011) Halophytes: Biology and economic potentials. *Karachi University Journal of Science* 39, 40–44.

Hammiche, V. and Maiza, K. (2006) Traditional medicine in Central Sahara: Pharmacopoeia of Tassili N'ajjer. *Journal of Ethnopharmacology* 105, 358–367.

Hasanuzzaman, M., Nahar, K., Alam, M.M., Bhowmik, P.C., Hossain, M., *et al.* (2014) Potential use of halophytes to remediate saline soils. *BioMed Research International* 2014, 1–12.

Hegedűs, R., Kerepeczki, É., Gál, D., Pekár, M., Bíróné, M.O., *et al.* (2010) Potential role of halophytic macrophytes in saline effluent treatment. *World Academy of Science, Engineering and Technology* 64, 273–277.

Herlt, A.J., Mander, L.N., Pongoh, E., Rumampuk, R.J. and Tarigan, P. (2002) Two major saponins from seeds of *Barringtonia asiatica*: putative antifeedants toward *Epilachna* sp. larvae. *Journal of Natural Products* 65, 115–120.

Hooper, D.U., Adair, E.C., Cardinale, B.J., Byrnes, J.E., Hungate, B.A. *et al.* (2012) A global synthesis reveals biodiversity loss as a major driver of ecosystem change. *Nature* 486, 105–108.

Hussein, S.R., Marzouk, M.M., Ibrahim, L.F., Kawashty, S.A. and Saleh, N.A. (2011) Flavonoids of *Zygophyllum album* L. f. and *Zygophyllum simplex* L. (Zygophyllaceae). *Biochemical Systematics and Ecology* 39, 778–780.

Isabelle, M., Lee, B.L., Lim, M.T., Koh, W.P., Huang, D. *et al.* (2010) Antioxidant activity and profiles of common fruits in Singapore. *Food Chemistry* 123, 77–84.

Jallali, I., Zaouali, Y., Missaoui, I., Smeoui, A., Abdelly, C. *et al.* (2014) Variability of antioxidant and antibacterial effects of essential oils and acetonic extracts of two edible halophytes: *Crithmum maritimum* L. and *Inula crithmoïdes* L. *Food Chemistry* 145, 1031–1038.

Jdey, A., Falleh, H., Jannet, S.B., Hammi, K.M., Dauvergne, X. *et al.* (2017) Phytochemical investigation and antioxidant, antibacterial and anti-tyrosinase performances of six medicinal halophytes. *South African Journal of Botany* 112, 508–514.

Jesus, J.M., Danko, A.S., Fiúza, A. and Borges, M.T. (2015) Phytoremediation of salt-affected soils: a review of processes, applicability, and the impact of climate change. *Environmental Science and Pollution Research* 22, 6511–6525.

Jlassi, A., Zorrig, W., Khouni, A.E., Lakhdar, A., Smaoui, A. *et al.* (2013) Phytodesalination of a moderately-salt-affected soil by *Sulla carnosa*. *International Journal of Phytoremediation* 15, 398–404.

Jordan, F.L., Robin-Abbott, M., Maier, R.M. and Glenn, E.P. (2002) A comparison of chelator-facilitated metal uptake by a halophyte and a glycophyte. *Environmental Toxicology and Chemistry* 21, 2698–2704.

Joseph, D., Chakraborty, K., Subin, C. and Vijayan, K. (2013) Halophytes of chenopodiaceae and aizoaceae from south-east coast of India as potential sources of essential nutrients and antioxidants. *Journal of Food and Nutrition Research* 1, 97–107.

Kadlec, R.H. and Wallace, S.D. (2009) *Treatment Wetlands*. 2nd edn. Lewis Publishers, New York.

Kang, S., Kim, D., Lee, B.H., Kim, M.R., Chiang, M., *et al.* (2011) Antioxidant properties and cytotoxic effects of fractions from glasswort (*Salicornia herbacea*) seed extracts on human intestinal cells. *Food Science and Biotechnology* 20, 115–122.

Kathiresan, K. and Bingham, B.L. (2001) Biology of mangrove and mangrove ecosystems. *Advances in Marine Biology* 40, 81–251.

Kaur, T., Hussain, K., Koul, S., Vishwakarma, R. and Vyas, D. (2013) Evaluation of nutritional and antioxidant status of *Lepidium latifolium* Linn.: A novel phytofood from Ladakh. *PLoS One* 8. doi: 10.1371/journal.pone.0069112.

Khan, M.A., Ansari, R., Ali, H., Gul, B. and Nielsen, B.L. (2009) *Panicum turgidum*, a potentially sustainable cattle feed alternative to maize for saline areas. *Agriculture, Ecosystems and Environment* 129, 542–546.

Khare, P., Chaudhary, S., Kumar, A., Yadav, G. and Thakur, N. (2013) A study on the standardization parameters of a halophytic plant (*Cressa cretica* L.). *Middle-East Journal of Science and Research* 15, 1472–1477.

Kim, R. (2017) Antioxidant effects of *Salicornia bigelovii* seed extracts. *Korean Society for Biotechnology and Bioengineering Journal* 32, 140–145.

Kong, C. and Seo, Y. (2012) Antiadipogenic activity of isohamnetin 3-O-β-D-glucopyranoside from *Salicornia herbacea*. *Immunopharmacology and Immunotoxicology* 34, 907–911.

Kong, N., Fang, S., Liu, Y. and Wang, J. (2014) Flavonoids from the halophyte *Apocynum venetum* and their antifouling activities against marine biofilm-derived bacteria. *Natural Product Research* 28, 37–41.

Korzeniowska, J. and Stanislawska-Glubiak, E. (2015) Phytoremediation potential of *Miscanthus*× *giganteus* and *Spartina pectinata* in soil contaminated with heavy metals. *Environmental Science and Pollution Research* 22, 11648–11657.

Koyro, H.W. (2006) Effect of salinity on growth, photosynthesis , water relations and solute composition of the potential cash crop halophyte *Plantago coronopus* (L.). *Environmental and Experimental Botany* 56, 136–146.

Koyro, H.W., Khan, M.A. and Lieth, H. (2011) Halophytic crops: A resource for the future to reduce the water crisis? *Emirates Journal of Food and Agriculture* 23, 1–16.

Koyro, H., Hussain, T., Huchzermeyer, B. and Khan, M.A. (2013) Photosynthetic and growth responses of a perennial halophytic grass *Panicum turgidum* to increasing NaCl concentrations. *Environmental and Exprerimental Botany* 91, 22–29.

Ksouri, R., Megdiche, W., Debez, A., Falleh, H., Grignon, C., *et al.* (2007) Salinity effects on polyphenol content and antioxidant activities in leaves of the halophyte *Cakile maritima*. *Plant Physiology and Biochemstry* 45, 244–249.

Ksouri, R., Falleh, H., Megdiche, W., Trabelsi, N. and Mhamdi, B. (2009) Antioxidant and antimicrobial activities of the edible medicinal halophyte *Tamarix gallica* L. and related polyphenolic constituents. *Food Chemistry and Toxicology* 47, 2083–2091.

Ksouri, R., Ksouri, W.M., Jallali, I., Debez, A. and Magné, C. (2012a) Medicinal halophytes: potent source of health promoting biomolecules with medical, nutraceutical and food applications. *Critical Review of Biotechnology* 32, 289–326.

Ksouri, R., Smaoui, A., Isoda, H. and Abdelly, C. (2012b) Utilization of halophyte species as new sources of bioactive substances. *Journal of Arid Land Studies* 22, 41–44.

Ksouri, W.M., Medini, F., Mkadmini, K., Legault, J. and Magné, C. (2013) LC-ESI-TOF-MS identification of bioactive secondary metabolites involved in the antioxidant, anti-inflammatory and anticancer activities of the edible halophyte *Zygophyllum album* Desf. *Food Chemistry* 139, 1073–1080.

Kumar, K.A., Patel, J., Choudhary, R.K. and Choudhary, R.K. (2010) Chemical composition and antimicrobial activity of the essential oil of *Desmostachya bipinnata* Linn. *International Journal of Phytomedicine* 2, 436–439.

Kumar, S., Rani, C. and Mangal, M. (2012) A critical review on *Salvadora persica*: An important medicinal plant of arid zone. *International Journal of Phytomedicine* 4, 292–303.

Kumari, A. and Parida, A.K. (2016) Metabolite profiling of the leaf extract reveals the antioxidant and nutraceuticals potential of the halophyte *Salvadora persica*. *RSC Advances* 6, 51629–51641.

Kumari, A. and Parida, A.K. (2018) Metabolomics and network analysis reveal the potential metabolites and biological pathways involved in salinity tolerance of the halophyte *Salvadora persica*. *Environmental and Experimental Botany* 148, 85–99. doi: 10.1016/j.envexpbot.2017.12.021.

Kumari, A., Das, P., Parida, A.K. and Agarwal, P.K. (2015) Proteomics, metabolomics and ionomics perspectives of salinity tolerance in halophytes. *Frontiers in Plant Science* 6. doi: 10.3389/fpls.2015.00537.

Kumari, A., Parida, A.K., Rangani, J. and Panda, A. (2017) Antioxidant activities, metabolic profiling, proximate analysis, mineral nutrient composition of *Salvadora persica* fruit unravel a potential functional food and a natural source of pharmaceuticals. *Frontier in Pharmacology* 8. doi:10.3389/fphar.2017.00061.

Ladeiro, B. (2012) Saline agriculture in the 21st century: Using salt contaminated resources to cope food requirements. *Journal of Botany* 2012, 1–7.

Lajnef, H.B., Pasini, F., Politowicz, J., Tlili, N., Khaldi, A., *et al.* (2017) Lipid characterization of *Eryngium maritimum* seeds grown in Tunisia. *Industrial Crops Products* 105, 47–52.

Lakshmi, K.P. and Rao, G.N. (2013) Antimicrobial activity of *Suaeda monoica*. (Forsst. ex Geml.) against human and plant pathogens. *Research Journal of Pharmaceutical, Biological and Chemical Sciences* 4, 680–685.

Li, C., Tang, B. and Jiang, S. (2017) Chemical composition and nutritional evaluation of the seeds of *Suaeda salsa*. (L.) Pall. *Journal of Food and Nutrition Research* 5, 22–26.

Li, J., Pan, L., Naman, C.B., Deng, Y. and Chai, H. (2014) Pyrrole alkaloids with potential cancer chemopreventive activity isolated from a goji berry-contaminated commercial sample of African mango. *Journal of Agricultural and Food Chemistry* 62, 5054–5060.

Liang, L., Liu, W., Sun, Y., Huo, X., Li, S. *et al.* (2016) Phytoremediation of heavy metal contaminated saline soils using halophytes : current progress and future perspectives. *Environmental Review* 999, 1–13.

Liebezeit, G. (2008) Ethnobotany and phytochemistry of plants dominant in salt marshes of the Lower Saxonian Wadden Sea, southern North Sea. *Senckenbergiana Maritima* 38, 1–30.

Lokhande, V.H. and Suprasanna, P. (2012) Prospects of halophytes in understanding and managing abiotic stress tolerance. In: Ahmad, P. and Prasad, M.N.V. (eds) Environmental Adaptations and Stress Tolerance of Plants in the Era of Climate Change. Springer, New York, pp. 29–56.

Lokhande, V.H., Gor, B.K., Desai, N.S., Nikam, T.D. and Suprasanna, P. (2013) *Sesuvium portulacastrum*, a plant for drought, salt stress, sand fixation, food and phytoremediation. A review. *Agronomy for Sustainable Development* 33, 329–348.

Lu, D., Zhang, M., Wang, S., Cai, J., Zhou, X., *et al.* (2010) Nutritional characterization and changes in quality of *Salicornia bigelovii* Torr. during storage. LWT - *Food Science and Technology* 43, 519–524.

Lutts, S. and Lefevre, I. (2015) How can we take advantage of halophyte properties to cope with heavy metal toxicity in salt-affected areas? *Annals of Botany* 115, 509–528.

Maciel, E., Leal, M.C., Lillebø, A.I., Domingues, P., Domingues, M.R. *et al.* (2016) Bioprospecting of marine macrophytes using MS-based lipidomics as a new approach. *Marine Drugs* 14, 49–77.

Magwa, M.L., Gundidza, M., Gweru, N. and Humphrey, G. (2006) Chemical composition and biological activities of essential oil from the leaves of *Sesuvium portulacastrum*. *Journal of Ethnopharmacology* 103, 85–89.

Mahmud, I., Islam, M.K., Saha, S., Barman, A.K., Rahman, M.M., *et al.* (2014) Pharmacological and ethnomedicinal overview of *Heritiera fomes:* Future prospects. *International Scholarly Research Notices* 2014, 938543. doi: 10.1155/2014/938543.

Manikandan, T., Neelakandan, T. and Rani, G.U. (2009) Antibacterial activity of *Salicornia brachiata*, a halophyte. *Journal of Phytology* 1, 441–443.

Manousaki, E. and Kalogerakis, N. (2011a) Halophytes—An emerging trend in phytoremediation. *International Journal of Phytoremediation* 13, 959–969.

Manousaki, E. and Kalogerakis, N. (2011b) Halophytes present new opportunities in phytoremediation of heavy metals and saline soils. *Industrial & Engineering Chemistry Research* 50, 656–660.

Manousaki, E., Kokkali, F. and Kalogerakis, N. (2009) Influence of salinity on lead and cadmium accumulation by the salt cedar (*Tamarix smyrnensis* Bunge). *Journal of Chemical Technology* and *Biotechnology* 84, 877–883.

Manousaki, E., Galanaki, K., Papadimitriou, L. and Kalogerakis, N. (2014) Metal phytoremediation by the halophyte *Limoniastrum monopetalum* (L.) Boiss, two contrasting ecotypes. *International Journal of Phytoremediation* 16, 755–769.

Marangoni, F. and Poli, A. (2010) Phytosterols and cardiovascular health. *Pharmacology Research* 61, 193–199.

Marcone M.F. (2003) *Batis maritima* (saltwort/beachwort): A nutritious, halophytic, seed bearing, perennial shrub for cultivation and recovery of otherwise unproductive agricultural land affected by salinity. *Food Research International* 36, 123–130.

Mardhiah, H.H., Ong, H.C., Masjuki, H.H., Lim, S. and Lee, H.V. (2017) A review on latest developments and future prospects of heterogeneous catalyst in biodiesel production from non-edible oils. *Renewable and Sustainable Energy Reviews* 67, 1225–1236.

Mariem, S., Hanen, F., Inès, J., Mejdi, S. and Riadh, K. (2014) Phenolic profile, biological activities and fraction analysis of the medicinal halophyte *Retama raetam*. *South African Journal of Botany* 94, 114–121.

Marina, L.U.Z. and Sci, M. (2013) Bioprospecting as a possible development mechanism for Colombia. *Acta Biológica Colombiana* 18, 19–30.

Marriboina, S., Sengupta, D., Kumar, S. and Reddy, A.R. (2017) Physiological and molecular insights into the high salinity tolerance of *Pongamia pinnata* (L.) Pierre, a potential biofuel tree species. *Plant Science* 258, 102–111.

Meot-Duros, L. and Magné, C. (2009) Antioxidant activity and phenol content of *Crithmum maritimum* L. leaves. *Plant Physiology and Biochemistry* 47, 37–41.

Milic, D.J., Lukovic, J., Ninkov, T., Zeremski-Skoric, L., Zoric, J. *et al.* (2012) Heavy metal content in halophytic plants from inland and maritime saline areas. *Central European Journal of Biology* 7, 307–317.

Mondal, S., Ghosh, D. and Ramakrishna, K. (2016) A complete profile on blind-your-eye Mangrove *Excoecaria agallocha* L. (Euphorbiaceae): Ethnobotany, phytochemistry, and pharmacological aspects. *Pharmacognosy Reviews* 10, 123–138.

Moser, B.R., Dien, B.S., Seliskar, D.M. and Gallagher, J.L. (2013) Seashore mallow (*Kosteletzkya pentacarpos*) as a salt-tolerant feed stock for production of biodiesel and ethanol. *Renewable Energy* 50, 833–839.

Muchate, N.S., Nikalje, G.C., Rajurkar, N.S., Suprasanna, P. and Nikam, T.D. (2016) Physiological responses of the halophyte *Sesuvium portulacastrum* to salt stress and their relevance for saline soil bio-reclamation. *Flora* 224, 96–105.

Mumby, P.J., Edwards, A.J., Arias-González, J.E., Lindeman, K.C., Blackwell, P.G., *et al.* (2004) Mangroves enhance the biomass of coral reef fish communities in the Caribbean. *Nature* 427, 533–536.

Murray, N.J., Ma, Z. and Fuller, R.A. (2015) Tidal flats of the Yellow Sea: A review of ecosystem status and anthropogenic threats. *Austral Ecology* 40, 472–481.

Nabet, N., Boudries, H., Chougui, N., Loupassaki, S., Souagui, S., *et al.* (2017) Biological activities and secondary compound composition from *Crithmum maritimum* aerial parts. *International Journal of Food Properties* 20, 1843–1855.

Nedjimi, B. and Daoud, Y. (2009) Cadmium accumulation in *Atriplex halimus* subsp. *schweinfurthii* and its influence on growth, proline, root hydraulic conductivity and nutrient uptake. *Flora - Morphology, Distribution, Functional Ecology of Plants* 204, 316–324.

Nelson, J.L. and Zavaleta, E.S. (2012) Salt marsh as a coastal filter for the oceans: Changes in function with experimental increases in nitrogen loading and sea-level rise. *PLoS One* 7, e38558. doi: 10.1371/journal.pone.0038558.

Nguir, A., Besbes, M., Jannet, H., Flamini, G. and Harzallah-Skhiri, F. (2011) Chemical composition, antioxidant and anti-acetylcholinesterase activities of Tunisian *Crithmum maritimum* L. essential oils. *Journal of Chemistry* 1, 173–179.

Nostro, A., Filocamo, A., Giovannini, A., Catania, S., Costa, C. *et al.* (2012) Antimicrobial activity and phenolic content of natural site and micropropagated *Limonium avei* (De Not.) Brullo & Erben plant extracts. *Natural Product Research* 26, 2132–2146.

Nsanganwimana, F., Pourrut, B., Mench, M. and Douay, F. (2014) Suitability of *Miscanthus* species for managing inorganic and organic contaminated land and restoring ecosystem services. A review. *Journal of Environmental Management* 143, 123–134.

Oosting, S.J., Mekoya, A., Fernandez-Rivera, S. and van der Zijpp, A.J. (2011) *Sesbania sesban* as a fodder tree in Ethiopian livestock farming systems. Feeding practices and farmers' perception of feeding effects on sheep performance. *Livestock Science* 139, 135–141.

Oueslati, S., Ksouri, R., Falleh, H., Pichette, A., Abdelly, C., *et al.* (2012) Phenolic content, antioxidant, anti-inflammatory and anticancer activities of the edible halophyte *Suaeda fruticosa* Forssk. *Food Chemistry* 132, 943–947.

Panda, A., Rangani, J., Kumari, A. and Parida, A.K. (2017) Efficient regulation of arsenic translocation to shoot tissue and modulation of phytochelatin levels and antioxidative defense system confers salinity and arsenic tolerance in the halophyte *Suaeda maritima*. *Environmental and Experimental Botany* 143, 149–171.

Pandey, K.B. and Rizvi, S.I. (2009) Plant polyphenols as dietary antioxidants in human health and disease. *Oxidative Medicine and Cellular Longevity* 2, 270–278.

Panta, S., Flowers, T., Lane, P., Doyle, R., Haros, G., *et al.* (2014) Halophyte agriculture: success stories. *Environmental and Experimental Botany* 107, 71–83.

Parida, A.K. and Jha, B. (2010) Salt tolerance mechanisms in mangroves: A review. *Trees* 24, 199–217.

Parida, A.K., Das, A.B., Sanada, Y. and Mohanty, P. (2004) Effects of salinity on biochemical components of the mangrove, *Aegiceras corniculatum*. *Aquatic Botany* 80, 77–87.

Parraga-Aguado, I., Gonzalez-Alcaraz, M.N., Lopez-Orenes, A., Ferrer-Ayala, M.A. and Conesa, H.M. (2016) Evaluation of the environmental plasticity in the xerohalophyte *Zygophyllum fabago* L. for the phytomanagement of mine tailings in semiarid areas. *Chemosphere* 161, 259–265.

Patel, S. (2016) *Salicornia* : Evaluating the halophytic extremophile as a food and a pharmaceutical candidate. *3 Biotech* 6, 1–10.

Pearlstein, S.L., Felger, R.S., Glenn, E.P., Harrington, J., Al-Ghanem, K.A. *et al.* (2012) Nipa. (*Distichlis palmeri*): A perennial grain crop for saltwater irrigation. *Journal of Arid Environments* 82, 60–70.

Petropoulos, S.A., Levizou, E., Ntatsi, G., Fernandes, Â., Petrotos, K., *et al.* (2017) Salinity effect on nutritional value, chemical composition and bioactive compounds content of *Cichorium spinosum* L. *Food Chemistry* 214, 129–136.

Pietta, P., Minoggio, M. and Bramati, L. (2003) Plant polyphenols: structure, occurrence and bioactivity. *Studies in Natural Products Chemistry* 28, 257–312.

Plouguerné, E., da Gama, B.A.P., Pereira, R.C. and Barreto-Bergter, E. (2014) Glycolipids from seaweeds and their potential biotechnological applications. *Frontiers in Cellular and Infection Microbiology* 4. doi: 10.3389/fcimb.2014.00174.

Qasim, M., Gulzar, S. and Khan, M.A. (2011) Halophytes as medicinal plants. In: Öztürk, M., Mermut, A.R. and Celik, A. (eds) *Urbanisation, Land Use, Land Degradation and Environment*. Daya Publishing House, New Delhi, pp. 330–343.

Qasim, M., Abideen, Z., Adnan, M.Y., Gulzar, S., Gul, B., *et al.* (2017) Antioxidant properties, phenolic composition, bioactive compounds and nutritive value of medicinal halophytes commonly used as herbal teas. *South African Journal of Botany* 110, 240–250.

Qian, D., Zhao, Y., Yang, G. and Huang, L. (2017) Systematic review of chemical constituents in the genus *Lycium* (Solanaceae). *Molecules* 22, 1–33.

Qureshi, R. and Bhatti, G.R. (2008) Ethnobotany of plants used by the Thari people of Nara Desert, Pakistan. *Fitoterapia* 79, 468–473.

Qureshi, R., Raza Bhatti, G. and Memon, R.A. (2010) Ethnomedicinal uses of herbs from northern part of Nara desert, Pakistan. *Pakistan Journal of Botany* 42, 839–851.

Rabhi, M., Hafsi, C., Lakhdar, A., Hajji, S., Barhoumi, Z., *et al.* (2009) Evaluation of the capacity of three halophytes to desalinize their rhizosphere as grown on saline soils under nonleaching conditions. *African Journal of Ecology* 47, 463–468.

Rabhi, M., Ferchichi, S., Jouini, J., Hamrouni, M.H., Koyro, H.W., *et al.* (2010) Phytodesalination of a salt-affected soil with the halophyte *Sesuvium portulacastrum* L. to arrange in advance the requirements for the successful growth of a glycophytic crop. *Bioresource Technology* 101, 6822–6828.

Radwan, H.M., Shams, K.A., Tawfik, W.A. and Soliman, A.M. (2008) Investigation of the glucosinolates and lipids constituents of *Cakile maritima* (Scope) growing in Egypt and their biological activity. *Journal of Medicine and Medical Sciences* 3, 182–187.

Rangani, J., Parida, A.K., Panda, A. and Kumari, A. (2016) Coordinated changes in antioxidative enzymes protect the photosynthetic machinery from salinity induced oxidative damage and confer salt tolerance in an extreme halophyte *Salvadora persica* L. *Frontiers in Plant Science* 7. doi.org/10.3389/fpls.2016.00050.

Ravindran, K.C., Venkatesan, K., Balakrishnan, V., Chellappan, K.P., Balasubramanian, T., *et al.* (2007) Restoration of saline land by halophytes for Indian soils. *Soil Biology and Biochemistry* 39, 2661–2664.

Rhee, M.H., Park, H. and Cho, J.Y. (2009) *Salicornia herbacea* : Botanical, chemical and pharmacological review of halophyte marsh plant. *Journal of Medicinal Plant Research* 3, 548–555.

Rodrigues, M.J., Gangadhar, K.N., Vizetto-Duarte, C., Wubshet, S.G., Nyberg, N.T., *et al.* (2014) Maritime halophyte species from southern Portugal as sources of bioactive molecules. *Marine Drugs* 12, 2228–2244.

Rozema, J. and Schat, H. (2012) Salt tolerance of halophytes, research questions reviewed in the perspective of saline agriculture. *Environmental and Experimental Botany* 92, 83–95.
Rozema, J., Muscolo, A. and Flowers, T. (2013) Sustainable cultivation and exploitation of halophyte crops in a salinising world. *Environmental and Experimental Botany* 92, 1–3.
Saïdana, D., Mahjoub, M.A., Boussaada, O., Chriaa, J., Chéraif, I., *et al.* (2008) Chemical composition and antimicrobial activity of volatile compounds of *Tamarix boveana* (Tamaricaceae). *Research in Microbiology* 163, 445–455.
Samiullah, Bano, A., Naz, R. and Yasmin, H. (2011) In vitro inhibition potential of *Lespedeza bicolor* Turcz against selected bacterial and fungal strains. *Journal* of *Medicinal Plants* 5, 3708–3714.
Samuelsen, A.B. (2000) The traditional uses, chemical constituents and biological activities of *Plantago major* L. A review. *Journal of Ethnopharmacology* 71, 1–21.
Sangwan, S., Rao, D.V. and Sharma, R.A. (2010) A review on *Pongamia pinnata* (L.) Pierre: A great versatile leguminous plant. *Natural Sciences* 8, 130–139.
Santos, E.S., Abreu, M.M., Peres, S., Magalhães, M.C.F., Leitão, S., *et al.* (2015) Potential of *Tamarix africana* and other halophyte species for phytostabilisation of contaminated salt marsh soils. *Journal of Soils and Sediments* 17, 1459–1473.
Santos Pedro, C.A., Gonçalves, S.C. and Ferreira, S.M.F. (2015) Phytoremediation of cadmium by the facultative halophyte plant *Bolboschoenus maritimus* (L.) Palla, at different salinities. *Environmental Science and Pollution Research* 22, 15598–15609.
Senatore, F., Napolitano, F. and Ozcan, M. (2000) Composition and antibacterial activity of the essential oil from *Crithmum maritimum* L. (Apiaceae) growing wild in Turkey. *Flavour and Fragrance Journal* 15, 186–189.
Shabala, S. (2013) Learning from halophytes: physiological basis and strategies to improve abiotic stress tolerance in crops. *Annals of Botany* 112, 1209–1221.
Shafi, M.S., Ashraf, M.Y. and Sarwar, G. (2001) Wild medicinal plants of Cholistan area of Pakistan. *Pakistan Journal of Biological* Science 4, 112–116.
Sharma, B., Salunke, R., Balomajumder, C., Daniel, S. and Roy, P. (2010) Anti-diabetic potential of alkaloid rich fraction from *Capparis decidua* on diabetic mice. *Journal of Ethnopharmacology* 127, 457–462.
Sharma, R., Wungrampha, S., Singh, V., Pareek, A. and Sharma, M.K. *et al.* (2016) Halophytes as bioenergy crops. *Frontiers in Plant Science* 7. doi: 10.3389/fpls.2016.01372.
Shelef, O., Gross, A. and Rachmilevitch, S. (2013) Role of plants in a constructed wetland: Current and new perspectives. *Water* 5, 405–419.
Sher, H., Al-Yemeni, M.N., Masrahi, Y.S. and Shah, H. (2010) Ethnomedicinal and ethnoecological evaluation of *Salvadora persica* L.: A threatened medicinal plant in Arabian Peninsula. *Journal of Medicinal Plant Research* 4, 1209–1215.
Shpigel, M., Ben-Ezra, D. Shauli, L., Sagi, M., Ventura, Y., *et al.* (2013) Constructed wetland with *Salicornia* as a biofilter for mariculture effluents. *Aquaculture* 412–413, 52–63.
Silva, Y.J.A., Silva, Y.J.A., Freire, M.B.G., Lopes, E.A., Santos, M.A., *et al.* (2016) *Atriplex nummularia* Lindl. as alternative for improving salt-affected soils conditions in semiarid environments: a field experiment. *Chilean Journal of Agricultural Research* 76, 343–348.
Simlai, A. and Roy, A. (2013) Biological activities and chemical constituents of some mangrove species from Sundarban estuary: An overview. *Pharmacognosy Reviews* 7, 170.
Simopoulos, A.P. (2016) An increase in the Omega-6/Omega-3 fatty acid ratio increases the risk for obesity. *Nutrients* 8, 1–17.
Sokolowska-Krzaczek, A., Skalicka-Wozniak, K. and Czubkowska, K. (2009) Variation of phenolic acids from herb and roots of *Salsola kali* L. *Acta Societatis Botanicorum Poloniae* 78, 197–201.
Stanley, O.D. (2008) Bio prospecting marine halophyte *Salicornia brachiata* for medical importance and salt encrusted land development. *Journal of Coastal Development* 11, 62–69.
Stöckigt, J., Sheludko, Y., Unger, M., Gerasimenko, I., Warzecha, H. *et al.* (2002) High-performance liquid chromatographic, capillary electrophoretic and capillary electrophoretic-electrospray ionisation mass spectrometric analysis of selected alkaloid groups. *Journal of Chromatography A* 967, 85–113.
Stuchlík, M. and Žák, S. (2002) Vegetable lipids as components of functional foods. *Biomedical Papers* 146, 3–10.
Suaire, F., Durickovic, I., Framont-Terrasse, L., Leblain, J.Y., De Rouck, A.C. *et al.* (2016) Phytoextraction of Na^+ and Cl^- by *Atriplex halimus* L. and *Atriplex hortensis* L.: A promising solution for remediation of road runoff contaminated with deicing salts. *Ecological Engineering* 94, 182–189.
Sundareshwar, P.V., Morris, J.T., Koepfler, E.K. and Fornwalt, B. (2003) Phosphorus limitation of coastal ecosystem processes. *Science* 299, 563–565.

Sunita, P., Jha, S. and Pattanayak, S. (2011) Pharmacognostic studies on leaf and stem of *Cressa cretica* Linn. *International Journal of Pharmaceutical Sciences and Research* 2, 849–855.

Sunita, P., Jha, S., Pattanayak, S.P. and Mishra, S.K. (2012) Antimicrobial activity of *Cressa cretica* L., a halophytic plant. *Journal of Scientific Research* 4, 203–212.

Szota, C., Farrell, C., Livesley, S.J. and Fletcher, T.D. (2015) Salt tolerant plants increase nitrogen removal from biofiltration systems affected by saline stormwater. *Water Research* 83, 195–204.

Taamalli, M., Ghabriche, R., Amari, T., Mnasri, M., Zolla, L., *et al.* (2014) Comparative study of Cd tolerance and accumulation potential between *Cakile maritima* L. (halophyte) and *Brassica juncea* L. *Ecological Engineering* 71, 623–627.

Taha, A. and Alsayed, H. (2000) Brine shrimp bioassay of ethanol extracts of *Sesuvium verrucosum*, *Salsola baryosma* and *Zygophyllum quatarense* medicinal plants from Bahrain. *Phyther Research* 14, 48–50.

Tiwari, P., Rahuji, N., Kumar, R., Lakshmi, V., Srivastava, M.N. *et al.* (2008) Search for antihyperglycemic activity in few marine flora and fauna. *Indian Journal of Science and Technology* 1(5), 1–5.

Trabelsi, N., Oueslati, S., Falleh, H., Waffo-Téguo, P., Papastamoulis, Y. *et al.* (2012) Isolation of powerful antioxidants from the medicinal halophyte *Limoniastrum guyonianum*. *Food Chemistry* 135, 1419–1424.

Tuan, N.Q., Lee, W., Oh. J., Kulkarni, R.R., Gény, C., *et al.* (2015) Flavanones and chromones from *Salicornia herbacea* mitigate septic lethality via restoration of vascular barrier integrity. *Journal of Agricultural and Food Chemistry* 63, 10121–10130.

Um, Y.R., Kong, C.S., Lee, J.I., Kim, Y.A., Nam, T.J. *et al.* (2010) Evaluation of chemical constituents from *Glehnia littoralis* for antiproliferative activity against HT-29 human colon cancer cells. *Process Biochemistry* 45, 114–119.

Van Oosten, M.J. and Maggio, A. (2015) Functional biology of halophytes in the phytoremediation of heavy metal contaminated soils. *Environmental and Experimental Botany* 111, 135–146.

Varli, S.N. and Şanlıer, N. (2016) Nutritional and health benefits of quinoa (*Chenopodium quinoa* Willd.). *Journal of Cereal Science* 69, 371–376.

Ventura, Y. and Sagi, M. (2013) Halophyte crop cultivation: The case for *Salicornia* and *Sarcocornia*. *Environmental and Experimental Botany* 92, 144–153.

Ventura, Y., Wuddineh, W.A., Myrzabayeva, M., Alikulov, Z., Khozin-Goldberg, I., *et al.* (2011) Effect of seawater concentration on the productivity and nutritional value of annual *Salicornia* and perennial *Sarcocornia* halophytes as leafy vegetable crops. *Scientia Horticulturae* 128, 189–196.

Ventura, Y., Eshel, A., Pasternak, D. and Sagi, M. (2015) The development of halophyte-based agriculture: Past and present. *Annals of Botany* 115, 529–540.

Vromman, D., Lefèvre, I., Šlejkovec, Z., Martínez, J.P., Vanhecke, N. *et al.* (2016) Salinity influences arsenic resistance in the xerohalophyte *Atriplex atacamensis* Phil. *Plant Physiology and Biochemistry* 103, 208–218.

Wang, H., Wang, X., You, J., Suo, Y., Zhao, X. *et al.* (2007) Comparative analysis of allantoin, quercetin, and 1-methyl-1,2,3,4-tetrahydro-β-carboline-3-carboxylic acid in *Nitraria tangutorum* Bobr. seed by HPLC-APCI-MS and CE. *Journal of Liquid Chromatography and Related Technologies* 30, 363–376.

Wang, H., Li, J., Tao, W., Zhang, X., Gao, X., *et al.* (2017) *Lycium ruthenicum* studies: Molecular biology, phytochemistry and pharmacology. *Food Chemistry* 240, 759–766.

Wang, H.L., Tian, C.Y., Jiang, L. and Wang, L. (2014) Remediation of heavy metals contaminated saline soils: A halophyte choice? *Environmental Science and Technology* 48, 21–22.

Wang, L., Baskin, J.M., Baskin, C.C., Cornelissen, J.H.C., Dong, M. and Huang, Z. (2012a) Seed dimorphism, nutrients and salinity differentially affect seed traits of the desert halophyte *Suaeda aralocaspica* via multiple maternal effects. *BMC Plant Biology* 12, 170.

Wang, L., Zhao, Z.Y., Zhang, K. and Tian, C.Y. (2012b) Oil content and fatty acid composition of dimorphic seeds of desert halophyte *Suaeda aralocaspica*. *African Journal of Agricultural Research* 7, 1910–1914.

Wang, W., Yan, Z., You, S., Zhang, Y., Chen, L. *et al.* (2011) Mangroves: obligate or facultative halophytes? A review. *Trees* 25, 953–963.

Webb, J.M., Quintã, R., Papadimitriou, S., Norman, L., Rigby, M., *et al.* (2012) Halophyte filter beds for treatment of saline wastewater from aquaculture. *Water Research* 46, 5102–5114.

Weber, D.J., Gul, B., Khan, A., Williams, T., Williams, N. *et al.* (2001) Composition of vegetable oil from seeds of native halophytes. In: McArthur, E.D., Fairbanks, D.J.C. *Proceedings of Shrubland Ecosystem Genetics and Biodiversity*, 13–15 June 2001, Provo, Utah, USA. US Department of Agriculture, Forest Service, Rocky Mountain Research Station, Ogden, Utah, USA, pp. 287–290.

Weber, D.J., Ansari, R., Gul, B. and Ajmal Khan, M. (2007) Potential of halophytes as source of edible oil. *Journal of Arid and Environment* 68, 315–321.

Wu, T., Lv, H., Wang, F. and Wang, Y. (2016) Characterization of polyphenols from *Lycium ruthenicum* fruit by UPLC-Q-TOF/MSE and their antioxidant activity in Caco-2 Cells. *Journal of Agricultural and Food Chemistry* 64, 2280–2288.

Wu, Y., Chung, A., Tam, N.F.Y., Pi, N. and Wong, M.H. (2008) Constructed mangrove wetland as secondary treatment system for municipal wastewater. *Ecological Engineering* 34, 137–146.

Yajun, B., Xiaojing, L. and Weiqiang, L. (2003) Primary analysis of four salt tolerant plants growing in Hai-He Plain, China. In: Lieth, H. and Mochtchenko, M. (eds) *Cash Crop Halophytes: Recent Studies.* Springer, Dordrecht, Netherlands, pp. 135–138.

Yao, L.H., Jiang, Y.M., Shi, J., Tomas-Barberan, F.A., Datta, N. *et al.* (2004) Flavonoids in food and their health benefits. *Plant Foods for Human Nutrition* 59(3), 113–122.

Yao, X. and Chen, G. (2007) Simultaneous determination of p-hydroxyacetophenone, chlorogenic acid, and caffeic acid in Herba *Artemisiae Scopariae* by capillary electrophoresis with electrochemical detection. *Analytical and Bioanalytical Chemistry* 388, 475–481.

Yu, X., Zhang, Y., Shao, R. and Xu, W. (2012) Study on antibacterial and antioxidant activities of *Salicornia herbacea* extracts. *Advanced Materials Research* 421, 47–50.

Zarrouk, M., El Almi, H., Youssef, N.B., Sleimi, N., Smaoui, A., *et al.* (2003) Lipid composition of seeds of local halophytes: *Cakile maritima, Zygophyllum album* and *Crithmum maritimum*. In: Lieth, H. and Mochtchenko, M. (eds) *Cash Crop Halophytes: Recent Studies*. Springer, Dordrecht, Netherlands, pp. 121–124.

Zhang, Y., and Xu, H. (2017) Recent progress in the chemistry and biology of limonoids. *RSC Advances* 7(56), 35191–35220.

Zheng, J., Li, H., Ding C., Suo, Y., Wang, L. *et al.* (2011) Anthocyanins composition and antioxidant activity of two major wild *Nitraria tangutorun* Bobr. variations from Qinghai–Tibet Plateau. *Food Research International* 44, 2041–2046.

Zhou, Z.Q., Fan, H.X., He, R.R., Xiao, J., Tsoi, B. *et al.* (2016) Lycibarbarspermidines A-O., new Dicaffeoylspermidine derivatives from Wolfberry, with activities against Alzheimer's disease and oxidation. *Journal of Agricultural and Food Chemistry* 64, 2223–2237.

20 Practical Uses of Halophytic Plants as Sources of Food and Fodder

Tiziana Centofanti[1,*] and Gary Bañuelos[2]
[1]*Central European University, Budapest, Hungary;* [2]*San Joaquin Valley Agricultural Sciences Center, Parlier, California, USA*

Abstract

Halophytes are plants that are adapted to saline soils in their natural habitats because they are salt tolerant. They are found in a range of environments with varied salinity and climatic conditions. These plant species can be irrigated with saline water and cultivated on saline soils that are unsuitable for commercial crops. Halophytes are rich in nutrients, such as antioxidants, fatty acids and amino acids, and many species have been used traditionally as herbs and vegetables, feed and fodder. Therefore, halophytes are considered one of the alternative solutions to problems related to food security, fresh water scarcity, salinization and diversification of diets for healthier nutrition. However, despite the promising future for multiple uses of halophytes, many problems related to halophyte cultivation for human and animal consumption and their commercialization have still not been tackled. This chapter is intended to provide an overview of the development of halophytes as vegetable, feed and fodder, and to highlight the importance of creating a demand in the marketplace for halophyte consumption.

Keywords: Alternative foods; Climate change; Marine ecosystem; Medicinal plants; Salinity

20.1 Introduction

20.1.1 Halophytes and the environment

Food production needs to increase by up to 70% by 2050 to match the food needs of a growing population that is expected to reach 9 million by 2050 (Panta *et al.*, 2014). This increase is a challenge because of two main problems we are facing today: land and water scarcity. Worldwide urbanization and land degradation are the main causes of land scarcity for agricultural production. Soil salinization is a major factor contributing to soil degradation. High levels of salts in soil cause salinity, and in a global context NaCl is the major cause (Munns and Tester, 2008; Rozema and Flowers, 2008). Only about 6% of the world's land surface is composed of naturally saline soils that have never been cultivated because of their high salinity. Irrigated agriculture, which produces 70% of crops, is the cause of soil degradation, called secondary salinization. It is estimated that about 20% of irrigated land (230 million ha) is seriously affected by salts (Boscaiu and Vicente, 2013). Irrigation water carries salts (dissolved ions) that accumulate in the soil with cycles of irrigation and cause the loss of more than 10 million ha of arable land every year

* Corresponding author e-mail: centofantit@spp.ceu.edu

(Owens, 2001). The problem of soil salinization has occurred since ancient times. Yield losses and crop failures were documented in old written records of early civilizations such as those in the Indus and Nile valleys. It is thought that the collapse of the Mesopotamian civilization was partly due to the similar phenomenon of soil salinization due to irrigation (Aronson, 1985; Boscaiu and Vicente, 2013). Civilizations have historically been shaped by soil degradation (including salinity) that restricted population numbers, limited where domestication could occur and forced relocation (Hasegawa, 2013). The Aral Sea Basin in central Asia represents one of the worst cases of salinization, where salinity and waterlogging affect up to 50% of the irrigated area (Kijne, 2005; Qadir *et al.*, 2009). In addition to soil salinity, freshwater aquifers are becoming increasingly salty because of sea-level rise and the increase in droughts. In human history, agriculture has expanded into new land at the loss of natural ecosystems. Today, we are aware of the consequences of this process (forest conversion), which lead to significant losses of ecosystem services, such as habitat necessary to maintain biodiversity, storage of carbon, flood mitigation, and soil and watershed protection. Hence, use of degraded lands has often been suggested as an alternative to the issue of land scarcity. Definitive data on the annual worldwide loss of farmland due to soil degradation and related causes are lacking. Gibbs and Salmon (2015) used four approaches (expert opinion, satellite observations, biophysical models and inventories of abandoned agricultural land) to assess degraded land at the global scale. They concluded that there is a wide disagreement in the spatial distribution of these degraded lands, but estimated that the world's total degraded area ranges from 1 billion ha to 6 billion ha. The disagreement stems from the uncertainty in categorizing an area as degraded, as this classification depends on a number of characteristics that are subjective and site- and time-specific.

To meet the future food demand, we need not only to develop alternative and more efficient technologies to increase yield, but also to enlarge the diversity of crops that are grown and consumed. Today, among the thousands of crops deemed edible, humans consume only about 20 crops on a large scale, of which the major four (corn, rice, wheat and soybean) represent staple foods worldwide. A diversification of our diets with inclusion of traditional or neglected crops that are rich in micronutrients, require less inputs (fertilizers, water, pesticides) for growth and can adapt to degraded environments (i.e. saline soils) could be a win–win situation to tackle the problem of food security and resources scarcity. In line with other authors (Panta *et al.*, 2014; Ventura *et al.*, 2014), we think that halophytes (plants adapted to saline soil, see section 20.1.2) may be a commercial, alternative plant group for the use of saline land and to ease pressure on the requirement for good-quality land and water (Panta *et al.*, 2014). Others have shown that halophytes can be grown to produce oilseeds, grains, forage, fuel, food, medicine, chemicals, timber and fibre (Kahn and Qaiser, 2006; Glenn *et al.*, 2013) or be used for soil or water conservation or remediation (Ravindran *et al.*, 2007; Zhao *et al.*, 2002). In addition, halophytes have been tested for the treatment of saline aquaculture effluent (Brown *et al.*, 1999) and used as landscaping ornamentals (Zia *et al.*, 2008).

However, despite the promising future for multiple uses of halophytes, many problems related to halophyte cultivation for human and animal consumption and their commercialization have still not been tackled. This chapter considers the development of halophytes as vegetable, feed and fodder and highlights the importance of creating market demand for halophyte consumption. We also analyse the environmental impacts and benefits of halophytes in bio-saline agriculture.

20.1.2 Characteristics of halophytic plants

Halophytes are plants that are adapted to saline soils in their natural habitats because they are salt tolerant. Halophytes are found in a range of environments with varied salinity and climatic conditions. They are trees, grasses and saltbushes, growing in salt marshes, saline deserts and coastal areas around the world (Epstein *et al.*, 1980). They represent only a small fraction of the total number of flowering plants (2600 out of 400,000) and only a small percentage of halophytes is domesticated and used as food and fodder. There is no broadly accepted definition of halophyte but the most commonly used is: 'a halophyte is any plant that can complete its life cycle and reproduce itself under conditions of soil-water salinity of 8–10 dS m^{-1} electrical conductivity (EC) (approximately 20% seawater) or more' (Aronson, 1985, 1989).

For a long time researchers have tried to identify the mechanisms of salt tolerance, but these have not yet been fully clarified. Indeed, the effect of salinity on growth varies among halophytes and many differences do exist among species in the balance of Na^+ and Cl^- in shoot tissues (Flowers and Colmer, 2008).

For example, it has not yet been clarified whether mechanisms for salt tolerance are similar between all halophytes and whether mechanisms for salt tolerance in halophytes are linked with other environmental variables such as aridity, flooding, etc. (Flowers and Colmer, 2008). There is clearly still much to learn about halophytes and the diversity of mechanisms that they employ to cope with salinity.

Salt tolerance comprises morphological, physiological and biochemical adaptations at the whole plant, tissue and cellular levels (Koyro *et al.*, 2011). These adaptations consist of complex polygenic traits controlled by a number of genes or groups of genes (Türkan and Demiral, 2009). Transferring salinity tolerance traits to glycophytes (less salt-tolerant crops) would be useful to obtain conventional crops that are adapted to saline environments, but is currently impractical due to complex polygenic traits. According to Rozema and Schats (2013), glycophytes would not accommodate the appropriate allelic variants at all the loci involved in halophyte salt tolerance.

Generally, in both glycophytes and halophytes, salt-tolerance mechanisms involve controlled uptake of salt ions in the root system, compartmentalization of toxic ions in the vacuole and synthesis of organic solutes in the cytoplasm to metabolically balance the osmotic potential of the Na^+ and Cl^- accumulated in the vacuole (Zhu, 2001; Flowers and Colmer, 2008; Shabala and Mackay, 2011). Sugars (e.g. sucrose), sugar alcohols (e.g. sorbitol), amino acids (e.g. proline), methylated proline-related compounds (e.g. methyl-proline), betaines (e.g. glycinebetaine) and methylated sulfonio compounds (e.g. dimethylsulfonioproprionate, DMSP) have been proposed to fulfill this function in halophytes (Flowers and Colmer, 2008). In addition, to counteract the oxidative damage caused by reactive oxygen species (ROS), halophytes produce antioxidative enzymes and non-enzymatic molecules with antioxidant properties – such as ascorbate, glutathione, alpha tocopherol, flavonoids, anthocyanins, carotenoids and polyphenolic compounds (Türkan and Demiral, 2009; Zhu, 2001). These antioxidants and organic solutes confer nutritional value to the halophytes and are ultimately responsible for the quality of the vegetable product (Maggio *et al.*, 2011). The enrichment of the edible plant part with antioxidant compounds, such as ascorbic acid (vitamin C), β-carotene and polyphenols, is highly desirable for the human diet. Halophytes have gained considerable attention and have been the subject of many research studies in the past 30 years because they are considered one of the alternative solutions to problems related to food security, freshwater scarcity, salinization and diversification of diets for healthier nutrition. Their potential as alternative crops is multi-faceted: (1) they can be irrigated with seawater; (2) they can grow in arid saline environments, e.g. deserts, marginal and degraded soils; and (3) they can contain nutraceutical products that may be beneficial to human health. In section 20.2 we discuss the role of halophytes as alternative crops for sustainable use of natural resources.

20.2 Halophytes as Food

20.2.1 Traditional wild halophytes

Many halophytic plant species have been used traditionally as herbs and vegetables, and a review of the history of their uses can be found in ethnobotanic reviews and literature (Davy *et al.*, 2001; Simopoulos, 2004; Guarrera *et al.*, 2006; Tardío *et al.*, 2006).

Sugar beet (*Beta vulgaris*) is the only major food crop derived from the halophyte wild progenitor *B. vulgaris* subsp. *maritima* (Aronson, 1985). Similarly, the date palm (*Phoenix dactylifera*) cultivated today seems to have derived from the wild halophytic population that grew in the Middle East. Other traditional food staples derived from halophytes that grow in ocean water are eelgrass (*Zostera maritima*) and Palmer's saltgrass

(*Distichlis palmeri*) from the Gulf of California. Eelgrass plant seeds have been used by the Seri Indians as gruel after being toasted and ground. The gruel was then mixed with fats or other fatty seeds oils (i.e. columnar cactus) to prepare a meal very rich in protein, fats and carbohydrates (Felger *et al.*, 1980; Aronson, 1985). *Distichlis palmeri* (Glenn *et al.*, 2013) is a highly salt-tolerant grass endemic to the delta of the Colorado River and in the northern Gulf of California, USA. A perennial species, its grains are similar to rice and wheat in size and nutritional value. The plant can grow in flooded conditions – like paddy rice – in seawater, and the grains were used by the Cocopa Indians to make bread (Felger, 1979). Another halophyte species that has been used since ancient times is the mangrove. Mangroves represent an abundant source of timber, firewood and charcoal, as well as tannins. The bark of mangrove trees from genera like *Excaecaria agallocha*, *Kandelia*, *Rhizophora*, *Ceriops*, *Sonnertia acida* and *Carapa* are cut and boiled in large vessels of water to extract the tannins that are then used for fishing and in the plywood industry (Aronson, 1985). Many species of mangroves have palatable foliage that can be used as animal feed. In particular, the foliage of *Avicennia marina*, *Sonneratia alba* and *Rhizophora mangle* has served as camel and cattle feed (Chapman, 1976). In Colombia, mangrove species including *Avicenna marina* and *A. germinans* are used as food (Lieth *et al.*, 2000). In different regions of India, young shoots and leaves of *Chenopodium album* and *Amaranthus* spp. are used as salads and vegetables, and raw fruits of *Capparis decidua* are used for pickles (Dagar, 2005; Rameshkumar and Eswaran, 2013). People in the Maldives regularly eat *Sonneratia caseolaris*, also known as mangrove apple or crabapple mangrove fruit, which are eaten raw and also used to make vinegar.

Sea spinach (*Tetragonia* sp.) was introduced to the culinary world in the 18th century. Its edible leaves are the halophyte version of spinach and they have been used pickled and cooked to fight scurvy. Sea kale (*Crambe maritima*) is another traditional halophyte vegetable precursor of a widely consumed commercial cabbage. Its shoots are served steamed or sauced like asparagus (Panta *et al.*, 2014).

Many of the plants described above have been collected from wild populations or cultivated in private backyard and kitchen gardens and sometimes sold at local farmers' markets (Wilson *et al.*, 2000). These halophytic species have great potential to be transformed into vegetable crops for saline agriculture. Growing interest over the past decade in cultivating crops under saline conditions has led to rediscovery of the potential of several promising halophytic plant species to be farmed as gourmet vegetables. Most are eaten as raw vegetables or in fresh salads, and some are also cooked or pickled. Apart from their palatability, these plants are – in general – rich in protein, antioxidant compounds and/or essential nutrients (minerals, vitamins, amino acids and/or fatty acids). However, halophytes are wild species and they can hardly compete with standard crops, which were domesticated thousands of years ago and improved over the centuries (Boscaiu and Vicente, 2013). Traits that are adaptive for wild plants, such as uneven germination, lack of seed retention (seed shattering) and toxic substances in tissues (saponins, tannins) are undesirable for use of halophytes as cultivated plants (Brown *et al.*, 2014). For example, the agronomic characteristics of *Distichlis palmeri* and its crop potential as grain still need to be identified to develop efficient breeding programmes (Boscaiu and Vicente, 2013). The history of traditional uses of halophytes in communities is important in overcoming the problem of domestication. Brown *et al.* (2014) discuss the problems related to domestication of halophytic plants for commercial production. One important aspect to consider is that the success of halophyte growth in saline agriculture is related to the environmental conditions. These species can be grown better in environments similar to those in which the plant occurs in nature.

To date, most halophytic plants have been tested for their yield and adaptability to saline agriculture. A summary of saline agriculture using domesticated halophytes is provided in section 20.2.2.

20.2.2 Saline agriculture

Saline agriculture is the cultivation of crops using seawater (approx. 40 dS m^{-1}) or brackish water. Seawater agriculture, as proposed by Glenn *et al.* (1998), is an extreme case of halophyte farming and allows the cultivation of only the most salt-tolerant halophytes. Instead, the use of brackish water represents a more viable option for commercial saline agriculture, because a higher number of halophyte crops can be grown in much lower salinity levels than seawater (Rozema and Flowers, 2008; Ventura and Sagi, 2013). Saline agriculture can be carried out on marginal land, deserts and degraded soils close to the sea. About 43% of the Earth's total land surface is arid or semi-arid, and seawater in the oceans makes up 97% of water on earth. Hence, saline agriculture could be an alternative, sustainable way of increasing food production without competing with conventional crops for increasingly scarce resources such as fertile land and good-quality irrigation water (Khan and Duke, 2001; Boscaiu and Vicente, 2013). Despite these positive environmental factors of saline agriculture, successful agronomic techniques must be developed for growing saline water-irrigated crops in a sustainable manner. These methods must also not contribute to further damage of natural environments (Khan and Duke, 2001). One primary consideration for sustainable saline agriculture is to choose species from climate regimes similar to those where the potential crop is being planted, as the candidate species will be better adapted to local environmental conditions (Zerai *et al.*, 2010; Brown *et al.*, 2014).

Optimization of irrigation quantity at each salinity level (brackish or seawater) is another important aspect for sustainable saline agriculture, to allow salt leaching below the root zone, and consequently avoiding soil salinization (O'Leary, 1988; Lieth, 2000). Multi-year long-term experiments with halophytes have been conducted in various parts of the world; and no decrease in yield, or negative impact on soil structure has been observed because of saline water irrigation. Sandy soils in coastal areas or inland sand dunes may be readily available for large-scale halophyte production without the risk of salt contamination occurring on fertile soils through leaching Ca^{2+}/Na^{+} exchange and subsequent clay dispersion. Likewise, underground freshwater contamination should be avoided by the existence of sufficiently deep water tables or adequate drainage. In addition, saline agriculture must justify the expense of pumping seawater for irrigation by producing useful crops at high yields (Glenn *et al.*, 1998). However, seawater agriculture should be cost-effective in desert regions – even though the yields may be smaller than in traditional agriculture – because it is cheaper to pump seawater at sea level than to pump freshwater from wells (Glenn *et al.*, 1998). Typical agricultural wells lift water from 20 m to as deep as 100 m, whereas the lift of typical coastal seawater wells is only 3–10 m. Furthermore, in some locations tides can be used to irrigate crops without the need for pumping (Glenn *et al.*, 1994; Brown *et al.*, 2014). In addition, coastal desert farms on sandy soils generally have unimpeded drainage back to the sea, thus avoiding groundwater salt contamination. Furthermore, coastal and inland salt desert aquifers often already have elevated concentrations of salts and so should not be damaged by seawater irrigation (Glenn *et al.*, 1998). Glenn *et al.* (1998) have shown that normal farm irrigation equipment can be modified to protect it from seawater damage. Successful long-term field trials have been carried out with the halophyte *Salicornia bigelovii* in saline agriculture for fodder production, as discussed in the section 20.4 below. Some halophytes have also been cultivated in saline agriculture as gourmet vegetables. The long history of halophyte vegetable cultivation and consumption has developed consumers' appreciation for halophytes as gourmet vegetables because of their salty taste and high nutritional value (Mudie *et al.*, 2005; Lu *et al.*, 2010). In section 20.2.3 we describe some of the most cultivated halophyte gourmet vegetables.

20.2.3 Halophytes as gourmet vegetables

20.2.3.1 Salicornia *and* Sarcocornia *spp.*

Perhaps the most cultivated and consumed halophyte vegetables are *Salicornia* and *Sarcocornia* spp., which have been studied for their potential as gourmet food, animal feed and oils for biodiesel. *Salicornia* and *Sarcocornia* species are distinguished by their growth habit and flower morphology (Kadereit *et al.*, 2006, 2007).

In coastal communities, both *Salicornia* and *Sarcocornia* have traditionally been used as vegetables in fresh salads for self-consumption or for sale in local markets. Different species of the *Salicornia* genus are suitable for vegetable production. These halophytes are rich in minerals, fatty acids and antioxidant compounds, such as polyphenols (Boscaiu and Vicente, 2013; Ventura and Sagi, 2013). In *Salicornia* and *Sarcocornia*, total polyphenols are high, representing 1.2 and 2.0 mg GAE g^{-1} fresh weight (FW), respectively (Ventura *et al.*, 2011). The lower limit of other, non-halophytic leafy vegetables rated as rich in phenolic compounds is > 0.5 mg GAE g^{-1} FW (Isabelle *et al.*, 2010). Leaves of *S. bigelovii* are a source of omega-3 polyunsaturated fatty acids, and antioxidant β-carotene, with quantities ranging from 4.7 (Ventura *et al.*, 2011) to 15.9 mg 100 g^{-1} FW (Lu *et al.*, 2010). These values are similar to those found in seaweeds (4.0 mg/100 g FW) and spinach (5.1 mg/100 g FW) (Isabelle *et al.*, 2010). The highly saline-adapted shoots of *S. bigelovii* contain ascorbic acid content of 6 mg 100 g^{-1} FW (Lu *et al.*, 2010). These values are in a similar range to those for non-halophyte leafy vegetables, such as spinach (7 mg 100 g^{-1}) and lettuce (< 2 mg 100 g^{-1}) (Proteggente *et al.*, 2002).

S. europaea has been grown in several trials and also in small-scale commercial cultivation using seawater in Ensenada, Mexico (Ventura and Sagi, 2013). *S. bigelovii* is also grown in subtropical climates, particularly in Mexico, where growers can supply European markets with *Salicornia* shoots in the off-season from September to June (OASE Foundation, 2009). Indeed, most of the halophyte vegetable production is sold as gourmet vegetables in Europe and the USA (Böer, 2006; Zerai *et al.*, 2010; Ventura *et al.*, 2011). The young shoots are sold in the market as 'samphire' or 'sea asparagus'. They are succulent and have a salty taste, which makes them suitable as vegetables (Ventura *et al.*, 2011). High-yield production of *Salicornia* under saline conditions is a necessary requirement for its cultivation to be commercially viable (O'Leary, 1988). In addition to irrigation with seawater, *Salicornia* has been tested in small lysimeter experiments for irrigation with hypersaline drainage water (Grattan *et al.*, 2008) or saline aquaculture effluent (Brown *et al.*, 1999). Glenn *et al.* (1998) proposed to combine shrimp farms with halophyte production. Seawater is used to grow the shrimps and the effluent water is then used to grow halophytes (Buhmann *et al.*, 2013; Turcios and Papenbrock, 2014). This water is also a source of nutrients. *Salicornia* and *Sarcocornia* spp. have proved to be efficient biofilters for the removal of nutrients from aquaculture effluents (Shpigel *et al.*, 2013). It is estimated that a production level of 20–30 kg m^{-2} $year^{-1}$ of fresh *Salicornia* biomass in the constructed wetlands can remove 1–3 g m^{-2} day^{-1} of nitrogen (Shpigel *et al.*, 2013).

Some of the problems encountered in halophyte vegetable production are related to harvest and germination. Only the young shoots are used in fresh market vegetable production, and these must be harvested manually, thus creating a labour-intensive element critical to halophyte crop production (Panta *et al.*, 2014). Often these halophytes have repeated harvests during the growing cycle. However, the disadvantage of labour-intensive manual harvest can be offset by the advantage of accumulating higher final yields per area than crops harvested once per season. In addition, high crop quality is obtained by harvesting only the young shoot tips. Ventura *et al.* (2011) obtained the highest yield (16 kg m^{-2}) during a 6-month harvest period, using the 3-week interval for *Salicornia*; this is when the harvest is done every three weeks for a total duration of 6 months and only the shoots of the Salicornia plant are harvested, hence the harvest can be repeated multiple times.These types of studies are crucial to assess the commercial feasibility of halophytes as gourmet vegetables.

Various factors affect germination rate: for example, salinity levels of irrigation water applied to the seeds, and temperature, both influence germination potential. *Salicornia* and *Sarcocornia* spp. can be germinated with irrigation water comprising concentrations of up to 75% seawater. Germination limits for *Salicornia* and Sarcocornia spp. occur only at extreme hypersaline conditions of double the seawater concentration. Temperature has a strong impact on the germination rate of *Salicornia* spp. Germination usually happens during the winter and spring months when the sediment salt concentration is lowest (Davy *et al.*, 2001, 2006).

20.2.3.2 Crithmum maritimum

Humans have consumed *C. maritimum* (marine fennel) for centuries, and to date it is still often gathered from the coastal salt marshes and inland saltpans of Europe (Franke, 1982; Wagenvoort *et al.*, 1989; Davy *et al.*, 2001; Simopoulos, 2004; Tardío *et al.*, 2006).

C. maritimum is a facultative halophyte and has been largely used for nutritional and medicinal purposes. The plant is rich in several bioactive substances, minerals, vitamin C, essential oils and other biomolecules (Atia *et al.*, 2009; Meot-Duros *et al.*, 2010). It grows in rocky coastal environments in the Mediterranean region, where it is often subjected to sea spray (Ben-Hamed *et al.*, 2004). Traditionally, *C. maritimum* has been collected and eaten for its antiscorbutic properties, owing to the vitamin C content. This halophyte has the potential to become a multipurpose cash crop (Ventura *et al.*, 2014). Oils extracted from the plant have shown the presence of high concentrations of fatty acids of the omega-3 and omega-6 series, which have beneficial effects against coronary heart diseases (Guil-Guerrero and Rodriguez-Garcia, 1999). The succulent leaves are consumed fresh or pickled as salty vegetables and its inflorescence, in the shape of an umbrella, can be used for ornamental purposes. As discussed for *Salicornia*, the germination potential of *C. maritimum* can hinder its cultivation for large-scale commercial purposes. Atia *et al.* (2006) reported that germination was significantly inhibited when NaCl concentrations exceeded 50 mM (approx. 5 dS m^{-1}).

20.2.3.3 Salsola soda

Salsola soda, more commonly known as 'agretti', is a halophyte native to the Mediterranean basin. It is a relatively small plant that grows to about 0.7 m on average in soils located within coastal regions that are at times saturated with salt water. Throughout history, the plant was a very important source of soda ash, as people would extract the ashes from *S. soda*. The plant is no longer grown for the use of its soda ash, but rather is farmed as a vegetable in Mediterranean countries such as Italy. Various *Salsola* species are traditionally used in folk medicine for the treatment of hypertension, constipation and inflammation (Tundis *et al.*, 2009). In this regard, alkaloid extracts from *Salsola* species have been evaluated for the treatment of Alzheimer's disease (Tundis *et al.* 2009). A preliminary study conducted by Centofanti and Bañuelos (2015) in central California, showed that *Salsola soda* can grow in saline (EC 2 >10 dS m^{-1}) and boron-laden soils (10 mg l^{-1}) of the San Joaquin Valley, California, and easily tolerates irrigation with saline and B-rich water (EC 3 dS m^{-1} and 4 mg B l^{-1}). In this poor-quality soil, under these growing conditions, the plant accumulated and removed high amounts of Na (80 g Na kg^{-1} dry weight (DW), B (100 mg B kg^{-1} DW) and Se (3–4 mg Se kg^{-1} DW) without showing any toxicity symptoms (Figs 20.1 and 20.2). Hence, *S. soda* showed promising potential as a plant species that can be irrigated and grown in saline and B-laden conditions and accumulate and harvest unwanted ions from the soil. We have no direct evidence of its ability to grow under wetland growing conditions, although the plant is commonly grown under salty marshy regions in coastal areas of Mediterranean Europe. The plant can, apparently, survive under continuous wet and saline growing conditions.

Currently, the authors of this chapter are investigating the potential of agretti as a vegetable crop in large-scale field trials (Fig. 20.3) at Red Rock Ranch, Five Points, central California. The goal of the study is to understand how it can be grown and used on a sustained basis for phyto-harvest of Na, B and Se field soil drip-irrigated with saline drainage water, as well as in wetland conditions with simulated drainage water representative of the salinity, B and Se levels found on the west side of the San Joaquin Valley in central California. In addition, consumers' preferences and the potential market demand for alternative vegetable crops such as this will be performed in these current studies through economic analysis and consumer surveys carried out at farmers' markets and retail centres. Currently, agretti produced in the field trials at Red Rock Ranch is sold at farmers' markets and served in local restaurants in Fresno, California (Fig. 20.4).

20.2.3.4 Minor gourmet vegetables

Inula crithmoides is consumed in salads or pickled in vinegar and is a dietary source of iodine. It also contains a high total lipid content with a significant portion of omega-3 fatty acids. The attractive yellow flowers of *I. crithmoides* make it a candidate for flowering pots and landscaping in arid saline areas such as Lebanon (Zurayk and Baalbaki, 1996).

Fig. 20.1. *Salsola soda* (agretti) grown on saline soil (Oxalis silty clay loam) with EC >10 dS m^{-1}, water soluble B concentration of 10 mg l^{-1}, water soluble Se concentration of 0.2 mg l^{-1} and water soluble Na concentration of 1470 mg l^{-1}. Plants were grown on the soil for 3 weeks and were 37 days old when the photograph was taken.

Fig. 20.2. *Salsola soda* (agretti) grown on non-saline soil (Hanford sandy loam) with EC <1 dS m^{-1}, water soluble B concentration of 0.12 mg l^{-1}, water soluble Se concentration of 0.002 mg l^{-1} and water soluble Na concentration of 97 mg l^{-1}. Plants were grown on the soil for 3 weeks and were 37 days old when the picture was taken.

Various minor halophytes used in fresh salads or pickled are sold as gourmet vegetables in local markets. Among these are *Aster tripolium*, *Batis maritima* and *Plantago coronopus*, which are rich in minerals and polyphenols, essential amino acids and tocopherol antioxidants (vitamin E); there are also high levels of vitamins A, C and K, and minerals. Perennial wall rocket (*Diplotaxis tenuifolia*) is another halophyte leafy vegetable commonly cultivated in many parts of the worlds (de Vos *et al*., 2013).

Three halophytic plant species, *Atriplex hortensis*, *A. triangularis* and *Tetragonia tetragonioides* (New Zealand spinach), are alternatives to spinach (Wilson *et al*., 2000; Słupski *et al*., 2010) owing to their similar chemical composition

Fig. 20.3. *Salsola soda* (agretti) grown in a farmer's field at Red Rock Ranch, central California.

Fig. 20.4. *Salsola soda* (agretti) grown in a farmer's field at Red Rock Ranch, central California, and served in a restaurant in Fresno, California.

(Carlsson and Clark, 1983) and their edible leaves. These halophytes are used for human consumption in the Netherlands, Belgium and Portugal (Lieth, 2000; Panta *et al.*, 2014). Yield of *A. triangularis* was 21.2 t ha^{-1} on a FW basis when grown under seawater irrigation (30 g NaCl l^{-1}) (Gallagher, 1985).

20.2.3.5 Nutritional aspects of gourmet vegetables

In section 20.2.3.4 we have listed the types of secondary metabolites and nutraceutical compounds that are naturally present in the edible parts of halophyte vegetables. However, it is possible to enhance the content of secondary metabolites by specific agronomic practices (Ventura *et al.*, 2011), including amount and salinity of irrigation water, plant fertilization, harvest time and cycle, and harvested plant material (young or old leaves) (Ventura *et al.*, 2010, 2011, 2013). Little is known about how to implement these agronomic techniques in large-scale halophyte production, because these studies have often been conducted in small field trials for research purposes, rather than for improvement of commercialization of halophytes.

Microelement deficiency may result from soil alkalization as a side effect of soil salinization (Grattan and Grieve, 1999). Ventura *et al.* (2013) obtained an improvement of the quality of leafy vegetable *A. tripolium* by iron fertilization because, during cultivation, a leaf chlorosis was indirectly induced by the high soil pH (> 8) that affected the product quality. This species has been tested for its potential as a halophytic cash crop in pilot projects in The Netherlands, Belgium and Portugal (Lieth and Mochtchenko, 2002; Geissler *et al.*, 2009).

It is often mentioned that halophyte vegetables have potential negative health impacts because of the high Na and Cl levels in the edible shoot. However, they may only have a minor impact on the plant's nutritional value, because a gourmet product is consumed in small quantities. In addition, adding little or no salt to the halophyte dishes during food preparation can counterbalance the high NaCl content (Ventura and Sagi, 2013). Many of those aspects related to the taste of halophytes have not been tested for consumers' preferences and attitudes, because yield has received much more attention than the effects of salinity on the quality parameters of the marketable product (Maggio *et al.*, 2011).

20.2.3.6 Quinoa

We discuss here the South American seed crop *Chenopodium quinoa* because some of the 2500 accessions can tolerate salinity levels present in seawater (approx. 40 dS m^{-1}). Quinoa can tolerate other abiotic stresses such as drought, frost and water stress (Boscaiu and Vicente, 2013), and it also withstands temperatures from −4°C to almost 40°C. Interest in this seed crop is increasing worldwide, owing to its exceptional nutritional quality (e.g. Vega-Gálvez *et al.*, 2010). *C. quinoa* is considered to contribute to global food security because of the exceptional nutritional quality of the seeds, which are rich in vitamins, minerals, essential amino acids and fatty acids (Adolf *et al.*, 2013). *C. quinoa* is considered a 'pseudocereal' as its seeds contain starch, proteins, essential amino acids (i.e. lysine) in the nucleus of the grain, unlike wheat and rice. In addition, quinoa seeds are gluten free (James, 2009). *C. quinoa* has a long history of cultivation, being grown in the Andean region for thousands of years. Because of the high nutritional quality of its seeds, and its adaptability to a wide array of climates and environmental conditions, it has been successfully grown – at a small scale – in the USA and Canada, several EU countries, Africa (Kenya) and Asia (the Himalaya and the plains of northern India) (Boscaiu and Vicente, 2013). Panta *et al.* (2014) reported that *C. quinoa* had a grain yield potential of up to 5.2 t ha^{-1} when grown in temperate environments in Argentina. Despite the high environmental adaptability and high nutritional value of *C. quinoa*, its market is still relatively small, and it is mainly sold in specialized shops at high prices.

20.2.3.7 Commercialization of halophytes as food: is there a market for halophytes as vegetables?

Halophytes have been studied for decades but the main focus of past research efforts has been understanding salt tolerance mechanisms. Large-scale commercial production of halophytes is still lacking. *C. quinoa* is one halophyte crop with a larger market. Commercial cultivation of halophytes requires the selection of superior genotypes as gourmet vegetables, and definition of the plant's reproducible growing conditions (Ventura and Sagi, 2013). Halophytes should also be produced at a cost that is competitive with conventional crops (Brown *et al.*, 2014). In addition, Brown *et al.* (2014) point out that the halophytic crop should not adversely affect water supplies or land that could support conventional agriculture. Some of the common objectives of traditional plant breeding and agricultural practices should be applied to the domestication of halophytes for large-scale commercialization. Some of the most important objectives are: selection of the best genotypes for particular agro-ecological conditions, reduction in toxic compounds or anti-nutrient content, increased yields, improvement of marketing characteristics (uniformity of the product with respect to taste, size, colour, etc.) and tailoring standard agricultural practices to particular species (Boscaiu and Vicente, 2013).

Besides the development of plant breeding strategies and agronomic practices for cultivation of halophytes, another crucial factor for successful commercialization is consumers' preferences toward halophyte products and their willingness to pay for them. There is very little knowledge about these important aspects, which are fundamental to the commercialization of halophytes because farmers will start investing in the crop only if there is marketing potential. Consumers' food choices depend on many factors

(Clark, 1998) but taste, flavour and freshness are considered the most important attributes in fresh food quality (Migliore *et al.*, 2015). Migliore *et al.* (2015) showed that 'healthiness' has an important impact on consumers' choice, especially with regard to prickly pear fruit. Some of the 'credence attributes' (i.e. environmental conservation and nutritional aspects) can represent added value in products obtained from halophytes, and should be studied in depth to understand their role in consumers' preferences towards halophytes. This knowledge could create a bigger market demand and trigger farmers' investment in saline agriculture. The company OceanDesertFood in The Netherlands represents one example of halophyte commercialization. Farmers in Mexico produce *Salicornia* and then sell it to OceanDesertFood. Jeannette Hoek, the company president, is creating a market demand for halophyte food, fodder and products (OASE Foundation, 2009). The company produces *Salicornia* crackers and seaweed chips, which are among its most popular products.

In general, the gourmet vegetable and herb market requires the marketed products to be fresh and visually appealing in respect of colour and packaging. The product should also have a particular taste and nutritional value cherished by its consumers (van der Voort *et al.*, 2007). In particular, some halophytes can meet the standards of being 'functional foods' for health enthusiasts, owing to the high levels of nutritionally valuable metabolites such as omega-3 fatty acids, phenolic compounds, antioxidants or minerals (Ares and Gámbaro, 2007; Del Giudice and Pascucci, 2010). Other factors (credence attributes) such as environmental conservation claims, support to local farming, organic or origin certification, and knowledge and trust of the science, are more difficult to evaluate and need further analysis (Moser *et al.*, 2011; Dentoni *et al.*, 2014).

20.3 Oilseed Production for Human Consumption and Biodiesel

Halophytes are grown commercially for the biofuels generated from the seeds and biomass. Growing halophytes for biofuels has an important environmental advantage because these plants can be gown on marginal saline land and irrigated with seawater, thus reducing the competition for land with agriculture. *Salicornia bigelovii* is probably the most widely grown halophyte for oilseed and biodiesel production. Global Seawater Inc., a Mexican company, supports commercial biodiesel production of *S. bigelovii* in farms in Mexico. Christiansen (2008) reported that 1 ha of *Salicornia* grown on coastal land can produce between 890 and 950 l of biodiesel. The Masdar Institute of Science and Technology started a seawater-irrigated cultivation of *Salicornia* in Abu Dhabi to produce biofuel for the aviation industry, with the support of Boeing, Ethiad Airways and UOP Honeywell (ICBA, 2011; Panta *et al.*, 2014). *Tamarix* spp. and *Euphorbia tirucalii* (a desert succulent from East Africa) are saline- and drought-tolerant plants that have been grown for biomass production to generate biofuel. *T. jordanis* is most suited for ethanol fermentation as it contains preferred characteristics such as high cellulose, and low hemicellulose and phenol content (Eshel *et al.*, 2010, 2011; Santi *et al.*, 2014). *T. aphylla* irrigated with reclaimed sewage (EC approx. 3 dS m^{-1}) or brine (EC approx. 7–10 dS m^{-1}) produced 52 and 26 t ha^{-1}, respectively (Ventura *et al.*, 2014). Further studies on the yield potential of halophytes for the production of biomass or oilseed for biofuels are needed.

In addition to biofuel, seeds from halophytes can also be used to produce edible oils. Panta *et al.* (2014) reported that among the 50 seed-bearing halophytic species that are potential sources of edible oil and proteins the most studied are: *S. bigelovii* (Glenn *et al.*, 1991), *Suaeda moquinii* (Weber *et al.*, 2001), *Kosteletzkya virginica* (Gallagher, 1985; He *et al.*, 2003), *S. aralocaspica* (Wang *et al.*, 2012), *Salvadora persica* (Rao *et al.*, 2004; Reddy *et al.*, 2008), *Batis maritima* (Marcone, 2003), *Crithmum maritimum* and *Zygophyllum album* (Zarrouk *et al.*, 2003), *Nitraria sibiria*, *S. salsa*, *Chenopodium glaucum* and *Descurainia sophia* (Yajun *et al.*, 2003). There have been many trials of *S. bigelovii* as an oilseed plant in various parts of the world (ICBA, 2007). Commercial *Salicornia* production trials have been carried out in the USA, Middle East (Jaradat, 2005; Abdal, 2009), India (Rameshkumar and Eswaran, 2013), Mexico (Grattan *et al.*, 2008) and Africa (Zerai *et al.*, 2010). *S. bigelovii* has shown high seed yield (2 t ha^{-1})

when cultivated under seawater irrigation (40 g NaCl l^{-1}) or at root zone salinities greater than 70 g l^{-1} of total dissolved salts (Glenn *et al.*, 1997, 1999). These yield values are equivalent or higher than freshwater-irrigated oilseed crops such as sunflower and soybean. With respect to the nutritional value of *S. bigelovii* oil, the protein content is about 35% and its properties are similar to those of safflower oil (Glenn *et al.*, 1991; Zerai *et al.*, 2010). The seeds of *Salicornia* are very small (only about 1 mg in weight); this makes retention at harvest difficult, with seed recoveries of only about 75% compared to 90% for most crops. However, *Salicornia* seeds have been successfully harvested using ordinary mechanical equipment (Panta *et al.*, 2014). A 25% larger seed size was reported in *Salicornia* plants that had been cultivated for several crop cycles and harvested with mechanical equipment, suggesting that at least this species is subject to improvement through mass selection (Zerai *et al.*, 2010). Other problems affect the success of commercial production of *Salicornia* as oilseed. At maturity, the plant tends to lie flat in the field (lodge) and the seeds may shatter before harvest. *Salicornia* must grow for 100 days at cool temperatures before flowering, to obtain high seed yields (Glenn *et al.*, 1998). Hence, commercial production of *Salicornia* is strictly bound to climatic regions of the subtropics that have cool winters and hot summers. This region is where large areas of coastal desert are found, thus facilitating the use of marginal land for halophyte production.

Other minor halophytes have been trialled to obtain data on oilseed yield and quality. The seeds of *Kosteletzkya virginica*, the perennial seashore mallow, contain 32% protein and 22% lipids (Gallagher, 1985). *Suaeda persica* can be cultivated in soils with EC values ranging from 25 to 65 dS m^{-1}, but the maximum seed yield is obtained at 25–35 dS m^{-1} because, at higher EC values (55–65 dS m^{-1}), the seed yield can decline by 40%–47% (Rao *et al.*, 2004). The seeds contain 40%–45% of oil rich in lauric (C12) and myristic (C14) acid, widely used in the cosmetic and pharmaceutical industries (Reddy *et al.*, 2008). Weber *et al.* (2007) reported that seeds of *S. fruticosa* could be used as a source of edible oil for human consumption owing to their 74% unsaturated fatty acid content.

20.4 Halophytes as Forage

In principle, animals could be fed all halophytes that are used as food for humans, either as vegetable crops (e.g. *Aster tripolium*) or grains (e.g. *C. quinoa*) (Boscaiu and Vicente, 2013). However, several factors such as biomass production, the voluntary feed intake by animals and the nutritional value of the product are still poorly known, and should be further analysed to make forage saline agriculture efficient (Norman *et al.*, 2013).

One of the most challenging agricultural problems in dry lands that have been degraded through over-grazing is to find enough forage for cattle, sheep and goat herds. Livestock such as sheep, camels and cattle thrive well on certain halophyte feeds (Swingler *et al.*, 1996; Khan and Ansari, 2008). Pigs and poultry are the most sensitive to feed composition and the least saline-tolerant farm animals (Masters *et al.*, 2007). Some of the most productive halophytes used for forage are shrubby species of *Salicornia*, *Suaeda*, and *Atriplex* from the family Chenopodiaceae. Average yield potential of these halophytes when irrigated with saline water is 10-20 t ha^{-1}; this is comparable to the productivity of conventional forage species under non-saline conditions (O'Leary *et al.*, 1985; Glenn *et al.*, 1999; Masters *et al.*, 2007). Other highly productive species are the salt grass *Distichlis* and the succulent viney ground covers such as *Batis* (Glenn *et al.*, 1998). *D. spicata* was examined and showed considerable promise for selection as a fodder crop for ruminants (Bustan *et al.*, 2005).

In 2011, the International Centre of Biosaline Agriculture (ICBA) started large-scale halophyte forage model farms to study the adaptability and yield potential of salt-tolerant shrubs, trees and grasses in the western regions of Abu Dhabi Emirates (ICBA, 2012). Other studies are currently ongoing in cooperation with ICBA and other research organizations in Saudi Arabia, Pakistan and Bangladesh for the development of biosaline fodder and forage production using *Sporobolus virginicus*, *D. spicata* and *Atriplex* spp. (Panta *et al.*, 2014). The most well-studied halophyte for forage is *Atriplex* spp. Many farmers around the world feed *Atriplex* saltbush to sheep, which creates a leaner, more hydrated lamb meat, with higher levels of vitamin E. Of all the

Atriplex species, *A. lentiformis* is an important forage halophyte because its leaves have a nutritional quality (rich N source up to 17% crude protein under 100% seawater) similar to that of alfalfa (Pasternak *et al*., 1985; Pasternak, 1990). *A. lentiformis* is a perennial, deep-rooted desert shrub, which can be grown with seawater irrigation, but its seeds need to be germinated in nurseries because they do not germinate under high-salt conditions (Boscaiu and Vicente, 2013). In addition to its salt tolerance, *A. lentiformis* is also highly tolerant to drought and extremely efficient in water use. However, the energy content of the seeds is lower (approximately 15 Mcal kg DW) than other conventional forages, and the seeds contain some antinutrients. For example, the seeds of *S. bigelovii* contain saponins, which can have deleterious effects on animal growth (Glenn *et al*., 1991, 2013. As a consequence of these negative attributes the animal intake of *Atriplex* spp. was low (0.3–0.4 kg dry matter day^{-1}) (Pasternak *et al*., 1985; Pasternak, 1990). The plant is a perennial but it is advisable to cultivate it as an annual because it tends to become woody with age (Glenn *et al*., 2013).

Forage halophytes are often used as a complementary feed source because of their high salt content, to avoid meat containing levels of Na that are too high and unhealthy. The use of halophytes as feed or fodder for livestock will depend on the specific animals to be fed, some animals being more sensitive to feed composition than others. For example, the seed meal remaining after extraction of oil from *S. bigelovii* seeds can be used as a protein supplement in fish and ruminant diets, but it cannot be used as a protein source in poultry feeds because of its high saponin content (Glenn *et al*., 2013). In dry lands of different climatic zones in Faisalabad, Peshawar, Bhawalpur and Karachi in Pakistan, forage halophytes (i.e. *Atriplex* and *Maireana* spp.) represent a viable additional source of forage (Hollington *et al*., 2001). In arid coastal regions where freshwater for crop irrigation is limited, the straw or seed meal of *S. bigelovii* fed to lambs turned out to be an acceptable feed substitute (Swingle *et al*., 1996). For example, *S. bigelovii* grown in seawater-irrigated agriculture in the eastern coastal region of the Arabian Peninsula can replace 25% of alfalfa in animal diets (e.g. sheep) (Abdal, 2009). Other halophytes are also successful as complementary feed for animals that tolerate a halophyte-based diet, such as goats and sheep. Feeding trials with goats and sheep using 70% *Sporobolus virginicus* and *Distichlis spicata* and 30% conventional feed (Rhodes grass, *Chloris gayana*), showed that animals performed better in meat–fat–bone ratio and body composition than animals fed on 100% conventional feed (ICBA, 2007). Similarly, 1-year-old cow calves fed only on halophytic grass (*Panicum turgidum*) in Pakistan were leaner than animals fed on maize (Khan and Ansari, 2008). Other halophytic grasses, such as *Leptochloa fusca* and *Spartina patens*, are grown in arid lands for forage production. Mangrove and other coastal halophytic species (i.e. *Terminalia catappa*, *Aeluropus lagopoides*, *Cynodon dactylon* and *Brachiaria mutica*) are fodder for cattle, camels and goats (Dagar, 2005).

To increase the palatability and voluntary feed intake in animals it is important to select for low-salt-accumulating varieties of halophytes (Ventura *et al*., 2014).

Animals that feed on a salty diet due to the presence of halophytes in the diet may increase their water consumption, which may in turn affect animal weight. In addition to higher animal requirements for freshwater when the diet is halophyte-based, all salt ingested by grazing animals needs to be processed through the kidneys and this may increase the energy requirement to metabolize the feed (Panta *et al*., 2014). Furthermore, levels of metabolizable energy come from the organic matter, which is usually lower in halophytes (O'Connell *et al*., 2006). Finally, the high salinity in the feed of halophyte-based diets limits the amount an animal can eat and dilutes the nutritional value, because salt has no calories but takes up space (Glenn *et al*., 1998). In open grazing situations, animals turn to halophytes only when the more palatable plants are absent.

20.5 Conclusion

In this chapter, we have provided an overview of the advantages and disadvantages of halophytes as alternative plants for production of vegetables, feed and fodder. One of the most important environmental benefits is that halophytes can grow on land that would be otherwise unused for agricultural production, thus reducing land competition for food production. In addition, halophytes can grow on extremely saline soils in dry lands and can be irrigated with seawater or brackish water. However, there are many problems associated with halophyte cultivation. First and most important, many halophytes need to be domesticated to reach yields, taste and agricultural practices comparable to conventional crops. Second, but no less important, is the creation of a market for successful commercialization of halophytes as vegetables and animal feed. Although there are many success stories of halophyte production and commercialization (Panta *et al.*, 2014), large-scale and long-term field studies are needed to attain profitable and consumer-acceptable products. Studies on consumers' acceptance of alternative crops are needed for halophytes to become a recognized crop that starts to generate economic profits (Ventura and Sagi, 2013). As in the case of quinoa, halophytes' high nutritional value can be a leverage to attract consumers' attention and develop large-base consumption, other than just being gourmet vegetables in high-end restaurants. Educational programmes and marketing campaigns can be targeted toward the consumption of halophytes, not only as functional foods, but also as environment-friendly and less resource-intense food.

Acknowledgements

Funds for part of the research reported in this chapter were provided by the California Department of Water Resources, Proposition 204 Water Reuse Program, Sacramento, Fresno, CA, USA.

References

Abdal, M.S. (2009) Salicornia production in Kuwait. *World Applied Science Journal* 6, 1033–1038.

Adolf, V.I., Jacobsen, S.E. and Shabala, S. (2013) Salt tolerance mechanisms in quinoa (*Chenopodium quinoa* Willd.). *Environmental and Experimental Botany* 92, 43–54.

Ares, G. and Gámbaro, A. (2007) Influence of gender, age and motives underlying food choice on perceived healthiness and willingness to try functional foods. *Appetite* 49, 148–158.

Aronson, J. (1985) Economic halophytes—a global review. In: Wickens, G.E., Goodin, J.R. and Field, D.V. (eds) *Plants for Arid Lands*. Springer, Dordrecht, Netherlands, pp. 177–188.

Aronson, J.A. (1989) *HALOPH: A Data Base of Salt Tolerant Plants of the World*. University of Arizona, Tucson, USA. Available at https://www.sussex.ac.uk/affiliates/halophytes/index.php?content=about (accessed 18 August 2018).

Atia, A., Ben Hamed, K., Debez, A. and Abdelly, C. (2006) Salt and seawater effect on the germination of *Crithmum maritimum*. In: Öztürk, M., Waisel, Y., Khan, M.A. and Görk, G. (eds) *Biosaline Agriculture and Salinity Tolerance in Plants*. BirkhauserVerlag, Basel, Switzerland, pp. 29–33.

Atia, A., Debez, A., Barhoumi, Z., Abdelly, C. and Smaoui, A. (2009) Histochemical localization of essential oils and bioactive substances in the seed coat of the halophyte *Crithmum maritimum* L. (Apiaceae). *Journal of Plant Biology* 52, 448–452. Available at: http://www.biodieselmagazine.com/articles/2600/sea-asparagus-can-be-oil-feedstock/, accessed 21 September 2017.

Ben-Hamed, K., Debez, A., Chibani, F. and Abdelly, C. (2004) Salt response of *Crithmum maritimum*, an oleaginous halophyte. *Tropical Ecology* 45, 151–159.

Böer, B. (2006) Halophyte research and development: what needs to be done next? In: Khan, M.A. and Weber D.J. (ed.) *Ecophysiology of High Salinity Tolerant Plants*. Springer Verlag, Berlin, pp. 397–399.

Boscaiu, M. and Vicente, O. (2013) Halophytic crops for a salinising world. *Bulletin UASVM Horticulture* 70(1), 1–9.

Brown, J.J., Glenn, E.P., Fitzsimmons, K.M. and Smith, S.E. (1999) Halophytes for the treatment of saline aquaculture effluent. *Aquaculture* 175, 255–268.

Brown, J.J., Glenn, E.P. and Smith, S.E. (2014) Feasibility of halophyte domestication for high-salinity agriculture. In: Khan, M.A., Böer, B., Öztürk, M., Al Abdessalaam, T.Z., Clüsener-Godt, M. and Gul, B. (eds) *Sabkha Ecosystems*. Springer, Dordrecht, Netherlands, pp. 73–80.

Buhmann, A. and Papenbrock, J. (2013) Biofiltering of aquaculture effluents by halophytic plants: basic principles, current uses and future perspectives. *Environmental and Experimental Botany* 92, 122–133.

Bustan, A., Pasternak, D., Pirogova, I., Durikov, M., deVries, T.T. *et al.* (2005) Evaluation of saltgrass as a fodder crop for saline environments. *Journal of the Science of Food and Agriculture* 85, 2077–2084.

Carlsson, R. and Clarke, E.M.W. (1983) *Atriplex hortensis* L. as a leafy vegetable, and as a leaf protein concentrate plant. *Plant Foods for Human Nutrition* 33, 127–133.

Centofanti, T. and Bañuelos, G. (2015) Evaluation of the halophyte *Salsola soda* as an alternative crop for saline soils high in selenium and boron. *Journal of Environmental Management* 157, 96–102.

Chapman, V.J. (1976) *Mangrove Vegetation*. J. Cramer, Vaduz, Liechtenstein.

Christiansen, R.C. (2008) Sea asparagus can be oil feedstock. *Biomass Magazine*. Available at http://www.biodieselmagazine.com/articles/2600/sea-asparagus-can-be-oil-feedstock/ (accessed 23 December 2018).

Clark, J.E. (1998) Taste and flavour: their importance in food choice and acceptance. *Proceedings of the Nutrition Society* 57, 639–643.

Dagar, J. (2005) Ecology, management and utilization of halophytes. In: Gupta, S.R., Matta, N.K., Aggarwal, A., Kohli, R.K. and Chawla, A.K. (eds) *Ecology and Environmental Management: Issues and Research Needs*. *Bulletin of the National Institute of Ecology* 15, 81–97.

Davy, A.J., Bishop, G.F. and Costa, C.S.B. (2001) *Salicornia* L. (*Salicornia pusilla* J. Woods, *S. ramosissima* J. Woods, *S. europaea* L., *S. obscura* P.W. Ball & Tutin, *S. nitens* P.W. Ball & Tutin, *S. fragilis* P.W. Ball & Tutin, and *S. dolichostachya* Moss). *Journal of Ecology* 89, 681–707.

Davy, A.J., Bishop, G.F., Mossman, H., Redondo-Gómez, S., Castillo, J.M., *et al.* (2006) Biological flora of the British Isles: *Sarcocornia perennis* (Miller) A.J. Scott. *Journal of Ecology* 94, 1035–1048.

de Vos, A.C., Broekman, R., Guerra, C.C.D., van Rijsselberghe, M. and Rozema, J. (2013) Developing and testing new halophyte crops: a case study of salt tolerance of two species of the Brassicaceae, *Diplotaxis tenuifolia* and *Cochlearia officinalis*. *Environmental and Experimental Botany* 92, 154–164.

Del Giudice, T. and Pascucci, S. (2010) The role of consumer acceptance in the food innovation process: young consumer perception of functional foods in Italy. *International Journal on Food System Dynamics* 2, 111–122.

Dentoni, D., Tonsor, G.T., Calantone, R. and Peterson, H.C. (2014) Disentangling direct and indirect effects of credence labels. *British Food Journal* 116, 931–51.

Epstein, E., Norlyn, J.D., Rush, D.W., Kingsbury, R.W., Kelley, D.B. *et al.* (1980) Saline culture of crops: a genetic approach. *Science* 210, 399–404.

Eshel, A., Zilberstein, A., Alekparov, C., Eilam, T., Oren, I., *et al.* (2010) Biomass production by desert halophytes: alleviating the pressure on food production. *Proceedings of the 5th IASME/WSEAS International Conference on Energy and Environment. Recent Advances in Energy and Environment.* Cambridge, UK, pp. 362–367.

Eshel, A., Oren, I., Alekparov, C., Eilam, T. and Zilberstein, A. (2011) Biomass production by desert halophytes: alleviating the pressure on the scarce resources of arable soil and fresh water. *European Journal of Plant Science and Biotechnology* 5, 48–53.

Felger, R.S. (1979) Ancient crops for the 21st century. In: Ritchie, G.A. (ed.) *New Agricultural Crops.* Westview, Boulder, CO, USA, pp. 5–20.

Felger, R.S., Moser, M.B. and Moser, E.W. (1980) Seagrasses in Seri Indian culture. In: Phillips, R.C. and McRoy, C.P. (eds) *Handbook of Seagrass Biology, an Ecosystem Perspective.* Garland STPM, New York, pp. 260–276.

Flowers, T.J. and Colmer, T.D. (2008) Salinity tolerance of halophytes. *New Phytologist* 179, 945–963.

Franke, W. (1982) Vitamin C in sea fennel (*Crithmum maritimum*), an edible wild plant. *Economic Botany* 3, 163–165.

Gallagher, J.L. (1985) Halophytic crops for cultivation at seawater salinity. *Plant and Soil* 89, 323–336.

Geissler, N., Hussin, S. and Koyro, H.W. (2009) Elevated atmospheric CO_2 concentration ameliorates effects of NaCl salinity on photosynthesis and leaf structure of *Aster tripolium* L. *Journal of Experimental Botany* 60, 137–151.

Gibbs, H.K. and Salmon, J.M. (2015) Mapping the world's degraded lands. *Applied Geography* 57, 12–21.

Glenn, E.P., O'Leary, J.W., Watson, M.C., Thompson, T.L. and Kuehl, R.O. (1991) *Salicornia bigelovii* Torr.: an oilseed halophyte for seawater irrigation. *Science* 251, 1065–1067.

Glenn, E.P., Olsen, M., Frye, R. and Moore, D. (1994) Use of halophytes to remove carbon from the atmosphere: Results of a demonstration experiment. Electric Power Research Institute, TR-103310, Research Report 8011–03, Palo Alto, CA, USA.

Glenn, E., Miyamoto, S., Moore, D., Brown, J.J., Thompson, T.L., *et al.* (1997) Water requirements for cultivating *Salicornia bigelovii* Torr. with seawater on sand in a coastal desert environment. *Journal of Arid Environments* 36, 711–730.

Glenn, E.P., Brown, J.J. and O'Leary, J.W. (1998) Irrigating crops with seawater. *Scientific American* 279, 76–81.

Glenn, E.P., Brown, J.J. and Blumwald, E. (1999) Salt tolerance and crop potential of halophytes. *Critical Reviews in Plant Sciences* 18, 227–255.

Glenn, E.P., Anday, T., Chaturvedi, R., Martinez-Garcia, R., Pearlstein, S., *et al.* (2013) Three halophytes for saline-water agriculture: An oilseed, a forage and a grain crop. *Environmental and Experimental Botany* 92, 110–121.

Grattan, S.R. and Grieve, C.M. (1999) Salinity – mineral nutrient relations in horticultural crops. *Scientia Horticulturae* 78, 127–157.

Grattan, S.R., Benes, S.E., Peters, D.W. and Diaz, F. (2008) Feasibility of irrigating pickleweed (*Salicornia bigelovii* Torr.) with hypersaline drainage water. *Journal of Environmental Quality* 37, S149–S156.

Guarrera, P.M., Salerno, G. and Caneva, G. (2006) Food, flavouring and feed plant traditions in the Tyrrhenian sector of Basilicata, Italy. *Journal of Ethnobiology and Ethnomedicine* 2, 1–6.

Guil-Guerrero, J.L. and Rodríguez-García, I. (1999) Lipids classes, fatty acids and carotenes of the leaves of six edible wild plants. *European Food Research and Technology* 209, 313–316.

Hasegawa, P.M. (2013) Sodium (Na^+) homeostasis and salt tolerance of plants. *Environmental and Experimental Botany* 92, 19–31.

He, Z., Ruana, C., Qin, P., Seliskar, D.M. and Gallagher, J.L. (2003) *Kosteletzkya virginica*, a halophytic species with potential for agroecotechnology in Jiangsu Province, China. *Ecological Engineering* 21, 271–276.

Hollington, P.A., Hussain, Z., Kahlown, M.A. and Abdullah, M. (2001) Success stories in saline agriculture in Pakistan: from research to production and development. *BAC Saline Agriculture Conference,* 19–21.

ICBA (2007) Annual Report 2006 (1426-27H). International Center for Biosaline Agriculture, Dubai, UAE. Available at http://www.biosaline.org/news, accessed 25 September 2017.

ICBA (2011) Biosalinity News. *Newsletter of the International Center for Biosaline Agriculture* 12, 6–7. Available at http://www.biosaline.org/news, accessed 25 September 2017.

ICBA (2012) Biosalinity News. *Newsletter of the International Center for Biosaline Agriculture* 13(1), 4. Available at http://www.biosaline.org/news, accessed 25 September 2017.

Isabelle, M., Lee, B.L., Lim, M.T., Koh, W.P., Huang, D. *et al.* (2010) Antioxidant activity and profiles of common vegetables in Singapore. *Food Chemistry* 120, 993–1003.

James, L.E.A. (2009) Quinoa (*Chenopodium quinoa* Willd.): composition, chemistry, nutritional, and functional properties. *Advances in Food and Nutrition Research* 58, 1–31.

Jaradat, A.A. (2005) Saline agriculture in the Arabian Peninsula: management of marginal lands and saline water resources. *Journal of Food, Agriculture & Environment* 2, 302–306.

Kadereit, G., Mucina, L. and Freitag, H. (2006) Phylogeny of Salicornioideae (Chenopodiaceae): diversification, biogeography, and evolutionary trends in leaf and flower morphology. *Taxon* 55, 617–642.

Kadereit, G., Ball, P., Beer, S., Mucina, L., Sokoloff, D., *et al.* (2007) A taxonomic nightmare comes true: phylogeny and biogeography of glassworts (*Salicornia* L. Chenopodiaceae). *Taxon* 56, 1143–1170.

Khan, M.A. and Ansari, R. (2008) Potential use of halophytes with emphasis on fodder production in coastal areas of Pakistan. In: Abdelly, C., Ozturk, M., Ashraf, M. and Grignon, C. (eds) *Biosaline Agriculture and High Salinity Tolerance*. Birkhauser Verlag, Basel, pp. 157–162.

Khan, M.A. and Duke, N.C. (2001) Halophytes – A resource for the future. *Wetlands Ecology and Management* 9, 455–456.

Khan, M.A., and Qaiser, M. (2006) Halophytes of Pakistan: characteristics, distribution and potential economic usages. In: Khan, M.A., Böer, B., Kust, G.S. and Barth, H.J. (eds) *Sabkha Ecosystems*. Springer, New York, pp. 129–153.

Kijne, J.W. (2005) Aral Sea basin initiative: towards a strategy for sustainable irrigated agriculture with feasible investment in drainage. Synthesis Report. IPTRID/FAO, Rome, p. 72.

Koyro, H.W., Khan, M.A. and Lieth, H. (2011) Halophytic crops: A resource for the future to reduce the water crisis? *Emirates Journal of Food and Agriculture* 23, 1–16.

Lieth, H. (2000) Cashcrop halophytes for future halophyte growers. EU concerted action project IC 18CT96-0055, final meeting, EXPO 2000. Institute of Environmental Systems Research, University of Osnabrück, Germany. ISSN 09336-3114, No. 20.

Lieth, H. and Mochtchenko, M. (2002) Halophyte uses in different climates. IV. In: *QUEST Conference for the Middle East and North Africa.* Backhuys Publishers, Dawhah, Qatar.

Lu, D., Zhang, M., Wang, S., Cai, J., Zhou, X. and Zhu, C. (2010) Nutritional characterization and changes in quality of *Salicornia bigelovii* Torr. during storage. *LWT – Food Science and Technology* 43, 519–524.

Maggio, A., De Pascale, S., Fagnano, M. and Barbieri, G. (2011) Saline agriculture in Mediterranean environments. *Italian Journal of Agronomy* 6, 36–43.

Marcone, M.F. (2003) *Batis maritima* (Saltwort/Beachwort): A nutritious, halophytic, seed bearings, perennial shrub for cultivation and recovery of otherwise unproductive agricultural land affected by salinity. *Food Research International* 36, 123–130.

Masters, D., Benes, S. and Norman, H. (2007) Biosaline agriculture for forage and livestock production. *Agriculture Ecosystem Environments* 119, 234–248.

Meot-Duros, L., Cérantola, S., Talarmin, H., Le, M.C., Le, F.G. and Magné, C. (2010) New antibacterial and cytotoxic activities of falcarindiol isolated in *Crithmum maritimum* L. leaf extract. *Food Chemical Toxicology* 48, 553–557.

Migliore, G., Galati, A., Romeo, P., Crescimanno, M. and Schifani, G. (2015) Quality attributes of cactus pear fruit and their role in consumer choice: The case of italian consumers. *British Food Journal* 117, 1637–1651.

Moser, R., Raffaelli, R. and Thilmany-McFadden, D. (2011) Consumer preferences for fruit and vegetables with credence-based attributes: A review. *International Food and Agribusiness Management Review* 14, 121–142.

Mudie, P.J., Greer, S., Brakel, J., Dickson, J.H., Schinkel, C., *et al.* (2005) Forensic palynology and ethnobotany of *Salicornia* species (Chenopodiaceae) in northwest Canada and Alaska. *Canadian Journal of Botany* 83, 111–123.

Munns, R. and Tester, M. (2008) Mechanisms of salinity tolerance. *Annual Review of Plant Biology* 59, 651–681.

Norman, H.C., Masters D.G. and Barrett-Lennard, E.G. (2013) Halophytes as forages in saline landscapes: Interactions between plant genotype and environment change their feeding value to ruminants. *Environmental and Experimental Botany* 92, 96–109.

O'Connell, M., Young, J. and Kingwell, R. (2006) The economic value of saltland pastures in a mixed farming system in Western Australia. *Agricultural Systems* 89, 371–389.

O'Leary, J.W. (1988) Saline environments and halophytic crops. In: Whitehead, E.E., Hutchinson, C.F., Timmerman, B.N. and Varady, R.G. (eds) *Arid Lands Today and Tomorrow: Proceedings of an International Research and Development Conference*. Westview Press, Boulder, CO, USA.

O'Leary, J.W., Glenn, E.P. and Watson, M.C. (1985) Agricultural production of halophytes irrigated with seawater. *Plant and Soil* 89, 311–321.

OASE Foundation (2009) Biosaline agriculture, Amsterdam, The Netherlands. Available at: http://www.oasefoundation.eu/, accessed 25 September 2017.

Owens, S. (2001) Salt of the earth. *EMBO Reports* 21, 877–879.

Panta, S., Flowers, T., Lane, P., Doyle, R., Haros, G. *et al.* (2014) Halophyte agriculture: Success stories. *Environmental and Experimental Botany* 107, 71–83.

Pasternak, D. (1990) Fodder production with saline water. Project Report January 1982–December 1989. BGUN-ARI-35–90. pp 173.

Pasternak, D., Danon, A., Aronson, J.A. and Benjamin, R.W. (1985) Developing the sea- water agriculture concept. *Plant and Soil* 89, 337–348.

Proteggente, A.R., Pannala, A.S., Paganga, G., van Buren, L., Wagner, E., *et al.* (2002) The antioxidant activity of regularly consumed fruit and vegetables reflects their phenolic and vitamin C composition. *Free Radical Research* 36, 217–233.

Qadir, M., Noble, A.D., Qureshi, A.S., Gupta, R.K., Yuldashev, T. *et al.* (2009) Salt-induced land and water degradation in the Aral Sea basin: A challenge to sustainable agriculture in Central Asia. *Natural Resources Forum* 33, 134–149.

Rameshkumar, S. and Eswaran, K. (2013) Ecology, utilization and coastal management of salt tolerant plants (halophytes and mangroves) of Mypad coastal regions, Andhra Pradesh India. *International Journal of Environmental Biology* 3, 1–8.

Rao, G.G., Nayak, A.K., Chinchmalatpure, A.R., Nath, A. and Babu, V.R. (2004) Growth and yield of *Salvadora persica* a facultative halophyte grown on saline black soil (vertic haplustept). *Arid Land Research and Management* 18, 51–61.

Ravindran, K., Venkatesan, K., Balakrishnan, V., Chellappan, K. and Balasubramanian, T. (2007) Restoration of saline land by halophytes for Indian soils. *Soil Biology and Biochemistry* 39, 2661–2664.

Reddy, M.P., Shah, M.T. and Patolia, J.S. (2008) *Salvadora persica*, a potential species for industrial oil production in semiarid saline and alkali soils. *Industrial Crops and Products* 28, 273–278.

Rozema, J. and Flowers, T.J. (2008) Crops for a salinized world. *Science* 322, 1478–1480.

Rozema, J. and Schats, H. (2013) Salt tolerance of halophytes, research questions reviewed in the perspective of saline agriculture. *Environmental and Experimental Botany* 92, 83–95.

Santi, G., D'Annibale, A., Eshel, A., Zilberstein, A., Crognale, S., *et al.* (2014) Ethanol production from xerophilic and salt-resistant *Tamarix jordanis* biomass. *Biomass and Bioenergy* 61, 73–81.

Shabala, S. and Mackay, A. (2011) Ion transport in halophytes. *Advances in Botanical Research* 57, 151–187.

Shpigel, M., Ben-Ezra, D., Shauli, L., Sagi, M., Ventura, Y., *et al.* (2013) Constructed wetland with *Salicornia* as a biofilter for mariculture effluents. *Aquaculture* 412, 52–63.

Simopoulos, A.P. (2004) Omega-3 fatty acids and antioxidants in edible wild plants. *Biological Research* 37, 263–277.

Słupski, J., Achrem-Achremowicz, J., Lisiewska, Z. and Korus, A. (2010) Effect of processing on the amino acid content of New Zealand spinach (*Tetragonia tetragonioides* Pall. Kuntze). *International Journal of Food Science and Technology* 45, 1682–1688.

Swingle, R., Glenn, P. and Squires, S. (1996) Growth performance of lambs fed mixed diets containing halophyte ingredients. *Animal Food Science Technology* 63, 137–148.

Tardío, J., Pardo De Santayana, M. and Morales, R. (2006) Ethnobotanical review of wild edible plants in Spain. *Botanical Journal of the Linnean Society* 152, 27–71.

Tundis, R., Menichini, F., Conforti, F., Loizzo, M.R., Bonesi, M., *et al.* (2009) A potential role of alkaloid extracts from *Salsola* species (Chenopodiaceae) in the treatment of Alzheimer's disease. *Journal of Enzyme Inhibition and Medicinal Chemistry* 24, 818–824.

Turcios, A.E. and Papenbrock, J. (2014) Sustainable treatment of aquaculture effluents – what can we learn from the past for the future? *Sustainability* 6, 836–856.

Türkan, I. and Demiral, T. (2009) Recent developments in understanding salinity tolerance. *Environmental and Experimental Botany* 67, 2–9.

van der Voort, M., Baricicova, V., Dandar, M., Grzegorzewska, M., Schoorlemmer, H., *et al.* (2007) Quality requirements for vegetables and fruit products in the European Union. Applied Plant Research, Research Unit Arable Farming and Vegetable Production. EU Access Report 06.

Vega-Gálvez, A., Miranda, M., Vergara, J., Uribe, E., Puente, L. *et al.* (2010) Nutrition facts and functional potential of quinoa (*Chenopodium quinoa* Willd.), an ancient Andean grain: a review. *Journal of the Science of Food and Agriculture* 90, 2541–2547.

Ventura, Y. and Sagi, M. (2013) Halophyte crop cultivation: the case for *Salicornia* and *Sarcocornia*. *Environmental and Experimental Botany* 92, 144–153.

Ventura, Y., Wuddineh, W.A., Ephrath, Y., Shpigel, M. and Sagi, M. (2010) Molybdenum as an essential element for improving total yield in seawater-grown *Salicornia europaea* L. *Scientia Horticulturae* 126, 395–401.

Ventura, Y., Wuddineh, W.A., Myrzabayeva, M., Alikulov, Z., Khozin-Goldberg, I., *et al.* (2011) Effect of seawater concentration on the productivity and nutritional value of annual *Salicornia* and perennial *Sarcocornia* halophytes as leafy vegetable crops. *Scientia Horticulturae* 128, 189–196.

Ventura, Y., Myrzabayeva, M., Alikulov, Z., Cohen, S., Shemer, Z. *et al.* (2013) The importance of iron supply during repetitive harvesting of *Aster tripolium*. *Functional Plant Biology* 40, 968–976.

Ventura, Y., Eshel, A., Pasternak, D. and Sagi, M. (2014) The development of halophyte-based agriculture: past and present. *Annals of Botany* 115, 529–540.

Wagenvoort, W.A., van de Vooren, J.G. and Brandenburg, W.A. (1989) Plant domestication and the development of sea starwort (*Aster tripolium* L.) as a new vegetable crop. *Acta Horticulturae* 242, 115–122.

Wang, L., Zhao, Z.Y., Zhang, K. and Tian, C.Y. (2012) Oil content and fatty acid composition of dimorphic seeds of desert halophyte *Suaeda aralocaspica*. *African Journal of Agricultural Research* 7, 1910–1914.

Weber, D.J., Gul, B., Khan, M.A., Williams, T., Wayman, P. *et al.* (2001) Comparison of vegetable oil from seeds of native halophytic shrubs. In: McArthur, E.D. and Fairbanks, D.J. (eds) *Proceeding of Shrubland Ecosystem Genetics and Biodiversity*. RMRS-P-21. Rocky Mountain Research Station, USDA Forest Service, Ogden, UT, USA, pp. 287–290.

Weber, D.J., Ansari, R., Gul, B. and Khan, M.A. (2007) Potential of halophytes as source of edible oil. *Journal of Arid Environments* 68, 315–321.

Wilson, C., Lesch, S.M. and Grieve, C.M. (2000) Growth stage modulates salinity tolerance of New Zealand spinach (*Tetragonia tetragonioides* Pall.) and red orach (*Atriplex hortensis* L.). *Annals of Botany* 85, 501–509.

Yajun, B., Xiaojing, L. and Weiqiang, L. (2003) Primary analysis of four salt tolerant plants growing in Hai-He plain, China. In: Leith, H. and Mochtchenko, M. (eds) *Cash Crop Halophytes: Recent Studies*. Kluwer Academic Publishers, London, pp. 135–138.

Zarrouk, M., El Almi, H., Ben Youssef, N., Sleimi, N., Smaoui, A., *et al.* (2003) Lipid composition of seeds of local halophytes: *Cakile maritima*, *Zygophyllum album* and *Crithmum maritimum*. In: Leith, H. and Mochtchenko, M. (eds) *Cash Crop Halophytes: Recent Studies.* Kluwer Academic Publishers, London, pp. 121–124.

Zerai, D.B., Glenn, E.P., Chatervedi, R., Lu, Z., Mamood, A.N. *et al.* (2010) Potential for the improvement of *Salicornia bigelovii* through selective breeding. *Ecological Engineering* 36, 730–739.

Zhao, K., Fan, H., Jiang, X. and Song, J. (2002) Improvement and utilization of saline soil by planting halophytes. *Chinese Journal of Applied Environmental Biology* 8, 31–35.

Zhu, J.K. (2001) Plant salt tolerance. *Trends in Plant Science* 6, 66–71.

Zia, S., Egan, T.P. and Khan, M.A. (2008) Growth and selective ion transport of *Limonium stocksii* Plumbaginaceae under saline conditions. *Pakistan Journal of Botany* 40, 697–709.

Zurayk, R. and Baalbaki, R. (1996) *Inula crithmoides*: a candidate plant for saline agriculture. *Arid Soil Research and Rehabilitation* 10, 213–223.

21 Use of Halophytes as Medicinal Plants: Phytochemical Diversity and Biological Activity

Zora Dajić Stevanović[1,*], Milan S. Stanković[2], Jovana Stanković[1], Pedja Janaćković[1] and Miroslava Stanković[3]

[1]University of Belgrade, Belgrade, Serbia; [2]University of Kragujevac, Kragujevac, Serbia; [3]Nuclear Facilities of Serbia, Vinča, Serbia

Abstract

This chapter reviews the basic phytochemical characteristics and related biological activity in a range of halophytic species, in addition to their ethnobotanical relevance and traditional use as medicinal plants. The specific features of saline habitats (the osmotic and toxic effects of salts) encourage investigation into halophyte biosynthesis and accumulation of various secondary metabolites produced during adaptation. Recently, therefore, halophytes have become objects of investigation for their qualitative and quantitative characterization and isolation of biologically active substances. Taxonomic and habitat affiliations, main active substances and biological activity have been reviewed and presented for more than 50 halophytic species from families such as: Acanthaceae, Aizoaceae, Amaranthaceae, Apiaceae, Asteraceae, Brassicaceae, Caryophyllaceae, Chenopodiaceae, Combretaceae, Convolvulaceae, Euphorbiaceae, Fabaceae, Frankeniaceae, Fumariaceae, Juncaceae, Lamiaceae, Malvaceae, Plantaginaceae, Plumbaginaceae, Primulaceae, Rhizophoraceae, Salvadoraceae, Saxifragaceae, Tamaricaceae and Zygophyllaceae. Among the most frequent secondary metabolites of halophytes, different types of polyphenols have been determined, including phenolic acids (e.g. chlorogenic acid, vanillic and gallic acids), aromatic acids or phenylpropanoids, flavonoids (such as catechin and isorhamnetin), flavonoid glycosides, isoflavone glycosides, sulfated flavonoids, tannins, stilbenes, curcuminoids and coumarins, in addition to saponins, fatty acids, polyacetylenes, essential oils, carotenoids, alkaloids and other metabolites. We discuss the biological activity of plant extracts and their chemical composition, including, in particular, the antioxidant, antimicrobial and anticancerogenic effects of halophytes.

Keywords: Antioxidants; Ethnobotany; Phytochemicals; Polyphenols; Salinity

21.1 Introduction

Medicinal plants have provided the essential resources for human health from ancient times to the present day. According to the World Health Organization (WHO), the majority of the world's population depends on traditional medicine for primary health care needs. Plants and their various remedy products can be used in prevention and alleviation of pre-ailment symptoms, problems associated with chronic conditions, for recovery facilitation and human well-being. Apart from empirically learned medicinal and pharmacological properties, the selection of

* Corresponding author e-mail: dajic@agrif.bg.ac.rs

medicinal plants is dependent on cognitive features, ecological factors and cultural history (Leonti, 2011).

Modern science searches for new biological effects and interactions of already accepted medicinal herbs, and also for new metabolites that might show prospective bioactive effects. Metabolomic studies in plants revealed tens of thousands of such compounds. Secondary metabolites generally show greater individuality and diversity in their molecular structure than primary metabolites. Certain compound classes also appear to be extraordinarily rich in secondary metabolites, such as the structurally diverse groups of alkaloids, phenolics, acetogenins and terpenoids (e.g. Wink, 2003). The wide molecular diversity of secondary metabolites throughout the plant kingdom represents an extremely rich biogenic resource for the discovery of novel drugs and for developing innovative drugs. Such drugs are sometimes hidden in less-studied plants, including salt-tolerant species – the halophytes (Dajic Stevanovic *et al.*, 2014a). As plants inhabiting extreme and ecologically diverse habitats, halophytes exhibit many morphological, anatomical and biochemical adaptations related to biosynthesis and accumulate a of range of metabolites involved in salinity tolerance. Some recent studies on halophytic secondary metabolites and related biological interactions pointed out that some highly salt-tolerant species, such as *Crithmum maritimum*, *Cakile maritima*, *Eryngium maritimum*, *Atriplex halimus* and *Mesembryanthemum crystallinum* are characterized by the presence of different phenolic compounds responsible for antioxidant effects (e.g. Ksouri *et al.*, 2007; Ksouri *et al.*, 2008; Meot-Duros *et al.*, 2008; Meot-Duros and Magne, 2009; Atia *et al.*, 2011; Pereira *et al.*, 2017a). It is known that halophytes evolved very efficient systems of radical oxygen scavenging as an adaptive mechanism linked to salt stress, including both enzymatic and non-enzymatic pathways, where the latter refers to biosynthesis of a range of secondary metabolites able to neutralize the harmful effects of reactive oxygen species (e.g. Dajic, 2006). Our results, performed on a range of halophytes collected from different types of salt-affected soils of the southern Balkans, indicated their solid capacity for free radical scavenging, which was related to the high content of total phenolics in some species studied (Stanković *et al.*, 2015). It has already been reported that numerous salt-tolerant species including *Salicornia herbacea*, *Reaumuria vermiculata*, *Suaeda fruticosa*, *Corydalis heterocarpa*, *Cressa cretica*, *Tribulus terrestris*, *Aloe barbadensis*, *Calotropis procera*, *Acanthus ilicifolius* and *Eryngium maritimum* exhibit certain antimicrobial, antiviral, anticancer and other favourable biological effects (e.g. Dagar and Singh, 2007; Rhee *et al.*, 2009; Oueslati *et al.*, 2012a; Wang *et al.*, 2013; Karker *et al.*, 2016; Kong *et al.*, 2016).

Halophytes have been recognized as a source of polyphenols, terpenes and many other secondary metabolites, and therefore are being researched for their bioactive effects and possible use as drugs and curatives. Halophytes have multiple uses for local inhabitants and are generally an underused resource (Menzel and Lieth, 1999). In a recent study of the flora and vegetation of south-eastern Europe it was noticed that, among over 50 medicinal and aromatic species capable of growing on salt-affected soils, the highest abundance and frequency was exhibited by *Achillea collina*, *Agropyrum repens*, *Artemisia monogyna*, *Chamomilla recutita*, *Taraxacum officinale*, *T. serotinum*, *Polygonum aviculare*, etc., as well as by several typical halophytes of the genus *Plantago* (Dajic Stevanovic *et al.*, 2016), which illustrates the point that saline habitats could be a source for wild collection of a range of medicinal herbs.

Halophytes have many commercial uses and potentials, for example as raw material for food or fodder, a source of oilseed with high nutritional value and biomass. In addition, their specialized metabolites can be used as pharmaceuticals, food additives and phyto-nutraceuticals (Buhmann and Papenbrock, 2013; Buhmann *et al.*, 2015). Therefore, halophytes represent an economically very important resource.

21.2 Ethnobotanical Records and Traditional Use of Halophytes as Medicinal Plants

There are several comprehensive research studies on traditional use of a number of halophytic species native to the Far East (e.g. India, Pakistan and China) (e.g. Dagar, 1995; Dagar and Singh, 2007; Ahmad and Husain, 2008; Qasim *et al.*,

2011, 2014; Zhao *et al.*, 2011). According to some interesting ethnobotanical reports, a considerable number of halophytic medicinal plants show strong biological activity for treating different diseases and disorders.

There is a list of halophytes used in traditional medicine worldwide (Ahmad and Husain, 2008). For example, *Artemisia scoparia* (Asteraceae) is a well-known diuretic plant used for the treatment of diarrhoea, also for lowering the blood pressure, protecting the liver, normalizing the function of the gallbladder and eliminating fever. It possesses antibacterial, anti-inflammatory and antitumour activity (Ksouri *et al.*, 2012). *Atriplex halimus* (Chenopodiaceae) is used to aid glycaemic control in diabetic patients (Day, 1990). It can also be used in the treatment of internal parasites in veterinary medicine (Bayoumi and El-Shaer, 1992). *Cakile maritima* (Brassicaceae) has purgative, antiscorbutic, digestive and diuretic properties. This species was used to treat oedema and jaundice, and to detoxify the lungs (Davy *et al.*, 2006; Ksouri *et al.*, 2012). *Chenopodium album* (Chenopodiaceae) is acknowledged in folk medicine for its antiparasitic, sedative, diuretic, hepatoprotective and laxative properties (Fournier, 1999). *Crithmum maritimum* (Apiaceae) is used in folk medicine because of its digestive, diuretic, purgative and antiscorbutic properties (Ksouri *et al.*, 2012). *Eryngium maritimum* (Apiaceae) is an exceptional plant with diuretic, diaphoretic, cystotonic and anthelmintic properties. It can also be used as a stimulant, a kidney stone inhibitor and aphrodisiac (Meot-Duros *et al.*, 2008). *Inula viscosa* (Asteraceae) has been used in Mediterranean folk medicine for centuries for its antiseptic, antipyretic, antiphlogistic activities, antidiabetes and anti-inflammatory effects (Lauro and Rolih, 1990; Ksouri *et al.*, 2012). In Jordanian folk medicine, *I. viscosa* is used in the treatment of anaemia, bronchitis, rheumatic pain and tuberculosis. The species also exhibits diuretic, expectorant and anthelmintic properties, and is prescribed for female sterility and the induction of abortion (Ksouri *et al.*, 2012). A leaf infusion from *Limoniastrum monopetalum* (Plumbaginaceae) is known in folk medicine for its antidysenteric properties (Ksouri *et al.*, 2012). *Nitraria tangutorum* (Zygophyllaceae) is used in traditional medicine to cure neurasthenia, indigestion and cold (Duan *et al.*, 1999). People in north-western China used to take its fruits in the treatment of abdominal pain, hypertension and indigestion (Ksouri *et al.*, 2012). *Salicornia herbacea* (Chenopodiaceae) is used as a vegetable and herb in coastal areas of Korea and as an ingredient of vinegar in Europe. Its applications in traditional medicine are predominantly limited to the treatment of diabetes, obesity, constipation and cancer (Bang *et al.*, 2002). *Suaeda fruticosa* (Amaranthaceae, including Chenopodiaceae) possesses antibacterial, hypolipidaemic and hypoglycaemic activities (Qasim *et al.*, 2011; Ksouri *et al.*, 2012). *Salsola kali* (Chenopodiaceae) is known for its hypotensive properties (Ksouri *et al.*, 2012). *Tribulus terrestris* (Zygophyllaceae) extracts may serve as an aphrodisiac and to increase the level of sexual hormones (Gauthaman and Ganesan, 2008). Extracts of this species have traditionally been used in the treatment of asthma, diabetes, hyperlipidaemia, kidney stones, hypertension, nephrotoxicity, spermatorrhoea, fungal infections, and skin and liver disorders (Qasim *et al.*, 2011; Ksouri *et al.*, 2012). *Tetraena alba* (syn. *Zygophyllum album*) (Zygophyllaceae) is frequently applied in the treatment of asthma, gout and rheumatism (Ksouri *et al.*, 2012). The facultative halophyte *Verbena officinalis* (Verbenaceae) has been used in Serbian folk medicine for its antiphlogistic and analgesic properties (Jarić *et al.*, 2007). It is also used in the treatment of rheumatism (Guarrera *et al.*, 2005). Moderately tolerant halophytes of saline steppes in Eurasian habitats, such as *Achillea collina*, *Mentha pulegium* and especially *Chamomilla recutita* are well-known, acknowledged herbs, frequently used in traditional Balkan medicine (Dajic Stevanovic *et al.*, 2014b, 2016).

21.3 Phytochemistry and Bioactivity of Halophytes

Many halophytic species have been studied for assessment of their chemical characterization and bioactivity of their major secondary metabolites. A range of halophytes of different phylogenetic position was reviewed, including species of the following plant families: Acanthaceae, Aizoaceae, Amaranthaceae, Apiaceae, Asteraceae, Brassicaceae, Caryophyllaceae, Chenopodiaceae, Combretaceae, Convolvulaceae, Euphorbiaceae,

Fabaceae, Frankeniaceae, Fumariaceae, Juncaceae, Lamiaceae, Malvaceae, Plantaginaceae, Plumbaginaceae, Primulaceae, Rhizophoraceae, Salvadoraceae, Saxifragaceae, Tamaricaceae and Zygophyllaceae (Table 21.1).

Our review of biological effects of bioactive molecules in halophytes (Table 21.1) showed that many plant species contain different bioactive compounds, such as phenols (polyphenols – *phenolic acids* e.g. chlorogenic acid, vanillic and gallic acids, aromatic acids; *phenylpropanoids* – hydroxycinnamates e.g. trans 3-hydroxycinnamic acid; *flavonoids*–flavanols e.g. catechin, isorhamnetin, flavonoid glycosides, isoflavone glycosides, sulfated flavonoids, tannins, stilbenes, curcuminoids; and *coumarins* e.g. p-coumaric acid, lignans, quinones, and others), saponins, fatty acids, polyacetylenes, essential oils, carotenoids, alkaloids, glucosinolates, lactones, aliphatic acids, alcohols and sterols, alkaloids, quinones, cyclitols, chalcones, cycloalkanes (e.g. cyclitol), diterpenes, triterpenes, steroids, furochromone, coumarins, lipopolysaccharides (LPS) (also known as lipoglycans and endotoxins), phenanthrenoids, waxes (alkanes and lipids) and phytoalexins (terpenoids, glycosteroids and alkaloids), many of which exhibit solid antioxidant activities, and various biological and therapeutic effects. Because of the great variability of secondary metabolites (the specialized compounds from halophytes), it could be projected that such compounds will be of great importance and future use in pharmacy, as functional foods, nutraceuticals and in technical implementations.

21.3.1 Antioxidant activity

A large percentage of the therapeutic effect of natural products is based on the presence of specialized metabolites with antioxidant activity. Harmful effects caused by free radicals can be prevented by the action of antioxidant substances which inhibit their activity. Antioxidants are natural biomolecules that partially or completely prevent the oxidation of substrates exposed to the oxidation process. The mechanism of the action of antioxidant substances is based on their role as donors of the electron or hydrogen atom, as well as the ability to destroy the hydroperoxide of the lipid molecules (Xu *et al.*, 2017). Oxidative stress is known as an additional phenomenon of stress impact on plants, as a consequence of hyperosmolarity in conditions of increased salinity, which results in the appearance of reactive oxygen molecules, mainly hydrogen peroxide, hydroxyl radicals and superoxide anions (Xiong *et al.*, 2002). Several reports comment on the antioxidant activity of medicinal halophytic plants. In previous studies concerning the quantitative qualitative composition of secondary metabolites and their biological activity, the antioxidant activity in *in vitro* conditions was examined for a number of coastal and inland halophytes, including *Limoniastrum guyonianum* (Trabelsi *et al.*, 2012; Bouzidi *et al.*, 2016), *L. monopetalum* (Ksouri *et al.*, 2008), *Cressa cretica* (Sunita *et al.*, 2011), *Suaeda fruticosa* (Oueslati *et al.*, 2012a; Oueslati *et al.*, 2014), *Eryngium maritimum*, *Crithmum maritimum*, *Cakile maritima* (Ksouri *et al.*, 2007; Ksouri *et al.*, 2008; Meot-Duros *et al.*, 2008; Atia *et al.*, 2011; Pereira *et al.*, 2017a), *Limonium tetragonum* (Lee *et al.*, 2011), *L. gmelinii* (Korul'kina *et al.*, 2004), *Mesembryanthemum crystallinum*, *M. edule*, *Salsola kali*, *Tamarix gallica* (Ksouri *et al.*, 2008; Ksouri *et al.*, 2009; Falleh *et al.*, 2011a; Falleh *et al.*, 2011b; Rodrigues *et al.*, 2014), *Atriplex halimus* (Benhammou *et al.*, 2009), *Reaumuria vermiculata* (Karker *et al.*, 2016), *Zygophyllum album* (Ksouri *et al.*, 2013), *Arthrocnemum macrostachyum*, *Halimione portulacoides*, *Juncus acutus* and *Plantago coronopus* (Behery *et al.*, 2013; Rodrigues *et al.*, 2014; Pereira *et al.*, 2017b), *Spergularia marina* (Kim *et al.*, 2014), *Statice gmelinii*, *Mentha pulegium*, *Camphorosma monspeliaca*, *Salicornia europaea*, *Suaeda maritima*, *Artemisia santonicum*, *Achillea collina* and *Aster tripolium* (Stanković *et al.*, 2015).

Testing the antioxidant capacity of plant extracts and simultaneous monitoring of the number of specialized metabolites indicates that polyphenol compounds are the main group of secondary metabolites, able to provide antioxidant activity in halophytes. In addition, their qualitative composition is highly dependent on various influences, of which the taxonomic affiliation of a species and ecological conditions of its habitat are among the most important. The often expressed relation between antioxidant activity and total phenol compound content in methanol extracts was confirmed in a study on a range of inland halophytes. *Mentha pulegium*,

Table 21.1. Taxonomic and habitat affiliation, phytochemicals and bioactivity of halophytes.

Family	Species	Active substance	Biological activity	Source
Acanthaceae	*Avicennia marina*	Reducing sugars polyphenols, flavonoids, tannins	Antiviral	Beula *et al.* (2012)
Aizoaceae	*Mesembryanthemum crystallinum*	Phenols	Antioxidative	Ksouri *et al.* (2008)
Aizoaceae	*Mesembryanthemum edule*	Procyanidins and propelargonidins, polyphenol gallic and salicylic acids and quercetin, all in the ethyl acetate extract	Antioxidative and antimicrobial	Ksouri *et al.* (2008); Falleh *et al.* (2009, 2011a, b, 2013); Rocha *et al.* (2017)
Amaranthaceae	*Salicornia herbacea*	Flavonoid glycosides, polyphenols, flavonoids, isorhamnetin, 3-O-β-D-glucopyranoside, triterpenoid saponins	Antioxidative	Kong *et al.* (2008); Kim *et al.*, (2009); Rhee *et al.* (2009); Kang *et al.* (2011); Kim *et al.* (2012)
Amaranthaceae	*Arthrocnemum macrostachyum*	Polyunsaturated fatty acids, phenolic compounds	Antioxidative	Rodrigues *et al.* (2014)
Amaranthaceae	*Salicornia europaea*	Phenol compounds	Antioxidative	Stanković *et al.* (2015)
Apiaceae	*Crithmum maritimum*	Phenols, falcarindiol, polyacetylene, essential oils	Antimicrobial	Amor *et al.* (2005); Meot-Duros *et al.* (2008, 2010); Atia *et al.* (2011); Jallali *et al.* (2014)
Apiaceae	*Daucus carota*	Chlorogenic acid and catechin	Antioxidative	Jdey *et al.* (2017)
Apiaceae	*Eryngium maritimum*	Phenols	Antimicrobial	Meot-Duros *et al.* (2008)
Asteraceae	*Cynara cardunculus*	Polyphenols, flavonoids, tannins	Antioxidative	Falleh *et al.* (2008)
Asteraceae	*Inula crithmoides*	Chlorogenic acid	Antioxidative	Jdey *et al.* (2017)
Asteraceae	*Pluchea lanceolata*	Phenols, flavonoids tannins, proanthocyanidins, carotenoids, alkaloids and saponins	Antioxidative	Qasim *et al.* (2017)
Asteraceae	*Artemisia santonicum*	Phenols	Antioxidative	Serkerov and Aleskerova (1984); Stanković *et al.* (2015)
Asteraceae	*Aster tripolium*	Phenol compounds	Antioxidative	Stanković *et al.* (2015)
Asteraceae	*Achillea collina*	Phenol compounds	Antioxidative	Stanković *et al.* (2015)
Asteraceae	*Artemisia campestris*	Phenolic compounds	Antimicrobial	Ksouri *et al.* (2015)
Brassicaceae	*Cakile maritima*	Phenols, phytoalexins, glucosinolates	Antioxidative	Ksouri *et al.* (2007); Meot-Duros *et al.* (2008)

Continued

Table 21.1. Continued.

Family	Species	Active substance	Biological activity	Source
Brassicaceae	*Raphanus raphanistrum*	p-coumaric acid	Antioxidative	Jdey *et al.* (2017)
Caryophyllaceae	*Spergularia marina*	Phenolic compounds	Antioxidative	Kim *et al.* (2014)
Chenopodiaceae	*Suaeda fruticosa*	Phenols, flavonoids, tannins, proanthocyanidins, carotenoids, alkaloids and saponins	Antioxidative	Oueslati *et al.* (2014); Qasim *et al.* (2017)
Chenopodiaceae	*Suaeda pruinosa*	Polyphenols	Antioxidative	Oueslati *et al.* (2012b)
Chenopodiaceae	*Suaeda maritima*	Polyphenols	Antioxidative	Oueslati *et al.* (2012b)
Chenopodiaceae	*Suaeda mollis*	Polyphenols	Antioxidative	Oueslati *et al.* (2012b)
Chenopodiaceae	*Atriplex inflata*	Hydroxyl fatty acid methyl ester, fatty acid, lactone	Antioxidative	Nejma *et al.* (2017)
Chenopodiaceae	*Salsola kali*	Polyphenols	Antioxidative	Ksouri *et al.* (2008)
Chenopodiaceae	*Halimione portulacoides*	Sulfated flavonoids, aliphatic acids, alcohols and sterols, triterpenic ketone	Antioxidative	Rodrigues *et al.* (2014)
Chenopodiaceae	*Suaeda monoica*	Alkaloids, terpenoids, steroids, tannins, quinones, saponins, flavonoids and phenols	Antimicrobial	Dinesh *et al.* (2016)
Chenopodiaceae	*Camphorosma monospeliaca*	Phenol compounds	Antioxidative	Stanković *et al.* (2015)
Chenopodiaceae	*Atriplex halimus*	Polyphenols	Antioxidative	Benhammou *et al.* (2009)
Combretaceae	*Luminitzera racemosa*	Cyclitols, sugars, tannins	Antimicrobial	Bandaranayake (2002); Chandrasekaran *et al.* (2009)
Combretaceae	*Luminitzera littorea*	Sugars	Antimicrobial	Bandaranayake (2002); Saad *et al.* (2011)
Convolvulaceae	*Cressa cretica*	Flavonoids (quercetin, rutin) scopoletin, umbelliferone, isoflavone glycoside	Antioxidative	Sunit *et al.* (2011)
Convolvulaceae	*Ipomoea pes-caprae*	Phenols, flavonoids, tannins, proanthocyanidins, carotenoids, alkaloids and saponins	Antioxidative	Qasim *et al.* (2017)
Euphorbiaceae	*Excoecaria agallocha*	Alkaloids, carotenoids, chalcones, cyclitol, diterpenes, polyphenols, saponins, steroids, sugars, tannins, triterpenes	Antimicrobial	Bandaranayake (2002); Chandrasekaran *et al.* (2009)
Fabaceae	*Acacia salicina*	Polyphenols, tannins	Antiviral	Bouhlel *et al.* (2007)
Frankeniaceae	*Frankenia laevis*	Chlorogenic acid and catechin	Antioxidative	Jdey *et al.* (2017)
Fumariaceae	*Corydalis heterocarpa*	Furochromone, coumarins	Anticancerogenic	Kong *et al.* (2016)

Continued

Table 21.1. Continued.

Family	Species	Active substance	Biological activity	Source
Juncaceae	*Juncus acutus*	Lipopolysaccharide, phenanthrenoids	Anticancerogenic and antioxidative	Behery *et al.* (2013); Rodrigues *et al.* (2014)
Lamiaceae	*Mentha pulegium*	Phenols	Antioxidative	Tahira *et al.* (2011); Stanković *et al.* (2015)
Malvaceae	*Thespesia populneoides*	Phenols, flavonoids, tannins, proanthocyanidins, carotenoids, alkaloids and saponins	Antioxidative	Qasim *et al.* (2017)
Plantaginaceae	*Plantago coronopus*	Phenolic compounds	Antioxidative	Pereira *et al.* (2017)
Plumbaginaceae	*Limonium tetragonum*	Flavonol glycosides	Antioxidative	Lee *et al.* (2011)
Plumbaginaceae	*Limonium delicatulum*	Phenolic, flavonoid and tannins	Antioxidative and antimicrobial	Medini *et al.* (2014a)
Plumbaginaceae	*Limonium densiflorum*	Myrctin, trans 3-hydroxycinnamic acid and isorhamnetin	Antiviral and antioxidative	Medini *et al.* (2014b, 2015)
Plumbaginaceae	*Limonium algarvense*	Phenolic compounds	Antioxidative	Rodriguesa *et al.* (2015)
Plumbaginaceae	*Limoniastrum monopetalum*	Vanillic and gallic acids	Antioxidative	Ksouri *et al.* (2008); Trabelsi *et al.* (2010)
Plumbaginaceae	*Limoniastrum guyonianum*	Polyphenols	Antioxidative and antimicrobial	Bouzidi *et al.* (2016)
Plumbaginaceae	*Statice gmelinii*	Phenolic acids and flavonoids	Antioxidative	Korul'kina *et al.* (2004)
Primulaceae	*Aegiceras corniculatum*	Benzoquinones, carotenoids, hydrolysable tannins, coumarins, flavonoids, minerals, polyphenols, proteins, sugars, saponins, triterpenes, triterpenes, glycosides	Antimicrobial	Bandaranayake (2002); Chandrasekaran *et al.* (2009)
Rhizophoraceae	*Rhizophora mucronata*	Acid polysaccharides	Antiviral	Premanathan *et al.* (1999)
Rhizophoraceae	*Ceriops decandra*	Carotenoids, flavonoids, lipids and waxes, polyphenols, proteins, steroids, tannins, triterpenes	Antimicrobial	Bandaranayake (2002); Chandrasekaran *et al.* (2009)
Rhizophoraceae	*Ceriops tagal*	Alkaloids, tannins, flavonoids, sugars	Antimicrobial	Arivuselvan *et al.* (2011)
Salvadoraceae	*Salvadora persica*	Phenols, flavonoids, tannins, proanthocyanidins, carotenoids, alkaloids and saponins	Antioxidative	Qasim *et al.* (2017)
Saxifragaceae	*Bergenia ligulata*	Polyphenols	Antiviral	Rajbhandari *et al.* (2001)

Continued

Table 21.1. Continued.

Family	Species	Active substance	Biological activity	Source
Tamaricaceae	*Tamarix gallica*	Phenolic compounds	Anticancerogenic and antioxidative	Boulaaba *et al.* (2013b)
Tamaricaceae	*Reaumuria vermiculata*	Phenolic compounds	Anticancerogenic	Karker *et al.* (2016)
Zygophyllaceae	*Zygophyllum album*	Triterpenes, flavonoids and sterols	Antioxidative	Ksouri *et al.* (2013)

Artemisia santonicum and especially *Statice gmelinii* (Fig. 21.1) exhibited the best antioxidant properties (Stanković *et al.*, 2015). Studies of major specialized metabolites in shoots of *Statice gmelinii* revealed the presence of gallic, syringic and ellagic acid, as well as quercetin, rutin, myricetin and its glycosides (Korul'kina *et al.*, 2004). In species of the genus *Artemisia*, apart from well-studied sesquiterpene lactones, including α-santonin and artemisin found in *A. santonicum* (Serkerov and Aleskerova, 1984), caffeic acid, chlorogenic acid and apigenin were reported as important active substances of its extracts, whereas the caffeic, rosmarinic and ferulic acid are major phenolics in *Mentha pulegium* (Ivanescu *et al.*, 2010; Tahira *et al.*, 2011).

21.3.2 Antimicrobial activity

In addition to antioxidant potential, many secondary metabolites affect the functional ability of bacterial cells and other microorganisms. The mechanism of antimicrobial action of phenolic acids is based on intensifying the external permeability of the membrane, resulting in loss of selective permeability and uncontrolled loss of important biomolecules. The mode of antimicrobial action of flavonoids is linked to interference or complete blocking of some functions of the cytoplasmic membrane, as well as some enzymes important for the DNA replication process of the bacteria. Terpenoids affect cell membrane disintegration; coumarins inhibit cellular respiration; while specialized metabolites of the tannin group disrupt the functioning of biomolecules of membranes and enzyme activity (Cowan, 1999; Cushnie and Lamb 2005; Lou *et al.*, 2011). Owing to the variability of the structure of secondary metabolites, predominantly determined by taxonomic and ecological variability, specialized metabolites of halophytes exhibit a different intensity of antimicrobial activity. After surveying the antimicrobial activity of specialized metabolites in halophytes, the microbicidal and microbiostatic effects were established for different plant extracts (obtained by application of organic solvents of different polarity), followed by essential oils, infusions and decoctions, as well as for isolated and purified single plant compounds.

Testing the antimicrobial activity of essential oils from the aerial parts of *Artemisia judaica*, *A. herba-alba* and *A. arborescens* from Libya against several pathogenic bacterial and fungal species showed that the best antimicrobial activity was obtained for *A. judaica* oil. Also, the effect of the tested oils was higher against Gram (+) than Gram (−) bacteria (Janaćković *et al.*, 2015). When examining the quantitative and qualitative composition of the phenolic compounds and the antimicrobial activity of *Tamarix gallica* extract, it was found that syringic acid, isoquercetin and catechin were the major phenolics of the shoot extract, whereas the strongest activity was recorded against *Micrococcus luteus* (Ksouri *et al.*, 2009). The antimicrobial activity of extracts of the species *Limoniastrum guyonianum* showed the antimicrobial potential of specialized metabolites against *Pseudomonas aeruginosa*, *Staphylococcus aureus*, *Candida glabrata*, *Candida krusei* and *Candida parapsilesis*, due to effects of the major active components, such as gallic, procatechuic and trans-cinnamic acids, methyl-4-hydroxybenzoate, n-propyl-3,4,5-trihydroxybenzoate, epicatechin, naringin and myricetin (Bouzidi *et al.*, 2016). For *Frankenia laevis*, the results of a similar study have showed that plant extracts rich in chlorogenic acid and catechin exhibited a broad spectrum of antibacterial activity against the four human pathogenic strains, mainly *Micrococcus luteus* and *Salmonella enterica* (Jdey *et al.*, 2017). A comparative study of antimicrobial

Fig. 21.1. *Limonium gmelinii*, syn. *Statice gmelinii* (Plumbaginaceae) in a saline habitat, south-eastern Serbia (Photo: Milan S. Stanković).

activity in relation to the polarity of extracts of *Eryngium maritimum*, *Crithmum maritimum* and *Cakile maritima* showed that almost all bacterial strains tested (11 in total) were inhibited, with the exception of *Listeria monocytogenes*. In addition, the nonpolar fractions (chloroformic) showed greater activity in relation to polar (methanolic) fractions (Meot-Duros *et al*., 2008).

Essential oils of halophytes, as well as of other species, show a different level of activity in relation to plant extracts. A comparative analysis of antimicrobial and antioxidant activities of essential oils and acetone extracts of *C. maritimum* and *Inula crithmoides* showed opposite results: the essential oils exhibited better antimicrobial activity compared to acetone extracts, while the acetone extracts showed better antioxidant activity (Jallali *et al*., 2014), which is related to the different composition of major secondary metabolites of essential oil (terpenes and terpenoid compounds) versus the acetone extract (phenolic compounds). Expressed antifungal activity was determined for essential oils of several species of *Achillea millefolium* aggregate (mainly *A. collina*) collected from saline habitats of different salinity levels, where the highest effects exhibited oils rich in chamazulene, trans-caryophyllene and β-pinene (Dajic Stevanovic *et al*., 2015a). In the case of effects of essential oils of different populations of the obligate halophyte *Artemisia santonicum*, the highest antimicrobial activity against *Staphylococcus aureus*, *Listeria monocytogenes* and *Salmonella typhimurium*, as well as antifungal activity on *Aspergillus fumigatus*, *A. niger* and *Penicillium funiculosum*, was related to high content of oxygenated monoterpenes, mainly 1,8 cineole (Dajic Stevanovic *et al*., 2015b). By examining the content and type of phenolic compounds, as well as the antioxidant and antimicrobial activity of the extracts of various polarities in the facultative halophyte *Artemisia campestris*, it has been shown that the diversity and content of specialized metabolites contribute significantly to the biological activity. The results of this analysis indicate that extracts of *A. campestris* are characterized by the presence of 39 different secondary metabolites, including coumarins, flavones, flavonols, phenolic acids, sesquiterpenes, luteolin-7-O-ruthinoside, rhamnetin, isorhamnetin, hydroxycoumarin, kaempferol rutinoside and di-O-caffeoylquinic acid isomers metabolites, which contribute to the significant antioxidant and antimicrobial activity of its extract (Ksouri *et al*., 2015). Our preliminary results relating to screening the biological activity of methanolic extracts of a number of halophytes from inland and coastal habitats of south-eastern Europe indicated that *Limonium vulgare*, *L. angustifolium* and *Artemisia maritima* possess metabolites with significant antimicrobial activity, especially against *Staphylococcus aureus*, *Bacillus cereus*, *Sarcina lutea*, *Klebsiella oxytoca*, *Pseudomonas aeruginosa*, *Salmonella enteritidis* and *Aspergillus brasiliensis*.

21.3.3 Anticancer activity

Plant secondary metabolites in *in vitro* and *in vivo* conditions can affect uncontrolled cell division resulting in neoplasia (i.e. local, atypical and uncontrolled tissue growth (tumour), benign or malignant). The mechanism of antitumour activity of specialized plant metabolites differs depending on the type of cell and tissue, as well as the type of specialized biomolecule. Some of them block the cell cycle in the G_2/M phase, in addition to their ability to induce apoptosis, inhibit angiogenesis or alter the activity of enzymes involved in cell cycle regulation, as well as transport through the cell membrane (Ren *et al.*, 2003; Duangmano *et al.*, 2010; Khoo *et al.*, 2010). In previous studies of the pharmacological potential of plants in various experimental models, it has been shown that specialized halophyte metabolites manifest a certain anticancer activity. The level of their anticancer activity varies depending on taxonomic affiliation and the type of the sample preparation.

Anticancer activity has been established for essential oils, the extracts of the aboveground plant parts using various solvents, as well as for isolated specialized plant metabolites (Ksouri *et al.*, 2012). During the examination of anticancer activity, the results of various *in vitro* treatments of cancer cells are usually accompanied with quantitative and qualitative chemical analysis in order to identify specialized metabolites of halophytes responsible for such activity. For example, for *Reaumuria vermiculata* it has been found that hexane and dichloromethane extracts showed significant anticancer activity against the A-549 lung carcinoma cells, and that myricetin is the main component of the investigated extracts (Karker *et al.*, 2016). By examining the biological potential of the halophytic species *Zygophyllum album*, it has been shown that triterpenes, flavonoids and sterols, whose presence is found in extracts of different polarities, act as anti-cancer compounds against human lung carcinoma (A-549) and colon adenocarcinoma (DLD-1) cells, in addition to antioxidant and anti-inflammatory effects (Ksouri *et al.*, 2013). In a comparative study of five maritime halophytic species from southern Portugal (Rodrigues *et al.*, 2014), high *in vitro* cytotoxic activity and selectivity was obtained with the ether extract of *Juncus acutus*. Juncunol was identified as the active compound and for the first time it was shown to be selectively cytotoxic to various human cancer cells *in vitro*. Species of the genus *Suaeda* were examined for the purpose of determining the chemical composition of extracts and biological activity. Due to proven healing and nutritional use, *S. fruticosa* has been tested in several studies. In addition to antioxidant and anti-inflammatory activity, the report by Oueslati *et al.* (2012a) indicated that the dichloromethane extract showed the highest anticancer activity against human lung carcinoma (A-549) and colon adenocarcinoma cell lines (DLD-1, Caco-2 and HT-29) with specificity against DLD-1. Studies with a similar concept were also carried out for species of the genus *Salicornia* (Patel, 2016). A study of the antiproliferative activity of *S. herbacea* ethyl acetate extract showed that phytol selectively inhibited HepG2 cells, whereas γ-linolenic acid and pentadecyl ferulate had potent antiproliferative activity against HepG2 and A549 cells (Wang *et al.*, 2013), as well as the anticancer potential of seed extracts on human HCT 116 and HT-29 colon cancer cells (Kang *et al.*, 2011). Halophytes of the genus *Limonium* are known for their use in folk medicine, pharmacy and nutrition. In a study of the species *L. franchetii* (Kong *et al.*, 2014), it has been found that, in addition to already identified metabolites, the species contains two new flavonoid glycosides, named quercetin-3-O-(2''-O-tigloyl)-α-rhamnopyranoside and quercetin-3-O-(3''-O-tigloyl)-α-L-rhamnopyranoside with anticancer potential against rat C6 glioma cell lines. In addition, a study of the chemical composition and biological activity of *L. densiflorum* extracts revealed that trans 3-hydroxycinnamic acid, myricetin and isorhamnetin were the major molecules present, and may be responsible for biological activities such as the highest anticancer activity against human plungous carcinoma A-549 and colon adenocarcinoma cell lines DLD-1 (Medini *et al.*, 2015). Experimental analyses of anticancer activity for specialized metabolites of some species also include the assays that illuminate the mechanism of their action. For the species *Tamarix gallica* and *Arthrocnemum indicum*, results revealed that their extracts decrease DNA synthesis and confirm their effect on Caco-2 cell proliferation, which were stopped at the G2/M phase, as well as cell-cycle-associated proteins (cyclin B1, p38, Erk1/2, Chk1 and Chk2), whose concentration was correlated with the changes in cell cycle distribution (Boulaaba *et al.*, 2013a, 2013b). *Corydalis heterocarpa* is a particularly interesting halophytic species owing to the presence of new secondary metabolites with anticancer potential. The heterocarpin showed significant cytotoxic effects against AGS (human gastric cancer), HT-29 (human colon cancer), HT-1080 (human fibrosarcoma) and MCF-7 (human breast carcinoma) cells by inducing apoptosis via regulated Bax-Bcl-2 ratio, overproduced caspases and suppressed XIAP (Kong *et al.*, 2016). New coumarin from this species identified as (2′S)-columbianetin-3′-sulfate showed potent inhibitory effect by induction of apoptosis through activating Bax, p53 and p21 expressions (Kim *et al.*, 2010). Concerning anti-cytogenetic effects in terms of preventing DNA damage caused by γ-radiation in human lymphocytes, frequently related to cancer development, the obligate halophyte *Atriplex littoralis* has recently been assessed for the presence and expressed solid radiation-mitigating activity of the spinacetin 3-O-β-D-glucopyranoside, arbutin and 4-hydroxybenzyl-β-D-glucopyranoside in addition to novel flavonoid glycosides, named atriplexins (Godjevac *et al.*, 2015).

21.4 Conclusion

Halophytes have been already recognized as a valid source of polyphenols and many other specialized metabolites. These natural bioactive substances are of vital importance in various fields of application. Halophytes have multiple uses for local inhabitants and are an underused resource (Menzel and Lieth, 1999). The use of numbers of different halophytes for many years, in the traditional medicine, food, perfumery and cosmetics industries, was later justified by numerous researches that confirmed their great diversity of ecological and biological activities, including antioxidant, antimicrobial, antiviral, anti-inflammatory and antiproliferative effects.

Recent studies have pointed out that, apart from widely studied biological activities of secondary metabolites of halophytes, there are promising effects of a range of phytochemicals in respect of the health-related inhibition of enzymes, as shown for acetylcholinesterase, butyrylcholinesterase, α-amylase, α-glucosidase and tyrosinase by halophyte extracts rich in quinic acid, p-coumaric acid and rhamnetin (Zengin *et al.*, 2018). Moreover, some halophytes have been evaluated for prevention and treatment of metabolic diseases, including the extract of *Mesembryanthemum crystallinum*, which exhibited positive effects on the treatment and prevention of obesity, as well as influence on adipocytes in terms of fat reduction (Kurosu and Kazuichi, 2011). Differentiation of adipocytes was also reported in the case of effects of the flavonoid glucopyranoside-rich extract of the halophyte *Salicornia europea* (Kong *et al.*, 2012). Molecular docking should be performed in future studies on the metabolite bioactivity of halophytic species, to better understand the molecular interactions of the enzymes and phytochemicals, as pioneered by Zengin *et al.* (2018).

It could be anticipated that a range of already identified and novel secondary halophyte metabolites will be evaluated for specific bioactive roles. As oxidative DNA damage is considered to be a pathogenic event in the induction of many diseases, a reduction in the rate of such damage by metabolites of halophytic species, acting as antioxidants, may lead to a reduced risk of disease (Godjevac *et al.*, 2015). A future focus on the antigenotoxic and anti-cytotoxic activity of bioactive molecules in halophytes might be expected.

The long-term objective should be to establish a solid ground for pharmacological application of secondary metabolites of halophytes. This should be achieved by testing the various bioactivity of extracts, essential oils and individual isolated compounds, and their roles in suppression and prevention of specific diseases or health-related problems, in *in vitro* conditions and finally in clinical studies (Buhmann and Papenbrock,

2013). There is a worldwide tendency to replace synthetic antioxidants with natural bioactive substances, particularly in pharmaceutical, agri-food and cosmetic industries, and therefore further research into the potential of halophytes will be of significant importance.

References

Ahmad, S.S. and Husain, S.Z. (2008) Ethno medicinal survey of plants from salt range (Kallar Kahar) of Pakistan. *Pakistan Journal of Botany* 40, 1005–1011.

Amor, B.N., Hamed, B.K., Debez, A., Grignon, C. and Abdelly, C. (2005) Physiological and antioxidant responses of the perennial halophyte *Crithmum maritimum* to salinity. *Plant Science* 168, 889–899.

Arivuselvan, N., Silambarasan, D., Govindan, T. and Kathiresan, K. (2011) Antibacterial activity of mangrove leaf and bark extracts against human pathogens. *Advances in Biological Research* 5(5), 251–254.

Atia, A., Barhoumi, Z., Mokded, R., Abdelly, C. and Smaoui, A. (2011) Environmental eco-physiology and economical potential of the halophyte *Crithmum maritimum* L. (Apiaceae). *Journal of Medicinal Plants Research* 5, 3564–3571.

Bandaranayake, W.M. (2002) Bioactivities, bioactive compounds and chemical constituents of mangrove plants. *Wetlands Ecology and Management* 10, 421–452.

Bang, M.A., Kim, H.A. and Cho, Y.J. (2002) Hypoglycemic and antioxidant effect of dietary hamcho powder in streptozotocin-induced diabetic rats. *Journal of the Korean Society of Food Science and Nutrition* 31, 840–846.

Bayoumi, M.T. and El-Shaer, H.M. (1992) Impact of halophytes on animal health and nutrition. In: Squires, V.R. and Ayoub, A.T. (eds) *Halophytes as a Resource for Livestock and for Rehabilitation of Degraded Lands*. Kluwer Academic Publishers, Dordrecht, The Netherlands, pp. 267.

Behery, F.A., Naeem, Z.E., Maatooq, G.T., Amer, M.M. and Ahmed, A.F. (2013) A novel antioxidant phenanthrenoid dimer from *Juncus acutus* L. *Natural Product Research* 2(2), 155–63.

Benhammou, N., Bekkara, F.A. and Kadifkova-Panovska, T. (2009) Antioxidant activity of methanolic extracts and some bioactive compounds of *Atriplex halimus*. *Comptes Rendus Chimie* 12, 1259–1266.

Beula, J.M., Gnanadesigan, M., Rajkumar, P.B., Ravikumar, S. and Anand, M. (2012) Antiviral, antioxidant and toxicological evaluation of mangrove plant from South East coast of India. *Asian Pacific Journal of Tropical Biomedicine* 2(1), S352–S357.

Bouhlel, I., Mansour, B.H., Limem, I., Sghaier, B.M., Mahmoud, A., *et al.* (2007) Screening of antimutagenicity via antioxidant activity in different extracts from the leaves of *Acacia salicina* from the center of Tunisia. *Environmental Toxicology and Pharmacology* 23, 56–63.

Boulaaba, M., Mkadmini, K., Tsolmon, S., Han, J., Smaoui, A., *et al.* (2013a) *In vitro* antiproliferative effect of *Arthrocnemum indicum* extracts on Caco-2 cancer cells through cell cycle control and related phenol LC-TOF-MS identification. *Evidence-Based Complementary and Alternative Medicine* 2013. doi: 10.1155/2013/529375.

Boulaaba, M., Tsolmon, S., Ksouri, R., Han, J., Kawada, K., *et al.* (2013b) Anticancer effect of *Tamarix gallica* extracts on human colon cancer cells involves Erk1/2 and p38 action on G2/M cell cycle arrest. *Cytotechnology* 65(6), 927–936.

Bouzidi, A., Benzarti, A., Arem, A.E., Mahfoudhi, A., Hammami, S., *et al.* (2016) Chemical composition, antioxidant and antimicrobial effects of Tunisian *Limoniastrum guyonianum* Durieu ex Boiss extracts. *Pakistan Journal Pharmaceutical Sciences* 29(4), 1299–1305.

Buhmann, A. and Papenbrock, J. (2013) An economic point of view of secondary compounds in halophytes. *Functional Plant Biology* 40, 952–967.

Buhmann, A.K., Waller, U., Wecker, B. and Papenbrock, J. (2015) Optimization of culturing conditions and selection of species for the use of halophytes as biofilter for nutrient rich saline water. *Agricultural Water Management* 149, 102–114.

Chandrasekaran, M., Kannathasan, K., Venkatesalu, V. and Prabhakar, K. (2009) Antibacterial activity of some salt marsh halophytes and mangrove plants against methicillin resistant *Staphylococcus aureus*. *World Journal of Microbiology and Biotechnology* 25, 155–160.

Cowan, M.M. (1999) Plant products as antimicrobial agents. *Clinical Microbiology Reviews* 12, 564–582.

Cushnie, T.P.T. and Lamb, A.J. (2005) Antimicrobial activity of flavonoids. *International Journal of Antimicrobial Agents* 26, 343–356.

Dagar, J.C. (1995) Ecology of halophytic vegetation in India: a review. *International Journal of Ecology and Environmental Sciences* 21, 273–296.

Dagar, J.C. and Singh, G. (2007) *Biodiversity of saline and waterlogged environments: documentation, utilization and management*. National Biodiversity Authority, Chennai, India, pp. 57–60.

Dajić, Z. (2006) Salt stress – salinity and tolerance mechanisms in plants. In: Madhava Rao, K.V., Raghavendra, A.S. and Jenardhan Reddy, K. (eds) *Physiology and Molecular Biology of Stress Tolerance in Plants*. Springer, Dordrecht, The Netherlands, pp. 41–99.

Dajic Stevanovic, Z., Janackovic, P. and Stankovic, M. (2014a) Are there still neglected medicinal plants beyond official and traditional consideration? In: Dajic Stevanovic, Z. and Ibraliu, A. (eds) Proceedings of the 8th Conference on Medicinal and Aromatic Plants of Southeast European Countries (8th CMAPSEEC), 19–22 May 2014. Albanian Academy of Science, Agricultural University of Tirana and AMAPSEEC, Durrës, Albania, pp. 13–23.

Dajić Stevanović, Z., Petrović, M. and Aćić, S. (2014b) Ethnobotanical knowledge and traditional use of plants in Serbia in relation to sustainable rural development. In: Pieroni, A. and Quave, C.L. (eds) *Ethnobotany and Biocultural Diversities in the Balkans*. Springer, New York, pp. 229–252.

Dajic Stevanovic, Z., Pljevljakusic, D., Ristic, M., Sostaric, I., Kresovic, M., *et al.* (2015a) Essential oil composition of *Achillea millefolium* agg. populations collected from saline habitats in Serbia. *Journal of Essential Oil Bearing Plants* 18(6), 1343–1352.

Dajić Stevanović, Z., Pljevljakušić, D., Soković, M., Glamočlija, J., Stanković, M., *et al.* (2015b) Essential oil of the halophyte *Artemisia santonicum*: tissue localization, chemical characterization and biological activity. The Second Mediterranean Symposium on Medicinal and Aromatic Plants (oral presentation), April 22–25 2015. Republic of Turkey Ministry of Forestry and Water Affairs General Directorate of Forestry, Ankara, Turkey. Abstract book, p. 73.

Dajić Stevanović, Z., Aćić, S., Luković, M., Zlatković, I., Vasin, J., *et al.* (2016) Classification of continental halophytic grassland vegetation of Southeastern Europe. *Phytocoenologia* 46, 317–331.

Davy, A.J., Scott, R. and Cordazzo, C.V. (2006) Biological flora of the British Isles: *Cakile maritima* Scop. *Journal of Ecology* 94, 695–711.

Day, C. (1990) Hypoglycaemic compounds from plants. In: Bailey, C.J. and Flatt, P.R. (eds) *New Antidiabetic Drugs*. Smith-Gordon, London, pp. 267.

Dinesh, P., Arunprabu, S. and Ramanathan, T. (2016) Phytoconstituents, antioxidant, antimicrobial and haemolytic activity of *Suaeda maritima* and *Suaeda monoica* a natural halophyte. *World Journal of Pharmacy and Pharmaceutical Sciences* 5(11), 1002–1013.

Duan, J.A., Williams, I.D., Che, C.T., Zhou, R.H. and Zhao, S.X. (1999) Tangutorine: A novel 13-carboline alkaloid from *Nitraria tangutorum*. *Tetrahedron Letters* 40, 2593–2596.

Duangmano, S., Dakeng, S., Jiratchariyakul, W., Suksamrarn, A., Smith, D.R., *et al.* (2010) Antiproliferative effects of cucurbitacin B in breast cancer cells: Down-regulation of the c-Myc/hTERT/Telomerase pathway and obstruction of the cell cycle. *International Journal of Molecular Sciences* 11, 5323–5338.

Falleh, H., Ksouri, R., Chaieb, K., Karray-Bouraoui, N., Trabelsi, N., *et al.* (2008) Phenolic composition of *Cynara cardunculus* L. organs, and their biological activities. *Comptes Rendus Biologies* 331, 372–379.

Falleh, H., Ksouri, R., Oueslati, S., Guyot, S., Magne, C., *et al.* (2009) Interspecific variability of antioxidant activities and phenolic composition in *Mesembryanthemum* genus. *Food and Chemical Toxicology* 47(9), 2308–2313.

Falleh, H., Ksouri, R., Medini, F., Guyot, S., Abdelly, C., *et al.* (2011a) Antioxidant activity and phenolic composition of the medicinal and edible halophyte *Mesembryanthemum edule* L. *Industrial Crops and Products* 34, 1066–1071.

Falleh, H., Oueslati, S., Guyot, S., Dali, B.A., Magné, C., *et al.* (2011b) LC/ESI-MS/MS characterisation of procyanidins and propelargonidins responsible for the strong antioxidant activity of the edible halophyte *Mesembryanthemum edule* L. *Food Chemistry* 127, 1732–1738.

Falleh, H., Trabelsi, N., Magné-Bonenfant, M., Le Floch, G., Abdelly, C., *et al.* (2013) Polyphenol content and biological activities of *Mesembryanthemum edule* organs after fractionation. *Industrial Crops and Products* 42, 145–152.

Fournier, P. (1999) *Plantes Medicinales*. CME, Luxembourg, pp. 369.

Gauthaman, K. and Ganesan, A.P. (2008) The hormonal effects of *Tribulus terrestris* and its role in the management of male erectile dysfunction – an evaluation using primates, rabbit and rat. *Phytomedicine* 15, 44–54.

Godjevac, D., Stanković, J., Novaković, M., Anđelković, B., Dajić-Stevanović, Z., *et al.* (2015) Phenolic compounds from *Atriplex littoralis* and their radiation-mitigating activity. *Journal of Natural Products* 78(9), 2198–2204.

Guarrera, P.M., Forti, G. and Marignoli, S. (2005) Ethnobotanical and ethnomedicinal uses of plants in the district of Acquapendente (Latium, Central Italy). *Journal of Ethnopharmacology* 96, 429–444.

Ivanescu, B., Vlase, L., Corciova, A. and Lazar, M.I. (2010) HPLC-DAD-MS study of polyphenols from *Artemisia absinthium*, *A. annua*, and *A. vulgaris*. *Chemistry of Natural Compounds* 46, 468–470.

Jallali, I., Zaouali, Y., Missaoui, I., Smeoui, A., Abdelly, C., *et al.* (2014) Variability of antioxidant and antibacterial effects of essential oils and acetonic extracts of two edible halophytes: *Crithmum maritimum* L. and *Inula crithmoides* L. *Food Chemistry* 15(145), 1031–1038.

Janaćković, P., Novaković, J., Soković, M., Vujisić, L., Abdulhmid, A., *et al.* (2015) Composition and antimicrobial activity of essential oils of *Artemisia judaica, A. herba-alba* and *A. arborescens* from Libya. *Archives of Biological Sciences* 67, 455–466.

Jarić, S., Popović, Z., Mačukanović-Jocić, M., Djurdjević, L., Mijatović, M., *et al.* (2007) An ethnobotanical study of the usage of wild medicinal herbs from Kopaonik Mountain (Central Serbia). *Journal of Ethnopharmaology* 111, 160–175.

Jdey, A., Falleh, H., Jannet, B.S., Mkadmini, H.K., Dauvergne, X., *et al.* (2017) Phytochemical investigation and antioxidant, antibacterial and anti-tyrosinase performances of six medicinal halophytes. *South African Journal of Botany* 112, 508–514.

Kang, S., Kim, D., Lee, B.H., Kim, M.R., Chiang, M., *et al.* (2011) Antioxidant properties and cytotoxic effects of fractions from glasswort (*Salicornia herbacea*) seed extracts on human intestinal cells. *Food Science and Biotechnology* 20, 115–122.

Karker, M., Falleh, H., Msaada, K., Smaoui, A., Abdelly, C., *et al.* (2016) Antioxidant, anti-inflammatory and anticancer activities of the medicinal halophyte *Reaumuria vermiculata*. *Experimental and Clinical Sciences Journal* 15, 297–307.

Khoo, B.Y., Chua, S.L. and Balaram, P. (2010) Apoptotic effects of chrysin in human cancer cell lines. *International Journal of Molecular Sciences* 11, 2188–2199.

Kim, A.Y., Kong, C., Yea, S.S. and Seo, Y. (2010) Constituents of *Corydalis heterocarpa* and their antiproliferative effects on human cancer cells. *Food and Chemical Toxicology* 48(2), 722–728.

Kim, A.Y., Kong, C.S., Lee, J.I., Kim, H., Park, H.Y., *et al.* (2012) Evaluation of novel antioxidant triterpenoid saponins from the halophyte *Salicornia herbacea*. *Bioorganic & Medicinal Chemistry Letters* 22, 4318–4322.

Kim, H.S., Kim, J.A., Karadeniz, F., Ahn, B.N. and Kong, C.S. (2014) Radical scavenging and anti-inflammatory effects of the halophyte *Spergularia marina* Griseb. *Zeitschrift Naturforschung C. Journal of Biosciences* 69(11–12), 425–433.

Kim, Y.A., Kong, C.S., Um, R.Y., Lim, Y.S., Yea, S.S., *et al.* (2009) Evaluation of *Salicornia herbacea* as a potential antioxidant and antiinflammatory agent. *Journal of Medicinal Food* 12(3), 661–668.

Kong, C.S., Kim, Y.A., Kim, M.M., Park, J.S., Kim, S.K., *et al.* (2008) Antioxidant activity and inhibition of MMP-9 by Isorhamnetin and Quercetin 3-O-β-D-Glucopyranosides isolated from *Salicornia herbacea* in HT1080 cells. *Food Science and Biotechnology* 17, 983–989.

Kong, C.S., Lee, J.I., Kim, Y.A., Kim, J.A., Bak, S.S., *et al.* (2012) Evaluation on anti-adipogenic activity of flavonoid glucopyranosides from *Salicornia herbacea*. *Process Biochemistry* 47, 1073–1078.

Kong, C.S., Kim, Y.A., Kim, H. and Seo, Y. (2016) Evaluation of a furochromone from the halophyte *Corydalis heterocarpa* for cytotoxic activity against human gastric cancer (AGS) cells. *Food & Function* 7(12), 4823–4829.

Kong, N.N., Fang, S.T., Wang, J.H., Wang, Z.H. and Xia, C.H. (2014) Two new flavonoid glycosides from the halophyte *Limonium franchetii*. *Journal of Asian Natural Products Research* 16(4), 370–375.

Korul'kina, L.M., Shul'ts, E.E., Zhusupova, G.E., Abilov, Zh.A., Erzhanov, K.B., *et al.* (2004) Biologically active compounds from *Limonium gmelinii* and *L. popovii* I. *Chemistry of Natural Compounds* 40, 465–471.

Ksouri, R., Megdiche,W., Debez, A., Falleh, H., Grignon, C., *et al.* (2007) Salinity effects on polyphenol content and antioxidant activities in leaves of the halophyte *Cakile maritima*. *Plant Physiology and Biochemistry* 45, 244–249.

Ksouri, R., Megdiche, W., Falleh, H., Trabelsi, N., Boulaaba, M., *et al.* (2008) Influence of biological, environmental and technical factors on phenolic content and antioxidant activities of Tunisian halophytes. *Compte Rendues de Biologies* 331, 865–873.

Ksouri, R., Falleh, H., Megdiche, W., Trabelsi, N., Mhamdi, B., *et al.* (2009) Antioxidant and antimicrobial activities of the edible medicinal halophyte *Tamarix gallica* L. and related polyphenolic constituents. *Food and Chemical Toxicology* 47(8), 2083–2091.

Ksouri, R., Ksouri, W.M., Jallali, I., Debez, A., Magné, C., *et al.* (2012) Medicinal halophytes: potent source of health promoting biomolecules with medical, nutraceutical and food applications. *Critical Reviews Biotechnology* 32(4), 289–326.

Ksouri, W.M., Medini, F., Mkadmini, K., Legault, J., Magné, C., *et al.* (2013) LC-ESI-TOF-MS identification of bioactive secondary metabolites involved in the antioxidant, anti-inflammatory and anticancer activities of the edible halophyte *Zygophyllum album* Desf. *Food Chemistry* 139(1–4), 1073–1080.

Ksouri, W.M., Trabelsi, N., Mkadmini, K., Bourgou, S., Noumi, A., *et al.* (2015) *Artemisia campestris* phenolic compounds have antioxidant and antimicrobial activity. *Industrial Crops and Products* 63, 104–113.

Kurosu, M. and Kazuichi, S. (2011) *Mesembryanthemum crystallinum* extract suppressed the early differentiation of mouse 3T3–L1 preadipocytes. *Journal of Natural Pharmaceuticals* 2, 184–189.

Lauro, L. and Rolih, C. (1990) Observations and research on an extract of *Inula viscosa* Ait. *Bollettino-Societa Italiana Biologia Sperimentale* (Napoli) 66, 829–834.

Lee, J., Kong, C.S., Jung, M., Wan Hong, J., Young, L.S., *et al.* (2011) Antioxidant activity of the halophyte *Limonium tetragonum* and its major active components. *Biotechnology and Bioprocess Engineering* 16, 992–999.

Leonti, M. (2011) The future is written: Impact of scripts on the cognition, selection, knowledge and transmission of medicinal plant use and its implications for ethnobotany and ethnopharmacology. *Journal of Ethnopharmacology* 134, 542–555.

Lou, Z., Wang, H., Zhu, S., Ma, C. and Wang, Z. (2011) Antibacterial activity and mechanism of action of chlorogenic acid. *Journal of Food Science* 76(6), 398–403.

Medini, F., Fellah, H., Ksouri, R. and Abdelly, C. (2014a) Total phenolic, flavonoid and tannin contents and antioxidant and antimicrobial activities of organic extracts of shoots of the plant *Limonium delicatulum*. *Journal of Taibah University for Science* 8, 216–224.

Medini, F., Legault, J., Pichette, A., Abdelly, C. and Ksouri, R. (2014b) Antiviral efficacy of *Limonium densiflorum* against HSV-1 and influenza viruses. *South African Journal of Botany* 92, 65–72.

Medini, F. Bourgou, S., Lalancette, G.K., Snoussi, M., Mkadmini, K., *et al.* (2015) Phytochemical analysis, antioxidant, anti-inflammatory, and anticancer activities of the halophyte *Limonium densiflorum* extracts on human cell lines and murine macrophages. *South African Journal of Botany* 99, 158–164.

Menzel, U. and Lieth, H. (1999) Halophyte database vers. 2.0. Halophyte uses in different climates I: Ecological and ecophysiological studies. In: Lieth, H., Moschenko, M., Lohman, M., Koyro, H.-W. and Hamdy, A. (eds) *Progress in Biometeorology*. Backhuys Publishers, Leiden, The Netherlands, pp. 77–88.

Meot-Duros, L. and Magne, C. (2009) Antioxidant activity and phenol content of *Crithmum maritimum* L. leaves. *Plant Physiology and Biochemistry* 47, 37–41.

Meot-Duros, L., Le Floch, G. and Magné, C. (2008) Radical scavenging, antioxidant and antimicrobial activities of halophytic species. *Journal of Ethnopharmacology* 116, 258–262.

Meot-Duros, L., Cérantola, S., Talarmin, H., Meur, L.C., Le Floch, G., *et al.* (2010) New antibacterial and cytotoxic activities of falcarindiol isolated in *Crithmum maritimum* L. leaf extract. *Food and Chemical Toxicology* 48, 553–557.

Nejma, B.A., Znati, M., Nguir, A., Daich, A., Othman, M., *et al.* (2017) Phytochemical and biological studies of *Atriplex inflata* f. Muell.: Isolation of secondary bioactive metabolites. *Journal of Pharmacy and Pharmacology* 69, 1064–1074.

Oueslati, S., Ksouri, R., Falleh, H., Pichette, A., Abdelly, C., *et al.* (2012a) Phenolic content, antioxidant, anti-inflammatory and anticancer activities of the edible halophyte *Suaeda fruticosa* Forsk. *Food Chemistry* 132, 943–947.

Oueslati, S., Trabelsi, N., Boulaaba, M., Legault, J., Abdelly, C., *et al.* (2012b) Evaluation of antioxidant activities of the edible and medicinal *Suaeda* species and related phenolic compounds. *Industrial Crops and Products* 36, 513–518.

Oueslati, S., Ksouri, R., Pichette, A., Lavoie, S., Girard-Lalancette, K., *et al.* (2014) A new flavonol glycoside from the medicinal halophyte *Suaeda fruticosa*. *Natural Product Research* 28(13), 960–966.

Patel, S. (2016) *Salicornia*: evaluating the halophytic extremophile as a food and a pharmaceutical candidate. *3 Biotech* 6(1), 104.

Pereira, C.G., Barreira, L., da Rosa Neng, N., Nogueira, J.M.F., Marques, C., *et al.* (2017a) Searching for new sources of innovative products for the food industry within halophyte aromatic plants: *In vitro* antioxidant activity and phenolic and mineral contents of infusions and decoctions of *Crithmum maritimum* L. *Food and Chemical Toxicology* 107(B), 581–589.

Pereira, C.G., Custódio, L., Rodrigues, M.J., Neng, N.R., Nogueira, J.M.F., *et al.* (2017b) Profiling of antioxidant potential and phytoconstituents of *Plantago coronopus*. *Brazilian Journal of Biology* 77(3), 632–641.

Premanathan, M., Kathiresan, K. and Nakashima, H. (1999) Mangrove halophytes: A source of antiviral substances. *South Pacific Study* 19, 1–2.

Qasim, M., Gulzar, S. and Khan, A. (2011) Halophytes as medicinal plants. In: Ozturk, M., Mermut, A.R. and Celik, A. (eds) *Urbanisation, Land Use, Land Degradation and Environment*. Daya Publishing House, Delhi, pp. 330–343.

Qasim, M., Abideen, Z., Adnan, M.Y., Ansari, R., Gul, B., *et al.* (2014) Traditional ethno-botanical uses of medicinal plants from coastal areas of Pakistan. *Journal of Coastal Life Medicine* 2(1), 22–30.

Qasim, M., Abideen, Z., Adnan, M.Y., Gulzar, S., Gul, B., *et al.* (2017) Antioxidant properties, phenolic composition, bioactive compounds and nutritive value of medicinal halophytes commonly used as herbal teas. *South African Journal of Botany* 110, 240–250.

Rajbhandari, M., Wegner, U., Julich, M., Schopke, T. and Mentel, R. (2001) Screening of Nepalese medicinal plants for antiviral activity. *Journal of Ethnopharmacology* 74(3), 251–255.

Ren, W., Qiao, Z., Wang, H., Zhu, L. and Zhang, L. (2003) Flavonoids: Promising anticancer agents. *Medicinal Research Reviews* 23(4), 519–534.

Rhee, M.H., Park, H.J. and Cho, J.Y. (2009) *Salicornia herbacea*: Botanical, chemical and pharmacological review of halophyte marsh plant. *Journal of Medicinal Plants Research* 3(8), 548–555.

Rocha, M.I., Rodrigues, M.J., Pereira, C., Pereira, H., da Silva, M.M., *et al.* (2017) Biochemical profile and *in vitro* neuroprotective properties of *Carpobrotus edulis* L., a medicinal and edible halophyte native to the coast of South Africa. *South African Journal of Botany* 111, 222–231.

Rodrigues, M.J., Gangadhar, K.N., Vizetto-Duarte, C., Wubshet, S.G., Nyberg, N.T., *et al.* (2014) Maritime halophyte species from southern Portugal as sources of bioactive molecules. *Marine Drugs* 12(4), 2228–2244.

Rodrigues, M.J., Soszynski, A., Martins, A., Rauter, P.A., Neng, R., *et al.* (2015) Unravelling the antioxidant potential and the phenolic composition of different anatomical organs of the marine halophyte *Limonium algarvense*. *Industrial Crops and Products* 77, 315–322.

Saad, S., Taher, M., Susanti, D., Qaralleh, H., Binti, N.A., *et al.* (2011) Antimicrobial activity of mangrove plant (*Lumnitzera littorea*). *Asian Pacific Journal of Tropical Medicine*, 523–525.

Serkerov, S.V. and Aleskerova, A.N. (1984) Sesquiterpene lactones of *Artemisia santonica*. *Chemistry of Natural Compounds* 20, 391–392.

Stanković, S.M., Petrović, M., Godjevac, D. and Stevanović Dajić, Z. (2015) Screening inland halophytes from the central Balkan for their antioxidant activity in relation to total phenolic compounds and flavonoids: Are there any prospective medicinal plants? *Journal of Arid Environments* 120, 26–32.

Sunita, P., Jha, S. and Pattanayak, S.P. (2011) Anti-inflammatory and *in vivo* antioxidant activities of *Cressa cretica* Linn., a halophytic plant. *Middle-East Journal of Scientific Research* 8, 129–140.

Tahira, R., Naeemullah, M., Akbar, F. and Masood, M.S. (2011) Major phenolic acids of local and exotic mint germplasm grown in Islamabad. *Pakistan Journal of Botany* 43, 151–154.

Trabelsi, N., Megdiche, W., Ksouri, R., Falleh, H., Oueslati, S., *et al.* (2010) Solvent effects on phenolic contents and biological activities of the halophyte *Limoniastrum monopetalum* leaves. *Food Science and Technology* 43, 632–639.

Trabelsi, N., Oueslati, S., Falleh, H., Waffo-Téguo, P., Papastamoulis, Y. *et al.* (2012) Isolation of powerful antioxidants from the medicinal halophyte *Limoniastrum guyonianum*. *Food Chemistry* 135, 1419–1424.

Wang, X., Zhang, M., Zhao, Y., Wang, H., Liu, T., *et al.* (2013) Pentadecyl ferulate, a potent antioxidant and antiproliferative agent from the halophyte *Salicornia herbacea*. *Food Chemistry* 141(3), 2066–2074.

Wink, M. (2003) Evolution of secondary metabolites from an ecological and molecular phylogenetic perspective. *Phytochemistry* 64, 3–19.

Xiong, L., Schumaker, K.S. and Zhu, J.K. (2002) Cell signaling during cold, drought and salt stress. *The Plant Cell* (Supplement 2002), 165–183.

Xu, D.P., Li, Y., Meng, X., Zhou, T., Zhou, Y., *et al.* (2017) Natural antioxidants in foods and medicinal plants: Extraction, assessment and resources. *International Journal of Molecular Sciences* 18(1), 96.

Zengin, G., Aumeeruddy-Elalfi, Z., Mollica, A., Yilmaz, M.A. and Mahomoodally, M.F. (2018) *In vitro* and in silico perspectives on biological and phytochemical profile of three halophyte species – A source of innovative phytopharmaceuticals from nature. *Phytomedicine* 38, 35–44.

Zhao, K., Song, J., Feng, G., Zhao, M. and Liu, J. (2011) Species, types, distribution, and economic potential of halophytes in China. *Plant and Soil* 342, 495–509.

22 Lipids in Halophytes: Stress Physiology Relevance and Potential Future Applications

Bernardo Duarte[1,*], Ana Rita Matos[1], João Carlos Marques[2] and Isabel Caçador[1]
[1]*University of Lisbon, Lisbon, Portugal;* [2]*University of Coimbra, Coimbra, Portugal*

Abstract

The present environmental conditions impose serious constraints on typical cultivar farming. Halophytes are known to be highly resistant to harsh environments, withstanding drought and salinity while maintaining positive biomass production rates. It is known that, in adapting to adverse conditions, plants suffer some lipid remodelling to maintain membrane stability and the efficiency of the metabolic processes associated with these systems. This thus opens a new door for potential biotechnological applications based in added-value fatty acids (FAs) produced by halophytes. These can be used in the food industry sector, as some halophytes show very interesting concentrations of essential omega 3 and 6 FAs, even under seawater irrigation schemes. Another physiological consequence of growing under stress conditions is the accumulation of storage lipids in the highly productive photosynthetic organs. Along with seeds, these are the major sinks of triacylglycerols (TAGs) found in halophytes. Under stress conditions, halophytes are able to maintain their productivity while increasing the TAG content in their photosynthetic leaves, with high levels of saturated FAs; this would be useful in biodiesel production. Halophytes can, therefore, be considered sources of added-value FAs with applications in both food production and biotechnology.

Keywords: Coastal ecosystems; Economic plants; Halophytes; Lipids; Medicinal plants

22.1 Unravelling the Potential of Halophytes as Sources of Lipids

Soil salinization is now a major concern in modern agriculture as it is one of the most important factors impairing plant productivity. This problem affects about 3.6 billion ha of the world's 5.2 billion ha of agricultural land (Zhang and Shi, 2013). Moreover, climate change (increasing drought frequency and intensity, global warming and sea level rise with consequent salt water intrusion in coastal soils) adds to these concerns (Duarte *et al.*, 2014c). Salt stress in glycophytes is normally associated with a reduction in leaf expansion, stomatal closure, low primary productivity and nutritional insufficiencies (Rahnama *et al.*, 2010; Tuteja, 2010; James *et al.*, 2011). In contrast, halophyte plants are extremely productive under saline irrigation (Flowers and Colmer, 2008). Classically, halophytes are able to subsist and reproduce under soil salinities of more than 200 mM NaCl (Flowers and Colmer, 2008). Obligatory halophytes such as *Suaeda maritima* and *Mesembryanthemum crystallinum* demand saline environments for optimal growth. Other species such as *Puccinellia maritima*

* Corresponding author e-mail: baduarte@fc.ul.pt

 Halophytes and Climate Change: Adaptive Mechanisms and Potential Uses (eds M. Hasanuzzaman, S. Shabala and M. Fujita) and *Thellungiella halophila*, despite having optimal growth without salt, have the ability to tolerate high NaCl concentrations and are therefore classified as facultative halophytes (Flowers, 1972; Gong *et al.*, 2005; Gao *et al.*, 2006; Agarie *et al.*, 2007; Wang *et al.*, 2007, 2009; Guo *et al.*, 2012). These adaptations to a saline environment result from a complex metabolic network, including several biochemical and physiological processes. As global soil salinization has increased in recent years, the issue of adaptation has acquired reinforced importance, and there is an urgent need to understand the biochemical mechanisms allowing salt toleration, particularly for economically relevant crops. Some halophytic species have also been found to be potentially high nutritive solutions for arid environments. Several halophytes have already been identified and are used commercially as food sources, such as *Aster tripolium* (Ventura *et al.*, 2013; Duarte *et al.*, 2017), *Chenopodium quinoa* (Becker *et al.*, 2016) and *Salicornia* sp. (Ventura and Sagi, 2013). Despite the high salt tolerance exhibited by halophytes, little attention has been given to them by the scientific community, and most studies focusing on salt tolerance have mainly involved glycophytes (salt-sensitive plant species) (Hameed and Khan, 2011). The use of halophytes as alternative crops would lead to a reduction in agricultural freshwater consumption, ideal for areas with low water availability (Hameed and Khan, 2011; Duarte *et al.*, 2017a). In addition to the physiological aspects of salt tolerance and potential application for food production, in recent years attention has also turned to added-value metabolite production from halophytes, such as their fatty acids (FAs) (Barreira *et al.*, 2017; Duarte *et al.*, 2017a).

Plant oils in the form of triacylglycerols (TAGs) are composed of three FAs esterified to a glycerol backbone (Yang and Benning, 2018). TAGs can be precursors of cell building blocks, such as sugars. Additionally, they can also be used in ATP production from TAG degradation followed by FA β-oxidation and sequestration of potentially harmful free FAs in TAGs together with other lipids in lipid droplets. Lipid droplets containing TAGs may have extra functions, acting as compartments for enzyme aggregates that facilitate metabolic channelling, and thus allowing the dynamic sequestration of proteins and protein transport, contributing in this way to cellular lipid-based signalization mechanisms (Yang and Benning, 2018). TAGs are also the largest carbon reserve in oilseeds, acting as carbon sources and providing essential energy to fuel early growth until the emergence of the seedling photoautotrophic organs. Prior to germination, TAGs sequestered in lipid droplets are hydrolysed, releasing FAs (Cui *et al.*, 2016). Seed TAGs accumulate in the embryo and endosperm, with around 10% of seed TAGs stored in the latter; they are the basis of sucrose production in the endosperm, with an important role in skotomorphogenesis (Penfield *et al.*, 2004). Although TAG levels are generally low in photosynthetic tissues, they can accumulate in the leaves of stressed plants (Yang and Benning, 2018). TAGs in leaf mesophyll cells are usually found in cytosolic lipid droplets, and also inside the chloroplasts as plastoglobules (Yang and Benning, 2018). They usually serve to transiently sequester free FAs that can be harmful to the cell (Fan *et al.*, 2013; Yang *et al.*, 2015). Free FAs are generated under stress conditions due to the activity of phospholipases A (Matos *et al.*, 2008) in membrane lipids during senescence or membrane remodelling (Yang and Benning, 2018). The number and size of the plastoglobules increase during senescence or in plants subjected to abiotic stresses (van Wijk and Kessler, 2017).

Halophytes are permanently stressed plants as they inhabit very harsh environments with high soil salinity (Duarte *et al.*, 2014c, 2017b; Becker *et al.*, 2016), submersion periods (Duarte *et al.*, 2014a,b), altered thermal conditions (Duarte *et al.*, 2015b,c, 2016; Becker *et al.*, 2016) and anthropogenic contaminants (Anjum *et al.*, 2016; Caçador *et al.*, 2009; Santos *et al.*, 2014). These features make halophytes potential TAG-producing organisms, along with a high value as cash crops, due to their large, high-quality lipid content and ability to grow in adverse environments.

22.2 Halophytes as Sources of High Nutritional Value FAs

Halophytes typically present high saturated FA profiles, with values several times above 50%, sometimes associated with their salt-stress mechanisms (Barreira *et al.*, 2017). This can turn some disregarded halophyte species into potentially interesting sources of FAs. Linolenic acid (C18:3) and linoleic acid (C18:2) are also very abundant in halophytes, being part of the omega-3 and omega-6 categories, respectively, and thus essential FAs for the primary consumers. *Spartina patens* is an invasive species that has already spread across several Mediterranean countries such as France and Spain (Duarte *et al.*, 2015a; Baumel *et al.*, 2016). There has been recent verification of an increased invasion of *S. patens* in the Portuguese marshes, competing for habitats normally occupied by endemic Chenopodeaceae species (like *Halimione portulacoides* and *Sarcocornia* sp.) (Duarte *et al.*, 2015a). *S. patens* is a salt-excreting grass with high biomass production in terms of both rate and net production (Duarte *et al.*, 2015a). Being an invasive species in Europe, its potential use as animal fodder would be beneficial in ecological terms. Upon exposure to different concentrations of NaCl, the leaves of this grass increase their content of 16:0 and 18:0 FAs, with a concomitant decrease in its content of 18:2 and 18:3 FAs (Fig. 22.1). Although this is a typical sign of stress, the plants exhibit a high degree of tolerance to these salt concentrations, maintaining photosynthetic efficiency at high values mainly through the use of proline as an osmocompatible solute, in this way preserving the cellular osmotic potential at physiological levels (Duarte *et al.*, 2015d).

Turning to the role of polyunsaturated FAs (PUFAs) in cardiac diseases: indexes of atherogenicity (IA) and thrombogenicity (IT) are commonly used to infer and predict the potential health benefits associated with the ingestion of a certain food. IA indicates the relationship between the main classes of saturated and unsaturated FAs. Since saturated FAs are considered to be pro-atherogenic and unsaturated FAs anti-atherogenic, a low IA ratio is recommended (Garaffo *et al.*,

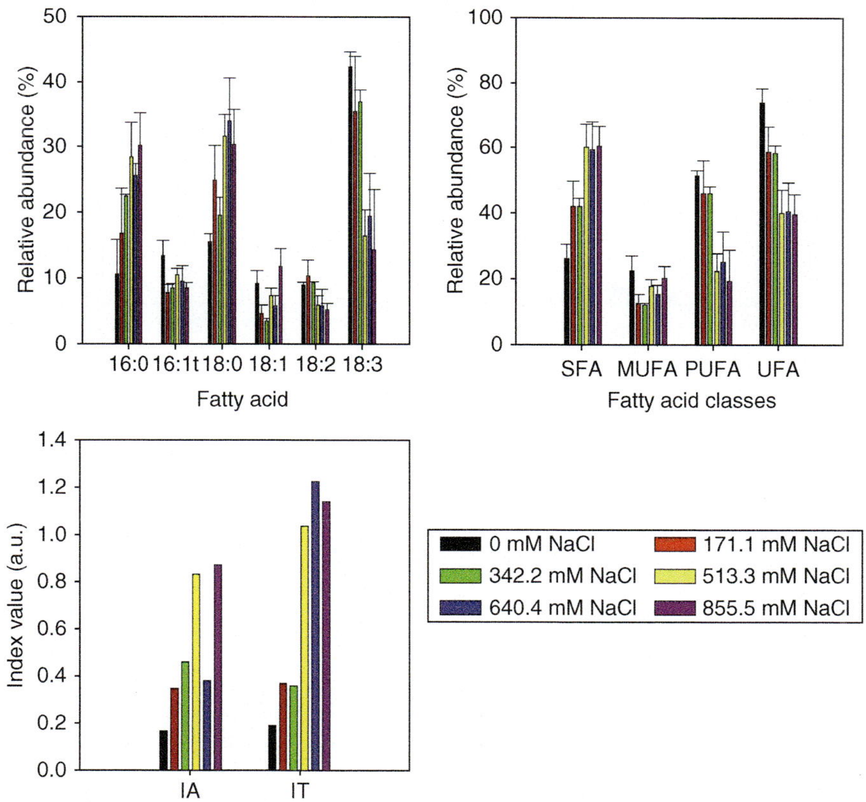

Fig. 22.1. Fatty acid profile, saturation classes (mean ± standard error) and nutritional health indexes in leaf tissues of *Spartina patens* along the tested NaCl concentrations. SFA, saturated fatty acids; MUFA, monounsaturated fatty acids; PUFA, polyunsaturated fatty acids; UFA, unsaturated fatty acids.

2011). The IT indicates the propensity to form masses in the blood vessels and is defined as the relationship between the pro-thrombogenic (saturated FA) and the anti-thrombogenic monounsaturated fatty acids (MUFA), n-3 and n-6 PUFA). Therefore, a low IT value is also desirable (Garaffo *et al.*, 2011). For example, tuna – a fish considered to have a high nutritional value – has an IA and IT of approximately 0.7 and 0.3, respectively (Garaffo *et al.*, 2011). Only under extreme salinities do the leaves of *S. patens* present higher values, making it suitable for harvesting in its normal habitat (salinity range from 0 to 513.3 mM), or even under saline agriculture cultivation when irrigated with seawater (salinity around 342.2–513.3 mM).

Aster tripolium is a promising potential cash crop halophyte with potential applications for human consumption and as a nutraceutical, as well as for forage and as an ornamental plant (Geissler *et al.*, 2009). In the Netherlands, Belgium, Portugal and Pakistan it is already under cultivation in pilot schemes (Geissler *et al.*, 2009). The FA profile of *A. tripolium* is typical from a 18:3 plant, as indicated by the absence of 16:3 FAs (Fig. 22.2). The total FA content in the leaves of *A. tripolium* exposed to a salinity gradient showed a decrease under low (0 mM NaCl) and very high salt exposure. Both 18:3 and 18:2 are included in the omega-3 and omega-6 categories, respectively, and are two of the most abundant FAs present in the leaves of this species. Some changes were also detected in the degree of unsaturation of FAs in the leaves. Saturated FAs showed a significant decrease under external salinities above 375 mM. Monounsaturated

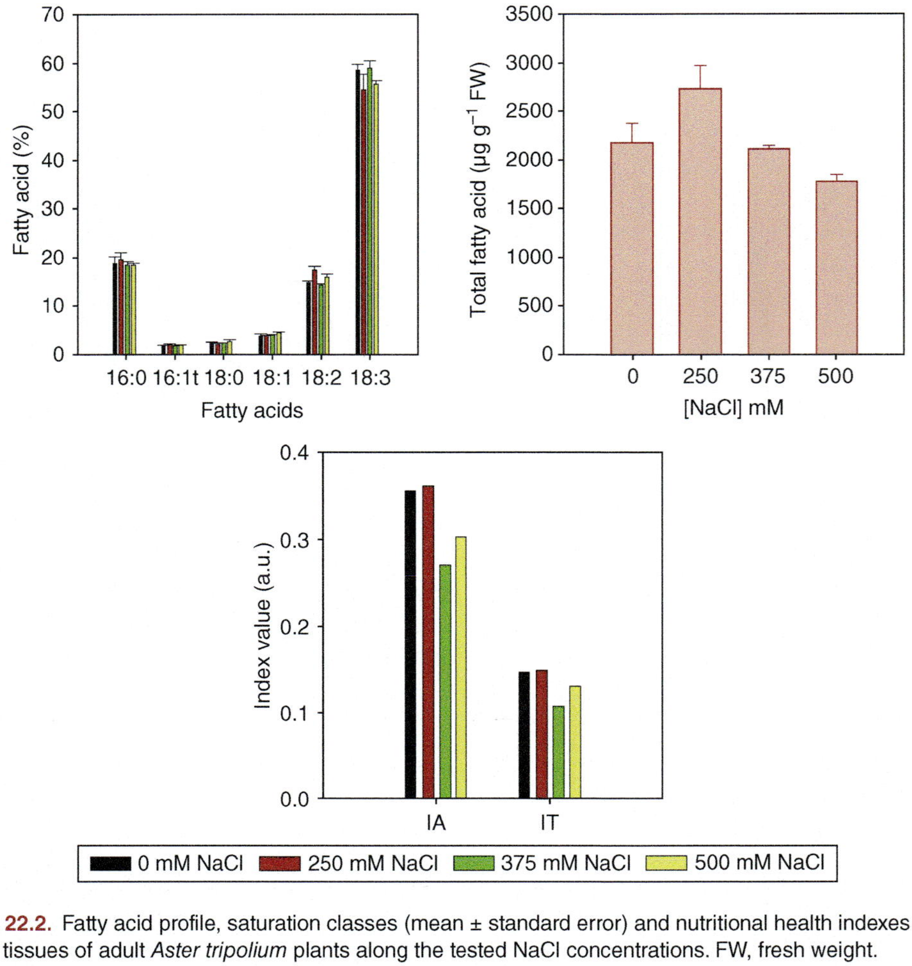

Fig. 22.2. Fatty acid profile, saturation classes (mean ± standard error) and nutritional health indexes in leaf tissues of adult *Aster tripolium* plants along the tested NaCl concentrations. FW, fresh weight.

FAs also showed a pronounced decrease in the plants exposed to 375 mM NaCl, whereas the PUFA concentration increased in these same individuals. As the plant is used for human and animal consumption, the FA profile of *A. tripolium* has acquired a reinforced and renewed importance (Duarte *et al.*, 2017). This plant showed high levels of omega-3 (18:3, 56% of the total FA content) and omega-6 (18:2, 15% of the total FA content). The abundance of linoleic acid (18:2), when compared to linolenic acid (18:3), may explain one aspect of halophyte adaptation during salt-stress conditions. At optimal external salinities, there is a shift in the 18:2/18:3 ratio, towards the linoleic acid. This can be due to membrane restructuring events, leading to the incorporation of lower amounts of polyunsaturated acids such as 18:3 (Ben Hamed *et al.*, 2005). In fact, the amount of omega-6 (18:2) found in *A. tripolium* plants is higher than most of the halophytes studied until now (Wu *et al.*, 1998; Ben Hamed *et al.*, 2005; Duarte *et al.*, 2017). Its low IA and IT values, and its FA profile, reinforce its potential use as a cash crop for salinized soils or for those with low freshwater supplies.

Other halophytes significantly change not only their physiology, but also their morphology, under salt stress. This is the case of the stem-succulent *Sarcocornia fruticosa*. Plants of this species show a typical reddish coloration on exposure to excessive salt concentrations, resulting from betacyanin accumulation (Duarte *et al.*, 2013, 2014c). This normally allows two ecotypes to be observable in the field, allowing fast identification of stressed individuals (Fig. 22.3). As the two ecotypes are green and red, and they have metabolic differences resulting from salt stress and betacyanin accumulation (Duarte *et al.*, 2013), we might also expect a change in FA composition of the edible photosynthetic stems. Salt-stressed plants present higher amounts of FAs on a fresh weight (FW) basis (Fig. 22.4). Red plants are richer in 16:0 while green individuals are most abundant in 18:2 FAs. As result the nutritional value of this species changes markedly according to its morphotype. With respect to IA and IT values, it is evident that the green plants are healthier for human consumption, with lower values than those observed in many other stem-succulent halophytes like *Arthrocnemum macrostachyum*,

Fig. 22.3. *Sarcocornia fruticosa* exhibiting the red coloration typical of betacyanin accumulation under salt-stress conditions (Photo by B. Duarte).

Sarcocornia perennis and *Salicornia ramosissima* (Barreira *et al.*, 2017). Nevertheless, the red morphotype can have other biotechnological applications, and these will be further discussed in the next section.

22.3 Halophytes as Potential Sources of Biodiesel

Most of the plant oils we consume are accumulated in seeds. World production from oilseed crops was around 100 billion kg of oil in 2011, and was valued at nearly US$120 billion (Bates *et al.*, 2013). The important uses of oilseed crops in recent years (including the explosion in oil biosynthesis investigations), their high value and growing demand are major reasons for this. Most of the FAs stored in seeds are in the form of triacylglycerol (TAG). Biodiesel is a mono-alkyl ester derived from oils, which has characteristics like petroleum-derived diesel. The conversion of edible or non-edible oil from oilseed plant species into biodiesel is relatively easy to perform, as it only involves the hydrolysis and separation of glycerol (Abideen *et al.*, 2012). This reaction consists of transforming triglycerides

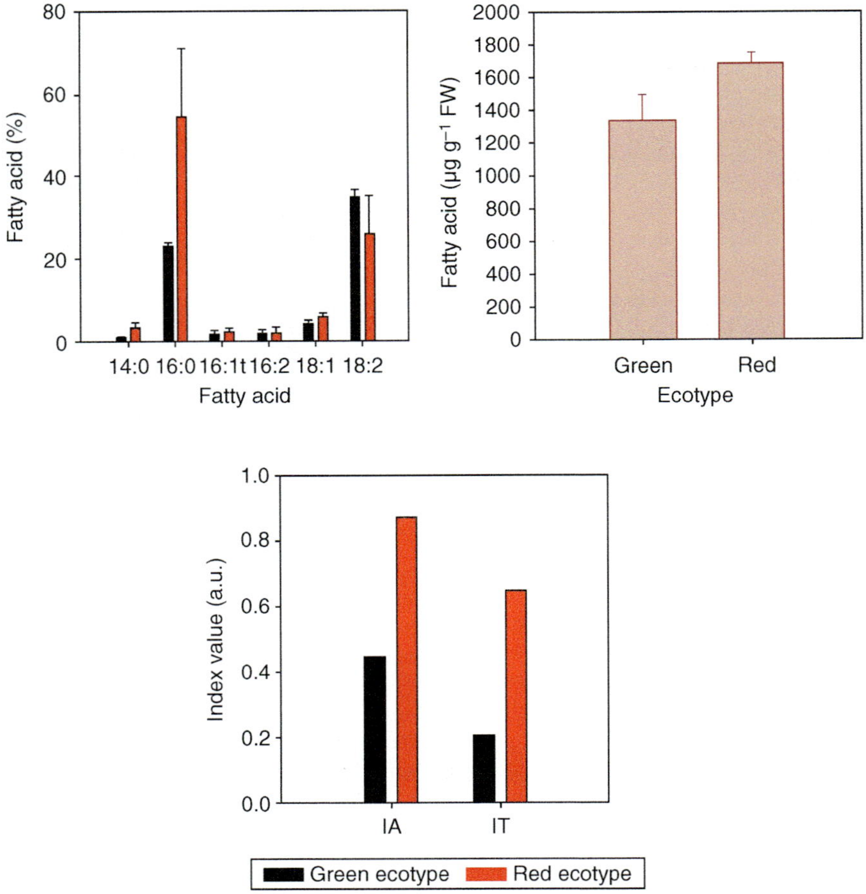

Fig. 22.4. Total fatty acids and profile (mean ± standard error) and nutritional health indexes in field-grown adult *Sarcocornia fruticosa* plants: green and red photosynthetic stems. FW, fresh weight.

into FA alkyl esters in the presence of an alcohol (such as methanol or ethanol) and a catalyst, with glycerol as a co-product (Hincapié *et al.*, 2011). Saturated FAs have higher melting points than unsaturated FA compounds. Oils with high percentages of saturated FA molecules crystallize at temperatures within the normal engine operation range (Nascimento *et al.*, 2013). Saturated FA methyl esters also produce biodiesel that has reduced air and noise emissions (Redel-Macías *et al.*, 2012). *Aster tripolium* appears to have higher total TAG concentrations on a seed FW basis than *S. fruticosa* or *C. quinoa* (Fig. 22.5). Nevertheless, TAGs derived from *A. tripolium* seed are richer in unsaturated FAs; the higher abundance of 16:0 and 18:0 in *S. fruticosa* makes it richer in saturated FAs and, thus, more suitable for biodiesel production purposes.

TAG accumulation in leaves is a typical response that can be found in stressed plants (Yang and Benning, 2018). *Aster tripolium* is an obligatory halophyte, having its optimum salinity around 250 mM (corresponding approximately to estuarine salinity, its natural habitat) (Duarte *et al.*, 2017b). Therefore, at this salinity, it is not expected that there would be large TAG accumulation in the leaves of individuals growing at optimal conditions. In fact, this was observed in the leaves of *A. tripolium* exposed to different levels of salinity, with high TAG values at NaCl concentrations below and above the

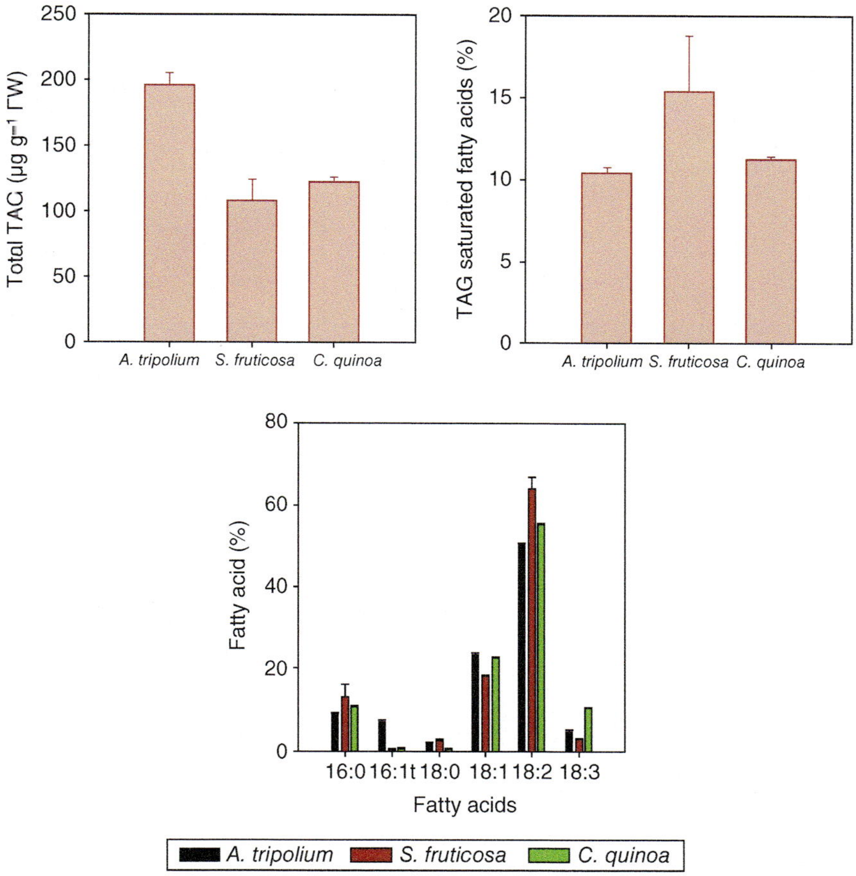

Fig. 22.5. TAG content and fatty acid composition in seeds of *Aster tripolium*, *Sarcocornia fruticosa* and *Chenopodium quinoa* (mean ± standard error). FW, fresh weight; TAG, triacylglycerol.

growth optimum (Fig. 22.6). Although in low amounts, the TAGs produced by leaves from individuals exposed to optimum salinities share similarly high saturation values to those from individuals exposed to extreme salinity. This high saturation is mostly due to increases in the 18:0 FA (stearic acid). The esterification of stearic acid is now suggested as a potential alternative catalyst for biodiesel preparation, using cheaper vegetable oil containing high amounts of free FAs (Soriano *et al.*, 2009). Comparing biomass production in terms of leaves and seeds, this species is suitable for biodiesel production because it produces high levels of leaf biomass.

However, not all stresses lead to the same TAG production and quality. Drought stress is considered to be one of the major constraints, along with soil salinization, to plant productivity. One of the known consequences of drought is the increasing number of plastoglobules in the chloroplast (Bréhélin and Kessler, 2008; Marchin *et al.*, 2017). These vesicles consist of an outer polar lipid monolayer containing neutral lipids such as prenylquinones and carotenoids (Bréhélin and Kessler, 2008). Severe drought has been reported to enhance TAG synthesis, which has been associated with stabilization of membranes during dehydration. Exposing *A. tripolium* to drought stress has significantly increased its TAG production in leaves, to levels much higher than those verified even under severe salinity stress (Fig. 22.7). Under drought stress it was possible to achieve TAG

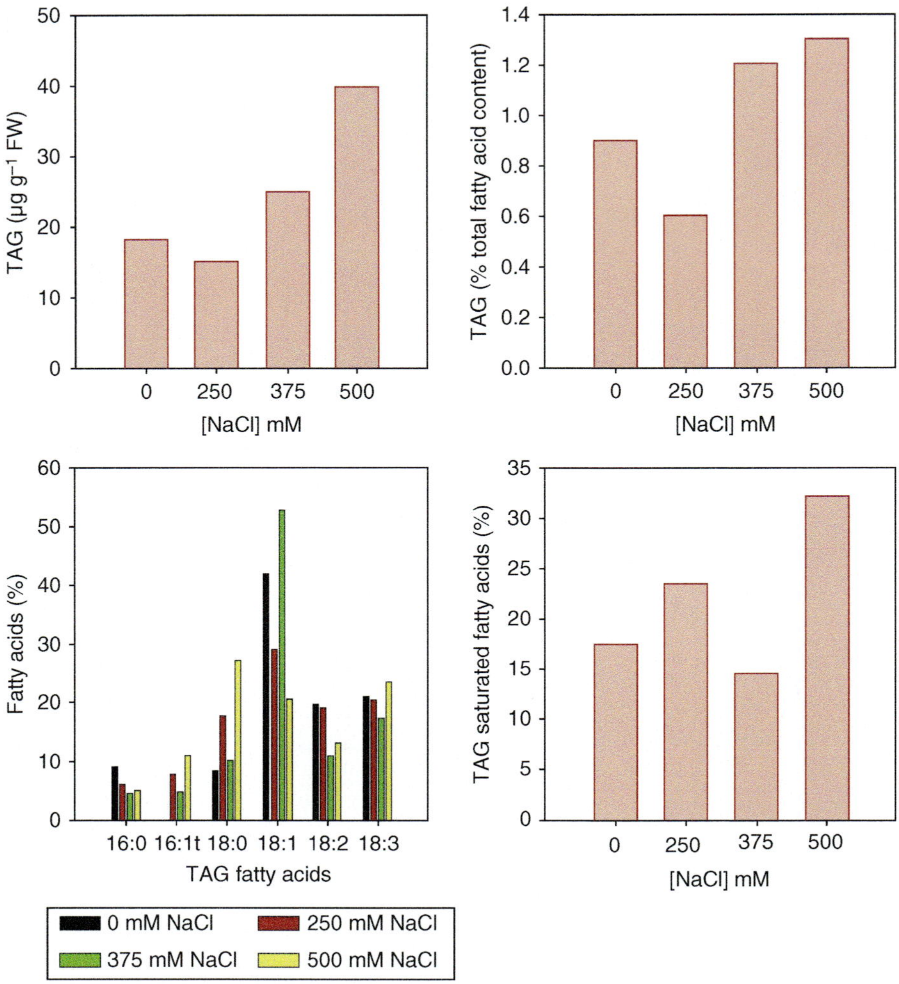

Fig. 22.6. TAG content and fatty acid composition (mean ± standard error) in leaf tissues of adult *Aster tripolium* plants along the tested NaCl concentrations. FW, fresh weight; TAG, triacylglycerol.

values of around 5% in all leaf lipid content. Nevertheless, the saturated FA abundance of TAG suffered a reduction, mainly due to a decrease in the 16:0 FA abundance under drought stress, indicating a membrane FA recycling to adjust to the stress conditions (Duarte *et al.*, 2017).

It is possible to observe the same tendency regarding TAG production in green and red *Sarcocornia* ecotypes (Fig. 22.8). The red ecotypes have a slightly higher TAG content, although a much lower saturated FA content. This is due to a sharp decrease in the 16:0 and 18:0 content in photosynthetic stems from the red ecotype. Although the *Sarcocornia* and *Salicornia* genera are often considered as potential new seed oil crops (Elsebaie *et al.*, 2013), their photosynthetic biomass is not as promising, even under stress, owing to the low percentage of saturated FAs.

22.4 Conclusion

Soil salinization, climate change and scarcity of freshwater resources have undoubtedly

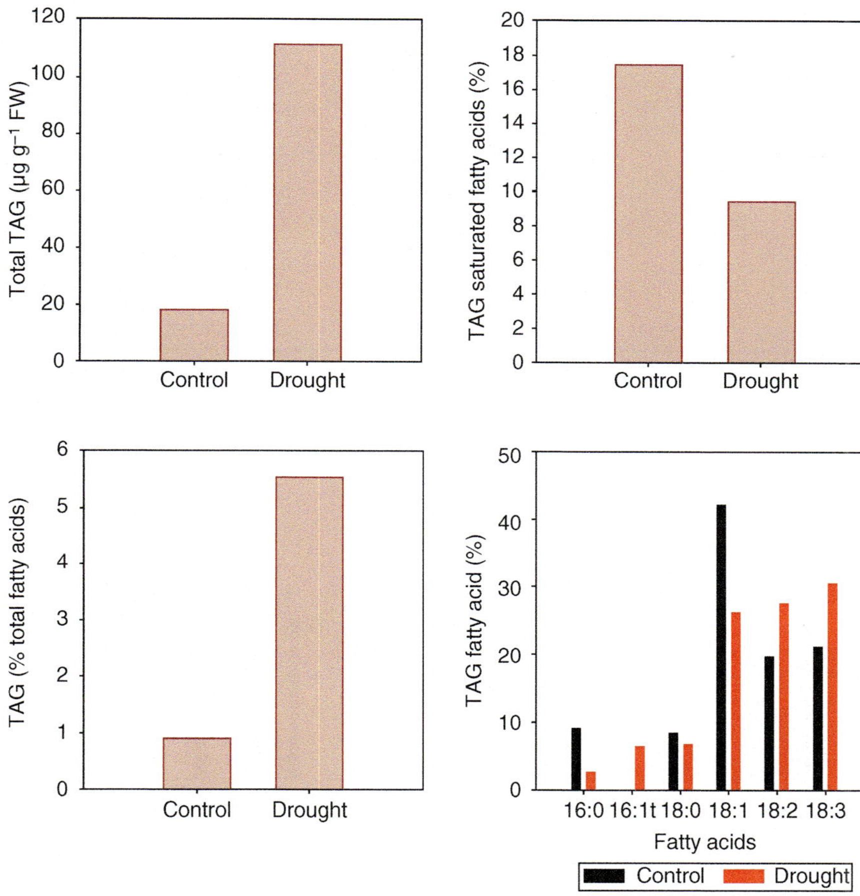

Fig. 22.7. TAG content and fatty acid composition (mean ± standard error) in leaf tissues of adult *Aster tripolium* plants under control conditions and drought stress. FW, fresh weight; TAG, triacylglycerol.

opened the door to saline agriculture and the use of halophytes as new crops. Nevertheless, beyond their physiological tolerance to salinity and harsh environments, it is important to secure quality food from halophytic biomass. Even in adverse conditions, halophytes have been proved to be able to maintain their nutritional value, with high omega-3 and -6 levels in the edible biomass. The seeds of some halophytes also have great potential in terms of biodiesel production, and they can grow in certain conditions – such as high soil salinity and drought – that impair the production of typical agronomic cultivars. Halophytes can, therefore, be considered sources of added-value FAs, with several applications in both food production and biotechnology.

Acknowledgements

The authors would like to thank the "Fundação para a Ciência e Tecnologia (FCT)" for funding the research in the Marine and Environmental Sciences Centre (MARE) throughout the project UID/MAR/04292/2013; and the Biosystems and Integrative Sciences Institute (BioISI) throughout the project UID/MULTI/04046/2013. The investigations by B. Duarte were supported by FCT through a postdoctoral grant (SFRH/BPD/115162/2016).

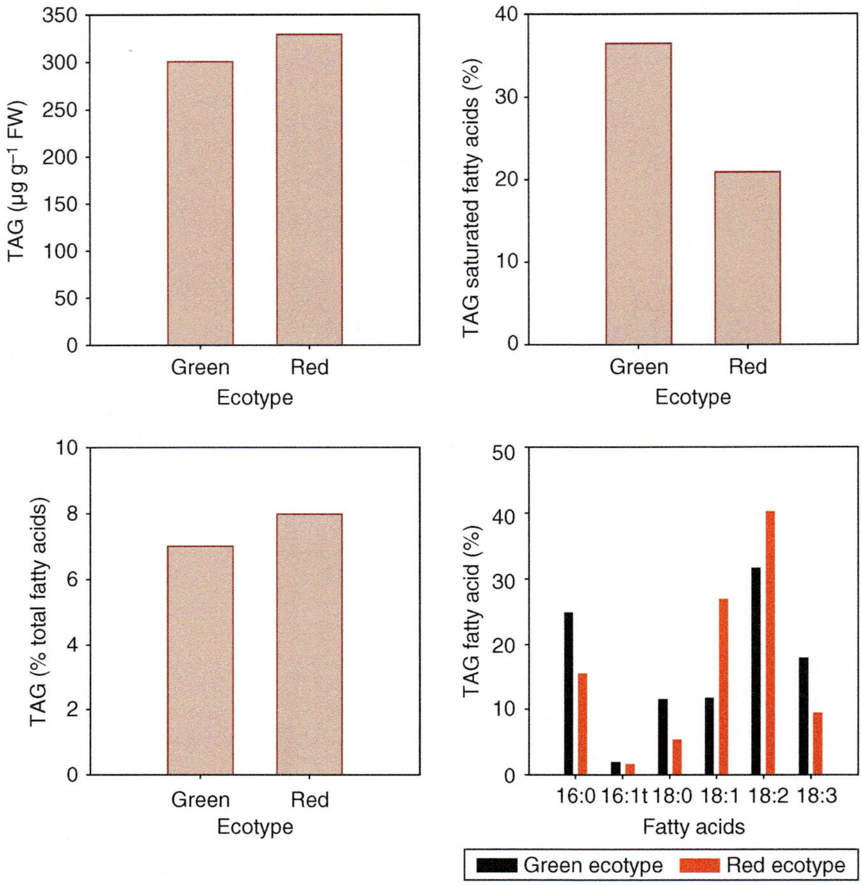

Fig. 22.8. TAG content and fatty acid composition in field-grown *Sarcocornia fruticosa* green and red photosynthetic stems. FW, fresh weight; TAG, triacylglycerol.

References

Abideen, Z., Ansari, R., Gul, B. and Khan, M.A. (2012) The place of halophytes in Pakistan's biofuel industry. *Biofuels* 3, 211–220.

Agarie, S., Shimoda, T., Shimizu, Y., Baumann, K., Sunagawa, H., Kondo, A., Ueno, O., Nakahara, T., Nose, A. and Cushman, J.C. (2007) Salt tolerance, salt accumulation, and ionic homeostasis in an epidermal bladder-cell-less mutant of the common ice plant *Mesembryanthemum crystallinum*. *Journal of Experimental Botany* 58, 1957–1967.

Anjum, N.A., Duarte, B., Caçador, I., Sleimi, N., Duarte, A.C. and Pereira, E. (2016) Biophysical and biochemical markers of metal/metalloid-impacts in salt marsh halophytes and their implications. *Frontiers in Environmental Science* 4. doi: 10.3389/fenvs.2016.00024.

Barreira, L., Resek, E., Rodrigues, M.J., Rocha, M.I., Pereira, H., Bandarra, N., da Silva, M.M., Varela, J. and Custódio, L. (2017) Halophytes: Gourmet food with nutritional health benefits? *Journal of Food Composition and Analysis* 59, 35–42.

Bates, P.D., Stymne, S. and Ohlrogge, J. (2013) Biochemical pathways in seed oil synthesis. *Current Opinion in Plant Biology* 16, 358–364.

Baumel, A., Rousseau-Gueutin, M., Sapienza-Bianchi, C., Gareil, A., Duong, N., Rousseau, H., Coriton, O., Amirouche, R., Sciandrello, S., Duarte, B., Caçador, I., Castillo, J.M. and Ainouche, M. (2016)

Spartina versicolor Fabre: another case of *Spartina* trans-Atlantic introduction? *Biological Invasions* 18, 2123–2135.

Becker, V.I., Goessling, J.W., Duarte, B., Caçador, I., Rosenqvist, E. *et al.* (2016) Combined effects of soil salinity and high temperature on photosynthesis and growth of quinoa plants (*Chenopodium quinoa*). *Functional Plant Biology* 44, 665–678.

Ben Hamed, K., Ben Youssef, N., Ranieri, A., Zarrouk, M. and Abdelly, C. (2005) Changes in content and fatty acid profiles of total lipids and sulfolipids in the halophyte *Crithmum maritimum* under salt stress. *Journal of Plant Physiology* 162, 599–602.

Bréhélin, C. and Kessler, F. (2008) The plastoglobule: A bag full of lipid biochemistry tricks. *Photochemistry and Photobiology* 84, 1388–1394.

Caçador, I., Caetano, M., Duarte, B. and Vale, C. (2009) Stock and losses of trace metals from salt marsh plants. *Marine Environmental Research* 67(2), 75–82.

Cui, S., Hayashi, Y., Otomo, M., Mano, S., Oikawa, K. *et al.* (2016) Sucrose production mediated by lipid metabolism suppresses the physical interaction of peroxisomes and oil bodies during germination of *Arabidopsis thaliana*. *Journal of Biological Chemistry* 291, 19734–19745.

Duarte, B., Santos, D., Marques, J.C. and Caçador, I. (2013) Ecophysiological adaptations of two halophytes to salt stress: Photosynthesis, PS II photochemistry and anti-oxidant feedback - Implications for resilience in climate change. *Plant Physiology and Biochemistry* 67, 178–188.

Duarte, B., Santos, D., Marques, J.C. and Caçador, I. (2014a) Biophysical probing of *Spartina maritima* photo-system II changes during prolonged tidal submersion periods. *Plant Physiology and Biochemistry* 77, 122–132.

Duarte, B., Santos, D., Silva, H., Marques, J.C., Caçador, I. *et al.* (2014b) Light-dark O_2 dynamics in submerged leaves of C_3 and C_4 halophytes under increased dissolved CO_2: Clues for saltmarsh response to climate change. *AoB Plants* 6. doi: 10.1093/aobpla/plu067.

Duarte, B., Sleimi, N. and Caçador, I. (2014c) Biophysical and biochemical constraints imposed by salt stress: Learning from halophytes. *Frontiers in Plant Science* 5. doi: 10.3389/fpls.2014.00746.

Duarte, B., Baeta, A., Rousseau-Gueutin, M., Ainouche, M., Marques, J.C. *et al.* (2015a) A tale of two *Spartinas*: Climatic, photobiological and isotopic insights on the fitness of non-indigenous versus native species. *Estuarine, Coastal and Shelf Science* 167, 178–190.

Duarte, B., Goessling, J.W., Marques, J.C. and Caçador, I. (2015b) Ecophysiological constraints of *Aster tripolium* under extreme thermal events impacts: Merging biophysical, biochemical and genetic insights. *Plant Physiology and Biochemistry* 97, 217–228.

Duarte, B., Santos, D., Marques, J.C. and Caçador, I. (2015c) Impact of heat and cold events on the energetic metabolism of the C_3 halophyte *Halimione portulacoides*. *Estuarine, Coastal and Shelf Science* 167, 166–177.

Duarte, B., Santos, D., Marques, J.C. and Caçador, I. (2015d) Ecophysiological constraints of two invasive plant species under a saline gradient: Halophytes versus glycophytes. *Estuarine, Coastal and Shelf Science* 167, 154–165.

Duarte, B., Marques, J.C. and Caçador, I. (2016) Ecophysiological response of native and invasive *Spartina* species to extreme temperature events in Mediterranean marshes. *Biological Invasions* 18, 2189–2205.

Duarte, B., Cabrita, M.T., Gameiro, C., Matos, A.R., Godinho, R. *et al.* (2017) Disentangling the photochemical salinity tolerance in *Aster tripolium* L.: connecting biophysical traits with changes in fatty acid composition. *Plant Biology* 19, 239–248.

Elsebaie, E.M., Elsanat, S.Y., Gouda, M.S. and Elnemr, K.M. (2013) Oil and fatty acids composition in glasswort (*Salicornia fruticosa*) seeds. *IOSR Journal of Applied Chemistry* 4, 06–09.

Fan, J., Yan, C. and Xu, C. (2013) Phospholipid: diacylglycerol acyltransferase-mediated triacylglycerol biosynthesis is crucial for protection against fatty acid-induced cell death in growing tissues of *Arabidopsis*. *The Plant Journal* 76, 930–942.

Flowers, T.J. (1972) Salt tolerance in *Suaeda maritima* (L.) Dum. *Journal of Experimental Botany* 23, 310–321.

Flowers, T.J. and Colmer, T.D. (2008) Salinity tolerance in halophytes. *New Phytologist* 179, 945–963.

Gao, F., Gao, Q., Duan, X.G., Yue, G.D., Yang, A.F. *et al.* (2006) Cloning of an H^+-PPase gene from *Thellungiella halophila* and its heterologous expression to improve tobacco salt tolerance. *Journal of Experimental Botany* 57, 3259–3270.

Garaffo, M.A., Vassallo-Agius, R., Nengas, Y., Lembo, E., Rando, R. *et al.* (2011) Fatty acids profile, atherogenic (IA) and thrombogenic (IT) health lipid indices, of raw roe of blue fin tuna and their salted product "Bottarga". *Food and Nutrition Sciences* 2, 736–743.

Geissler, N., Hussin, S. and Koyro, H.W. (2009) Interactive effects of NaCl salinity and elevated atmospheric CO_2 concentration on growth, photosynthesis, water relations and chemical composition of the potential cash crop halophyte *Aster tripolium* L. *Environmental and Experimental Botany* 65, 220–231.

Gong, Q., Li, P., Ma, S., Indu Rupassara, S. and Bohnert, H.J. (2005) Salinity stress adaptation competence in the extremophile *Thellungiella halophila* in comparison with its relative *Arabidopsis thaliana*. *The Plant Journal* 44, 826–839.

Guo, Q., Wang, P., Ma, Q., Zhang, J.L., Bao, A.K. *et al.* (2012) Selective transport capacity for K^+ over Na^+ is linked to the expression levels of *PtSOS1* in halophyte *Puccinellia tenuiflora*. *Functional Plant Biology* 39, 1047–1057.

Hameed, A. and Khan, M.A. (2011) Halophytes: Biology and economic potentials. *Karachi University Journal of Science* 39, 40–44.

Hincapié, G., Mondragón, F. and López, D. (2011) Conventional and in situ transesterification of castor seed oil for biodiesel production. *Fuel* 90, 1618–1623.

James, R.A., Blake, C., Byrt, C.S. and Munns, R. (2011) Major genes for Na^+ exclusion, *Nax1* and *Nax2* (wheat HKT1;4 and HKT1;5), decrease Na^+ accumulation in bread wheat leaves under saline and waterlogged conditions. *Journal of Experimental Botany* 62, 2939–2947.

Marchin, R.M., Turnbull, T.L., Deheinzelin, A.I. and Adams, M.A. (2017) Does triacylglycerol (TAG) serve a photoprotective function in plant leaves? An examination of leaf lipids under shading and drought. *Physiologia Plantarum* 161, 400–413.

Matos, A.R., Gigon, A., Laffray, D., Pêtres, S., Zuily-Fodil, Y. *et al.* (2008) Effects of progressive drought stress on the expression of patatin-like lipid acyl hydrolase genes in *Arabidopsis* leaves. *Physiologia Plantarum* 134, 110–120.

Nascimento, I.A., Marques, S.S.I., Cabanelas, I.T.D., Pereira, S.A., Druzian, J.I. *et al.* (2013) Screening microalgae strains for biodiesel production: Lipid productivity and estimation of fuel quality based on fatty acids profiles as selective criteria. *Bioenergy Research* 6, 1–13.

Penfield, S., Rylott, E.L., Gilday, A.D., Graham, S., Larson, T.R. *et al.* (2004) Reserve mobilization in the *Arabidopsis* endosperm fuels hypocotyl elongation in the dark, is independent of abscisic acid, and requires *phosphoenolpyruvate carboxykinase1*. *The Plant Cell* 16, 2705–2718.

Rahnama, A., James, R.A., Poustini, K. and Munns, R. (2010) Stomatal conductance as a screen for osmotic stress tolerance in durum wheat growing in saline soil. *Functional Plant Biology* 37, 255–263.

Redel-Macías, M.D., Pinzi, S., Ruz, M.F., Cubero-Atienza, A.J. and Dorado, M.P. (2012) Biodiesel from saturated and monounsaturated fatty acid methyl esters and their influence over noise and air pollution. *Fuel* 97, 751–756.

Santos, D., Duarte, B. and Caçador, I. (2014) Unveiling Zn hyperaccumulation in *Juncus acutus*: Implications on the electronic energy fluxes and on oxidative stress with emphasis on non-functional Zn-chlorophylls. *Journal of Photochemistry and Photobiology B: Biology* 140, 228–239.

Soriano, N.U., Venditti, R. and Argyropoulos, D.S. (2009) Biodiesel synthesis via homogeneous lewis acid-catalyzed transesterification. *Fuel* 88, 560–565.

Tuteja, N. (2010) Cold, salinity, and drought stress. In: Hirt, H. (ed.) *Plant Stress Biology: from Genomics to Systems Biology*. Wiley-VCH, Weinheim, Germany, pp. 137–159.

van Wijk, K.J. and Kessler, F. (2017) Plastoglobuli: Plastid microcompartments with integrated functions in metabolism, plastid developmental transitions, and environmental adaptation. *Annual Review of Plant Biology* 68, 253–289.

Ventura, Y. and Sagi, M. (2013) Halophyte crop cultivation: The case for *Salicornia* and *Sarcocornia*. *Environmental and Experimental Botany* 92, 144–153.

Ventura, Y., Myrzabayeva, M., Alikulov, Z., Cohen, S., Shemer, Z. *et al.* (2013) The importance of iron supply during repetitive harvesting of *Aster tripolium*. *Functional Plant Biology* 40, 968–976.

Wang, C.M., Zhang, J.L., Liu, X.S., Li, Z., Wu, G.Q. *et al.* (2009) *Puccinellia tenuiflora* maintains a low Na^+ level under salinity by limiting unidirectional Na^+ influx resulting in a high selectivity for K^+ over Na^+. *Plant, Cell and Environment* 32, 486–496.

Wang, S.M., Zhang, J.L. and Flowers, T.J. (2007) Low-affinity Na^+ uptake in the halophyte *Suaeda maritima*. *Plant Physiology* 145, 559–71.

Wu, J., Seliskar, D.M. and Gallagher, J.L. (1998) Stress tolerance in the marsh plant *Spartina patens*: Impact of NaCl on growth and root plasm membrane lipid composition. *Physiologia Plantarum* 102, 307–317.

Yang, Y. and Benning, C. (2018) Functions of triacylglycerols during plant development and stress. *Current Opinion in Biotechnology* 49, 191–198.

Yang, Y., Munz, J., Cass, C., Zienkiewicz, A., Kong, Q. *et al.* (2015) Ectopic expression of *WRI1* affects fatty acid homeostasis in *Brachypodium distachyon* vegetative tissues. *Plant Physiology* 169. doi: 10.1104/pp.15.01236.

Zhang, J.L. and Shi, H. (2013) Physiological and molecular mechanisms of plant salt tolerance. *Photosynthesis Research* 115, 1–22.

Index

Note: bold page numbers indicate figures; italic page numbers indicate tables.